Filippus S. Roux

Functional Phase Space Methods

Also of Interest

Quantum Information Theory
Concepts and Methods
Joseph M. Renes, 2022
ISBN 978-3-11-057024-3, e-ISBN (PDF) 978-3-11-057025-0

Quantum Technologies
For Engineers
Rainer Müller, Franziska Greinert, 2024
ISBN 978-3-11-071744-0, e-ISBN (PDF) 978-3-11-071745-7

Classical Mechanics
Hiqmet Kamberaj, 2021
ISBN 978-3-11-075581-7, e-ISBN (PDF) 978-3-11-075582-4

Quantum Mechanics
Guo-Ping Zhang, Mingsu Si, Thomas F. George, 2024
ISBN 978-3-11-067212-1, e-ISBN (PDF) 978-3-11-067215-2

Data Science
Time Complexity, Inferential Uncertainty, and Spacekime Analytics
Ivo D. Dinov, Milen Velchev Velev, 2021
ISBN 978-3-11-069780-3, e-ISBN (PDF) 978-3-11-069782-7

Quantum Mechanics
An Introduction to the Physical Background and Mathematical Structure
Gregory Naber, 2021
ISBN 978-3-11-075161-1, e-ISBN (PDF) 978-3-11-075194-9

Filippus S. Roux

Functional Phase Space Methods

Quantum Optics in All Degrees of Freedom

DE GRUYTER

Mathematics Subject Classification 2020
Primary: 81S30; Secondary: 81V80

Author
Dr. Filippus S. Roux
School of Chemistry and Physics
University of KwaZulu-Natal
Durban 4000
South Africa
rouxf@ukzn.ac.za

ISBN 978-3-11-144531-1
e-ISBN (PDF) 978-3-11-144534-2
e-ISBN (EPUB) 978-3-11-144535-9

Library of Congress Control Number: 2025935123

Bibliographic information published by the Deutsche Nationalbibliothek
The Deutsche Nationalbibliothek lists this publication in the Deutsche Nationalbibliografie; detailed bibliographic data are available on the Internet at http://dnb.dnb.de.

Cover image: The artistic rendition of an anisotropically scaled version of the Wigner function for a two-photon subtracted squeezed vacuum state; created by the author.
Typesetting: VTeX UAB, Lithuania

www.degruyter.com
Questions about General Product Safety Regulation:
productsafety@degruyterbrill.com

To the memory of my mother
Martha Johanna Roux (née Potgieter)

Contents

Part IV: **Epilogue**

Preface

Purpose of the book

The aim of this book is to provide a comprehensive discussion of a functional phase space formalism (*Wigner functional theory*) that is powerful enough to model any quantum optical system regardless of how complex it may be. In practical implementations of such quantum optical systems, all the degrees of freedom can potentially play a role. The quantum nature of such a system implies that the *particle-number* degree of freedom plays a significant role. At the same time, any physical implementation involves physical devices and components that introduce physical dimension parameters, affecting the *spatiotemporal* degrees of freedom in the system. In addition to these continuous degrees of freedom, there are also discrete internal degrees of freedom in the form of spin or polarization.

The functional phase space formalism presented here is powerful in the sense that it allows the modelling of such systems without any constraints. In other words, the continuous degrees of freedom are represented as infinite-dimensional degrees of freedom without the need to discretize and truncate them.

The Wigner functional formalism has been used to analyse various sophisticated quantum systems. However, it is not widely known. Hence, the need for a textbook that provides a complete description of the formalism, together with discussions of various applications. The main purpose of this book is to achieve this goal. The book presents the Wigner functional formalism to students and professionals in engineering and physics who are studying and working on the development and design of highly sophisticated photonic quantum information systems, providing them with the tools to perform the calculations that are necessary for technical work in this field, together with an understanding of the underlying principles.

Structure of the book

The material in the book is divided into three parts, preceded by an introductory chapter, and followed by a summary and outlook chapter at the end. The first part provides the background material that is necessary to make the book self-contained. The background information includes chapters on classical optics and basic quantum optics. The intention is to present the material at such a level that the mathematical background of a student with a bachelors from either physics or electrical engineering would suffice. It is nevertheless not the intention to address introductory quantum optics too exhaustively. While introductory quantum optics is presented comprehensively, it is not the intention to write a textbook on quantum optics. The focus is on that which serves as background for the rest of the book.

https://doi.org/10.1515/9783111445342-203

The second part provides the core material of the book, dealing with the presentation of functional phase space and the Wigner functional theory. In this part, the formal aspects of the functional phase space formalism is derived and developed. It is used to present the functional equivalents of standard quantum optics concepts. At the end of the second part, functional representations of physical systems that are commonly used to perform measurements in quantum optics and quantum information systems are discussed.

In the third part of the book, the application of parametric down-conversion is addressed. A full description of the process in terms of all the degrees of freedom is presented. Since parametric down-conversion is such a complex nonlinear process in which spatiotemporal and spin degrees of freedom combine with the particle-number degree of freedom, it is a challenging process to model. As such, it provides an excellent demonstration of the power of the functional phase space formalism. A variety of applications involving parametric down-conversion are addressed.

How to use the book

The book can be used in different ways. Its main intention is for graduate study, where the material in this book can be used for a graduate course, or perhaps more than one. It is not a good idea to cover all the material in just one course. The ideal course would use the material in the second part for a course on functional phase space methods, requiring knowledge of quantum optics as a prerequisite. The prerequisite material is provided in the first part. Some selection of material from the third part can be included in such a course to provide a demonstration of the methods. Alternatively, the third part of the book can serve as a separate specialist course.

For the individual interested in this field, the book can serve as a source for self-study. In this case, the person should work through the book from the start to at least the end of the second part. Several worked examples and exercises are provided to aid such self-study.

The book can also be used as a reference for professionals in the fields of engineering and physics, provided that they already have some familiarity with phase space methods. A comprehensive index is provided for keywords, which are emphasized in the text for easy identification.

Filippus S. Roux

Acknowledgments

The work presented in the book grew out of an extended period of research, which started years ago with the study of theoretical quantum optics in terms of discrete and continuous variable formalisms, but also benefited from interactions with experimental work in classical and quantum optics. In this regard, various discussions and interactions with individuals aided in one way or another the development of the material presented here. These individuals include (in no particular order) Thomas Konrad, Andrew Forbes, Andreas Buchleitner, Giacomo Sorelli, Tobias Brünner, Miles Padgett, Robert W. Boyd, Sandeep K. Goyal, Yingwen Zhang, Thomas Wellens, Vyacheslav N. Shatokhin, Shashi Prabhakar, Ebrahim Karimi, Adam Vallés, Micheal V. Berry, Alexander I. Lvovsky, and many more. Some individuals also contributed in ways other than technical discussions by, for example, serving as hosts during visits. I am sincerely grateful to all these people for their part in this process.

https://doi.org/10.1515/9783111445342-204

1 Introduction

In physics, a formalism is always provided in the context of physical scenarios. It is the purpose of the formalism to model these scenarios. As such, the formalism enables mathematical descriptions of physical systems within this context allowing them to be analysed and leading to predictions of their behaviour.

Functional phase space and the *Wigner functionals* defined on it represent such a formalism. The context for this formalism is *quantum optics* [1–4]. In what follows, we demonstrate that this context is rich in terms of all the degrees of freedom it represents. We show that a complete description of some physical systems in quantum optics requires a functional representation.

Quantum optics is the part of quantum physics associated with light. While *quantum physics* represents the physical scenarios in which we can observe quantum phenomena, the theoretical description of these scenarios requires *quantum theory*. It includes various formalisms in terms of which these physical quantum scenarios can be modelled. The first such quantum formalism is *quantum mechanics* [5–7]. Other quantum formalisms include: *quantum field theory* [8–12] in the context of particle physics; and the *Moyal formalism* [13]. Physical systems in quantum optics are mostly modelled in terms of quantum mechanics, with some inspiration taken from quantum field theory and the Moyal formalism.

In recent years, there has been a significant growth of research in quantum optics. The reason for this growth can be traced back to the development of new technology providing a broader context for quantum optics. To illustrate how functional formalism can enhance research in quantum optics, we first discuss this broader context.

1.1 Quantum information technology

The formulation of quantum mechanics is roughly 100 years old. It led to the development of revolutionary technologies, including semiconductor devices and lasers. These technologies were developed during what is now called the *First Quantum Revolution*. A better understanding of the quantum properties of nature currently produces a new wave of technological advances. This new development in quantum technology is called the *Second Quantum Revolution* [14] due to its expected disruptive nature.

Any process of technological development naturally goes through various stages [15]. It usually starts with an understanding of the underlying physics, which is followed by the proof-of-principle demonstrations of early ideas. Then preliminary implementations are designed and produced, becoming gradually more sophisticated. Eventually maturity is reached where the products are provided to a growing market.

An understanding of *quantum information science* [16, 17] represents the necessary underlying physics for the Second Quantum Revolution. Currently, there are already

https://doi.org/10.1515/9783111445342-001

initial stage engineered systems implementing quantum information technologies. Yet, there is a long road ahead leading to maturity.

The physical implementations of quantum information systems range from trapped ions and atoms, through various solid state technologies, to photonic systems. Each of these physical technologies represents its own challenges and techniques. In this book, we focus exclusively on photonic technologies, incorporating quantum optics.

The increase in sophistication of the systems calls for an improvement in the formalism with which such increasingly complex systems are analysed, and in terms of which they are designed. For quantum technology, the initial *Dirac formalism of quantum mechanics* was augmented by the *second quantization* formalism of *quantum field theory* [10]. The development of functions on phase space, such as Wigner functions, broadened the range of approaches to represent quantum states and quantum operations, leading to the *Moyal formalism* [13, 18, 19].

Among the formalisms in quantum optics, one that is currently widely used is the so-called *continuous variable formalism* [20, 21] in which quantum states are expressed as phase space distributions in terms of operators inherited from quantum field theory (second quantization). It represents the *particle-number* (photon-number) degrees of freedom as continuous degrees of freedom, hence the name. On the other hand, the *spatiotemporal degrees of freedom* are given as a finite number of discrete modes. The latter restriction can impose a limitation in its capability to analyse complex systems. For example, the process of *parametric down-conversion* [22], which is widely used in photonic quantum information systems, involves the spatiotemporal properties as infinitely many degrees of freedom. When it comes to the design of physical systems based on parametric down-conversion, for example, to prepare exotic photonic states, such limitations in the formalism can be detrimental for systems that must operate in the real-world environment. Such applications thus call for a more powerful formalism.

Recently, an extension of the Moyal formalism has emerged in which the two-dimensional phase space is converted to an infinite-dimensional phase space on which distributions (e. g., Wigner functions) are represented by *functionals* (functions of functions). This functional Moyal formalism incorporates both the photon-number degrees of freedom and the spatiotemporal degrees of freedom as *continuous degrees of freedom*. The *Wigner functional formalism* [23–25] does not impose any restrictions, such as truncations, on the modelling of physical systems. Therefore, it is powerful enough to model any sophisticated practical system relevant to photonic quantum information technology.

This functional formalism is built on various concepts found in a range of topics, ranging from mathematics, optics, quantum physics, and even particle physics. A successful presentation of Wigner functional theory thus necessitates a thorough introduction of all this background information. While mathematics is introduced as it is needed during the course of the discussions, the topics of classical optics and quantum optics are discussed in the following chapters. Among these topics, the concepts associated with quantum physics present a special challenge [26]. While we can provide a cold formal discussion of the basic formalism of quantum mechanics and quantum optics, such concepts may be difficult to absorb due to their abstract and often counter-intuitive nature. For this reason, we spend the rest of this introductory chapter to provide some discussion of the concepts in quantum physics.

1.2 Quantum physics

Humanity's confrontation with quantum phenomena started with the work of Max Planck around 1900, when he introduced the concept of quantized radiation to explain the blackbody radiation spectrum. A complete mathematical formulation with which this quantum behaviour could be modelled reached maturity only several years later, roughly at the time of the 5th Solvay conference in 1927. The resulting formulation of *quantum mechanics* succeeded in providing a formalism to model quantum physics within the context of the experimental observations at that time, allowing one to study microscopic systems in which such quantum phenomena are found.

The problem with these microscopic systems is that it is difficult to extract information from them through experiments. Today we understand that these restrictions in gaining information from such systems is a fundamental property of nature. The success of quantum mechanics is thus partly due to its capability to separate that which can be known (observed) from what cannot. The former became part of the subject of quantum physics in the context of the *scientific method*, and the latter was relegated to the domain of philosophy under the topic of *interpretations of quantum mechanics*.

For a long time after the development of quantum mechanics, it was considered inappropriate (even career-limiting) for a physicist to ponder the interpretations of quantum mechanics. However, that situation gradually changed through the activities of various people, notably the theoretical work of John S. Bell [27]. Ironically, these investigations into the interpretation of quantum mechanics led to an understanding of something that can be measured, namely *entanglement* [28–35]. This concept was demonstrated by the experimental work of John Clauser [36] and Alain Aspect [37].

The concept of *entanglement* is one of those ideas that challenges our intuition. It also seems to allow communication faster than the speed of light, in defiance of Albert Einstein's theory of special relativity. However, it was then shown that faster-than-light communication is not possible thanks to the *no-cloning theorem* [38], which shows that unknown quantum states cannot be cloned. In this way, our understanding of the quantum world increased.

The increased understanding represented by notions such as entanglement and the no-cloning theorem, led to the development of a new topic related to quantum physics called *quantum information science*. It spawned ideas of how to perform secure communication (*quantum communication*) [39], high performance computing (*quantum com-*

puting) [40], and more accurate measurements (*quantum metrology*) [41, 42], in what we now call *quantum information technologies* [16].

Despite this increase in our understanding of the quantum world, it is still widely regarded as “mysterious” and “weird.” Even Richard Feynman famously claimed that “nobody understands quantum mechanics” [26]. Part of the problem is that there are many different views of the underlying mechanisms in quantum physics that cannot be directly assessed through scientific observations.

As much as we need to separate science from that which is not science, it helps to have a bigger picture in mind when we think about the fundamental properties of our universe. Unfortunately, when it comes to that which cannot be distinguished through experimental observations, the bigger picture held by different people may vary significantly. It is not the intention of this book to provide any lengthy discussion of the *interpretations of quantum mechanics*. When it is necessary to provide some underlying interpretation for the sake of clarity, it is advisable to keep such a representation as simple as possible. In that sense, one can invoke *Occam’s Razor*, which selects the simplest possible explanation. Such a *minimalist interpretation* is used as the underlying interpretation for the discussions to follow.

1.2.1 Physics as a science

What do we mean when we refer to quantum mechanics? Is there a distinction between *quantum mechanics* and *quantum physics*? How is quantum mechanics related to *quantum field theory* and *quantum information science*? How does quantum physics differ from *classical physics*?

Before we can address these questions, we first need to step outside of physics and take a bird’s eye view of our activities to understand quantum phenomena in the physical world. We need to distinguish between that which is physics and that which is not physics. So what is physics?

Physics is the study of the physical world with the aid of the *scientific method*. It means that whenever we think we have some understanding of how the physical world works we need to test that understanding. The way to test the understanding is to use it to predict how the physical world will behave in a specific situation and then see if it really happens like that.

Our understanding of the physical world is often expressed in terms of some *mathematical formulation*. We can use this formulation to compute predicted outcomes of a proposed experiment. Then we perform the experiment and record the outcomes. If the predicted outcomes agree with the experimentally observed outcomes, we gain some confidence in our understanding. If an understanding has passed enough such tests, we can call it a *scientific theory*.

Mathematics is a very versatile tool in terms of which we can express our understanding of the physical world. In itself, mathematics is not a science, because we do not use comparisons between theory and experiments to determine whether something in mathematics is correct. Instead, mathematics is founded on the rules of logic that are used to construct logical proofs of *theorems*. The theorems are the logical consequences of sets of *axioms*—assumptions about the properties of mathematical concepts. When we apply mathematics in a physical scenario, we use the set of axioms that is valid for that physical scenario. The mathematical theorems that follow from that set of axioms are then valid for that physical scenario thanks to the rules of logic. These theorems then lead to methods that can be used to calculate predicted results for experiments in that physical scenario.

One needs to realize that the mathematical formulation in terms of which an understanding is formulated is always just a *model* of the physical world. There could be different successful mathematical formulations for one and the same scenario in the physical world. To quote Feynman: "every theoretical physicist who is any good knows six or seven different theoretical representations for exactly the same physics" [26]. Moreover, there is nothing to prevent a formalism, invented for a specific physical scenario, to be used in a different physical scenario where it is equally valid. Another situation that can occur is that some of the solutions obtained from the mathematical formulation for a given physical scenario are not valid in that scenario. Such solutions are called *spurious solutions*. In view of these considerations, it is clearly possible to be misled when properties or solutions of the mathematical formulation are confused for properties of the physical scenario. We need to distinguish between the *physics* and the *formalism* and never confuse the two.

1.2.2 Comparison with classical physics

When we contrast quantum physics with classical physics, we need to keep in mind that there are different theories that make up *classical physics*. The formalism of quantum mechanics evolved out of different aspects taken from these different classical theories. Perhaps one may think that quantum mechanics just corresponds to *classical mechanics* [43]. However, as shown in Figure 1.1, quantum mechanics also contains various concepts taken from *classical field theory*, as used in optics [44] and electromagnetism [45], as well as *classical statistical physics* [46], as used in thermodynamics.

The notion of a *classical particle* corresponds to our intuitive understanding of a particle as informed by our experience. Such a classical particle usually has a fixed size that is maintained for as long as it exists. The basic kinematics and dynamics of such particles are described by Newton's laws in classical mechanics [43]. A good example of a system governing the behaviour of such a particle is the *classical harmonic oscillator*. In such a system, the *state* of the particle is given by its position and velocity (or momentum). Position and momentum are called *canonical variables*. For a one-dimensional system, we can represent this state on a two-dimensional plane with position represented along one direction and the velocity (or momentum) orthogonal to

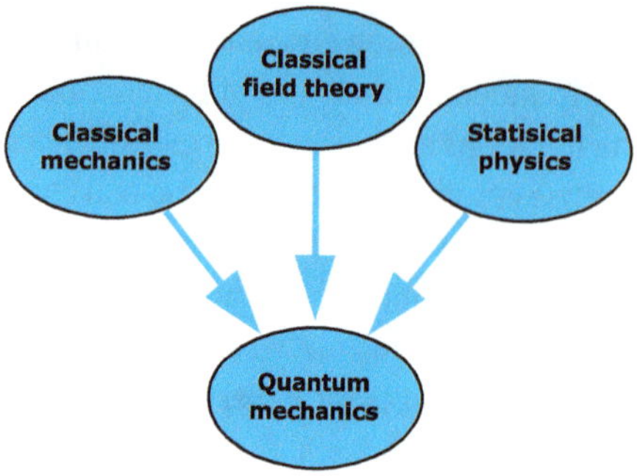

Figure 1.1: Classical theories that contributed to the formulation of quantum mechanics.

that. Such a plane is called the (classical) *phase space*. The evolution of the state of the particle is given by a continuous trajectory of points on phase space as a function of time.

A different context in classical physics that we can consider for comparison with quantum physics is that of *classical fields*. The phenomenon of *interference* and the associated concept of *superposition*, which play significant roles in quantum mechanics, already appear in some form in *classical field theory*. We encounter such classical fields in electromagnetism with their dynamics governed by *Maxwell's equations*. For the current context, the electromagnetic field includes light as a classical optical field obeying a *wave equation*, namely the *Helmholtz equation*, derived from Maxwell's equations under appropriate conditions.

Statistical physics contributed the concept of *probability* to the formulation of quantum theory. However, it had to be developed further. Due to the phenomenon of *interference*, probability needs a substructure. In classical field theory, where interference is often found, the intensity of a field is provided with a substructure in the form of a complex amplitude, which introduces the concept of a *phase* to allow the manifestation of interference. To merge this idea with probability, the notion of a *complex probability amplitude* is introduced. It serves as the required substructure for *classical probability*. As an implication, the *conservation of probability* requires that the "energy" of the complex probability amplitude must be normalized. The equivalent requirement in *classical field theory* is merely that the energy is finite.

1.2.3 Planck and de Broglie

Max Planck introduced a relationship between energy E and frequency ν (or angular frequency ω) to explain the *blackbody radiation spectrum*. This linear relationship

$$E = h\nu = \hbar\omega, \tag{1.1}$$

contains the *Planck constant* denoted by h (or the *reduced Planck constant,* which is denoted by $\hbar = h/2\pi$).[a] It indicates that the amount of energy exchanged during an *interaction* is proportional to the frequency.

Planck's relationship was later augmented by Louis de Broglie with a similar relationship between the *wavelength* λ (or *wavenumber* $k = 2\pi/\lambda$) and the momentum p. However, the de Broglie relationship relates two *vectorial* quantities (the *momentum vector* $\mathbf{p}$ to the *wavevector or propagation vector* $\mathbf{k}$) and is represented as

$$\mathbf{p} = \hbar\mathbf{k}, \tag{1.2}$$

with a proportionality constant given by the reduced Planck constant. It shows that the momentum exchanged during an *interaction* is quantized in the same way as energy.

The Planck and the de Broglie relationships have subsequently received widespread experimental support, establishing them as cornerstones of the fundamental quantum property of nature. The idea of quantized systems has led to the formulation of models for micro systems, ranging from atoms to various more complex systems.

Caution is necessary though. It may be tempting to associate any form of discreteness with "quantum properties" even when the same discreteness can be reproduced in classical scenarios. Striving to understand what quantum physics is, one also *needs to understand what it is not.* To avoid such traps, we regard quantum physics to be exclusively represented by the Planck relationship (1.1) and the de Broglie relationship (1.2). In their absence, what we are dealing with may just be artefacts of the formalism. Such exclusivity must be able to explain the numerous quantum phenomena that have been observed. In what follows, we explain the basis of our approach and how the large variety of quantum phenomena is a consequence of a simple principle.

The relationships in (1.1) and (1.2) relate quantities that are associated with classical particles (energy and momentum), as in classical mechanics, to quantities that are only valid for waves (frequency and wavenumber), which are associated with *classical field theory.* Inadvertently, these relationships imply that fundamental interactions always manifest as *interactions among waves or fields.* In the current discussion these terms are used interchangeably. Fields are discussed more thoroughly in Section 1.4.1.

Ironically, the quantized aspect of fundamental interactions creates the impression that the notion of particles (corresponding to our intuitive concept of a classical particle) is ideal to explain the quantization, in contrast to the implied wave nature of the relationships. This dichotomy has led to the idea of a *particle-wave duality.* However, our discussions show that the properties of interactions make the use of (classical) particles to explain the quantization unnecessary.

---

a The Planck constant is currently defined as an exact quantity with the dimensions of *action*, given in SI units by

$$h = 6.62607015 \times 10^{-34} \text{ Joule-seconds.}$$

1.2.4 Heisenberg uncertainty

The de Broglie relationship has another consequence. While position and momentum are *canonical variables* in the context of classical mechanics, the de Broglie relationship converts them into the *Fourier variables* of position and wavenumber. Therefore, quantum physics inherits the properties of Fourier analysis, which are added to the properties of canonical variables in the context of classical mechanics.

To understand the implication, we remind ourselves that the Fourier relationship between a function (of position) and its spectrum (a function of the wavenumber) carries the following property: the width of the spectrum is inversely proportional to the width of the smallest feature as a function of position. The fact that the de Broglie relationship converts the canonical variables into Fourier variables implies that quantum systems inherit this Fourier property. The state of a quantum particle cannot represent arbitrary fixed values of both position and wavenumber. Instead, such a quantum state can be represented in terms of a single function of both the position and the wavenumber. As a result of this Fourier property, the smallest support of such a function is an area having a lower bound with a numerical value on the order of 1. Hence, $\Delta x \Delta k \gtrsim 1$. Multiplying both sides by the *reduced Planck constant* to convert the wavenumber width into a momentum width, we find that the minimum area is proportional to the reduced Planck constant

$$\Delta x \Delta p \geq \frac{1}{2}\hbar. \tag{1.3}$$

Given a function of three-dimensional space and time, representing the *configuration space* distribution of a quantum particle, we find that the width of the function gives an *uncertainty* when the position of the particle is measured. When the same quantum particle's momentum is measured, the width of the associated spectrum on the *Fourier domain* serves as the corresponding uncertainty. The Fourier relationship between the configuration space function and its spectrum then leads to the relationship between these uncertainties given in (1.3). This relationship is called the *Heisenberg uncertainty principle* in honour of Werner Heisenberg who interpreted the widths as *uncertainties* of measurable quantities.

1.3 Quantized interactions

Thanks to Einstein's understanding of the photoelectric effect, the relationships in (1.1) and (1.2) were eventually interpreted to imply that all waves that take part in these interactions do so in terms of discrete entities. It is tempting to regard these discrete entities as particles. However, it would be misleading to think of them as *classical particles*. Instead we use a different concept for these quantized entities.

Each of these entities can be understood as a *quantized amount of an interacting field* (or just a *quantum*) that carries the full complement of all the degrees of freedom of the field (apart from the amplitude or intensity). A mental picture that we can use for this situation is depicted in Figure 1.2. You dip a hypothetical measuring cup into an incoming field to retrieve a fixed measured amount of the "substance" of that field and pour that measured amount into the interaction. After the interaction, you catch each output from the interaction as a single measure in that cup and pour it into its associated outgoing field. The amount that this measuring cup contains has the units of *action* and is given by one h (the *Planck constant*). This quantized amount of action can be produced by multiplying the energy in a quantum by one period of its oscillation (the inverse of its frequency) or by multiplying the momentum of a quantum by its wavelength. The incoming and outgoing fields can consist of multiples of these measured amounts or quanta, each of which can have a different spatiotemporal distribution and different *internal degrees of freedom* like spin or polarization.

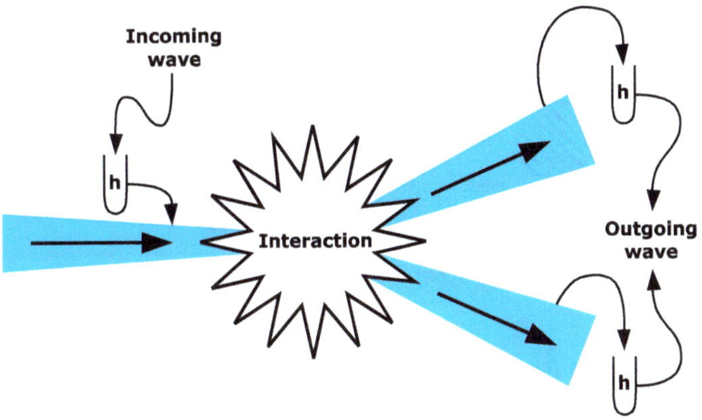

Figure 1.2: Quantized nature of interactions.

To make things even more interesting, the concept of *superposition* comes to play along. We already know from *classical field theory* that any field can be expanded as a superposition in terms of a linear combination of other fields. Such a classical superposition is depicted symbolically in the top part of Figure 1.3. The different fields in the linear combination can be a *basis*—a special set of fields in terms of which all possible fields can be expressed as linear combinations. Such a basis is not unique: the elements of one basis can be expressed as linear combinations of the elements of any other basis.

The situation becomes a bit more complicated in quantum physics. Usually, the process of an interaction is under-constrained. In other words, the initial conditions of the interaction do not produce a unique solution for the outgoing quanta. In general, there is a continuous variation of the possible solutions, each having a certain *probability amplitude* to occur. What then happens is that the interaction produces all the possible combinations of outgoing field quanta in superposition. Since interactions often produce more than one outgoing field, a superposition of all the different solutions involves all the different combinations of outgoing quanta that satisfy the constraints of the interaction. In other words, the different terms in the linear combination each consists of a combination of multiple quanta. Each combination is multiplied by its probability amplitude, serving as the coefficients in the linear combination. Such a quantum superposition of multiple quanta is shown in the bottom part of Figure 1.3.

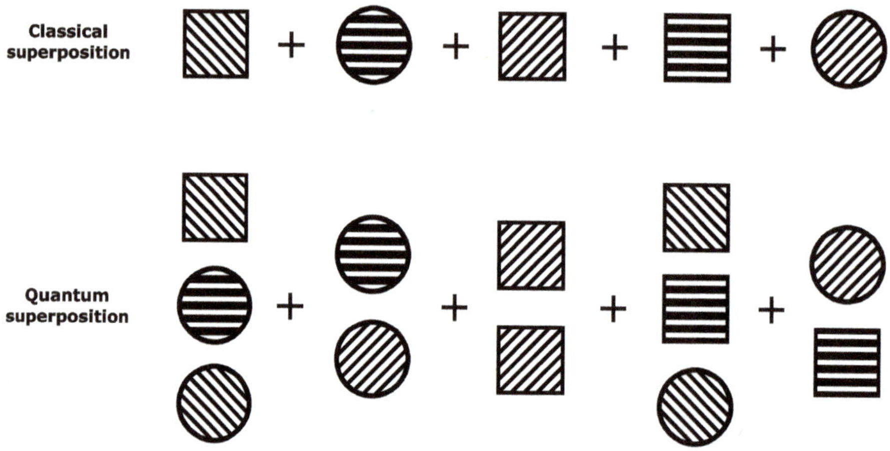

Figure 1.3: Comparison of classical and quantum superpositions.

Hence, the quantization effect of interactions implies that the elements of any basis in terms of which the outgoing state is expressed can consist of multiple measures or quanta, each having all the degrees of freedom. It then follows as an inevitable consequence of the quantization effect of interactions that the most general state that can exist in the physical world can consist of such superpositions of multiples of quanta of the field, as shown in Figure 1.3. This situation does not exist in *classical field theory*.

These superpositions among multiple fields often exhibit properties that are associated with quantum phenomena, such as *entanglement*. The quantum properties of states thus follow as logical consequences of the quantization effect of interactions.

1.3.1 Fundamental principle of quantum physics

Due to the fundamental role that the quantization effect of *interactions* plays in the understanding of quantum physics, we can propose it as a fundamental *principle of quantum physics*. The exact meaning of the quantized nature of fundamental interactions, as provided by the preceding discussion, is that the *action* is quantized in units of the *Planck constant*.

Being a *physics principle*, the statement of such a principle should not make any reference to any mathematical concepts, such as those introduced by the formalism in terms of which quantum physics is modelled. It is therefore not a *postulate* as one would find in mathematics.

Interactions are also *localized*. Often the interactions involve atoms or molecules, especially when measurements are involved. Such interactions are naturally localized due to the small sizes of these particles. So, when a detection is observed at a specific point on a screen, for example, it is because the wave being measured interacted with a single atom or molecule at that point in the screen. Even if the wave is spread out over the entire screen, it can be expanded in terms of the *measurement basis* suitable for the detection by the individual atoms or molecules in the screen, leading to a superposition of all the possible localized detections, each multiplied by the probability amplitude for that detection. The final measurement may eventually only be represented by one of the terms in the superposition.[b]

When no mediating particles (such as the atoms or molecules in a measurement system) are involved, the different fields that take part in an interaction always interact at a point in spacetime. Such a point is called an *event*. Quantum field theory is based on such pointwise interactions, which are then integrated over all spacetime.[c]

The concept of localization is combined with the quantized nature into a simple yet comprehensive statement of a fundamental principle of quantum physics. It reads

fundamental interactions are quantized and localized.

This principle forms the complete description of the physical characteristic of nature that leads to quantum phenomena in the context of bosonic fields such as light. When fermions are also considered, other principles, such as the Pauli exclusion principle, need to be added to formulate their unique characteristics.

1.3.2 Quantum particles

Clearly, the notion of a quantum or a quantized measure of a field contributing to an interaction does not fit well with the notion of a classical particle. Therefore, to avoid any confusion, we need to emphasize that, whenever the term "particle" is used, it does

b This picture of the measurement process leads to the *measurement problem* in the context of the interpretations of quantum mechanics, which is beyond the scope of our interest.

c The pointwise nature of these interactions can lead to divergences in the calculation of scattering amplitudes that require a process of *renormalization* to obtain finite results for predictions.

not refer to the same thing as a particle in the traditional classical sense, having a fixed size and moving on some classical trajectory. Instead, the term particle henceforth refers to a *quantum particle*, which is a quantum of the field, carrying the full set of degrees of freedom of that field. A quantum state may consist of multiple quantum particles (field quanta). Each such quantum behaves like a wave and obeys a *wave equation* as its *equation of motion*. The quantum state as a whole obeys a *quantum evolution equation*, such as the *Schrödinger equation*. When we need to talk about a particle in the traditional sense, we specifically call it a *classical particle*.

This notion of a quantum particle as a field quantum is specifically valid for photons. They are regarded as such quantum particles without any notions traditionally associated with classical particles. As the material discussed in this book is focussed on physical scenarios involving photons, we do not need to concern ourselves with any other (fundamental or composite) quantum particles. However, it is worth mentioning that this clear distinction between quantum and classical particles can become blurred when considering particles in other quantum scenarios. *Composite particles* that are built up from fundamental particles, such as protons, neutrons, nuclei, atoms, etc., have finite sizes associated with them. Such a finite size is a concept that we intuitively associate with classical particles. The sizes of nucleons (protons and neutrons) are introduced by a fundamental process called *confinement* [47]. It is the result of a phase transition caused by the highly nonlinear behaviour of the force among quarks, as described by *quantum chromodynamics* (QCD). This phase transition introduces a fundamental scale, called the *the QCD scale*, which then determines the sizes of these nucleons, found in the nuclei of atoms. The sizes of atoms are determined by the charged potentials of these nuclei. Due to the finite scales associated with these particle sizes, such composite particles can under certain conditions behave according to our notions of classical particles. However, they can still form superpositions in the sense of quantum particles. Therefore, their complex probability amplitudes can still behave like waves.

1.4 Quantum states

Based on the preceding discussion, it follows that the properties of quantum states are determined by the properties of quantum interactions. Now we can discuss the notion of a *quantum state* more carefully. At the same time, we address some of the formal representations of quantum states and how they are related.

Previously, we mentioned the concept of the *state* of a classical particle. Ignoring any internal properties, it can be represented as a point on a classical phase space in terms of its position and velocity. As such, we can distinguish between the physical existence of the classical particle and the state in which it finds itself. Such a classical particle has a unique identity that distinguishes it from any other classical particle. Different classical particles generally cannot have the same state because they cannot occupy the same position in space at the same time.

A *quantum state*, on the other hand, is a bit more complicated. It is difficult to think of such a state as being distinct from the physical existence of the quantum particles, because a quantum state also incorporates the particle-number degrees of freedom into its description. Even when a quantum state consists of multiple quanta, we cannot associate a unique identity to each of these quanta. Bosonic particles can have the same properties, making them *indistinguishable*. It means that they can occupy the same position at the same time, as found in a *Bose–Einstein condensate*, for example. Fermions, on the other hand, obey the *Pauli exclusion principle*, which means that two fermions can never have exactly the same properties.

There are many formal ways to represent quantum states. The *Schrödinger equation* (formulated by Erwin Schrödinger), describes the evolution of a *wavefunction*, representing a quantum state. For a single quantum, a wavefunction is a complex-valued function. It can be a (configuration space) function of the spatial coordinates and time $\psi(\mathbf{x}, t)$. However, we can perform a Fourier transform to compute its spectrum, representing the *Fourier domain wavefunction* as a function of the Fourier domain coordinates. A wavefunction is interpreted as a *probability amplitude*. The squared modulus of a wavefunction for a single quantum represents the probability distribution for the detection of that quantum. The *conservation of probability* requires that the wavefunction is normalized so that its squared modulus integrates to 1 ($\|\psi\|^2 = 1$). The wavefunctions for multiple quanta can be functions of multiple sets of coordinates. The situation becomes more complicated for the general case.

Paul Dirac invented the *Dirac bra-ket notation* [48], representing the quantum state as a *state vector* $|\psi\rangle$, leaving the choice of coordinates unspecified. To recover the wavefunction for a single quantum in a given coordinate system, one computes the *inner products* of the state vector with the chosen coordinate basis vectors: $\psi(\mathbf{x}, t) = \langle\mathbf{x}|\psi(t)\rangle$. It emphasizes the use of inner products, implying that these state vectors are elements of a *Hilbert space* (a complete inner product space) [49]. The state vector is normalized $\langle\psi|\psi\rangle = 1$ to ensure the conservation of probability. Such a state vector can be represented as a *density operator*, which allows one to generalize the concept to include both pure states and mixed states.

In phase space, states (both pure and mixed) can be represented by different kinds of functions, such as the *Wigner function* introduced by Eugene Wigner. Any density operator can be associated (or converted into) a Wigner function. In general, these phase space representations can have negative values. So, they are not true probability distributions. Other kinds of phase space functions include the *Glauber–Sudarshan P-function* and the *Husimi Q-function*. All of them are functions of the phase space variables. Operators can also be represented by functions on phase space. Products of operators are represented by *Moyal star products* for their functions. In fact, all of quantum mechanics can be done as calculations on phase space in terms of the *Moyal formalism* without the use of any operators.

Regardless of the formalism that is chosen to represent the quantum state, they all represent the same physical thing. The physical properties of quantum states can be very different. If the quantum state represents a single quantum, the formal representation of the state contains *one full set of degrees of freedom* to describe the state. For example, a single photon has a spatial mode and a frequency spectrum. These properties can be generalized into a combined spatiotemporal spectral function. It also has *spin* or *internal degrees of freedom* in terms of which its state of polarization is defined. In general, the polarization can be mixed with the spatiotemporal function. The combination of these spatiotemporal and spin degrees of freedom defines the *full set* of degrees of freedom of a single-photon state.

The superposition of different single-photon states is still a single photon. The evolution of such a single quantum of an optical field in the absence of interactions obeys the same equations of motion as the classical field. For example, an optical field can be considered either as a classical field obeying the Helmholtz equation, or it can be considered as the wavefunction (probability amplitude) for a single photon, also obeying this wave equation.

This correspondence, which is dictated by the fact that one and the same nature incorporates both the classical and the quantum viewpoints, helps to identify one of the true differences between classical physics and quantum physics. A classical field and the state of a single quantum particle have the same number of degrees of freedom: one full set. However, this correspondence is no longer true when the quantum state consists of multiple quanta.

When a quantum state consists of more than one particle or quantum, then the situation becomes more complicated. Unless the different quanta are described by exactly the same spatiotemporal and spin properties, each one of them needs to be described in terms of its own full set of degrees of freedom. Such a quantum state cannot be compared to a classical field any more. The description of each quantum in terms of its degrees of freedom represents a solution of the equations of motion. In the most general case, the state thus needs as many full sets of degrees of freedom as the number of quanta it contains. To make matters worse, the state may also consist of a superposition of terms each of which consists of multiple quanta. These terms may have different numbers of quanta, so that the number of quanta in the state is not fixed. In that case, the most general description of the spatiotemporal and spin properties of the state is extremely complicated.

If this discussion created the impression that each quantum in a state must be regarded as an isolated entity with a unique identity, then that would be an incorrect impression. The description of a state in terms of a specific superposition of terms with multiple quanta is not unique. One can perform a transformation on the degrees of freedom of the quanta in the different terms that would recombine them in different ways, giving a different representation of the same state.

Fortunately, we seldom encounter a state that is so complicated. Often, all the quanta in the state carry the exact same description and can be defined in terms of just one function that parametrizes the spatiotemporal degrees of freedom of the entire state, even though it consists of multiple quanta. Some states can be represented by a finite number of terms in a superposition with the same number of particles in each term. It then needs only a few functions in its definition even if the number of quantum in the state is very large. Nevertheless, it is useful to have a formalism that is powerful enough to handle any possible situation.

The role of interactions in quantum physics now also becomes clearer. Such interactions can change the number of quanta and, therefore, also the number of degrees of freedom. When a state evolves in the presence of interactions, the number of quanta does not in general remain constant, neither does the number of functions in terms of which it needs to be described. Such a complicated process cannot be described by classical theories.

1.4.1 Fields

The term *field* is often encountered in the context of quantum theory, especially in the case of quantum *field* theory, as the name indicates. However, the physical concept associated with this term can represent different things.

In *classical field theories*, such as classical optics or electromagnetism, a field is a function of spacetime. It can represent the electric field as a vector field indicating the magnitude and direction of the force that a charged particle at a specific spacetime point experiences. Such a function is a solution of the equations of motion. One such function suffices to describe the state of the classical field for a specific case.

In quantum physics, the situation is more complicated due to the fact that quantum states can consist of multiple quanta. Clearly, the concept of a field as it exists in classical scenarios cannot be directly applied to quantum scenarios. We need to decide what aspect of the classical notion of a field we want to transfer to the quantum case. To help us with this decision, we can take a look at the different ways that this term is used in existing formulations of quantum theory. Many of these usages are related to the formalism in terms of which quantum physics is modelled.

In quantum field theory, according to the approach of *second quantization*, theories are formulated in terms of *field operators*. In these cases, the second quantization approach regards a field as an operator-valued function of spacetime. These field operators are then used to model the dynamics by producing or destroying so-called *field excitations* associated with point-like locations in spacetime.[d] A related notion derived from this formalism is the idea that a field is a resource permeating all space and time from which particles can be produced when created or into which they are absorbed when annihilated. The term *excitation* suggests that particles or quanta are like ripples

d The reason we state it as being "associated with" a point-like location is because it would be incorrect to assume that such an excitation appears at an isolated point in spacetime. The equation of motion quickly spreads this excitation over all space.

on this resource field. Note that these concepts are specifically associated with an operator approach.

An alternative view emerges when we consider the formulation of quantum field theory in terms of path-integrals. In this case, the field operators are replaced by functions or "paths," and the path-integral becomes a functional integral over all such functions. These paths are functions of spacetime, akin to the fields found in classical field theories. However, it is not necessarily assumed that they are solutions of the equations of motion, because such equations of motion can be highly nonlinear in an interacting theory. Instead, these functions are all the possible functions that can be constructed as arbitrary superpositions of plane waves. The path-integral thus produces the *complex probability amplitude* for specific configurations of interactions from the superpositions of all such functions in the presence of such configurations.

The formalism that we develop here does not use operators in its final manifestation. Therefore, we do not end up with field operators. Our approach is more closely related to the path-integral approach, because we also represent the formalism in terms of functional integrals. In our case though, we are not investigating fundamental dynamics with these functional integrals. Instead, we are interested in the formulation of states on which certain unitary processes are performed and which are then subjected to measurements. So, the notion of fields are specifically related to how these states are represented. In this sense, we can employ the classical scenario's notion of a field as a solution of an equation of motion. To be more precise, we refer to the solutions of the *equations of motion* as *fields*. When interactions are introduced, we assume that each such interaction maps the space of all such solutions back onto itself.

In the case where a state consists of a single quantum, it is perfectly parameterized by a single field. However, a general state is not just a single field. An arbitrary quantum state is a *functional* of fields. It corresponds to the explanation provided by Steven Weinberg,[e] stating: "In a relativistic theory, the wavefunction is a functional of these fields, not a function of particle coordinates." In the formalism that we develop here, an arbitrary state is defined on a functional phase space. Each point in the functional phase space is represented by a field. The independent variables in terms of which these states are defined are therefore *field variables*.

Bibliography

[1] R. Loudon. *Quantum Theory of Light*. Oxford University Press, Oxford, 1973.
[2] L. Mandel and E. Wolf. *Introductory Quantum Optics*. Cambridge University Press, New York, 1995.
[3] D. F. Walls and G. J. Milburn. *Quantum Optics*. Springer-Verlag, Berlyn, 1995.

e Talk presented at the conference "Historical and Philosophical Reflections on the Foundations of Quantum Field Theory," at Boston University, March 1996.

[4] C. C. Gerry and P. L. Knight. *Optical Coherence and Quantum Optics*. Cambridge University Press, New York, 2005.
[5] J. J. Sakurai. *Modern Quantum Mechanics*. Addison-Wesley Publishing Company, Reading, Massachusetts, USA, 1994.
[6] D. J. Griffiths. *Introduction to Quantum Mechanics*. Prentice Hall, Englewood Cliffs, 1995.
[7] R. Shankar. *Principles of Quantum Mechanics*. Plenum, New York, USA, 1980.
[8] C. Itzykson and J.-B. Zuber. *Quantum Field Theory*. McGraw-Hill, New York, USA, 1985.
[9] F. Mandl and G. Shaw. *Quantum Field Theory*. John Wiley & Sons, New York, USA, 1984.
[10] M. E. Peskin and D. V. Schroeder. *An Introduction to Quantum Field Theory*. Addison-Wesley Publishing Company, Reading, Massachusetts, USA, 1995.
[11] S. Weinberg. *The Quantum Theory of Fields, Volume I*. Cambridge University Press, New York, 1995.
[12] S. Weinberg. *The Quantum Theory of Fields, Volume II*. Cambridge University Press, New York, 1996.
[13] T. L. Curtright and C. K. Zachos. Quantum mechanics in phase space. *Asia Pac. Phys. Newsl.*, 1:37–46, 2012.
[14] J. P. Dowling and G. J. Milburn. Quantum technology: the second quantum revolution. *Philos. Trans. R. Soc. Lond. A*, 361:1655–1674, 2003.
[15] R. U. Ayres. Barriers and breakthroughs: an -expanding frontiers- model of the technology-industry life cycle. *Technovation*, 7:87–115, 1988.
[16] M. A. Nielsen and I. L. Chuang. *Quantum Computation and Quantum Information*. Cambridge University Press, Cambridge, England, 2000.
[17] C. Weedbrook, S. Pirandola, R. García-Patrón, N. J. Cerf, T. C. Ralph, J. H. Shapiro, and S. Lloyd. Gaussian quantum information. *Rev. Mod. Phys.*, 84:621–669, 2012.
[18] H. J. Groenewold. On the principles of elementary quantum mechanics. *Physica*, 12:405–460, 1946.
[19] J. E. Moyal. Quantum mechanics as a statistical theory. *Math. Proc. Camb. Philos. Soc.*, 45:99–124, 1949.
[20] S. L. Braunstein and P. Van Loock. Quantum information with continuous variables. *Rev. Mod. Phys.*, 77:513–577, 2005.
[21] G. Adesso, S. Ragy, and A. R. Lee. Continuous variable quantum information: Gaussian states and beyond. *Open Syst. Inf. Dyn.*, 21:1440001, 2014.
[22] C. K. Hong and L. Mandel. Theory of parametric frequency down conversion of light. *Phys. Rev. A*, 31:2409, 1985.
[23] S. Mrowczynski and B. Mueller. Wigner functional approach to quantum field dynamics. *Phys. Rev. D*, 50:7542–7552, 1994.
[24] F. S. Roux. Combining spatiotemporal and particle-number degrees of freedom. *Phys. Rev. A*, 98:043841, 2018.
[25] F. S. Roux. Erratum: Combining spatiotemporal and particle-number degrees of freedom [Phys. Rev. A 98, 043841 (2018)]. *Phys. Rev. A*, 101:019903(E), 2018.
[26] R. P. Feynman. *The Character of Physical Law*. Penguin Books, London, England, 1965.
[27] J. S. Bell. *Speakable and Unspeakable in Quantum Mechanics*. Cambridge University Press, Cambridge, USA, 1993.
[28] S. Hill and W. K. Wootters. Entanglement of a pair of quantum bits. *Phys. Rev. Lett.*, 78:5022–5025, 1997.
[29] W. K. Wootters. Entanglement of formation of an arbitrary state of two qubits. *Phys. Rev. Lett.*, 80:2245–2248, 1998.
[30] A. V. Thapliyal. Multipartite pure-state entanglement. *Phys. Rev. A*, 59:3336–3342, 1999.
[31] A. Mair, A. Vaziri, G. Weihs, and A. Zeilinger. Entanglement of the orbital angular momentum states of photons. *Nature*, 412:313–316, 2001.
[32] W. P. Bowen, N. Treps, R. Schnabel, and P. K. Lam. Experimental demonstration of continuous variable polarization entanglement. *Phys. Rev. Lett.*, 89:253601, 2002.
[33] N. Akopian. Entangled photon pairs from semiconductor quantum dots. *Phys. Rev. Lett.*, 96:130501, 2006.

[34] R. Horodecki, P. Horodecki, M. Horodecki, and K. Horodecki. Quantum entanglement. *Rev. Mod. Phys.*, 81:865–942, 2009.

[35] P. J. Shadbolt, M. R. Verde, A. Peruzzo, A. Politi, A. Laing, M. Lobino, J. C. F. Matthews, M. G. Thompson, and J. L. O'Brien. Generating, manipulating and measuring entanglement and mixture with a reconfigurable photonic circuit. *Nat. Photonics*, 6:45–59, 2012.

[36] S. J. Freedman and J. F. Clauser. Experimental test of local hidden-variable theories. *Phys. Rev. Lett.*, 28:938–941, 1972.

[37] A. Aspect, P. Grangier, and G. Roger. Experimental realization of Einstein–Podolsky–Rosen–Bohm Gedankenexperiment. *Phys. Rev. Lett.*, 49:91–94, 1982.

[38] W. Wootters and W. Zurek. A single quantum cannot be cloned. *Nature*, 299:802–803, 1982.

[39] N. Gisin and R. Thew. Quantum communication. *Nat. Photonics*, 1:165–171, 2007.

[40] A. Steane. Quantum computing. *Rep. Prog. Phys.*, 61:117–173, 1997.

[41] V. Giovannetti, S. Lloyd, and L. Maccone. Advances in quantum metrology. *Nat. Photonics*, 5:222–229, 2011.

[42] E. Polino, M. Valeri, N. Spagnolo, and F. Sciarrino. Photonic quantum metrology. *AVS Quantum Sci.*, 2:024703, 2020.

[43] H. Goldstein. *Classical Mechanics, 2nd ed.* Addison-Wesley Publishing Company, Reading, Massachusetts, USA, 1980.

[44] B. E. A. Saleh and M. C. Teich. *Fundamentals of Photonics*. John Wiley & Sons, New York, USA, 1991.

[45] J. D. Jackson. *Classical Electrodynamics, 2nd ed.* John Wiley & Sons, New York, USA, 1975.

[46] F. Mandl. *Statistical Physics, 2nd ed.* John Wiley & Sons, New York, USA, 1988.

[47] M. Shifman. Understanding confinement in QCD: elements of a big picture. *Int. J. Mod. Phys. A*, 25:4015–4031, 2010.

[48] P. A. M. Dirac. A new notation for quantum mechanics. *Math. Proc. Camb. Philos. Soc.*, 35:416–418, 1939.

[49] E. Kreyszig. *Introductory Functional Analysis with Applications*. John Wiley & Sons, New York, USA, 1978.

Part I: **Background material**

2 Classical optics

A thorough understanding of quantum optics is inevitably built on an adequate understanding of classical optics, which in turn requires an understanding of classical electromagnetic theory. Here, we review the pertinent aspects of *classical field theory*, in the context of classical electromagnetic theory and the classical optics that follows from it, starting with a discussion of linear systems. This chapter thus provides a self-contained introduction to most of the background information not directly related to quantum theory. It includes various topics of classical optics that is used in subsequent chapters. Apart from the derivations of the Helmholtz equation and the paraxial wave equation starting from Maxwell's equations, and discussions of the solutions of these equations (plane waves and Gaussian beams, respectively), it also includes discussions of *free space propagation*, *Fourier optical systems* [1, 2], *polarization* [3], and *optical angular momentum*. The latter is provided for scenarios where continuous variable analyses incorporate more sophisticated applications of *structured light* [4]. Some of these topics cover more material than is strictly necessary for the material discussed in subsequent chapters (such as optical angular momentum), and can therefore be skipped.

For the purpose of the subsequent discussions, various mathematical concepts are introduced from a number of different topics, such as *functional analysis* [5] and *Lie group theory* [6], albeit without too much formal mathematics. We start with a discussion of linear systems, which is more general than optical systems. It provides a deeper context for the general approach with which the physical systems are modelled in both classical optics and quantum optics.

2.1 Linear systems

Much of our understanding of the physical world follows from being able to model some processes in terms of *linear systems*; it is much easier to solve linear equations than nonlinear equations. A linear system can be described as a system where the output obtained from a linear combination of inputs is given by the same linear combination of the individual outputs. It means that

$$\mathcal{L}\{af_1 + bf_2\} = a\mathcal{L}\{f_1\} + b\mathcal{L}\{f_2\}, \tag{2.1}$$

where $\mathcal{L}\{\cdot\}$ represents the linear process, a and b are constants and f_1 and f_2 are the inputs. The models for such linear systems are often represented in terms of linear algebra [7]. Due to the continuous nature of the systems that we discuss here, we use integral representations instead.

https://doi.org/10.1515/9783111445342-003

A general (one-dimensional) linear process (or linear *transformation*) can be represented as a *superposition integral*[a]

$$g(x) = \int K(x,x')f(x')\,\mathrm{d}x' \triangleq \mathcal{L}\{f\}, \tag{2.2}$$

where $f(x)$ is the *input function* and $g(x)$ is the *output function* or *transformed function*. The linear process is described by the *bilinear kernel function* $K(x,x')$,[b] which is also called a *point-spread function*. The integral is a definite integral with the integration extending over the entire domain of the functions, that is, from $-\infty$ to ∞. For such integrals, we do not show the integration boundaries.

Sometimes, we need to use the complex conjugate of the output function. In such a case, the kernel is both transposed and complex conjugated. The combination of these two changes represents the *Hermitian adjoint*, denoted as $K(x,x') \rightarrow K^{\dagger}(x',x)$. The transformation producing the complex conjugate of the output function is thus expressed as

$$g^*(x) = \int f^*(x')K^{\dagger}(x',x)\,\mathrm{d}x' \triangleq \mathcal{L}^{\dagger}\{f^*\}. \tag{2.3}$$

2.1.1 Orthogonality

It is often helpful to consider the *eigenfunctions* of such a linear process. Such eigenfunctions are represented in terms of an eigenvalue equation of the form

$$\mathcal{L}\{\phi_n\} = \lambda_n \phi_n, \tag{2.4}$$

where $\phi_n(x)$ and λ_n are the *eigenfunctions* and their associated *eigenvalues*, respectively, and n is an index that labels the different eigenfunctions and eigenvalues. In some cases, the eigenfunctions can also be labelled by a continuous parameter. For the moment, we assume that it is a discrete index.

An *orthogonal* set of eigenfunctions satisfy an *orthogonality condition*, expressed in terms of an *inner product*. The concept of *orthogonality* needs the concept of an *inner product*, which then implies that the functions are elements of an *inner product space* or *Hilbert space*.[c] We define the inner product as

$$\langle \phi_m, \phi_n \rangle \triangleq \int \phi_m^*(x)\phi_n(x)\,\mathrm{d}x = \delta_{m,n}, \tag{2.5}$$

---

a In mathematics, it is also called a *Hilbert–Schmidt integral operator*.

b The term "bilinear" means that the kernel is linear with respect to contractions on either side of it.

c A Hilbert space is a *complete* inner product space, in the sense that all convergent sequences converge to functions that are elements of the space.

where x is a generic (spatial) variable ($x \in \mathbb{R}$), and $\delta_{m,n}$ is the *Kronecker delta function*

$$\delta_{m,n} = \begin{cases} 1 & m = n, \\ 0 & m \neq n. \end{cases} \tag{2.6}$$

Note that the variable with respect to which the integration is performed to evaluate the inner product is not shown inside the notation $\langle \cdot, \cdot \rangle$.

For (2.5), it is assumed that the eigenfunctions ϕ_n are *normalized* in the sense that

$$\|\phi_n\|^2 \triangleq \int |\phi_n(x)|^2 \, \mathrm{d}x \equiv \langle \phi_n, \phi_n \rangle = 1. \tag{2.7}$$

Here, $\|\phi_n\|$ represents the *Euclidean norm* (or the L^2-norm), which is equivalent to the square root of the inner product of the eigenfunction with itself. Note that the variable with respect to which the integration is performed to evaluate the norm is not shown inside the notation $\| \cdot \|$. A set of orthogonal normalized functions is called *orthonormal*.

The inner product in (2.5) contains the complex conjugate of one of the eigenfunctions. The complex conjugated function serves as the *dual* in the inner product.

The function on which the inner product is applied, can be a transformed version of another function, as shown in (2.2). In such a case, we have

$$\langle \phi_m, g \rangle = \int \phi_m^*(x) g(x) \, \mathrm{d}x = \int \phi_m^*(x) K(x, x') f(x') \, \mathrm{d}x' \, \mathrm{d}x. \tag{2.8}$$

When there are multiple integrals all extending over the entire domain, they share the same integral sign. We can also have the situation where the dual is the transformed version of another function. In such a case, the kernel is replaced by its Hermitian adjoint. The inner product is then given by

$$\langle g, \phi_n \rangle = \int g^*(x) \phi_n(x) \, \mathrm{d}x = \int f^*(x') K^\dagger(x', x) \phi_n(x) \, \mathrm{d}x' \, \mathrm{d}x. \tag{2.9}$$

Note that $\langle \phi_m, g \rangle = \langle g, \phi_m \rangle^*$.

2.1.1.1 Bases

The set of functions that can be produced by linear combinations of a given set of functions $\mathcal{B}$ is called the *span* of $\mathcal{B}$; or, one can say that $\mathcal{B}$ *spans* the former set of functions. If $\mathcal{B}$ spans the entire space of possible input functions, it is said to be *complete*. In that case, the set of functions can be used to *resolve the identity*:

$$\sum_n \phi_n(x) \phi_n^*(x') = \delta(x - x'), \tag{2.10}$$

where $\delta(\cdot)$ denotes the *Dirac delta function*, in terms of which the identity operation on the space of functions is expressed.

A *complete orthonormal* set of functions can serve as a *basis* in terms of which the input function can be represented by a linear combination

$$f(x) = \sum_n C_n \phi_n(x). \tag{2.11}$$

The *coefficients* C_n are complex numbers obtained by computing the inner products between the input function and the basis functions

$$C_n = \langle \phi_n, f \rangle = \int \phi_n^*(x) f(x)\, dx. \tag{2.12}$$

The benefit of such an expansion is that it makes the calculation of the output from a linear process much easier. Consider the case where the set of functions (or basis) is the set of all the eigenfunctions of a linear process. When we substitute the input function, expanded in terms of the eigenfunctions, into the linear process, we get

$$g = \mathcal{L}\{f\} = \sum_n C_n \mathcal{L}\{\phi_n\} = \sum_n C_n \lambda_n \phi_n, \tag{2.13}$$

based on (2.4). So, the output function is given by a similar expansion in terms of the same basis functions in which the coefficients are modified by being multiplied by the eigenvalues: $C_n \to C_n' = C_n \lambda_n$.

Note that the orthonormal basis functions are discrete functions, that is, they are *countable*. A Hilbert space with a countable basis is a *separable Hilbert space*.

2.1.1.2 Dirac delta function

The Dirac delta function is a distribution with the property that

$$\int \delta(x' - x) f(x')\, dx' = f(x), \tag{2.14}$$

for any function $f(x)$. It implies that

$$\int \delta(x)\, dx = 1. \tag{2.15}$$

Unlike many other functions (such as the exponential function or the trigonometric functions), the argument of a Dirac delta function is not in general dimensionless. Changing the variable $x \to ay$, where a is a real constant, we get

$$\delta(ay) = \frac{1}{|a|} \delta(y). \tag{2.16}$$

It thus follows that the Dirac delta function carries the inverse of the dimensions (units) of its argument.

Considered as a function, $\delta(x) = 0$ for all $x \neq 0$, and at zero $\delta(0)$ is undefined (infinite). The latter makes the Dirac delta function awkward to use as a function; it can produce divergent results in calculations. To avoid such issues, we can use a *limit process* where the Dirac delta function is replaced by a well-defined normalized function that approaches the Dirac delta function in a suitable limit. A good example of such a function is the normalized Gaussian function. It follows that

$$\delta(x) = \lim_{\epsilon \to 0} \frac{1}{\epsilon\sqrt{\pi}} \exp\left(-\frac{x^2}{\epsilon^2}\right). \tag{2.17}$$

2.1.1.3 Exponential functions

An often encountered set of orthogonal functions are expressed as *exponential functions*. For one dimension, such an exponential function is given by

$$\phi(x; k) = \exp(\mathrm{i}xk). \tag{2.18}$$

Here, k is a real-valued *parameter* that labels the different basis functions, taking over the role of the index n. The orthogonality of these exponential functions is expressed as a *continuous orthogonality condition*

$$\langle \phi(k), \phi(k') \rangle \triangleq \int \exp(-\mathrm{i}xk) \exp(\mathrm{i}xk')\,\mathrm{d}x = 2\pi\delta(k - k'). \tag{2.19}$$

where it now produces the Dirac delta function, instead of the Kronecker delta function. Equivalently, we can consider the integration over the parameter k, leading to

$$\int \exp(-\mathrm{i}xk) \exp(\mathrm{i}x'k)\,\mathrm{d}k = 2\pi\delta(x - x'). \tag{2.20}$$

It indicates the *completeness* of the set of orthogonal functions.

Due to the properties of the Dirac delta function, the exponential basis functions are not *square integrable*, because

$$\|\phi(k)\|^2 = \langle \phi(k), \phi(k) \rangle = 2\pi\delta(0), \tag{2.21}$$

which is a *divergent constant*. In other words, they are *not normalizable*. Moreover, because they are parameterized by a continuous variable they are *not countable*. So, although the inner product defined in (2.5) can be used for the set of square integrable functions to define a separable Hilbert space, the set of exponential functions do not belong to such a Hilbert space.

Nevertheless, the exponential functions represent a *continuous complete orthogonal basis* (though not orthonormal). The continuous nature of the parameter that distinguishes the different elements of this basis, makes them quite useful in analyses.

2.1.2 Fourier analysis

The exponential functions and the concept of orthogonality lead to *Fourier analysis*, which plays a significant role in the mathematical formulation of quantum theory. The *Fourier transform* and the *inverse Fourier transform* are generically defined as

$$\begin{aligned} F(k) = \mathcal{F}\{f\} &\triangleq \int \exp(-\mathrm{i}xk) f(x)\,\mathrm{d}x \equiv \langle \phi(k), f \rangle\,, \\ f(x) = \mathcal{F}^{-1}\{F\} &\triangleq \int F(k)\exp(\mathrm{i}xk)\,\frac{\mathrm{d}k}{2\pi}\,, \end{aligned} \tag{2.22}$$

respectively, where $f(x)$ is the function in *configuration space* and $F(k)$ is the *spectrum* or the *coefficient function* in the *Fourier domain*. The variable with respect to which the integration is evaluated is not shown in the arguments of $\mathcal{F}\{\cdot\}$ and $\mathcal{F}^{-1}\{\cdot\}$. It is assumed that x is a spatial coordinate, carrying the units of a distance, while the units of the Fourier domain variable k is an inverse distance.

The Fourier transform is the inner product between the function and an exponential basis function, the latter serving as the *Fourier kernel function*. It produces the spectrum with which the function can be expanded in terms of the exponential functions, which then represents the inverse Fourier transform.

The signs in the arguments of the Fourier kernel functions in (2.22) follow the *physics sign convention* when x represents a spatial coordinate. (The opposite convention is used by the electrical engineering community.) The Fourier transform with respect to a temporal coordinate has the opposite sign.

For a well-defined Fourier transform, a function and its spectrum must be *square integrable* functions. Stated in terms of *Parseval's theorem* or *Plancherel's theorem* [8],

$$\|f(x)\|^2 \triangleq \int |f(x)|^2\,\mathrm{d}\xi = \|F(k)\|^2 \triangleq \int |F(k)|^2\,\frac{\mathrm{d}k}{2\pi} < \infty. \tag{2.23}$$

Since the exponential basis functions do not satisfy this condition, they do not belong to the set of square integrable functions. It leads to the awkward situation where the basis is not included in the set of functions that are expanded in terms of the basis, as mentioned above.[d] Nevertheless, it does not prevent Fourier analysis from being widely used and forming the foundation for the formulation of quantum mechanics.

2.1.2.1 Kernels in the Fourier domain

The superposition integral in (2.2) can be converted to the Fourier domain, thanks to the linearity of the expression. The Fourier domain superposition integral reads

d There are formal ways to treat this issue, but that is beyond the scope of our discussion.

$$G(k) = \int \widetilde{K}(k,k')F(k')\,\frac{dk'}{2\pi}, \tag{2.24}$$

where

$$\begin{aligned} G(k) &= \mathcal{F}\{g\} = \int \exp(-ixk)g(x)\,dx, \\ F(k) &= \mathcal{F}\{f\} = \int \exp(-ixk)f(x)\,dx, \end{aligned} \tag{2.25}$$

are the *spectra* of the output and input functions, respectively, and $\widetilde{K}(k,k')$ is a *Fourier domain kernel function.*

Exercise 2.1. Show that the Fourier domain kernel function is given by

$$\widetilde{K}(k,k') = \int K(x,x')\exp(-ixk)\exp(ix'k')\,dx\,dx', \tag{2.26}$$

in terms of the point-spread function $K(x,x')$.

2.1.3 Shift invariance

Often a linear process is invariant with respect to shifts in the independent variable (position). In other words, one can perform the same process at arbitrary locations of the input function and the output function would be the same apart from having the same shift. Such a *shift invariant* linear process is represented by a *convolution integral*

$$g(x) = \int h(x-x')f(x')\,dx' = \int h(x')f(x-x')\,dx' \triangleq h * f, \tag{2.27}$$

where $*$ represents the convolution process. (When $*$ is used to represent the convolution, the variables of the functions are not shown.) The kernel $h(x)$ is now a one-dimensional function that is convolved with the input function, and is called the *impulse response* because it is produced as the output (response) function when the input is a Dirac delta function (or impulse function)

$$h * \delta = \int h(x-x')\delta(x')\,dx = h(x). \tag{2.28}$$

The Fourier transform of the convolution of two functions is the product of their spectra. Therefore, the shift invariant linear process is given by a product

$$G(k) = H(k)F(k), \tag{2.29}$$

in the Fourier domain, where $H(k) = \mathcal{F}\{h(x)\}$ is called the *transfer function.*

Exercise 2.2. Show that, if $g = h * f$, then $G(k) = H(k)F(k)$, where $G(k)$, $H(k)$, and $F(k)$ are the Fourier transforms of $g(x)$, $h(x)$, and $f(x)$, respectively.

Hint: Express one of the functions in the convolution in terms of its inverse Fourier transform and change the order of integration.

2.1.4 Transmission function

We can also have the opposite situation, where the output function is given by a product of the input function and another function

$$g(x) = t(x)f(x). \tag{2.30}$$

We call $t(x)$ a *transmission function*, based on the fact that they operate through such products. In this case, the Fourier domain process is given by a convolution

$$G(k) = \int T(k - k')F(k') \frac{\mathrm{d}k'}{2\pi} = T * F, \tag{2.31}$$

where $T(k)$ is the spectrum of $t(x)$, serving as a *convolution kernel*. Several optical elements are conveniently represented in terms of transmission functions, including apertures, thin lenses, thin holograms, and diffraction gratings. In a lossless scenario, the transmission function is a *phase-only* element. Such optical components include thin lenses and phase-only *diffractive optical elements*. We can also have transmission functions that are not phase-only elements and therefore introduce loss. In such cases, they still lead to convolution kernels, but such scenarios do not represent lossless optics.

2.1.5 Unitarity

Energy is conserved in physical systems. For a *lossless* system, the energy of the output function is therefore equal to the energy of the input function, which is proportional to the squared magnitude of the function. It leads to $\|G(k)\|^2 = \|F(k)\|^2$, where these norms represent the Euclidean norm, defined in (2.23). For the general linear transformation given in (2.24), it implies that

$$\begin{aligned}\int |G(k)|^2\,\mathrm{d}k &= \int F^*(k_1)\widetilde{K}^\dagger(k_1,k)\widetilde{K}(k,k_2)F(k_2)\,\frac{\mathrm{d}k_2}{2\pi}\,\frac{\mathrm{d}k_1}{2\pi}\,\frac{\mathrm{d}k}{2\pi} \\ &= \int |F(k)|^2\,\frac{\mathrm{d}k}{2\pi}.\end{aligned} \tag{2.32}$$

The integrated product of the two kernels produces the Dirac delta function

$$\int \widetilde{K}^\dagger(k_1,k)\widetilde{K}(k,k_2)\,\frac{\mathrm{d}k}{2\pi} = 2\pi\delta(k_1 - k_2). \tag{2.33}$$

Such a process (or kernel) is said to be *unitary*. It maintains the magnitude (Euclidean norm) of a function. Based on (2.23), the Fourier transform is a unitary process. The same applies for the inverse Fourier transform.

2.1.6 Multidimensional systems

So far, we have discussed linear systems in the context of a one-dimensional continuous variable and its Fourier domain. For a linear *optical* system, the number of dimensions increases to four, comprising a propagation direction (denoted by z), a two-dimensional transverse plane (denoted by x and y) and a time dependence t. In the Fourier domain, the optical field is represented by an *angular spectrum*. The Fourier domain is however only three-dimensional. (The Fourier kernel is a plane wave that satisfies the wave equation, as discussed in Section 2.3.) It is either represented in terms of a three-dimensional *propagation vector* or *wavevector* $\mathbf{k}$, with *angular frequency* ω fixed by the dispersion relation $\omega = c|\mathbf{k}|$, or it is represented in terms of a two-dimensional *transverse propagation vector* $\mathbf{K}$ and the angular frequency ω, with k_z being fixed by the dispersion relation. (The latter set is called *optical beam variables* and is discussed in Section 2.4.2.) In either case, the Fourier integrals are expressed as three-dimensional integrals with one of the four spacetime coordinates x, y, z, t set as a fixed value. (See Sections 2.4.3 and 2.4.4.)

Bold symbols denote vectors. For example, the wavevector is given by a three-dimensional vector

$$\mathbf{k} \triangleq k_x\vec{x} + k_y\vec{y} + k_z\vec{z}, \tag{2.34}$$

in terms of the orthogonal *unit vectors* $\vec{x}$, $\vec{y}$, and $\vec{z}$ in three-dimensional space. Often, we consider two-dimensional vectors, such as the transverse propagation vector, which is given by

$$\mathbf{K} \triangleq k_x\vec{x} + k_y\vec{y}. \tag{2.35}$$

The magnitude of a vector is denoted by

$$|\mathbf{k}| \triangleq \sqrt{\mathbf{k}\cdot\mathbf{k}} = \sqrt{k_x^2 + k_y^2 + k_z^2}, \tag{2.36}$$

where $\cdot$ is the *dot-product* between vectors. The three-dimensional version of the superposition integral in (2.24) thus reads

$$G(\mathbf{k}) = \int K(\mathbf{k}, \mathbf{k}')F(\mathbf{k}')\,\frac{\mathrm{d}^3k'}{(2\pi)^3}, \tag{2.37}$$

where $G(\mathbf{k})$ and $F(\mathbf{k}')$ are the *angular spectra* of the output and input functions, respectively. The integration over the wavenumber becomes an integral over the three-dimensional wavevectors.

In its most general form, the angular spectrum as a function of **k**, represents an optical field that propagates in all directions. For practical linear optical systems, optical fields are assumed to propagate predominantly in a particular direction, even in those cases where the fields are *non-paraxial* (the *paraxial condition* is discussion in Section 2.3.5). Therefore, the angular spectra are assumed to be nonzero only in half of the full three-dimensional space represented by the wavevectors. Assuming light propagates predominantly in the z-direction, this *beaming condition* is enforced by

$$F(\mathbf{k}) \to F(\mathbf{k})\theta(k_z), \tag{2.38}$$

where the *Heaviside function* $\theta(\cdot)$ is defined as

$$\theta(x) = \begin{cases} 1 & \text{for } x > 0, \\ 0 & \text{for } x \leq 0. \end{cases} \tag{2.39}$$

In such cases, it may be more convenient to represent the angular spectra in terms of optical beam variables. Their benefit is that they facilitate the inclusion of *evanescent waves*, which are generated by physical devices (such as transmission functions) that impose boundary conditions with feature sizes smaller than the wavelength.

Note that a system may be separately shift invariant (or not) in time and/or space. Usually, we can assume that the system is shift invariant (or translation invariant) in time. If not, it implies that the energy in the system is not conserved. (It follows from *Noether's theorem* [9], which states that continuous global symmetries are associated with conserved quantities. In the case of the *conservation of energy*, the associated symmetry is time-translation invariance.) In nonlinear systems or systems with absorption, the energy may not be conserved, depending on how the system is modelled. Moreover, some systems such as sources or detectors have finite durations during which light is emitted or finite integration times for detection. Such systems are usually modelled by functions that are not shift invariant.

Most of the time, when we are not dealing with sources or detectors, the systems are modelled in such a way that any energy which is converted into other forms (due to absorption, for example) is taken into account. Therefore, systems are usually assumed to be shift invariant (or translation invariant) in time. It leads to a description of the system in terms of a pointwise multiplication of the angular spectrum by a *temporal transfer function*. The general case, with the spatial degrees of freedom included, can be represented in optical beam variables by a two-dimensional integral

$$G(\mathbf{K}, \omega) = \int K(\mathbf{K}, \mathbf{K}', \omega) F(\mathbf{K}', \omega) \frac{d^2k'}{(2\pi)^2}. \tag{2.40}$$

For a transmission function, it becomes

$$G(\mathbf{K}, \omega) = \int T(\mathbf{K} - \mathbf{K}', \omega) F(\mathbf{K}', \omega) \frac{d^2k'}{(2\pi)^2}, \tag{2.41}$$

where $T(\mathbf{K}-\mathbf{K}',\omega)$ is the *convolution kernel*. Since we often encounter them, we briefly discuss such convolution kernels here.

2.1.7 Convolution kernel

In three-dimensions, a general *convolution kernel* has the form

$$K(\mathbf{k}_1,\mathbf{k}_2) = T(\mathbf{k}_1-\mathbf{k}_2). \tag{2.42}$$

It is essentially a function of only one wavevector.

A convolution kernel is the Fourier transform of a *transmission function*, which is a two-dimensional spatial function. Therefore, it is usually more convenient to express such convolution kernels in *optical beam variables*, (see Section 2.4.2 below). It allows us to express the temporal properties separately. If the temporal part is also represented as a convolution kernel (as, e. g., with detector kernels), we can model it as

$$\begin{aligned} T(\mathbf{K}_1-\mathbf{K}_2,\omega_1-\omega_2) = & \int h(\mathbf{X})\exp[-\mathrm{i}\mathbf{x}\cdot(\mathbf{k}_1-\mathbf{k}_2)]\,\mathrm{d}^2x \\ & \times \int d(t)\exp[\mathrm{i}(\omega_1-\omega_2)t]\,\mathrm{d}t. \end{aligned} \tag{2.43}$$

Here, $h(\mathbf{X})$ is the two-dimensional *transmission function* and $d(t)$ is a temporal function that can be used to model the duration of a process.

2.2 Maxwell's equations

James Clerk Maxwell unified all classical electrical and magnetic phenomena in terms of a set of equations, which carries his name. Among their many implications, these equations show that light is an electromagnetic wave. Maxwell's equations form the foundation upon which our knowledge of classical optics is built. It is also incorporated into *quantum electrodynamics* [10], which represents our best theory of the electromagnetic force coupled to charged particles.

In SI units, Maxwell's equations are given by

$$\nabla\times\mathbf{E} = -\partial_t\mathbf{B}, \tag{2.44}$$

$$\nabla\times\mathbf{H} = \partial_t\mathbf{D}+\mathbf{J}, \tag{2.45}$$

$$\nabla\cdot\mathbf{D} = \rho, \tag{2.46}$$

$$\nabla\cdot\mathbf{B} = 0, \tag{2.47}$$

where **E** is the *electric field strength*, **H** is the *magnetic field strength*, **D** is the *electric flux density*, **B** is the *magnetic flux density*, **J** is the *electric current density*, and ρ is the *electric* *charge density*. They are all real-valued quantities. The partial derivative is represented by the simpler notation

$$\partial_t \triangleq \frac{\partial}{\partial t}. \tag{2.48}$$

Again the bold symbols denote vectors, such as

$$\mathbf{E} \triangleq E_x\vec{x} + E_y\vec{y} + E_z\vec{z}, \tag{2.49}$$

for example. Using the (Einstein) *summation convention*, where repeated symbolic indices imply a summation over all the values of those indices, we can also represent a vector in a component-based notation as

$$\mathbf{E} = E_a\vec{x}_a, \tag{2.50}$$

without the summation symbol. Here, $\vec{x}_a$ represents the orthogonal unit vectors in three-dimensional space, with a ranging over either $\{x, y, z\}$ or $\{1, 2, 3\}$. A quantity with such indices is called a *tensor*. With only one index, the tensor represents a *vector*. Those with two indices are *matrices*. The component-based notation can be used for expressions with arbitrary tensors, and is referred to as *tensor notation*.

2.2.1 Constituent relations

There are additional relationships that exist among **E**, **H**, **D**, and **B** that are caused by the properties of the medium in which these fields exist. In vacuum, these *constituent relationships* are given by

$$\mathbf{D} = \epsilon_0\mathbf{E} \quad \text{and} \quad \mathbf{B} = \mu_0\mathbf{H}, \tag{2.51}$$

where

$$\mu_0 = 4\pi \times 10^{-7}\,\mathrm{H\,m^{-1}} \tag{2.52}$$

is the *magnetic permeability of vacuum*, and

$$\epsilon_0 = (\mu_0 c^2)^{-1} = 8.854187817 \times 10^{-12}\,\mathrm{F\,m^{-1}} \tag{2.53}$$

is the *electric permittivity of vacuum*, with

$$c = 299792458\,\mathrm{m\,s^{-1}} \approx 3 \times 10^8\,\mathrm{m\,s^{-1}}, \tag{2.54}$$

being the *speed of light*. It is an exact quantity; the meter is currently defined as the distance travelled by light in vacuum in $(299792458)^{-1}$ s.

2.2.2 Properties of the medium

The propagation of light or electromagnetic waves through a medium depends on the properties of the medium. Here, we consider the effects of the properties of the medium on the electromagnetic field that exist in such a medium.

In general, $\mathbf{J} = \sigma\mathbf{E}$, where σ is the *electric conductivity*. For $\sigma = 0$, the medium does not support currents ($\mathbf{J} = 0$), and thus acts as an insulator. A medium is said to be *source free* when $\mathbf{J} = 0$ and $\rho = 0$. For $\sigma \neq 0$, there is absorption in the medium, often represented in terms of an *absorption coefficient*.

In a general medium, the flux density can be an arbitrary function of the field strength, as well as a function of the spatial and temporal derivatives of the field strength. In that case, the medium is (temporally or spatially) *dispersive*; the effect is nonlocal in space and/or time. For a temporally dispersive medium, the constituent properties (permittivity or permeability) are frequency dependent. It can also give rise to phenomena such as the *magneto-optic* effect [11].

If the flux densities **D** or **B** do not depend on the temporal or spatial derivatives of the field strengths **E** or **H**, the medium is *non-dispersive*. In other words, the reaction of the medium to the field strength is purely local.

When the flux density only depends on the field strength and not on its derivatives, we can expand **D** as

$$\begin{aligned} D_a = \epsilon_0 \left\{\delta_{ab} + [\chi_e^{(1)}]_{ab}\right\} E_b + \epsilon_0 [\chi_e^{(2)}]_{abc} E_b E_c \\ + \epsilon_0 [\chi_e^{(3)}]_{abcd} E_b E_c E_d + \cdots, \end{aligned} \tag{2.55}$$

and **B** as

$$\begin{aligned} B_a = \mu_0 \left\{\delta_{ab} + [\chi_m^{(1)}]_{ab}\right\} H_b + \epsilon_0 [\chi_m^{(2)}]_{abc} H_b H_c \\ + \epsilon_0 [\chi_m^{(3)}]_{abcd} H_b H_c H_d + \cdots, \end{aligned} \tag{2.56}$$

in tensor notation. Here, δ_{ab} is an identity matrix, $\chi_e^{(n)}$ is the n-th order *electric susceptibility*, and $\chi_m^{(n)}$ is the n-th order *magnetic susceptibility*. The higher-order terms with more than one factor of **E** or **H** are associated with *nonlinear media*. They give rise to nonlinear optical phenomena such as the *electro-optic effect*, *four-wave mixing*, *second-harmonic generation*, and many others [11]. The nonlinear optical process of *parametric down-conversion* is discussed in Chapter 8, with some applications in Chapter 9.

For a *linear medium*, we have

$$D_a = \epsilon_0 \left\{\delta_{ab} + [\chi_e^{(1)}]_{ab}\right\} E_b = \epsilon_{ab} E_b, \tag{2.57}$$

$$B_a = \mu_0 \left\{\delta_{ab} + [\chi_m^{(1)}]_{ab}\right\} H_b = \mu_{ab} H_b \tag{2.58}$$

where ϵ_{ab} and μ_{ab} represent permittivity and permeability matrices, respectively. All other higher-order susceptibilities are zero.

If a linear medium is also *isotropic*, the properties of the medium does not depend on the direction of propagation or the directions of **E** or **H**. Then

$$[\chi_e^{(1)}]_{ab} = \chi_e^{(1)}\,\delta_{ab}. \tag{2.59}$$

Similarly for $\chi_m^{(1)}$. The permittivity of such a medium is defined as

$$\epsilon \triangleq \epsilon_0\left[1+\chi_e^{(1)}\right] = \epsilon_0\epsilon_r = \epsilon_0 n^2, \tag{2.60}$$

where ϵ_r is the *dielectric constant* or *relative permittivity*, and n is the *refractive index* of the medium. When the medium has a nonzero conductivity ($\sigma \neq 0$), associated with absorption, the dielectric constant (or the refractive index) is a complex-valued quantity.

Anisotropic media (discussed in Section 2.9.4) give rise to *polarization effects*. They include *dichroism* where the absorption coefficient depends on the orientation of the electric field, and *retardance* and *walk-off* where the refractive index depends on the orientation of the electric field.

A medium is *homogeneous* when the properties of the medium are constant as functions of space. The implication is that

$$\nabla\epsilon = \nabla\mu = 0. \tag{2.61}$$

Examples of inhomogeneous media are *gradient index media* (discussed in Section 2.7.4), *volume holograms*, *Bragg gratings*, and *photonic crystals*.

A medium is called *non-magnetic* when $\mu = \mu_0$. It implies that $\mathbf{B} = \mu_0\mathbf{H}$.

2.3 Equation of motion

For the moment, we assume that the medium is source-free, linear, isotropic, non-dispersive, homogeneous, and non-magnetic. Under such conditions, Maxwell's equations can be expressed purely in terms of the field strengths and without the source terms

$$\nabla\times\mathbf{E} = -\mu_0\partial_t\mathbf{H}, \tag{2.62}$$

$$\nabla\times\mathbf{H} = \epsilon\partial_t\mathbf{E}, \tag{2.63}$$

$$\nabla\cdot\mathbf{E} = 0, \tag{2.64}$$

$$\nabla\cdot\mathbf{H} = 0. \tag{2.65}$$

Eventually, some of these constraints are relaxed to discuss specific phenomena that are of interest to us.

2.3.1 Wave equation

Take the curl of (2.62) and eliminate **H** with (2.63). Then use the identity

$$\nabla \times \nabla \times \mathbf{A} = \nabla(\nabla \cdot \mathbf{A}) - \nabla^2 \mathbf{A}, \tag{2.66}$$

and (2.64) to obtain

$$\nabla^2 \mathbf{E} - \mu_0 \epsilon \partial_t^2 \mathbf{E} = \nabla^2 \mathbf{E} - \frac{1}{v^2} \partial_t^2 \mathbf{E} = 0. \tag{2.67}$$

An equivalent equation can be derived for **H**. The equation in (2.67) has the form of a *wave equation* with a *phase velocity*

$$v = \frac{1}{\sqrt{\mu_0 \epsilon}}. \tag{2.68}$$

In vacuum, electromagnetic fields propagate at the speed of light, so that

$$c = \frac{1}{\sqrt{\mu_0 \epsilon_0}}, \tag{2.69}$$

which is numerically given in (2.54). From (2.60), the *refractive index* is the speed of light in vacuum divided by the speed of light in the medium

$$n = \sqrt{\epsilon_r} = \sqrt{\frac{\epsilon}{\epsilon_0}} = \frac{c}{v}. \tag{2.70}$$

2.3.2 Monochromatic fields

The electromagnetic field sometimes has a narrow spectrum. For such cases, it is often assumed that the field has a single fixed frequency. Under this *monochromatic* assumption, the electric and the magnetic fields can be expressed as

$$\begin{aligned} \mathbf{E}(\mathbf{x}, t) &= \operatorname{Re}\{\widetilde{\mathbf{E}}(\mathbf{x}) \exp(-i\omega t)\}, \\ \mathbf{H}(\mathbf{x}, t) &= \operatorname{Re}\{\widetilde{\mathbf{H}}(\mathbf{x}) \exp(-i\omega t)\}, \end{aligned} \tag{2.71}$$

where $\operatorname{Re}\{\cdot\}$ represents the real part of the argument, and $\widetilde{\mathbf{E}}(\mathbf{x})$ and $\widetilde{\mathbf{H}}(\mathbf{x})$ are time-independent complex-valued *phasor* fields. Using these phasor fields, we can express quantities in a time-independent fashion. We can thus use the complex phasor fields (or their complex-conjugates) to perform linear calculations and postpone the process of taking the real part until after the calculations. The choice of which form represents the phasor field and which one represents its complex-conjugate is determined by the *physics sign convention*. Hence, $\mathbf{E}(\mathbf{x}, t) \to \widetilde{\mathbf{E}}(\mathbf{x}) \exp(-i\omega t)$ and similar for **H**.

The (source-free) Maxwell equations for the monochromatic electric and magnetic phasor fields are

$$\nabla \times \tilde{\mathbf{E}}(\mathbf{x}) = \mathrm{i}\omega\mu_0\tilde{\mathbf{H}}(\mathbf{x}), \quad \nabla \cdot \tilde{\mathbf{E}}(\mathbf{x}) = 0,$$
$$\nabla \times \tilde{\mathbf{H}}(\mathbf{x}) = -\mathrm{i}\omega\epsilon\tilde{\mathbf{E}}(\mathbf{x}), \quad \nabla \cdot \tilde{\mathbf{H}}(\mathbf{x}) = 0. \tag{2.72}$$

The time dependence dropped out, and the equations now contain the *angular frequency* ω as an explicit parameter.

The monochromatic property allows us to remove all time dependencies and temporal derivatives. As a result, we can express the electric (magnetic) field directly in terms of spatial derivatives of the magnetic (electric) field by

$$\tilde{\mathbf{E}} = \mathrm{i}\frac{\nabla \times \tilde{\mathbf{H}}}{\omega\epsilon}, \tag{2.73}$$

$$\tilde{\mathbf{H}} = -\mathrm{i}\frac{\nabla \times \tilde{\mathbf{E}}}{\omega\mu_0}. \tag{2.74}$$

The implication is that we can remove one of these fields in favour of the other. For instance, using (2.74) to replace the magnetic field, we can formulate all expressions purely in terms of the electric field.

2.3.3 Helmholtz equation

When the monochromatic Maxwell equations in (2.72) are used to derive a wave equation for the monochromatic electric (or magnetic) phasor fields, as was done to obtain (2.67), the result is

$$\nabla^2\tilde{\mathbf{E}} + \frac{\omega^2}{v^2}\tilde{\mathbf{E}} = 0. \tag{2.75}$$

It is a *monochromatic* wave equation due to the explicit appearance of ω. Alternatively, we can substitute $\mathbf{E}(\mathbf{x}, t) \to \tilde{\mathbf{E}}(\mathbf{x}) \exp(-\mathrm{i}\omega t)$ into (2.67) to obtain the same result.

The phase velocity v (or c in vacuum) can be expressed in terms of the *wavelength* λ and the *frequency* ν, or in terms of angular frequency ω and the *wavenumber* k by

$$v = \lambda\nu = \frac{\omega}{k}. \tag{2.76}$$

Therefore,

$$\frac{\omega}{v} = k = \frac{2\pi}{\lambda}. \tag{2.77}$$

The wavenumber is the magnitude of the wavevector, as represented in (2.36). In a dielectric medium, it is related to the *wavenumber in vacuum* k_0 by

$$k = nk_0, \tag{2.78}$$

where n is the refractive index.

The *Helmholtz equation* follows by substituting (2.77) into (2.75):

$$\nabla^2\tilde{\mathbf{E}} + k^2\tilde{\mathbf{E}} = 0. \tag{2.79}$$

Being expressed in terms of $\tilde{\mathbf{E}}$, the expression reminds us that the electric field is represented as a complex-valued phasor field. However, one often finds that it is simply expressed in terms of the electric field $\mathbf{E}$, under the assumption that it does not create any confusion. Therefore, we henceforth remove the tilde. At times, we may also include the time-dependent phase factor in the complex-valued fields.

2.3.4 Scalar fields

The wave equation in (2.67) and the Helmholtz equation in (2.79) both operate on each component of the vector field independently. So, we can express these equations as *scalar* equations for the individual components. The vector field is simply the superposition of the different components, each satisfying the *scalar Helmholtz equation*

$$\nabla^2 f + k^2 f = 0, \tag{2.80}$$

where f represents the scalar field.

When the electric field is uniformly polarized (has a constant polarization vector), it can be expressed as a *scalar* field times a constant polarization vector $\mathbf{E}(\mathbf{x}) = f(\mathbf{x})\vec{\eta}$. The *polarization vector* $\vec{\eta}$ can then be factored out and removed. The remaining scalar field $f(\mathbf{x})$ is a solution of the scalar Helmholtz equation and gives rise to *scalar diffraction theory* (see Section 2.6 below).

When the polarization varies as a function of position in the electric field, the field can be expressed in terms of two uniformly polarized fields

$$\mathbf{E}(\mathbf{x}) = f_1(\mathbf{x})\vec{\eta}_1 + f_2(\mathbf{x})\vec{\eta}_2, \tag{2.81}$$

such that $\vec{\eta}_1$ and $\vec{\eta}_2$ are constant polarization vectors that are mutually orthogonal $\vec{\eta}_1^* \cdot \vec{\eta}_2 = 0$. The two scalar fields $f_1(\mathbf{x})$ and $f_2(\mathbf{x})$ are also orthogonal in the sense that

$$\langle f_1, f_2 \rangle \triangleq \int f_1^*(\mathbf{x}) f_2(\mathbf{x})\, d^3x = 0. \tag{2.82}$$

Each of the two terms in the decomposition satisfies an independent scalar Helmholtz equation. The decomposition in (2.81) corresponds to a *Schmidt decomposition* of the vector field, which is formally equivalent to the *Schmidt decomposition* of quantum states discussed in Section 4.3.3.

In general, we retain the polarization of the field so that we can include those cases where the polarization is not uniform. However, all such cases can be understood in terms of solutions of the scalar Helmholtz equation.

2.3.5 Paraxial condition

The situation where the light propagates in a beam with a relatively small *beam divergence angle* (defined below) is quite common in optical applications. In such a case, all the light propagates close to the *optical axis*, which is defined to be the z-axis. Such an optical field is referred to as a *paraxial* field.

Here, we focus on the paraxial condition in configuration space as it affects the equation of motion. In Section 2.6.1, we return to the paraxial condition applied in the Fourier domain in the context of free space propagation.

The beam divergence angle depends on the wavelength of the light. Therefore, to ensure that the paraxial condition is satisfied, we assume that the light is monochromatic. As a result, the *monochromatic condition* is a prerequisite for the paraxial condition.

For a general optical field that satisfies the paraxial condition and propagates in the z-direction, we can remove a factor of $\exp(ikz)$ from the electric and magnetic phasor fields. Although it is similar to the way the temporal phase factor is separated and removed under the monochromatic assumption, we do not completely remove the z-dependence in the remaining fields under the paraxial condition. Hence,

$$\begin{aligned} \mathbf{E}(\mathbf{x}) &= \bar{\mathbf{E}}(\mathbf{x})\exp(ikz), \\ \mathbf{H}(\mathbf{x}) &= \bar{\mathbf{H}}(\mathbf{x})\exp(ikz), \end{aligned} \tag{2.83}$$

where $\bar{\mathbf{E}}$ and $\bar{\mathbf{H}}$ are the remaining parts of the phasor fields after the z-directed plane wave is removed. These remaining parts of the phasor fields are slow-varying in z.

2.3.5.1 Paraxial expansion

The paraxial condition leads to simplifications in the expressions of the relevant equations. The simplifications can be implemented simply by discarding any second-order derivatives in z of the fields, thanks to their slow-varying nature in z. It is referred to as the *paraxial approximation*.

Here, we do the paraxial expansion of expressions more systematically as an asymptotic expansion, where the expansion parameter is a dimensionless quantity that is small provided that the paraxial condition is well satisfied. The appropriate expansion parameter for this purpose is the *beam divergence angle* given by

$$\Theta \triangleq \frac{\lambda}{\pi d}, \tag{2.84}$$

where λ is the wavelength and d is a scale parameter on the transverse plane. Often, d is the beam radius w_0 of a Gaussian beam. In the general case, we can use the inverse of the size of the angular spectrum of the paraxial beam to define the transverse scale.[e]

While d is the characteristic scale on the transverse plane, the scale that is associated with the longitudinal direction (z-direction) is the *Rayleigh range*

$$z_R \triangleq \frac{\pi d^2}{\lambda} \equiv \frac{1}{2}kd^2. \tag{2.85}$$

The beam divergence angle can be written as the ratio of the transverse and longitudinal scales $\Theta = d/z_R$.

To perform the expansion, we express the fields as functions of *normalized coordinates* that are defined by

$$u = \frac{x}{d}, \quad v = \frac{y}{d}, \quad \text{and} \quad w = \frac{z}{z_R}. \tag{2.86}$$

The electric phasor field, for example, then becomes

$$\begin{aligned} \mathbf{E}(x,y,z) &= \bar{\mathbf{E}}(x,y,z)\exp(ikz) \\ &\to \mathbf{E}'[u(x),v(y),w(z)]\exp(ikz). \end{aligned} \tag{2.87}$$

We can substitute these fields into any monochromatic expression. The dimension parameters are then combined into dimensionless factors that contain powers of the beam divergence angle. The expression is then expanded in terms of the beam divergence angle. Higher-order terms are discarded based on their suppression by factors of the beam divergence angle. After converting the truncated result back to the original coordinates (x,y,z), we obtain the relevant expression under the paraxial approximation.

2.3.5.2 Paraxial wave equation

The paraxial expansion of the Helmholtz equation, given in (2.79), leads to the *paraxial wave equation*. We derive it with the aid of the systematic approach. For this purpose, we first separate the differential operator ∇ into a transverse part ∇_{xy} and a longitudinal part and express them in terms of normalized coordinates. It becomes

$$\nabla = \nabla_{xy} + \vec{z}\partial_z = \frac{1}{d}\nabla_{uv} + \frac{\vec{z}}{z_R}\partial_w. \tag{2.88}$$

Substituting (2.87), (2.88), and $k = 2z_R/d^2$ into (2.79), we obtain

---

e One way to compute the transverse scale is to take the Fourier transform of the two-dimensional complex function of the cross-section of the beam to obtain the angular spectrum. The inverse of the square root of the second centralized moment of the modulus square of this spectrum gives an estimate of the scale on the transverse plane.

$$\begin{aligned}0 &= \left(\frac{1}{d^2}\nabla_{uv}^2 + \frac{1}{z_R^2}\partial_w^2 + k^2\right)\mathbf{E}'(u,v,w)\exp(ikz) \\ &= \left[\nabla_{uv}^2\mathbf{E}'(u,v,w) + \mathrm{i}4\partial_w\mathbf{E}'(u,v,w) \right. \\ &\quad \left. + \Theta^2\partial_w^2\mathbf{E}'(u,v,w)\right]\frac{\exp(ikz)}{d^2}. \end{aligned} \tag{2.89}$$

The second derivative in w is suppressed by two factors of Θ. Therefore, we drop this Θ^2-term. The remaining terms produce an equation in normalized coordinates. Converted back into the original coordinates, it becomes the *paraxial wave equation*:

$$\nabla_{xy}^2\mathbf{E}(\mathbf{x}) + \mathrm{i}2k\partial_z\mathbf{E}(\mathbf{x}) = 0. \tag{2.90}$$

Here, we removed the bar from the slow-varying electric phasor field to express it as $\mathbf{E}(\mathbf{x})$, assuming that there is no confusion.

The same arguments presented in Section 2.3.4 can be applied here to express the paraxial wave equation for scalar fields. So, we can remove a constant polarization vector from the slow-varying electric phasor field in the paraxial wave equation to obtain the *scalar paraxial wave equation* that reads

$$\nabla_{xy}^2 f(\mathbf{x}) + \mathrm{i}2k\partial_z f(\mathbf{x}) = 0. \tag{2.91}$$

2.3.5.3 Paraxial expansion of the Maxwell equations

The paraxial expansion of the Maxwell equations provides direct relationships between the transverse parts of the electric and magnetic fields and shows how the z-components are given in terms of the transverse parts of these fields, but are suppressed by one factor of Θ. Therefore, knowledge of the transverse electric field determines the complete electromagnetic field in the paraxial limit.

Exercise 2.3. Show that the paraxial expansion of the source-free Maxwell equations in vacuum leads to

$$\begin{aligned} H_x(\mathbf{x}) &= -\frac{1}{\eta_0}E_y(\mathbf{x}), \quad H_y(\mathbf{x}) = \frac{1}{\eta_0}E_x(\mathbf{x}), \\ E_z(\mathbf{x}) &= \frac{\mathrm{i}}{k}\nabla_{xy}\cdot\mathbf{E}(\mathbf{x}), \quad H_z(\mathbf{x}) = \frac{\mathrm{i}}{k}\nabla_{xy}\cdot\mathbf{H}(\mathbf{x}), \end{aligned} \tag{2.92}$$

where

$$\eta_0 = \sqrt{\frac{\mu_0}{\epsilon_0}}, \tag{2.93}$$

is the *intrinsic impedance of vacuum*.

Hint: drop all those terms that are suppressed by a factor of Θ^2 with respect to the leading-order term and separate the result into separate components.

2.4 Solutions of the Helmholtz equation

Different solutions of (2.80) represent different modes or basis functions. A *mode* is a solution that can exist independently, satisfying all relevant boundary conditions without having to excite additional fields. The solutions are also called *basis functions* because we can express an arbitrary field in terms of a linear combination of them.

A general approach to obtain solutions for the scalar Helmholtz equation in (2.80) is through the *separation of variables*, using the following steps:

1. Select a suitable coordinate system $\{a, b, c\}$ and express the differential operator ∇^2 in (2.80) in terms of this coordinate system.
2. Substitute the scalar field as a product of functions, one for each coordinate $f(a, b, c) \to A(a)B(b)C(c)$, into (2.80) for the chosen coordinates.
3. Separate the wave equation into decoupled differential equations for the respective coordinates.
4. Solve the differential equations to find expressions for the solutions.
5. Apply the boundary conditions (where applicable) to find the allowed solutions.

2.4.1 Cartesian coordinates: Plane waves

The solutions of the scalar Helmholtz equation in Cartesian coordinates are the *plane waves*, shown in Figure 2.1. They are given by

$$f(\mathbf{x}; \mathbf{k}) = \exp(\mathrm{i}\mathbf{k} \cdot \mathbf{x}), \tag{2.94}$$

where $\mathbf{k}$ is the wavevector (or propagation vector) that denotes the direction of propagation of the wave, and

$$\mathbf{x} \triangleq x\vec{x} + y\vec{y} + z\vec{z}, \tag{2.95}$$

is the *position vector* pointing from the origin of the coordinate system to the point denoted by (x, y, z). The relationship between the wavenumber and the angular frequency is given by the *dispersion relation*, which reads

$$\omega = v|\mathbf{k}| = v\sqrt{k_x^2 + k_y^2 + k_z^2}, \tag{2.96}$$

in a dielectric medium, where v is the phase velocity. In vacuum, v is replaced by c. Thanks to (2.36), we can express the wavevector as $\mathbf{k} = k\vec{k}$, where $\vec{k}$ is a unit vector that points in the direction of propagation.

The plane wave solution of the Helmholtz equation can be combined with the appropriate time dependent factor $\exp(-\mathrm{i}\omega t)$ and a constant unit polarization vector $\vec{\eta}$, representing its polarization, to get an expression for the electric field given by

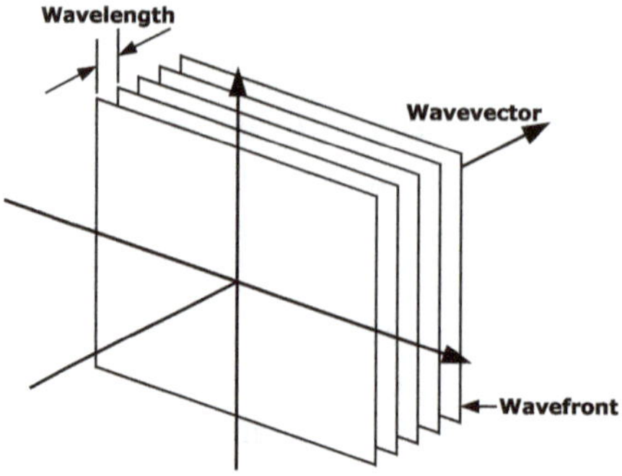

Figure 2.1: Plane wave.

$$\mathbf{E}_{\text{pw}}(\mathbf{x}, t; \mathbf{k}) = \vec{\eta} \exp(-i\omega t + i\mathbf{k} \cdot \mathbf{x}). \tag{2.97}$$

It could also be multiplied by a constant amplitude E_0 with the correct units for an electric field. However, since the plane wave is not normalizable, it does not represent a physical electric field. Nevertheless, we can use the plane wave solutions in (2.97) to define physical electric fields in terms of wavevector integrals, as discussed below.

The constant unit polarization vector $\vec{\eta}$ is orthogonal to the direction of propagation $\vec{\eta} \cdot \vec{k} = 0$. Polarization is discussed in more detail in Section 2.9. The real-valued electric field is obtained by taking the real part of the complex-valued field, as in (2.71).

Note that we could also have expressed the plane waves as

$$\mathbf{E}^*_{\text{pw}}(\mathbf{x}, t; \mathbf{k}) = \vec{\eta}^* \exp(i\omega t - i\mathbf{k} \cdot \mathbf{x}). \tag{2.98}$$

The two expressions in (2.97) and (2.98) are equivalent: they produce the same real-valued electric field. Here, we use the *physics sign convention* by choosing (2.97), according to which *phase decreases with time.*

2.4.2 Optical beam variables

In the following discussions on the orthogonality and *completeness of plane waves* (Sections 2.4.3 and 2.4.4), and the angular spectra (Section 2.4.5), we distinguish between two ways to reduce the number of spacetime dimensions to match the number of Fourier dimensions. The wavevector integrals found in these sections run over only three of the four quantities ω, k_x, k_y, and k_z. By specifying any three of these quantities, we obtain the fourth from the *dispersion relation* (2.96).

One way is to fix the time coordinate. This *fixed-time condition* leads to integrals where the three integration variables are k_x, k_y, and k_z, covering the entire three-dimensional space of propagating plane waves. Another way is to fix one of the three spatial coordinates. Without loss of generality, we assume that z is fixed and call it the *fixed-z condition.* It separates the longitudinal direction (i. e., the propagation direction) from the transverse directions.

The fixed-z condition is provided in anticipation of the *paraxial approximation* applied on the Fourier domain. The spectrum of plane waves is restricted to a half-sphere, centred on the general direction of propagation. In other words, we only need half of the space of all plane waves, including only those whose wavevectors point at an angle of less than 90° relative to the general propagation direction. For convenience, we define the general propagation direction of the optical beam as the z-axis (hence, the fixed-z condition). The plane waves that form part of the optical beam are those that have positive k_z-components.

For an optical beam, it is usually more convenient to specify ω, k_x, and k_y and thereby fix k_z via the dispersion relation. In other words, we prefer to integrate over ω instead of k_z. The transverse wavevector components are combined into a two-dimensional transverse wavevector, denoted by $\mathbf{K}$, as opposed to the three-dimensional wavevector, denoted by $\mathbf{k}$. The Fourier domain variables $\{\omega, k_x, k_y\}$ are called the *optical beam variables* and the z-component is given by

$$k_z = \sqrt{k^2 - k_x^2 - k_y^2} = \sqrt{k^2 - |\mathbf{K}|^2}. \tag{2.99}$$

For propagating plane waves, $k_x^2 + k_y^2 < k^2$. Therefore, if we only consider propagating plane waves, the integrations over k_x and k_y are restricted to within a circular area with a radius of the wavenumber k. However, we can also include the *evanescent field*, in which case the integrations over k_x and k_y are extended to infinity. Evanescent waves, with $k_x^2 + k_y^2 > k^2$, have imaginary values for $k_z = \mathrm{i}\beta$, leading to a decaying z-dependence given by $\exp(-\beta z)$. Evanescent waves are excited by boundary conditions with features smaller than the wavelength. They decay along the direction of propagation and do not represent propagating fields.

Exercise 2.4. Show that the evanescent wave

$$g(x, y, z) = \exp(\mathrm{i}k_x x + \mathrm{i}k_y y - az) \tag{2.100}$$

is a solution of the Helmholtz equation. What is the condition for a to be a real-valued constant?

The optical beam variables have the benefit that they naturally include the evanescent part of the field. Usually, the evanescent field is not relevant, because it decays away relatively quickly along the propagation direction, but there may be cases where it plays a role. As a result, we can maintain the integral over the transverse wavevectors all the way to infinity, even when we are not specifically interested in the evanescent field.

2.4.3 Orthogonality of plane waves

Different wavevectors represent different plane waves that are mutually orthogonal functions based on the inner product discussed in Section 2.1.1. These wavevectors fill a three-dimensional space, but the plane waves are functions in four-dimensional configuration space. Due to the reduced number of degrees of freedom associated with the wavevectors, the domain over which the *orthogonality condition* is defined also needs to be restricted so that the number of degrees of freedom match. In the following discussions, we consider two ways to reduce the configuration space dimensions. For this purpose, we use the expressions for the plane waves in (2.97), so that $\mathbf{E} = \mathbf{E}_{\text{pw}}$.

2.4.3.1 Fixed time

Here, the temporal coordinates of the two plane waves are set equal and the inner product is evaluated over the remaining three-dimensional space. Thus, we obtain

$$\begin{aligned}\langle \mathbf{E}_1, \mathbf{E}_2 \rangle_t &= \int \mathbf{E}^*_{\text{pw}}(\mathbf{x}, t; \mathbf{k}_1) \cdot \mathbf{E}_{\text{pw}}(\mathbf{x}, t; \mathbf{k}_2)\, \mathrm{d}^3x \\ &= \vec{\eta}_1^* \cdot \vec{\eta}_2 \int \exp(\mathrm{i}\omega_1 t - \mathrm{i}\omega_2 t - \mathrm{i}\mathbf{k}_1 \cdot \mathbf{x} + \mathrm{i}\mathbf{k}_2 \cdot \mathbf{x})\, \mathrm{d}^3x \\ &= \vec{\eta}_1^* \cdot \vec{\eta}_2 (2\pi)^3 \delta(\mathbf{k}_1 - \mathbf{k}_2). \end{aligned} \tag{2.101}$$

Note that $\delta(\mathbf{k}_1 - \mathbf{k}_2)$ represents the product of three Dirac delta functions for the three respective components of the vector in the argument. As a general rule, when a Dirac delta function contains a vectorial quantity in its argument, it represents a product of Dirac delta functions for the different components of the vectorial quantity.

2.4.3.2 Fixed plane

In this case, the z-coordinates of the two plane waves are set equal and the inner product is evaluated over the two-dimensional transverse plane (perpendicular to the z-direction) and time:

$$\begin{aligned}\langle \mathbf{E}_1, \mathbf{E}_2 \rangle_z &= \int \mathbf{E}^*_{\text{pw}}(\mathbf{x}, t; \mathbf{K}_1, \omega_1) \cdot \mathbf{E}_{\text{pw}}(\mathbf{x}, t; \mathbf{K}_2, \omega_2)\, \mathrm{d}^2x\, \mathrm{d}t \\ &= \vec{\eta}_1^* \cdot \vec{\eta}_2 \int \exp(\mathrm{i}\omega_1 t - \mathrm{i}\omega_2 t - \mathrm{i}\mathbf{k}_1 \cdot \mathbf{x} + \mathrm{i}\mathbf{k}_2 \cdot \mathbf{x})\, \mathrm{d}^2x\, \mathrm{d}t \\ &= \vec{\eta}_1^* \cdot \vec{\eta}_2 (2\pi)^3 \delta(\mathbf{K}_1 - \mathbf{K}_2) \delta(\omega_1 - \omega_2). \end{aligned} \tag{2.102}$$

Here, we use *optical beam variables* (discussed in Section 2.4.2) to separate the longitudinal direction from the transverse plane, with $\mathbf{K}$ being the two-dimensional transverse part of the wavevector. It implies that $\delta(\mathbf{K}_1 - \mathbf{K}_2)$ represents the product of two Dirac delta functions for the two respective components of the vector in the argument.

2.4.4 Completeness of plane waves

The plane waves obey *completeness conditions*, provided that the domain within which the completeness condition is defined is appropriately restricted. Here, we consider the completeness conditions that correspond to the two orthogonality conditions. The plane waves are treated as scalar fields, ignoring the polarization. (Polarization has its own completeness condition, as shown in Section 2.9.)

2.4.4.1 Fixed time

When the time is set equal, the time dependent terms in the exponents of the plane waves cancel. The resulting completeness condition is restricted to the set of all three-dimensional spatial fields:

$$\int \exp(-\mathrm{i}\mathbf{k}\cdot\mathbf{x}_1 + \mathrm{i}\mathbf{k}\cdot\mathbf{x}_2)\,\frac{\mathrm{d}^3k}{(2\pi)^3} = \delta(\mathbf{x}_1 - \mathbf{x}_2). \tag{2.103}$$

2.4.4.2 Fixed plane

When the z-coordinates are set equal, the z-dependent terms in the exponents of the plane waves cancel and the completeness condition becomes

$$\begin{aligned}&\int \exp(\mathrm{i}\omega t_1 - \mathrm{i}\mathbf{K}\cdot\mathbf{X}_1 - \mathrm{i}\omega t_2 + \mathrm{i}\mathbf{K}\cdot\mathbf{X}_2)\,\mathrm{d}_\mathrm{b}k\\ &= \delta(\mathbf{X}_1 - \mathbf{X}_2)\delta(t_1 - t_2),\end{aligned} \tag{2.104}$$

where $\mathbf{X} = x\vec{x} + y\vec{y}$ is the two-dimensional transverse part of the position vector, and the *integration measure for the optical beam variables* is represented by

$$\mathrm{d}_\mathrm{b}k \triangleq \frac{\mathrm{d}^2k\mathrm{d}\omega}{(2\pi)^3}. \tag{2.105}$$

This completeness condition is restricted to the set of all fields that depend on two spatial coordinates and time.

2.4.5 Angular spectrum

The orthogonality of plane waves leads naturally to a Fourier representation for optical fields. It gives rise to the notion of *Fourier optics* [2]. The plane waves can be used to represent any propagating electric field. The form of these plane wave expansions depend on the way in which the domain is restricted. Again, we could consider the two possibilities—fixed time or fixed z—leading to inequivalent angular spectra due to the difference in the integration measures. However, the nature of the typical applications where angular spectra are used, makes it more sensible to use the fixed-z condition, as expressed in terms of optical beam variables. Therefore, we define the angular spectrum of an optical field specifically under the fixed-z condition.

The plane wave expansion of an arbitrary electric phasor field, corresponding to the fixed-z orthogonality condition, is expressed in terms of optical beam variables by

$$\mathbf{E}(\mathbf{x},t) = \sum_s \int F_s(\mathbf{K},\omega)\vec{\eta}_s \exp(-\mathrm{i}\omega t + \mathrm{i}\mathbf{k}\cdot\mathbf{x})\,\mathrm{d}_\mathrm{b}k, \tag{2.106}$$

where we included the time dependence, $F_s(\mathbf{K},\omega)$ is the (fixed-z) *angular spectrum*, with spin index $s = 1, 2$, and $\vec{\eta}_s^*$ is the polarization vectors for the respective spins. The angular spectrum is given by the Fourier transform of the two spin components of the electric phasor field with respect to time and the transverse coordinates

$$F_s(\mathbf{K},\omega) = \int \vec{\eta}_s^* \cdot \mathbf{E}(\mathbf{x},t)\exp(\mathrm{i}\omega t - \mathrm{i}\mathbf{k}\cdot\mathbf{x})\,\mathrm{d}^2x\,\mathrm{d}t, \tag{2.107}$$

where the value of z is fixed, and $\vec{\eta}_s^*$ is the associated polarization vector. The angular spectrum $F_s(\mathbf{k})$ has the dimensions of an electric field multiplied by an area and time.

Having fixed the value of z, the angular spectrum is referenced to the transverse plane (the *reference plane*) defined by that z-value. When the propagation of the optical field is computed with the aid of such an angular spectrum, we need to use the plane waves that are defined relative to this reference plane.

2.5 Solutions of the paraxial wave equation

The solutions of the scalar paraxial wave equation in (2.91) are *polynomial Gaussian beams*—Gaussian functions multiplied by bivariate complex polynomials. They include different sets of complete orthonormal bases, such as the *Laguerre–Gauss* modes or the *Hermite–Gauss* modes.

2.5.1 Gaussian beam

First, we consider the simple *Gaussian beam*. It is without doubt the most widely used optical beam in optical experiments and photonic systems, both classical and quantum. The expression of the scalar Gaussian beam is given by

$$g(x,y,z) = \frac{N}{z_R + \mathrm{i}z}\exp\left[-\frac{k(x^2+y^2)}{2(z_R+\mathrm{i}z)}\right], \tag{2.108}$$

where N is a normalization constant, k is the wavenumber, and z_R is the *Rayleigh range*, given in (2.85). Here, we show the spatial coordinates individually instead of combining them into **x** or **X**. The propagation distance z is defined relative to the waist where the beam has its smallest width.

Exercise 2.5. Compute the normalization constant N in (2.108) so that

$$\int |g(x,y,z)|^2 \, dxdy = 1. \tag{2.109}$$

The transverse scale of a Gaussian beam can be determined by setting $z = 0$ where the waist is located and assessing the width of the resulting two-dimensional Gaussian function. For $z = 0$, (2.108) becomes

$$g(x,y,0) = \frac{N}{z_R} \exp\left[-\frac{k(x^2+y^2)}{2z_R}\right] = \frac{N}{z_R} \exp\left(-\frac{x^2+y^2}{w_0^2}\right), \tag{2.110}$$

where w_0 is the beam radius at the waist ($z = 0$). It is defined as the width of the Gaussian function such that the *intensity* (optical power per unit area) at $r = w_0$ is $\exp(-2) = 1/e^2$ times the peak intensity. Hence,

$$z_R = \frac{1}{2} k w_0^2 = \frac{\pi w_0^2}{\lambda}, \tag{2.111}$$

which corresponds to (2.85) for $d = w_0$. The Rayleigh range is a scale parameter that determines the form of the Gaussian beam as a function of z. For $z \ll z_R$, the beam is effectively collimated—the wavefront is (almost) flat and the beam radius remains approximately constant. For $z \gg z_R$, the beam is similar to a diverging spherical wave with a radius of curvature $R \approx z$. In Figure 2.2, a diagram of the typical structure of a Gaussian beam is shown.

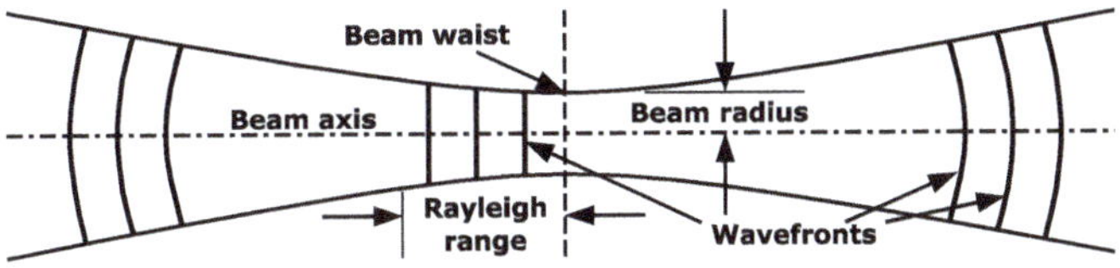

Figure 2.2: Gaussian beam.

Expressed in terms of the radial coordinate $r = \sqrt{x^2+y^2}$, the Gaussian beam is a rotationally symmetric function; it does not depend on the azimuthal coordinate ϕ. At $z = 0$, the function is real valued. It implies that the phase is constant over the xy-plane at $z = 0$. The shape of the beam profile is a Gaussian function at all values of z. The smallest beam radius exists at the *waist* of the beam where $z = 0$. The shape of the Gaussian beam is symmetrical upon inverting the z-axis (i. e., for $z \to -z$). Therefore, the width of the beam at z is equal to the width of the beam at $-z$, and the same applies for the radius of curvature of the wavefront apart from a sign change.

For $z \neq 0$, the function becomes complex valued and the phase is not constant over the transverse plane anymore. Separating the amplitude and phase of the function in (2.108), we can express the Gaussian beam as the product of a real-valued Gaussian amplitude function and a parabolic phase factor

$$g(x,y,z) = \frac{N}{z_R + \mathrm{i}z} \exp\left[-\frac{x^2+y^2}{w^2(z)}\right] \exp\left[\frac{\mathrm{i}\pi(x^2+y^2)}{\lambda R(z)}\right], \tag{2.112}$$

where

$$w(z) \triangleq w_0 \left(1 + \frac{z^2}{z_R^2}\right)^{1/2} \tag{2.113}$$

is the $1/e^2$ radius of the beam as a function of z, and

$$R(z) \triangleq z\left(1 + \frac{z_R^2}{z^2}\right) \tag{2.114}$$

is the *radius of curvature* of the parabolic wavefront as a function of z. The parabolic phase function with the radius of curvature R follows from the *paraxial approximation* of a spherical wave

$$\exp[\mathrm{i}k(R - \sqrt{R^2 - r^2})] \approx \exp\left(\frac{\mathrm{i}\pi r^2}{\lambda R}\right). \tag{2.115}$$

The sign of the radius of curvature indicates whether the beam is diverging (positive) or converging (negative). At $z = 0$, the radius of curvature is infinite, which means that the wavefront is flat. It decreases to $2z_R$ at $z = z_R$ and then increases again gradually. The smallest radius of curvature is found at $z = \pm z_R$, with a value of $R_{\mathrm{min}} = \pm 2z_R$.

The complex-valued factor $(z_R + \mathrm{i}z)^{-1}$ in (2.108) can be written as

$$\frac{1}{z_R + \mathrm{i}z} = \left(z^2 + z_R^2\right)^{-1/2} \exp\left[-\mathrm{i}\arctan\left(\frac{z}{z_R}\right)\right]. \tag{2.116}$$

It produces an additional z-dependent phase shift

$$\gamma(z) \triangleq \arctan\left(\frac{z}{z_R}\right), \tag{2.117}$$

which is called the *Gouy phase*.

2.5.2 Polynomial Gaussian modes

The Gaussian function in (2.108) is not the only solution of the scalar paraxial wave equation in (2.91). There are sets of solutions (modes) representing complete orthonormal bases in terms of which arbitrary paraxial beams can be expressed. The different sets of modes depend on the coordinate system that is used when solving (2.91).

For Cartesian coordinates, we obtain the *Hermite–Gauss modes*. Expressed in terms of the normalized coordinates defined in (2.86), they are given by

$$\begin{aligned}\mathrm{HG}_{mn} = {} & N_{mn}^{(\mathrm{HG})}\frac{(1-\mathrm{i}w)^{(m+n)/2}}{(1+\mathrm{i}w)^{1+(m+n)/2}}H_m\left(\frac{\sqrt{2}u}{\sqrt{1+w^2}}\right)H_n\left(\frac{\sqrt{2}v}{\sqrt{1+w^2}}\right)\\ & \times\exp\left[\frac{-(u^2+v^2)}{1+\mathrm{i}w}\right],\end{aligned} \tag{2.118}$$

where m and n are integers representing *mode indices*, $N_{mn}^{(\mathrm{HG})}$ is a normalization constant, and H_m and H_n represent the *Hermite polynomials* [12].

In cylindrical coordinates, the solutions of the paraxial wave equation are the *Laguerre–Gauss modes*. They are given by

$$\begin{aligned}\mathrm{LG}_{p\ell} = {} & N_{p\ell}^{(\mathrm{LG})}\frac{(1-\mathrm{i}w)^{p}}{(1+\mathrm{i}w)^{p+|\ell|+1}}\rho^{|\ell|}\exp(\mathrm{i}\ell\phi)\\ & \times L_p^{|\ell|}\left(\frac{2\rho^2}{1+w^2}\right)\exp\left(\frac{-\rho^2}{1+\mathrm{i}w}\right),\end{aligned} \tag{2.119}$$

where p is a non-negative integer called the *radial index*, ℓ is a signed integer called the *azimuthal index*, $\rho^2 = u^2+v^2$ is a normalized radial coordinate, and $L_p^{|\ell|}(\cdot)$ represents the *associate Laguerre polynomials* [12]. The normalization constant is given by

$$N_{p\ell}^{(\mathrm{LG})} = \left[\frac{p!2^{|\ell|+1}}{\pi(p+|\ell|)!}\right]^{1/2}. \tag{2.120}$$

The *configuration space* intensity functions of the Laguerre–Gauss modes are rotationally symmetric. The modes are eigenfunctions of the rotation operation (see Section 2.8). Therefore, they are associated with fixed amounts of *orbital angular momentum* that is proportional to the azimuthal index ℓ [13]. The order of the polynomial factor of the Laguerre–Gauss modes is given by $n = 2p + |\ell|$. The Gouy phase is given by the phase of the w-dependent Gouy phase factors $1 \pm \mathrm{i}w$ in (2.119), as shown in (2.116).

2.5.2.1 Optical vortices

The Laguerre–Gauss modes have *phase singularities* of order ℓ on the axis of the beam. The phase is undefined on the axis for $\ell \neq 0$. These singularities are *essential singularities* in that all phase values are produced around them. In other words, any contour around such a phase singularity contains all phase values continuously varying from 0 to 2π. The phase can cycle multiple times through 0 to 2π for one trip around the singularity. The phase increases either in a clockwise or anticlockwise direction around the singularity. All these possibilities are represented in terms of the *topological charge* of the phase singularities. The topological charge of a phase singularity is computed by an *index integral*, given by

$$2\pi\nu = \oint_C \nabla\theta(x,y) \cdot \mathrm{d}\hat{s}, \tag{2.121}$$

where ν is a signed integer representing the topological charge, C is a closed contour enclosing the phase singularity, $\theta(x,y)$ is the phase function on the two-dimensional plane, and $\mathrm{d}\hat{s}$ is a line element tangential to the contour. For the Laguerre–Gauss modes, the topological charge is equal to the azimuthal index ℓ. The mode index ℓ is often referred to as the topological charge. Such a phase singularity is also called an *optical vortex* due to the associated vortex structure in the region of the phase singularity in a complex optical field. Optical vortices can be found in any complex optical field [14].

2.5.3 Generating functions for polynomial Gaussian modes

There are well-known generating functions for the polynomials and special functions upon which the solutions of the Helmholtz equation and the paraxial wave equation are based [12]. In Appendix A, we provide a general discussion of generating functions. Here, we discuss the generating functions for the Hermite–Gauss modes and the Laguerre–Gauss modes. Due to their Gaussian form, these generating functions are much easier to use in calculations than their explicit expressions in terms of the polynomials.

2.5.3.1 Generating function for Hermite–Gauss modes

The *generating function for the Hermite polynomials* $H_n(x)$ is given by [12]

$$\mathcal{H}_0(\eta) = \exp(2x\eta - \eta^2) = \sum_{n=0}^{\infty} \frac{\eta^n}{n!} H_n(x), \tag{2.122}$$

where η is the generating parameter. Individual Hermite polynomials are extracted by

$$\partial_\eta^n \mathcal{H}_0(\eta)\big|_{\eta=0} = H_n(x). \tag{2.123}$$

We combine two such generating functions with different generating parameters μ and ν for the normalized coordinates u and v, respectively, by replacing the respective generating parameters by

$$\eta \to \mu \frac{1 - iw}{\sqrt{1 + w^2}} \quad \text{and} \quad \eta \to \nu \frac{1 - iw}{\sqrt{1 + w^2}}, \tag{2.124}$$

and the coordinates (arguments of the Hermite polynomials) by

$$x \to \frac{\sqrt{2}u}{\sqrt{1 + w^2}} \quad \text{and} \quad x \to \frac{\sqrt{2}v}{\sqrt{1 + w^2}}. \tag{2.125}$$

Finally, we append all the other factors that make up the Hermite–Gauss modes, but remove the normalization constant because they depend on the mode indices in ways that do not conform to the summation that produces the generating function. The resulting expression of the *generating function for the Hermite–Gauss modes* in terms of normalized coordinates then reads

$$\mathcal{H}(\mu, \nu) = \frac{1}{(1 + iw)} \exp\left[\frac{2\sqrt{2}(u\mu + v\nu)}{1 + iw} - \frac{1 - iw}{1 + iw}(\mu^2 + \nu^2) - \frac{u^2 + v^2}{1 + iw}\right]. \tag{2.126}$$

Without the normalization constants, the individual functions are not normalized. The generating function in (2.126) can be used to compute a generating function for the inverse squares of the normalization constants. We compute the inner product in terms of the normalized coordinates between two generating functions with different generating parameters. The result is

$$\langle \mathcal{H}(\mu', \nu'), \mathcal{H}(\mu, \nu) \rangle = \frac{\pi}{2} \exp(2\mu'\mu + 2\nu'\nu). \tag{2.127}$$

Since the result is a function of the products $\mu'\mu$ and $\nu'\nu$, we conclude that the Hermite–Gauss modes are mutually orthogonal, as expected. The normalization constants are obtained by expanding the result of the inner product in powers of $\mu'\mu$ and $\nu'\nu$, respectively. The coefficients of the terms in the expansion are the inverses of the squared normalization constants for the associated modes. When the modes are expressed in terms of $\{x, y, z\}$ instead of the normalized coordinates, additional factors of the beam width w_0 appear.

Exercise 2.6. Show that the normalization constants for the Hermite–Gauss modes that are produced by the generating function in (2.127) are given by

$$N_{m,n} = \sqrt{\frac{2}{\pi 2^{m+n} m! n!}}. \tag{2.128}$$

Exercise 2.7. Show that the generating function for the Hermite–Gauss modes in the Fourier domain reads

$$\mathcal{H}(k_x, k_y; \mu, \nu) = \pi w_0 \exp\left[-i\sqrt{2}(k_x\mu + k_y\nu)w_0 - \frac{z_R + iz}{2k}(k_x^2 + ky^2) + \mu^2 + \nu^2\right], \tag{2.129}$$

Hint: Computing the two-dimensional Fourier transform of (2.126).

2.5.3.2 Generating function for Laguerre–Gauss modes

The *generating function for the Laguerre–Gauss modes* is based on the *generating function for the associated Laguerre polynomials*, which is given by [12]

$$\mathcal{L}_0(\nu;\ell) = \frac{1}{(1-\nu)^{1+\ell}} \exp\left(-\frac{x\nu}{1-\nu}\right) = \sum_{p=0}^{\infty} \nu^p L_p^{\ell}(x), \tag{2.130}$$

for positive integers ℓ. After transforming this generating function appropriately and combining it with the necessary factors in a way similar to how it is done for the Hermite–Gauss modes, we obtain the generating function for the Laguerre–Gauss modes in normalized coordinates, given by

$$\mathcal{L}(\mu,\nu,\sigma) = \frac{1}{(1-\nu)+\mathrm{i}(1+\nu)w} \exp\left[\frac{(u+\mathrm{i}\sigma v)\mu - (u^2+v^2)(1+\nu)}{(1-\nu)+\mathrm{i}(1+\nu)w}\right], \tag{2.131}$$

where $\sigma = \pm 1$ is the sign of ℓ. Again, the modal functions produced by this generating function are unnormalized. The Laguerre–Gauss modes are obtained by calculating

$$\mathrm{LG}_{p\ell}(u,v,w) = N_{p\ell}^{(\mathrm{LG})} \frac{1}{p!} \partial_\nu^p \partial_\mu^{|\ell|} \mathcal{L}(\mu,\nu,\sigma)\Big|_{\mu,\nu=0}, \tag{2.132}$$

and specifying the sign of ℓ through σ.

When we test the *orthogonality* of the Laguerre–Gauss modes with the aid of their generating function (keeping the sign in the two generating functions the same so that $\sigma^2 = 1$), in terms of the normalized coordinates, we get

$$\langle \mathcal{L}(\mu',\nu',\sigma), \mathcal{L}(\mu,\nu,\sigma)\rangle = \frac{\pi}{2(1-\nu'\nu)} \exp\left(\frac{1}{2}\frac{\mu'\mu}{1-\nu'\nu}\right). \tag{2.133}$$

Again, the fact that the result is a function only of the products $\mu'\mu$ and $\nu'\nu$ indicates the expected mutually orthogonality of the Laguerre–Gauss modes. The normalization constants in (2.120) can be obtained from this result.

Exercise 2.8. Show that the normalization constants for the Laguerre–Gauss modes, as given in (2.120), can be obtained from (2.133).

Exercise 2.9. Compute the two-dimensional Fourier transform of (2.131) after converting it to the original coordinates, to show that the generating function for the transverse Fourier domain Laguerre–Gauss modes is given by

$$\mathcal{L}(k_x,k_y,z;\mu,\nu,\sigma) = \frac{1}{1+\nu} \exp\left[-\mathrm{i}\frac{(k_x+\mathrm{i}\sigma k_y)w_0\mu}{2(1+\nu)} - \frac{1}{4}|\mathbf{K}|^2 w_0^2\left(\frac{1-\nu}{1+\nu} + \mathrm{i}\frac{z}{z_R}\right)\right], \tag{2.134}$$

where w_0 is the beam radius and z_R is the Rayleigh range.

2.6 Free space propagation and diffraction

One can say that any *deflection* experienced by a beam of light that is not due to *reflection* or *refraction* is the result of *diffraction*. According to this definition, deflection is the collective noun for reflection, refraction, and diffraction. However, diffraction differs from the other two in an essential way. Reflection and refraction are not wave phenomena—they can be explained in terms of geometrical optics. Diffraction, on the other hand, is a result of the wave properties of light.

Often one may think of diffraction purely as the result of the sharp edges of an obstacle in the path of a light beam. However, although such edges do affect the diffraction of an optical beam, the process of diffraction is what happens to the beam of light as it propagates. An optical beam has a given configuration space amplitude function in some initial plane. After propagating some distance, the amplitude function has changed. This change is the result of the diffraction process. The effect of any obstacle is to set up boundary conditions, which in turn affects the amplitude function of the electric field in the immediate vicinity of the obstacle. Diffraction is what happens when this electric field in the immediate vicinity of the obstacle propagates further. Here, we are not interested in how the electric field in the immediate vicinity of the obstacle is produced. Instead, our focus is on the propagation of an optical beam over some distance, given an initial amplitude function, which may have been imposed by some boundary conditions.

The generic free space optical system is shown in Figure 2.3. The general direction of propagation is denoted by the *z*-axis. We specify a plane at $z = z_1$ (often $z_1 = 0$), which is called the *input plane*, and another plane at $z = z_2 > z_1$, which is called the *output plane*. (If we want to consider *backward propagation*, we would have $z_2 < z_1$.)

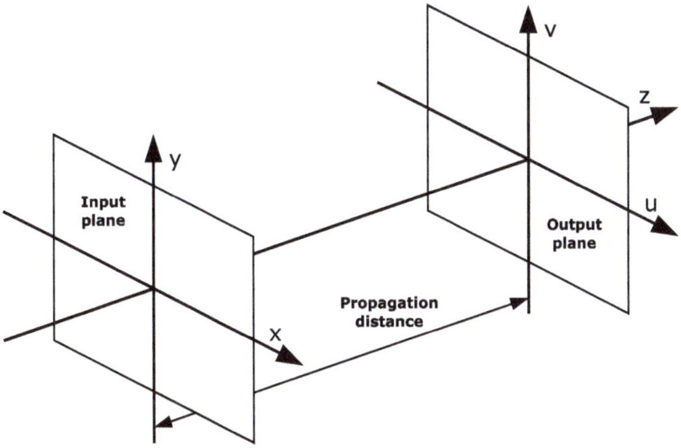

Figure 2.3: Free space system associated with beam propagation.

For a given two-dimensional complex-valued function $f_{\mathrm{in}}(x,y)$ in the input plane (the *input function*), representing the scalar electric field of an optical beam in that plane, we compute the complex-valued function $f_{\mathrm{out}}(u,v)$ for the scalar electric field in the output plane (the *output function*). This process defines the scalar propagation of a beam of light from z_1 to z_2. The input and output functions are both parts of a continuous three-dimensional complex-valued function $f(\mathbf{x})$ of the scalar electric field. In the input plane $f_{\mathrm{in}}(x,y) = f(x,y,z_1)$ and in the output plane $f_{\mathrm{out}}(u,v) = f(u,v,z_2)$. The propagation process reveals the rest of the function $f(\mathbf{x})$.

Based on the discussion about linear systems, we know that if an optical beam has a specific angular spectrum in a plane at $z = z_1$, then this beam would have the same angular spectrum in any another plane at another value of z, provided that the *reference frame* remains the same. Therefore, fixing the value of z to that of the input plane, we compute the angular spectrum as in (2.107) for the input function given by

$$f_{\mathrm{in}}(x,y,t) = \vec{\eta}^* \cdot \mathbf{E}(\mathbf{x},t)\big|_{z=z_1}, \tag{2.135}$$

where $\vec{\eta}$ is a constant state of polarization. The scalar electric phasor field of the optical beam at any other value of z, with the time dependence included, is then given by the inverse Fourier transform

$$f(\mathbf{x},t) = \int F(\mathbf{K},\omega)\exp(-\mathrm{i}\omega t + \mathrm{i}\mathbf{k}\cdot\mathbf{x})\,\mathrm{d}_{\mathrm{b}}k, \tag{2.136}$$

which is equivalent to (2.106) for scalar fields. Under the *monochromatic approximation*, it becomes a two-dimensional integral without the time dependence. In the rest of this section, we employ the monochromatic approximation.

If the same angular spectrum is used to reconstruct the beam at different values of z, why would the beam ever look different at different values of z? The key lies in the z-dependent part of the phase function in the integrand. It is represented as

$$\exp[\mathrm{i}zk_z(\mathbf{K})] = \exp\left(\mathrm{i}z\sqrt{k^2 - |\mathbf{K}|^2}\right) \triangleq \Psi(\mathbf{K},z), \tag{2.137}$$

where the wavenumber k is a fixed parameter under the monochromatic approximation. For propagation from z_1 to z_2, the reconstructed beams at z_1 and z_2 differ in their expressions for the integrand by a factor of $\Psi(\mathbf{K},\Delta z)$, where $\Delta z = z_2 - z_1$. This phase factor indicates how the phase relationship among the different plane waves changes as the beam propagates along the z-direction. This change in the phase is a result of the fact that, although the plane waves all have the same wavelength (in three dimensions), they do not propagate in the same direction. We call this phase factor the *propagation phase factor*. The three-dimensional plane waves consist of two-dimensional *slices* of the plane waves times the propagation phase factor

$$\exp[\mathrm{i}\mathbf{K}\cdot\mathbf{X} + \mathrm{i}k_z(\mathbf{K})\Delta z] = \exp(\mathrm{i}\mathbf{K}\cdot\mathbf{X})\Psi(\mathbf{K},\Delta z), \tag{2.138}$$

which implies that the propagation phase factor $\Psi(\mathbf{K}, \Delta z)$ relates the two-dimensional Fourier transforms at different values of z to each other. For $\Delta z = 0$, the propagation phase factor equals unity $\Psi(\mathbf{K}, 0) = 1$. Therefore, it has no effect when evaluated at the *reference plane* (input plane).

To obtain the output function, we reconstruct it in the output plane from its angular spectrum. This reconstruction process is done in two steps. First, the angular spectrum of the input function is multiplied by the propagation phase factor, using the appropriate propagation distance $\Delta z = z_2 - z_1$. Then we compute the inverse Fourier transform of this result. The entire propagation process is expressed as

$$f(\mathbf{X}, z_2) = \mathcal{F}^{-1}\left\{\mathcal{F}\{f(z_1)\}\,\Psi(\mathbf{K}, z_2 - z_1)\right\}. \tag{2.139}$$

It represents a *rigorous propagation procedure for scalar optical fields*. The two integrations of the respective Fourier transformations can be interchanged, leading to

$$f(\mathbf{X}, z_2) = f(\mathbf{X}, z_1) * \mathcal{F}^{-1}\{\Psi(z_2 - z_1)\} = \int f(\mathbf{X}', z_1) G(\mathbf{X} - \mathbf{X}', \Delta z)\, \mathrm{d}^2 x', \tag{2.140}$$

where

$$G(\mathbf{X}, \Delta z) = \int \exp[\mathrm{i}\mathbf{K} \cdot \mathbf{X} + \mathrm{i}k_z(\mathbf{K})\Delta z]\, \frac{\mathrm{d}^2 k}{(2\pi)^2}. \tag{2.141}$$

The propagation process can therefore be expressed as a *convolution* between the input function $f(\mathbf{X}, z_1)$ and $G(\mathbf{X}, \Delta z)$, which represents the *shift invariant* propagation process in free space. The latter is the *propagation kernel*. It acts like an *impulse response* of free space propagation. The propagation kernel is the inverse Fourier transform of the propagation phase factor

$$G(\mathbf{X}, \Delta z) = \mathcal{F}^{-1}\{\Psi(\Delta z)\}. \tag{2.142}$$

The integral for the propagation kernel in (2.141) produces a Dirac delta function for $\Delta z = 0$, as it should. For $\Delta z > 0$, it is difficult to evaluate. By solving the integral numerically, one can see that it looks like a spherical wave radiating outward from the origin, tapering to zero on the sides where the waves curve toward the input plane. This kernel function is reminiscent of the spherical wave used in the *Huygens principle*.

Exercise 2.10. Show that the propagation of a plane wave reproduces the same plane wave together with a phase shift.

Exercise 2.11. Show that backward propagation is consistent with forward propagation: If we propagate an input function to some distance z, and then propagate it backward to the input plane it reproduces the original input function.

Exercise 2.12. Show that the propagation kernel is unitary, in the context of Section 2.1.5.

2.6.1 Fresnel propagation

Often a beam of light propagates predominantly in a specific direction. The angular spectrum of such a beam is narrow compared to the circular region of propagating plane waves. For such cases, the wavevectors of all the plane waves in the beam lie close to the axis of the beam. It is the *paraxial condition* that we encountered in Section 2.3.5. In the context of scalar propagation theory, it is also called the *Fresnel condition.*

Under the paraxial condition, the magnitude of the transverse wavevector is small compared to the wavenumber

$$|\mathbf{K}| \ll k = \frac{2\pi}{\lambda}. \tag{2.143}$$

As a consequence, the expression for $k_z(\mathbf{K})$ can be approximated as

$$k_z(\mathbf{K}) = \sqrt{k^2 - |\mathbf{K}|^2} \approx k - \frac{|\mathbf{K}|^2}{2k}. \tag{2.144}$$

The propagation phase factor becomes the *paraxial propagation phase factor*, given by

$$\Psi(\mathbf{K}, z) = \exp[\mathrm{i}k_z(\mathbf{K})z] \approx \exp(\mathrm{i}kz)\exp\left(-\mathrm{i}\frac{z|\mathbf{K}|^2}{2k}\right). \tag{2.145}$$

The first exponential function represents the *phase advance* due to propagation over a distance z. It is the phase factor that is removed in (2.83) to get the slow-varying fields. The second exponential is a parabolic phase factor on the Fourier domain.

Upon substituting the paraxial propagation phase factor into the propagation kernel in (2.142) and evaluating the inverse Fourier integral, we get a paraxial approximation of the propagation kernel.[f] The integral can now be evaluated, leading to

$$\begin{aligned} G(\mathbf{X}, z) &\approx \exp(\mathrm{i}kz) \int \exp\left(\mathrm{i}\mathbf{K}\cdot\mathbf{X} - \mathrm{i}\frac{z|\mathbf{K}|^2}{2k}\right) \frac{\mathrm{d}^2k}{(2\pi)^2} \\ &= \frac{-\mathrm{i}\exp(\mathrm{i}kz)}{\lambda z} \exp\left(\mathrm{i}\frac{\pi|\mathbf{X}|^2}{\lambda z}\right). \end{aligned} \tag{2.146}$$

Next, we substitute this paraxial approximation of the propagation kernel back into the propagation integral in (2.140). The result

$$f(\mathbf{X}, z) = \frac{-\mathrm{i}\exp(\mathrm{i}kz)}{\lambda z} \int f(\mathbf{X}') \exp\left(\mathrm{i}\frac{\pi}{\lambda z}|\mathbf{X} - \mathbf{X}'|^2\right) \mathrm{d}^2x', \tag{2.147}$$

f Strictly speaking, the integral is not well-defined for such phase factors because they are not finite-energy functions. To make the calculation more rigorous, we can multiply the integrand by $\exp(-d|\mathbf{X}|^2)$. Then, after evaluating the integral, we take the limit $d \to 0$.

is the *Fresnel diffraction integral*. The paraxial approximation of the propagation kernel in (2.146) is called the *Fresnel kernel*.

Exercise 2.13. Provide a detailed calculation of the expression for the paraxial approximation of the propagation kernel given in (2.146).

Hint: Add $-d|\mathbf{K}|^2$ in the exponent and set $d = 0$ after the integration.

Exercise 2.14. Show that the Fresnel kernel in (2.146) is unitary, in the context of Section 2.1.5.

A different version of the Fresnel diffraction integral is obtained by expanding the argument of the exponent in the Fresnel kernel. It leads to

$$\begin{aligned} f(\mathbf{X}, z) &= \frac{-\mathrm{i}\exp(\mathrm{i}kz)}{\lambda z} \exp\left(\mathrm{i}\frac{\pi|\mathbf{X}|^2}{\lambda z}\right) \\ &\quad \times \int f(\mathbf{X}') \exp\left(\mathrm{i}\frac{\pi|\mathbf{X}'|^2}{\lambda z}\right) \exp\left(-\mathrm{i}\frac{2\pi}{\lambda z}\mathbf{X}\cdot\mathbf{X}'\right) \mathrm{d}^2x' \\ &= \frac{-\mathrm{i}\exp(\mathrm{i}kz)}{\lambda z} Q(\mathbf{X}, z)\mathcal{F}\{fQ(z)\}, \end{aligned} \tag{2.148}$$

where we define the parabolic phase factors as

$$Q(\mathbf{X}, z) \triangleq \exp\left(\mathrm{i}\frac{\pi|\mathbf{X}|^2}{\lambda z}\right) = \exp\left(\mathrm{i}\frac{k|\mathbf{X}|^2}{2z}\right). \tag{2.149}$$

The Fourier transform in (2.148), as represented by $\mathcal{F}\{\cdot\}$ here, is evaluated with the Fourier variables given by

$$\mathbf{K}' = \frac{k\mathbf{X}'}{z}. \tag{2.150}$$

According to (2.148), the Fresnel diffraction process can be expressed in terms of a Fourier transform. The input function is first multiplied by a parabolic phase factor. Then it is Fourier transformed, and the result is again multiplied by a parabolic phase factor. The rigorous propagation procedure requires two Fourier transforms whereas the Fresnel procedure only requires one.

Due to the paraxial approximation, Fresnel diffraction is not as accurate as the rigorous computations. In practice, the rigorous computation is only used for short propagation distances where Fresnel diffraction breaks down. For longer distances, Fresnel diffraction is accurate enough.

2.6.2 Fraunhofer approximation

When the propagation distance z becomes sufficiently large compared to the size of the input plane (input aperture), the parabolic phase factor under the integral can be neglected. For

$$z \gg \frac{\pi|\mathbf{X}|}{\lambda}, \tag{2.151}$$

we can assume that

$$\exp\left(\mathrm{i}\frac{\pi|\mathbf{X}|^2}{\lambda z}\right) \approx 1. \tag{2.152}$$

This situation is the *Fraunhofer approximation*, which is only valid at very large distances (in the *far field*). The condition is often stated as

$$z \gg \frac{D^2}{\lambda}, \tag{2.153}$$

where D is the size of the input function. Under this approximation, the Fresnel diffraction integral reduces to the *Fraunhofer diffraction integral*

$$f(\mathbf{X}, z) = \frac{-\mathrm{i}\exp(\mathrm{i}kz)}{\lambda z}\exp\left(\mathrm{i}\frac{\pi|\mathbf{X}|^2}{\lambda z}\right)\int f(\mathbf{X}')\exp\left(-\mathrm{i}\frac{2\pi}{\lambda z}\mathbf{X}\cdot\mathbf{X}'\right)\,\mathrm{d}^2x'. \tag{2.154}$$

Apart from the parabolic phase factor, the output function is the Fourier transform of the input function. If we are only interested in the intensity of the output function, we only need to compute the modulus square of the Fourier transform of the input function.

2.7 Lens systems

A *thin lens* can be represented as a *phase-only transmission function*

$$t_{\text{lens}}(\mathbf{X}) = \exp[\mathrm{i}\theta(\mathbf{X})], \tag{2.155}$$

provided that any optical beam traversing it is paraxial and can pass through the lens without being clipped by the edge of the lens. Physical lenses have finite sizes and light that passes beyond the edge of the lens is generally considered to be lost. Therefore, if there is a possibility that the traversing beam can be clipped at the edge of a lens, the transmission function can be modified by multiplying the phase-only transmission function by an *aperture function*

$$t_{\text{lens}}(\mathbf{X}) = A(\mathbf{X})\exp[\mathrm{i}\theta(\mathbf{X})]. \tag{2.156}$$

The aperture function $A(\mathbf{X})$ is a binary function equal to 1 within the region of the lens and equal to 0 beyond the edge where the light is lost. Usually, we assume that the optical beam can pass through the lens without being clipped. Therefore, we treat the lens as a phase-only transmission function.

Under the *paraxial approximation*, the phase-only transmission function of a positive[g] lens is represented by a parabolic phase factor given by

$$t_{\text{lens}}(\mathbf{X}) = \exp\left(-\mathrm{i}\frac{kr^2}{2f}\right) \equiv Q^*(\mathbf{X}, f), \tag{2.157}$$

where f is the *focal length* of the lens, k is the wavenumber and $r = |\mathbf{X}|$. We also express it in terms of (2.149).

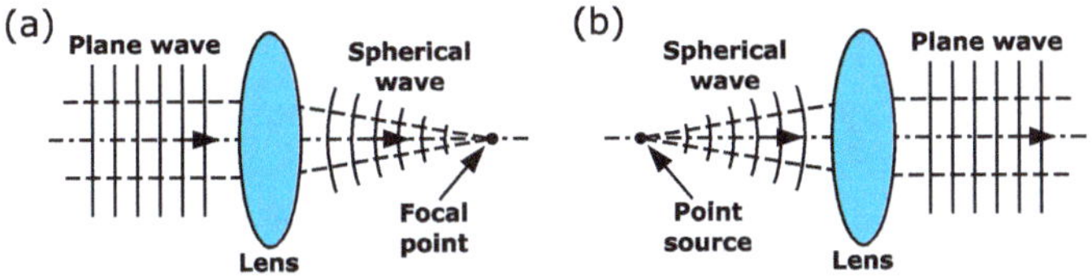

Figure 2.4: Positive lens (a) focusing a plane wave and (b) collimating a spherical wave.

A positive thin lens is an optical element that focuses light. When a plane wave is incident on a lens, as shown in Figure 2.4(a), it focuses the light, causing it to *converge* to a *focal point* located at a distance of one *focal length* behind the lens. A spherical wave diverging from a point source located at one focal length in front of a lens is *collimated* by the lens, forming a plane wave behind the lens, as shown in Figure 2.4(b). The *front-focal plane* (*back-focal plane*) is defined as the transverse plane located at a distance of one focal length in front of (behind) a lens.

Strictly speaking, a physical beam passing through a physical lens is neither an exact plane wave, nor an exact spherical wave. A better description is given in terms of Gaussian beams, discussed in Section 2.5.1.

Exercise 2.15. Use the Fresnel diffraction integral in (2.148) to show that the transmission function of the positive thin lens in (2.157) leads to a focal point at $z = f$.

Exercise 2.16. Show that a tilted plane wave passing through the lens produces a corresponding shift for the focal point at $z = f$.

---

g A negative lens increases divergence.

2.7.1 Fourier transformation by a 2*f* system

Considering an arbitrary beam passing through a lens, we can express it in terms of its plane wave expansion. Each plane wave passing through the lens forms a focal point in the back-focal plane. If the plane wave passes obliquely through the lens, then the focal point is shifted to the side by a distance given by the angle of incidence. Since the amplitude of each plane wave in the beam is given by the function value of the input function's angular spectrum for the wavevector of that plane wave, the output function in the back-focal plane is given by the angular spectrum of the input function as a function of the output plane coordinates instead of the wavevectors. Similar to (2.150), the Fourier domain variables are related to the output plane coordinates by

$$\mathbf{K} = \frac{k\mathbf{U}}{f}. \tag{2.158}$$

Hence, the lens breaks the beam up into its Fourier components, thus performing a Fourier transform on the input function of the beam.

Here, we provide a quantitative analysis to show that the lens performs a Fourier transform on the input function located in the front-focal plane. For this purpose, we consider an arbitrary complex-valued input function $g_0(x,y)$ in the front-focal plane. To compute the function in the back-focal plane, we perform a three-step process: first, we compute the Fresnel propagation from the front-focal plane to the lens plane; then we multiply the result with the lens transmission function; and finally we compute the Fresnel propagation of the last result from the lens plane to the back-focal plane.

The Fresnel transform of the input function for the first step is performed over a propagation distance $z = f$. Represented in terms of (2.148), the result is

$$g_1(\mathbf{X}) = \frac{-\mathrm{i}\exp(\mathrm{i}kf)}{\lambda f} Q(\mathbf{X},f)\mathcal{F}\{g_0 Q(f)\}, \tag{2.159}$$

where $g_1(\mathbf{X})$ is the complex-valued function in the lens plane right in front of the lens, and $Q(\mathbf{X},f)$ is given by (2.149) with $z = f$.

The next step is to multiply $g_1(\mathbf{X})$ with the transmission function of the lens, which is given by $Q^*(\mathbf{X},f)$ according to (2.157). As a result, it removes the factor of $Q(\mathbf{X},f)$,

$$g_1'(\mathbf{X}) = g_1(\mathbf{X})Q^*(\mathbf{X},f) = \frac{-\mathrm{i}\exp(\mathrm{i}kf)}{\lambda f}\mathcal{F}\{g_0 Q(f)\}. \tag{2.160}$$

The final step is to apply another Fresnel transform, again with $z = f$, on the result after the lens. It produces

$$\begin{aligned} g_2(\mathbf{U}) &= \frac{-\mathrm{i}\exp(\mathrm{i}kf)}{\lambda f} Q(\mathbf{U},f)\mathcal{F}\{g_1' Q(f)\} \\ &= \frac{-\exp(\mathrm{i}2kf)}{\lambda^2 f^2} Q(\mathbf{U},f)\mathcal{F}\{\mathcal{F}\{g_0 Q(f)\}Q(f)\}, \end{aligned} \tag{2.161}$$

where $\mathbf{U} = u\vec{x} + v\vec{y}$ is the two-dimensional position vector on the output plane (the back-focal plane of the lens). The Fourier transform of the parabolic phase factor produces a complex conjugated parabolic phase factor in terms of the output coordinates

$$\begin{aligned}\mathcal{F}\{Q(f)\} &= \int \exp\left(-\mathrm{i}\frac{2\pi}{\lambda f}\mathbf{U}\cdot\mathbf{X}\right)\exp\left(\mathrm{i}\frac{\pi}{\lambda f}|\mathbf{X}|^2\right)\,\mathrm{d}^2x \\ &= \mathrm{i}\lambda f \exp\left(-\mathrm{i}\frac{\pi}{\lambda f}|\mathbf{U}|^2\right) \equiv \mathrm{i}\lambda f Q^*(\mathbf{U},f).\end{aligned} \tag{2.162}$$

The integral in (2.161) is evaluated by changing the order of integration and then expanding the resulting parabolic phase factor. It then becomes

$$\begin{aligned}&\mathcal{F}\{\mathcal{F}\{g_0 Q(f)\}Q(f)\} \\ &= \int g_0(\mathbf{X}')Q(\mathbf{X}',f)\int \exp\left[-\mathrm{i}\frac{2\pi}{\lambda f}(\mathbf{X}'+\mathbf{U})\cdot\mathbf{X}\right]Q(\mathbf{X},f)\,\mathrm{d}^2x\,\mathrm{d}^2x' \\ &= \int g_0(\mathbf{X}')Q(\mathbf{X}',f)\mathrm{i}\lambda f Q^*(\mathbf{X}'+\mathbf{U},f)\,\mathrm{d}^2x' \\ &= \mathrm{i}\lambda f Q^*(\mathbf{U},f)\int g_0(\mathbf{X}')\exp\left(-\mathrm{i}\frac{2\pi}{\lambda f}\mathbf{U}\cdot\mathbf{X}'\right)\,\mathrm{d}^2x'.\end{aligned} \tag{2.163}$$

Substituted back into (2.161), it produces the output given by the Fourier transform

$$g_2(\mathbf{U}) = \frac{-\mathrm{i}\exp(\mathrm{i}2kf)}{\lambda f}\int g_0(\mathbf{X}')\exp\left(-\mathrm{i}\frac{2\pi}{\lambda f}\mathbf{U}\cdot\mathbf{X}'\right)\,\mathrm{d}^2x' \tag{2.164}$$

$$\equiv \frac{-\mathrm{i}\exp(\mathrm{i}2kf)}{\lambda f}\mathcal{F}\{g_0(\mathbf{X}')\} \equiv \frac{-\mathrm{i}\exp(\mathrm{i}2kf)}{\lambda f}G\left(\frac{2\pi\mathbf{U}}{\lambda f}\right),$$

where $G(\mathbf{K})$ is the angular spectrum of the input function. Apart from the complex constant factor in front, the result in the output plane (back-focal plane) is the Fourier transform of the complex function in the input plane (front-focal plane). This simple optical system, shown in Figure 2.5, is called a $2f$ system, because it stretches over a distance of twice the focal length. It produces the same result as Fraunhofer diffraction in the far field. Therefore, the output plane of the $2f$ system is often called the "far field."

Exercise 2.17. Show that a Gaussian beam with its waist in the input plane (front-focal plane) of a $2f$ system produces another Gaussian beam with its waist in the output plane (back-focal plane), and that the output Gaussian beam waist radius w_{out} is related to the input Gaussian beam waist radius w_{in} by

$$w_{out} = \frac{\lambda f}{\pi w_{in}}. \tag{2.165}$$

Exercise 2.18. Find an expression for the function in the back-focal plane of a thin lens if an input function $g(x,y)$ is placed directly against the front of the lens. Show that the intensity of the output function corresponds to the intensity of the Fourier transform of the input function.

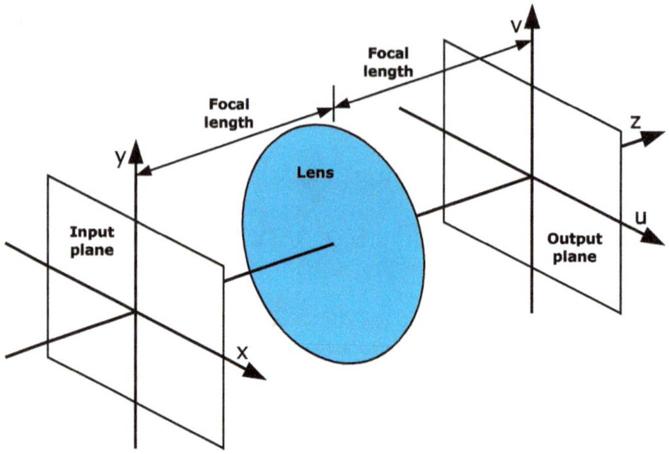

Figure 2.5: Fourier transforming $2f$ system.

2.7.2 Imaging properties of a lens and the $4f$ system

A well-known property of a lens is that it can be used to form an image of an object. To see how this imaging process works, we can repeat the calculation of the previous section with arbitrary distances on either side of the lens. We leave it as an exercise.

Exercise 2.19. Repeat the calculation in Section 2.7.1, but instead of propagating over distances of a focal length before and after the lens, compute the propagation over arbitrary distances z_1 and z_2. Show that when

$$\frac{1}{z_1}+\frac{1}{z_2}=\frac{1}{f}, \tag{2.166}$$

the output function is a 180° rotated and scaled version of the input function times a parabolic phase factor, and that the scaling factor or *magnification* is

$$M=\frac{z_2}{z_1}. \tag{2.167}$$

When a single lens is used for imaging, the complex-valued output function is modulated by a parabolic phase factor. A way to obtain an image without this parabolic phase modulation is to cascade two $2f$ systems, thus producing a $4f$ system, as shown in Figure 2.6. The reason why two consecutive $2f$ systems produce an image is because two consecutive Fourier transforms reproduce the original input function, albeit with a negative argument, which corresponds to a 180° rotation. The calculation is similar to what we had in (2.163), but without the Q's. Two consecutive Fourier transformations, as performed by two $2f$ systems, produce

$$
\begin{aligned}
g_2(\mathbf{U}) &= \frac{-\exp(\mathrm{i}4kf)}{\lambda^2 f^2}\mathcal{F}\{\mathcal{F}\{g_0\}\} \\
&= \frac{-\exp(\mathrm{i}4kf)}{\lambda^2 f^2}\int g_0(\mathbf{X}')\int \exp\left[-\mathrm{i}\frac{2\pi}{\lambda f}(\mathbf{X}'+\mathbf{U})\cdot\mathbf{X}\right]\mathrm{d}^2x\,\mathrm{d}^2x' \\
&= \frac{-\exp(\mathrm{i}4kf)}{\lambda^2 f^2}\int g_0(\mathbf{X}')\delta\left(\frac{\mathbf{X}'}{\lambda f}+\frac{\mathbf{U}}{\lambda f}\right)\mathrm{d}^2x' \\
&= -\exp(\mathrm{i}4kf)g_0(-\mathbf{U}). \qquad (2.168)
\end{aligned}
$$

The constant global phase factor can be discarded, leaving the 180° rotation due to the minus sign in the argument. Here, we have assumed that the focal lengths of the two $2f$ systems are the same. When they are different, the output function is magnified.

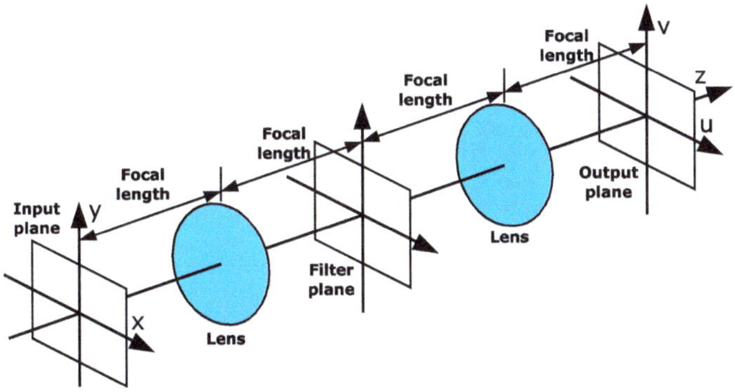

Figure 2.6: Imaging $4f$ system.

Exercise 2.20. Show that a $4f$ system consisting of two $2f$ systems with different focal lengths f_1 and f_2 produce a magnified image with a magnification factor:

$$M = \frac{f_2}{f_1}. \qquad (2.169)$$

2.7.3 Resolution

In the previous two sections, it is assumed that the lens performs an ideal Fourier transformation. In a physical lens system, there are various physical properties of the system that introduce unwanted effects on the process. The most prominent effect is a reduction in the *resolution*, caused by the finite size of the lens. It produces a *point-spread function* that distorts the Fourier transformed function in the back-focal plane of the lens. This distortion only happens when the beam that passes through the lens is broader than the lens diameter causing it to be clipped by the edge of the lens. Often, this effect is modelled as being shift invariant. The point-spread function then becomes an impulse response, which is convolved with the Fourier transformed function in the back-focal plane of the lens. This impulse response is the Fourier transform of the aperture function that determines the size of the lens. For a circular aperture, it has the form

$$h(r) = \frac{D}{2r} J_1\left(\frac{\pi D r}{\lambda f}\right), \tag{2.170}$$

which is called the *Airy pattern*, where $J_1(\cdot)$ is the first-order Bessel function of the first kind [12], r is the radial coordinate in the output plane, λ is the wavelength, f is the focal length of the lens, and D is the diameter of the lens. The impulse response has a transverse size that is inversely proportional to the size of the lens, setting a limit on the resolution (the smallest feature size of the function) that can be obtained in the back-focal plane of the lens. For a circular aperture, the resolution in the output plane is given by a distance of

$$\Delta d = \frac{1.22\lambda f}{D}, \tag{2.171}$$

based on the first null of the Airy pattern.

Exercise 2.21. Compute the Airy pattern in (2.170).

Another effect of the lens is produced by distortions of the wavefront caused by *lens aberrations* [1, 2]. High quality lenses have less aberrations, but even the best lenses have some aberrations. These aberrations can also contribute to the reduction of resolution in the output plane. However, optical systems are often designed in such a way that the effect of the lens apertures on the resolution dominates. In such a case, the system is said to be *diffraction limited*, and the effect of aberrations can be ignored. In all optical systems considered here, we assume diffraction limited conditions.

2.7.4 Gradient index lens

A *gradient index (GRIN) medium* is a versatile medium that can serve as a complete lens system. The construction of a *GRIN lens* is a cylindrical dielectric medium with a radially varying refractive index. The function of the refractive index is

$$n(r) = n_0\sqrt{1 - Ar^2} \approx n_0 - \frac{1}{2}n_0 A r^2, \tag{2.172}$$

where n_0 is the maximum refractive index found on the symmetry axis (z-axis) of the GRIN lens, A is a *gradient constant*, and the squared radial distance is $r^2 = x^2 + y^2$. The approximation assumes that A times the squared radius of the GRIN lens is small. It implies *paraxial* propagation in the GRIN lens. Depending on the length of a GRIN lens, it can act as a $2f$ system, as a $4f$ system, or as a lens with a virtually unlimited range of focal lengths.

When an optical beam enters such a GRIN lens, propagating in the z-direction, it gradually turns into a converging beam, eventually forming a focal point (or Gaussian waist). Beyond the focal point the beam diverges again, but the GRIN medium again turns this diverging beam into a converging beam to form another focal point some distance later. This process is repeated over and over again, producing a periodic pattern along the z-direction. The period for this pattern is the same, regardless of the mode of the optical beam. It is a characteristic property of the GRIN lens, referred to as the *pitch* and given by

$$P = \frac{2\pi}{\sqrt{A}}. \tag{2.173}$$

Within a distance of one pitch, there are two focal points.

A GRIN lens with the length of a quarter pitch acts like a $2f$ system. The optical field on the back surface of such a GRIN lens is the two-dimensional Fourier transform of the field entering the GRIN lens at its front surface. A GRIN lens with the length of half a pitch thus acts like a $4f$ system, with the optical field entering the front surface being imaged onto the back surface with a 180° rotation around the z-axis.

The paraxial wave equation for a GRIN lens medium is given by

$$\nabla_{x,y}^2 f(\mathbf{x}) + i2n_0 k \partial_z f(\mathbf{x}) - \frac{4r^2}{w_m^4} f(\mathbf{x}) = 0, \tag{2.174}$$

where $r^2 = x^2 + y^2$, and

$$w_m^2 = \frac{2}{n_0 k \sqrt{A}}. \tag{2.175}$$

The first two terms are equivalent to those of the free space paraxial wave equation in (2.91), the third term is due to the inhomogeneous medium. As a result, the solutions are expected to be expressed in terms of Gaussian functions. Indeed, the equivalent of the Gaussian solution for (2.91) is

$$G_0(\mathbf{x}) = \sqrt{\frac{2}{\pi}} \frac{w}{g_1(z)} \exp\left[-\frac{g_2(z)}{g_1(z) w_m^2} |\mathbf{X}|^2\right], \tag{2.176}$$

where w is the mode size at the front surface for $z = 0$, and

$$\begin{aligned} g_1(z) &= w^2 \cos(z\sqrt{A}) + i w_m^2 \sin(z\sqrt{A}), \\ g_2(z) &= w_m^2 \cos(z\sqrt{A}) + i w^2 \sin(z\sqrt{A}). \end{aligned} \tag{2.177}$$

After a quarter pitch, $z = \frac{1}{4}P = \pi/2\sqrt{A}$ (a $2f$ system), the mode size changes to

$$w' = \frac{w_m^2}{w}. \tag{2.178}$$

So, it oscillates between w and w' during propagation. For $w = w_m$, the beam size remains constant. Thus, w_m represents a characteristic mode size.

When the initial beam enters the GRIN lens at an angle or is shifted to the side, the beam inside the GRIN lens propagates in a sinusoidal fashion, oscillating from side to side. The solution for this situation is

$$G_{\text{oss}}(\mathbf{x}) = \sqrt{\frac{2}{\pi}}\frac{w}{g_1(z)} \exp\left[-\frac{\left|g_2(z)\mathbf{X} - w_m^2\mathbf{X}_0\right|^2}{g_1(z)g_2(z)w_m^2} - \mathrm{i}\frac{\sin(z\sqrt{A})}{g_2(z)}|\mathbf{X}_0|^2\right], \tag{2.179}$$

where $\mathbf{X}_0$ represents the maximum transverse displacement that occurs at $z = 0$, for the assumed initial conditions. At a quarter pitch, the beam is located on the symmetry axis, with a phase tilt given by $\exp(-\mathrm{i}2\mathbf{X}\cdot\mathbf{X}_0/w_m^2)$. In Figure 2.7, this oscillation is shown for the case when two oppositely displaced beams enter the GRIN lens. Their oscillations cause them to pass through each other periodically, producing interference. This scenario is analogous to the evolution of the *marginal probability distribution* of a Schrödinger cat state, which is considered in Section 6.1.4.

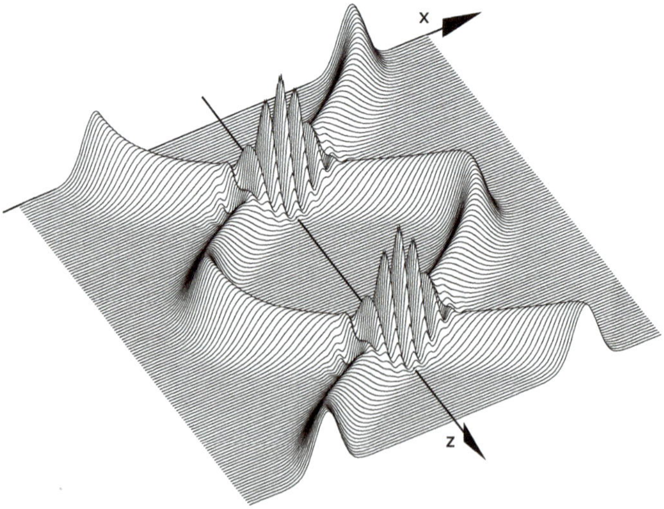

Figure 2.7: Oscillations for two displaced Gaussian beams are shown as the evolution of a one-dimensional cross-section of the two beams along x as a function of z for one full period of oscillation. It shows the interference between the two beams when they pass through each other.

There are also discrete solutions for the paraxial wave equation in (2.174). One such set of solutions are related to the Hermite–Gauss modes, discussed in Section 2.5.2. In the GRIN lens medium, these Hermite–Gauss modes are

$$G_{m,n}(\mathbf{x}) = \frac{N_{m,n}}{g_1(z)}\left[\frac{v(z)}{g_1(z)}\right]^{n+m} H_m\left[\frac{x}{v(z)}\right] H_n\left[\frac{y}{v(z)}\right]\exp\left[-\frac{g_2(z)|\mathbf{X}|^2}{g_1(z)w_m^2}\right], \tag{2.180}$$

where $N_{m,n}$ is a normalization constant, $g_1(z)$ and $g_2(z)$ are given in (2.177), H_m and H_n are Hermite polynomials [12], and

$$v(z) = \frac{1}{\sqrt{2}w}\left[w^4\cos^2(z\sqrt{A}) + w_m^4\sin^2(z\sqrt{A})\right]^{1/2}. \tag{2.181}$$

If $w = w_m$, these modes become

$$G_{m,n}(\mathbf{x}) = N_{m,n}H_m\left(\frac{\sqrt{2}x}{w_m}\right)H_n\left(\frac{\sqrt{2}y}{w_m}\right)\exp\left[-\frac{|\mathbf{X}|^2}{w_m^2} - \mathrm{i}(1+n+m)z\sqrt{A}\right]. \tag{2.182}$$

The z-dependence becomes an exponential phase factor.

Apart from the practical usefulness of such GRIN lenses, they serve as a physical analogue of a harmonic oscillator for waves or fields. With an appropriate redefinition of variables and a rearrangement of the terms in its equation of motion under the paraxial approximation in (2.174), we find it to be formally equivalent to the *Schrödinger equation* of the *quantum harmonic oscillator*, discussed in Section 3.6.6. By studying the solutions of (2.174), we can obtain an idea of the physical nature of such a system.

2.8 Angular momentum

The *angular momentum* in an optical field can take on two forms, one being *spin angular momentum*, associated with the state of polarization of the optical field, and the other is *orbital angular momentum* that is associated with the spatial degrees of freedom of the optical field.

According to *Noether's theorem* [9], a continuous symmetry always has a conserved quantity associated with it. It thus follows that a study of angular momentum in optical fields is closely related to rotation symmetry: if the system through which an optical field propagates is rotationally symmetric, the angular momentum in that field is conserved. For this reason, we first discuss three-dimensional rotations in the mathematical language of *Lie group theory* [6]. However, it is not intended as a comprehensive treatment of the topic of Lie group theory.

2.8.1 Three-dimensional rotations

From a mathematical point of view, the set of all three-dimensional rotations forms a *Lie group*, denoted by SO(3). Each of these rotations in the set can be represented as a matrix. However, the nature of such a set of matrices (i. e., their *representation*) depends on the nature of the quantities on which they operate. There are different *irreducible representations* of a Lie group. In the case of the SO(3) group, they are referred to as the *spin* representations. For the three-dimensional case considered here, the representation is called the *spin-1* ("spin one") representation. The *spin-0* representation is a *singlet* giving a 1-dimensional "matrix," which remains invariant under transformations. The Lie group SO(3) only has integer spin representations. However, SU(2) is another Lie group, which is the *universal cover group*[h] of SO(3) and which also contains half-integer spin representations, in addition to the integer spin representations. Based on the *spin-statistics theorem* [15], the half-integer spin representations are specifically associated with fermion fields. However, that does not mean that half-integer spin representations do not appear in other scenarios. Although we only consider photon fields (which are boson fields) and, therefore, should only need the integer spin representations, the states of polarization of coherent paraxial optical beams (discussed in Section 2.9) are conveniently represented by the spin-$\frac{1}{2}$ ("spin-half") representation of SU(2). It thus follows that the properties of symmetries, as presented in terms of Lie group theory, are relevant in both classical and quantum physics. Nevertheless, only a little bit of Lie group theory is needed here.

The Lie groups SO(3) and SU(2) have the same *Lie algebra*, consisting of the same *generators* of the group. An arbitrary element of a Lie group is expressed as an exponential function with a linear combination of the generators in its argument. In two dimensions (associated with its spin-$\frac{1}{2}$ representation), the generators of SU(2) are the *Pauli matrices*

$$\sigma_x = \begin{bmatrix} 0 & 1 \\ 1 & 0 \end{bmatrix}, \quad \sigma_y = \begin{bmatrix} 0 & -\mathrm{i} \\ \mathrm{i} & 0 \end{bmatrix}, \quad \sigma_z = \begin{bmatrix} 1 & 0 \\ 0 & -1 \end{bmatrix}. \tag{2.183}$$

In three dimensions (for the spin-1 representation), the generators are given by

$$J_x = \begin{bmatrix} 0 & 0 & 0 \\ 0 & 0 & -\mathrm{i} \\ 0 & \mathrm{i} & 0 \end{bmatrix}, \quad J_y = \begin{bmatrix} 0 & 0 & \mathrm{i} \\ 0 & 0 & 0 \\ -\mathrm{i} & 0 & 0 \end{bmatrix}, \quad J_z = \begin{bmatrix} 0 & -\mathrm{i} & 0 \\ \mathrm{i} & 0 & 0 \\ 0 & 0 & 0 \end{bmatrix}. \tag{2.184}$$

Both these sets of generators obey a commutation relation that defines the Lie algebra. In tensor notation, it is given by

h The space of all SU(2) transformations contains all the SO(3) transformations twice.

$$[T_a, T_b] \triangleq T_a T_b - T_b T_a = \mathrm{i}\varepsilon_{abc} T_c, \tag{2.185}$$

where T_a represents the generators (either σ_a or J_a), and ε_{abc} is the *totally antisymmetric tensor* with $\varepsilon_{123} = 1$. For a general Lie algebra, ε_{abc} is replaced by the *structure constants* f_{abc} that define the properties of the Lie algebra.

The elements of the Lie group for a specific irreducible representation are constructed from the generators in that irreducible representation by

$$U(\mathbf{g}) = \exp(\mathbf{g} \cdot \mathbf{T}), \tag{2.186}$$

where **g** represents a vector of coefficients and **T** a vector of generator matrices. For the SU(2) group, it becomes the rotation matrix

$$R(\mathbf{a}) = \exp(a_x \sigma_x + a_y \sigma_y + a_z \sigma_z) = \exp(\mathbf{a} \cdot \vec{\sigma}), \tag{2.187}$$

in two dimensions, or

$$R(\mathbf{a}) = \exp(a_x J_x + a_y J_y + a_z J_z) = \exp\left(\mathbf{a} \cdot \vec{J}\right), \tag{2.188}$$

in three dimensions, where **a** is the coefficient vector, and $\vec{\sigma}$ and $\vec{J}$ are the generators in the respective number of dimensions expressed as vectors.

When we apply a rotation to an electromagnetic field, the rotation matrix operates on the indices of the field vector and in its argument, leading to

$$\mathbf{E}(\mathbf{x}) \to R\mathbf{E}\left(R^{-1}\mathbf{x}\right), \tag{2.189}$$

where R represents the 3×3 rotation matrix. The inverse is applied in the argument. When applied to a differential equation (such as the Helmholtz equation), the rotation matrix also operates on the derivatives with respect to the spatial coordinates.

2.8.2 Angular momentum conservation

The conserved quantity that is related to rotation is obtained as the zeroth component of the associated *Noether current*. For the electromagnetic field, it is given by the *angular momentum vector*

$$\mathbf{M} = \mathbf{x} \times \mathbf{S}, \tag{2.190}$$

where **x** is the position vector, and **S** is the *Poynting vector*, given by

$$\mathbf{S} = \frac{1}{c}\mathbf{E} \times \mathbf{H}, \tag{2.191}$$

in terms of the electric and magnetic fields **E** and **H** and the speed of light c.

When dealing with the angular momentum of massive particles, we can distinguish between their *intrinsic* angular momentum (or spin angular momentum) and their *extrinsic* angular momentum (or orbital angular momentum) by considering their angular momentum in their *rest frames*. Here, we treat them as *classical particles*—dimensionless points travelling on classical trajectories. The orbital angular momentum requires that a particle has a nonzero momentum. As a result, there is no orbital angular momentum for a particle in its rest frame.

Alternatively, we can consider the component of the angular momentum along the direction of motion. The orbital angular momentum of a particle is always perpendicular to its motion. Therefore, the angular momentum along the direction of its motion can only come from the intrinsic part. The component of the angular momentum along its direction of motion is called the *helicity* of the particle. The helicity is computed as

$$\text{helicity} = \frac{\mathbf{p} \cdot \vec{J}}{|\mathbf{p}|}, \tag{2.192}$$

where $\mathbf{p}$ is the momentum of the particle and $\vec{J}$ is a vector of angular momentum operations. Expressed as matrices, these operations are the generators given in (2.184).

We can use (2.192) to study the helicity of an electromagnetic field. Having zero mass, the electromagnetic field always propagates at the speed of light and does not have a rest frame. Therefore, helicity is the obvious choice.

An electromagnetic field is not like a classical particle, which is often treated as a dimensionless point. The electromagnetic field, on the other hand, is an extended field. Therefore, even if we consider the angular momentum along the direction of propagation, we can still observe orbital angular momentum. In this case, it is called *intrinsic orbital angular momentum*. The spin angular momentum is the part of the intrinsic angular momentum that is not orbital angular momentum. For the electromagnetic field, the spin angular momentum is represented in terms of its state of polarization. However, for a general electromagnetic field, it is not so easy to distinguish between spin angular momentum and orbital angular momentum.

2.8.3 Angular momentum of paraxial fields

Under the *paraxial approximation*, the three-dimensional rotation invariance of the electromagnetic field becomes a one-dimensional rotation invariance. The only rotation invariance that remains is with respect to rotations around the propagation axis (also called the *optical axis*).

Instead of the SO(3) Lie group, we now have the SO(2) Lie group, which is equivalent to the U(1) Lie group. It is an *Abelian group* (the group transformations commute with one another). In contrast, the three-dimensional rotations form a *non-Abelian group* because they do not commute with one another.

In the case of the helicity in (2.192), the momentum vector is given by the optical axis (z-axis). So, the helicity becomes

$$\text{helicity} = \vec{z} \cdot \vec{J} = J_z. \tag{2.193}$$

For the intrinsic (spin) part of the helicity h_{spin}, represented as a matrix, we have $h_{\text{spin}} = J_z$, where J_z is given in (2.184). Since we are only interested in the transverse dimensions, we can restrict J_z to a two-dimensional matrix, in which case it becomes the Pauli matrix σ_y given in (2.183), which is associated with the spin-$\frac{1}{2}$ representation of SU(2). The polarization eigenvectors of σ_y are the left- and right-hand circular states of polarization (see Section 2.9). A paraxial optical field can therefore be represented as a superposition of left- and right-hand circular states of polarized light.

Here, σ_y is the only generator for the two-dimensional representation of the SO(2) group. In other words, all the two-dimensional rotations around the optical axis can be constructed with σ_y. Therefore, the rotation matrix is represented by

$$R_{\text{spin}}(\alpha) = \exp\left(\alpha\sigma_y\right), \tag{2.194}$$

where α is the rotation angle.

Considering the orbital part of the helicity h_{orb}, we find

$$h_{\text{orb}} = \mathrm{i}\left(\mathbf{x}_{\perp} \times \nabla_{xy}\right) \cdot \vec{z} = \mathrm{i}\left(x\partial_y - y\partial_x\right) \triangleq L_z. \tag{2.195}$$

In cylindrical coordinates, it becomes

$$h_{\text{orb}} = L_z = \mathrm{i}\partial_{\phi}. \tag{2.196}$$

The eigenfunctions of the orbital helicity operation are those modes for which the azimuthal ϕ-dependence is completely given by a factor of the form $\exp(\mathrm{i}\ell\phi)$ where ℓ is the *azimuthal index* (a signed integer), which we encountered before in the context of the Laguerre–Gauss modes in Section 2.5.2. It then follows that the orbital angular momentum in these modes is proportional to ℓ, which also represents the topological charge of the phase singularity on the axis of these modes. Since the ϕ-dependence of a Laguerre–Gauss mode is completely given by $\exp(\mathrm{i}\ell\phi)$, it is an eigenfunction of the orbital helicity operation. The Laguerre–Gauss modes are therefore called *orbital angular momentum* (OAM) modes. However, they are not the only OAM modes. Other examples include the *Bessel modes*, which are solutions of the Helmholtz equation in cylindrical coordinates.

2.9 Polarization

When an electromagnetic field propagates through a source-free, isotropic, homogeneous medium, both the electric field and the magnetic field are orthogonal to the direction of propagation. The orientation of the electric field vector can change as a function of time and propagation distance. The orientation of the electric field vector and its temporal evolution on the transverse plane defines the *state of polarization* of the electromagnetic field. Such a transverse plane is best defined for a *paraxial* beam. Therefore, we assume for the sake of the discussion on polarization that we are dealing with a paraxial optical beam propagating in the z-direction.

2.9.1 States of polarization

Based on their qualitative properties, as determined by the nature of the electric field vector and how it changes in time and as a function of the propagation distance, the different states of polarization of a coherent paraxial optical field can be divided into different kinds. Generically, the electric field vector of a plane wave propagating in the z-direction is given by (2.97), where $\vec{\eta}$ is a complex unit vector, called the *polarization vector* representing the state of polarization.

When the orientation of the electric field vector remains constant (apart from the oscillation of the field), both as a function of time and along the propagation direction, the light is said to be *linearly polarized*. The polarization vector of a linearly polarized plane wave propagating in the z-direction is represented by

$$\vec{\eta} = \cos(\alpha)\vec{x} + \sin(\alpha)\vec{y}, \tag{2.197}$$

where α is the *inclination angle*, representing the orientation of the electric field vector on the two-dimensional transverse plane. Two inclination angles that differ by 180° (π rad) represent the same linear state of polarization.

When the orientation of the electric field vector rotates as a function of time and propagation distance, but retains a constant magnitude, the light is said to be *circularly polarized*. Looking in the propagation direction, one observes a right- (left-) circularly polarized electric field rotating clockwise (anticlockwise) as a function of time.[i] A circularly polarized plane wave propagating in the z-direction is represented by a complex polarization vector given by

$$\vec{\eta} = \frac{1}{\sqrt{2}}(\vec{x} \pm i\vec{y}). \tag{2.198}$$

The positive (negative) sign represents the left-handed (right-handed) circularly polarized plane wave (using the *physics sign convention*).

The most general state of polarization is called *elliptical polarization*. It is represented by a polarization vector

i This definition of the handedness is the IEEE definition used by the electrical engineering community. Many optics textbooks use the opposite convention.

$$\vec{\eta} = \eta_x \vec{x} + \eta_y \vec{y}, \tag{2.199}$$

where η_x and η_y are complex-valued constants obeying $|\eta_x|^2 + |\eta_y|^2 = 1$. In general, the magnitude of the electric field changes (oscillates) and the orientation of the vector rotates. The tip of the electric field vector moves along an ellipse as it oscillates. Such an ellipse has a major axis and a minor axis. The major axis indicates the orientation of the state of polarization. The motion along the ellipse can either be right-handed or left-handed. While linear polarization has an orientation but no handedness and circular polarization has a handedness but no orientation, general elliptical polarization has both an orientation and a handedness. The linear and circular states of polarization are special cases of elliptical states of polarization.

An optical beam can also be *partially polarized* or completely *unpolarized* if they are not perfectly coherent. In such cases, the optical beam is represented by an *incoherent sum* of pure states of polarization for coherent optical beams.

Exercise 2.22. Find the real-valued expression for a left-circularly polarized plane wave and explain how this shows that the electric field vector rotates anticlockwise.

2.9.2 Poincaré sphere

The fact that the states of polarization of a coherent paraxial beam are expressed by two-dimensional unit vectors, implies that the different states of polarization can be represented by the points on the surface of a sphere, called the *Poincaré sphere*, as shown in Figure 2.8. Imagine the sphere is like the earth. At the north and south poles we get the right- and left-hand circular states of polarization, respectively. On the equator, we find all the linear states of polarization with their inclination angles gradually changing as one moves along the equator so that geometrically opposite points represent orthogonal polarizations. As a result, the inclination angle α is equal to half the azimuthal coordinate ϕ (i. e., $\alpha = \frac{1}{2}\phi$).

Between the equator and the poles, we find all the remaining elliptical states of polarization, becoming more circular closer to the poles and more linear closer to the equator. The orientations of the elliptical states of polarization are given by half the azimuthal angle and their helicity is determined by the hemisphere (upper or lower) in which they fall. Each point on the Poincaré sphere represents a unique state of polarization and all possible states of polarization of a coherent paraxial beam are represented by points on the Poincaré sphere.

Formally, the concept of a Poincaré sphere is equivalent to the *Bloch sphere*, used in the context of two-dimensional quantum systems. Both are obtained due to the spin-half representation of the SU(2) Lie group. For the states of polarization, the spin-half representation is used to represent the two-dimensional polarization vectors. Two-dimensional SU(2) transformations represent the polarization transformations that can be implemented with various optical polarization components.

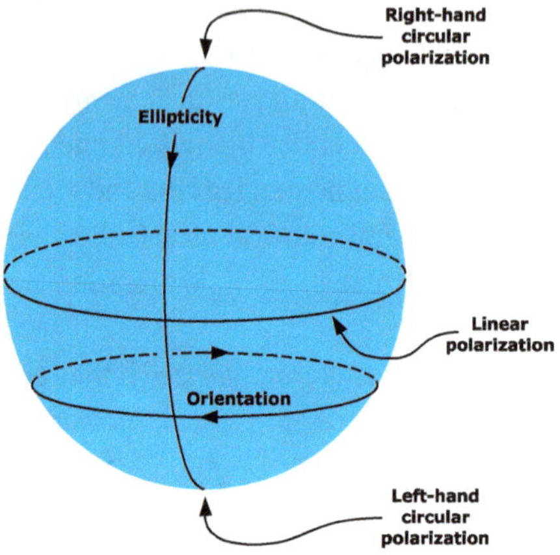

Figure 2.8: Poincaré sphere.

2.9.3 Orthogonal states of polarization

The states of polarization can be represented as a linear combination of the x- and y-linear states of polarization with complex-valued coefficients. The x- and y-linear states of polarization serve as the polarization basis vectors. However, any pair of orthogonal states of polarization can be used as a basis. In other words, if $\vec{\eta}_1$ and $\vec{\eta}_2$ are the two elements of the polarization basis, then

$$\vec{\eta}_1^* \cdot \vec{\eta}_2 = 0. \tag{2.200}$$

Any two geometrically opposite points on the Poincaré sphere represent orthogonal states of polarization. For example, the two states of circular polarization are orthogonal to each other, as can be verified by applying (2.198) in (2.200).

Since the electric field vector is orthogonal to the wavevector

$$\mathbf{k} \cdot \vec{E}_0 = 0, \tag{2.201}$$

the orientation of the electric field vector is restricted by the direction of the wavevector, but not by its magnitude (the wavenumber). The amplitude of the electric field reads

$$\vec{E}_0 = \vec{\eta}(\vec{k}) E_0, \tag{2.202}$$

where $\vec{\eta}(\vec{k})$ is the polarization vector, which is a (dimensionless) unit vector, and E_0 is a constant (scalar) amplitude that carries the dimensions of an electric field. A pair of polarization basis vectors $\vec{\eta}_s(\vec{k})$ with $s = 1, 2$, in terms of which any state of polarization can be represented, always have the following properties:

$$|\vec{\eta}_s(\vec{k})| = 1,$$
$$\vec{\eta}_r^*(\vec{k}) \cdot \vec{\eta}_s(\vec{k}) = \delta_{r,s},$$
$$\vec{\eta}_1^*(\vec{k}) \cdot \vec{k} = \vec{\eta}_2^*(\vec{k}) \cdot \vec{k} = 0, \tag{2.203}$$

where $\delta_{r,s}$ represent the 2×2 identity matrix (the Kronecker delta function). The properties show why the polarization basis vectors depend on $\vec{k}$. For real-valued polarization vectors, it follows that

$$\vec{\eta}_1(\vec{k}) \times \vec{\eta}_2(\vec{k}) = \vec{k}. \tag{2.204}$$

The *completeness condition* for these three vectors implies that

$$\vec{\eta}_1^*(\vec{k}) \otimes \vec{\eta}_1(\vec{k}) + \vec{\eta}_2^*(\vec{k}) \otimes \vec{\eta}_2(\vec{k}) + \vec{k} \otimes \vec{k} = \mathbb{1}_3, \tag{2.205}$$

where $\otimes$ is the *tensor product* and $\mathbb{1}_3$ is the identity of the three-dimensional space.

2.9.4 Anisotropic media

As mentioned in Section 2.2.2, anisotropic dielectric media can affect the state of polarization of light propagating through it. In fact, it can even affect the way the light is propagating through the medium. Here, we only consider such effects in a linear medium. The dielectric tensor of a linear medium can be written in terms of the first-order electric susceptibility as

$$\epsilon_{ab} = \delta_{ab}\epsilon_0 + [\chi_e^{(1)}]_{ab}\epsilon_0. \tag{2.206}$$

It is a 3×3 Hermitian matrix, with positive real eigenvalues, representing three dielectric constants along three orthogonal *principle axes*. In diagonalized form, the *dielectric tensor* is represented as

$$\epsilon_{ab} = \sum_{i=1}^{3} \vec{m}_a^{(i)} \epsilon_i \vec{m}_b^{(i)}, \tag{2.207}$$

where $\vec{m}_a^{(i)}$ represents the unit vectors along the three principal axes and ϵ_i denotes the three dielectric constants associated with these principal axes. If all three dielectric constants are the same (degenerate), the medium is isotropic and does not affect the state of polarization of light propagating through it. If they are all different (non-degenerate), the medium is referred to as a *biaxial* crystal or medium, and if one dielectric constant differs from the other two, it is called a *uniaxial* or *birefringent* crystal or medium. For uniaxial crystals, the principal axis with the unique dielectric constant (the one that is different from the other two) is called the *optic axis* of the crystal or simply the *crystal axis*. The refractive index (the square root of the dielectric constant) along the optic axis is the *extraordinary refractive index* denoted by n_e, while the other refractive index is the *ordinary refractive index* denoted by n_o.

The refractive index that an electric field experiences when propagating through an anisotropic dielectric media is determined by the direction of the electric field vector. We see this effect from the constituent relation that relates the electric field to the electric flux density. It reads

$$\mathbf{D} = \sum_{i=1}^{3} \epsilon_i \vec{m}^{(i)} \vec{m}^{(i)} \cdot \mathbf{E} = \sum_{i=1}^{3} \epsilon_i \mathbf{E}_i, \tag{2.208}$$

where $\mathbf{E}_i$ is the projections of the electric field along the three principal axes. If the electric field is polarized along one of these principle axes, the projections along the other two would be zero, and the electric field would only experience the dielectric constant (refractive index) associated with the principle axis along which it is polarized.

For a birefringent (uniaxial) medium, the light separates into two beams (or waves) propagating independently. The polarization vector of the *extraordinary wave* lies in the plane defined by the propagation axis and the crystal axis (assuming they are not parallel). The *ordinary wave's* polarization vector is perpendicular to this plane. Unless the propagation axis is perpendicular or parallel to the crystal axis, the extraordinary wave would experiences *walk-off*, as discussed below.

2.9.4.1 Retardance

Consider a plane wave propagating through a uniaxial crystal. When the propagation vector is parallel to the optic axis nothing strange happens because the wave only experiences the ordinary refractive index. For a propagation vector that is perpendicular to the optic axis, the wave generally decomposes into two waves, one having its polarization along the optic axis (the *extraordinary polarization*) and the other with its polarization perpendicular to it (the *ordinary polarization*). These waves propagate at different velocities through the crystal due to the different refractive indices. As a result, the state of polarization of the combined wave changes continuously during propagation. This effect is called *retardance*. An optical element made from a uniaxial crystal with a specific thickness and with its optic axis perpendicular to the wavevector of a normally incident wave is called a *wave plate*.

When the incident wave is linearly polarized either along or perpendicular to the optic axis, its state of polarization remains unchanged. These two cases are called the *normal modes*. If the incident wave is linearly polarized at an angle lying between these two cases, for example, 45° with respect to the optic axis of a wave plate, then the state of polarization would change, as shown in Figure 2.9. It becomes more elliptical, with a particular helicity (say right-handed), but with the same orientation as it propagates through the wave plate. At some point, the state of polarization is exactly (right-handed) circular. A wave plate with such a thickness is called a *quarter wave plate*. Beyond this point, the state of polarization becomes elliptical again, with the same helicity but with an orthogonal orientation with respect to the orientation of the input state of polarization. It becomes more elliptical until the state of polarization is linear along in the orthogonal orientation. A wave plate with this thickness is called a *half wave plate*. From this point onward, the states of polarization go through the reversed order of states back to the original input state of polarization, but now these states of polarization have the opposite helicity. A wave plate with a thickness that reproduces the input state of polarization is called a *full wave plate*.

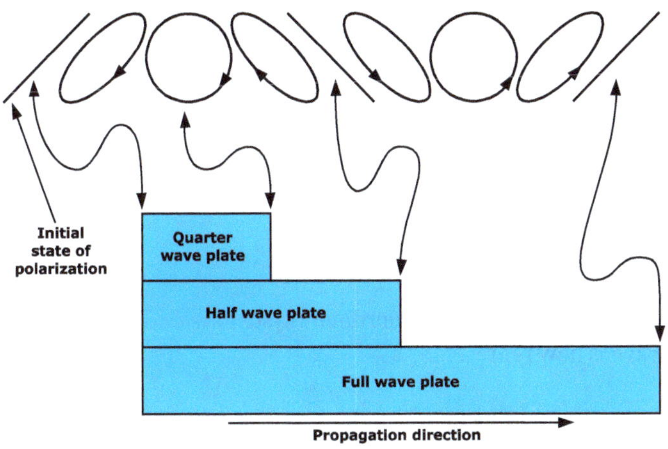

Figure 2.9: The states of polarization due to propagation through a wave plate.

One can determine the effect of a wave plate with the aid of the Poincaré sphere. The two normal modes that are unaffected by the wave plate are represented by geometrical opposite points on the equator of the Poincaré sphere. The wave plate performs a rotation on the states on the Poincaré sphere about a rotation axis that passes through the two normal modes. The rotation angle is given by the thickness of the wave plate. For a quarter wave plate, the rotation angle is only 90°, for a half wave plate it is 180°, and for a full wave plate it is 360°.

2.9.4.2 Walk-off

When the propagation axis of the incident wave is neither orthogonal, nor perpendicular to the crystal axis, an additional phenomenon occurs. While the ordinary polarized beam propagates directly forward through the crystal, the extraordinary polarized beam shifts to the side during propagation. This *walk-off* effect depends on the orientation of the crystal axis.

To derive the *walk-off* effect, we consider the situation where the electromagnetic field is a plane wave with a propagation vector

$$\mathbf{k} = k\vec{k} = \frac{n\omega}{c}\vec{k}, \tag{2.209}$$

where n is the relevant refractive index. The plane wave propagates through the birefringent medium with a *crystal axis* given by the unit vector $\vec{e}_X$, which is neither perpendicular nor parallel to $\vec{k}$. The angle between them is given by $0 < \theta_X < \pi/2$, so that

$$\vec{k} \cdot \vec{e}_X = \cos(\theta_X). \tag{2.210}$$

The *crystal plane* is uniquely defined as the plane parallel to both $\vec{k}$ and $\vec{e}_X$. However, since they are not perpendicular, we define another unit vector $\vec{p}$ that lies in the crystal plane and is orthogonal to $\vec{k}$. The crystal axis can then be expressed as

$$\vec{e}_X = \cos(\theta_X)\vec{k} + \sin(\theta_X)\vec{p}. \tag{2.211}$$

The unit vector perpendicular to the crystal plane is denoted as $\vec{s}$, so that

$$\vec{k} \times \vec{e}_X = \sin(\theta_X)\vec{k} \times \vec{p} = \sin(\theta_X)\vec{s}. \tag{2.212}$$

Thus, $\vec{p}$, $\vec{s}$, and $\vec{k}$ represent the unit vectors of a coordinate system.

The generic plane wave is now represented as

$$\mathbf{E}(\mathbf{x}, t) = \mathbf{E}_0 \exp(-i\omega t + i\mathbf{k} \cdot \mathbf{x}), \tag{2.213}$$

where $\mathbf{E}_0 = E_p\vec{p} + E_s\vec{s} + E_k\vec{k}$ is a constant vector for the electric field. Due to the anisotropic medium, it is not in general perpendicular to the wavevector anymore. Similar vector fields are defined for $\mathbf{H}(\mathbf{x}, t)$ and $\mathbf{D}(\mathbf{x}, t)$, but $\mathbf{B}(\mathbf{x}, t) = \mu_0\mathbf{H}(\mathbf{x}, t)$. When substituted into the source-free Maxwell equations, they produce

$$\mathbf{k} \times \mathbf{E}_0 = \mu_0\omega\mathbf{H}_0, \quad \mathbf{k} \times \mathbf{H}_0 = -\omega\mathbf{D}_0, \quad \mathbf{k} \cdot \mathbf{D}_0 = 0, \quad \mathbf{k} \cdot \mathbf{H}_0 = 0. \tag{2.214}$$

The last two follow from the first two when they are dotted with **k**. The first two equations then lead to

$$\mathbf{D}_0 = -\frac{\mathbf{k} \times (\mathbf{k} \times \mathbf{E}_0)}{\mu_0\omega^2} = -n^2\epsilon_0 \left[\vec{k} \times (\vec{k} \times \mathbf{E}_0)\right], \tag{2.215}$$

where n is the appropriate refractive index, depending on the components of $\mathbf{E}_0$. Using

$$\vec{k} \times (\vec{k} \times \mathbf{E}_0) = (\vec{k} \cdot \mathbf{E}_0)\vec{k} - \mathbf{E}_0, \tag{2.216}$$

we then get

$$\mathbf{D}_0 = n^2\epsilon_0\left[\mathbf{E}_0 - (\vec{k}\cdot\mathbf{E}_0)\vec{k}\right]. \tag{2.217}$$

The right-hand side removes the component along $\vec{k}$, leaving only the transverse part of $\mathbf{E}_0$ equated to $\mathbf{D}_0$. We already know from the third equation in (2.214) that $\mathbf{D}_0$ is orthogonal to $\vec{k}$. For the remaining transverse components, we can use the constituent relation for the electric field and electric flux to obtain three separate equations for the three components given by

$$\begin{aligned}
D_p &= \epsilon_0\left[n_e^2\cos^2(\theta_X) + n_o^2\sin^2(\theta_X)\right]E_p + \epsilon_0(n_e^2 - n_o^2)\sin(\theta_X)\cos(\theta_X)E_k,\\
D_s &= \epsilon_0 n_o^2 E_s,\\
0 &= \epsilon_0(n_e^2 - n_o^2)\sin(\theta_X)\cos(\theta_X)E_p + \epsilon_0\left[n_e^2\sin^2(\theta_X) + n_o^2\cos^2(\theta_X)\right]E_k,
\end{aligned} \tag{2.218}$$

The two nonzero components of $\mathbf{D}_0$ can be replaced in terms of the components of $\mathbf{E}_0$ with the aid of (2.217). Along the $\vec{s}$ direction, we have $n = n_o$, representing the *ordinary polarization*. For the *extraordinary polarization*, with the electric field vector lying in the crystal plane, the remaining equations lead to the *effective refractive index* $n = n_{\text{eff}}$ (see Appendix B).

To satisfy the last equation in (2.218), the components of the electric field with the extraordinary polarization are given by

$$\begin{aligned}
E_p &= \frac{|\mathbf{E}_0|\left(n_o^2\cos^2(\theta_X) + n_e^2\sin^2(\theta_X)\right)}{\sqrt{n_o^4\cos^2(\theta_X) + n_e^4\sin^2(\theta_X)}} = \frac{|\mathbf{E}_0|n_o n_e}{n_{\text{eff}}\sqrt{n_o^2 + n_e^2 - n_{\text{eff}}^2}},\\
E_k &= \frac{|\mathbf{E}_0|\cos(\theta_X)\sin(\theta_X)\left(n_o^2 - n_e^2\right)}{\sqrt{n_o^4\cos^2(\theta_X) + n_e^4\sin^2(\theta_X)}} = \frac{|\mathbf{E}_0|\sqrt{n_o^2 - n_{\text{eff}}^2}\sqrt{n_{\text{eff}}^2 - n_e^2}}{n_{\text{eff}}\sqrt{n_o^2 + n_e^2 - n_{\text{eff}}^2}},
\end{aligned} \tag{2.219}$$

where we replaced the angle dependence in favour of the effective refractive index in the final expressions. We see that, according to the expression for E_k, the electric field for the extraordinary polarization has a component along the direction of the wavevector. The associated magnetic field vector points in the $\vec{s}$ direction, perpendicular to the crystal plane. As a result, the Poynting vector (2.191) is not parallel to the propagation vector. Instead, there is an angle between the two vectors, causing the field to be laterally shifted. This shift, produced by the Poynting vector relative to the propagation direction, is referred to as the *walk-off*. The cosine of the *walk-off angle* is given by

$$\begin{aligned}
\cos(\theta_{\text{wo}}) &= \frac{n_o^2\sin^2(\theta_X) + n_e^2\cos^2(\theta_X)}{\sqrt{n_o^4\sin^2(\theta_X) + n_e^4\cos^2(\theta_X)}}\\
&= \frac{n_o^2\sin(\theta_X)}{\sqrt{n_o^4\sin^2(\theta_X) + \left(n_o^2 - n_{\text{eff}}^2\right)^2\cos^2(\theta_X)}}.
\end{aligned} \tag{2.220}$$

Since the walk-off angle is fairly small, we can express it approximately as

$$\theta_{\mathrm{wo}} \approx \frac{\left|n_{\mathrm{o}}^2 - n_{\mathrm{eff}}^2\right|}{n_{\mathrm{o}}^2 \tan(\theta_X)}. \tag{2.221}$$

The refractive indices n_{o} and n_{e} are in general dispersive, and thus depend on the wavelength of the light. See Appendix B.2 for more detail.

We could have picked a convenient direction (such as the z-axis) for the wavevector of the plane wave in the above calculations. However, since we use these results later and since the physical beam is never a simple plane wave, we use generic directions $\vec{p}$, $\vec{s}$, and $\vec{k}$ in the calculation. Eventually, we need to integrate over the wavevector. Therefore, we need to determine how the other vectors depend on the wavevector.

Having determined the basic process and learned how the different vectors are formed, we can formulate a procedure to define all the vectors. As input, we take the crystal axis $\vec{e}_X$ and the wavevector $\mathbf{k}$. They are not orthogonal, but they define the crystal plane as given in (2.211). It then follows that

$$\vec{p} = \frac{\vec{e}_X - \cos(\theta_X)\vec{k}}{\sin(\theta_X)} = \frac{k\vec{e}_X - \cos(\theta_X)\mathbf{k}}{k\sin(\theta_X)}. \tag{2.222}$$

The unit vector perpendicular to the crystal plane $\vec{s}$ is then given by

$$\vec{s} = \vec{k} \times \vec{p} = \frac{\vec{k} \times \vec{e}_X}{\sin(\theta_X)} = \frac{\mathbf{k} \times \vec{e}_X}{k\sin(\theta_X)}. \tag{2.223}$$

Above, we found that $\vec{s}$ represents the polarization direction of both $\mathbf{E}$ and $\mathbf{D}$ for the ordinary polarized field. We can represent them as

$$\mathbf{D}_{\mathrm{o}} = D_{\mathrm{o}}\vec{s} \quad \text{and} \quad \mathbf{E}_{\mathrm{o}} = E_{\mathrm{o}}\vec{s}, \tag{2.224}$$

where D_{o} and E_{o} are scalar fields.

For the extraordinary polarized field, we saw that $\mathbf{D}$ is parallel to $\vec{p}$; it lies in the crystal plane and is perpendicular to $\mathbf{k}$. Therefore, it reads

$$\mathbf{D}_{\mathrm{e}} = D_{\mathrm{e}}\vec{p}, \tag{2.225}$$

where D_{e} is the associated scalar field. To obtain $\mathbf{E}$ for the extraordinary polarization, we scale the different parts of $\mathbf{D}_{\mathrm{e}}$ that are parallel and perpendicular to the crystal axis by their respective inverse dielectric constants. Hence,

$$\begin{aligned}\mathbf{E}_{\mathrm{e}} &= \frac{(\mathbf{D}_{\mathrm{e}} \cdot \vec{e}_X)\vec{e}_X}{\epsilon_0 n_{\mathrm{e}}^2} + \frac{\mathbf{D}_{\mathrm{e}} - (\mathbf{D}_{\mathrm{e}} \cdot \vec{e}_X)\vec{e}_X}{\epsilon_0 n_{\mathrm{o}}^2} \\ &= \frac{[n_{\mathrm{o}}^2 \sin^2(\theta_X) + n_{\mathrm{e}}^2 \cos^2(\theta_X)]D_{\mathrm{e}}\vec{p}}{\epsilon_0 n_{\mathrm{o}}^2 n_{\mathrm{e}}^2} + \frac{(n_{\mathrm{o}}^2 - n_{\mathrm{e}}^2)\sin(\theta_X)\cos(\theta_X)D_{\mathrm{e}}\vec{k}}{\epsilon_0 n_{\mathrm{o}}^2 n_{\mathrm{e}}^2}.\end{aligned} \tag{2.226}$$

We also found that the component of the extraordinary electric field perpendicular to **k** is related to D_e by the effective refractive index given in (B.1). So, we can define a scalar electric field for the extraordinary polarization as

$$D_e = \epsilon_0 n_{\text{eff}}^2 E_e = \frac{\epsilon_0 n_o^2 n_e^2 E_e}{n_o^2 \sin^2(\theta_X) + n_e^2 \cos^2(\theta_X)}. \tag{2.227}$$

The electric vector field for the extraordinary polarization is therefore given by

$$\mathbf{E}_e = E_e \vec{p} + \frac{(n_o^2 - n_e^2)\sin(\theta_X)\cos(\theta_X)}{n_o^2 \sin^2(\theta_X) + n_e^2 \cos^2(\theta_X)} E_e \vec{k}. \tag{2.228}$$

The component along $\vec{k}$ is in general small because the difference in values of n_o and n_e is much smaller than their respective values, as seen in Appendix B.2.

In summary, the vectors and vector fields are

$$\vec{p} = \frac{k\vec{e}_X - \cos(\theta_X)\mathbf{k}}{k\sin(\theta_X)}, \quad \vec{s} = \frac{\mathbf{k} \times \vec{e}_X}{k\sin(\theta_X)}, \quad \vec{k} = \frac{\mathbf{k}}{k},$$
$$\mathbf{D}_o = D_o \vec{s}, \quad \mathbf{E}_o = E_o \vec{s}, \quad \mathbf{D}_e = D_e \vec{p}, \quad \mathbf{E}_e = E_e \vec{p} + \Delta_X E_e \vec{k}, \tag{2.229}$$

where

$$\Delta_X = \frac{(n_o^2 - n_e^2)\sin(\theta_X)\cos(\theta_X)}{n_o^2 \sin^2(\theta_X) + n_e^2 \cos^2(\theta_X)}. \tag{2.230}$$

2.10 Coherence

A characteristic property of light is its ability to produce *interference*. It is the key feature that identifies light as a wave with wave properties. However, light can lose its ability to produce interference under certain circumstances. The conditions under which light can produce interference is governed by a property called *coherence* [16].

In the analyses that we considered thus far in this chapter, we have effectively assumed that the optical field is perfectly coherent. The fields are presented as well-defined deterministic functions that can be precisely evaluated everywhere. In the physical world, it does not exactly work like that. Physical sources of light in general produce ensembles of such deterministic fields without well-defined relative phases, leading to a *stochastic optical field*. As a result, each field in such an ensemble can only interfere with itself. The extent to which the whole ensemble can produce a visible interference pattern is then determined by whether the different fields in the ensemble would produce the same interference pattern. When this is not possible, the combined interference pattern becomes blurred and the stochastic optical field is said to be *incoherent*. In general, there is a variation of conditions leading to different degrees of *partial coherence.*

Obviously, the study of coherence is closely linked to the phenomenon of *optical interference.* It is an indication that there exist correlations among different parts of an optical field. Therefore, *optical interference* measurements are measurements of *first-order correlations*[j] in an optical field.

First-order correlations are obtained by observing the intensity (using a single detector) of a superposition of different optical fields. For scalar fields, it is given by

$$I = |f_1 + f_2|^2 = |f_1|^2 + |f_2|^2 + f_1 f_2^* + f_1^* f_2, \tag{2.231}$$

where f_1 and f_2 are complex-valued scalar fields (see Section 2.3.4), representing the two overlapping optical fields. The intensity is detected at a point $\mathbf{x}$ and at a time t (ignoring the fact that physical detectors have finite areas and finite integration times).

Often, the two functions are different parts of the same field that are made to overlap with the aid of an *interferometer.* In such a situation, we can write the observed intensity related to the superposition of different points in an optical field in terms of the field at the two points. Then $f_1 \to f(\mathbf{x}_1, t_1)$ and $f_2 \to f(\mathbf{x}_2, t_2)$. The last two terms in (2.231) can then be combined into a *correlation function* (or the real part of such a correlation function). In the case of stochastic optical fields, we also need to perform an ensemble averaging to represent the observed intensity. Therefore, we define the *first-order correlation function* as

$$\Gamma^{(1)}(\mathbf{x}_1, t_1, \mathbf{x}_2, t_2) = \langle f^*(\mathbf{x}_1, t_1) f(\mathbf{x}_2, t_2) \rangle , \tag{2.232}$$

where $\langle \cdot \rangle$ denotes the ensemble average. When the two spacetime points coincide, the first-order correlation function becomes the *intensity* at that point:

$$\Gamma^{(1)}(\mathbf{x}, t, \mathbf{x}, t) = \left\langle |f(\mathbf{x}, t)|^2 \right\rangle \equiv I(\mathbf{x}, t). \tag{2.233}$$

The concept of *coherence* can be viewed as *the ability of a field to produce interference.* For this reason, *first-order coherence* is based on the observation of interference fringes in first-order correlation experiments. It forms the foundation of classical coherence theory. The *visibility* V of the interference fringes is defined by

$$V = \frac{I_{\max} - I_{\min}}{I_{\max} + I_{\min}}, \tag{2.234}$$

as demonstrated in Figure 2.10. The *normalized first-order correlation function* or *complex degree of coherence* is defined as

j Note that what we refer to as "first-order correlations" are sometimes referred to as "second-order correlations" and "second-order coherence" in other literature [16].

$$g^{(1)}(\mathbf{x}_1, t_1, \mathbf{x}_2, t_2) = \frac{\langle f^*(\mathbf{x}_1, t_1) f(\mathbf{x}_2, t_2) \rangle}{\sqrt{\langle |f(\mathbf{x}_1, t_1)|^2 \rangle \langle |f(\mathbf{x}_2, t_2)|^2 \rangle}}. \tag{2.235}$$

The maximum value for $g^{(1)}$ is obtained when the two spacetime points coincide. Then $g^{(1)}(\mathbf{x}, t, \mathbf{x}, t) = 1$. Alternatively, the magnitude of $g^{(1)}$ is smaller than 1.

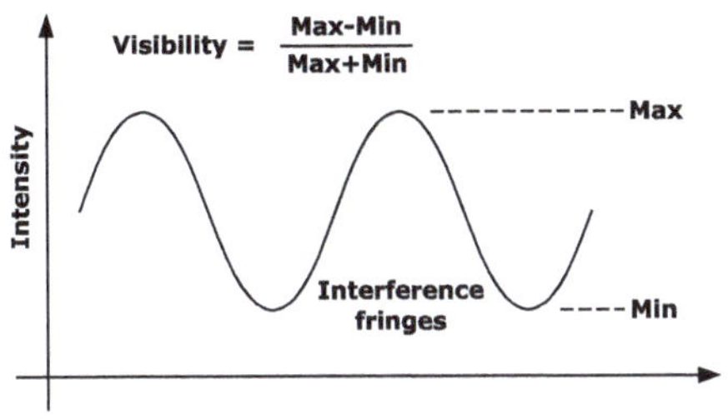

Figure 2.10: Definition of visibility.

2.10.1 Optical interferometry

Typical examples of first-order correlations are those obtained from measurements using *optical interferometry*. Such measurements can be used to determine the coherence of an optical source.

Different types of optical interferometer setups can be represented in the same way. The *Mach–Zehnder interferometer*, *Michelson–Morley interferometer*, and the *Sagnac interferometer* all share similar constructions. (See Figure 2.11.) In all these cases, the input beam is divided between two paths by a beamsplitter. In these paths, one may encounter different operations, introducing relative phase modulations. The two beams are then recombined with another beamsplitter (which may be the same physical optical device). Both output ports produce interference, but usually only one output port is observed.

The Mach–Zehnder interferometer separates the two beams along different paths and recombines them with a second beamsplitter. It is relatively easy to align and is therefore often used in optical experiments and systems.

The Michelson–Morley interferometer redirects the two beams back toward the same beamsplitter that separated the two beams. They enter the same ports from which they emerged to be recombined. This interferometer is famous for being the setup that Michelson and Morley used to determine that the aether does not exist.

In Figure 2.11, we show the Sagnac interferometer as being constructed in terms of an optical fibre. However, the same setup can be made using bulk optics components. After being separated by the beamsplitter (fibre coupler), the two beams follow opposite paths to enter the same beamsplitter at the opposite ports from which the emerged, to be recombined toward the observation plane. The fibre optical implementation of Sagnac interferometers are often used to implement *optical gyroscopes*.

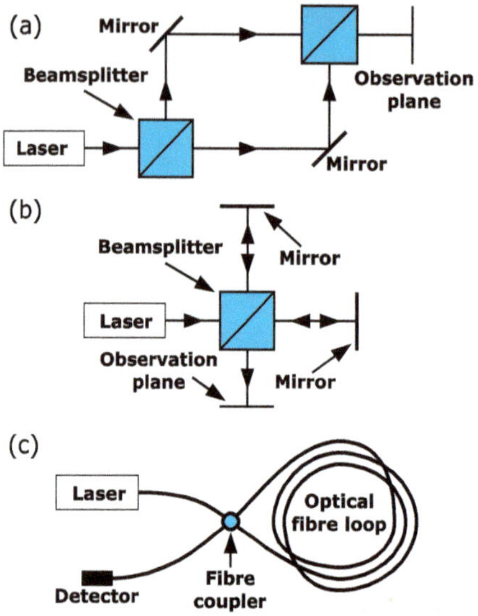

Figure 2.11: Optical interferometer setups (a) Mach–Zehnder, (b) Michelson–Morley, and (c) Sagnac (in optical fibre).

In Section 7.5, the operation of these interferometers are generically formulated in terms of quantum optics. Therefore, we do not discuss these interferometers further in the classical context.

Bibliography

[1] M. Born and E. Wolf. *Principles of Optics, 6th ed.* Pergamon Press, New York, 1980.
[2] J. W. Goodman. *Introduction to Fourier Optics, 2nd ed.* McGraw-Hill, New York, USA, 1996.
[3] B. E. A. Saleh and M. C. Teich. *Fundamentals of Photonics*. John Wiley & Sons, New York, USA, 1991.
[4] A. Forbes, M. De Oliveira, and M. R. Dennis. Structured light. *Nat. Photonics*, 15:253–262, 2021.
[5] E. Kreyszig. *Introductory Functional Analysis with Applications*. John Wiley & Sons, New York, USA, 1978.
[6] W.-K. Tung. *Group Theory in Physics*. World Scientific, Singapore, 1985.
[7] G. Strang. *Linear Algebra and its Applications, 3rd ed.* Harcourt Brace Jovanovich, Inc., Orlando, USA, 1988.
[8] M. Plancherel. Contribution à l'etude de la representation d'une fonction arbitraire par les integrales définies. *Rend. Circ. Mat. Palermo*, 30:298–335, 1910.
[9] E. Noether. Invariante Variationsprobleme. *Nachr. Ges. Wiss. Gött., Math.-Phys. Kl.*, 1918:235–257, 1918.
[10] F. Mandl and G. Shaw. *Quantum Field Theory*. John Wiley & Sons, New York, USA, 1984.
[11] Y. R. Shen. *The Principles of Nonlinear Optics*. John Wiley & Sons, New York, USA, 2003.

[12] M. Abramowitz and I. A. Stegun. *Handbook of Mathematical Functions*. Dover, Toronto, 1972.
[13] L. Allen, M. W. Beijersbergen, R. J. C. Spreeuw, and J. P. Woerdman. Orbital angular momentum of light and the transformation of Laguerre–Gaussian laser mode. *Phys. Rev. A*, 45:8185–8189, 1992.
[14] J. F. Nye and M. V. Berry. Dislocations in wave trains. *Proc. R. Soc. Lond. A*, 336:165–190, 1974.
[15] W. Pauli. The connection between spin and statistics. *Phys. Rev.*, 58:716–722, 1940.
[16] L. Mandel and E. Wolf. *Introductory Quantum Optics*. Cambridge University Press, New York, 1995.

3 Quantum optics

Our understanding of quantum physics is expressed in the language of quantum formalisms, such as *quantum mechanics* [1–3], *quantum field theory* [4–8], or the *Moyal formalism* [9]. For light, the formalism is called *quantum optics* [10–13].

In this chapter, we introduce the formalism of quantum optics in the context of the *particle-number* (photon-number) degree of freedom only, ignoring the other degrees of freedom (spin and spatiotemporal degrees of freedom). The other degrees of freedom are combined with the particle-number degree of freedom in Chapter 4, starting with the quantization of the electromagnetic field in Section 4.1. The role of the other degrees of freedom become progressively more significant, eventually leading to the functional formulation presented in Chapter 5.

The reason why we only consider the *particle-number degree of freedom* here, is because the concepts from quantum physics that sets them apart from classical theories are exclusively associated with the particle-number degree of freedom. (The validity of this statement is demonstrated over the course of the discussions.) From a purely formal point of view, we simply ignore the other degrees of freedom. However, one can view it conceptually as a situation where the other degrees of freedom are fixed to specific values (a specific state of polarization and a specific spatiotemporal mode).

It is not the intention of the discussion in this chapter to be an exhaustive treatise of the topic of quantum optics. Instead, we provide a development of the formalism to a level where it can serve as suitable background information that is adequate for the discussions in subsequent chapters. The development in this chapter is predominantly based on quantum mechanics, with some inspiration from quantum field theory. Other aspects associated with quantum field theory are considered in Section 4.1.

While quantum mechanics is first and foremost a *formalism*, the aim, as with any field in physics, is not just a method to perform calculations but also to increase our *understanding* of quantum physics. However, there is always the danger that such an understanding could involve misleading notions, especially in quantum physics where direct observations are often not possible. To avoid such a situation, the development of the formalism that we follow here emphasizes the difference between physics and formalism. The idea is that, to understand quantum physics, *one also needs to understand what it is not*. For this reason, the formalism of quantum optics is presented in stages, emphasizing its development due to inspirations from quantum physics, without imposing quantum physics. In other words, we postpone the introduction of concrete concepts from quantum physics until that point where they become relevant in the context of the physical scenario. Prior to that point, the discussion of the formalism may often refer to its relevance in the quantum context, but does not impose the quantum context as an exclusive application. Eventually, after concepts from quantum physics are imposed, these concepts reveal the pertinent aspects of nature not represented by classical theories.

https://doi.org/10.1515/9783111445342-004

3.1 Operator formalism

Quantum mechanics is a formalism based on operators. It may create the idea that operators are directly associated with quantum physics. However, one neither needs operators to formulate a successful theory of quantum physics, as the Moyal formalism demonstrates, nor does it mean that one cannot use operators to formulate a successful classical theory.

Here, we start by developing an operator formalism, which can be used for quantum physics, while it can equally well be used for *classical field theory*. It allows us to introduce the notions of a *quadrature operator* and a *ladder operator*, which play significant roles in the formulation of quantum theory, even if their derivation does not require quantum physics.

3.1.1 Fourier theory

One of the implications of the Planck and the de Broglie relationships (1.1) and (1.2) is that there exists a Fourier relationship between *canonical variables*. Fourier theory is a widely applicable mathematical topic, not exclusively associated with quantum physics. Here, we start with Fourier theory, as discussed in Section 2.1.2, without invoking the de Broglie relationship that would convert the Fourier domain variable (the wavenumber) to a momentum.

For the one-dimensional case, consider two functions $\psi(\xi)$ and $\tilde{\psi}(\zeta)$ in the *configuration space* and the *Fourier domain*, respectively, related by Fourier integrals

$$\begin{aligned}\psi(\xi) &= \int \tilde{\psi}(\zeta)\exp(\mathrm{i}\xi\zeta)\,\frac{\mathrm{d}\zeta}{2\pi},\\ \tilde{\psi}(\zeta) &= \int \psi(\xi)\exp(-\mathrm{i}\xi\zeta)\,\mathrm{d}\xi.\end{aligned} \tag{3.1}$$

Here ξ is the configuration space variable and ζ is the Fourier domain variable. For the *physics sign convention* used in these definitions, it is assumed that ξ is a spatial coordinate. However, these coordinates are eventually identified with the *particle-number degree of freedom* that has nothing to do with the spatial degrees of freedom. In the quantum context, we can interpret the two functions as representations of the state of a quantum system in terms of *wavefunctions* in the respective domains. However, there is nothing in the expressions at this stage that indicates that these functions are quantum states. The same functions can be used to represent classical fields. Here, we just refer to them as functions (or as *spectra* for the functions in the Fourier domain).

For convenience in notation, and since it needs to be done eventually anyway, we convert the coordinates ξ and ζ to dimensionless variables. Based on the fact that the argument of the Fourier kernel must be dimensionless, ξ and ζ carry mutually inverse dimensions. To make the coordinate variables dimensionless, we assume that there is a dimension parameter Λ with the same units as ξ. Then we define the dimensional coordinate variables as

$$q \triangleq \Lambda^{-1}\xi \quad \text{and} \quad p \triangleq \Lambda\zeta. \tag{3.2}$$

Although we use the traditional notation for the *canonical variables*, we must point out that q and p are not physical position and momentum. To avoid confusion, we simply refer to them as *quadrature variables* in anticipation of their eventual role. The Fourier integrals are now represented by

$$\begin{aligned} \psi(q) &= \int \tilde{\psi}(p)\exp(\mathrm{i}qp)\,\frac{\mathrm{d}p}{2\pi}, \\ \tilde{\psi}(p) &= \int \psi(q)\exp(-\mathrm{i}qp)\,\mathrm{d}q, \end{aligned} \tag{3.3}$$

in terms of the quadrature variables.

It may seem from (3.2) that we may be free to rescale q and p by picking a different Λ leading to an exactly equivalent formulation. However, when we introduce the dynamics, the value of Λ is fixed by the requirement that the free (non-interacting) theory *conserves particle number*.

Fourier theory imposes conditions on the functions that are related by (3.3). Moreover, when such functions are used to model a physical situation, the physics of the scenario also imposes conditions. For the sake of the development of a general formalism that can in principle be used for any physical scenario, it may be helpful to have a general description of these functions. Without going into too much detail about the mathematical properties of these functions, we state that these pairs of functions must both have finite Euclidean norms (defined in Section 2.1.1) and be infinitely differentiable.

3.1.2 Dirac notation

Before we proceed with our discussion, let us briefly pause to discuss some notation. Here, we introduce the Dirac notation, density operators, and operator traces. Although it is associated with quantum theory, the notation itself does not impose quantum theory. This notation facilitates the introduction of coordinate bases when we return to our discussion in the next section.

The two functions that depend on q and p, respectively, contain the same information because they are directly related by Fourier transformations. Paul Dirac proposed a way to represent the "state" of a particle or a *quantum system* without specifying the coordinates in terms of which it is expressed. Using abstract notation, he represents this information as a *state vector*, without indicating the coordinates.

States in quantum mechanics are often expressed in terms of *Dirac notation* [14], where the state of a system is denoted by $|\psi\rangle$, which is called a *ket*. All the possible states that such a system can have are collected into a set to form a *vector space*. It means that the different elements of this set (the state vectors) can be combined in linear combinations to form new elements belonging to the same set.

The state vectors of a quantum system are all *normalized*. It is a requirement that follows from the statistical nature of quantum theory. The measurements that are calculated from a state vector are associated with different probabilities. These probabilities must add up to 1. This requirement is often referred to as the *conservation of probability*.

The normalization of state vectors imposes a restriction on the coefficients in a linear combination in terms of which state vectors can be expressed. The fact that the elements of the vector space are normalized also means that it is (a subset of) a *normed space* with an associated *norm*.[a]

The norm of a state vector is calculated by taking the *inner product* of the state vector with itself, leading to the Euclidean norm of (2.7). It means that the vector space also has an inner product associated with it, and thus forms a *Hilbert space*. The definition of a Hilbert space does not imply that the elements are all normalized. The normalization of quantum state vectors is an additional property. However, we can assume that all the elements in the Hilbert space are *normalizable*, and are related to quantum state vectors via a normalization constant. With this understanding, we can call it the *Hilbert space of quantum states*.

In Dirac notation, the inner product of the state vector with itself is

$$\langle\psi|\psi\rangle \equiv \|\psi\|^2 (= 1), \tag{3.4}$$

where $\|\psi\|$ represents the Euclidean norm of the state vector. The inner product indicates that there exists a *dual* for every ket $|\psi\rangle$. It is denoted by $\langle\psi|$ and is called a *bra*. The inner product between different state vectors produces a complex constant $\langle\psi|\psi'\rangle = \eta$ with $|\eta| < 1$. The opposite inner product (with the ket and bra interchanged) then produces the complex conjugate $\langle\psi'|\psi\rangle = \eta^*$.

The elements of a vector space can all be represented in terms of linear combinations of a linearly independent subset of elements that is called a *basis*, as discussed in Section 2.1.1. It is often assumed that the elements of the basis are also elements of the vector space. In the case of the space of all quantum state vectors, the elements of such a basis are normalized. For a separable Hilbert space, such basis elements are discrete and countable. The number of elements in the basis represents the *dimension* of the vector space.

In the quantum context, the kets and bras represent state vectors that are normalized according to (3.4). However, at this point, they can also be used to represent classical state vectors, but then the normalization condition is relaxed and it is simply required that $\langle\psi|\psi\rangle \equiv \|\psi\|^2 < \infty$.

---

a If the normed space is *complete*—that is, when all convergent sequences of the elements converge to elements inside the set—it is called a *Banach space*.

Exercise 3.1. For a state vector $|\psi\rangle$, expanded as

$$|\psi\rangle = \sum_n |\phi_n\rangle C_n, \tag{3.5}$$

in terms of a complete orthogonal basis $|\phi_n\rangle$, where C_n represents the complex coefficients, show that the condition for its normalization is that

$$\sum_n |C_n|^2 = 1. \tag{3.6}$$

3.1.2.1 Density operator

When a quantum state is represented as a wavefunction $\psi(\mathbf{x}, t)$ or a state vector $|\psi\rangle$, it implies that there is no doubt about the description of the state. In such a case, a *physical system* is said to be in a *pure state*, meaning that the system has a definite state. The normalization of a pure quantum state $\langle\psi|\psi\rangle = 1$ ensures *conservation of probability*, which is necessary for the statistical nature of observations.

In practice, the exact description of the state may not be so certain. Sometimes, the system does not have a definite state, but rather has varying probabilities to be in different states. Then we need to work with a statistical ensemble of different state vectors. In such a case, the system is said to be in a *mixed state*. For this purpose, the state is represented as a *density operator*. For a pure state, the density operator is given by

$$\hat{\rho} = |\psi\rangle \langle\psi| . \tag{3.7}$$

When $\hat{\rho}$ is the density operator of a pure state, it then follows that $\hat{\rho}^2 = \hat{\rho}$.

For a mixed state, the density operator becomes a *convex sum* of different pure states, each multiplied by a *probability*. The sum is called *convex* because the coefficients are probabilities that must add up to 1. The density operator of a mixed state is given by

$$\hat{\rho} = \sum_n |\psi_n\rangle P_n \langle\psi_n| , \tag{3.8}$$

where $|\psi_n\rangle$ represents different pure states that are combined to form the mixed state and P_n denotes the probabilities for the system to be in the different pure states. For a given mixed state, the representation in terms of a given set of pure states with their associated probabilities is not in general unique.

3.1.2.2 Hermitian adjoint and Hermitian operators

The concept of a *Hermitian adjoint* follows from the inner product. If we have a state vector $|\psi\rangle$ with an associated dual $\langle\psi|$ and we apply and operator $\hat{A}$ on $|\psi\rangle$ to get another (unnormalized) state $|\psi'\rangle = \hat{A}|\psi\rangle$ with an associated (unnormalized) dual $\langle\psi'|$, then the inner product between the transformed state and the original dual is

$$\langle\psi|\psi'\rangle = \langle\psi|\hat{A}|\psi\rangle \triangleq \eta. \tag{3.9}$$

On the other hand,

$$\langle\psi'|\psi\rangle = \eta^* \neq \langle\psi|\hat{A}|\psi\rangle\,. \tag{3.10}$$

So, what would be the equivalent operator that we need to insert between $\langle\psi|$ and $|\psi\rangle$ to obtain η^*? It is defined as the *Hermitian adjoint* of $\hat{A}$, represented by $\hat{A}^\dagger$, so that

$$\langle\psi'|\psi\rangle = \langle\psi|\hat{A}^\dagger|\psi\rangle = \eta^*. \tag{3.11}$$

Hermitian operators, which are equal to their Hermitian adjoints $\hat{A} = \hat{A}^\dagger$, play a significant role in quantum formalism. The eigenvalues of such operators are real valued, which can thus represent the numerical values of a measurable quantity. As a result, measurable quantities are represented by Hermitian operators in quantum formalism, and are called *observables*. When the eigenvalues of a Hermitian operator are all different (non-degenerate), their associated *eigenvectors* are *mutually orthogonal*. When two or more eigenvalues are the same (degenerate), their associated eigenvectors are *linearly independent*. The set of eigenvectors associated with degenerate eigenvalues are not unique, because any linear combination of such eigenvectors is also an eigenvector. However, it is always possible to obtain a set of eigenvectors for degenerate eigenvalues that are mutually orthogonal. Such a set of mutually orthogonal eigenvectors can be obtained from a degenerate set of linearly independent eigenvectors by a process called *Graham–Schmidt orthogonalization* [15]. Therefore, we can always assume that the eigenvectors of a Hermitian operator (an observable) are mutually orthogonal regardless of whether the eigenvalues are degenerate or not.

The density operator is *Hermitian*

$$\hat{\rho}^\dagger = \hat{\rho}. \tag{3.12}$$

The eigenvalues of a density operator are not only real valued, but also non-negative (positive or zero). It means that

$$\langle\phi|\hat{\rho}|\phi\rangle \geq 0, \tag{3.13}$$

for any state $|\phi\rangle$.

3.1.2.3 Operator trace

The normalization of the state, expressed in terms of the density operator, is represented in terms of a trace

$$\text{tr}\{\hat{\rho}\} = 1. \tag{3.14}$$

Here, the trace represents an *operator trace*. It is evaluated by inserting the identity $\mathbb{1} = \sum_n |\phi_n\rangle\langle\phi_n|$ resolved in terms of a completed basis of the Hilbert space $|\phi_n\rangle$. For an arbitrary operator $\hat{A}$, it becomes

$$\mathrm{tr}\{\hat{A}\} = \sum_n \langle\phi_n|\hat{A}|\phi_n\rangle . \tag{3.15}$$

The operator trace is distinguished from other traces in that it contains operators (denoted by a caret ˆ unless it is the identity operator $\mathbb{1}$) in its argument.

Exercise 3.2. Show that, if a pure state $|\psi\rangle$ is expanded in terms of a complete orthogonal basis as in (3.5), then the trace of its density operator is normalized.

Hint: Use (3.6).

Since mixed states must also *conserve probability*, all density operators (pure or mixed), must be normalized. The normalization of the mixed state, as represented by (3.8), requires that

$$\mathrm{tr}\{\hat{\rho}\} = \sum_n \langle\psi_n|\psi_n\rangle P_n = \sum_n P_n = 1. \tag{3.16}$$

It implies that the probabilities add up to 1.

If $\hat{\rho}$ is not a pure state, then $\hat{\rho}^2 \neq \hat{\rho}$. The *purity* of the state is defined as

$$\text{purity} \triangleq \mathrm{tr}\{\hat{\rho}^2\}. \tag{3.17}$$

3.1.2.4 Expectation values

With states being represented by *density operators* and *observables* being represented by Hermitian operators, we can now express the *expectation value* $\langle A\rangle$ of a quantity associated with the observable $\hat{A}$ for a state $\hat{\rho}$ in terms of the operator trace

$$\langle A\rangle \triangleq \mathrm{tr}\{\hat{\rho}\hat{A}\}. \tag{3.18}$$

For a pure state $|\psi\rangle$, it becomes

$$\langle A\rangle = \langle\psi|\hat{A}|\psi\rangle . \tag{3.19}$$

3.1.3 Coordinate bases

As discussed in Section 2.1.1, there are situations where it is convenient to represent the elements of a vector space in terms of a basis that does not lie inside that vector space. Those bases are not discrete and their elements are neither normalizable nor countable.

Here, we consider such a situation. The functions in (3.3) can be obtained from a state vector $|\psi\rangle$ with the aid of *coordinate bases*, denoted by $|q\rangle$ and $|p\rangle$ (with their associated duals $\langle q|$ and $\langle p|$) for the configuration space and its Fourier domain, respectively. These bases are assumed to be orthogonal and complete. The *orthogonality conditions* are expressed by

$$\langle q|q'\rangle = \delta(q-q') \quad \text{and} \quad \langle p|p'\rangle = 2\pi\delta(p-p'), \tag{3.20}$$

and the *completeness conditions* are

$$\int |q\rangle \langle q| \, \mathrm{d}q = \mathbb{1} \quad \text{and} \quad \int |p\rangle \langle p| \, \frac{\mathrm{d}p}{2\pi} = \mathbb{1}, \tag{3.21}$$

where $\mathbb{1}$ is the identity operator on the Hilbert space. The orthogonality conditions clearly show that the elements of these coordinate bases are not (and cannot be) normalized. They are also not countable and are not elements of the Hilbert space. Therefore, *they cannot and do not represent state vectors of quantum systems* in the quantum context. Moreover, the elements of these coordinate bases cannot represent classical fields because their magnitudes are not finite. However, we can use them to represent such state vectors in terms of expansions in these bases, either as

$$|\psi\rangle = \int |q\rangle \, \psi(q) \, \mathrm{d}q \quad \text{or as} \quad |\psi\rangle = \int |p\rangle \, \tilde{\psi}(p) \, \frac{\mathrm{d}p}{2\pi}, \tag{3.22}$$

where the functions $\psi(q)$ and $\tilde{\psi}(p)$ serve as *coefficient functions*. It follows that these functions are obtained from the state vectors via inner products

$$\psi(q) = \langle q|\psi\rangle \quad \text{and} \quad \tilde{\psi}(p) = \langle p|\psi\rangle, \tag{3.23}$$

respectively. The normalization of the state vector in the quantum context implies that the coefficient functions are normalized, similar to what we have in (2.23):

$$\|\psi\|^2 \triangleq \int |\psi(q)|^2 \, \mathrm{d}q = \|\tilde{\psi}\|^2 \triangleq \int |\tilde{\psi}(p)|^2 \, \frac{\mathrm{d}p}{2\pi} = 1. \tag{3.24}$$

In the context of *classical field theory*, on the other hand, it is just required that these magnitudes are finite.

The fact that the two expressions in (3.22) represent exactly the same state is part of a more general property that we can call *unitary equivalence* or *unitary invariance*. Different coordinate bases can be related by unitary transformations. Assume that $|u\rangle$ and $|v\rangle$ represent the elements of two different coordinate bases. Then we can convert one into the other by

$$|v\rangle = \int |u\rangle \, U(u,v) \, \mathrm{d}u, \tag{3.25}$$

where $U(u,v)$ is a unitary kernel, as defined in (2.1.5). It then follows that, for a state given in terms one such coordinate basis

$$|\psi\rangle = \int |u\rangle\,\psi(u)\,\mathrm{d}u, \tag{3.26}$$

we can introduce arbitrary unitary operators to convert the coordinate basis to another coordinate basis

$$|\psi\rangle = \int |u\rangle\,U(u,v)U^{\dagger}(v,u')\psi(u')\,\mathrm{d}u'\,\mathrm{d}u\,\mathrm{d}v = \int |v\rangle\,\psi'(v)\,\mathrm{d}v, \tag{3.27}$$

where the coefficient function has now been transformed to

$$\psi'(v) = \int U^{\dagger}(v,u')\psi(u')\,\mathrm{d}u'. \tag{3.28}$$

Since there are infinitely many such unitary transformations, it means that there are infinitely many different ways to represent the same state vector in terms of different coordinate bases, all related by unitary transformations. The Fourier kernel is an example of such a unitary kernel.

When we insert the identity, resolved in terms of a coordinate basis, as in (3.21), into the inner product for the function in (3.23) in the opposite basis, and use the other relationship in (3.23) to replace the resulting inner product in the integrand, we get

$$\tilde{\psi}(p) = \int \langle p|q\rangle\psi(q)\,\mathrm{d}q, \tag{3.29}$$

or the similar expression for $\psi(q)$ in terms of $\tilde{\psi}(p)$. Comparing these expressions with (3.3), we see that

$$\langle q|p\rangle = \exp(ipq) \quad \text{and} \quad \langle p|q\rangle = \exp(-ipq). \tag{3.30}$$

It reveals the Fourier relationship between the two bases: the elements of one quadrature basis can be expressed in terms of the other, as

$$\begin{aligned} |p\rangle &= \int |q\rangle\,\exp(iqp)\,\mathrm{d}q, \\ |q\rangle &= \int |p\rangle\,\exp(-iqp)\,\frac{\mathrm{d}p}{2\pi}. \end{aligned} \tag{3.31}$$

It also means that $|\langle q|p\rangle| = 1$, which implies that the two bases are *mutually unbiased*: when the elements of one basis are expanded in terms of the other, none of the elements in the expansion dominates.

3.1.4 Quadrature operators

Now we define operators (observables) for the respective coordinates. These operators, which are called *quadrature operators*, are given by

$$\hat{q} \triangleq \int |q\rangle\, q\, \langle q|\; \mathrm{d}q \quad \text{and} \quad \hat{p} \triangleq \int |p\rangle\, p\, \langle p|\; \frac{\mathrm{d}p}{2\pi}. \tag{3.32}$$

They are Hermitian: $\hat{q}^{\dagger} = \hat{q}$ and $\hat{p}^{\dagger} = \hat{p}$ (see the discussion on Hermitian operators in Section 3.1.2), and can thus be used to compute the *expectation values*, as in (3.18), of the respective coordinates.

Note that there is no *Planck constant* in the definition of the Fourier domain operator $\hat{p}$. It is *not the momentum operator*. At this point, the operator $\hat{p}$ is analogous to an operator for the expectation value of the wavenumber and not for the momentum.

3.1.4.1 Properties of quadrature operators

In the current context, the eigenvalues of the quadrature operators are the real-valued quadrature variables, which are associated with the dimensionless (normalized) coordinates of the configuration space and the Fourier domain, respectively. Their associated *eigenvectors* $|q\rangle$ and $|p\rangle$ thus form orthogonal bases *by construction*. The quadrature operators are thus also Hermitian by construction. Their eigenvalue equations are

$$\hat{q}\,|q\rangle = |q\rangle\, q \quad \text{and} \quad \hat{p}\,|p\rangle = |p\rangle\, p. \tag{3.33}$$

Holomorphic functions[b] of the respective quadrature operators are converted to the same holomorphic functions of the corresponding quadrature variables when they are applied to the eigenvectors:

$$\begin{aligned} f(\hat{q})\,|q\rangle &= |q\rangle\, f(q) \\ g(\hat{p})\,|p\rangle &= |p\rangle\, g(p). \end{aligned} \tag{3.34}$$

We can use the definitions of the quadrature bases in (3.31) to define *shift* or *translation operators*, given by

---

b A function is holomorphic on a domain if it is differentiable at every point on that domain, which means that is can be expanded as a power series of the variables of that domain.

$$
\begin{aligned}
|q-q_0\rangle &= \int |p\rangle \exp[-\mathrm{i}(q-q_0)p]\,\frac{\mathrm{d}p}{2\pi} \\
&= \exp(\mathrm{i}q_0\hat{p}) \int |p\rangle \exp(-\mathrm{i}qp)\,\frac{\mathrm{d}p}{2\pi} = \exp(\mathrm{i}q_0\hat{p})\ |q\rangle \\
|p-p_0\rangle &= \int |q\rangle \exp[\mathrm{i}q(p-p_0)]\,\mathrm{d}q \\
&= \exp(-\mathrm{i}\hat{q}p_0) \int |q\rangle \exp(\mathrm{i}qp)\,\mathrm{d}q = \exp(-\mathrm{i}\hat{q}p_0)\,|p\rangle\,.
\end{aligned} \tag{3.35}
$$

Here, we used (3.34) to *pull* the exponential function *back through* the eigenvector to obtain the definition of these shift operators in the form of exponentiated quadrature operators. The fact that a shift in the basis element is related to such an exponentiated operator in the form of an operator-valued phase factor follows from the associated Fourier theorem [16] and is therefore a consequence of the Fourier analysis on which this development is based. It also follows that the quadrature operators are the *generators of translation* in the opposite quadrature variables. These shift operators are also called *Weyl operators*.

The quadrature operators can be expressed in terms of the opposite bases by applying identity operators resolved in the opposite bases to them. For example,

$$
\begin{aligned}
\hat{p} = \mathbb{1}\hat{p}\mathbb{1} &= \int |q\rangle \exp(\mathrm{i}pq)p\exp(-\mathrm{i}pq')\,\langle q'|\,\frac{\mathrm{d}p}{2\pi}\,\mathrm{d}q'\,\mathrm{d}q \\
&= \int |q\rangle\,(-\mathrm{i}\partial_q)\exp(\mathrm{i}pq)\exp(-\mathrm{i}pq')\,\langle q'|\,\frac{\mathrm{d}p}{2\pi}\,\mathrm{d}q'\,\mathrm{d}q \\
&= \int |q\rangle\,(-\mathrm{i}\partial_q)\,\delta(q-q')\,\langle q'|\,\mathrm{d}q'\,\mathrm{d}q \\
&= \int |q\rangle\,(-\mathrm{i}\partial_q)\,\langle q|\,\mathrm{d}q.
\end{aligned} \tag{3.36}
$$

In a similar way, we obtain

$$
\hat{q} = \int |p\rangle\,(\mathrm{i}\partial_p)\,\langle p|\,\frac{\mathrm{d}p}{2\pi}. \tag{3.37}
$$

3.1.5 Noncommuting operators

The two quadrature operators $\hat{q}$ and $\hat{p}$ do not commute with each other. We can demonstrated this fact by using the representations of these operators in terms of the same basis, with $\hat{q}$ given in (3.32) and $\hat{p}$ as represented in (3.36). Their commutation applied to an arbitrary state vector $|\psi\rangle$ produces

$$
\begin{aligned}
[\hat{q},\hat{p}]\,|\psi\rangle &= \int |q'\rangle\,q'\langle q'|q\rangle\,(-\mathrm{i}\partial_q)\,\langle q|\psi\rangle\,\mathrm{d}q'\,\mathrm{d}q - \int |q\rangle\,(-\mathrm{i}\partial_q)\,\langle q|q'\rangle q'\langle q'|\psi\rangle\,\mathrm{d}q'\,\mathrm{d}q \\
&= \int |q\rangle\,q\,(-\mathrm{i}\partial_q)\,\psi(q)\,\mathrm{d}q - \int |q\rangle\,(-\mathrm{i}\partial_q)\,q\psi(q)\,\mathrm{d}q = \mathrm{i}\int |q\rangle\,\psi(q)\,\mathrm{d}q = \mathrm{i}\,|\psi\rangle\,.
\end{aligned} \tag{3.38}
$$

It thus follows that

$$[\hat{q}, \hat{p}] = \mathrm{i}\mathbb{1}. \tag{3.39}$$

The shift or Weyl operators derived in (3.35) provide an alternative representation of the *commutation relation for quadrature operators*. For this purpose, we need the *Baker–Campbell–Hausdorff formula* to combine exponentiated operators. It is given by

$$\exp(\hat{X})\exp(\hat{Y}) = \exp\left(\hat{X} + \hat{Y} + \frac{1}{2}[\hat{X}, \hat{Y}]\right), \tag{3.40}$$

for the case when $[\hat{X}, [\hat{X}, \hat{Y}]] = [\hat{Y}, [\hat{X}, \hat{Y}]] = 0$. The Baker–Campbell–Hausdorff formula can also be expressed in the opposite form, as

$$\exp\left(\hat{X} + \hat{Y}\right) = \exp\left(-\frac{1}{2}[\hat{X}, \hat{Y}]\right)\exp(\hat{X})\exp(\hat{Y}), \tag{3.41}$$

to separate exponentiated operators, still assuming $[\hat{X}, [\hat{X}, \hat{Y}]] = [\hat{Y}, [\hat{X}, \hat{Y}]] = 0$.

Exercise 3.3. Use the Baker–Campbell–Hausdorff formulas in (3.40) and (3.41) to show that by interchanging the order of the product of the two Weyl operators, one produces a phase factor

$$\exp(-\mathrm{i}\hat{q}p_0)\exp(\mathrm{i}q_0\hat{p}) = \exp(\mathrm{i}q_0p_0)\exp(\mathrm{i}q_0\hat{p})\exp(-\mathrm{i}\hat{q}p_0). \tag{3.42}$$

A linear transformation of these quadrature operators to another set that also obeys a commutation relation of the form given in (3.39), as given by

$$\hat{q} \to \hat{q}' = A_{qq}\hat{q} + A_{qp}\hat{p} \quad \text{and} \quad \hat{p} \to \hat{p}' = A_{pq}\hat{q} + A_{pp}\hat{p}, \tag{3.43}$$

leads to the condition that the coefficients must obey:

$$A_{qq}A_{pp} - A_{qp}A_{pq} = 1. \tag{3.44}$$

In Section 3.5, we show that this condition is consistent with the requirement for a *symplectic transformation* as associated with the *canonical transformations* of phase space variables. It follows that a *commutation relation* of the form (3.39) between two operators identifies them as candidates for phase space variables. The *Poisson bracket* provides an analogous mathematical structure in classical theory [23] without operators.

The resulting non-commutation has nothing to do with quantum physics *per se*. It is derived here for the operators defined in terms of coordinate bases that are related by Fourier transforms. Therefore, the non-commutation is an unambiguous consequence of the Fourier relationship. The result we arrive at here is equally valid in the quantum context and in the context of a *classical field theory* formulated in terms of operators.

Consider, for example, classical *scalar diffraction theory* as presented in Section 2.6, in which Fourier theory plays a significant role. If one defines operators for the expectation values of the coordinates in the configuration space and the Fourier domain, respectively, these operators would exhibit the same non-commutation found in (3.39). It thus follows that the non-commutation of operators cannot be used as an indication of quantum physics.

3.1.6 Ladder operators

The formulation of ladder operators is motivated by the form of the Hamiltonian of the harmonic oscillator, which is considered later. However, nothing prevents us from introducing the ladder operators in terms of the quadrature operators at this point.

The ladder operators are defined as

$$\hat{a} \triangleq \frac{1}{\sqrt{2}}(\hat{q} + \mathrm{i}\hat{p}) \quad \text{and} \quad \hat{a}^{\dagger} \triangleq \frac{1}{\sqrt{2}}(\hat{q} - \mathrm{i}\hat{p}) . \tag{3.45}$$

The quadrature operators are therefore given by

$$\hat{q} = \frac{1}{\sqrt{2}}(\hat{a} + \hat{a}^{\dagger}) \quad \text{and} \quad \hat{p} = -\mathrm{i}\frac{1}{\sqrt{2}}(\hat{a} - \hat{a}^{\dagger}), \tag{3.46}$$

in terms of the ladder operators. The reason for calling them *ladder operators* becomes clear later. Based on the commutation relation of the quadrature operators in (3.39), it follows that

$$[\hat{a}, \hat{a}^{\dagger}] = \mathbb{1}, \tag{3.47}$$

representing a *commutation relation for ladder operators*.

Exercise 3.4. Show that $[\hat{a}^n, \hat{a}^{\dagger}] = n\hat{a}^{n-1}$, and that

$$[\hat{a}^n, \hat{a}^{\dagger m}] = \sum_{p=1}^{m} \frac{n!m!}{p!(n-p)!(m-p)!} \hat{a}^{\dagger(m-p)} \hat{a}^{n-p} . \tag{3.48}$$

for $n > m$.

Hint: use the identity $[\hat{a}, \hat{b}\hat{c}] = [\hat{a}, \hat{b}]\hat{c} + \hat{b}[\hat{a}, \hat{c}]$.

3.1.7 Number operator and the Fock states

The ladder operators are not *Hermitian* and (unlike the quadrature operators) do not represent observables. However, they can be used to construct Hermitian operators that do represent observables. A simple yet versatile example is the operator $\hat{a}^{\dagger}\hat{a}$, which is called the *number operator* for reasons provided below. For an arbitrary state vector $|\psi\rangle$, the *expectation value of the number operator* is always positive

$$\langle\psi|\,\hat{a}^{\dagger}\hat{a}\,|\psi\rangle \geq 0, \tag{3.49}$$

because it represents the squared magnitude of the unnormalized vector $|\psi'\rangle = \hat{a}\,|\psi\rangle$, and such a squared magnitude is always non-negative. The eigenvalue λ_n associated with an *eigenvector* $|\phi_n\rangle$ of $\hat{a}^{\dagger}\hat{a}$, through

$$\hat{a}^{\dagger}\hat{a}\,|\phi_n\rangle = |\phi_n\rangle\,\lambda_n, \tag{3.50}$$

must also be non-negative. Moreover, from (3.47) we get

$$[\hat{a}, \hat{a}^{\dagger}\hat{a}] = \hat{a} \quad \text{and} \quad [\hat{a}^{\dagger}\hat{a}, \hat{a}^{\dagger}] = \hat{a}^{\dagger}. \tag{3.51}$$

It then follows that, if $|\phi_n\rangle$ is an eigenvector, then both $\hat{a}\,|\phi_n\rangle$ and $\hat{a}^{\dagger}\,|\phi_n\rangle$ must also be eigenvectors, because

$$\begin{aligned}\hat{a}^{\dagger}\hat{a}\hat{a}\,|\phi_n\rangle &= \hat{a}\,|\phi_n\rangle\,(\lambda_n - 1)\,,\\ \hat{a}^{\dagger}\hat{a}\hat{a}^{\dagger}\,|\phi_n\rangle &= \hat{a}^{\dagger}\,|\phi_n\rangle\,(\lambda_n + 1)\,.\end{aligned} \tag{3.52}$$

Now we see why $\hat{a}^{\dagger}$ and $\hat{a}$ are called ladder operators: they respectively increase and decrease the eigenvalues in steps of 1. Since $\hat{a}^{\dagger}$ increases the value, we call it the *creation operator*, and since $\hat{a}$ decreases the value, it is called the *annihilation operator*. Most of the time, we refer to these two operators collectively as the *ladder operators*. Since the eigenvalues must all be non-negative, there must be an eigenvector $|\phi_0\rangle$ with the smallest eigenvalue. When $\hat{a}$ is applied to this eigenvector, we must impose that $\hat{a}\,|\phi_0\rangle = 0$. It then follows that $\hat{a}^{\dagger}\hat{a}\,|\phi_0\rangle = 0$, which means that $\lambda_0 = 0$. We call this eigenvector the *vacuum state* $|\phi_0\rangle = |\text{vac}\rangle$. The other eigenvectors are produced by applying $\hat{a}^{\dagger}$ repeatedly and normalizing the result. These eigenvectors are called the *Fock states* and we denote them by $|n\rangle$. The *number operator* is denoted by

$$\hat{n} \triangleq \hat{a}^{\dagger}\hat{a}. \tag{3.53}$$

Since the lowest eigenvalue is zero, the other eigenvalues are the positive integers $\lambda_n = n$. The normalization of the Fock states can be determined by noting that

$$\langle n|n\rangle = \mathcal{N}^2\,\langle\text{vac}|\,\hat{a}^{n}\hat{a}^{\dagger n}\,|\text{vac}\rangle = 1, \tag{3.54}$$

where $\mathcal{N}$ is the normalization constant. To compute the overlap, we convert the operators to *normal order*, which means that all the annihilation operators are placed on the left-hand side of the creation operators. It is done with the aid of the commutation relation in (3.48) for $m = n$:

$$\left[\hat{a}^n, \hat{a}^{\dagger n}\right] = n!\,, \tag{3.55}$$

which leads to the definition of the Fock states as

$$|n\rangle \triangleq \frac{\hat{a}^{\dagger n}}{\sqrt{n!}}\,|\text{vac}\rangle \quad \text{and} \quad \langle n| \triangleq \langle \text{vac}|\,\frac{\hat{a}^n}{\sqrt{n!}}. \tag{3.56}$$

Exercise 3.5. Show that the number operator expressed in terms of the quadrature operators is

$$\hat{n} = \hat{a}^\dagger\hat{a} = \frac{1}{2}(\hat{q}^2 + \hat{p}^2 - \mathbb{1}), \tag{3.57}$$

where $\mathbb{1}$ comes from the commutation relation (3.39).

Exercise 3.6. Use (3.48), to show that

$$[\hat{a}^m, \hat{n}] = m\hat{a}^m \quad \text{and} \quad \left[\hat{n}, \hat{a}^{\dagger m}\right] = m\hat{a}^{\dagger m}. \tag{3.58}$$

In the quantum context, the notion of the *particle-number degree of freedom* naturally leads to the idea of a quantum state that contains a well-defined integer number of particles, as represented by the Fock states. The number of particles n is referred to as its *occupation number*. We found that such a Fock state is an eigenvector of the number operator

$$\hat{n}\,|n\rangle = |n\rangle\,n. \tag{3.59}$$

Considering the particle-number degree of freedom only, we find that the Fock states form a complete orthonormal basis. They obey the *orthogonality condition*

$$\langle m|n\rangle = \delta_{m,n}, \tag{3.60}$$

where $\delta_{m,n}$ is the Kronecker delta, define in (2.6). By implication, these Fock states are dimensionless. They also satisfy the *completeness condition*

$$\sum_n |n\rangle\,\langle n| = \mathbb{1}, \tag{3.61}$$

where $\mathbb{1}$ is the identity operator. When the ladder operators are applied once to the Fock states, we get

$$\begin{aligned} \hat{a}\,|n\rangle &= |n-1\rangle\,\sqrt{n}, \\ \hat{a}^\dagger\,|n\rangle &= |n+1\rangle\,\sqrt{n+1}. \end{aligned} \tag{3.62}$$

The factors of $\sqrt{n}$ and $\sqrt{n+1}$ can lead to a *bosonic enhancement* for processes in quantum physics.

Thanks to their orthogonality and completeness, the Fock basis can be used to expand any state in terms of its particle-number degree of freedom, as shown in (3.5), but with $|\phi_n\rangle \to |n\rangle$, so that

$$|\psi\rangle = \sum_{n=0}^{\infty} |n\rangle\, C_n. \tag{3.63}$$

The complex coefficients C_n are obtained by computing the overlap

$$C_n = \langle n|\psi\rangle. \tag{3.64}$$

and the normalization implies (3.6).

3.1.8 Fock state coefficient functions

One can compute the coefficient functions of the Fock states by evaluating the overlap $\psi_n(q) = \langle q|n\rangle$. They are used to expand the Fock states in terms of the q-basis

$$|n\rangle = \int |q\rangle\, \langle q|n\rangle\, \mathrm{d}q = \int |q\rangle\, \psi_n(q)\, \mathrm{d}q. \tag{3.65}$$

In the quantum context, these Fock state coefficient functions represent the *wavefunctions* of the Fock states. Their complex conjugates also represent the coefficient functions when we expand a quadrature basis element in terms of the Fock states:

$$|q\rangle = \sum_n |n\rangle\, \langle n|q\rangle = \sum_n |n\rangle\, \psi_n^*(q). \tag{3.66}$$

To compute the coefficient functions, we can express the Fock states in terms of the ladder operators, as in (3.56), which are in turn expressed in terms of the quadrature operators $\hat{q}$ and $\hat{p}$. In the q-basis, the latter two operators convert the overlap into a differential equation with solutions based on the *Hermite polynomials*, which we encountered before in Chapter 2. The solutions of this differential equation represent the coefficient functions.

Here, we provide a different approach to compute the Fock state coefficient function, using generating functions as discussed in Appendix A. When we express the Fock state in the overlap for the coefficient function in terms of ladder operators, using (3.56), we have

$$\psi_n^*(q) = \langle n|q\rangle = \frac{1}{\sqrt{n!}}\, \langle \mathrm{vac}|\, \hat{a}^n\, |q\rangle\,. \tag{3.67}$$

The products of *annihilation operators* can be converted into a generating function for such products, as given by

$$\sum_n \frac{\eta^n}{\sqrt{2^n n!}}\psi_n^*(q) = \langle \text{vac}| \exp\left(\frac{1}{\sqrt{2}}\eta\hat{a}\right)|q\rangle , \tag{3.68}$$

where η is the generating parameter and where we introduced convenient constants in anticipation of our goal. Next, we exploit the fact that

$$\langle \text{vac}| \exp\left(K\hat{a}^\dagger\right) = \langle \text{vac}| \tag{3.69}$$

for an arbitrary constant K, to insert exponentiated creation operators:

$$\sum_n \frac{\eta^n}{\sqrt{2^n n!}}\psi_n^*(q) = \langle \text{vac}| \exp\left(\frac{1}{\sqrt{2}}\eta\hat{a}^\dagger\right) \exp\left(\frac{1}{\sqrt{2}}\eta\hat{a}\right)|q\rangle . \tag{3.70}$$

The exponentiated ladder operators can now be combined using the *Baker–Campbell–Hausdorff formula* in (3.40), as applicable here. Hence,

$$\begin{aligned}\sum_n \frac{\eta^n}{\sqrt{2^n n!}}\psi_n^*(q) &= \langle \text{vac}| \exp\left(\frac{1}{\sqrt{2}}\eta\hat{a}^\dagger + \frac{1}{\sqrt{2}}\eta\hat{a} - \frac{1}{4}\eta^2\right)|q\rangle \\ &= \langle \text{vac}| \exp\left(\eta\hat{q} - \frac{1}{4}\eta^2\right)|q\rangle = \langle \text{vac}|q\rangle \exp\left(\eta q - \frac{1}{4}\eta^2\right) \\ &= \psi_0^*(q) \exp\left(\eta q - \frac{1}{4}\eta^2\right),\end{aligned} \tag{3.71}$$

where we combined the ladder operators into a quadrature operator, according to the definitions in (3.46); used the eigenvalue equations in (3.34) to *push* the exponentiated quadrature operator *forward through* the basis element to produce a function of the quadrature variable; and finally defined $\langle \text{vac}|q\rangle \triangleq \psi_0^*(q)$ as (the complex conjugate of) the zeroth-order coefficient function. This result can be interpreted in terms of the *generating function for Hermite polynomials*, given in (2.122), with the appropriate replacements. It thus follows that

$$\sum_n \frac{\eta^n}{\sqrt{2^n n!}}\psi_n^*(q) = \psi_0^*(q)\sum_{n=0}^{\infty} \frac{\eta^n}{2^n n!}H_n(q). \tag{3.72}$$

The coefficient functions are then given by

$$\psi_n^*(q) = \frac{\psi_0^*(q)}{\sqrt{2^n n!}}H_n(q), \tag{3.73}$$

up to the zeroth-order coefficient function $\psi_0^*(q)$, which is yet to be determined.

To find the expressions for the zeroth-order coefficient function, we consider the inner product between coefficient functions of arbitrary orders

$$\int \psi_m^*(q)\psi_n(q)\, dq = \int \frac{\psi_0^*(q)\psi_0(q)}{\sqrt{2^m m! 2^n n!}}H_m(q)H_n(q)\, dq. \tag{3.74}$$

Comparing this expression with the *orthogonality condition for Hermite polynomials* [17], given by

$$\int H_m(x)H_n(x)\exp(-x^2)\,\mathrm{d}x = \sqrt{\pi}2^n n!\,\delta_{m,n}, \tag{3.75}$$

where $\delta_{m,n}$ is the Kronecker delta, we see that, for

$$|\psi_0(q)|^2 = \frac{1}{\sqrt{\pi}}\exp(-q^2), \tag{3.76}$$

the coefficient functions obey the orthogonality condition

$$\int \psi_m^*(q)\psi_n(q)\,\mathrm{d}q = \delta_{m,n}. \tag{3.77}$$

It follows from (3.65), based on the orthogonality of the Fock states in (3.60). Assuming that the zeroth-order coefficient function is real-valued, we obtain

$$\psi_0(q) = \pi^{-1/4}\exp\left(-\frac{1}{2}q^2\right). \tag{3.78}$$

So, the general expression for the coefficient functions is

$$\psi_n(q) = \frac{1}{\pi^{1/4}\sqrt{2^n n!}}H_n(q)\exp\left(-\frac{1}{2}q^2\right). \tag{3.79}$$

These coefficient functions are the one-dimensional versions of the Hermite–Gauss modes that we found in (2.118) at the waist ($z = 0$), with an appropriate scaling of the coordinate ($q \to \sqrt{2}u$). The shapes of the coefficient functions for $n = 1, 2, 3$ are shown in Figure 3.1.

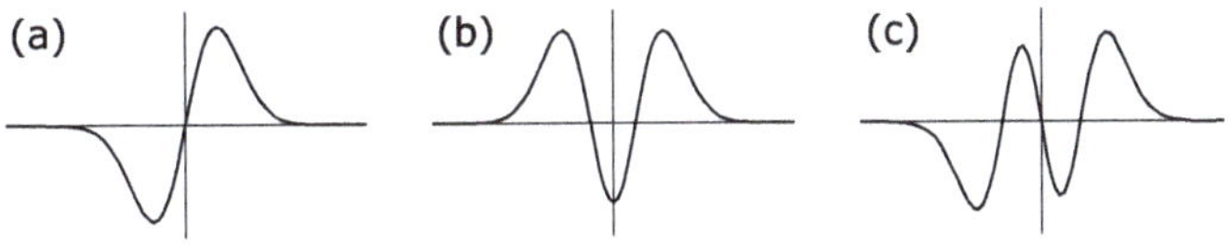

Figure 3.1: The shapes of the three Hermite–Gauss modes: (a) $n = 1$, (b) $n = 2$, (c) $n = 3$.

Exercise 3.7. Follow the same procedure for the p-basis to show that

$$\tilde{\psi}_n(p) = \frac{(-\mathrm{i})^n\sqrt{2}\pi^{1/4}}{\sqrt{2^n n!}}H_n(p)\exp\left(-\frac{1}{2}p^2\right). \tag{3.80}$$

Although $\psi_n(q)$ is real valued, $\tilde{\psi}_n(p)$ is not. However, since they are related by the Fourier transform, we have $\tilde{\psi}_n^*(p) = \tilde{\psi}_n(-p)$.

3.1.8.1 Orthogonality of the Fock basis

Given that the number operator is Hermitian and all its eigenvalues (the occupation numbers) are non-degenerate, we know that its eigenvectors (the Fock states) are orthogonal, as assumed in (3.60). If we are not convinced of that yet, we can use the overlap between the Fock states and the quadrature bases to show that they are orthogonal. For this purpose, we express the Fock states in terms of their expansions in the quadrature bases. Focussing on the q-basis, we have (3.65). The inner product between the Fock states then leads to

$$\begin{aligned}\langle m|n\rangle &= \int \langle q|q'\rangle \psi_m^*(q)\psi_n(q')\,\mathrm{d}q\,\mathrm{d}q' = \int \psi_m^*(q)\psi_n(q)\,\mathrm{d}q \\ &= \int \frac{H_m(q)H_n(q)}{\sqrt{\pi 2^m m! 2^n n!}} \exp\left(-q^2\right)\mathrm{d}q = \delta_{m,n},\end{aligned} \tag{3.81}$$

where we used (3.75). The result is the *orthogonality condition* for Fock states in terms of only the particle-number degree of freedom.

3.1.8.2 Completeness of the Fock basis

For the *completeness of the Fock basis*, we again consider their expansion in terms of the quadrature basis. Here, we form the summation

$$\begin{aligned}\sum_{n=0}^{\infty} |n\rangle\langle n| &= \sum_{n=0}^{\infty} \int |q_1\rangle \psi_n(q_1)\psi_n^*(q_2)\langle q_2|\,\mathrm{d}q_1\,\mathrm{d}q_2 \\ &= \sum_{n=0}^{\infty} \int |q_1\rangle \frac{H_n(q_1)H_n(q_2)}{\sqrt{\pi}2^n n!} \exp\left(-\frac{1}{2}q_1^2 - \frac{1}{2}q_2^2\right)\langle q_2|\,\mathrm{d}q_1\,\mathrm{d}q_2.\end{aligned} \tag{3.82}$$

To evaluate this summation, we use *Mehler's formula* [18]:

$$\sum_{n=0}^{\infty} \frac{c^n}{2^n n!} H_n(x)H_n(y) = \frac{1}{\sqrt{1-c^2}} \exp\left[\frac{2cxy}{1-c^2} - \frac{(x^2+y^2)c^2}{1-c^2}\right], \tag{3.83}$$

where c is an arbitrary constant. The case in (3.82) with $c = 1$, produces a singularity due to the denominators $1 - c^2$. Therefore, we consider the limit $c \to 1$. It leads to

$$\begin{aligned}\sum_{n=0}^{\infty} |n\rangle\langle n| &= \lim_{c\to 1} \int |q_1\rangle \frac{1}{\sqrt{\pi}\sqrt{1-c^2}} \exp\left[\frac{2cq_1q_2}{1-c^2} - \frac{(q_1^2+q_2^2)c^2}{1-c^2}\right] \\ &\quad \times \exp\left(-\frac{1}{2}q_1^2 - \frac{1}{2}q_2^2\right)\langle q_2|\,\mathrm{d}q_1\,\mathrm{d}q_2 \\ &= \lim_{c\to 1} \int |q_1\rangle \frac{1}{\sqrt{\pi(1-c^2)}} \exp\left[-\frac{(q_1-q_2)^2}{1-c^2}\right] \\ &\quad \times \exp\left(\frac{1}{2}q_1^2 + \frac{1}{2}q_2^2 - \frac{2q_1q_2}{1+c}\right)\langle q_2|\,\mathrm{d}q_1\,\mathrm{d}q_2.\end{aligned} \tag{3.84}$$

The singular part of the expression now represents a limit process for the Dirac delta function, as in (2.17). Here, it gives

$$\lim_{c\to 1} \frac{1}{\sqrt{\pi(1-c^2)}} \exp\left[-\frac{(q_1-q_2)^2}{1-c^2}\right] = \delta(q_1-q_2). \tag{3.85}$$

The remaining terms in the exponent vanish when the argument of the Dirac delta is zero. Hence, we obtain the completeness condition

$$\sum_{n=0}^{\infty} |n\rangle\langle n| = \int |q_1\rangle\,\delta(q_1-q_2)\,\langle q_2|\;dq_1\,dq_2 = \int |q\rangle\langle q|\;dq = \mathbb{1}, \tag{3.86}$$

where we used (3.21). The result confirms the completeness of the Fock states, as demonstrated here in terms of the particle-number degree of freedom only.

3.1.8.3 Inner products between quadrature bases

In (3.30), we obtained expressions for the inner products between elements from the two quadrature bases by comparing the expressions for state vectors in terms of these bases. Here, we can obtain the same expressions in a different way, by using the orthogonality of the Fock states together with Mehler's formula. Expanding the quadrature bases in terms of Fock states, we have

$$|q\rangle = \sum_{n=0}^{\infty} |n\rangle\,\psi_n^*(q) \quad \text{and} \quad |p\rangle = \sum_{m=0}^{\infty} |m\rangle\,\tilde{\psi}_m^*(p). \tag{3.87}$$

The inner product between them gives

$$\begin{aligned}\langle q|p\rangle &= \sum_{m,n=0}^{\infty} \psi_n(q)\langle n|m\rangle\tilde{\psi}_m^*(p) = \sum_{n=0}^{\infty} \psi_n(q)\tilde{\psi}_n^*(p)\\ &= \sum_{n=0}^{\infty} \frac{(\mathrm{i})^n\sqrt{2}}{2^n n!} H_n(q)H_n(p)\exp\left(-\frac{1}{2}q^2-\frac{1}{2}p^2\right),\end{aligned} \tag{3.88}$$

where we substituted the coefficient functions from (3.79) and (3.80) in the last expression. Using Mehler's formula in (3.83) with $c = \mathrm{i}$, we reproduce

$$\langle q|p\rangle = \exp(\mathrm{i}pq), \tag{3.89}$$

as given in (3.30).

3.1.9 Projection operators

A general state vector can consist of a superposition of different numbers of particles. It is often useful to select the part with a specific number of particles from such a state. The *projection operators for the number of particles* serves this purpose. It is defined in terms of the Fock states by

$$\hat{P}_n \triangleq |n\rangle \langle n| . \tag{3.90}$$

These operators have a number of significant properties. The normalization of the Fock states ($\langle n|n\rangle = 1$) implies that the projection operators are *idempotent*:

$$\hat{P}_m \hat{P}_n = \delta_{m,n} \hat{P}_n. \tag{3.91}$$

From the completeness of the Fock basis, it follows that the sum of all the projection operators for the number of particles is the identity

$$\sum_{n=0}^{\infty} \hat{P}_n = \sum_{n=0}^{\infty} |n\rangle \langle n| = \mathbb{1}. \tag{3.92}$$

3.1.10 Discussion

Let us summarize what we have achieved so far. We managed to derive the ladder operators and Fock states without using any concept from quantum physics. Evidently, the discrete nature of the ladder operators simply follows from Fourier theory and does not require quantum physics. It implies that one cannot regard the presence of ladder operators in any formalism as a conclusive indication of quantum physics. Nevertheless, we can use them to *model* the discrete nature of quantum physics in a purely formal way.

How does Fourier theory lead to the discrete property of the ladder operators? We can argue that it follows from the discreteness of the Hermite modes in (3.79). For this purpose, it is worth noting that the coefficient functions of the Fock states, which are represented by these Hermite modes, are *eigenfunctions of the Fourier transform.*

Exercise 3.8. Use the one-dimensional generating function for the Hermite–Gauss modes

$$\mathcal{H}(\mu) = \exp\left(2q\mu - \mu^2 - \frac{1}{2}q^2\right), \tag{3.93}$$

as obtained from (2.126) at $z = 0$, to show that the Fourier transform of this generating function reproduces the same generating function in terms of p, but with $\mu \to -i\mu$. Use the expansion of the resulting generating function to identify the associated eigenvalues as given by $(-i)^n$.

So, while the identity operator can be resolved in terms of the Fock states, as shown in (3.61), and the number operator can be diagonalized in terms of the Fock states to be represented by

$$\hat{n} = \hat{a}^{\dagger}\hat{a} = \sum_{n=0}^{\infty} |n\rangle\, n\, \langle n| \,, \tag{3.94}$$

the Fourier transform, expressed as an operator, can also be diagonalized in terms of the Fock states. As an operator, it can be represented by

$$\hat{\mathcal{F}} = \sum_{n=0}^{\infty} |n\rangle\, (-\mathrm{i})^{n}\, \langle n| \,. \tag{3.95}$$

It produces an output state with a coefficient function given by the Fourier transform of the input state's coefficient function when expressed in the same basis. The inverse Fourier transform is likewise represented as an operator by

$$\hat{\mathcal{F}}^{-1} = \sum_{n=0}^{\infty} |n\rangle\, \mathrm{i}^{n}\, \langle n| \,. \tag{3.96}$$

In the next chapter, we discuss the quantization of the electromagnetic field as it is done in quantum field theory. For the purpose of this quantization process, we start with the ladder operators obeying their commutation relations without invoking quantum physics explicitly. These ladder operators suffice to model the discreteness of the quantized electromagnetic field, without the need for the Planck and the de Broglie relationships. Although the expressions of the quantized electromagnetic fields are in the form of Fourier transforms, these Fourier transforms do not operate on the particle-number degree of freedom. So, it seems that the discreteness is now invoked by the ladder operators without the presence of particle-number Fourier theory.

It turns out that we can invert the order of the derivation in this section to show that the particle-number Fourier relationship follows from the ladder operators. All we need to do is to assume that the quadrature operators that are defined in terms of these ladder operators satisfy eigenvalue equations with continuous eigenvalues. It thus allows us to derive the Fock states as *eigenstates*[c] of the number operator in terms of the Hermite modes, which then leads to the overlap between the quadrature bases being given by the Fourier kernel.

---

c The term *eigenstate* represents an *eigenvector* that is normalized. In general, eigenvectors cannot always be normalized.

3.2 Coherent states

In Section 3.1, we developed the concept of a quantum state, even though we did not specifically introduce any quantum physics. In the process, we arrived at the set of Fock states, which is a complete orthogonal basis for states in the Hilbert space. We also obtained the quadrature bases, which are complete orthogonal bases. While the elements in the quadrature bases cannot be normalized and thus do not represent states in the Hilbert space, they can be used to express such states in terms of expansions.

Here, we discuss another set of states called the *coherent states*. They are not orthogonal, but they can be used as a complete basis. Considered as a quantum state, a laser beam is often represented as a coherent state.

3.2.1 Displacement operator

There are different defining properties for coherent states. One of them is that coherent states are *displaced vacuum states*. While the Fock states are produced from the vacuum state by multiple applications of the creation operator, the coherent states are produced from the vacuum state with the aid of a *displacement operator*. It is given by

$$\hat{D}(\alpha) = \exp(\alpha\hat{a}^{\dagger} - \alpha^{*}\hat{a}), \tag{3.97}$$

where α is an arbitrary (dimensionless) complex-valued parameter.

We often encounter operators expressed as exponentiated operators. They are formally defined in terms of the expansion of an exponential function, given as

$$\exp(\hat{O}) = \mathbb{1} + \hat{O} + \frac{1}{2!}\hat{O}\hat{O} + \frac{1}{3!}\hat{O}\hat{O}\hat{O}\ldots, \tag{3.98}$$

with the leading order term represented by the identity operator.[d]

When a displacement operator is applied to the vacuum state it produces a coherent state. As expansions in terms of the Fock states, the coherent states are given by

$$\hat{D}(\alpha)\,|\text{vac}\rangle = |\alpha\rangle = \exp\left(-\frac{1}{2}|\alpha|^2\right)\sum_{n=0}^{\infty}|n\rangle\,\frac{\alpha^n}{\sqrt{n!}}. \tag{3.99}$$

For $\alpha = 0$, we get the vacuum state $\hat{D}(0)\,|\text{vac}\rangle = \mathbb{1}\,|\text{vac}\rangle = |\text{vac}\rangle$. Although the vacuum state is a Fock state, it is also an element in the set of all coherent states.

---

d This expansion may run into convergence issues when $\hat{O}$ is an unbounded operator. However, provided that we restrict the domain to physical states for which the application of unbounded operators produce well-defined finite results, the expansion is convergent.

Exercise 3.9. Use the Baker–Campbell–Hausdorff identity in (3.41), to express the displacement operator in normal order and use this result to derive the expansion in (3.99).

3.2.2 Unitary operator

We briefly pause to discuss an special property of displacement operators. It is a property that is often found in quantum theory. In terms of the operator formalism, this property is represented by a special class of operators.

Unitary operators are operators whose Hermitian adjoints are equal to their inverses $\hat{U}^{\dagger} = \hat{U}^{-1}$, so that $\hat{U}^{\dagger}\hat{U} = \hat{U}\hat{U}^{\dagger} = \mathbb{1}$. If we can diagonalize such an operator, we would find that its eigenvalues always have a magnitude of unity $|\lambda_n| = 1$. (Compare this definition with that of *Hermitian operators* that are equal to their Hermitian adjoints $\hat{A}^{\dagger} = \hat{A}$ and whose eigenvalues are real.) Unitary operators play a significant role in quantum theory: they maintain the normalization of states and, therefore, also the probabilities computed from them. If $\hat{U}|\psi\rangle = |\psi'\rangle$ and $\langle\psi|\psi\rangle = 1$, then

$$\langle\psi'|\psi'\rangle = \langle\psi|\hat{U}^{\dagger}\hat{U}|\psi\rangle = \langle\psi|\psi\rangle = 1. \tag{3.100}$$

based on this property, we can say that all physical processes in quantum physics are unitary, that is, they can be modelled by unitary operators.

The displacement operators are *unitary operators*. They maintaining the normalization of the state, because $\hat{D}^{\dagger}(\alpha)\hat{D}(\alpha) = \hat{D}(\alpha)\hat{D}^{\dagger}(\alpha) = \mathbb{1}$. So, the exponent of a displacement operator $\alpha\hat{a}^{\dagger} - \alpha^*\hat{a}$ is anti-Hermitian. It is the only anti-Hermitian linear combination of the ladder operators with α as a complex-valued free parameter.

3.2.3 Eigenstates of the annihilation operator

Another defining property of coherent states is that they are *eigenstates of the annihilation operator*

$$\hat{a}|\alpha\rangle = |\alpha\rangle\alpha. \tag{3.101}$$

The Hermitian adjoint of this equation gives

$$\langle\alpha|\hat{a}^{\dagger} = \alpha^*\langle\alpha|. \tag{3.102}$$

One can use the identity

$$\exp(\hat{X})\hat{Y}\exp(-\hat{X}) = \hat{Y} + \left[\hat{X},\hat{Y}\right] + \frac{1}{2!}\left[\hat{X},\left[\hat{X},\hat{Y}\right]\right] + \frac{1}{3!}\left[\hat{X},\left[\hat{X},\left[\hat{X},\hat{Y}\right]\right]\right] + \cdots, \tag{3.103}$$

to show that coherent states are eigenstates of the annihilation operator. We leave it as an exercise.

Exercise 3.10. Compute the commutation $[\hat{a}, \hat{D}(\alpha)]$ with the aid of the identity in (3.103) to show that coherent states are eigenstates of the annihilation operator.

3.2.4 Statistical properties of coherent states

Unlike the Fock states, a coherent state does not have a fixed number of particles because it is given by a superposition of different Fock states. However, we can compute the *average number of particles* in a coherent state as an *expectation value* given by

$$\langle n \rangle = \langle \alpha | \hat{n} | \alpha \rangle = \langle \alpha | \hat{a}^\dagger \hat{a} | \alpha \rangle = |\alpha|^2. \tag{3.104}$$

The *probability* for a particular number of particles in the coherent state is given by the expectation value of the project operator for the number of particles defined in (3.90)

$$P_n = \langle \alpha | \hat{P}_n | \alpha \rangle = |\langle n | \alpha \rangle|^2 = \exp\left(-|\alpha|^2\right) \frac{|\alpha|^{2n}}{n!}. \tag{3.105}$$

The result is a *Poisson distribution*, which means that the observation of one particle in this state is independent of the observation of any other particle. Using (3.104), we can express it as

$$P_n = \exp\left(-\langle n \rangle\right) \frac{\langle n \rangle^n}{n!}, \tag{3.106}$$

in terms of the average number of particles in the state. The Poisson distributions for different values of $\langle n \rangle$ are shown in Figure 3.2.

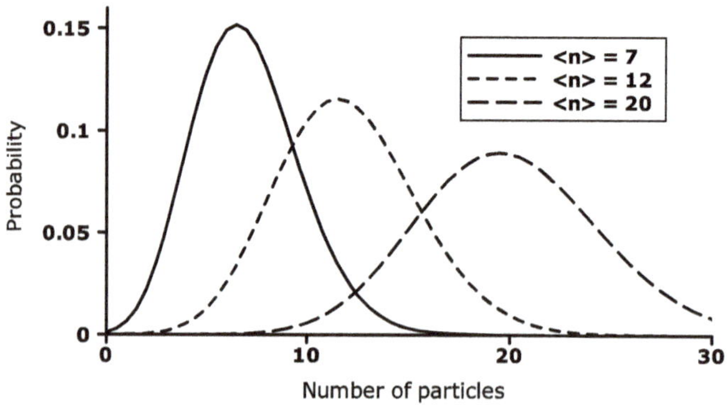

Figure 3.2: Poisson distributions for $\langle n \rangle = 7, 12, 20$.

One can form a generating function (discussed in Appendix A) for the probabilities in the Poisson distribution by multiplying (3.106) by J^n and summing the result over n from 0 to infinity. The resulting generating function

$$\mathcal{G}_{\text{Poisson}} = \exp\left(J\langle n\rangle - \langle n\rangle\right), \tag{3.107}$$

can be used to compute the moments of the distribution. The average is readily produced by the first moment

$$\partial_J \mathcal{G}_{\text{Poisson}}\big|_{J=1} = \langle n\rangle. \tag{3.108}$$

The second moment is given by

$$\langle n^2\rangle = \partial_J \left(J\partial_J \mathcal{G}_{\text{Poisson}}\right)\big|_{J=1} = \langle n\rangle + \langle n\rangle^2. \tag{3.109}$$

The variance is the second moment minus the square of the average, which gives

$$\langle n^2\rangle - \langle n\rangle^2 \equiv \langle n\rangle. \tag{3.110}$$

Thus, we find that the variance of a Poisson distribution is equal to its average.

3.2.5 Non-orthogonality and over-completeness

Coherent states are not mutually orthogonal. The inner product between two coherent states with different parameters α and β gives

$$\begin{aligned}\langle\alpha|\beta\rangle &= \exp\left(-\frac{1}{2}|\alpha|^2 - \frac{1}{2}|\beta|^2\right)\sum_n \frac{\alpha^{*n}\beta^n}{n!} \\ &= \exp\left(-\frac{1}{2}|\alpha|^2 - \frac{1}{2}|\beta|^2 + \alpha^*\beta\right).\end{aligned} \tag{3.111}$$

To show that the coherent states can resolve the identity, we consider

$$\int |\alpha\rangle\langle\alpha|\,\mathrm{d}^2\alpha = \int_0^\infty\int_0^{2\pi} \exp(-r^2)\sum_{mn} r^{m+n}\exp[\mathrm{i}(m-n)\phi]\frac{|m\rangle\langle n|}{\sqrt{m!}\sqrt{n!}}\,\mathrm{d}\phi\, r\mathrm{d}r, \tag{3.112}$$

where the parameter of the coherent state becomes a complex-valued integration variable. We replace $\alpha \to r\exp(\mathrm{i}\phi)$ and treat the amplitude and phase as cylindrical coordinates. The integral over ϕ produces a Kronecker delta

$$\int_0^{2\pi} \exp[\mathrm{i}(m-n)\phi]\,\mathrm{d}\phi = 2\pi\delta_{m,n}. \tag{3.113}$$

The expression thus becomes

$$\int |\alpha\rangle\langle\alpha|\, d^2\alpha = 2\pi\sum_n \frac{|n\rangle\langle n|}{n!}\int_0^\infty r^{2n}\exp(-r^2)\, r dr$$
$$= 2\pi\sum_n \frac{|n\rangle\langle n|}{n!}\frac{n!}{2} = \pi\sum_n |n\rangle\langle n| = \pi\mathbb{1}. \tag{3.114}$$

The integral over r is readily evaluated and in the end we used (3.61). So, we find that the identity operator is resolved in terms of the coherent states as

$$\mathbb{1} = \frac{1}{\pi}\int |\alpha\rangle\langle\alpha|\, d^2\alpha. \tag{3.115}$$

Although coherent states can be used to resolve the identity, they are not linearly independent. We can expand any coherent state in terms of all the other coherent states. Such an expansion is obtained by applying the identity resolved in terms of all the coherent state on an arbitrary coherent states

$$|\beta\rangle = \mathbb{1}\,|\beta\rangle = \frac{1}{\pi}\int |\alpha\rangle\langle\alpha|\beta\rangle\, d^2\alpha$$
$$= \frac{1}{\pi}\int |\alpha\rangle \exp\left(-\frac{1}{2}|\alpha|^2 - \frac{1}{2}|\beta|^2 + \alpha^*\beta\right) d^2\alpha. \tag{3.116}$$

It thus follows that the coherent states are *over complete.*

3.2.6 Relationship with quadrature variables

The expectation values of the quadrature operators $\hat{q}$ and $\hat{p}$ for a given coherent state are readily calculated as

$$\langle\alpha|\,\hat{q}\,|\alpha\rangle = \frac{1}{\sqrt{2}}\langle\alpha|\,(\hat{a}+\hat{a}^\dagger)\,|\alpha\rangle = \frac{1}{\sqrt{2}}(\alpha+\alpha^*) = \sqrt{2}\,\mathrm{Re}\{\alpha\},$$
$$\langle\alpha|\,\hat{p}\,|\alpha\rangle = \frac{-i}{\sqrt{2}}\langle\alpha|\,(\hat{a}-\hat{a}^\dagger)\,|\alpha\rangle = \frac{-i}{\sqrt{2}}(\alpha-\alpha^*) = \sqrt{2}\,\mathrm{Im}\{\alpha\}. \tag{3.117}$$

For this reason, we now associate the real and imaginary parts of α with the quadrature variables as $q = \sqrt{2}\,\mathrm{Re}\{\alpha\}$ and $p = \sqrt{2}\,\mathrm{Im}\{\alpha\}$. The coherent state parameter treated as a complex variable are then given by

$$\alpha = \frac{1}{\sqrt{2}}(q + ip) \quad \text{and} \quad \alpha^* = \frac{1}{\sqrt{2}}(q - ip), \tag{3.118}$$

in terms of the quadrature variables. When α is the parameter that identifies a specific coherent state, we can use (3.118) to represent it in terms of real-valued parameters that are related to the quadrature variables.

Substituting (3.118) together with (3.45) into the displacement operator in (3.97), we obtain an expression for the displacement operator in terms of quadrature variables:

$$\hat{D}(\alpha) = \exp(\mathrm{i}p\hat{q} - \mathrm{i}q\hat{p}). \tag{3.119}$$

If we use the Baker–Campbell–Hausdorff identity in (3.41) to separate the two terms in the exponent in (3.119), we recover the two shift operators in (3.35). Hence, the displacement produced by the displacement operator is related to the shifts produced by the shift operators in (3.35).

3.2.7 Coherent states in term of quadrature bases

The coherent states can be expressed in terms of the quadrature bases by

$$\begin{aligned}
|\alpha\rangle &= \int |q\rangle \langle q|\alpha\rangle \,\mathrm{d}q = \int |q\rangle \exp\left(-\frac{1}{2}|\alpha|^2\right) \sum_n \langle q|n\rangle \frac{\alpha^n}{\sqrt{n!}} \,\mathrm{d}q, \\
|\alpha\rangle &= \int |p\rangle \langle p|\alpha\rangle \frac{\mathrm{d}p}{2\pi} = \int |p\rangle \exp\left(-\frac{1}{2}|\alpha|^2\right) \sum_n \langle p|n\rangle \frac{\alpha^n}{\sqrt{n!}} \frac{\mathrm{d}p}{2\pi},
\end{aligned} \tag{3.120}$$

where we used (3.99). Focussing only on the q-basis, we substitute the coefficient functions from (3.79) to obtain

$$\langle q|\alpha\rangle = \exp\left(-\frac{1}{2}|\alpha|^2\right) \sum_n \langle q|n\rangle \frac{\alpha^n}{\sqrt{n!}} = \exp\left(-\frac{1}{2}|\alpha|^2 - \frac{1}{2}q^2\right) \sum_n \frac{\alpha^n H_n(q)}{\pi^{1/4}\sqrt{2^n}n!}. \tag{3.121}$$

The generating function for the Hermite polynomials in (2.122) helps us to evaluate the summation. After we replace $x \to q$ and $\eta \to \alpha/\sqrt{2}$ in (2.122), the result reads

$$\begin{aligned}
\langle q|\alpha\rangle &= \pi^{-1/4} \exp\left(-\frac{1}{2}|\alpha|^2 - \frac{1}{2}q^2 + \sqrt{2}q\alpha - \frac{1}{2}\alpha^2\right) \\
&= \pi^{-1/4} \exp\left[-\frac{1}{2}(q - q_0)^2 + \mathrm{i}(q - \frac{1}{2}q_0)p_0\right],
\end{aligned} \tag{3.122}$$

where we replaced the parameter of the coherent state by real-valued parameters

$$\alpha \to \frac{1}{\sqrt{2}}(q_0 + \mathrm{i}p_0). \tag{3.123}$$

We can now express the coherent state in terms of the q-basis as

$$|\alpha\rangle = \int |q\rangle \pi^{-1/4} \exp\left[-\frac{1}{2}(q - q_0)^2 + \mathrm{i}qp_0\right] \mathrm{d}q, \tag{3.124}$$

where we discarded the global phase factor. A similar result is obtained when we express the coherent state in terms of the p-basis.

Exercise 3.11. Show that the inner product between a p-basis element and a coherent state is

$$\langle p|\alpha\rangle = \sqrt{2}\pi^{1/4}\exp\left[-\frac{1}{2}(p-p_0)^2 - iq_0(p-\frac{1}{2}p_0)\right] \tag{3.125}$$

in terms of (3.123).

Expanding the coherent state in terms of the quadrature bases, we find that the coefficient functions are Gaussian functions of the quadrature variable. For the coefficient function in the q-basis, the two parameters q_0 and p_0 represent a shift and the frequency of a phase modulation, respectively. For an expansion in terms of the p-basis, the roles of q_0 and p_0 are interchanged, as expected for the Fourier domain function. The Gaussian nature of these coefficient functions is discussed in more detail in Section 3.4 below.

3.3 Variance and uncertainty

A concept that is closely associated with quantum physics is the concept of *uncertainty*. Here, we discuss this concept in the context of the current formalism. It requires some general concepts from statistical analysis [19], together with the abstract notion of noncommuting operators. Therefore, we do not need to introduce any quantum physics yet.

We start with the variance of an *observable* $\hat{A}$, as defined by

$$\sigma_A^2 = \left\langle \Delta A^2 \right\rangle \triangleq \left\langle A^2 \right\rangle - \langle A\rangle^2 , \tag{3.126}$$

where

$$\left\langle A^2 \right\rangle = \operatorname{tr}\{\hat{\rho}\hat{A}^2\} \quad \text{and} \quad \langle A\rangle = \operatorname{tr}\{\hat{\rho}\hat{A}\}. \tag{3.127}$$

The square root of the variance is the *standard deviation* denoted by σ_A. It represents the *uncertainty* ΔA of the quantity associated with the observable $\hat{A}$.

Consider, for example, the uncertainty in the number of particles in a state. The variance in the particle-number in, for instance, coherence states is readily calculated by using their eigenvalue equations

$$\sigma_n^2 \triangleq \langle\alpha|\,\Delta\hat{n}^2\,|\alpha\rangle = \langle\alpha|\,\hat{n}^2\,|\alpha\rangle - \langle\alpha|\,\hat{n}\,|\alpha\rangle^2 = |\alpha|^2. \tag{3.128}$$

It implies that the variance in the *occupation number* equals the expectation value of the occupation number $\langle n\rangle$, as found for the *Poisson distribution* in Section 3.2.4. The uncertainty (or standard deviation) is then given by $\Delta n = |\alpha|$.

For a Fock state, the uncertainty in n is exactly zero. Since the Fock states are the eigenstates of the number operator with the occupation number n as the associated eigenvalues, there is no uncertainty in their occupation numbers.

3.3.1 Schrödinger uncertainty relation

For a more general understanding of uncertainty as it appears in quantum physics, we follow the work of Robertson [20] and Schrödinger [21] to develop a general representation of uncertainty relationships [22]. For this purpose, we consider the uncertainty in the context of noncommuting operators.

Using (3.126), we express the variance for an observable $\hat{A}$ as

$$\sigma_A^2 = \left\langle \Delta A^2 \right\rangle = \langle\psi| \left(\hat{A} - \langle A\rangle\right)^2 |\psi\rangle \,. \tag{3.129}$$

If we define an unnormalized state vector,

$$|f\rangle \triangleq \left(\hat{A} - \langle A\rangle\right) |\psi\rangle \,, \tag{3.130}$$

then

$$\sigma_A^2 = \langle f|f\rangle. \tag{3.131}$$

In the same way, we can define another unnormalized state

$$|g\rangle \equiv \left(\hat{B} - \langle B\rangle\right) |\psi\rangle \,, \tag{3.132}$$

for another observable $\hat{B}$, so that

$$\sigma_B^2 = \langle g|g\rangle. \tag{3.133}$$

The two observables (Hermitian operators) $\hat{A}$ and $\hat{B}$, associated with arbitrary measurable quantities, do not in general commute; they may not have the same *eigenvectors*.

It follows from the *Cauchy–Schwarz inequality* that

$$\sigma_A^2\sigma_B^2 = \langle f|f\rangle\langle g|g\rangle \geq |\langle f|g\rangle|^2. \tag{3.134}$$

Considering the right-hand side of the inequality, we have

$$\begin{aligned} |\langle f|g\rangle|^2 &= \langle f|g\rangle\langle g|f\rangle \\ &= \langle\psi| \left(\hat{A} - \langle A\rangle\right)\left(\hat{B} - \langle B\rangle\right) |\psi\rangle \, \langle\psi| \left(\hat{B} - \langle B\rangle\right)\left(\hat{A} - \langle A\rangle\right) |\psi\rangle \\ &= \left(\left\langle \hat{A}\hat{B}\right\rangle - \langle A\rangle\langle B\rangle\right)\left(\left\langle \hat{B}\hat{A}\right\rangle - \langle B\rangle\langle A\rangle\right). \end{aligned} \tag{3.135}$$

The products of operators can be written in terms of the commutator and the anticommutator, as follows

$$\hat{A}\hat{B} = \frac{1}{2}\{\hat{A}, \hat{B}\} + \frac{1}{2}[\hat{A}, \hat{B}],$$

$$\hat{B}\hat{A} = \frac{1}{2}\{\hat{A},\hat{B}\} - \frac{1}{2}[\hat{A},\hat{B}], \tag{3.136}$$

where $\{\hat{A},\hat{B}\} = \hat{A}\hat{B} + \hat{B}\hat{A}$ and $[\hat{A},\hat{B}] = \hat{A}\hat{B} - \hat{B}\hat{A}$. Hence,

$$|\langle f|g\rangle|^2 = \left(\frac{1}{2}\left\langle\{\hat{A},\hat{B}\}\right\rangle - \langle A\rangle\langle B\rangle\right)^2 - \left(\frac{1}{2}\left\langle[\hat{A},\hat{B}]\right\rangle\right)^2. \tag{3.137}$$

The inequality becomes

$$\sigma_A^2\sigma_B^2 \geq \left(\frac{1}{2}\left\langle\{\hat{A},\hat{B}\}\right\rangle - \langle A\rangle\langle B\rangle\right)^2 + \left(\frac{1}{\mathrm{i}2}\left\langle[\hat{A},\hat{B}]\right\rangle\right)^2, \tag{3.138}$$

where we changed the last term because the commutator often produces an imaginary-valued quantity. The expression in (3.138) is called the *Schrödinger uncertainty relation*. This relationship relates the product of variances to the commutation and anticommutation relations. In terms of density operators, the uncertainty relation is expressed as

$$\sigma_A^2\sigma_B^2 \geq \left(\frac{1}{2}\operatorname{tr}\{\{\hat{A},\hat{B}\}\hat{\rho}\} - \operatorname{tr}\{\hat{A}\hat{\rho}\}\operatorname{tr}\{\hat{B}\hat{\rho}\}\right)^2 + \left(\frac{1}{\mathrm{i}2}\operatorname{tr}\{[\hat{A},\hat{B}]\hat{\rho}\}\right)^2. \tag{3.139}$$

For a *minimum uncertainty state*, the inequality becomes an equality.

3.3.2 Covariance

The first term on the right-hand side of (3.138) represents the *covariance* of the two quantities that are associated with the two observables. In the context of *stochastic processes* [19], the *covariance* is defined for a *joint probability density function* $f_{XY}(X,Y)$ of two random variable X and Y as

$$\begin{aligned} V(X,Y) &\triangleq \langle(X - \langle X\rangle)(Y - \langle Y\rangle)\rangle \\ &= \int (X - \langle X\rangle)(Y - \langle Y\rangle) f_{XY}(X,Y)\,\mathrm{d}X\,\mathrm{d}Y. \end{aligned} \tag{3.140}$$

When we expand the expectation value, it becomes

$$V(X,Y) = \langle XY\rangle - \langle X\rangle\langle Y\rangle. \tag{3.141}$$

In terms of the operator formalism, where the role of the random variables are taken over by Hermitian operators (observables), the expression for two noncommuting operators $\hat{A}$ and $\hat{B}$ becomes

$$V(A,B) \triangleq \frac{1}{2}\left\langle\{\hat{A},\hat{B}\}\right\rangle - \langle A\rangle\langle B\rangle. \tag{3.142}$$

3.3.3 Heisenberg uncertainty

If the two observables are uncorrelated, the covariance is zero. Then the first term on the right-hand side of (3.138) drops away, leading to the *Robertson inequality*, given by

$$\sigma_A^2\sigma_B^2 \geq \left(\frac{1}{\mathrm{i}2}\left\langle[\hat{A},\hat{B}]\right\rangle\right)^2 . \tag{3.143}$$

When the commutator of two observables produces an imaginary constant, such as $[\hat{A},\hat{B}] = \mathrm{i}C$, where C is a positive real-valued constant, the uncertainty relation reduces to the familiar *Heisenberg uncertainty relation*

$$\sigma_A\sigma_B \geq \frac{1}{2}C. \tag{3.144}$$

At this point, we notice that the uncertainty relationships that we obtained here follow directly from the commutation properties of Hermitian operators. For two such operators to commute, their eigenvectors must be the same (or can be made to be the same in the case of degeneracies). Therefore, any arbitrary pair of Hermitian operators are unlike to be commuting operators. When classical scenarios are modelled in terms of operators, it is quite likely to have situations where those operators do not commute. In such cases, the same uncertainty relations that we obtained here would apply. What makes the quantum scenarios different is that a quantum state often comprises a single quantum (single photon) and only allows a single measurement. The uncertainties in the measurement results in such cases are severely affected by these uncertainty relations. In classical scenarios where one does not have such discrete measurement results, the uncertainty relations, though still applicable, do not have such a severe impact.

3.3.4 Uncertainty in quadrature

Since the quadrature operators are Hermitian, they represent observables. The variances in quadrature, expressed in terms of the ladder operators, are given by

$$\begin{aligned}
\sigma_q^2 &= \mathrm{tr}\{\hat{\rho}\Delta\hat{q}^2\} = \left\langle\hat{q}^2\right\rangle - \langle\hat{q}\rangle^2 \\
&= \frac{1}{2} + \langle\hat{n}\rangle + \frac{1}{2}\left(\left\langle\hat{a}^2\right\rangle + \left\langle\hat{a}^{\dagger 2}\right\rangle\right) - \frac{1}{2}\left(\langle\hat{a}\rangle + \left\langle\hat{a}^{\dagger}\right\rangle\right)^2 , \\
\sigma_p^2 &= \mathrm{tr}\{\hat{\rho}\Delta\hat{p}^2\} = \left\langle\hat{p}^2\right\rangle - \langle\hat{p}\rangle^2 \\
&= \frac{1}{2} + \langle\hat{n}\rangle - \frac{1}{2}\left(\left\langle\hat{a}^2\right\rangle + \left\langle\hat{a}^{\dagger 2}\right\rangle\right) + \frac{1}{2}\left(\langle\hat{a}\rangle - \left\langle\hat{a}^{\dagger}\right\rangle\right)^2 .
\end{aligned} \tag{3.145}$$

Exercise 3.12. Show that the variances in quadrature for the Fock states are given by

$$\langle n|\,\Delta\hat{q}^2\,|n\rangle = \langle n|\,\Delta\hat{p}^2\,|n\rangle = \frac{1}{2} + n. \tag{3.146}$$

Based on the Robertson inequality in (3.143), states for which the covariance in quadrature is zero, have an uncertainty relationship given by

$$\sigma_q\sigma_p \geq \frac{1}{\mathrm{i}2}\,\langle[\hat{q},\hat{p}]\rangle = \frac{1}{2}. \tag{3.147}$$

It represents a lower bound for the product of the uncertainties in the respective quadratures. When the relationship becomes an equality for a given state, then that state is called a *minimum uncertainty state.*

Exercise 3.13. Show that the coherent states are *minimum uncertainty states* by computing their quadrature variances to obtain

$$\langle \alpha|\,\Delta\hat{q}^2\,|\alpha\rangle = \langle \alpha|\,\Delta\hat{p}^2\,|\alpha\rangle = \frac{1}{2}. \tag{3.148}$$

More generally, the uncertainty relationship for the quadrature operators, given in term of (3.139), reads

$$\sigma_q^2\sigma_p^2 \geq \left(\frac{1}{2}\,\langle\{\hat{q},\hat{p}\}\rangle - \langle q\rangle\langle p\rangle\right)^2 + \frac{1}{4}. \tag{3.149}$$

In terms of ladder operators, the anticommutator is

$$\{\hat{q},\hat{p}\} = \mathrm{i}\left(\hat{a}^{\dagger 2} - \hat{a}^2\right). \tag{3.150}$$

For Fock states, it readily follows that

$$\langle n|\,\{\hat{q},\hat{p}\}\,|n\rangle = \langle n|\,\hat{q}\,|n\rangle = \langle n|\,\hat{p}\,|n\rangle = 0. \tag{3.151}$$

As a result, the covariance in the quadratures is zero for Fock states. For coherent states, the individual terms are not zero. Instead, they are

$$\begin{aligned}\langle \alpha|\,\{\hat{q},\hat{p}\}\,|\alpha\rangle &= \mathrm{i}\left(\alpha^{*2} - \alpha^2\right),\\ \langle \alpha|\,\hat{q}\,|\alpha\rangle &= \frac{1}{\sqrt{2}}\left(\alpha + \alpha^*\right),\\ \langle \alpha|\,\hat{p}\,|\alpha\rangle &= -\mathrm{i}\frac{1}{\sqrt{2}}\left(\alpha - \alpha^*\right).\end{aligned} \tag{3.152}$$

When combined into the expression for the covariance, these results again produce zero. So, in both these cases the first term of the right-hand side of the uncertainty relation in (3.149) becomes zero. As a result, the uncertainty relationships for these cases simplify to the Heisenberg uncertainty relation

$$\sigma_q \sigma_p \geq \frac{1}{2}. \tag{3.153}$$

3.4 Gaussian states

At the end of Section 3.2, we saw that a coherent state is represented by a *Gaussian coefficient function* in terms of the quadrature bases. Since pure quantum states with such Gaussian coefficient functions are often encountered, we discuss them in more detail here. In general, a state is considered to be a Gaussian state due to the Gaussian nature of its phase space function, which is discussed in Section 3.5. However, one can show that, if the coefficient function (or the wavefunction) of a pure state has a Gaussian shape, then its representation in terms of a *Wigner function* on phase space also has a Gaussian shape. As a gentle introduction to Gaussian states, we consider such states here expressed in terms of the q-basis.

A general *pure Gaussian state* is given in terms of (3.22) by

$$|\psi\rangle = \int |q\rangle \exp(-aq^2 + bq + c)\, dq, \tag{3.154}$$

where a, b and c are complex-valued constants with $\operatorname{Re}\{a\} > 0$. The imaginary part of c produces a global phase that can be discarded and its real part can be converted into a normalization constant to ensure that the state is normalized. For the sake of this discussion, inspired by (3.124), we parameterize the Gaussian coefficient function with four real-valued parameters so that the state is expressed by

$$|\psi\rangle = \int |q\rangle \mathcal{N}_q \exp\left[-\frac{(q-q_0)^2}{2(w^2 + iR^2)} + ip_0 q\right] dq. \tag{3.155}$$

Here q_0 is a shift along the q-direction, w is a positive constant for the width of the Gaussian function, p_0 defines a phase tilt, and R parameterizes a phase curvature introducing a quadratic phase variation. The width and the phase curvature are combined into a complex width parameter. Constant global phase factors are discarded.

Exercise 3.14. Show that the normalization constant in (3.155) is given in terms of w and R by

$$\mathcal{N}_q = \frac{\sqrt{w}}{\pi^{1/4}(w^4 + R^4)^{1/4}}. \tag{3.156}$$

The expression of the state in (3.155) can be converted into the p-basis by applying an identity operator resolved in terms of the p-basis. The result is a Fourier transformation:

$$|\psi\rangle = \int |p\rangle \exp(-\mathrm{i}pq)\mathcal{N}_q \exp\left[-\frac{(q-q_0)^2}{2(w^2+\mathrm{i}R^2)} + \mathrm{i}p_0 q\right] \mathrm{d}q \frac{\mathrm{d}p}{2\pi}$$
$$= \int |p\rangle \mathcal{N}_p \exp\left[-\frac{1}{2}\left(w^2+\mathrm{i}R^2\right)(p-p_0)^2 - \mathrm{i}pq_0\right] \frac{\mathrm{d}p}{2\pi}. \quad (3.157)$$

Again, we discarded a constant global phase factor. The resulting function is still Gaussian, but the parameters take on different roles: the complex width parameter is inverted; the function is now shifted by p_0 and it has a phase tilt given by q_0. So, we see that p_0 and q_0 interchanged their roles and the width becomes inverted.

Exercise 3.15. Show that the normalization constant in (3.157) is given by

$$\mathcal{N}_p = \sqrt{2w\sqrt{\pi}}. \quad (3.158)$$

For $w = 1$ and $R = 0$, the width of the two expressions in (3.155) and (3.157) are the same, representing the coherent states in Section 3.2. For $w \neq 1$, we say that the state is *squeezed*. The width of the coefficient function in one quadrature basis is larger than its equivalent in the other quadrature basis. When $R \neq 0$, the squeezing becomes twisted between p and q. It can be understood as a rotation introduced on phase space, which is discussed in Section 3.5.

3.4.1 Moments of Gaussian states

The location and width of a Gaussian state is determined by the *moments* of that state. Using the definitions of the quadrature operators given in (3.32), we compute the expectation values of the quadrature operators (the first moment) for the Gaussian state. The results are

$$\langle q\rangle = \langle\psi|\hat{q}|\psi\rangle = q_0 \quad \text{and} \quad \langle p\rangle = \langle\psi|\hat{p}|\psi\rangle = p_0. \quad (3.159)$$

For the second moments, we compute the expectation values of the squares of the quadrature operators to obtain

$$\left\langle q^2\right\rangle = \langle\psi|\hat{q}^2|\psi\rangle = q_0^2 + \frac{w^2}{2} + \frac{R^4}{2w^2},$$
$$\left\langle p^2\right\rangle = \langle\psi|\hat{p}^2|\psi\rangle = p_0^2 + \frac{1}{2w^2}. \quad (3.160)$$

The variances are given in terms of the first and second moments by

$$\sigma_q^2 = \left\langle q^2 \right\rangle - \langle q \rangle^2 = \frac{w^2}{2} + \frac{R^4}{2w^2},$$
$$\sigma_p^2 = \left\langle p^2 \right\rangle - \langle p \rangle^2 = \frac{1}{2w^2}, \tag{3.161}$$

so that

$$\sigma_p \sigma_q = \frac{1}{2}\sqrt{1 + \frac{R^4}{w^4}}. \tag{3.162}$$

For $R = 0$, it becomes

$$\sigma_p \sigma_q = \frac{1}{2}, \tag{3.163}$$

regardless of the value of w. For $w = 1$, we then have the symmetric situation (coherent state) for which $\sigma_p = \sigma_q$. The average value and standard deviation in quadrature for a Gaussian state is demonstrated in Figure 3.3.

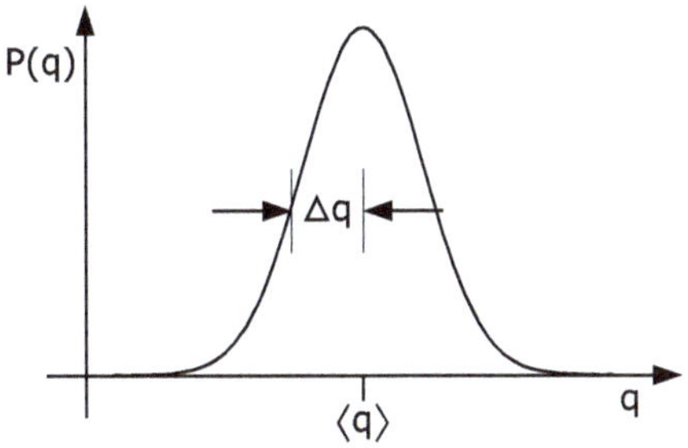

Figure 3.3: Average value $\langle q \rangle$ and standard deviation Δq in the quadrature for a Gaussian state.

Does the value that we obtain in (3.162) represent the smallest value for the uncertainty? To answer this question, we need to relate the widths of the Gaussian states for the different quadrature basis to the uncertainty. For this purpose, we compute the right-hand side of the uncertainty relation for quadrature operators obtained in (3.149). This calculation requires the expectation values for $\hat{q}\hat{p}$ and $\hat{p}\hat{q}$. They can be readily computed by using the expression in terms of the q- and p-basis for the bras and kets, depending on which quadrature variable sits closest to it. The results are

$$\langle \psi | \hat{q}\hat{p} | \psi \rangle = q_0 p_0 + \frac{R^2}{2w^2} + \frac{\mathrm{i}}{2},$$
$$\langle \psi | \hat{p}\hat{q} | \psi \rangle = q_0 p_0 + \frac{R^2}{2w^2} - \frac{\mathrm{i}}{2}. \tag{3.164}$$

Hence,

$$\langle[\hat{q},\hat{p}]\rangle = \langle\psi|\,[\hat{q},\hat{p}]\,|\psi\rangle = \mathrm{i},$$
$$\langle\{\hat{q},\hat{p}\}\rangle = \langle\psi|\,\{\hat{q},\hat{p}\}\,|\psi\rangle = 2q_0p_0 + \frac{R^2}{w^2}, \tag{3.165}$$

where the first relation is expected from (3.39). Substituting these expressions and (3.159) into (3.149), we obtain

$$\sigma_q^2\sigma_p^2 \geq \frac{R^4}{4w^4} + \frac{1}{4}. \tag{3.166}$$

Comparing it with (3.162), we conclude that a *pure Gaussian state* is a *minimum uncertainty state*. Hence, for $R \neq 0$, the covariance in the state is nonzero.

For *mixed Gaussian states*, the situation becomes a bit more complicated. In that case, it is better to consider these states in terms of their Wigner functions, which are discussed in Section 3.5.

We need to remember that the quantity in (3.162) is related to an uncertainty in the quadrature variables q and p, and is a consequence of the Fourier relationship between q and p. It does not represent an uncertainty in particle number. For the uncertainty in particle number, we must use the number operator $\hat{n}$ as in (3.128), instead of the quadrature operators.

3.4.2 Squeezing

The width of the coefficient function of a state can be changed with the aid of a *squeezing operator*. It is a unitary operator because the normalization of the state needs to be maintained. The squeezing operator is expressed in terms of ladder operators by

$$\hat{S} = \exp\left(\frac{1}{2}\hat{a}^2\xi^* - \frac{1}{2}\hat{a}^{\dagger 2}\xi\right), \tag{3.167}$$

where ξ is a complex-valued parameter called the *squeezing parameter*.

To understand the squeezing process of the operator, we investigate the effect of the squeezing operator on the moments that we calculated above in Section 3.4.1. The squeezed version of a state $|\psi\rangle$ is obtained by applying the squeezing operator on that state $\hat{S}\,|\psi\rangle$. So, the effect of the squeezing process on the expectation value of an *observable* $\hat{O}$ can be determined by comparing $\langle\psi|\,\hat{S}^\dagger\hat{O}\hat{S}\,|\psi\rangle$ to $\langle\psi|\,\hat{O}\,|\psi\rangle$. For a general state, the effect of the squeezing process is then revealed by computing the transformation produced by $\hat{S}^\dagger\hat{O}\hat{S}$. For this purpose, we use (3.103) to compute such transformations where $\hat{O}$ is given by different powers of the quadrature operators. To facilitate the calculation, we use (3.39) and (3.45) to express the squeezing operator with a positive real-valued squeezing parameter as

$$\hat{S} = \exp\left(-\frac{1}{2}\xi\right)\exp\left(\mathrm{i}\xi\hat{p}\hat{q}\right) = \exp\left(\frac{1}{2}\xi\right)\exp\left(\mathrm{i}\xi\hat{q}\hat{p}\right). \tag{3.168}$$

The results, for a positive real-valued squeezing parameter, are

$$\begin{aligned} \hat{S}^{\dagger}\hat{q}\hat{S} &= \exp(\xi)\hat{q}, \quad \hat{S}^{\dagger}\hat{q}^{2}\hat{S} = \exp(2\xi)\hat{q}^{2}, \\ \hat{S}^{\dagger}\hat{p}\hat{S} &= \exp(-\xi)\hat{q}, \quad \hat{S}^{\dagger}\hat{p}^{2}\hat{S} = \exp(-2\xi)\hat{q}^{2}. \end{aligned} \tag{3.169}$$

The transformations of $\hat{q}$ and $\hat{q}^2$ are represented by enhancements involving factors of $\exp(\xi)$, while $\hat{p}$ and $\hat{p}^2$ are suppressed by factors of $\exp(-\xi)$. If ξ is negative, the roles are interchanged. For a general complex-valued squeezing parameter, the phase of ξ determines the orientation of the squeezing process on the two-dimensional qp plane (i. e., phase space).

3.4.2.1 Effect of squeezing on location and width

The first moments give the location of a state. The squeezing process leads to the transformation of this location given by

$$\langle q\rangle \rightarrow \exp(\xi)\langle q\rangle \quad \text{and} \quad \langle p\rangle \rightarrow \exp(-\xi)\langle p\rangle. \tag{3.170}$$

The variance, which is determined by the first and second moments, represents the width of the state as a function of the quadrature variable. For the respective quadrature variables, the variances are transformed by the squeezing process according to

$$\sigma_q^2 \rightarrow \exp(2\xi)\sigma_q^2 \quad \text{and} \quad \sigma_p^2 \rightarrow \exp(-2\xi)\sigma_p^2. \tag{3.171}$$

If the original width of the state was $w = 1$ (as in the case of the coherent states), then the width after squeezing is given by $w = \exp(\xi)$, as a function of q, or by $w = \exp(-\xi)$, as a function of p. In summary, the effect of the squeezing process is to increase or reduce the width of the coefficient function and to change the state's location when it is not located at the origin.

Exercise 3.16. Show that when the squeezing process is applied to a vacuum state, it does not change the state's location, but it changes the width of the state according to (3.171).

3.4.2.2 Effect of squeezing on occupation number

The expectation value of the number operator gives the average number of particles in a state. The effect of the squeezing process on the average number of particles in a state can also be determined with the aid of (3.103). For a positive real-valued squeezing parameter, the result is

$$\begin{aligned} \hat{S}^{\dagger}\hat{n}\hat{S} &= \hat{S}^{\dagger}\hat{a}^{\dagger}\hat{a}\hat{S} \\ &= \cosh(2\xi)\hat{n} + \sinh^{2}(\xi) - \frac{1}{2}\sinh(2\xi)\left(\hat{a}^{\dagger 2} + \hat{a}^{2}\right). \end{aligned} \tag{3.172}$$

The effect depends on the initial state. When it operators on the vacuum state, we get

$$\langle \mathrm{vac}| \hat{S}^\dagger \hat{n} \hat{S} |\mathrm{vac}\rangle = \sinh^2(\xi). \tag{3.173}$$

It represents a function that starts from zero for $\xi = 0$ (the original vacuum state contains no particles) and grows with increasing magnitude of the squeezing parameter. Hence, the average number of particles in a squeezed vacuum state is not zero.

Exercise 3.17. Show that the average number of particles in an arbitrary squeezed Fock state is

$$\langle n| \hat{S}^\dagger \hat{n} \hat{S} |n\rangle = n + (1 + 2n) \sinh^2(\xi), \tag{3.174}$$

which is always larger than for the unsqueezed Fock states.

Exercise 3.18. Compute the average number of particles and the uncertainty in the number of particles for a squeezed coherent state when the parameter of the coherent state is real valued. Show that the number of particles initially decreases and then increases as the squeezing is increased.

3.4.3 Bogoliubov transformations

The squeezing process has a close relationship with *Bogoliubov transformations*. They are defined as linear transformations of the ladder operators that maintain the commutation relations. A Bogoliubov transformation is given by

$$\begin{aligned} \hat{a} &\to u\hat{a} + v\hat{a}^\dagger \triangleq \hat{b}, \\ \hat{a}^\dagger &\to v^*\hat{a} + u^*\hat{a}^\dagger \triangleq \hat{b}^\dagger, \end{aligned} \tag{3.175}$$

where $\hat{b}$ and $\hat{b}^\dagger$ are called *Bogoliubov operators* and u and v are complex *Bogoliubov coefficients*. If $[\hat{a}, \hat{a}^\dagger] = 1$, then the Bogoliubov operators would obey the same *commutation relation*, provided that

$$[\hat{b}, \hat{b}^\dagger] = [u\hat{a} + v\hat{a}^\dagger, v^*\hat{a} + u^*\hat{a}^\dagger] = |u|^2 - |v|^2 = 1, \tag{3.176}$$

leading to a condition that the Bogoliubov coefficients must satisfy. The *inverse Bogoliubov transformation* is given by

$$\begin{aligned} \hat{b} &\to u^*\hat{b} - v\hat{b}^\dagger \equiv \hat{a}, \\ \hat{b}^\dagger &\to u\hat{b}^\dagger - v^*\hat{b} \equiv \hat{a}^\dagger. \end{aligned} \tag{3.177}$$

When we apply the inverse Bogoliubov transformation to (3.175), it produces

$$\hat{b} = (|u|^2 - |v|^2)\hat{b} + (vu - uv)\hat{b}^\dagger,$$

$$\hat{b}^{\dagger} = (v^* u^* - u^* v^*)\hat{b} + (|u|^2 - |v|^2)\hat{b}^{\dagger}. \tag{3.178}$$

The right-hand sides reproduce the left-hand sides, provided that the last equality in (3.176) is satisfied. For this reason, the Bogoliubov coefficients can be parameterized in terms of hyperbolic trigonometric functions

$$u = \exp(\mathrm{i}\phi_u)\cosh(\zeta) \quad \text{and} \quad v = \exp(\mathrm{i}\phi_v)\sinh(\zeta), \tag{3.179}$$

where ζ, ϕ_u, and ϕ_v are real-valued parameters.

Although the Bogoliubov operators and the quadrature operators are both linear combinations of the ladder operators, there is no pair of Bogoliubov operators that correspond to the quadrature operators. For any finite value of ζ, the magnitude of one Bogoliubov coefficient is always strictly larger than that of the other. For quadrature operators, the coefficients are of equal magnitude.

3.4.4 Bogoliubov eigenstates

We introduced the Bogoliubov transformation in the previous section with the statement that it is related to the squeezing process. To make that statement more precise, we now show that the eigenstates of the Bogoliubov operators are *displaced squeezed vacuum states*.

If we did not know what the eigenstates of the Bogoliubov operators were, we could have looked for unitary operators $\hat{U}_b$ that would produce the eigenstates from the vacuum. The eigenvalue equation of the Bogoliubov operator would then lead to

$$\hat{b}\,|b\rangle = \hat{b}\hat{U}_b\,|\mathrm{vac}\rangle = |b\rangle\, b = \hat{U}_b\,|\mathrm{vac}\rangle\, b, \tag{3.180}$$

so that

$$\hat{U}_b^{\dagger}\hat{b}\hat{U}_b\,|\mathrm{vac}\rangle = |\mathrm{vac}\rangle\, b, \tag{3.181}$$

where b is the eigenvalue. Therefore, to demonstrate that displaced squeezed vacuum states are the eigenstates that we are looking for, we can substitute in $\hat{U}_b = \hat{D}\hat{S}$ and show that it satisfies the equation. The eigenvalue equation then reads

$$\hat{S}^{\dagger}(\xi)\hat{D}^{\dagger}(\alpha_0)\hat{b}\hat{D}(\alpha_0)\hat{S}(\xi)\,|\mathrm{vac}\rangle = |\mathrm{vac}\rangle\, b. \tag{3.182}$$

We can use (3.103) to evaluate the operator product on the left-hand side.

Considering only the displacement operators, as defined in (3.97), we get

$$\hat{D}^{\dagger}(\alpha_0)\hat{b}\hat{D}(\alpha_0) = \exp(\alpha_0^*\hat{a} - \alpha_0\hat{a}^{\dagger})(u\hat{a} + v\hat{a}^{\dagger})\exp(\alpha_0\hat{a}^{\dagger} - \alpha_0^*\hat{a}), \tag{3.183}$$

with the aid of (3.97). Using the commutator,

$$[\alpha_0^*\hat{a} - \alpha_0\hat{a}^\dagger, u\hat{a} + v\hat{a}^\dagger] = \alpha_0 u + \alpha_0^* v \triangleq b_0, \tag{3.184}$$

we get

$$\hat{D}^\dagger(\alpha_0)\hat{b}\hat{D}(\alpha_0) = \hat{b} + b_0. \tag{3.185}$$

Next, we apply the squeezing operators, as defined in (3.167), leading to

$$\hat{S}^\dagger(\xi)\hat{D}^\dagger(\alpha_0)\hat{b}\hat{D}(\alpha_0)\hat{S}(\xi) = \hat{S}^\dagger(\xi)\hat{b}\hat{S}(\xi) + b_0. \tag{3.186}$$

Considering the first term on the right-hand side, we get

$$\hat{S}^\dagger(\xi)\hat{b}\hat{S}(\xi) = \exp\left(\frac{1}{2}\hat{a}^{\dagger 2}\xi - \frac{1}{2}\hat{a}^2\xi^*\right)(u\hat{a} + v\hat{a}^\dagger)\exp\left(\frac{1}{2}\hat{a}^2\xi^* - \frac{1}{2}\hat{a}^{\dagger 2}\xi\right). \tag{3.187}$$

It requires the commutator

$$\left[\frac{1}{2}\hat{a}^{\dagger 2}\xi - \frac{1}{2}\hat{a}^2\xi^*, u\hat{a} + v\hat{a}^\dagger\right] = -u\xi\hat{a}^\dagger - v\xi^*\hat{a}. \tag{3.188}$$

The result has the form of a Bogoliubov operator with coefficients $u \to -v\xi^*$ and $v \to -u\xi$. However, it is not a Bogoliubov operator because the coefficients do not satisfy (3.176), but it still produces the same commutation relation with the necessary replacements. Therefore, we can use the commutation relation repeatedly to evaluate the expansion. The resulting expression obtained from (3.103) does not terminate, but the terms are summable, leading to

$$\begin{aligned}\hat{S}^\dagger(\xi)\hat{b}\hat{S}(\xi) = {} & [u\cosh(|\xi|) - v\exp(-\mathrm{i}\theta)\sinh(|\xi|)]\hat{a} \\ & + [v\cosh(|\xi|) - u\exp(\mathrm{i}\theta)\sinh(|\xi|)]\hat{a}^\dagger,\end{aligned} \tag{3.189}$$

where $\xi = \exp(\mathrm{i}\theta)|\xi|$. The term with the annihilation operator $\hat{a}$ is removed when we apply the operator product to the vacuum state, as in (3.182). To remove the creation operator, it is required that

$$u = \Phi\cosh(|\xi|) \quad \text{and} \quad v = \Phi\exp(\mathrm{i}\theta)\sinh(|\xi|), \tag{3.190}$$

where $\Phi = \exp(\mathrm{i}\phi_0)$ is an unknown common phase factor. Comparing these expressions to (3.179), we see that $\zeta \to |\xi|$, $\phi_u \to \phi_0$, and $\phi_v \to \theta + \phi_0$. It gives a relationship between the Bogoliubov coefficients u and v and the squeezing parameter $\xi = \exp(\mathrm{i}\theta)|\xi|$. The eigenvalue equation thus becomes

$$(u\hat{a} + v\hat{a}^\dagger)\,|\alpha_0;\xi\rangle = |\alpha_0;\xi\rangle\,(u\alpha_0 + v\alpha_0^*), \tag{3.191}$$

where $|\alpha_0;\xi\rangle \triangleq \hat{D}(\alpha_0)\hat{S}(\xi)\,|\text{vac}\rangle$ and u and v are related to ξ through (3.190). Note that, while the Bogoliubov operator depends on ξ, it does not depend on the displacement α_0.

Therefore, α_0 parameterizes all the eigenstates for a given Bogoliubov operator with a squeezing parameter ξ.

3.5 Phase space

Obviously, the concept of phase space plays a central role in the topic of this book. Thus far, we have provided only a few brief general comments about phase space. Here, a more thorough introduction is provided for the concept of a phase space, with some discussion of its properties. However, this discussion only considers the *particle-number degree of freedom*, as applicable for the current chapter. When we incorporate the spatiotemporal degrees of freedom in Chapter 4 and beyond, we end up with a *functional* phase space, which is discussed further there.

3.5.1 Classical phase space

The notion of a phase space originated in classical mechanics [23]. In its simplest form it is a two-dimensional plane with position and momentum representing the two perpendicular axes. Each point on this plane represents a unique state of a classical particle as determined by its position and momentum. The dynamics of the system in which the particle finds itself causes its state to change continuously. As a result, the changing state of the particle produces a trajectory on phase space, depicting the position and momentum of the particle as functions of time. An example of such a simple mechanical system is the classical harmonic oscillator, which we discuss in Section 3.5.2.

First, we review classical dynamics briefly. It is useful in preparation for the subsequent discussion of quantum dynamics. We introduce the concept of a Hamiltonian and some of its properties, and show how it is related to the Lagrangian. This review is not exhaustive. For more thorough introductions on the topic, one should consult books on *classical mechanics* [23].

3.5.1.1 Hamiltonian and Lagrangian

The dynamics of a system can often be represented by a set of equations of motion for all the quantities in the system. However, interactions may present some challenges. Such equations of motion can be derived from either a *Lagrangian* or a *Hamiltonian* representing the dynamics of the system.

The Hamiltonian represents the energy in the system as the sum of the kinetic energy and the potential energy $H = K + V$, where K is the kinetic energy and V is the potential energy. The Lagrangian, on the other hand, is the difference between the kinetic energy and the potential energy $L = K - V$. There are more differences between the Lagrangian and the Hamiltonian.

The Lagrangian depends on the quantities of interest and their derivatives. For a one-dimensional system, we have the quantity of interest $q(t)$ and its derivative

$$\dot{q}(t) \triangleq \frac{\mathrm{d}q(t)}{\mathrm{d}t}. \tag{3.192}$$

The derivative provides the rate of change as a function of the relevant coordinate parameter, which is time t in this case.

The Hamiltonian also depends on the quantity of interest, given by $q(t)$ for the one-dimensional system, but in place of its time derivative, the Hamiltonian has a dependence on the *conjugate momentum*. For a one-dimensional system, we represent it as $p(t)$. It is obtained from the Lagrangian by

$$p(t) \triangleq \frac{\partial L}{\partial \dot{q}(t)}. \tag{3.193}$$

In the context of classical mechanics, the quantities of interest are usually the position coordinates of the *configuration space*. Therefore, the conjugate variables are their associated momenta.

Since the Lagrangian and the Hamiltonian of a system represent the same dynamics, they are related. This relationship is used to convert the dependencies on some quantities (e. g., the time derivative of the quantity of interest) into dependencies on other quantities (e. g., the quantity that is conjugate to the quantity of interest). This transformation is accomplished by the *Legendre transform*. For a one-dimensional system, it is given by

$$H(q,p) = p\dot{q} - L(q,\dot{q}). \tag{3.194}$$

It now follows that H only depends on $\dot{q}$ through p. As a result, the partial derivative of H with respect to $\dot{q}$ is zero. We can interchange $H(q,p)$ and $L(q,\dot{q})$ to get the inverse Legendre transform, which recovers the Lagrangian from the Hamiltonian.

The two-dimensional space defined by q and p is called a *phase space*. Therefore, the Hamiltonian is a function defined on phase space. The evolution of systems due to the dynamics as described by the Hamiltonian are thus represented by trajectories on phase space, parameterized by the evolution parameter t. However, the initial state of a system may also be represented by a *probability distribution* on phase space. In such a case, the Hamiltonian governs the evolution of such a distribution on phase space as a function of time, as shown below.

3.5.1.2 Action principle

The *action* of a system is given by the integral of the Lagrangian over the evolution parameter, time. In the current context, the action is expressed as

$$S[q] = \int_a^b L(q, \dot{q})\,\mathrm{d}t, \tag{3.195}$$

with a and b being the initial and final values of the evolution parameter. The square brackets indicate that the action is a *functional* of $q(t)$, representing the quantity of interest as a function of the evolution parameter.

The evolution of a system is governed by the *variation* of the action. Stated differently, we can say that the solution for the evolution of a system is found where the action is *variational*, that is, where its variation vanishes.

We define the variation of q as $q + \delta q$, where δq is a function of t that is zero at the endpoints $\delta q(t = a) = \delta q(t = b) = 0$. The variation of the action is

$$\delta S = S[q + \delta q] - S[q]. \tag{3.196}$$

It leads to

$$\delta S = \int_a^b \frac{\partial L}{\partial q}\delta q + \frac{\partial L}{\partial \dot{q}}\delta \dot{q}\,\mathrm{d}t = \int_a^b \left[\frac{\partial L}{\partial q} - \frac{\mathrm{d}}{\mathrm{d}t}\left(\frac{\partial L}{\partial \dot{q}}\right)\right]\delta q\,\mathrm{d}t, \tag{3.197}$$

where we performed partial integration on the second term. For the action to be variational ($\delta S = 0$), it then follows that

$$\frac{\partial L}{\partial q} - \frac{\mathrm{d}}{\mathrm{d}t}\left(\frac{\partial L}{\partial \dot{q}}\right) = 0, \tag{3.198}$$

which is called the *Euler–Lagrange equation*. It produces the equation of motion for the system in the form of a second-order differential equation for $q(t)$.

3.5.1.3 Hamiltonian equations of motion

We can also obtain equations of motion from the Hamiltonian by considering the differentials for the two sides of the equation in (3.194). It gives

$$\begin{aligned}\frac{\partial H}{\partial q}\mathrm{d}q + \frac{\partial H}{\partial p}\mathrm{d}p = p\mathrm{d}\dot{q} + \dot{q}\mathrm{d}p - \frac{\partial L}{\partial q}\mathrm{d}q - \frac{\partial L}{\partial \dot{q}}\mathrm{d}\dot{q} &= \dot{q}\mathrm{d}p - \frac{\mathrm{d}}{\mathrm{d}t}\left(\frac{\partial L}{\partial \dot{q}}\right)\mathrm{d}q \\ &= \dot{q}\mathrm{d}p - \dot{p}\mathrm{d}q,\end{aligned} \tag{3.199}$$

where we used (3.193) and (3.198). Comparing the differentials we get two equations for the Hamiltonian

$$\frac{\partial H}{\partial q} = -\dot{p} \quad \text{and} \quad \frac{\partial H}{\partial p} = \dot{q}. \tag{3.200}$$

Now we have two first-order differential equations for the phase space variables.

Any measurement applied to the system can be expressed by a function of the phase space variables $A(q,p)$, representing the distribution of outcomes from such a measurement. The evolution of such a function as a function of time is then indirectly determined by the functions $q(t)$ and $p(t)$. We can represent it as

$$\frac{\mathrm{d}A}{\mathrm{d}t} = \frac{\partial A}{\partial q}\dot{q} + \frac{\partial A}{\partial p}\dot{p} = \frac{\partial A}{\partial q}\frac{\partial H}{\partial p} - \frac{\partial A}{\partial p}\frac{\partial H}{\partial q} \triangleq \{H,A\}, \tag{3.201}$$

assuming A only depends on t via q and p, and where we substitute in (3.200). The final expression represents the *Poisson bracket*, defined by

$$\{f,g\} \triangleq \frac{\partial f}{\partial p}\frac{\partial g}{\partial q} - \frac{\partial g}{\partial p}\frac{\partial f}{\partial q}, \tag{3.202}$$

where $f(q,p)$ and $g(q,p)$ are any smooth functions on phase space.

The implication of (3.201) is that the Hamiltonian is the *generator of time evolution*. The *evolution equation* in (3.201) is valid for any phase space distribution. For example, by replacing A by either q or p, we recover the two equations of motion in (3.200).

3.5.1.4 Symplectic geometry of phase space

The two variables that represent the coordinates of phase space q and p are generically called *canonical variables*. One can transform these canonical variables to different sets of canonical variables $q'(q,p)$ and $p'(q,p)$, provided that the new canonical variables satisfy equations of motion similar in form to those in (3.200). This condition implies that the differentials of these canonical variables are related by a *canonical transformation*. As a matrix-vector equation, it reads

$$\mathrm{d}\mathbf{q}' = M\mathrm{d}\mathbf{q}, \tag{3.203}$$

where $\mathrm{d}\mathbf{q}' = [\mathrm{d}q' \; \mathrm{d}p']^T$, $\mathrm{d}\mathbf{q} = [\mathrm{d}q \; \mathrm{d}p]^T$, and

$$M = \begin{bmatrix} \partial_q q' & \partial_p q' \\ \partial_q p' & \partial_p p' \end{bmatrix}, \tag{3.204}$$

is the *canonical transformation matrix*. The implication is that the canonical transformation matrix satisfies an equation

$$MJM^T = J, \tag{3.205}$$

where

$$J = \begin{bmatrix} 0 & \mathbf{1} \\ -\mathbf{1} & 0 \end{bmatrix}. \tag{3.206}$$

The expression in (3.205) is the *symplectic condition*. Hence, the canonical transformation matrix M is a *symplectic matrix*. In other words, it is an element of a *symplectic group*. All canonical transformations thus satisfy the symplectic condition. It shows that the phase space with coordinates q and p has a *symplectic geometry*.

Exercise 3.19. Show that when the canonical transformation matrix is a 2×2 matrix, as defined by the transformation in (3.43), the symplectic condition in (3.205) leads to (3.44).

The symplectic property of phase space has significant consequences. All the dynamics of a system represented on phase space are governed by symplectic transformations. Poisson brackets are invariant under such symplectic canonical transformations. Stated more precisely, the points on a trajectory representing the dynamics of a particle on phase space are related by symplectic canonical transformations. Another property (Liouville's theorem) is that any finite area on phase space retains its size under canonical transformations. It can be related to the dynamics of a system. If the state of a system is initially found within an area of phase space then, at some later time, it is still found inside the associated area obtained from the canonical transformations of the initial area, as determined by the dynamics.

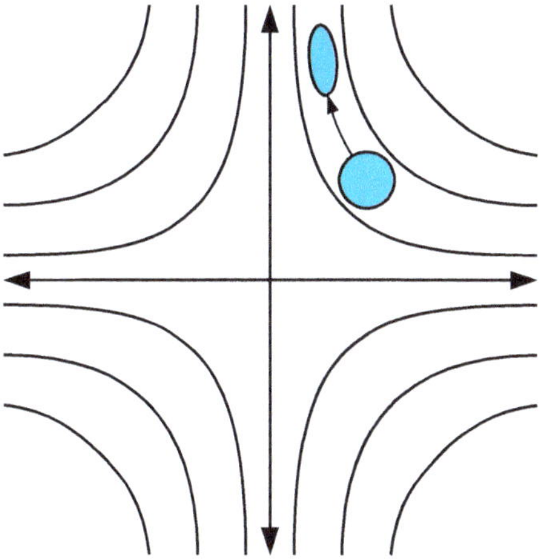

Figure 3.4: Squeezing process on phase space. The hyperbolic curves represent lines along which points are transformed during the squeezing process.

When the real-valued variables q and p are converted to complex-valued variables α and α^*, as in (3.118), the symplectic transformation of q and p is converted into a *Bogoliubov transformation* of α and α^*. The Bogoliubov transformation is obtained from (3.175) by replacing the ladder operators by complex variables $\hat{a} \rightarrow \alpha$ and $\hat{a}^\dagger \rightarrow \alpha^*$.

(See Section 5.7.4 for more detail.) The Bogoliubov coefficients are given in terms of the canonical transformation matrix elements.

Exercise 3.20. Show that when the phase space variables q and p are expressed in terms of the complex-value variables α and α^*, the equation in (3.205) leads to the condition in (3.176) for the associated Bogoliubov coefficients.

The effect of the Bogoliubov transformation associated with squeezing on phase space is demonstrated in Figure 3.4. Any function on the phase space is transformed along the hyperbolic contours. As a result, an isotropic function becomes squeezed, as demonstrated by the example in Figure 3.4.

3.5.2 Classical harmonic oscillator

As an example of a mechanical system and its representation on classical phase space, we consider the classical harmonic oscillator here. It is a system in which a particle or object with a mass m oscillates along one direction in space. For such an oscillation to occur, the particle must experience a *restoring force* whenever it is displaced away from an equilibrium position pushing it back toward the equilibrium position. This restoring force is proportional to the displacement. The equation of motion is obtained in terms of Newton's second law

$$F = -c_s x(t) = ma = m\partial_t^2 x(t), \tag{3.207}$$

where c_s is the *string constant*. The oscillations occur with an angular frequency of

$$\omega_m = \sqrt{\frac{c_s}{m}}. \tag{3.208}$$

The energy in the system, which is represented by the Hamiltonian as the sum of the kinetic and potential energy, is given by

$$H = \frac{1}{2}m\,[\partial_t x(t)]^2 + \frac{1}{2}c_s x^2(t) = \frac{1}{2}mv^2(t) + \frac{1}{2}c_s x^2(t), \tag{3.209}$$

where $v(t) = \dot{x}(t)$ is the velocity. We can replace the string constant in favour of the angular frequency $c_s \to m\omega_m^2$, and the velocity by $v(t) \to p(t)/m$ in terms of the momentum, to represent the Hamiltonian as

$$H = \frac{1}{2m}p^2(t) + \frac{1}{2}m\omega_m^2 x^2(t). \tag{3.210}$$

We use p to represent the momentum here, but it should not be confused with the quadrature variable for which we use the same symbol.

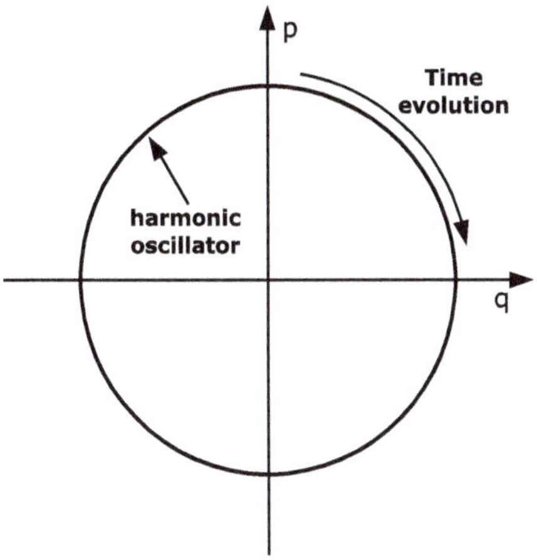

Figure 3.5: Phase space trajectory of the harmonic oscillator.

By replacing $m\omega_m x(t) \rightarrow q(t)$ (to give q the same dimensions as momentum), we obtain a simpler way to represent the dynamics. Since the energy in the system E is a constant, the Hamiltonian leads to the relationship

$$2mE = p(t)^2 + q(t)^2. \tag{3.211}$$

It describes a circle with a squared radius $R^2 = 2mE$ on a two-dimensional *phase space* with axes given by p and q, as shown in Figure 3.5. They represent the phase space variables (even though they carry the units of momentum in the current situation). Each point on the phase space represents a *state* of the system, as determined by $q(t)$ and $p(t)$. The point moves as a function of time t, as governed by the dynamics. For the harmonic oscillator, the dynamics produces a circular motion with time, as indicated in Figure 3.5.

3.5.3 Quantum phase space

3.5.3.1 Phase space for quantum kinematics

When considering the representation of quantum states on phase space, we notice that there are a few differences compared to the situation with a classical phase space. As a way to add quantum physics to the notion of a phase space for classical particles, we can see what happens when we simply apply the de Broglie relationship to the phase space for the kinematics of the classical particles. The de Broglie relationship converts the classical phase space, represented by the position and the momentum, into a phase space represented by the position and the *wavenumber*. At the same time, the classical particle (or the statistical distribution of states for the classical particle) is also converted into a *field* in the context of quantum physics. The quantities of interest represented on a quantum phase space are now *functions* instead of the coordinates of points. (The coordinates become operators in quantum mechanics.) Moreover, the probability interpretation as used in the phase space of classical mechanics does not remain the same.

3.5.3.2 Phase space for quantum optics

In a certain sense, the conversion of a kinematic system into a quantum system simply by introducing the de Broglie relationship is not natural. There does not really exist any physical quantum system in which the particle-number degree of freedom is represented by a position variable.

It is much more natural for the *particle-number degree of freedom* to correspond to the strength of a field. Such scenarios include those considered in quantum optics. The *canonical phase space variables* are now the quadrature variables q and p representing the particle-number degree of freedom. They do not represent the spatiotemporal (or spin) degrees of freedom. However, the spin and spatiotemporal degrees of freedom can be incorporated into the quadrature variables q and p, in such a way that these variables encapsulate all the degrees of freedom of the system. In such a case, the phase space becomes infinite dimensional. The inclusion of the other degrees of freedom are discussed in more detail in Chapters 4 and 5.

While the coefficient function (or wavefunction) that represents a state is either a function of q or a function of p, the function that represents the state on phase space is a function of both these variables. We have already seen that, for a given coefficient function represented as a function of one of these variables, its equivalent in terms of the other variable is completely determined via the Fourier transform. By implication, there are equivalent constraints on the functions that represent states on phase space.

So far, we just referred to "the function on phase space" without being too specific about how we define it and what we call it. The reason is that the representation of a specific state by a function on phase space is not unique; there are different ways to represent a state on phase space. The most common phase space representations are the *Wigner function* [24], the *Husimi Q-function* [25], and the *Glauber–Sudarshan P-function* [26, 27]. Although we mostly focus on the Wigner function, we also introduce the other two. Here, we briefly discuss Wigner functions in the context of the particle-number degree of freedom only. In Chapter 5, the *Wigner functional*, which incorporates the other degrees of freedom as well, is discussed in more detail.

3.5.4 Wigner function

The *Wigner function* for an arbitrary quantum state, as given by its density operator $\hat{\rho}$, is defined in terms of the q-basis by [24]

$$W(q,p) = \text{Wigner}\{\hat{\rho}\} \triangleq \int \left\langle q + \frac{1}{2}x \middle| \hat{\rho} \middle| q - \frac{1}{2}x \right\rangle \exp(-\mathrm{i}xp)\, \mathrm{d}x. \tag{3.212}$$

The Wigner function is thus partially defined in terms of a Fourier transform. If the density operator is pure $\hat{\rho} = |\psi\rangle \langle\psi|$, one gets

$$W(q,p) = \text{Wigner}\{|\psi\rangle \langle\psi|\} = \int \psi\left(q + \frac{1}{2}x\right) \psi^{*}\left(q - \frac{1}{2}x\right) \exp(-\mathrm{i}xp)\, \mathrm{d}x, \tag{3.213}$$

where $\psi(q) = \langle q|\psi\rangle$ is the coefficient function or wavefunction of the state in terms of the q-basis.

Note that the definition of the Wigner function in (3.212) implies that it is a *linear* operation on the density operator. The Wigner function of a linear combination of density operators is therefore given by the same linear combination of the Wigner functions of the individual operators. As a result, a mixed state, as represented in (3.8), can be expressed by

$$W(q,p) = \text{Wigner}\{\hat{\rho}\} = \sum_{n} P_n \,\text{Wigner}\{|\psi_n\rangle \langle\psi_n|\}, \tag{3.214}$$

in terms of the Wigner functions of pure states.

Wigner functions are not restricted to density operators. One can compute the Wigner function for any arbitrary operator. As an example, the Wigner function for the quadrature operator $\hat{q}$ is readily computed as

$$\begin{aligned} \text{Wigner}\{\hat{q}\} &= \int \left\langle q + \frac{1}{2}x \right| \hat{q} \left| q - \frac{1}{2}x \right\rangle \exp(-\mathrm{i}xp)\, \mathrm{d}x \\ &= \int \delta(x) \left(q - \frac{1}{2}x\right) \exp(-\mathrm{i}xp)\, \mathrm{d}x = q. \end{aligned} \tag{3.215}$$

To obtain the Wigner function for $\hat{p}$, we insert an identity operator, resolved in terms of the p-basis (3.21),

$$\begin{aligned} \text{Wigner}\{\hat{p}\} &= \int \left\langle q + \frac{1}{2}x \right| \hat{p} \left| q - \frac{1}{2}x \right\rangle \exp(-\mathrm{i}xp)\, \mathrm{d}x \\ &= \int \left\langle q + \frac{1}{2}x \middle| p' \right\rangle p' \left\langle p' \middle| q - \frac{1}{2}x \right\rangle \exp(-\mathrm{i}xp) \frac{\mathrm{d}p'}{2\pi}\, \mathrm{d}x \\ &= \int \exp[\mathrm{i}(p' - p)x] p' \frac{\mathrm{d}p'}{2\pi}\, \mathrm{d}x = \int 2\pi\delta(p' - p)p' \frac{\mathrm{d}p'}{2\pi} = p. \end{aligned} \tag{3.216}$$

Exercise 3.21. Show that the Wigner function for any operator that is expressed as a *holomorphic function* of either $\hat{q}$ or $\hat{p}$ is given by the same holomorphic function of q or p, respectively,

$$\text{Wigner}\{h(\hat{q})\} = h(q) \quad \text{and} \quad \text{Wigner}\{h(\hat{p})\} = h(p). \tag{3.217}$$

Thanks to the linearity of (3.212), it then follows that the Wigner functions for the ladder operators are directly obtained from those for $\hat{q}$ and $\hat{p}$ via (3.45). They are

$$\text{Wigner}\{\hat{a}\} = \frac{1}{\sqrt{2}}(q + ip) \equiv \alpha,$$
$$\text{Wigner}\{\hat{a}^\dagger\} = \frac{1}{\sqrt{2}}(q - ip) \equiv \alpha^*, \tag{3.218}$$

as can be verified by direct calculations.

There is much more that can be said about Wigner functions. However, we postpone such discussions to Chapter 5, after having incorporated the other degrees of freedom.

3.6 Quantum dynamics

In Section 3.1, we derived an operator formalism without specifically using any concepts from quantum physics, other than the implied Fourier relationship. It led us all the way to the definition of quadrature operators, ladder operators, and even the Fock states. Then we proceeded to discuss various states, such as the coherent states and general Gaussian states, which include squeezed states.

At no point during these discussions did the *Planck constant* show up. It is not because we used *natural units* in which the Planck constant is set equal to 1. As a rule, *we do not set the Planck constant equal to 1*, but we may sometimes "hide" it by combining it with other dimension parameters for simplicity. In such cases, the Planck constant is still present in terms of the definition of the dimensionless parameter.

The reason why the Planck constant does not appear in the preceding discussions is because the formalism thus far developed does not concern interactions. Therefore, the Planck and the de Broglie relationships that are associated with interactions, and which relate quantities via the Planck constant, do not explicitly enter these discussions. In a sense, the operator formalism developed so far can equally well be used in classical scenarios (assuming a context can be identified in which such a formulation would be useful). There is nothing in the context of classical physics that prohibits such a formulation. (However, some of the states we discussed do not have classical equivalents.)

This situation changes with the discussion of *quantum dynamics*. When dynamics includes interactions, it requires the introduction of the Planck and the de Broglie relationships, which are specifically associated with such interactions, as explained in Chapter 1. Therefore, the Planck constant makes an appearance.

However, in cases where the dynamics is not specifically associated with interactions, the Planck constant can usually be removed (cancelled) from the equations through some redefinitions of quantities, without changing the physics being modelled. Such examples lead to equations that may equally well be used to model classical scenarios, at least formally. Any reformulation of the equations for a given physical scenario that does not affect any observable predictions obtained from these equations carries no physical significance. Therefore, if we can remove the Planck constant without changing the observable predictions of such equations, as demonstrated below, then the Planck constant does not play any physically significant role in these equations.

3.6.1 Schrödinger equation

Quantum dynamics, as represented by a time evolution, is based on Planck's relationship $E = \hbar\omega$. We again start with Fourier analysis (see Section 2.1.2). Here, we consider it in terms of the temporal degree of freedom. In the time domain (discarding all other dimensions), we have

$$\begin{aligned} \psi(t) &= \int \tilde{\psi}(\omega)\exp(-\mathrm{i}\omega t)\,\frac{\mathrm{d}\omega}{2\pi}, \\ \tilde{\psi}(\omega) &= \int \psi(t)\exp(\mathrm{i}\omega t)\,\mathrm{d}t. \end{aligned} \tag{3.219}$$

The expectation value of the *energy* E is then given by

$$\begin{aligned} \langle E\rangle &= \int \hbar\omega|\tilde{\psi}(\omega)|^2\,\frac{\mathrm{d}\omega}{2\pi} \\ &= \int \left[\int \psi^*(t)\exp(-\mathrm{i}\omega t)\,\mathrm{d}t\right]\hbar\omega\tilde{\psi}(\omega)\,\frac{\mathrm{d}\omega}{2\pi} \\ &= \int \psi^*(t)\int \exp(-\mathrm{i}\omega t)\hbar\omega\tilde{\psi}(\omega)\,\frac{\mathrm{d}\omega}{2\pi}\,\mathrm{d}t \\ &= \int \psi^*(t)\,(\mathrm{i}\hbar\partial_t)\int \exp(-\mathrm{i}\omega t)\tilde{\psi}(\omega)\,\frac{\mathrm{d}\omega}{2\pi}\,\mathrm{d}t \\ &= \int \psi^*(t)\,(\mathrm{i}\hbar\partial_t)\,\psi(t)\,\mathrm{d}t. \end{aligned} \tag{3.220}$$

We see that the energy is extracted from $\psi(t)$ with the aid of a time derivative and the *Planck constant*:

$$E \to \mathrm{i}\hbar\partial_t. \tag{3.221}$$

When we replace the energy by the Hamiltonian, represented in terms differential operators, and allow it to operate on $\psi(t)$, we get

$$\mathrm{i}\hbar\partial_t\psi(t) = H\psi(t), \tag{3.222}$$

which is called the *Schrödinger equation* and $\psi(t)$ now serves as the *wavefunction* representing the quantum state. Note that the sign is determined by the *physics sign convention* used for the Fourier relationships in Section 2.1.2. Although the Hamiltonian *operates* on the state, we do not represent it as an *operator* $\hat{H}$. The latter notation is reserved for *q-number* operators defined on the Hilbert space that operate on quantum state vectors represented by Dirac vectors. Here, the Hamiltonian is a *c-number* operation expressed in terms of derivatives.

3.6.2 Spatial evolution

In quantum mechanics, the evolution of a system is usually given by its evolution in time. In quantum field theory, which is used to model particle physics, the situation becomes more complicated with the evolution in *spacetime*, because there are multiple *paths* that connect two points in spacetime. The situation is similar in quantum optics, but often it is still simply given by evolution in time. Nevertheless, there are scenarios where the system is *stationary* in the sense that the evolution in time is simply given by a phase rotation represented by $\exp(-i\omega t)$ that may be integrated over frequency, while the evolution along a spatial direction is more complicated.

Consider, for example, an optical beam propagating through a lens. While the temporal evolution is trivial, the spatial evolution requires the propagation of the beam through the transmission function of the lens. In such cases, it is simpler to represent the evolution in terms of a *propagation operator* instead of trying to force the temporal evolution as generated by the Hamiltonian to mimic the spatial evolution. We encounter such scenarios in later chapters. Here, we develop the basic formalism in terms of only the particle-number degree of freedom mainly in the context of temporal evolution. For this purpose, we introduce a time dependence that leads to a formulation of the dynamics. The corresponding expressions for the spatial evolution, which is provided without derivation, follow in a similar way.

For an equivalent derivation for the evolution along a spatial direction, we introduce a dependence on z, instead of a time dependence, and consider the momentum instead of the energy. By analogy, the *momentum* is extracted from $\psi(z)$ with a z-derivative times the Planck constant:

$$\text{momentum} \rightarrow -i\hbar\partial_z . \tag{3.223}$$

It is equivalent to the derivation for the quadrature operator $\hat{p}$ in (3.36). Note that there is a change in the sign compared to the temporal case. The momentum is then replaced by a *momentum operator* or *propagation operator* P, again representing a c-number operation expressed in terms of derivatives, leading to a *spatial evolution equation* that is analogous to the Schrödinger equation:

$$-i\hbar\partial_z\psi(z) = P\psi(z). \tag{3.224}$$

3.6.3 Evolving state vectors

Next, we incorporate the dynamics into the formalism that we developed in Section 3.1. For this purpose, we combine the temporal degree of freedom with the particle-number degree of freedom that we used for the operator formalism in Section 3.1 by assuming that those functions also have time dependencies. The particle-number degree of freedom is denoted by the dimensionless quadrature variables q and p. The resulting functions now represent *wavefunctions* in the quantum context. They obey Schrödinger equations of the form

$$i\hbar\partial_t\psi(q,t) = H\psi(q,t) \quad \text{and} \quad i\hbar\partial_t\tilde{\psi}(p,t) = H\tilde{\psi}(p,t). \tag{3.225}$$

The c-number Hamiltonian H is defined in terms of the quadrature variable on which the wavefunction depends.

We multiply the first of these equations by the coordinate basis $|q\rangle$ and integrated over q to form the ket of the state,

$$|\psi(t)\rangle = \int |q\rangle\, \psi(q,t)\, \mathrm{d}q. \tag{3.226}$$

Then we end up with

$$i\hbar\partial_t\, |\psi(t)\rangle = \hat{H}\, |\psi(t)\rangle\,, \tag{3.227}$$

where we converted the c-number Hamiltonian to a q-number Hamiltonian operator defined on the Hilbert space of the state vectors. In effect, we "pulled" the c-number Hamiltonian "through" the ket to convert it to a q-number operator $|q\rangle\, H \to \hat{H}\, |q\rangle$. The resulting equation represents the Schrödinger equation for the state vector. Overlapping this equation by $\langle p|$ on the left, we reproduce the second equation in (3.225). Here, the q-number Hamiltonian operator is "pulled through" the bra to convert it back into a c-number Hamiltonian defined in terms of p. Hence $\langle p|\, \hat{H} \to H\, \langle p|$.

When the state is represented by a density operator, we can decompose it in terms of bras and kets as in (3.8). After applying (3.227) and its Hermitian adjoint on the bras and kets, we reconstitute the density operator. The resulting evolution equation reads

$$i\hbar\partial_t\hat{\rho}(t) = [\hat{H}, \hat{\rho}(t)]. \tag{3.228}$$

It is called the *von Neumann equation*. For spatial evolution, it reads

$$-i\hbar\partial_z\hat{\rho}(z) = [\hat{P}, \hat{\rho}(z)]. \tag{3.229}$$

3.6.4 Unitary evolution and the Heisenberg equation

The evolution of a state vector $|\psi(t)\rangle$ as a function of time can be represented by a unitary operator applied to an initial state vector defined at some *reference time* (e. g., $t = 0$),

$$|\psi(t)\rangle = \hat{U}(t)\, |\psi(0)\rangle\,. \tag{3.230}$$

Since $\hat{U}^{\dagger}(t)\hat{U}(t) = \mathbb{1}$, the state vector maintains its normalization for all time, as required in the quantum context. When the state vector in the Schrödinger equation is represented in terms of the unitary evolution operator, it leads to an evolution equation for the unitary evolution operator given by

$$\mathrm{i}\hbar\partial_t\hat{U}(t) = \hat{H}\hat{U}(t). \tag{3.231}$$

Its solution is an expression for the unitary evolution operator in terms of the Hamiltonian, given by

$$\hat{U}(t) = \exp_T\left(\frac{-\mathrm{i}}{\hbar}\int_0^t \hat{H}\,\mathrm{d}t\right), \tag{3.232}$$

where $\exp_T(\cdot)$ is a *time-ordered exponential function* in which all the terms in its expansion have the products of the argument in order of time increasing from right to left. The way in which the unitary evolution operator depends on the Hamiltonian shows that the latter is the *generator of time evolution.*

The expectation value of an *observable* such as the quadrature operator $\hat{q}$ becomes a function of time

$$\langle q(t)\rangle = \langle\psi(t)|\,\hat{q}\,|\psi(t)\rangle\,, \tag{3.233}$$

when it is computed for an evolving state. The evolving state can be expressed in terms of unitary evolution operators, which can be absorbed into the observable, so that

$$\langle q(t)\rangle = \langle\psi(0)|\,\hat{q}(t)\,|\psi(0)\rangle\,, \tag{3.234}$$

where

$$\hat{q}(t) = \hat{U}^{\dagger}(t)\hat{q}\hat{U}(t). \tag{3.235}$$

It represents a different, but equally valid representation of the process, which is called the *Heisenberg picture.* The previous situation where the states evolve as functions of time is called the *Schrödinger picture.* The observable now obeys the *Heisenberg equation,* given in terms of a commutator by

$$\mathrm{i}\hbar\partial_t\hat{q}(t) = [\hat{q}(t),\hat{H}], \tag{3.236}$$

which follows from the time derivative of (3.235), together with replacements using (3.231) and its Hermitian adjoint.

In a similar way, the evolution along a spatial direction leads to a *quantum propagation equation for operators* (a spatial version of the Heisenberg equation). It reads

$$-\mathrm{i}\hbar\partial_z\hat{q}(z) = [\hat{q}(z),\hat{P}], \tag{3.237}$$

where $\hat{P}$ is the (q-number) *momentum operator* or the *propagation operator*. It is the *generator of spatial evolution.*

3.6.5 Unitary temporal evolution

The time evolution of states and operators are given by their unitary evolution in the Schrödinger and Heisenberg pictures, respectively. For states given by their density operators in the Schrödinger pictures, we have

$$\hat{\rho}(t) = \hat{U}(t)\hat{\rho}(0)\hat{U}^{\dagger}(t), \tag{3.238}$$

and for operators (or observables) in the Heisenberg pictures, we get

$$\hat{A}(t) = \hat{U}^{\dagger}(t)\hat{A}(0)\hat{U}(t), \tag{3.239}$$

where the unitary evolution operator is given by (3.232).

In the context of a free field theory, the unitary evolution operator contains the *interaction-free Hamiltonian*, as expressed in terms of ladder operators in (3.262) below. The unitary evolution operator can therefore be written as

$$\hat{U}(t) = \exp(-\mathrm{i}\hat{\omega}t), \tag{3.240}$$

in terms of a *frequency operator* $\hat{\omega}$ given by

$$\hat{\omega} = \frac{1}{2}\omega(\hat{a}^{\dagger}\hat{a} + \hat{a}\hat{a}^{\dagger}). \tag{3.241}$$

The *Planck constant* cancels for a free field theory, leaving the expression without it.

The unitary evolution operator expressed in terms of the frequency operator can be used to see how states evolve in time for a free field theory. The effect of the frequency operator on the Fock states is

$$\hat{\omega}\,|n\rangle = |n\rangle\left(n + \frac{1}{2}\right)\omega. \tag{3.242}$$

Therefore,

$$\hat{U}(t)\,|n\rangle = \exp(-\mathrm{i}\hat{\omega}t)\,|n\rangle = |n\rangle \exp\left[-\mathrm{i}\left(n + \frac{1}{2}\right)\omega t\right]. \tag{3.243}$$

Exercise 3.22. Show that when the unitary evolution operator for a free field theory is applied to a coherent state it produces

$$\hat{U}(t)\,|\alpha\rangle = \exp\left(-\mathrm{i}\frac{1}{2}\omega t\right)|\exp(-\mathrm{i}\omega t)\alpha\rangle\,. \tag{3.244}$$

Hint: use the expansion of the coherent state in terms of the Fock states given in (3.99).

Although the ladder operators are not Hermitian, they are used to construct *observables*. As a result, ladder operators also evolve as functions of time in the Heisenberg picture. Inserting them into the Heisenberg equation, we get

$$\begin{aligned} \mathrm{i}\hbar\partial_t\hat{a}(t) &= [\hat{a}(t),\hat{H},] = \hbar\omega\hat{a}(t), \\ \mathrm{i}\hbar\partial_t\hat{a}^{\dagger}(t) &= [\hat{a}^{\dagger}(t),\hat{H}] = -\hbar\omega\hat{a}^{\dagger}(t), \end{aligned} \tag{3.245}$$

leading to their free temporal evolution. Note that the Planck constant cancelled, as expected for a free field theory. The solutions are

$$\hat{a}(t) = \hat{a}\exp(-\mathrm{i}\omega t) \quad \text{and} \quad \hat{a}^{\dagger}(t) = \hat{a}^{\dagger}\exp(\mathrm{i}\omega t), \tag{3.246}$$

which relate the ladder operators in the Heisenberg picture to those in the Schrödinger picture. The transformations of the observables are then obtained from the transformation of the individual ladder operators in the definitions of the observables.

We can also obtain the solutions in (3.246) with the aid of (3.239). In this case, we use (3.103), together with the commutation relations

$$[\hat{a},\hat{\omega}] = \omega\hat{a} \quad \text{and} \quad [\hat{\omega},\hat{a}^{\dagger}] = \omega\hat{a}^{\dagger}. \tag{3.247}$$

Exercise 3.23. Show that the solutions in (3.246) are obtained from

$$\hat{a}(t) = \hat{U}^{\dagger}(t)\hat{a}\hat{U}(t) \quad \text{and} \quad \hat{a}^{\dagger}(t) = \hat{U}^{\dagger}(t)\hat{a}^{\dagger}\hat{U}(t). \tag{3.248}$$

with the aid of (3.103), (3.240), and (3.247).

3.6.6 Quantum harmonic oscillator

The quantum formalism that we developed here can now be used to formulate theories that model quantum physics. Often such quantum theories are based on classical theories through a process called *quantization*. Starting with a classical theory represented by a Hamiltonian, one would replace the classical canonical variables by operators obeying suitable commutation relations.

A simple example of such a process is provided by the quantization of the *classical harmonic oscillator*, which we discussed in Section 3.5.2. In this section, we consider the quantization of the harmonic oscillator with the aid of the formalism developed in Section 3.1. While it serves to demonstrate aspects of the quantum formalism, it does not provide the required incorporation of the spatiotemporal degrees of freedom as needed later. A quantization procedure that incorporates the spatiotemporal degrees of freedom is demonstrated in Section 4.1, where we use it to quantize the electromagnetic field.

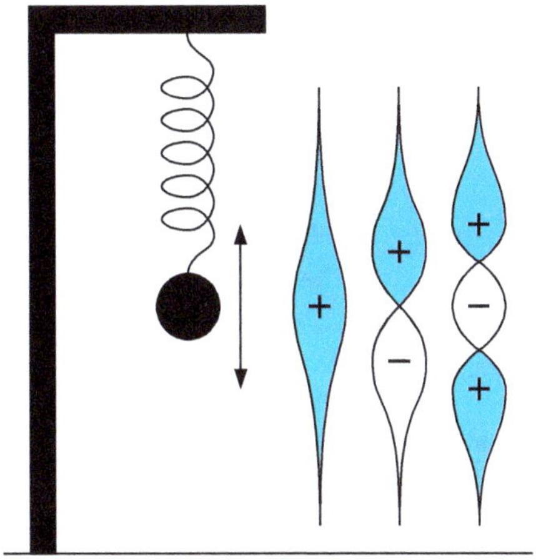

Figure 3.6: Physical harmonic oscillator with the wavefunctions of the first three modes of the associated quantum harmonic oscillator.

The proposal made by de Broglie that fundamental particles can be treated as waves, leads us to imagine (as a formal exercise) what would happen when the particle in the classical harmonic oscillator is replaced by a wave. This situation is depicted in Figure 3.6. The wave is given in the *Schrödinger picture* by a normalized complex-valued *wavefunction* $\psi(x,t)$, evolving as a function of time. Normalized in terms of (3.24), this wavefunction leads to a ***probability distribution*** given by its modulus square. Observable quantities such as the position, momentum, and energy of the wave can thus be represented in terms of their expectation values. For example, the expectation value for the position of the particle obtained from $\psi(x,t)$ is given by

$$\langle x(t)\rangle = \int x|\psi(x,t)|^2 \,\mathrm{d}x. \tag{3.249}$$

Since we are not integrating over time, the expectation value of the position can change as a function of time, even though the normalization of $\psi(x,t)$ is constant with time.

The expectation value for the momentum of the particle is given by an equivalent probability distribution as a function of the momentum evolving as a function of time. The de Broglie relationship, which relates the momentum to the wavenumber for the one-dimensional case, implies that such a probability distribution for the momentum, is readily obtained from the configuration space wavefunction $\psi(x,t)$ with the aid of a Fourier transform, similar to what we did in Section 3.1. The spectral function thus obtained as a function of the wavenumber, serves as the Fourier domain wavefunction, from which the required probability distribution for the momentum is obtained as the modulus square. As a result, we have

$$\begin{aligned} \psi(x,t) &= \int \tilde{\psi}(k,t)\exp(ikx)\,\frac{dk}{2\pi}, \\ \tilde{\psi}(k,t) &= \int \psi(x,t)\exp(-ikx)\,dx, \end{aligned} \tag{3.250}$$

where $\psi(x,t)$ and $\tilde{\psi}(k,t)$ represent the *configuration space wavefunction* and *Fourier domain wavefunction*, respectively, both represented as functions of time. It allows us to express the expectation values for momentum in an equivalent way to what we did for the position, and then to derive an expression that gives the expectation values for momentum in terms of the configuration space wavefunction $\psi(x,t)$.

The expectation value for momentum becomes

$$\begin{aligned} \langle p(t)\rangle &= \hbar \int k|\tilde{\psi}(k,t)|^2\,\frac{dk}{2\pi} \\ &= \int \psi^*(x,t)(-i\hbar\partial_x)\psi(x,t)\,dx, \end{aligned} \tag{3.251}$$

in a way similar to the derivation in (3.36), but here we obtain a quantity associated with momentum by incorporating the *Planck constant*. We see that the momentum is extracted from $\psi(x,t)$ with a differential operator $p \to -i\hbar\partial_x$. It extracts the wavenumber and converts it to momentum via the de Broglie relationship. The energy is expressed with a temporal differential operator $E \to i\hbar\partial_t$, as shown in Section 3.6.

Using these replacements, we can obtain an equation for the evolution of $\psi(x,t)$ from the Hamiltonian. We multiply the Hamiltonian on the right by the wavefunction $\psi(x,t)$ and then perform the replacements. The resulting equation

$$i\hbar\partial_t\psi(x,t) = -\frac{\hbar^2}{2m}\partial_x^2\psi(x,t) + \frac{1}{2}m\omega_m^2x^2\psi(x,t), \tag{3.252}$$

is the *Schrödinger equation* for the *quantum harmonic oscillator*.

In this case, we do not have a string constant. Instead, the frequency ω_m is related to the mass via the relation

$$\omega_m = \frac{mc^2}{\hbar}. \tag{3.253}$$

We use this relationship to replace the mass in terms of ω_m to get an equation in terms of wave parameters, given by

$$\begin{aligned} i\hbar\partial_t\psi(x,t) &= -\frac{1}{2}\hbar\omega_m\frac{c^2}{\omega_m^2}\partial_x^2\psi(x,t) + \frac{1}{2}\hbar\omega_m\frac{\omega_m^2}{c^2}x^2\psi(x,t) \\ &= -\frac{1}{2}\hbar\omega_m\left[\frac{1}{k_m^2}\partial_x^2\psi(x,t) - k_m^2x^2\psi(x,t)\right], \end{aligned} \tag{3.254}$$

where $k_m = \omega_m/c$. It allows us to express the equation in dimensionless variables without any free parameters as

$$i\partial_\tau\psi(\xi,\tau) = -\frac{1}{2}\partial_\xi^2\psi(\xi,\tau) + \frac{1}{2}\xi^2\psi(\xi,\tau), \tag{3.255}$$

where we cancelled Planck's constant on both sides, and defined $\tau = \omega_m t$ and $\xi = k_m x$. While (3.255) is a dimensionless version of the Schrödinger equation for the quantum harmonic oscillator, it still represents the same (hypothetical) physical scenario. Therefore, the solutions of (3.255) would be formally the same as those obtained for (3.252) apart from some rescaling to recover dimension parameters.

The solutions of (3.255) are readily obtained when we recognize it as being formally equivalent to a one-dimensional version of the *paraxial wave equation in a GRIN lens medium*, given in (2.174). Since we have already obtained solutions for (2.174), we can identify their one-dimensional versions as solutions of (3.255).

In this way, we obtain *discrete solutions* of (3.255) from the one-dimensional version of (2.180), with the necessary replacements of variables and parameters. Often associated with the quantum aspect of the Schrödinger equation for the quantum harmonic oscillator, these discrete solutions are interpreted as Fock states, similar to those discussed in Section 3.1.7.

The discrete solutions are not the only solutions of the Schrödinger equation in (3.255). Through its formal equivalence to the one-dimensional version of (2.174), we find that (3.255) also have *oscillating solutions*, given by the one-dimensional version of the solutions for the GRIN lens medium in (2.179). We express them as

$$\psi(\xi,\tau) = \frac{\sqrt{W}}{\pi^{1/4}\sqrt{\gamma_1(\tau)}}\exp\left[-\frac{\xi^2\gamma_2(\tau) - 2\xi\xi_0 + \xi_0^2\cos(\tau)}{2\gamma_1(\tau)}\right], \tag{3.256}$$

in terms of the dimensionless coordinates of (3.255), where

$$\gamma_1(\tau) = W^2\cos(\tau) + i\sin(\tau) \quad \text{and} \quad \gamma_2(\tau) = \cos(\tau) + iW^2\sin(\tau), \tag{3.257}$$

with the width W and the initial position ξ_0 being dimensionless free parameters. A third free parameter is the initial time, which is here assumed to be zero. At $\tau = 0$, the function becomes

$$\psi(\xi,0) = \frac{1}{\pi^{1/4}\sqrt{W}}\exp\left[-\frac{(\xi-\xi_0)^2}{2W^2}\right], \tag{3.258}$$

representing a displaced Gaussian function. As τ increases, the function oscillates along the ξ-axis and its width changes periodically with the oscillations. Such an oscillation is shown in Figure 2.7 for the superposition of two Gaussian functions, which serves as a classical analogue of a Schrödinger cat state (see Section 6.1.4).

3.6.6.1 Operator representation

While we found that, even after replacing the particle in the classical harmonic oscillator by a field, we still effectively have a classical scenario, we can now proceed to *quantize* this classical Hamiltonian as a purely formal exercise. In effect, what we do is to obtain a q-number *Hamiltonian operator* that is equivalent to the c-number Hamiltonian given on the right-hand of (3.254). This process is equivalent to the process we used to derive (3.227), where we represented it by $|q\rangle H \to \hat{H} |q\rangle$. Here $|q\rangle$ is an element of a coordinate basis (a quadrature basis) and the c-number *Hamiltonian* is

$$H = \frac{1}{2}\hbar\omega_m \left(-\partial_\xi^2 + \xi^2\right). \tag{3.259}$$

For this purpose, we retain the factor of $\hbar\omega_m$ in anticipation of the resulting quantized Hamiltonian and to have the correct units. The conversion process is done by representing the wavefunction as $\psi(\xi, \tau) \to \psi(q, t) = \langle q|\psi(t)\rangle$ in which the dimensionless coordinate is represented by the quadrature variable $\xi \to q$. Then we multiply the expression by $|q\rangle$ on the left and integrate over q. The right-hand side then reads

$$\int |q\rangle H\psi(q, t)\, \mathrm{d}q = \frac{1}{2}\hbar\omega_m \int |q\rangle \left(-\partial_q^2 + q^2\right) \psi(q, t)\, \mathrm{d}q. \tag{3.260}$$

The conversion is now perform by replacing $q \to \hat{q}$ and $-\mathrm{i}\partial_q \to \hat{p}$ while pulling the Hamiltonian through $|q\rangle$. Note that $\hat{p}$ is *not* the momentum operator, because the derivative does not come with the *Planck constant*. So, we end up with

$$\int \hat{H} |q\rangle \langle q|\psi(t)\rangle\, \mathrm{d}q = \hat{H} |\psi(t)\rangle = \frac{1}{2}\hbar\omega_m \left(\hat{p}^2 + \hat{q}^2\right) |\psi(t)\rangle . \tag{3.261}$$

We used the completeness of the quadrature basis in (3.21) to convert the integral over q to an identity operator. As a final step, we express the quadrature operators in terms of ladder operators, as in (3.46). The resulting Hamiltonian operator is given by

$$\hat{H} = \frac{1}{2}\hbar\omega_m \left(\hat{a}^\dagger\hat{a} + \hat{a}\hat{a}^\dagger\right) = \hbar\omega_m \left(\hat{a}^\dagger\hat{a} + \frac{1}{2}\right), \tag{3.262}$$

where we used the *commutation relation for ladder operators* in (3.47) to obtain the last expression. Although the expression for the Hamiltonian operator is derived here for the harmonic oscillator, it is more generally valid in that the Hamiltonian of most free field theories can be expressed in terms of ladder operators in this way. In case there are additional degrees of freedom, the expression needs to be integrated or summed over them. In Section 4.1, we obtain a similar expression for the Hamiltonian, apart from the incorporation of the spatiotemporal and spin degrees of freedom.

The definition of the dimensionless variable ξ involves specific dimension parameters. In view of the apparent scale invariance found in (3.2), one can ask what would happen if we rescale ξ by an arbitrary numerical factor. It follows that the two terms in the resulting Hamiltonian equivalent to (3.259) would have different numerical factors so that when we perform the quantization, the resulting expression in terms of ladder operators would not have had the form given in (3.262). It would not commute with the number operators, and thus not *conserve particle number*.

The Hamiltonian operator given in terms of the ladder operators is easily shown to commute with the number operator in (3.53). As a result, they share the same *eigenvectors*, namely the Fock states. In the case of the Hamiltonian, the eigenvalues are energies

$$E = n\hbar\omega_m \tag{3.263}$$

where n is the *occupation number*. Moreover, similar to the case with the number operator, the commutations of the Hamiltonian with the ladder operators give

$$[\hat{a}, \hat{H}] = \hbar\omega_m\hat{a} \quad \text{and} \quad [\hat{H}, \hat{a}^\dagger] = \hbar\omega_m\hat{a}^\dagger . \tag{3.264}$$

These results provide the justification for the terminology applied to the ladder operators. If $|\psi\rangle$ is an eigenvector of the Hamiltonian so that

$$\hat{H}\,|\psi\rangle = |\psi\rangle\, E, \tag{3.265}$$

where the eigenvalue E is the energy of the state, then (3.264) implies that

$$\begin{aligned}\hat{H}\hat{a}\,|\psi\rangle &= \left(\hat{a}\hat{H} - \hbar\omega_m\hat{a}\right)|\psi\rangle = \hat{a}\,|\psi\rangle\,(E - \hbar\omega_m)\,, \\ \hat{H}\hat{a}^\dagger\,|\psi\rangle &= \left(\hat{a}^\dagger\hat{H} + \hbar\omega_m\hat{a}^\dagger\right)|\psi\rangle = \hat{a}^\dagger\,|\psi\rangle\,(E + \hbar\omega_m)\,.\end{aligned} \tag{3.266}$$

By implication, when the ladder operators $\hat{a}$ and $\hat{a}^\dagger$ are applied to an *eigenvector* of the Hamiltonian (the Fock states), they produce new eigenvectors with energies that are respectively reduced and increased by $\hbar\omega_m$. In effect, $\hat{a}$ removes a quantum of energy and $\hat{a}^\dagger$ adds a quantum of energy. It is similar to the situation that we have with the number operator. The implication is that, in addition to changing the occupation number, these ladder operators also change the energy in the state. *Each quantum (or particle), as distinguished in terms of the occupation number, is associated with a quantum of energy.*

While this last observation is perhaps the single most significant concept that we learn from this analysis of the quantum harmonic oscillator, we can also obtain the same result from the quantization of the electromagnetic field. Here, the result is cleaner because the spatiotemporal and spin degrees of freedom are not included.

3.7 Discussion

In this chapter, we developed the basic formalism with which quantum optics is modelled in terms of the particle-number (or photon-number) degree of freedom. This formalism is used as the basis for the incorporation of the other degrees of freedom (spin and spatiotemporal degrees of freedom) in the next chapter, to develop a more comprehensive formalism used in subsequent chapters.

A large part of the formalism is derived without the explicit application of quantum physics, as captured in the relationships of Planck (1.1) and de Broglie (1.2). In this way, we obtained the definitions of the quadrature operators and the ladder operators. We also formulated definitions of various states such as Fock states, coherent states, and squeezed states, and discussed their properties. As a necessary part of the background, we also discussed the concept of phase space, albeit postponing some pertinent aspects until Chapter 5, where we have a *functional* phase space, after incorporating the spatiotemporal and spin degrees of freedom.

Quantum physics enters the development (via the Planck and the de Broglie relationships) when we introduce dynamics as represented by the evolution equations such as the Schrödinger and Heisenberg equations. It culminates in a discussion of the quantum harmonic oscillator.

The irony in our discussion of the quantum harmonic oscillator would not have escaped the attentive reader. Here, we have the much heralded prime example of a quantum system, widely used (often as a tacit analogy) to model more complicated quantum systems. And yet, we discuss it here in analogy with a purely *classical* system, the GRIN lens. So, what then is so "quantum" about the quantum harmonic oscillator?

3.7.1 Critical review

In the spirit of our endeavour to understand quantum physics by also understanding *what it is not*, a critical review of the analysis of the quantum harmonic oscillator is in order. We separate this critical review into four parts, based on specific notions that can lead to the miss-identification of exclusively quantum properties.

3.7.1.1 Particles and wavefunctions

The first notion is easily dealt with. Despite de Broglie's suggestion that a quantum system is represented by a scenario where a (classical) particle is replaced by a wavefunction, it does not mean that such a replacement produces a system that exclusively describes a quantum system. The formal equivalence between the quantum harmonic oscillator of Section 3.6.6 and the GRIN lens medium of Section 2.7.4 serves as a counterexample. Although such evolution equations for wavefunctions are often successfully used to describe the evolution in quantum systems, their formal representations can equally well represent *evolution equations of fields* in a *classical field theory* context.

3.7.1.2 Discrete solutions

Discrete solutions are often found in bounded potentials both in classical and quantum scenarios. Examples of such bounded potentials with discrete solutions in the classical context include the modes in optical fibres and the modes in laser cavities. As a result, the discreteness of solutions of any equation modelling a system cannot be taken as an indication that the system in question is a quantum system. For the case of the quantum harmonic oscillator, we found that, in addition to its discrete solutions, it also has a continuous set of oscillating solutions, in correspondence to the solutions of a (classical) GRIN lens medium.

The case of the atom needs special consideration. Here, we have discrete energy levels corresponding to the discrete solutions of a confining potential, separated into orbitals with the incorporation of spin. These orbitals are occupied by single electrons, which may give the misleading idea that it is the discrete solutions of the confining potential that leads to this quantization in terms of distinct particles. Instead this separation is a consequence of the fermion character of electrons, obeying the *Pauli exclusion principle*. If instead, we could have populated the orbitals of an atom with charged bosons, they could all have occupied the lowest energy state (as in a *Bose–Einstein condensate*). In fact, such bosons can occupy any solution (continuous or discrete) of the Schrödinger equation for the atom. It is therefore not so much the discreteness of a particular set of solutions that is responsible for the quantum properties of an atom, but to a larger extent rather Pauli's exclusion principle.

3.7.1.3 Operators and non-commutation

Operators form an essential part of quantum mechanics, which is used to model quantum physics. In contrast, one does not often find operators in classical theories. Yet there does not exist any physical reason why operators cannot also be used to model classical scenarios. Moreover, the Moyal formalism demonstrates that one can model quantum physics successfully without the use of operators. Therefore, the use of operators in a theory that models a particular physical scenario cannot be regarded as an exclusive indication of quantum physics.

The operators used in quantum mechanics often exhibit non-commutation, leading to *canonical commutation relations*. In Section 3.1, it is shown that noncommuting operators can appear simply due to the Fourier relationship. Such Fourier relationships also exist in classical theories, as for example found in *scalar diffraction theory* discussed in Section 2.6. An significant difference is that, in quantum physics the non-commutation appears between operators representing *canonical variables* (those associated with the particle-number degree of freedom), because the Planck and the de Broglie relationships convert these canonical variables into Fourier variables, which does not happen in classical theories. So, it means that non-commutation among arbitrary operators is not an exclusive indication of quantum physics, but non-commutation among operators representing canonical variables does indeed indicate a quantum scenario.

3.7.1.4 Planck constant

Perhaps the most pernicious notion of quantum exclusivity is the one associated with the *Planck constant*. While the valid appearance of the Planck constant in the description of a system can serve as a reliable proxy for the identification of quantum physics, the question of whether it should be present in a given description is not a trivial matter.

Consider, for example, the case of the quantum harmonic oscillator. Clearly, the Planck constant has been removed from both sides of the equation in (3.254) to produce (3.255). What does it mean? It follows from the fact that the potential in the harmonic oscillator does not include any *quantum interactions*. Despite the presence of the quadratic potential, it still represents a free system without interactions. It demonstrates that one cannot simply represent quantum physics by introducing the Planck constant without a valid physical reason, because it can be cancelled from the equations with suitable redefinitions if it does not belong there. One can say that if such a removal of the Planck constant from an equation is possible, it should not have been there in the first place.

3.7.1.5 Quantum interaction

What then constitutes a valid representation of quantum physics? It is when the scenario involves a *quantum interaction*. In such a case, the Planck constant appears in a way that cannot be cancelled through a simple redefinition. Such interactions are responsible for the existence of quantum states that cannot be produced by classical processes.

Due to its significance in the formulation of quantum theory, the Planck constant and its role need to be discussed in more detail. We do so in the next section.

3.7.2 The role of the Planck constant

The *Planck constant* is exclusively associated with quantum phenomena. Therefore, it is often used as a proxy for identifying quantum physics. Unfortunately, such a practice does not preclude the possibility to incorporate the Planck constant into expressions and scenarios where it does not really belong or is not strictly necessary. It is therefore necessary to consider expressions in quantum theory carefully and understand why the Planck constant makes an appearance in them. If one can remove the Planck constant from an expression with the aid of an innocuous transformation or redefinition (one that does not change the physical predictions of the expression), then the Planck constant does not have any physical significance in that expression. In such cases, it is best to remove Planck's constant lest it leads to erroneous conclusions.

The pertinent question is, what is the role of the Planck constant in quantum physics? The dynamics of a quantum system can either be represented by a *Hamiltonian* (often used in quantum mechanics) or a *Lagrangian* (the preference in quantum field theory). The Hamiltonian appears in the exponent of the *unitary evolution operator*, integrated over time and divided by the Planck constant to obtain a dimensionless exponent. In a similar way, the Lagrangian appears in the exponent inside the *path integral*, integrated to form an *action* and also divided by the Planck constant to give a dimensionless exponent. The terms in the Hamiltonian are composed of two or more canonical variables. Those terms with more than two canonical variables represent interactions. In the Lagrangian, the role of these canonical variables are taken over by fields and their derivatives. Each term in the Lagrangian contains two or more such fields and those consisting of more than two fields are interaction terms.

We can attempt to remove the Planck constant from the path integral through a *field redefinition*. Since the terms have at least two fields, we can pull a factor of the square root of the Planck constant from every field. Those terms with only two fields produce one factor of the Planck constant to be cancelled by the Planck constant with which the action is divided making the exponent dimensionless. However, all the interaction terms produce extra factors of the square root of the Planck constant that are not cancelled. An analogous redefinition of canonical variables in the Hamiltonian produces a similar situation. So, we see that the Planck constant is irrevocably associated with interactions.

If the Planck constant is set to zero after such a field redefinition, all the interaction terms would be removed, leaving a non-interacting (free) theory. Such a theory is the same for classical and quantum scenarios. Hence, it is the interactions and not the presence of the Planck constant *per se* that indicates the difference between classical and quantum physics. That is why we specifically associated the Planck constant with the exchange of energy and momentum during *interactions* when we introduced it and formulated the principle of quantum physics as a statement about the properties of *fundamental interactions*.

There is another way to arrive at this conclusion. We saw in the discussion of the quantum harmonic oscillator that the commutations between the Hamiltonian and the ladder operators produce factors of the quantized energy $\hbar\omega_m$. Earlier in Section 3.1.7, we saw that the commutations between the number operator and the ladder operators simply changes the occupation numbers. While we used the correspondence between these results to relate the quantized energy with individual quanta, these two sets of commutation relationships exemplify different contexts in the formulation of quantum physics. In the case where the commutations involve the Hamiltonian, we are specifically dealing with the dynamics of the quantum system. It is in this context where we expect to find the interactions and therefore these commutations produce factors of the Planck constant. On the other hand, those commutations that involve the number operator do not produce factors of the Planck constant. The relevant context in this case is the definition of states, which does not involve interactions and, therefore, does not require the presence of the Planck constant.

When quantum states are presented in terms of some mathematical formalism, each individual particle represents a single quantum excitation, measured in units of action with an amount given by the Planck constant. However, unless the formalism specifically represents these quanta in units of energy or momentum, the Planck constant does not appear in the representation of the state.

Moreover, in discussions about the formalism for photonic states and measurements performed on them, fundamental interactions do not play a significant role, unless nonlinear media, such as those mediating parametric down-conversion, are involved, or if the physical process of the measurement, which involves fundamental interactions, needs to be modelled. Therefore, the Planck constant is not expected to make an appearance in such discussions when nonlinear media or fundamental interactions are not involved.

In our development of the formalism here, we make an effort to point out what happened to the Planck constant; never setting it equal to 1. When it does not show up, it simply is not there, unless it is incorporated into some combination of dimension parameters.

Bibliography

[1] J. J. Sakurai. *Modern Quantum Mechanics*. Addison-Wesley Publishing Company, Reading, Massachusetts, USA, 1994.

[2] D. J. Griffiths. *Introduction to Quantum Mechanics*. Prentice Hall, Englewood Cliffs, 1995.

[3] R. Shankar. *Principles of Quantum Mechanics*. Plenum, New York, USA, 1980.

[4] C. Itzykson and J.-B. Zuber. *Quantum Field Theory*. McGraw-Hill, New York, USA, 1985.

[5] F. Mandl and G. Shaw. *Quantum Field Theory*. John Wiley & Sons, New York, USA, 1984.

[6] M. E. Peskin and D. V. Schroeder. *An Introduction to Quantum Field Theory*. Addison-Wesley Publishing Company, Reading, Massachusetts, USA, 1995.

[7] S. Weinberg. *The Quantum Theory of Fields, Volume I*. Cambridge University Press, New York, 1995.

[8] S. Weinberg. *The Quantum Theory of Fields, Volume II*. Cambridge University Press, New York, 1996.

[9] T. L. Curtright and C. K. Zachos. Quantum mechanics in phase space. *Asia Pac. Phys. Newsl.*, 1:37–46, 2012.

[10] R. Loudon. *Quantum Theory of Light*. Oxford University Press, Oxford, 1973.

[11] L. Mandel and E. Wolf. *Introductory Quantum Optics*. Cambridge University Press, New York, 1995.

[12] D. F. Walls and G. J. Milburn. *Quantum Optics*. Springer, Berlin, 1995.

[13] C. C. Gerry and P. L. Knight. *Optical Coherence and Quantum Optics*. Cambridge University Press, New York, 2005.

[14] P. A. M. Dirac. A new notation for quantum mechanics. *Math. Proc. Camb. Philos. Soc.*, 35:416–418, 1939.

[15] M. A. Nielsen and I. L. Chuang. *Quantum Computation and Quantum Information*. Cambridge University Press, Cambridge, England, 2000.

[16] J. W. Goodman. *Introduction to Fourier Optics, 2nd ed.* McGraw-Hill, New York, USA, 1996.

[17] M. Abramowitz and I. A. Stegun. *Handbook of Mathematical Functions*. Dover, Toronto, 1972.

[18] F. G. Mehler. Ueber die Entwicklung einer Function von beliebig vielen Variablen nach Laplaceschen Functionen höherer Ordnung. *J. Reine Angew. Math.*, 1866:161–176, 1866.

[19] A. Papoulis. *Probability, Random Variables, and Stochastic Processes*. McGraw-Hill, New York, USA, 1984.

[20] H. P. Robertson. The uncertainty principle. *Phys. Rev.*, 34:163–164, 1929.

[21] E. Schrödinger. Zum Heisenbergschen Unschärfeprinzip. *Ber. Kgl. Akad. Wiss. Berlin*, 24:296–303, 1930.

[22] E. Schrödinger. The statistical interpretation of quantum mechanism. *Rev. Mod. Phys.*, 42:358–381, 1970.

[23] H. Goldstein. *Classical Mechanics, 2nd ed.* Addison-Wesley Publishing Company, Reading, Massachusetts, USA, 1980.

[24] E. Wigner. On the quantum correction for thermodynamic equilibrium. *Phys. Rev.*, 40:749–759, 1932.

[25] K. Husimi. Some formal properties of the density matrix. *Nippon Sugaku-Buturigakkwai Kizi Dai 3 Ki*, 22:264–314, 1940.
[26] E. C. G. Sudarshan. Equivalence of semiclassical and quantum mechanical descriptions of statistical light beams. *Phys. Rev. Lett.*, 10:277–279, 1963.
[27] R. J. Glauber. Coherent and incoherent states of the radiation field. *Phys. Rev.*, 131:2766–2788, 1963.

Part II: **Functional phase space formalism**

4 Combining all degrees of freedom

The formulation of quantum theory in term of the particle-number degree of freedom in Chapter 3 ignores the spatiotemporal and spin degrees of freedom that are associated with photons, or at best, it freezes these degrees of freedom in terms of a single fixed mode with a fixed spin. Here, we consider the incorporation of the *spatiotemporal* and *spin* degrees of freedom into the formalism developed for the particle-number degree of freedom in Chapter 3. The infinite-dimensional nature of the spatiotemporal degrees of freedom leads to subtleties in their incorporation that need to be addressed with care.

Our primary goal is a formalism that is powerful enough to model any quantum optical scenario without restrictions in a way that is convenient enough to allow calculations with which an arbitrary physical scenario thus modelled can be analysed. The requirement for a powerful formalism implies that it incorporates all degrees of freedom without constraint. The convenience of the formalism is determined by the choices made in its development.

With only the particle-number degree of freedom, the formalism of Chapter 3 is one-dimensional. By incorporating the other degrees of freedom, we end up with an infinite-dimensional formalism. To some extent, the incorporation of the spatiotemporal and spin degrees of freedom is accomplished with the definition of the ladder operators in Section 4.1. It allows us to model dynamics in a way similar to how it is done in *quantum field theory*, [1–5] which has been used to formulate theories for the fundamental forces. These theories are combined to form the *standard model of particle physics* [1]. They have subsequently demonstrated remarkable agreements with experiment results [9].

Quantum field theory is not so much developed from a state-based approach as a formulation in terms of a Hilbert space, as is the case with quantum mechanics. Instead, it uses the operator algebra presented by field operators, incorporating ladder operators that are generalized to include the other degrees of freedom.[a] Furthermore, its success has largely been obtained from *perturbative calculations*. In that context, the fields are all effectively quantized as *free fields* and the interactions are added perturbatively. Multiple interactions are combined and integrated as superpositions to produce the *complex probability amplitudes* for fundamental processes found in particle physics.

Here, we take a leave from quantum field theory's book in our approach to derive the required formalism. For this reason, we review the *quantization of the electromagnetic field*, as it is done in *quantum electrodynamics* [1, 3, 5].

However, we do not consider scenarios as complex as the fundamental dynamics found in particle physics, and thus do not need the full toolkit provided by quantum field theory. Instead, our aim is to obtain a formalism for the descriptions of states, and

---

a Formally, there is always some space on which the operators are defined, and in the context of quantum physics, this space is a Hilbert space, but in theoretical particle physics this aspect does not play a central role.

https://doi.org/10.1515/9783111445342-006

to model the processes and measurements that are applied to them. It is not the aim to develop a formalism for fundamental dynamics, which is already provided by quantum field theory. We need a formalism that provides a powerful capability to formulate arbitrary exotic states with arbitrary exotic measurements applied to them, for which quantum field theory is not well suited. Our development eventually leads to a *functional phase space formalism*.

The incorporation of the additional degrees of freedom together with the particle-number degree of freedom into an exhaustive representation of arbitrary states presents its own challenges. The ideal is to formulate a complete orthogonal basis that incorporates all degrees of freedom in terms of which arbitrary states can be expressed conveniently. The convenience in the use of continuous variables for the particle-number degree of freedom has led to the distinction between formalisms employing *discrete variables* versus *continuous variables* [6–8].

A characteristic aspect of these bases is how they represent the spatiotemporal degrees of freedom. While the spin degree of freedom is naturally discrete due to its two-dimensional nature, spatiotemporal degrees of freedom are naturally *continuous degrees of freedom* in an infinite-dimensional space. If the spatiotemporal degrees of freedom are represented in terms of discrete modes, we run into the same situation that we have with the discrete variable particle-number degree of freedom. When using discrete modes, we need to represent an arbitrary state as a summation over all such modes. Practical calculations would then typically require the separate evaluation of each term in such a summation, which implies that it needs to be truncated at some point, leading to inaccuracies. Moreover, these infinite summations rarely lead to closed form expressions. It is more desirable to have a formalism in which states are represented in terms of integrals that can be evaluated without truncations and are more prone to closed-form solutions. Our aim here is thus to obtain a formalism in which not only the particle-number degree of freedom, but also the spatiotemporal degrees of freedom, are represented in terms of continuous variables. In this chapter, we consider a few options by investigating the properties of different sets of states in terms of which these degrees of freedom are incorporated.

An obvious candidate is the set of all Fock states, generalized to incorporate spatiotemporal and spin degrees of freedom. A formalism based on Fock states represents states as summations over discrete indices, both for the occupation number and for a discrete modal basis. It leads to truncations in practical calculations and is therefore not considered to be convenient for our purposes.

Another candidate is the set of all coherent states, specified in terms of parameter functions that determine their spatiotemporal and spin properties. Although they do not form an orthogonal basis, coherent states are often used in analyses due to their continuous representation of the particle-number degree of freedom.

We also consider *quadrature bases*. Although they are complete orthogonal bases, representing the particle-number degree of freedom as a continuous variable in the respective bases, they are not normalizable. The elements of these bases are not individually suitable to represent states, but they can be used as bases representing states in terms of linear combinations.

Eventually, having formulated a powerful representation of states, we must also be able to add interactions. For this purpose, the fundamental interactions are assumed to be perturbative. However, in quantum optics, such fundamental interaction can be enhanced thanks to the bosonic nature of light. The effective coupling strength can become strong, making the traditional perturbative approach invalid. We deal with this challenge when we encounter it in Chapter 8.

Our discussions on the inclusion of the other degrees of freedom with the particle-number degree of freedom kicks off with the quantization of the electromagnetic field in Section 4.1. It starts with the discussion of Lorentz covariant quantization of the electromagnetic field in vacuum. In Section 4.1.2, we consider Lorentz covariance and how to use it to formulate the orthogonality conditions for the single-photon bases and the commutation relations for the ladder operators. These definitions are used for the definition of the Lorentz covariant quantization of the electromagnetic field in vacuum. It is readily converted to expressions in terms of optical beam variables. The discussion about the quantization of the electromagnetic field serves as the first step in the process to develop the formalism by incorporating the spatiotemporal degrees of freedom with the particle-number degree of freedom. During this first step, we get the opportunity to address some issues that are relevant to this development.

4.1 Quantization of the electromagnetic field

One way to model a physical scenario is to start with a classical description of the scenario and then *quantize* it. Often the term "quantization" is used as being synonymous with "replacing quantities by operators." However, the concept of *quantization* should be understood as a process whereby classical fields are replaced by fields consisting of quanta. Due to the physical interpretation of the *commutation relations for ladder operators* (see Section 3.6.6), they are ideal for the quantization of *classical field theories*.

There are various formal quantization procedures. Often such procedures lead to some mathematical difficulties. Such issues are not our concern here. The canonical quantization approach used in quantum field theory [1] suffices to provide us with reliable models of the physical scenarios that we wish to investigate. In this approach, the theory is quantized as a *free field theory*. By implication, the interaction terms are not present during the quantization process. They are added as a *modelling* of the dynamics, based on physical considerations of the scenario under investigation. The success of this modelling process is determined in a scientific way by comparing its predictions with experimental results. The field operators in terms of which these additional interaction terms are modelled are exactly those obtained from the quantization of the *free* theory. It thus leads to the *interaction picture* [1]. In this way, the quantization mechanism associated with these field operators, namely the creation or annihilation of quanta, is exactly the same regardless of whether the theory has interactions or not. A quantum of a specific field is physically the same thing regardless of the presence and nature of interactions. In the context of quantum optics, we prefer to formulate the theory and its interactions in terms of ladder operators instead of field operators.

Our starting point in this chapter is to specify an algebra of operators based on a generalization of the ladder operators derived in Chapter 3. Here, we generalize these ladder operators to incorporate the spin and spatiotemporal degrees of freedom. They can then be used in the quantization of the electromagnetic field. We start this process by considering the ladder operators as the assumed basis on which a quantum formalism for optical fields is developed.

Since the ladder operators derived in Chapter 3 only represent the particle-number degree of freedom, they are (or can be made to be) dimensionless. One way to incorporate the spatiotemporal degrees of freedom is to define them in terms of a set of complete orthonormal modal functions. Although such a discrete set of modal functions can in principle be defined for an infinite space, it is often done by considering the field in a finite volume over a finite period of time under the assumption that one can take a *continuum limit* where the finite volume grows to an infinite size and the finite period grows to an infinite duration, tacitly assuming that such a limit is well-defined. Introducing the spatiotemporal degrees of freedom in terms of such a set of discrete modes, we obtain a *commutation relation for ladder operators* given as $[\hat{a}_m, \hat{a}_n^\dagger] = \delta_{m,n}$, where the modes are indexed by m or n and $\delta_{m,n}$ is the Kronecker delta, defined in (2.6). The spin degree of freedom can also be added, leading to additional spin indices. Thus we end up with a countable infinite set of distinct ladder operators. Such ladder operators would still be dimensionless.

Any formalism derived from a set of distinct ladder operators leads to expressions in terms of summations with infinitely many terms. As explained before, we prefer to avoid such a situation because infinite summations rarely produce closed-form solutions and practical calculations tend to require truncations, requiring approximations that are often uncontrolled. (An approximation is called *controlled* when the validity of the approximation can be quantified in terms of experimental parameters associated with specific experimental conditions.) Hence, a formalism based on a discrete set of functions is generally not convenient for practical calculations.

Instead, it is preferred that the spatiotemporal degrees of freedom are incorporated as *continuous degrees of freedom*, leading to integral expressions. Such integral expressions can to a large extent be evaluated without truncations, often with closed-form expressions as results.[b] In such cases, the spatiotemporal basis functions are distinguished by continuously varying parameters. A typical example of such functions is the *plane*

b Of course, there are various issues associated with the use of such integral expressions, especially in the context of quantum field theory. Fortunately, in quantum optics, many of those issues are avoided. The rest are dealt with as they are encountered.

waves in an infinite three-dimensional space, as considered in Section 2.4.1. These plane waves are distinguished by their *wavevectors*, as well as the spin degree of freedom (unless we are considering scalar optical fields) represented by the *optical state of polarization*. The wavevectors represent continuous degrees of freedom. Although mutually orthogonal, these plane waves are not normalizable. Moreover, the ladder operators are not dimensionless in this case.

4.1.1 Ladder operators

For the quantum description of light, we need a set of *ladder operators*, including a creation operator $\hat{a}^{\dagger}$ and an annihilation operator $\hat{a}$ that, respectively, creates or removes individual photons. The ladder operators that we obtained in Section 3.1.6 ($\hat{a}^{\dagger}$ and $\hat{a}$) are generalized to include spatiotemporal and spin degrees of freedom, so that they become $\hat{a}_s^{\dagger}(\mathbf{k})$ and $\hat{a}_s(\mathbf{k})$, where $\mathbf{k}$ is the three-dimensional wavevector and the subscript s represents a spin index. For a given wavevector and spin, the ladder operators are associated with a plane wave that satisfies the Helmholtz equation of (2.79), serving as the free-field equation of motion. Therefore, the angular frequency ω is fixed via the dispersion relation in (2.96). In this way, these operators become operator-valued functions of the wavevector for each of the two spins (orthogonal states of polarization).

The (non-vanishing) *commutation relation* for these operators, which generalizes (3.47), has the form

$$\left[\hat{a}_r(\mathbf{k}_1), \hat{a}_s^{\dagger}(\mathbf{k}_2)\right] = C(\mathbf{k}_1)\, \delta_{r,s}\delta(\mathbf{k}_1 - \mathbf{k}_2), \tag{4.1}$$

where $\delta(\mathbf{k}_1 - \mathbf{k}_2)$ represents a product of three Dirac delta functions, and $C(\mathbf{k}_1)$ is a function of the wavevector that we include for reasons that will become clear shortly. The right-hand side is also proportional to $\mathbb{1}$. However, we do not show this identity operator explicitly together with the spatiotemporal identity kernel, unless necessary within the context of the discussion. Due to its argument, $\delta(\cdot)$ has the units of [distance^3]. The dimensions of the ladder operators also depend on the dimensions of $C(\mathbf{k}_1)$.

Based on (4.1), these operators commute whenever $\mathbf{k}_1 \neq \mathbf{k}_2$ or $r \neq s$. For $\mathbf{k}_1 = \mathbf{k}_2$ and $r = s$, on the other hand, the result becomes divergent. This situation is not problematic because, in any representation of a physical state or a physical process, these operators are always found inside integrals over the wavevectors, so that the resulting Dirac delta functions produce well-defined results.

If we want to obtain a formulation that is valid for optical fields, which are relativistic fields, then the commutation relation in (4.1) should look the same in all reference frames. Therefore, the way in which $C(\mathbf{k}_1)$ depends on $\mathbf{k}_1$ is determined by the requirement for *Lorentz covariance*. We briefly digress to show that the Dirac delta function in (4.1) can only be Lorentz covariant if it is multiplied by the angular frequency ω.

4.1.2 Lorentz covariance and Dirac delta functions

4.1.2.1 Fourier domain Dirac delta function

How can a Dirac delta function for the wavevector be made Lorentz covariant? Consider a boost in the z-direction, given by

$$k_z' = \gamma\left(k_z + \frac{1}{c}\beta\omega\right), \tag{4.2}$$

$$\omega' = \gamma\left(\omega + c\beta k_z\right), \tag{4.3}$$

where $\beta = v/c$ and $\gamma = (1-\beta^2)^{-1/2}$, with v being the velocity and c being the speed of light. The transformation of a Dirac delta function under such a boost is

$$\delta\left(k_z - k_{z1}\right) = \delta\left(k_z' - k_{z1}'\right)\frac{\mathrm{d}k_z'}{\mathrm{d}k_z}. \tag{4.4}$$

The derivative in (4.4) becomes

$$\frac{\mathrm{d}k_z'}{\mathrm{d}k_z} = \gamma\left(1 + \frac{\beta}{c}\frac{\mathrm{d}\omega}{\mathrm{d}k_z}\right) = \gamma\left(1 + \frac{c\beta k_z}{\omega}\right) = \frac{\omega'}{\omega}, \tag{4.5}$$

where we first used (2.96) and then (4.3). The results in (4.4) and (4.5) lead to the covariant expression in terms of the Dirac delta function [1] that retains its form after a Lorentz transformation provided that it is multiplied by the *angular frequency*

$$\omega\delta\left(k_z - k_{z1}\right) \to \omega'\delta\left(k_z' - k_{z1}'\right). \tag{4.6}$$

4.1.2.2 Position space Dirac delta functions

It is instructive to see what happens when a Lorentz boost is applied to a Dirac delta function in *configuration space*. A boost in the z-direction gives

$$t' = \gamma\left(t + \frac{1}{c}\beta z\right) \quad \text{and} \quad z' = \gamma\left(z + c\beta t\right). \tag{4.7}$$

In this case, there are Dirac delta functions for both z and t. The transformation of the Dirac delta functions under such a boost gives

$$\delta\left(t - t_1\right)\delta\left(z - z_1\right) \to \delta\left(t' - t_1'\right)\frac{\mathrm{d}t'}{\mathrm{d}t}\delta\left(z' - z_1'\right)\frac{\mathrm{d}z'}{\mathrm{d}z}. \tag{4.8}$$

These derivatives produce

$$\frac{\mathrm{d}t'}{\mathrm{d}t} = \gamma\left(1 + \frac{\beta}{c}\frac{\mathrm{d}z}{\mathrm{d}t}\right) \quad \text{and} \quad \frac{\mathrm{d}z'}{\mathrm{d}z} = \gamma\left(1 + c\beta\frac{\mathrm{d}t}{\mathrm{d}z}\right). \tag{4.9}$$

However, the coordinates are all independent of each other. Therefore,

$$\frac{\mathrm{d}z}{\mathrm{d}t} = \frac{\mathrm{d}t}{\mathrm{d}z} = 0, \tag{4.10}$$

which leads to

$$\frac{\mathrm{d}t'}{\mathrm{d}t} = \frac{\mathrm{d}z'}{\mathrm{d}z} = \gamma. \tag{4.11}$$

Substituting (4.11) into (4.8), we obtain

$$\delta(t - t_1)\,\delta(z - z_1) \to \gamma^2 \delta(t' - t_1')\,\delta(z' - z_1'). \tag{4.12}$$

The resulting scaling factor suggests that we can compensate for it with the aid of an infinitesimal four-volume $V = \mathrm{d}t\,\mathrm{d}x\,\mathrm{d}y\,\mathrm{d}z$ because such an infinitesimal four-volume transforms as $V \to V/\gamma^2$. The infinitesimal volume element is the same as the integration measure, reminding us that Dirac delta functions are only well-defined under integrals. By including the integration measure, we obtain an expression that is Lorentz covariant.

4.1.2.3 Invariant integration measure

In view of the above discussion, would not the same then be true for the Fourier domain case considered previously? Indeed, from (4.5) we get

$$\frac{\mathrm{d}k_z'}{\omega'} = \frac{\mathrm{d}k_z}{\omega}. \tag{4.13}$$

So, if we multiply the Dirac delta function by $\mathrm{d}k_z$, the result is invariant. However, since there is no integration over ω, we can place ω explicitly in the expressions to obtain invariant expressions even without the integration measure. The benefit of this approach is that we can use (4.13) to define integrals as Lorentz invariant expressions even when they do not contain Dirac delta functions.

4.1.3 Lorentz covariant quantization

Hence, a *Lorentz covariant commutation relation* must have the form

$$\left[\hat{a}_r(\mathbf{k}_1), \hat{a}_s^\dagger(\mathbf{k}_2)\right] = C_0\,\omega_1 \delta_{r,s} \delta(\mathbf{k}_1 - \mathbf{k}_2), \tag{4.14}$$

where C_0 is a constant that is independent of the wavevectors. The ladder operators can only be dimensionless if C_0 contains the necessary dimension parameters to remove those from the rest of the expression. Such dimension parameters could be conversion parameters like c, or they could be scale parameters. However, apart from the speed of light, which would not render the expression dimensionless, there are no fundamental dimension parameters that could be used in this situation. Nor do we want this definition to be scale dependent. It implies that the constant C_0 needs to be dimensionless (apart from possible factors of c). Hence, a Lorentz covariant definition of the ladder operators implies that they carry dimensions, which in this case (without any factors of c) are given by [distance$^{3/2}$ time$^{-1/2}$]. Based on these considerations and the derivation in Section 4.1.2, the ladder operators defined in Section 4.1.1 obey *Lorentz covariant commutation relations* given by

$$\begin{aligned}
[\hat{a}_r(\mathbf{k}_1), \hat{a}_s(\mathbf{k}_2)] &= \left[\hat{a}_r^{\dagger}(\mathbf{k}_1), \hat{a}_s^{\dagger}(\mathbf{k}_2)\right] = 0, \\
\left[\hat{a}_r(\mathbf{k}_1), \hat{a}_s^{\dagger}(\mathbf{k}_2)\right] &= (2\pi)^3 \omega_1 \delta_{r,s} \delta(\mathbf{k}_1 - \mathbf{k}_2).
\end{aligned} \tag{4.15}$$

When the quantized electromagnetic field is considered in the context of physical experiments, we often encounter the notion of a *paraxial optical beam*, as discussed in Section 2.3.5. It leads to the requirement to formulate the quantized electromagnetic field in terms of *optical beam variables*, presented in Section 2.4.2. Such a formulation breaks Lorentz invariance, because it requires a specific choice for the direction of the beam axis. In Chapter 8, we encounter a situation where the electromagnetic field propagates through a dielectric medium. Such a medium also breaks Lorentz invariance. In the end, although the fundamental theory is Lorentz covariant, it may not always be convenient to represent the theory in its Lorentz covariant form. Therefore, different representations for the quantized electromagnetic field are provided, including expressions in terms of optical beam variables that are formally consistent with the Lorentz covariant quantization of the electromagnetic field.

4.1.3.1 Quantization in the Coulomb gauge

To quantize a real-valued field, we replace it by a *field operator* consisting of a linear combination of the ladder operators, respectively multiplied by their associated plane wave solutions of the equations of motion and the complex conjugates of these solutions. As such, the superposition of the two terms produces a Hermitian operator. It is multiplied by an overall constant consisting of dimension parameters, such as Planck's constant, and which depends on the commutation relation of the ladder operators. In effect, the quantization process replaces the angular spectra in the classical expansion by the ladder operators.

Expressions for the quantized electromagnetic field in a fully Lorentz covariant manner are obtained from a generic expression for the quantized *gauge field* $\hat{A}^{\mu}$, which is a four-vector (consisting of the *magnetic vector potential* and the *scalar potential*), expressed in tensor notation, where $\mu = 0, 1, 2, 3$ represents the spatiotemporal components associated with $\{ct, x, y, z\}$, respectively. The quantized gauge field in the Heisenberg picture of the free theory has the form

$$\begin{aligned}
\hat{A}^{\mu}(\mathbf{x}, t) = \int \mathcal{N}_A(\omega_{\mathbf{k}}) \Big[&\eta_s^{\mu}(\vec{k}) \hat{a}_s(\mathbf{k}) \exp(-\mathrm{i}\omega_{\mathbf{k}} t + \mathrm{i}\mathbf{k} \cdot \mathbf{x}) \\
&+ \eta_s^{\mu\dagger}(\vec{k}) \hat{a}_s^{\dagger}(\mathbf{k}) \exp(\mathrm{i}\omega_{\mathbf{k}} t - \mathrm{i}\mathbf{k} \cdot \mathbf{x}) \Big] \frac{\mathrm{d}^3 k}{(2\pi)^3},
\end{aligned} \tag{4.16}$$

where the prefactor $\mathcal{N}_A$, consisting of dimension parameters, is constant apart from its possible dependence on the *angular frequency*, defined via the dispersion relation $\omega_{\mathbf{k}} \triangleq c|\mathbf{k}|$; and $\eta_s^{\mu}(\vec{k})$ is the four-vector for the spin, which depends on $\vec{k}$, the direction of the wavevector $\mathbf{k}$. The summation over the spin indices s is imposed by the summation convention, stipulating that all repeated spin indices are summed over.

Next, the *Coulomb gauge condition* is employed, which sets $\hat{A}^0 = 0$ and $\nabla\cdot\hat{\mathbf{A}} = 0$. The quantized gauge field then becomes the *quantized magnetic vector potential*

$$\hat{\mathbf{A}}(\mathbf{x},t) = \int \mathcal{N}_A(\omega_{\mathbf{k}}) \left[\vec{\eta}_s(\vec{k})\hat{a}_s(\mathbf{k}) \exp(-\mathrm{i}\omega_{\mathbf{k}}t + \mathrm{i}\mathbf{k}\cdot\mathbf{x}) + \vec{\eta}_s^{\,*}(\vec{k})\hat{a}_s^{\dagger}(\mathbf{k}) \exp(\mathrm{i}\omega_{\mathbf{k}}t - \mathrm{i}\mathbf{k}\cdot\mathbf{x})\right] \frac{\mathrm{d}^3k}{(2\pi)^3}. \tag{4.17}$$

The gauge condition enforces $\mathbf{k}\cdot\vec{\eta}_s = 0$, as found with classical electromagnetic fields. As a result, there are only two polarization basis vectors $\vec{\eta}_s$ with $s = 1,2$, having the properties given in (2.203) and (2.205).

The form of the expressions for the quantized electric and magnetic fields follow from that of the gauge field (or the magnetic vector potential) in the Coulomb gauge. They are given by

$$\begin{aligned}
\hat{\mathbf{E}} &= -\partial_t\hat{\mathbf{A}} = \mathrm{i}\int \mathcal{N}_A(\omega_{\mathbf{k}})\omega_{\mathbf{k}} \left[\vec{\eta}_s\hat{a}_s(\mathbf{k}) \exp(-\mathrm{i}\omega_{\mathbf{k}}t + \mathrm{i}\mathbf{k}\cdot\mathbf{x}) - \vec{\eta}_s^{\,*}\hat{a}_s^{\dagger}(\mathbf{k}) \exp(\mathrm{i}\omega_{\mathbf{k}}t - \mathrm{i}\mathbf{k}\cdot\mathbf{x})\right] \frac{\mathrm{d}^3k}{(2\pi)^3}, \\
\hat{\mathbf{H}} &= \frac{1}{\mu_0}\nabla\times\hat{\mathbf{A}} = \mathrm{i}\frac{1}{\mu_0}\int \mathcal{N}_A(\omega_{\mathbf{k}}) \left[(\mathbf{k}\times\vec{\eta}_s)\hat{a}_s(\mathbf{k}) \exp(-\mathrm{i}\omega_{\mathbf{k}}t + \mathrm{i}\mathbf{k}\cdot\mathbf{x}) - (\mathbf{k}\times\vec{\eta}_s^{\,*})\hat{a}_s^{\dagger}(\mathbf{k}) \exp(\mathrm{i}\omega_{\mathbf{k}}t - \mathrm{i}\mathbf{k}\cdot\mathbf{x})\right] \frac{\mathrm{d}^3k}{(2\pi)^3}.
\end{aligned} \tag{4.18}$$

These expressions are independent of the gauge fixing condition, since they are always *gauge invariant*. It is only the relationships between the gauge field and the electric and magnetic fields that depend on the gauge fixing condition.

4.1.3.2 Determining the constant

To determine $\mathcal{N}_A$, we first obtain an expression for the quantized Hamiltonian, equivalent to (3.262). Substituting the expressions for the quantized electric and magnetic fields in (4.18) into the expression for the classical *Hamiltonian* of the free electromagnetic field in vacuum, which is given by

$$H = \frac{1}{2}\int \left(\epsilon_0|\mathbf{E}|^2 + \mu_0|\mathbf{H}|^2\right) \mathrm{d}^3x, \tag{4.19}$$

we obtain

$$\hat{H} = \frac{1}{\mu_0} \int \mathcal{N}_A^2(\omega_{\mathbf{k}}) k^2 \left[\hat{a}_s(\mathbf{k}) \hat{a}_s^\dagger(\mathbf{k}) + \hat{a}_s^\dagger(\mathbf{k}) \hat{a}_s(\mathbf{k}) \right] \frac{d^3k}{(2\pi)^3}. \tag{4.20}$$

Exercise 4.1. Derive (4.20) from (4.18) and (4.19).

We now use the relationships that we obtained in (3.264) for the *particle-number degree of freedom* only, and generalize them to incorporate the *spatiotemporal degrees of freedom*. The resulting relationships are

$$[\hat{a}_s(\mathbf{k}), \hat{H}] = \hbar\omega_{\mathbf{k}} \hat{a}_s(\mathbf{k}) \quad \text{and} \quad [\hat{H}, \hat{a}_r^\dagger(\mathbf{k})] = \hbar\omega_{\mathbf{k}} \hat{a}_r^\dagger(\mathbf{k}), \tag{4.21}$$

providing us with the defining property of the ladder operators: *they respectively create and annihilate quanta of energy.* These relationships are significant in that they represent the quantization of the energy associated with individual quanta, as a generalization of the situation discussed in Section 3.6.6, with the spin and spatiotemporal degrees of freedom included here. We then use these relationships to constrain the value of $\mathcal{N}_A$ so that the ladder operators perform their role in the quantization process correctly.

The Lorentz covariant commutation relations of the ladder operators are provided in (4.15). However, to be general, accommodating situations where the Lorentz covariant formulation is not convenient, it is assumed here that the non-vanishing *generic commutation relation* can have any (potentially angular frequency dependent) factor C. So, it becomes

$$\left[\hat{a}_s(\mathbf{k}_1), \hat{a}_r^\dagger(\mathbf{k}_2) \right] = (2\pi)^3 C(\omega_1) \delta_{s,r} \delta(\mathbf{k}_1 - \mathbf{k}_2). \tag{4.22}$$

Substituting the Hamiltonian in (4.20) into the second commutator in (4.21) and using (4.22), we obtain

$$\begin{aligned}
[\hat{H}, \hat{a}_r^\dagger(\mathbf{k}')] &= \frac{1}{\mu_0} \int \mathcal{N}_A^2(\omega_{\mathbf{k}}) k^2 \left\{ [\hat{a}_s(\mathbf{k}), \hat{a}_r^\dagger(\mathbf{k}')] \hat{a}_s^\dagger(\mathbf{k}) + \hat{a}_s^\dagger(\mathbf{k}) [\hat{a}_s(\mathbf{k}), \hat{a}_r^\dagger(\mathbf{k}')] \right\} \frac{d^3k}{(2\pi)^3} \\
&= \frac{2}{\mu_0} k'^2 \mathcal{N}_A^2(\omega'_{\mathbf{k}}) C(\omega'_{\mathbf{k}}) \hat{a}_r^\dagger(\mathbf{k}') = \hbar\omega'_{\mathbf{k}} \hat{a}_r^\dagger(\mathbf{k}').
\end{aligned} \tag{4.23}$$

The result provides a relationship between the two unknown quantities

$$\mathcal{N}_A(\omega_{\mathbf{k}}) = \sqrt{\frac{\hbar}{2\epsilon_0 \omega_{\mathbf{k}} C(\omega_{\mathbf{k}})}}. \tag{4.24}$$

The prefactor in the quantum Hamiltonian is therefore determined by the factor C in the commutation relation. The *quantum Hamiltonian* then reads

$$\hat{H} = \frac{1}{2} \int \omega_{\mathbf{k}} \hbar \left[\hat{a}_s(\mathbf{k}) \hat{a}_s^\dagger(\mathbf{k}) + \hat{a}_s^\dagger(\mathbf{k}) \hat{a}_s(\mathbf{k}) \right] \frac{d^3k}{(2\pi)^3 C(\omega_{\mathbf{k}})}. \tag{4.25}$$

It shows how the choice of C affects the expression of the quantum Hamiltonian. It also affects the expression of the magnetic vector potential

$$\begin{aligned}\hat{\mathbf{A}}(\mathbf{x},t) = \int \sqrt{\frac{\omega_{\mathbf{k}}\hbar}{2\epsilon_0 C(\omega_{\mathbf{k}})}}\, & [\vec{\eta}_s\hat{a}_s(\mathbf{k})\exp(-\mathrm{i}\omega_{\mathbf{k}}t + \mathrm{i}\mathbf{k}\cdot\mathbf{x}) \\ & + \vec{\eta}_s^*\hat{a}_s^\dagger(\mathbf{k})\exp(\mathrm{i}\omega_{\mathbf{k}}t - \mathrm{i}\mathbf{k}\cdot\mathbf{x})\big]\,\frac{\mathrm{d}^3k}{(2\pi)^3\omega_{\mathbf{k}}}.\end{aligned} \tag{4.26}$$

4.1.3.3 Lorentz covariant quantized fields

For the Lorentz covariant case, we set $C = \omega_{\mathbf{k}}$ so that (4.22) matches (4.15), leading to

$$\mathcal{N}_A(\omega_{\mathbf{k}}) = \frac{1}{\omega_{\mathbf{k}}}\sqrt{\frac{\hbar}{2\epsilon_0}}. \tag{4.27}$$

Substituting it into the expression for the *Hamiltonian*, we obtain

$$\hat{H} = \frac{\hbar}{2}\int \hat{a}_s(\mathbf{k})\hat{a}_s^\dagger(\mathbf{k}) + \hat{a}_s^\dagger(\mathbf{k})\hat{a}_s(\mathbf{k})\,\frac{\mathrm{d}^3k}{(2\pi)^3}. \tag{4.28}$$

Although it does not explicitly contain the angular frequency, the commutation relation of the ladder operators would produce it where necessary. The expressions for the magnetic vector potential, the electric field and the magnetic field, based on (4.27), are

$$\begin{aligned}\hat{\mathbf{A}}(\mathbf{x},t) = \int \sqrt{\frac{\hbar}{2\epsilon_0}}\, & [\vec{\eta}_s\hat{a}_s(\mathbf{k})\exp(-\mathrm{i}\omega_{\mathbf{k}}t + \mathrm{i}\mathbf{k}\cdot\mathbf{x}) \\ & + \vec{\eta}_s^*\hat{a}_s^\dagger(\mathbf{k})\exp(\mathrm{i}\omega_{\mathbf{k}}t - \mathrm{i}\mathbf{k}\cdot\mathbf{x})\big]\,\frac{\mathrm{d}^3k}{(2\pi)^3\omega_{\mathbf{k}}},\end{aligned} \tag{4.29}$$

$$\begin{aligned}\hat{\mathbf{E}}(\mathbf{x},t) = \mathrm{i}\int \sqrt{\frac{\hbar}{2\epsilon_0}}\, & [\vec{\eta}_s\hat{a}_s(\mathbf{k})\exp(-\mathrm{i}\omega_{\mathbf{k}}t + \mathrm{i}\mathbf{k}\cdot\mathbf{x}) \\ & - \vec{\eta}_s^*\hat{a}_s^\dagger(\mathbf{k})\exp(\mathrm{i}\omega_{\mathbf{k}}t - \mathrm{i}\mathbf{k}\cdot\mathbf{x})\big]\,\frac{\mathrm{d}^3k}{(2\pi)^3},\end{aligned} \tag{4.30}$$

$$\begin{aligned}\hat{\mathbf{H}}(\mathbf{x},t) = \mathrm{i}\int \sqrt{\frac{\hbar}{2\mu_0}}\, & \big[(\vec{k}\times\vec{\eta}_s)\hat{a}_s(\mathbf{k})\exp(-\mathrm{i}\omega_{\mathbf{k}}t + \mathrm{i}\mathbf{k}\cdot\mathbf{x}) \\ & - (\vec{k}\times\vec{\eta}_s^*)\hat{a}_s^\dagger(\mathbf{k})\exp(\mathrm{i}\omega_{\mathbf{k}}t - \mathrm{i}\mathbf{k}\cdot\mathbf{x})\big]\,\frac{\mathrm{d}^3k}{(2\pi)^3}.\end{aligned} \tag{4.31}$$

These expressions represent the Lorentz covariant fields in vacuum, without any approximations, apart from the fact that the gauge field is expressed in the Coulomb gauge, leading to the magnetic vector potential. However, it should be pointed out that the expressions in (4.30) and (4.31) are not individually Lorentz covariant. A Lorentz transformation causes the transverse components of the electric and magnetic fields to mix. For the *non-Lorentz covariant commutation relations*, where $C = 1$, we can simply replace $\hbar \to \hbar\omega_{\mathbf{k}}$ in these expressions.

In the context of quantum field theory, the gauge field (magnetic vector potential) is used to construct local *observables*. For this purpose, the gauge field is separated into two terms $\hat{\mathbf{A}} = \hat{\mathbf{A}}^{(+)} + \hat{\mathbf{A}}^{(-)}$, associated with the annihilation and creation operators, respectively. They obey commutation relations similar to those of the ladder operators.

Exercise 4.2. Use the commutation relations for the ladder operators in (4.15) to show that the *equal-time* commutation relation for the quantized magnetic vector potential is

$$\left[\hat{\mathbf{A}}^{(+)}(\mathbf{x}_1,t),\hat{\mathbf{A}}^{(-)}(\mathbf{x}_2,t)\right] = \frac{\hbar}{2\epsilon_0}\delta(\mathbf{x}_1-\mathbf{x}_2). \tag{4.32}$$

When we perform the same exercise with the electric field, we obtain

$$\left[\hat{\mathbf{E}}^{(+)}(\mathbf{x}_1,t),\hat{\mathbf{E}}^{(-)}(\mathbf{x}_2,t)\right] = \frac{\hbar}{2\epsilon_0}\int \omega_{\mathbf{k}}\exp[i\mathbf{k}\cdot(\mathbf{x}_1-\mathbf{x}_2)]\,\frac{d^3k}{(2\pi)^3}, \tag{4.33}$$

which does not produce a Dirac delta function. As with the quantized magnetic vector potential, the commutation becomes divergent for $\mathbf{x}_1 = \mathbf{x}_2$. However, it does not evaluate to zero for $\mathbf{x}_1 \neq \mathbf{x}_2$. As a result, one should not interpret the field operators for the electromagnetic field as being *local* operators in spacetime. Nor should one base any conclusions about the properties of states on the properties of these field operators without careful consideration. In our development of the formalism, we avoid such issues as far as possible by using the ladder operators, instead of the field operators.

4.1.4 Vectorial quadrature operators

The ladder operators in the expressions for the quantized electromagnetic fields in (4.29) to (4.31) can be expressed in terms of *vectorial quadrature operators*. For this purpose, we define them by combining the spin vectors with the ladder operators (and summing over the spin):

$$\begin{aligned}
\hat{\mathbf{q}}(\mathbf{k}) &= \frac{1}{\sqrt{2}}\left[\vec{\eta}_s\hat{a}_s(\mathbf{k}) + \vec{\eta}_s^*\hat{a}_s^{\dagger}(\mathbf{k})\right],\\
\hat{\mathbf{p}}(\mathbf{k}) &= \frac{-i}{\sqrt{2}}\left[\vec{\eta}_s\hat{a}_s(\mathbf{k}) - \vec{\eta}_s^*\hat{a}_s^{\dagger}(\mathbf{k})\right].
\end{aligned} \tag{4.34}$$

Hence,

$$\begin{aligned}
\vec{\eta}_s\hat{a}_s(\mathbf{k}) &= \frac{1}{\sqrt{2}}\left[\hat{\mathbf{q}}(\mathbf{k}) + i\hat{\mathbf{p}}(\mathbf{k})\right],\\
\vec{\eta}_s^*\hat{a}_s^{\dagger}(\mathbf{k}) &= \frac{1}{\sqrt{2}}\left[\hat{\mathbf{q}}(\mathbf{k}) - i\hat{\mathbf{p}}(\mathbf{k})\right].
\end{aligned} \tag{4.35}$$

The resulting expression for the quantized magnetic vector potential is

$$\hat{\mathbf{A}}(\mathbf{x},t) = \sqrt{\frac{\hbar}{\epsilon_0}} \int [\hat{\mathbf{q}}(\mathbf{k}) \cos(\omega_{\mathbf{k}} t - \mathbf{k}\cdot\mathbf{x}) + \hat{\mathbf{p}}(\mathbf{k}) \sin(\omega_{\mathbf{k}} t - \mathbf{k}\cdot\mathbf{x})] \frac{\mathrm{d}^3 k}{(2\pi)^3 \omega}. \tag{4.36}$$

The equivalent expressions for the electric and the magnetic fields are

$$\hat{\mathbf{E}}(\mathbf{x},t) = \sqrt{\frac{\hbar}{\epsilon_0}} \int [\hat{\mathbf{q}}(\mathbf{k}) \sin(\omega_{\mathbf{k}} t - \mathbf{k}\cdot\mathbf{x}) - \hat{\mathbf{p}}(\mathbf{k}) \cos(\omega_{\mathbf{k}} t - \mathbf{k}\cdot\mathbf{x})] \frac{\mathrm{d}^3 k}{(2\pi)^3},$$
$$\hat{\mathbf{H}}(\mathbf{x},t) = \sqrt{\frac{\hbar}{\mu_0}} \int \left[\vec{k} \times \hat{\mathbf{q}}(\mathbf{k}) \sin(\omega_{\mathbf{k}} t - \mathbf{k}\cdot\mathbf{x}) - \vec{k} \times \hat{\mathbf{p}}(\mathbf{k}) \cos(\omega_{\mathbf{k}} t - \mathbf{k}\cdot\mathbf{x})\right] \frac{\mathrm{d}^3 k}{(2\pi)^3}. \tag{4.37}$$

Here, we find that the coefficient functions are real valued, because the quadrature operators are Hermitian.

4.1.5 Quantization in optical beam variables

The development of a formalism for the quantization of an optical field in which the *paraxial approximation* can be imposed consistently requires a few conditions. One of these conditions is the *monochromatic assumption*, which is discussed in Section 2.3. It requires that the formalism imposes restrictions on the temporal spectrum.

The wavevector integrals found in Section 4.1.3 run over the three wavevector components k_x, k_y, and k_z. Specifying these three components, we obtain the angular frequency $\omega_{\mathbf{k}}$ from the dispersion relation (2.96). It corresponds to the fixed-time orthogonality condition, discussed in Sections 2.4.3 to 2.4.5.

In Section 2.4.2, we introduced the *optical beam variables* as the integration variables k_x, k_y, and ω, fixing k_z via the dispersion relation. The spectra are thus functions of the two-dimensional transverse wavevector $\mathbf{K}$ and the angular frequency ω, as opposed to the three-dimensional wavevector $\mathbf{k}$. The definitions of quantized fields can be converted into equivalent definitions in terms of these optical beam variables by changing the integration variables in the Fourier domain integrals from an integration over k_z to an integration over ω, provided that we satisfy the *optical beaming condition* that restricts plane waves to have $k_z > 0$. The purpose of the optical beaming condition is to allow the implementations of the monochromatic approximation and paraxial approximation in a consistent manner. Using the vacuum dispersion relation (2.96), we convert one of the integration variables (in vacuum) as follows:

$$\mathrm{d}k_z = \frac{\partial k_z}{\partial \omega} \mathrm{d}\omega = \frac{\omega \mathrm{d}\omega}{c^2 k_z}. \tag{4.38}$$

Therefore, the measure for the Fourier domain integrals is converted into

$$\frac{d^3k}{(2\pi)^3} \to \frac{\omega}{c^2k_z}\frac{d^2k d\omega}{(2\pi)^3}. \tag{4.39}$$

Using the properties of Dirac delta functions, together with (4.38), we have

$$\delta(k_{z1} - k_{z2}) = \delta(\omega_1 - \omega_2)\left[\frac{dk_z(\omega)}{d\omega}\right]^{-1} = \delta(\omega_1 - \omega_2)\frac{c^2k_{z1}}{\omega_1}. \tag{4.40}$$

It helps with converting the orthogonality conditions and commutation relations to optical beam variables.

Additional factors of the speed of light in vacuum c appear together with a factor of k_z in both (4.39) and (4.40). These factors of c will appear in all the orthogonality conditions, commutation relations, and Fourier domain integrals, unless we absorb them into the definitions of the basis elements and the ladder operators. For this reason, we relate the ladder operators expressed in terms of the optical beam variables to the original ladder operators by a factor of c,

$$\hat{a}_s^\dagger(\mathbf{k}) = c\hat{a}_s^\dagger(\mathbf{K},\omega) \quad \text{and} \quad \hat{a}_s(\mathbf{k}) = c\hat{a}_s(\mathbf{K},\omega). \tag{4.41}$$

Using (4.40) and (4.41), we convert the generic commutation relation for the ladder operators in (4.22) into *one in terms of optical beam variables*. It reads

$$\left[\hat{a}_r(\mathbf{K},\omega), \hat{a}_s^\dagger(\mathbf{K}',\omega')\right] = (2\pi)^3 C'\delta_{s,r}\delta(\mathbf{K}-\mathbf{K}')\delta(\omega-\omega'), \tag{4.42}$$

where

$$C' = \frac{k_z}{\omega}C(\omega). \tag{4.43}$$

The expression for the Hamiltonian in terms of optical beam variables is obtained from (4.25) with the appropriate substitutions, becoming

$$\hat{H} = \frac{1}{2}\int \omega\hbar\left[\hat{a}_s(\mathbf{K},\omega)\hat{a}_s^\dagger(\mathbf{K},\omega) + \hat{a}_s^\dagger(\mathbf{K},\omega)\hat{a}_s(\mathbf{K},\omega)\right]\frac{d^2k\, d\omega}{(2\pi)^3C'}. \tag{4.44}$$

The expressions for the quantized electromagnetic fields in terms of optical beam variables, with ladder operators obeying the generic commutation relation in (4.42), are obtained by applying (4.39) and (4.41) to the magnetic vector potential in (4.26) and using (4.18). We also replace $c \to 1/\sqrt{\epsilon_0\mu_0}$. The results are

$$\begin{aligned}\hat{\mathbf{A}}(\mathbf{x},t) = \int \sqrt{\frac{\mu_0 k_z\hbar}{2C'}}\,&[\vec{\eta}_s\hat{a}_s(\mathbf{K},\omega)\exp(-i\omega t + i\mathbf{k}\cdot\mathbf{x}) \\ &+ \vec{\eta}_s^*\hat{a}_s^\dagger(\mathbf{K},\omega)\exp(i\omega t - i\mathbf{k}\cdot\mathbf{x})]\,\frac{d^2k\, d\omega}{(2\pi)^3k_z},\end{aligned} \tag{4.45}$$

$$\hat{\mathbf{E}}(\mathbf{x},t) = \mathrm{i}\int \sqrt{\frac{\mu_0 k_z \hbar}{2C'}}\,[\vec{\eta}_s \hat{a}_s(\mathbf{K},\omega)\exp(-\mathrm{i}\omega t + \mathrm{i}\mathbf{k}\cdot\mathbf{x})$$

$$- \vec{\eta}_s^{\,*} \hat{a}_s^\dagger(\mathbf{K},\omega)\exp(\mathrm{i}\omega t - \mathrm{i}\mathbf{k}\cdot\mathbf{x})\big]\,\frac{\omega\,\mathrm{d}\omega\,\mathrm{d}^2k}{(2\pi)^3 k_z}, \tag{4.46}$$

$$\hat{\mathbf{H}}(\mathbf{x},t) = \mathrm{i}\int \sqrt{\frac{\epsilon_0 k_z \hbar}{2C'}}\,\big[(\vec{k}\times\vec{\eta}_s)\hat{a}_s(\mathbf{K},\omega)\exp(-\mathrm{i}\omega t + \mathrm{i}\mathbf{k}\cdot\mathbf{x})$$

$$- (\vec{k}\times\vec{\eta}_s^{\,*})\hat{a}_s^\dagger(\mathbf{K},\omega)\exp(\mathrm{i}\omega t - \mathrm{i}\mathbf{k}\cdot\mathbf{x})\big]\,\frac{\omega\,\mathrm{d}\omega\,\mathrm{d}^2k}{(2\pi)^3 k_z}. \tag{4.47}$$

Although the choice of propagation direction breaks the three-dimensional rotation invariance, and consequently also Lorentz invariance, we can express these fields in a way that is consistent with the original Lorentz covariant formalism by using $C' = k_z$. So, we do not need to maintain explicit Lorentz covariance, provided that we follow a consistent approach to obtain the final expressions of these quantities in terms of optical beam variables. Alternatively, it may be more convenient in certain scenarios where consistency with the Lorentz covariant formulation serves no purpose, to use a non-Lorentz covariant formulation by setting $C' = 1$.

4.2 Fock states

4.2.1 Number operator

Fock states are eigenvectors of the *number operator*, with the *occupation* number n being the eigenvalue. The appropriate number operator that incorporates all the degrees of freedom is a generalization of the one in (3.53) in which all these degrees of freedom are either integrated or summed. Such a number operator is expressed with the aid of the summation convention as

$$\hat{n} \triangleq \int \hat{a}_s^\dagger(\mathbf{k})\hat{a}_s(\mathbf{k})\,\frac{\mathrm{d}^3k}{(2\pi)^3\omega_{\mathbf{k}}}. \tag{4.48}$$

Here, we used the *Lorentz covariant integration measure* to define the wavevector integral. To simplify notation, we henceforth represent this measure by

$$\mathrm{d}_\omega k_m \triangleq \frac{\mathrm{d}^3k_m}{(2\pi)^3\omega_m}, \tag{4.49}$$

where m labels different integration variables, and $\omega_m = c|\mathbf{k}_m|$.

The commutation relations of the number operator with the ladder operators are obtained from those in (4.15), leading to

$$[\hat{a}_s(\mathbf{k}),\hat{n}] = \hat{a}_s(\mathbf{k}) \quad \text{and} \quad \big[\hat{n},\hat{a}_s^\dagger(\mathbf{k})\big] = \hat{a}_s^\dagger(\mathbf{k}). \tag{4.50}$$

When the ladder operators are raised to a power m in these commutations, they are reproduced with a factor of m, similar to what we found in (3.58).

4.2.2 Notation for contractions

The current development becomes progressively more complex, eventually involving the expressions of functionals. However, since these functionals are often in Gaussian form, one can exploit this property to simplify the notation. The Gaussian form implies an exponential function with an argument consisting of integrals over some degrees of freedom. Typically, these integrations run over all three-dimensional wavevectors. The exponent in these functionals are products of functions of the wavevectors, which are integrated. There may be multiple wavevectors that are being integrated. However, a particular wavevector normally appears exactly twice as arguments of functions.

At this point, we introduce a notation that is henceforth used to simplify expressions. Whenever we have two quantities (operators, functions, kernels, etc.) that depend on the same wavevector and carry the same spin indices, which are respectively integrated and summed, then it is regarded as a type of *contraction*, similar to the contraction of indices associated with matrix-matrix multiplication. Therefore, we denote such a wavevector integral and spin summation by a bilinear operation denoted by $\diamond$. We call it a $\diamond$-*contraction*. The number operator in (4.48) can thus be represented as

$$\hat{n} \triangleq \hat{a}^{\dagger} \diamond \hat{a}, \tag{4.51}$$

where we remove the spin indices and the wavevector dependencies from the contracted quantities. An inner product between two functions is likewise represented (using the summation convention) by

$$F^* \diamond G \triangleq \int F_s^*(\mathbf{k}) G_s(\mathbf{k}) \, \mathrm{d}_\omega k \equiv \langle F, G \rangle. \tag{4.52}$$

The $\diamond$-contraction notation is not suitable for cases where three or more quantities have the same wavevector or spin that is being integrated or summed. If there is a *kernel function* involved, we have

$$F^* \diamond K \diamond G \triangleq \int F_r^*(\mathbf{k}) K_{r,s}(\mathbf{k}, \mathbf{k}') G_s(\mathbf{k}') \, \mathrm{d}_\omega k \, \mathrm{d}_\omega k'. \tag{4.53}$$

To maintain the same dimensions as a direct contraction between two functions, the dimensions of the kernel $K_{r,s}(\mathbf{k}, \mathbf{k}')$ must be the inverse of the dimensions of the integration measure for the relevant $\diamond$-contraction.

Note that the integration measure in terms of which the $\diamond$-contraction is defined here is the Lorentz covariant measure of (4.49). However, when we discuss expressions where a different measure is used, we still use the $\diamond$-contraction with the understanding that it involves the relevant integration measure within the context.

We also define an *identity kernel* denoted by

$$\mathbf{1}_{s,r}(\mathbf{k}_1, \mathbf{k}_2) \triangleq (2\pi)^3 \omega_1 \delta_{s,r} \delta(\mathbf{k}_1 - \mathbf{k}_2), \tag{4.54}$$

in the context of Lorentz covariance. It removes $\diamond$-contractions: $\mathbf{1} \diamond G = G \diamond \mathbf{1} = G$. The same notation for the identity can be used even when the context is not Lorentz covariant. Comparing the expression of the identity kernel with the *commutation relation for ladder operators* in (4.15), we see that the latter produces the identity kernel

$$\left[\hat{a}_s(\mathbf{k}_1), \hat{a}_r^\dagger(\mathbf{k}_2)\right] = \mathbf{1}_{s,r}(\mathbf{k}_1, \mathbf{k}_2). \tag{4.55}$$

We usually follow a *generalized normal-order convention* when writing down such $\diamond$-contractions: complex conjugate or Hermitian adjoint quantities are placed toward the left-hand side of quantities that are not complex conjugated or the Hermitian adjoint. However, this convention cannot always be followed.

4.2.3 Fixed-spectrum ladder operators

Using the $\diamond$-contraction notation, we define *fixed-spectrum ladder operators* as

$$\hat{a}_F^\dagger \triangleq \hat{a}^\dagger \diamond F \quad \text{and} \quad \hat{a}_F \triangleq F^* \diamond \hat{a}, \tag{4.56}$$

in terms of a normalized complex-valued spectral function $F_s(\mathbf{k})$. This spectral function is generally called the *parameter function* (or the *angular spectrum* in analogy to the case of Section 2.4.5). The *commutation relation between fixed-spectrum ladder operators* with different spectral functions then leads to

$$\left[\hat{a}_F, \hat{a}_G^\dagger\right] = F^* \diamond \left[\hat{a}, \hat{a}^\dagger\right] \diamond G = F^* \diamond \mathbf{1} \diamond G = F^* \diamond G \equiv \langle F, G \rangle. \tag{4.57}$$

The commutation between two arbitrary fixed-spectrum ladder operators produces the inner product between their parameter functions. This calculation could be expressed in terms of the integrals, but the resulting expressions are much more cumbersome. For the same angular spectrum, the commutation becomes $[\hat{a}_F, \hat{a}_F^\dagger] = F^* \diamond F \equiv \|F\|^2 = 1$, showing that the parameter functions are normalized.

4.2.4 Coordinate bases

Applying the ladder operators defined in Section 4.1.1 once to the vacuum state, we obtain a Fourier domain coordinate basis in the form of a *wavevector basis*. These elements and their duals are respectively represented by kets and bras that read

$$|\mathbf{k}, s\rangle \triangleq \hat{a}_s^\dagger(\mathbf{k})\,|\mathrm{vac}\rangle \quad \text{and} \quad \langle \mathbf{k}, s| \triangleq \langle \mathrm{vac}|\,\hat{a}_s(\mathbf{k}). \tag{4.58}$$

Although we represent them as kets and bras, they are not valid quantum states because they are not normalizable.

The *orthogonality condition for this wavevector basis* is readily derived from the commutation relation

$$\langle \mathbf{k}, r|\mathbf{k}', s\rangle = \langle \mathrm{vac}|\left[\hat{a}(\mathbf{k}, r), \hat{a}_s^\dagger(\mathbf{k}', s)\right]|\mathrm{vac}\rangle = \mathbf{1}_{s,r}(\mathbf{k}, \mathbf{k}'). \tag{4.59}$$

We can also formulate a completeness relation, but since the elements of the wavevector basis are all produced from the vacuum through a single application of the ladder operators, they can only produce an identity operator for single-photon states. As such, they only resolve the *single-photon projection operator*.

Exercise 4.3. Use the commutation relations in (4.55) to show that

$$\begin{aligned} \hat{a}_r(\mathbf{k}')\,|\mathbf{k}, s\rangle &= |\mathrm{vac}\rangle\,\mathbf{1}_{r,s}(\mathbf{k}, \mathbf{k}'), \\ \langle \mathbf{k}, s|\,\hat{a}_r^\dagger(\mathbf{k}') &= \mathbf{1}_{r,s}(\mathbf{k}, \mathbf{k}')\,\langle \mathrm{vac}|\,. \end{aligned} \tag{4.60}$$

The wavevector basis can be used to represent an arbitrary *single-photon state*

$$\begin{aligned} |1_F\rangle &= \sum_s \int |\mathbf{k}, s\rangle\, F_s(\mathbf{k})\, \mathrm{d}_\omega k = \hat{a}_F^\dagger\,|\mathrm{vac}\rangle\,, \\ \langle 1_F| &= \sum_s \int F_s^*(\mathbf{k})\,\langle \mathbf{k}, s|\, \mathrm{d}_\omega k = \langle \mathrm{vac}|\,\hat{a}_F, \end{aligned} \tag{4.61}$$

in terms of the parameter function $F_s(\mathbf{k})$, which now also serves as the *coefficient function* of the single-photon state. The subscript F indicates that the single-photon state is parameterized by the same spectral function $F_s(\mathbf{k})$ that parametrizes the fixed-spectrum ladder operators in (4.56). It also follows that the *single-photon Fourier domain wavefunction* is given by the same spectral function $F_s(\mathbf{k}) = \langle \mathbf{k}, s|1_F\rangle$. In the single-photon context, the *parameter function, angular spectrum, coefficient function*, and *Fourier domain wavefunction*, are all synonymous. The normalization of the single-photon states implies the equivalent normalization of the wavefunction

$$\begin{aligned} \langle 1_F|1_F\rangle &= \sum_{r,s} \int F_s^*(\mathbf{k})\langle \mathbf{k}, r|\mathbf{k}', s\rangle F_s(\mathbf{k}')\, \mathrm{d}_\omega k\, \mathrm{d}_\omega k' \\ &= \sum_s \int |F_s(\mathbf{k})|^2\, \mathrm{d}_\omega k \equiv F^* \diamond F \equiv \|F\|^2 = 1, \end{aligned} \tag{4.62}$$

which is equivalent to the normalization for the parameter function imposed by the commutation relation of the fixed-spectrum ladder operators.

We could also define a *configuration space coordinate basis* with elements denoted by $|\mathbf{x}, t\rangle$ with appropriate orthogonality and completeness conditions. However, we do not use such a configuration space coordinate basis here. The wavevector basis suffices to represent the *spatiotemporal degrees of freedom* of any quantum state that we may wish to investigate.

Although these bases are analogues to the coordinate bases defined in Chapter 3, we must remember that those bases represent the *particle-number degree of freedom*. They are only associated with the position and momentum degrees of freedom in the context of the (hypothetical) quantum harmonic oscillator. Here, the coordinate bases are associated with the spatiotemporal degrees of freedom as represented by the wavevectors in the Fourier domain. Their particle-number degree of freedom is fixed to that of a single photon. The wavevectors are *not* associated with the particle-number degree of freedom in any way. For this reason, there is no reason to invoke the de Broglie relationship to convert the wavevector basis into a momentum basis.

We reiterate that the elements of these coordinate bases cannot be normalized and can therefore not individually represent quantum states. In the case of the configuration space coordinate basis, there is another reason why the elements cannot be quantum states. The state of a quantum system (particle) needs to exist for all time (assuming no interactions). The elements of the configuration space coordinate basis only exist (are only nonzero) at one spacetime point.

When states have more than one photon, we can still use the Fourier domain coordinate basis to represent them in terms of *multiphoton wavefunctions* for fixed numbers of photons. For example, a general two-photon state can be represented with the aid of the wavevector basis as

$$|2_F\rangle = \sum_{r,s} \int |\mathbf{k}_1, r\rangle\, |\mathbf{k}_2, s\rangle\, F_{r,s}(\mathbf{k}_1, \mathbf{k}_2)\, \mathrm{d}_\omega k_1\, \mathrm{d}_\omega k_2, \tag{4.63}$$

where $F_{r,s}(\mathbf{k}_1, \mathbf{k}_2)$ is the *two-photon wavefunction*. However, in the general case, a state can consist of a superposition of different numbers of photons. The representation of such a state in terms of the wavevector basis becomes cumbersome.

4.2.5 Fixed-spectrum Fock states

In a *fixed-spectrum state*, all the photons in the state share the exact same parameter function. Using the *fixed-spectrum ladder operators*, defined in (4.56), we express the *fixed-spectrum Fock states* by

$$|n_F\rangle \triangleq \frac{1}{\sqrt{n!}} \hat{a}_F^{\dagger n}\, |\mathrm{vac}\rangle \quad \text{and} \quad \langle n_F| \triangleq \frac{1}{\sqrt{n!}} \langle \mathrm{vac}|\, \hat{a}_F^{n}, \tag{4.64}$$

This notation corresponds to what we used in (4.61) and (4.63). Here, n is the occupation number (the number of photons in the state) and the subscript F represents the (fixed) parameter function in terms of which *all* the photons are parameterized.

Exercise 4.4. Show that the inner product between Fock states with different spectra gives

$$\langle m_F|n_G\rangle = \delta_{m,n} (F^* \diamond G)^n . \tag{4.65}$$

Ladder operators applied to arbitrary fixed-spectrum Fock states produce

$$\begin{aligned}\hat{a}_r(\mathbf{k}')\,|n_F\rangle &= |(n-1)_F\rangle\, F_r(\mathbf{k}')\sqrt{n},\\ \hat{a}_r^\dagger(\mathbf{k}')\,|n_F\rangle &= |n_F\rangle\,|\mathbf{k}',r\rangle .\end{aligned} \tag{4.66}$$

Contracting these expressions with a factor of the same parameter function, we obtain

$$\begin{aligned}\hat{a}_F\,|n_F\rangle &= |(n-1)_F\rangle\,\sqrt{n},\\ \hat{a}_F^\dagger\,|n_F\rangle &= |(n+1)_F\rangle\,\sqrt{n+1}.\end{aligned} \tag{4.67}$$

These expressions are similar to the cases for the particle-number degree of freedom only, provided in (3.62) with the *bosonic enhancement* factors $\sqrt{n}$ and $\sqrt{n+1}$. Applying the number operator on the fixed-spectrum Fock states, we get

$$\hat{n}\,|n_F\rangle = \frac{1}{\sqrt{n!}}[\hat{n},\hat{a}_F^{\dagger n}]\,|\text{vac}\rangle = \frac{n\hat{a}_F^{\dagger n}}{\sqrt{n!}}\,|\text{vac}\rangle = |n_F\rangle\, n. \tag{4.68}$$

Hence, the *fixed-spectrum Fock states* are *eigenvectors* of the *number operator*.

A general fixed-spectrum state can be expressed in terms of the fixed-spectrum Fock states, all with the same spectrum F, by

$$|\psi_F\rangle = \sum_{n=0}^{\infty} |n_F\rangle\, C_n. \tag{4.69}$$

The complex coefficients C_n maintain the normalization of the state in that

$$\sum_{n=0}^{\infty} |C_n|^2 = 1. \tag{4.70}$$

4.2.6 Discrete basis Fock states

The inner product between Fock states with different parameter functions (4.65) suggests that one can define fixed-spectrum Fock states that form a basis for both the particle-number degree of freedom and the spatiotemporal degrees of freedom, by using a set of orthonormal spectral modes for the parameter functions. Let us assume that $F_s(\mathbf{k}) \rightarrow M_{ms}(\mathbf{k})$, where $M_{ms}(\mathbf{k})$ represents the modal spectra of a discrete set of orthonormal modal functions, indexed by the integer m (the spin is still denoted by s). The orthogonality condition for such a set of orthonormal modal functions is

$$\langle M_m, M_n \rangle \equiv M_m^* \diamond M_n = \delta_{m,n}, \tag{4.71}$$

and the *completeness condition* is

$$\sum_m M_{mr}(\mathbf{k}_1) M_{ms}^*(\mathbf{k}_2) = \mathbf{1}_{r,s}(\mathbf{k}_1, \mathbf{k}_2). \tag{4.72}$$

In terms of this modal basis, we have single-photon states, given by

$$\begin{aligned} |1_m\rangle &= \sum_s \int |\mathbf{k}, s\rangle \, M_{ms}(\mathbf{k}) \, \mathrm{d}_\omega k, \\ \langle 1_m| &= \sum_s \int M_{ms}^*(\mathbf{k}) \, \langle \mathbf{k}, s| \, \mathrm{d}_\omega k, \end{aligned} \tag{4.73}$$

where the subscript now indicates the different modes by their mode index. We also have ladder operators associated with the discrete modes

$$\hat{a}_m^\dagger \triangleq \hat{a}^\dagger \diamond M_m \quad \text{and} \quad \hat{a}_m \triangleq M_m^* \diamond \hat{a}, \tag{4.74}$$

as in (4.56). The discrete Fock states are now defined as

$$|n_m\rangle \triangleq \frac{\hat{a}_m^{\dagger n}}{\sqrt{n!}} |\text{vac}\rangle \quad \text{and} \quad \langle n_m| \triangleq \langle \text{vac}| \frac{\hat{a}_m^n}{\sqrt{n!}}. \tag{4.75}$$

The ladder operators obey the *commutation relation*

$$[\hat{a}_p, \hat{a}_m^\dagger] = \delta_{p,m}, \tag{4.76}$$

and the Fock states satisfy an *orthogonality condition* that reads

$$\langle m_u | n_v \rangle = \delta_{m,n} \delta_{u,v}, \tag{4.77}$$

where m and n are occupation numbers and u and v are mode indices.

4.2.6.1 Completeness

The point of defining the discrete Fock states is to formulate a complete orthonormal basis for the entire Hilbert space of all quantum optical states. The reason why we are interested in the concept of a complete basis is because we often want to employ the ability of such a complete basis to resolve the identity in our calculations. The orthogonality is readily demonstrated with (4.77), but the *completeness* is more complex. The discrete Fock states, as defined in (4.75), do not produce a complete basis. This fact can be easily seen by trying to expand the state

$$|\psi\rangle = |1_m\rangle\,|1_n\rangle\,, \tag{4.78}$$

where $m \neq n$, in terms of these discrete Fock states. All the inner products between $|\psi\rangle$ and the Fock states in (4.75) are zero. As a result, such a state cannot be represented in terms of the discrete Fock basis.

This example demonstrates that a complete orthogonal basis for both particle-number and the spatiotemporal degrees of freedom needs to allow different photons in a state to have different parameter functions. The general form for such a basis element composed from the discrete Fock states is represented by an infinite tensor product of discrete Fock states

$$|m, n, \ldots\rangle = |m_1\rangle \otimes |n_2\rangle \otimes \cdots\,, \tag{4.79}$$

where the different slots in the ket on the left, corresponding to the subscripts on the right, represent the different discrete modes, and the integers m, n, etc. represent the occupation numbers associated with the respective discrete modes. In general, an element in the basis can consist of an arbitrary selection from all such modes, hence the infinite number of factors in the tensor product. However, a physical state (with a finite energy) can only contain a finite number of photons. Therefore, only a finite number of the occupation numbers would be nonzero.

The formulation of a discrete basis in terms of tensor products of Fock states implies a *stratification* of the Hilbert space. If $\mathcal{H}$ represents the Hilbert space of all quantum optical states, then the tensor product structure of an element of the Fock basis implies that $\mathcal{H}$ is stratified into slices

$$\mathcal{H} = \mathcal{H}_1 \otimes \mathcal{H}_2 \otimes \mathcal{H}_3 \otimes \cdots, \tag{4.80}$$

where each $\mathcal{H}_m$ is a slice of the Hilbert space for a fixed spatiotemporal mode representing the parameter function of the discrete Fock states for that slice. Each slice is equivalent to a particle-number degree of freedom only Hilbert space, as discussed in Chapter 3, in which the spatiotemporal degrees of freedom are fixed to just one mode.

These tensor product basis elements obey an orthogonality condition represented in terms of infinite products of Kronecker deltas for the number of particles in the factors associated with corresponding modes. Such an *orthogonality condition* has the form

$$\langle u, v, \ldots | m, n, \ldots\rangle = \delta_{u,m}\delta_{v,n}\cdots, \tag{4.81}$$

where the ... on the right-hand side represents the rest of the product of Kronecker deltas for all the slots in the states.

There is a bit of an issue with the vacuum state. To include cases where a given mode is not represented in a state, the ket associated with that mode must have a zero occupation number. It leads to the situation where a separate vacuum state is associated with each spatiotemporal mode, which is orthogonal to the vacuum states associated with other modes. This situation is a purely formal construction and does not reflect real physics. There is only one vacuum state in the physical world and it does not carry any spatiotemporal information.

A revealing aspect of these discrete Fock states is to identify the operators for which they serve as *eigenvectors*. It can be readily shown that they are eigenvectors of the number operator, in the same way it is done in (4.68) for general fixed-spectrum Fock states. However, they are not eigenvectors of the *interaction-free Hamiltonian* in (4.28). The reason is that the *angular frequency* becomes a continuous parameter that distinguishes different eigenvectors. Therefore, since they depend on a continuous parameter, the eigenvectors of the Hamiltonian are not a discrete countable set. A consequence is that the eigenvectors of the Hamiltonian in (4.28) cannot be normalized. (One can try to force normalizability only to end up with states that do not satisfy a proper orthogonality condition). These unnormalized eigenvectors are therefore not elements in the Hilbert space, and cannot be used to *stratify* the Hilbert space.

One can try to circumvent this issue by considering the spatiotemporal degrees of freedom in a finite volume, thus leading to discrete normalizable eigenvectors for the Hamiltonian. However, such an approach assumes that there is a valid continuum limit, as explained in Section 4.1, which is clearly not the case because the eigenvectors of the Hamiltonian in the continuum are not normalizable elements of the Hilbert space.

Since there are an infinite number of different spatiotemporal modes, the number of photons in such a state would be infinite if the occupation numbers associated with all those modes are nonzero. As mentioned before, a state with a finite number of photons can only have a finite number of spatiotemporal modes with nonzero occupation numbers. So, the vast majority of these basis elements are associated with unphysical states if we do not restrict the number of nonzero occupation numbers to be finite. Moreover, including these unphysical states, we end up with a basis that is uncountable (the infinite sequence of occupation numbers can be compared with the infinite sequence of digits in the decimal representation of the real numbers). It leads to a Hilbert space that is not separable. For a separable Hilbert space, we need to discard all such unphysical states and retain only a countable subset of those with finite numbers of photons.

Regardless of the issue with the unphysical states, a basis represented by infinitely many tensor products is extremely cumbersome and, therefore, not convenient for the purpose of calculations. However, their discrete nature is more suitable for the purpose of formal discussions, because they are normalized quantum states and can thus represent a basis for a Hilbert space. We do not consider them any further here, but we return to consider their completeness in Section 4.4.3, after having discussed fixed-spectrum coherent states.

4.3 Finite number of discrete degrees of freedom

Despite the benefits that one reaps from using continuous variables, there are some concepts that are easier to explain in terms of a finite number of discrete degrees of freedom. In this case, these degrees of freedom apply to both the particle-number degree of freedom and the spatiotemporal degrees of freedom, noting that spin is already restricted to a two-dimensional space. While the particle-number degree of freedom is represented by a fixed number of photons in the states, the other degrees of freedom are represented by different modes, which are in general considered to be different spatiotemporal modes with different states of polarization.

The purpose of this section is not to find a suitable basis for our purposes, but to discuss some concepts that are needed later. Having discussed the somewhat related concept of Fock states in the previous section, it is a natural point to include the discussion of these concepts.

4.3.1 Qubits

The simplest non-trivial quantum states are those in a two-dimensional Hilbert space represented by a basis consisting of only two elements. These two-dimensional states are called *qubits* in the context of *quantum information science* [10]. The two elements are denoted by $|0\rangle$ and $|1\rangle$. All the states in this Hilbert space can be represented as

$$|\Psi\rangle = |0\rangle\,\alpha + |1\rangle\,\beta, \tag{4.82}$$

where α and β are complex coefficients. The requirement for normalization implies that $|\alpha|^2 + |\beta|^2 = 1$. The values of α and β that satisfy this condition can be parameterized as

$$\begin{aligned} \alpha &= \exp\left(-\mathrm{i}\frac{1}{2}\phi\right)\cos\left(\frac{1}{2}\theta\right), \\ \beta &= \exp\left(\mathrm{i}\frac{1}{2}\phi\right)\sin\left(\frac{1}{2}\theta\right), \end{aligned} \tag{4.83}$$

in terms of two angles $0 \leq \theta \leq \pi$ and $0 \leq \phi < 2\pi$. As such, θ and ϕ represent the angular coordinates for points on the two-dimensional surface of a sphere, which is called the *Bloch sphere*. The poles of the sphere, where $\theta = 0$ and $\theta = \pi$, represent the states $|0\rangle$ and $|1\rangle$, respectively. The other points on the Bloch sphere are given by superpositions in terms of the complex coefficients α and β.

The similarity between the Bloch sphere and the *Poincaré sphere*, discussed in Section 2.9.2, indicates that the same mathematical structure describes both these scenarios. Indeed they both follow from the two-dimensional representations of the SU(2) Lie group. We can always remove a global phase from these states, because two states with different global phases represent the same state. Therefore, we can factor out a U(1) Lie group, which represents the global phase. The factor group SU(2)/U(1) has the topology of a sphere. The states of polarization of a classical optical field is a good analogue of a qubit system. It is therefore no surprise that polarization is often used in optical implementations of quantum information systems to represent qubits.

4.3.2 Bipartite states

Often in quantum information systems, the same quantum state is subjected to two or more measurements. Since photons are destroyed by measurements, a quantum optical state needs to contain two or more photons for a successful measurement in such a case. A quantum optical state consisting of exactly two photons is often called a *biphoton* state or a *photon pair*. However, it is not in general possible to be sure that a quantum optical state has exactly two photons. For two measurements, the state is divided into two parts each of which is directed toward one of two measurement setups. If the two setups both succeed in measuring a photon, then the complete state must have had at least two photons that were separated into the two parts.

Any state that is thus separated into two parts with at least one photon in each part is called a *bipartite state*. Often the two parts (or *subsystems*) are labelled by A and B, respectively. The density operator for such a bipartite state can be expressed in the measurement bases of the two measurement setups as

$$\hat{\rho} = \sum_{m,n,p,q} |m\rangle_A \, |p\rangle_B \; \rho_{m,p,n,q} \; \langle n|_A \, \langle q|_B \, , \tag{4.84}$$

where $|m\rangle_A$ and $|p\rangle_B$ (and their associated duals $\langle n|_A$ and $\langle q|_B$) are (measurement) basis elements for the A and B subsystems, respectively, and $\rho_{m,p,n,q}$ represents the complex coefficients of the expansion in terms of these bases. Here, it is assumed that the bases are finite dimensional. The complex coefficients have four indices, which implies a *fourth-rank tensor*. These elements can be rearranged to represent a *density matrix* with all possible combinations of m, p representing the rows and all possible combinations of n, q representing the columns. Such a density matrix is a square $NM \times NM$ matrix, where N and M represent the dimensions of the A and B subsystems, respectively.

4.3.2.1 Partial trace

The partial trace of a bipartite state is an *operator trace* (see Section 3.1.2) performed on only one of the two subsystems of its density operator. The result is a density operator for the state of the remaining subsystem only and is called the *reduced density operator*.

For example, the partial trace over the B subsystem applied to the density operator in (4.84) produces

$$\hat{\rho}_A = \text{tr}_B\{\hat{\rho}\} = \sum_{m,n} |m\rangle_A \, \rho'_{m,n} \, \langle n|_A \,, \tag{4.85}$$

where

$$\rho'_{m,n} = \sum_p \rho_{m,p,n,p}. \tag{4.86}$$

By analogy the matrix with elements $\rho'_{m,n}$ is called the *reduced density matrix*.

4.3.3 Schmidt decomposition

The density operator of a pure bipartite state can be represented by $\hat{\rho} = |\Psi\rangle\,\langle\Psi|$, where

$$|\Psi\rangle = \sum_{m,p} |m\rangle_A \, |p\rangle_B \; c_{m,p}, \tag{4.87}$$

with $c_{m,p}$ representing complex-valued coefficients. Using unitary operators, discussed in Section 3.2.2, we can transform the expression of the pure bipartite state into any orthonormal basis. This transformation can be done separately for the two subsystems, so that they do not need to have the same orthonormal basis.

There always exist two special orthonormal bases for the two respective subsystems of a pure bipartite state that would diagonalize the coefficients. The resulting expansion

$$|\Psi\rangle = \sum_m |\xi_m\rangle \, |\zeta_m\rangle \, \lambda_m, \tag{4.88}$$

runs over only one index m. It is called the *Schmidt decomposition* of the pure bipartite state, in which λ_m represents the *Schmidt coefficients*. They are real valued and non-negative. The normalization of the state implies that

$$\langle\Psi|\Psi\rangle = \sum_m \lambda_m^2 = 1. \tag{4.89}$$

The number of nonzero Schmidt coefficients represents the *Schmidt rank* of the state. The elements $\{|\xi_m\rangle\}$ and $\{|\zeta_m\rangle\}$ are the *Schmidt bases* associated with the respective subsystems. They are both orthonormal:

$$\langle\xi_m|\xi_n\rangle = \langle\zeta_m|\zeta_n\rangle = \delta_{m,n}. \tag{4.90}$$

4.3.3.1 Schmidt number

The partial trace of the density operator for a pure state can be calculated in terms of the Schmidt decomposition. It gives

$$\hat{\rho}_A = \text{tr}_B\{\hat{\rho}\} = \sum_{nm} |\xi_n\rangle \langle\zeta_m|\zeta_n\rangle\lambda_n\lambda_m \langle\xi_m| = \sum_n |\xi_n\rangle \lambda_n^2 \langle\xi_n| . \tag{4.91}$$

The purity of the resulting reduced density operator, which reads

$$\text{tr}\{\hat{\rho}_A^2\} = \sum_n \lambda_n^4, \tag{4.92}$$

represents a useful quantity. Consider, for example, a pure state with N Schmidt coefficients that are all equal. The requirement for normalization implies that

$$\lambda_n = \frac{1}{\sqrt{N}} \quad \text{for all } n. \tag{4.93}$$

The purity of the reduced density operator for such a state is

$$\text{tr}\{\hat{\rho}_A^2\} = \sum_n \frac{1}{N^2} = \frac{1}{N}. \tag{4.94}$$

It is therefore equal to the inverse of the dimension of the state, which represents the lower bound for the purity of a finite-dimensional state.

Exercise 4.5. Show that the smallest value for the purity of a finite-dimensional density operator is equal to the inverse of the number of dimensions.

We can generalize this concept by defining the *Schmidt number* for an arbitrary state as the inverse of the purity of its reduced density operator

$$K = \frac{1}{\text{tr}\{\hat{\rho}_A^2\}}. \tag{4.95}$$

It gives an indication of the effective number of Schmidt coefficients for a given pure bipartite state. The Schmidt number is in general smaller than the Schmidt rank. They are equal only when the values of all the nonzero Schmidt coefficients are the same.

4.3.4 Entanglement

One of the most enigmatic properties found in quantum systems is *entanglement*. Such systems exhibit correlations among remote observations of the system that defies any notions of a local realistic explanation for the fundamental properties of nature.

Although there are classical optical fields where the expression formally represents the same situation, the physical implication differs from what is found in the quantum case. In the classical case, the correlations are always associated with different properties (different degrees of freedom) of a single local measurement. For this reason, the classical manifestation is preferentially referred to as *classical non-separability*, instead of "classical entanglement," as sometimes found in the literature. It ensures that the term *entanglement* is reserved for the quantum version, making the use of the phrase "quantum entanglement" superfluous.

4.3.4.1 EPR paradox and Bell's inequality

Historically, the concept of entanglement emerged out of the original work by Einstein, Podolsky, and Rosen (EPR) [11]. They attempted to show that quantum mechanics is incomplete through an argument based on a hypothetical experiment in which a particle decays into two particles which are respectively measured to obtain position and momentum information about the original particle. It was argued that the individual measurements would be able to provide information with sufficient accuracy, so that, after being combined via *conservation of momentum*, the information about the position and momentum of the original particle would violate the Heisenberg uncertainty principle.

This work led Erwin Schrödinger to introduce the concept of *entanglement* [12]. It follows that the EPR argument fails because of the entanglement between the two particles, as imposed by the conservation of momentum.

To put this situation on a firmer foundation, John S. Bell derived an inequality for such EPR measurements. In the derivation, he used two seemingly innocuous assumptions: *nature only allows local interactions* and *nature has a unique reality*. They combine into an assumed condition for nature referred to as *local realism*. The *Bell inequality* must be satisfied by any EPR measurements if these assumptions are true. The *inequality*, in a form provided by Clauser, Horn, Shimony, and Holt (CHSH) [13], is given by

$$\langle \hat{R}_A \hat{U}_B \rangle + \langle \hat{S}_A \hat{U}_B \rangle + \langle \hat{S}_A \hat{V}_B \rangle - \langle \hat{R}_A \hat{V}_B \rangle \leq 2, \tag{4.96}$$

where $\hat{R}_A$ and $\hat{S}_A$ are two noncommuting *observables* measured in the A subsystem and $\hat{U}_B$ and $\hat{V}_B$ are two noncommuting observables measured in the B subsystem. The expectation values are obtained for the joint measurements of the two observables in their respective subsystems.

The CHSH inequality can be violated if the two subsystems are entangled, because an entangled state does not satisfy the assumption of a unique reality. Therefore, the violation of this inequality indicates that the bipartite state is entangled. In 1982, Alain Aspect performed an experiment using the polarization of light from a parametric down-converted source demonstrating the *violation of the CHSH inequality* [14].

4.3.4.2 Bell states

To explain what entanglement is, we consider a simple example: a bipartite system consisting of qubits. In other words, each of the two subsystems is a two-dimensional Hilbert space. The combined space is four-dimensional. An obvious set of four basis elements would be

$$\{|0\rangle_A\,|0\rangle_B\,,|0\rangle_A\,|1\rangle_B\,,|1\rangle_A\,|0\rangle_B\,,|1\rangle_A\,|1\rangle_B\}. \tag{4.97}$$

However, we can also consider sums and differences of these elements:

$$\begin{aligned}
\Phi_+ &= \frac{1}{\sqrt{2}}\left(|0\rangle_A\,|0\rangle_B + |1\rangle_A\,|1\rangle_B\right),\\
\Phi_- &= \frac{1}{\sqrt{2}}\left(|0\rangle_A\,|0\rangle_B - |1\rangle_A\,|1\rangle_B\right),\\
\Psi_+ &= \frac{1}{\sqrt{2}}\left(|0\rangle_A\,|1\rangle_B + |1\rangle_A\,|0\rangle_B\right),\\
\Psi_- &= \frac{1}{\sqrt{2}}\left(|0\rangle_A\,|1\rangle_B - |1\rangle_A\,|0\rangle_B\right).
\end{aligned} \tag{4.98}$$

These four states are called the *Bell states*. All of them are *maximally entangled*.

4.3.4.3 Separability

Imagine that one of the Bell states represents a state on which measurements are performed to determine the identity of the element ($|0\rangle$ or $|1\rangle$) in the respective subsystems. Consider, for example, Φ_+. If the outcome of the measurement in the A system is $|0\rangle$ then the outcome in the B system would also be $|0\rangle$. On the other hand, if the outcome of the measurement in the A system is $|1\rangle$, then the outcome in the B system would be $|1\rangle$. As a result, there is a correlation in the observed measurements. In fact, if we knew that we are measuring Φ_+, then we only have to measure one of the two systems and thus know what the measurement in the other system will give. The same situation applies for any of the other Bell states.

In contrast, consider the case

$$|\psi\rangle = \frac{1}{\sqrt{2}}\left(\Phi_+ + \Psi_+\right) = \frac{1}{2}\left(|0\rangle_A + |1\rangle_A\right)\left(|0\rangle_B + |1\rangle_B\right). \tag{4.99}$$

In this case, a specific outcome of the measurement at A would be completely uncorrelated to the outcome of the measurement at B.

The key difference between the two cases is that $|\psi\rangle$ in (4.99) can be factorized into a tensor product of the state for subsystem A and the state for subsystem B. If a state can be factorized in this way, it is said to be *separable*. On the other hand, if a state cannot be factorized, such as the Bell states, then it is called *non-separable* or *entangled*.

The Bell states are all pure states. The notion of separability can be generalized for mixed states. In that case, a state is said to be separable if it can be written as a linear combination (or convex sum) of separable states

$$\hat{\rho} = \sum_n P_n \hat{\rho}_{An} \otimes \hat{\rho}_{Bn}. \tag{4.100}$$

Here, ⊗ represents the tensor product and P_n is the coefficients of the expansion, indicated as probabilities. A mixed state that cannot be written as such a combination of separable pure states is entangled.

Although we discussed the property of entanglement here in the context of a two-dimensional Hilbert space, the concept of separability can be extended to arbitrary numbers of dimensions. It can also be generalized to more than two subsystems, leading to the concept of *multiparty entanglement*.

4.3.4.4 Quantifying entanglement

While the notion of separability provides some conceptual understanding of what entanglement means, it does not tell us how to know whether an arbitrary state is entangled or not. Therefore, we need to have some tests that we can apply to a state to determine whether it is separable or entangled. We may also want to know how entangled the state is. In other words, we need ways to quantify the entanglement in a state.

For a pure state, the number of nonzero Schmidt coefficients indicates whether a state is entangled. When there is only one nonzero Schmidt coefficient, the state is separable. With more than one nonzero Schmidt coefficient, the state is entangled. The amount of entanglement in a pure state can therefore be quantified in terms of the Schmidt number, as defined in (4.95).

For mixed states, the situation is significantly more complicated. To quantify entanglement in a mixed state, we generally have to perform a complicated optimization calculation. Fortunately, there is a direct way to calculate the amount of entanglement in a bipartite qubit state in terms of concurrence.

4.3.4.5 Concurrence

The concept of *concurrence* extends to arbitrary dimensions. For qubits (i. e., when the subsystems of a bipartite state are both two-dimensional, or when one is two-dimensional and the other three-dimensional), there is a simpler way to compute it.

The concurrence for such a system is given by

$$C = \max\{0, \lambda_1 - \lambda_2 - \lambda_3 - \lambda_4\}, \tag{4.101}$$

where the λ's (labelled from large to small $\lambda_1 > \lambda_2 > \lambda_3 > \lambda_4$) are the eigenvalues of

$$R = \sqrt{\sqrt{\rho}\tilde{\rho}\sqrt{\rho}}. \tag{4.102}$$

Here, ρ is the density matrix of the state, and

$$\tilde{\rho} = [\sigma_y \otimes \sigma_y]\, \rho^* \, [\sigma_y \otimes \sigma_y]\,, \tag{4.103}$$

with σ_y being the Pauli y-matrix, given in (2.183).

4.3.4.6 Entanglement witnesses

Often, it is not necessary to quantify the amount of entanglement. We may only want to know whether or not a state is entangled. For this purpose, we can use *entanglement witnesses*. One of the most often used examples of an entanglement witness is to determine whether a state violates the Bell (or CHSH) inequality.

4.4 Fixed-spectrum coherent state

The coherent states are discussed in terms of the particle-number degree of freedom in Section 3.2. Here, we incorporate the spatiotemporal degrees of freedom into those coherent states. To do so, we represent the Fock states in the expression for the coherent state in (3.99) as fixed-spectrum Fock states. The resulting *fixed-spectrum coherent states* are then given by

$$\begin{aligned} |\alpha_F\rangle &= \exp\left(-\frac{1}{2}|\alpha_0|^2\right) \sum_{n=0}^{\infty} \frac{\alpha_0^n}{\sqrt{n!}} \, |n_F\rangle \\ &= \exp\left(-\frac{1}{2}|\alpha_0|^2\right) \sum_{n=0}^{\infty} \frac{\alpha_0^n \hat{a}_F^{\dagger n}}{n!} \, |\text{vac}\rangle\,, \end{aligned} \tag{4.104}$$

where α_0 is an arbitrary complex parameter. In the last expression, we used (4.64) to express the fixed-spectrum Fock states in terms of fixed-spectrum creation operators. The subscript F indicates the normalized parameter function $F_s(\mathbf{k})$, to distinguish these coherent states from those in Section 3.2.

Combining the complex parameter α_0 with the normalized parameter function $F_s(\mathbf{k})$, we define a complex finite-energy parameter function

$$\alpha_s(\mathbf{k}) \triangleq \alpha_0 F_s(\mathbf{k}), \tag{4.105}$$

and then redefine the *fixed-spectrum ladder operators* as

$$\hat{a}_\alpha^\dagger \triangleq \hat{a}^\dagger \diamond \alpha \quad \text{and} \quad \hat{a}_\alpha \triangleq \alpha^\dagger \diamond \hat{a}, \tag{4.106}$$

in the context of the coherent states. The commutation relation for these fixed-spectrum ladder operators, equivalent of (4.57), is

$$\left[\hat{a}_\alpha, \hat{a}_\beta^\dagger\right] = \alpha^* \diamond \left[\hat{a}, \hat{a}_s^\dagger\right] \diamond \beta = \alpha^* \diamond \beta. \tag{4.107}$$

As a result, the *commutation between ladder operators* with the same parameter function produces the magnitude square of the parameter function. The expression for the coherent state in terms of this notation becomes

$$|\alpha\rangle = \exp\left(-\frac{1}{2}\|\alpha\|^2\right)\sum_{n=0}^{\infty}\frac{\hat{a}_{\alpha}^{\dagger n}}{n!}|\text{vac}\rangle\,, \tag{4.108}$$

allowing α to parameterize the state in terms of the complex finite energy parameter function $\alpha_s(\mathbf{k})$, where

$$\|\alpha\|^2 \triangleq \sum_s \int |\alpha_s(\mathbf{k})|^2 \, \mathrm{d}_\omega k = |\alpha_0|^2 \sum_s \int |F_s(\mathbf{k})|^2 \, \mathrm{d}_\omega k = |\alpha_0|^2. \tag{4.109}$$

By combining the complex parameter α_0 with the normalized parameter function $F_s(\mathbf{k})$, we obtain a parameter function with a finite magnitude. In the process, we also combine the *particle-number degree of freedom* represented by α_0 with the *spatiotemporal* and *spin* degrees of freedom as represented by $F_s(\mathbf{k})$. Henceforth, we use such unnormalized complex-valued parameter functions to parametrize the coherent states. The magnitude of such a parameter function $\|\alpha\|^2$ represents the particle-number degree of freedom (the average number of photons in the coherent state), while the normalized shape function $\alpha_s(\mathbf{k})/\|\alpha\|$ represents the spatiotemporal and spin degrees of freedom. The subscript F is removed for the coherent state unless it is necessary for clarity.

4.4.1 Fixed-spectrum displacement operator

An alternative representation of the fixed-spectrum coherent states is given in terms of the *fixed-spectrum displacement operator*, which is defined as

$$\hat{D}[\alpha] \triangleq \exp\left(\hat{a}^\dagger \diamond \alpha - \alpha^* \diamond \hat{a}\right) \equiv \exp\left(\hat{a}_\alpha^\dagger - \hat{a}_\alpha\right). \tag{4.110}$$

The notation for the displacement operator $\hat{D}[\alpha]$ uses square brackets for its argument to indicate that it is parameterized by a function instead of a constant. Therefore, it becomes a *functional* of the parameter function $\alpha_s(\mathbf{k})$.

Neither fixed-spectrum displacement operators, nor fixed-spectrum coherent states depend explicitly on the wavevector. Therefore, the eigenvalue equation of such fixed-spectrum coherent states has the form given in the following exercise.

Exercise 4.6. Show that, when the annihilation operator is applied to the coherent state, it produces

$$\hat{a}_s(\mathbf{k})\,|\alpha\rangle = |\alpha\rangle\,\alpha_s(\mathbf{k}). \tag{4.111}$$

The equation in (4.111) shows that the coherent states are the eigenstates of the annihilation operator. The resulting eigenvalue equation has an interesting form. To appreciate the situation, consider for a moment what we would have had if the coherent state explicitly depended on the wavevector. The corresponding eigenvalue equation

$$\hat{a}_s(\mathbf{k})\,|\alpha(\mathbf{k})\rangle = |\alpha(\mathbf{k})\rangle\,\alpha_s(\mathbf{k}), \tag{4.112}$$

would in that case be applicable for a specific wavevector. The function value of the parameter function for that wavevector would be the eigenvalue associated with the eigenstate for that wavevector.

By contrast, the eigenvalue equation in (4.111) implies that the entire function $\alpha_s(\mathbf{k})$ serves as the "eigenvalue." In other words, each function $\alpha_s(\mathbf{k})$ has a unique eigenstate associated with it. That eigenstate is the coherent state parameterized by that function. Since there is a very large number of such functions, there is an equally large number of fixed-spectrum coherent states.

It follows from (4.111) that, for any *holomorphic functional* of the annihilation operator $f[\hat{a}]$,[c] we have

$$f[\hat{a}]\,|\alpha\rangle = |\alpha\rangle\, f[\alpha], \tag{4.113}$$

where α denotes the eigenvalue function or parameter function. An equivalent equation is obtained for any holomorphic functional of the creation operator $f[\hat{a}^\dagger]$ by computing the adjoint of (4.113).

4.4.2 Non-orthogonality

The inner product between different fixed-spectrum coherent states can be calculated with the aid of (3.48) and (4.107). For two coherent states $|\alpha\rangle$ and $|\beta\rangle$ that are respectively parameterized by $\alpha_s(\mathbf{k})$ and $\beta_s(\mathbf{k})$, we have

$$\begin{aligned}\langle\alpha|\beta\rangle &= \exp\left(-\frac{1}{2}\|\alpha\|^2 - \frac{1}{2}\|\beta\|^2\right)\sum_{m,n=0}^{\infty}\frac{1}{n!m!}\langle\text{vac}|\left[\hat{a}_\alpha^m, \hat{a}_\beta^{\dagger n}\right]|\text{vac}\rangle \\ &= \exp\left(-\frac{1}{2}\|\alpha\|^2 - \frac{1}{2}\|\beta\|^2\right)\sum_{n=0}^{\infty}\frac{1}{n!}(\alpha^*\diamond\beta)^n \\ &= \exp\left(-\frac{1}{2}\|\alpha\|^2 - \frac{1}{2}\|\beta\|^2 + \alpha^*\diamond\beta\right).\end{aligned} \tag{4.114}$$

c For example, such a holomorphic functional can be a polynomial with terms consisting of arbitrary numbers of annihilation operators contracted on a coefficient function

$$f[\hat{a}] = \sum_n \int \hat{a}_r(\mathbf{k}_1)\hat{a}_s(\mathbf{k}_2)\cdots\hat{a}_t(\mathbf{k}_n)C_{r,s,\dots,t}(\mathbf{k}_1,\mathbf{k}_2,\dots,\mathbf{k}_n)\,\mathrm{d}_\omega k_1\,\mathrm{d}_\omega k_2\cdots\mathrm{d}_\omega k_n.$$

Even when $\langle \alpha, \beta \rangle = \alpha^* \diamond \beta = 0$, we still have $\langle \alpha | \beta \rangle \neq 0$. As a result, as we found in (3.111), fixed-spectrum coherent states still do not represent an orthogonal set of states.

For the completeness of the fixed-spectrum coherent states, we wait until Chapter 6, after we have discussed Wigner functionals and functional integration in Chapter 5. It makes the calculation much easier.

4.4.3 Complete orthogonal Fock basis

The expression for the overlap between fixed-spectrum coherent states provides us with a way to reconsider the *completeness of the discrete basis Fock states* introduced in Section 4.2.6. First, we show that the separate Fock states of a discrete basis do not produce a complete basis. For this case, the completeness assumes an operator of the form

$$\hat{L} = \sum_{m,n} |n_m\rangle \langle n_m| = \sum_{m,n} \frac{1}{n!} \hat{a}_m^{\dagger n} |\text{vac}\rangle \langle \text{vac}| \hat{a}_m^n, \tag{4.115}$$

where n is the occupation number (particle-number degree of freedom) and m is the modal index (spatiotemporal degrees of freedom). If $\hat{L}$ is the identity $\mathbb{1}$, then we should get

$$\langle \alpha_1 | \mathbb{1} | \alpha_2 \rangle = \langle \alpha_1 | \alpha_2 \rangle = \exp\left(-\frac{1}{2} \|\alpha_1\|^2 - \frac{1}{2} \|\alpha_2\|^2 + \alpha_1^* \diamond \alpha_2 \right), \tag{4.116}$$

where $|\alpha_1\rangle$ and $|\alpha_2\rangle$ are arbitrary fixed-spectrum coherent states. For convenience, we expand the coherent state parameter functions in terms of the normalised parameter functions of the discrete Fock basis. They are thus given by

$$\alpha_1(\mathbf{k}) = \sum_m A_m F_m(\mathbf{k}) \quad \text{and} \quad \alpha_2(\mathbf{k}) = \sum_m B_m F_m(\mathbf{k}), \tag{4.117}$$

so that

$$A_m = \langle F_m, \alpha_1 \rangle = F_m^* \diamond \alpha_1 \quad \text{and} \quad B_m = \langle F_m, \alpha_2 \rangle = F_m^* \diamond \alpha_2. \tag{4.118}$$

What we find is that

$$\begin{aligned}
\langle \alpha_1 | \hat{L} | \alpha_2 \rangle &= \sum_{m,n} \frac{1}{n!} \langle \alpha_1 | \hat{a}_m^{\dagger n} |\text{vac}\rangle \langle \text{vac}| \hat{a}_m^n | \alpha_2 \rangle \\
&= \exp\left(-\frac{1}{2} \|\alpha_1\|^2 - \frac{1}{2} \|\alpha_2\|^2 \right) \sum_{m,n} \frac{1}{n!} (\alpha_1^* \diamond F_m)^n (F_m^* \diamond \alpha_2)^n \\
&= \exp\left(-\frac{1}{2} \|\alpha_1\|^2 - \frac{1}{2} \|\alpha_2\|^2 \right) \sum_{m,n} \frac{1}{n!} A_m^{*n} B_m^n \\
&= \exp\left(-\frac{1}{2} \|\alpha_1\|^2 - \frac{1}{2} \|\alpha_2\|^2 \right) \sum_m \exp\left(A_m^* B_m \right).
\end{aligned} \tag{4.119}$$

Clearly,

$$\sum_m \exp\left(A_m^* B_m\right) \neq \exp\left(\alpha_1^* \diamond \alpha_2\right) = \exp\left(\sum_m A_m^* B_m\right). \tag{4.120}$$

It confirms the observation in Section 4.2.6, namely that the basis of individual discrete Fock states is not complete.

Next, we consider the case where the basis consists of all possible tensor products of discrete Fock basis elements. In Section 4.2.6, we found that most of these states are unphysical in that they consist of infinitely many photons. However, to consider their completeness, we retain these unphysical states.

It becomes notationally cumbersome to represent a generic element of the basis. One way to represent them is

$$|\{n\}\rangle = \bigotimes_m \frac{1}{\sqrt{n(m)!}} \hat{a}_m^{\dagger n(m)} |\mathrm{vac}\rangle , \tag{4.121}$$

where $\{n\} = \{n(1), n(2), \ldots\}$ is the set of occupation numbers for all the modes, with $n(m)$ representing the *occupation number* of the m-th mode. The operator for the completeness is then given by

$$\hat{L}' = \bigotimes_m \left(\sum_n \frac{1}{n!} \hat{a}_m^{\dagger n} |\mathrm{vac}\rangle \langle \mathrm{vac}| \hat{a}_m^n\right). \tag{4.122}$$

Although the notation calls for different factors with operators that are separately applied to the vacuum state, it is understood to represent the same vacuum. Following the same process using the overlaps with fixed-spectrum coherent states, we obtain

$$\begin{aligned}
\langle \alpha_1 | \hat{L}' | \alpha_2 \rangle &= \langle \alpha_1 | \bigotimes_m \left(\sum_n \frac{1}{n!} \hat{a}_m^{\dagger n} |\mathrm{vac}\rangle \langle \mathrm{vac}| \hat{a}_m^n\right) |\alpha_2\rangle \\
&= \exp\left(-\frac{1}{2}\|\alpha_1\|^2 - \frac{1}{2}\|\alpha_2\|^2\right) \prod_m \left(\sum_n \frac{1}{n!} A_m^{*n} B_m^n\right) \\
&= \exp\left(-\frac{1}{2}\|\alpha_1\|^2 - \frac{1}{2}\|\alpha_2\|^2 + \sum_m A_m^* B_m\right) \\
&= \exp\left(-\frac{1}{2}\|\alpha_1\|^2 - \frac{1}{2}\|\alpha_2\|^2 + \alpha_1^* \diamond \alpha_2\right) \equiv \langle \alpha_1 | \alpha_2 \rangle .
\end{aligned} \tag{4.123}$$

Hence, $\hat{L}' \equiv \mathbb{1}$, which means that the set of all tensor products of discrete Fock states is a complete orthonormal basis, provided that one considers all possible tensor products of the individual Fock bases for all the modes (including the unphysical ones).

In view of the discussion in Section 4.2.6, we make the following observations:
(a) The discrete tensor product Fock basis is a complete orthogonal basis for the Hilbert space of all quantum optical states.
(b) The elements of this basis are normalized and as such are elements of the Hilbert space of all quantum optical states.
(c) The majority of the basis elements represent unphysical states with an infinite number of photons. One can exclude such unphysical states from the basis, but then the completeness as demonstrated above may not apply anymore.
(d) The elements of the Fock basis are *not* eigenvectors of the free Hamiltonian (4.28) in which all the spatiotemporal degrees of freedom are incorporated.
(e) Together with the unphysical states, the basis contains an uncountable infinite number of elements, implying that the Hilbert space is *not separable*. However, with the unphysical states excluded, the resulting Hilbert space of all physical quantum optical states is *separable*.

The separability of the Hilbert space is significant because many theorems on which quantum mechanics is based assume that the Hilbert space is separable. Despite the significance of this basis, it is not convenient to use in calculations for our purposes. Therefore, we do not concern ourselves further with this issue.

4.5 Quadrature bases

The next candidate for a suitable basis in terms of which the states in the Hilbert space of all quantum optical states can be expanded are the *quadrature eigenvectors*. However, there are different ways to incorporate the spatiotemporal degrees of freedom into the quadrature eigenvectors. One way is to proceed with the fixed-spectrum approach by expanding the particle-number quadrature eigenvectors of Section 3.1.8 in terms of the fixed-spectrum Fock states. Unfortunately, the resulting bases do not have the properties we are looking for, as shown in the next section. Another approach, which is discussed subsequently, is to solve the suitable eigenvalue equations.

4.5.1 Fixed-spectrum quadrature bases

For the incorporation of the spatiotemporal and spin degrees of freedom into the quadrature bases elements, we can consider a similar procedure to what we used for the coherent states in Section 4.4. Such *fixed-spectrum quadrature bases* are defined through their expansions in terms of *fixed-spectrum Fock states* of Section 4.2.5, leading to generalizations of the equivalent expansions for the particle-number degree of freedom only case given by (3.66). The elements in these bases thus become

$$|q_F\rangle = \sum_n |n_F\rangle\, \psi_n^*(q) \quad \text{and} \quad |p_F\rangle = \sum_n |n_F\rangle\, \widetilde{\psi}_n^*(p). \tag{4.124}$$

The particle-number coefficient functions $\psi_n^*(q)$ and $\widetilde{\psi}_n^*(p)$ are not affected: by performing the same procedure that was followed in Section 3.1.8 to compute the coefficient function, we end up with the same result.

Unfortunately, these fixed-spectrum quadrature bases suffer from an undesirable property relating to the inner products between two different fixed-spectrum quadrature basis elements. We already know that when the spectra of two Fock states are the same, their inner product is the same as the case where we ignore the spatiotemporal degrees of freedom $\langle m_F|n_F\rangle = \delta_{m,n}$. Therefore, the inner product between q-basis elements with the same spectra or between p-basis elements with the same spectra (indicated by the same subscript F) remain the same as in cases without the spatiotemporal degrees of freedom, so that

$$\langle q_F|q'_F\rangle = \delta(q-q') \quad \text{and} \quad \langle p_F|p'_F\rangle = 2\pi\delta(p-p'). \tag{4.125}$$

Inner products between q- and p-basis elements with the same spectrum also produce

$$\langle q_F|p_F\rangle = \exp(iqp). \tag{4.126}$$

Provided that the Fock state parameter functions of the bases are the same, the elements of the quadrature bases are orthogonal for different values of the quadrature variables.

However, the inner product between two fixed-spectrum quadrature basis elements with different Fock state parameter functions is not zero. Based on (4.65), the inner product between two q-basis elements with different spectra becomes

$$\langle q_F|q'_G\rangle = \sum_{mn}\langle m_F|n_G\rangle\psi_m(q)\psi_n^*(q') = \sum_n \mu^n\psi_n(q)\psi_n^*(q'), \tag{4.127}$$

where $\mu = F^* \diamond G$. Using (3.79) and Mehler's formula (3.83), we obtain an expression for this inner product that reads

$$\langle q_F|q'_G\rangle = \frac{\pi^{-1/2}}{\sqrt{1-\mu^2}}\exp\left[-\frac{(q-q')^2\mu}{1-\mu^2}\right]\exp\left[-\frac{(1-\mu)(q^2+q'^2)}{2(1+\mu)}\right]. \tag{4.128}$$

For $G = F$, we have $\mu = 1$ and the inner product, when properly treated as a limit process, produces the required *orthogonality condition* in (4.125). On the other hand, if F and G are orthogonal, we have $\mu = 0$. Then

$$\langle q_F|q'_G\rangle = \frac{1}{\sqrt{\pi}}\exp\left[-\frac{1}{2}\left(q^2+q'^2\right)\right] \equiv \psi_0(q)\psi_0^*(q') \neq 0. \tag{4.129}$$

As a result, the quadrature bases elements lose their orthogonality. Even using a discrete modal basis for the spatiotemporal degrees of freedom, we still do not obtain an orthogonal quadrature basis for both particle-number and spatiotemporal degrees of freedom. Therefore, the concept of a *stratified* Hilbert space does not work with such fixed-spectrum quadrature bases defined in terms of orthogonal spatiotemporal modes.

4.5.2 Quadrature bases in all degrees of freedom

Next, we use a different approach to incorporation of the *spatiotemporal* and *spin* degrees of freedom into the *quadrature bases* elements. For this purpose, we consider the *eigenvectors* of the *quadrature operators* in terms of all degrees of freedom. First, we define these quadrature operators with all the degrees of freedom incorporated.

Given the ladder operators that are defined with all degrees of freedom in Section 4.1.1, we define the associated quadrature operators as

$$\begin{aligned} \hat{q}_s(\mathbf{k}) &\triangleq \frac{1}{\sqrt{2}}\left[\hat{a}_s(\mathbf{k}) + \hat{a}_s^{\dagger}(\mathbf{k})\right], \\ \hat{p}_s(\mathbf{k}) &\triangleq -\mathrm{i}\frac{1}{\sqrt{2}}\left[\hat{a}_s(\mathbf{k}) - \hat{a}_s^{\dagger}(\mathbf{k})\right], \end{aligned} \tag{4.130}$$

in analogy to how it is done in (3.46) for the particle-number degree of freedom only. It also means that

$$\begin{aligned} \hat{a}_s(\mathbf{k}) &= \frac{1}{\sqrt{2}}\left[\hat{q}_s(\mathbf{k}) + \mathrm{i}\hat{p}_s(\mathbf{k})\right], \\ \hat{a}_s^{\dagger}(\mathbf{k}) &= \frac{1}{\sqrt{2}}\left[\hat{q}_s(\mathbf{k}) - \mathrm{i}\hat{p}_s(\mathbf{k})\right]. \end{aligned} \tag{4.131}$$

Based on (4.15), these quadrature operators obey Lorentz covariant *commutation relations* that read

$$\begin{aligned} &[\hat{q}_s(\mathbf{k}_1), \hat{q}_r(\mathbf{k}_2)] = [\hat{p}_s(\mathbf{k}_1), \hat{p}_r(\mathbf{k}_2)] = 0, \\ &[\hat{q}_s(\mathbf{k}_1), \hat{p}_r(\mathbf{k}_2)] = \mathrm{i}(2\pi)^3\omega_1\delta_{s,r}\delta(\mathbf{k}_1 - \mathbf{k}_2) \equiv \mathrm{i}\mathbf{1}_{s,r}(\mathbf{k}_1, \mathbf{k}_2). \end{aligned} \tag{4.132}$$

The quadrature operators defined in (4.130) are clearly Hermitian. Therefore, we expect them to have real-valued eigenvalues. It suggests that we can formulate eigenvalue equations of the form

$$\hat{q}_s(\mathbf{k})\,|q\rangle = |q\rangle\, q_s(\mathbf{k}) \quad \text{and} \quad \hat{p}_s(\mathbf{k})\,|p\rangle = |p\rangle\, p_s(\mathbf{k}), \tag{4.133}$$

where $q_s(\mathbf{k})$ and $p_s(\mathbf{k})$ are real-valued eigenvalue functions. Here, we have a situation that is similar to what we have for the fixed-spectrum coherent states with the eigenvalue equation in (4.111) for the annihilation operator. The eigenvalue is itself a function of the wavevector and carries a spin index. If one allows for all possible such functions, even if one imposes some restrictions on the set of functions, this is still a truly vast space. Since each eigenvalue function is associated with a unique eigenvector, the sets of eigenvectors $|q\rangle$ and $|p\rangle$ are equally as large.

The set of all eigenvalue functions for any of these quadrature operators forms a *functional space*. Any quantity defined as a mapping from this functional space (as its *functional domain*) to the set of complex numbers, becomes a *functional* of the eigenvalue functions acting as independent real-valued field variables. A *field variable* is a generalization of an independent variable—each element in the functional domain of a functional is itself a function, which we call a *field*. For example, the generalization of (3.23) in the current quantum context with all degrees of freedom included leads to

$$\psi[q] = \langle q|\psi\rangle \quad \text{and} \quad \tilde{\psi}[p] = \langle p|\psi\rangle, \tag{4.134}$$

where $\psi[q]$ and $\tilde{\psi}[p]$ are the *wave functionals* of the state $|\psi\rangle$ in terms of the two respective *quadrature field variables*.

A direct result of the eigenvalue equations in (4.133) is that any operator that is a *holomorphic functional* of either of the quadrature operators produces an equivalent *holomorphic functional* with the quadrature eigenvalue function as independent quadrature field variable when applied to an eigenvector of the corresponding quadrature operator

$$\begin{aligned} h[\hat{q}]\,|q\rangle &= |q\rangle\, h[q], \\ h[\hat{p}]\,|p\rangle &= |p\rangle\, h[p]. \end{aligned} \tag{4.135}$$

These relationships, which are generalizations of those in (3.34), can also be applied in the reverse direction where holomorphic functionals of the quadrature field variables are *pulled back* through the eigenvectors to produce operators in terms of the associated quadrature operators.

As a first attempt to solve (4.133), we consider the *fixed-spectrum quadrature bases*, discussed in Section 4.5.1, to see if they can satisfy these eigenvalue equations. When we substitute such a basis element into the eigenvalue equation for $\hat{q}$, we get

$$\begin{aligned} \hat{q}_s(\mathbf{k})\,|q_F\rangle &= \frac{1}{\sqrt{2}}\sum_n \left[\hat{a}_s(\mathbf{k}) + \hat{a}_s^{\dagger}(\mathbf{k})\right] |n_F\rangle\, \psi_n^*(q) \\ &= \frac{1}{\sqrt{2}}\sum_n \left[|(n-1)_F\rangle\, F_s(\mathbf{k})\sqrt{n} + |n_F\rangle\,|\mathbf{k},s\rangle\right] \psi_n^*(q) \\ &= \frac{1}{\sqrt{2}}\sum_n |n_F\rangle \left[F_s(\mathbf{k})\sqrt{n+1}\psi_{n+1}^*(q) + |\mathbf{k},s\rangle\,\psi_n^*(q)\right]. \end{aligned} \tag{4.136}$$

The result does not reproduce the original basis element. Therefore, the fixed-spectrum quadrature bases elements are not eigenvectors of the eigenvalue equation in (4.133).

To find the eigenvectors, we define them in terms of operators that *render* them from the vacuum state $|q\rangle = \hat{\Xi}\,|\text{vac}\rangle$. The eigenvalue equation then reads

$$\hat{q}_s(\mathbf{k})\hat{\Xi}\,|\text{vac}\rangle = \hat{\Xi}\,|\text{vac}\rangle\, q_s(\mathbf{k}). \tag{4.137}$$

Hence, assuming $\hat{\Xi}$ is invertible, we have

$$\hat{\Xi}^{-1}\hat{q}_s(\mathbf{k})\hat{\Xi}\,|\text{vac}\rangle = |\text{vac}\rangle\, q_s(\mathbf{k}). \tag{4.138}$$

An ansatz for the *rendering operator* of the *q-eigenvectors* is given by

$$\hat{\Xi} = \exp(\hat{a}^\dagger \diamond G + c\hat{a}^\dagger \diamond \hat{a}^\dagger), \tag{4.139}$$

where G denotes an unknown parameter function and c is an unknown constant. The inverse rendering operator is given by

$$\hat{\Xi}^{-1} = \exp(-\hat{a}^\dagger \diamond G - c\hat{a}^\dagger \diamond \hat{a}^\dagger). \tag{4.140}$$

The equation then becomes

$$\exp(-\hat{a}^\dagger \diamond G - c\hat{a}^\dagger \diamond \hat{a}^\dagger)\hat{q}_s(\mathbf{k}) \exp(\hat{a}^\dagger \diamond G + c\hat{a}^\dagger \diamond \hat{a}^\dagger)\,|\text{vac}\rangle = |\text{vac}\rangle\, q_s(\mathbf{k}), \tag{4.141}$$

which is of the form suitable for the identity in (3.103). Without the vacuum state, the condition that we impose is that the right-hand side must be proportional to the identity operator. However, since it operates on a vacuum state, the result can also contain a term with an annihilation operator.

First, we need the commutation relation

$$[-\hat{a}^\dagger \diamond G - c\hat{a}^\dagger \diamond \hat{a}^\dagger, \hat{q}_s(\mathbf{k})] = \frac{1}{\sqrt{2}}G_s(\mathbf{k}) + \sqrt{2}c\hat{a}_s^\dagger(\mathbf{k}). \tag{4.142}$$

The identity in (3.103) times $\sqrt{2}$ then gives

$$\begin{aligned}&\sqrt{2}\exp(-\hat{a}^\dagger \diamond G - c\hat{a}^\dagger \diamond \hat{a}^\dagger)\hat{q}_s(\mathbf{k}) \exp(\hat{a}^\dagger \diamond G + c\hat{a}^\dagger \diamond \hat{a}^\dagger)\\ &= \hat{a}_s(\mathbf{k}) + \hat{a}_s^\dagger(\mathbf{k}) + G_s(\mathbf{k}) + 2c\hat{a}_s^\dagger(\mathbf{k}).\end{aligned} \tag{4.143}$$

We can set $c = -\frac{1}{2}$ to remove $\hat{a}^\dagger$. The annihilation operator $\hat{a}$ is removed when the expression is applied to the vacuum. The remaining parameter function is then related to the eigenvalue function by

$$G_s(\mathbf{k}) = \sqrt{2}q_s(\mathbf{k}). \tag{4.144}$$

It represents the parameter function labelling the different eigenvectors. The eigenvectors produced by this rendering operator are not normalizable, as expected. However, we can multiply it by a suitable prefactor for convenience. The rendering operator for the q-eigenvectors thus becomes

$$\hat{\Xi}_q = Q_0 \exp\left(\sqrt{2}\hat{a}^\dagger \diamond q - \frac{1}{2}\hat{a}^\dagger \diamond \hat{a}^\dagger\right), \tag{4.145}$$

where Q_0 is an unknown prefactor.

Exercise 4.7. Show that the equivalent rendering operator for the p-eigenvectors is

$$\hat{\Xi}_p = P_0 \exp\left(\mathrm{i}\sqrt{2}\hat{a}^\dagger \diamond p + \frac{1}{2}\hat{a}^\dagger \diamond \hat{a}^\dagger\right), \tag{4.146}$$

where P_0 is another unknown prefactor.

The quadrature eigenvectors are thus given by

$$\begin{aligned}
|q\rangle &= \hat{\Xi}_q\,|\mathrm{vac}\rangle = Q_0 \exp\left(\sqrt{2}\hat{a}^\dagger \diamond q - \frac{1}{2}\hat{a}^\dagger \diamond \hat{a}^\dagger\right)|\mathrm{vac}\rangle\,, \\
|p\rangle &= \hat{\Xi}_p\,|\mathrm{vac}\rangle = P_0 \exp\left(\mathrm{i}\sqrt{2}\hat{a}^\dagger \diamond p + \frac{1}{2}\hat{a}^\dagger \diamond \hat{a}^\dagger\right)|\mathrm{vac}\rangle\,.
\end{aligned} \tag{4.147}$$

Since the quadrature operators are Hermitian, we expect to find that their eigenvectors are orthogonal for different eigenvalue functions. In the next section, we calculate the inner products between such eigenvectors in detail.

4.5.3 Inner products

To determine whether the quadrature eigenvectors are orthogonal, we need to calculate $\langle q|q'\rangle$ and $\langle p|p'\rangle$, and at the same time it would be useful to calculate $\langle q|p\rangle$. For this purpose, we represent the eigenvectors in terms of their rendering operators. The inner products are then represented by

$$\begin{aligned}
\langle q|q'\rangle &= \langle \mathrm{vac}|\,\hat{\Xi}_q[q]\hat{\Xi}_q^\dagger[q']\,|\mathrm{vac}\rangle\,, \\
\langle p|p'\rangle &= \langle \mathrm{vac}|\,\hat{\Xi}_p[p]\hat{\Xi}_p^\dagger[p']\,|\mathrm{vac}\rangle\,, \\
\langle q|p\rangle &= \langle \mathrm{vac}|\,\hat{\Xi}_q[q]\hat{\Xi}_p^\dagger[p]\,|\mathrm{vac}\rangle\,,
\end{aligned} \tag{4.148}$$

where we show the functional dependence of the rendering operators on the parameter functions.

Since the two terms in the exponent of each rendering operator commute, we can separate them into a product of exponentiated operators. The different overlaps in (4.148) are generically represented by

$$\begin{aligned}
\langle g|h\rangle = &\ \langle \mathrm{vac}|\,G_0^* H_0 \exp\left(\sqrt{2}c_1 g \diamond \hat{a}\right)\exp\left(c_2\hat{R}\right) \\
&\times \exp\left(\sqrt{2}c_3\hat{a}^\dagger \diamond h\right)\exp\left(c_4\hat{R}^\dagger\right)|\mathrm{vac}\rangle\,,
\end{aligned} \tag{4.149}$$

where g and h are real-valued parameter functions, G_0 and H_0 are the unknown complex prefactors for the two eigenvectors, c_1, c_2, c_3, and c_4 are constants that determine whether it is a q- or p-eigenvector, and for convenience we defined

$$\hat{R} \triangleq \frac{1}{2}\hat{a} \diamond \hat{a} \quad \text{and} \quad \hat{R}^{\dagger} \triangleq \frac{1}{2}\hat{a}^{\dagger} \diamond \hat{a}^{\dagger}. \tag{4.150}$$

To evaluate (4.149), we perform commutations rearranging the exponentiated operators in normal order. However, in the process a more complex expression is produced with new operators generated due to commutation among the current ones. Therefore, we first need to consider the algebra of all such operators.

4.5.3.1 Quadrature operator algebra

Here, we derived the commutation relations that are associated with the operators that appear in (4.149) and all those that they can generate. These commutation relations are based on those for the ladder operators $\hat{a}_s(\mathbf{k})$ and $\hat{a}_r^{\dagger}(\mathbf{k})$ in (4.15), which is represented as in (4.55).

Most of the *commutation relations* in the *algebra* can be computed without much effort. For example,

$$\begin{gathered}
\left[\hat{a}_s(\mathbf{k}), \hat{a}^{\dagger} \diamond g\right] = g_s(\mathbf{k}), \quad \left[\hat{R}, \hat{a}_s^{\dagger}(\mathbf{k})\right] = \hat{a}_s(\mathbf{k}), \quad \left[\hat{R}, \hat{a}^{\dagger} \diamond g\right] = g \diamond \hat{a}, \\
\left[g^* \diamond \hat{a}, \hat{a}_s^{\dagger}(\mathbf{k})\right] = g_s^*(\mathbf{k}), \quad \left[\hat{a}_s(\mathbf{k}), \hat{R}^{\dagger}\right] = \hat{a}_s^{\dagger}(\mathbf{k}), \quad \left[g^* \diamond \hat{a}, \hat{R}^{\dagger}\right] = \hat{a}^{\dagger} \diamond g^*, \\
\left[g^* \diamond \hat{a}, \hat{a}^{\dagger} \diamond g'\right] = g^* \diamond g'.
\end{gathered} \tag{4.151}$$

One commutation relation that needs special mention is the following:

$$\left[\hat{R}, \hat{R}^{\dagger}\right] = \frac{1}{4}\left[\hat{a} \diamond \hat{a}, \hat{a}^{\dagger} \diamond \hat{a}^{\dagger}\right] = \frac{1}{2}\left(\hat{a}^{\dagger} \diamond \hat{a} + \hat{a} \diamond \hat{a}^{\dagger}\right) \triangleq \hat{s}, \tag{4.152}$$

where $\hat{s}$ is a *symmetrized number operator*. It can also be written in terms of the normal-ordered number operator, defined in (4.48), and a divergent constant

$$\hat{s} = \hat{n} + \frac{1}{2}\Omega, \tag{4.153}$$

where the *divergent constant* Ω is given by

$$\Omega \triangleq 2\int \delta(0)\, \mathrm{d}^3k. \tag{4.154}$$

The factor of 2 comes from the two spin states. This divergent constant represents the *cardinality of the Fourier space*. It is the cardinality of countable infinity, often denoted by $\aleph_0$, which follows from the fact that a basis with a countably infinite number of basis functions is necessary to represent all functions with finite norm, as required by Fourier analysis.

Note that, due to the commutation relation defined in (4.15), the factors of $(2\pi)^3\omega$ cancel and, therefore, do not appear in (4.154). Both $\delta(\mathbf{k}=0)$ and $\int \mathrm{d}^3k$ represent divergent constants with the same cardinality, but they carry different dimensions. The dimensions of the former is [distance^3] whereas that of the latter is [distance^{-3}]. So, the combination Ω is dimensionless. The commutation relations involving the symmetrized number operator are

$$
\begin{aligned}
&[\hat{a}_s(\mathbf{k}),\hat{s}] = \hat{a}_s(\mathbf{k}), \quad [g^*\diamond\hat{a},\hat{s}] = g^*\diamond\hat{a}, \quad \left[\hat{R},\hat{s}\right] = 2\hat{R}, \\
&\left[\hat{s},\hat{a}_s^\dagger(\mathbf{k})\right] = \hat{a}_s^\dagger(\mathbf{k}), \quad \left[\hat{s},\hat{a}^\dagger\diamond g\right] = \hat{a}^\dagger\diamond g, \quad \left[\hat{s},\hat{R}^\dagger\right] = 2\hat{R}^\dagger.
\end{aligned} \tag{4.155}
$$

The resulting algebra closes.

We also need the commutators of the exponentiated operators. For this purpose, we use the identity in (3.103). The results can be readily converted to expressions for the commutations. These results are

$$
\begin{aligned}
\exp(cg^*\diamond\hat{a})\hat{a}^\dagger\diamond h\exp(-cg^*\diamond\hat{a}) &= \hat{a}^\dagger\diamond h + cg^*\diamond h, \\
\exp(cg^*\diamond\hat{a})\hat{R}^\dagger\exp(-cg^*\diamond\hat{a}) &= \hat{R}^\dagger + c\hat{a}^\dagger\diamond g^* + \frac{1}{2}c^2g^*\diamond g^*, \\
\exp(cg^*\diamond\hat{a})\hat{s}\exp(-cg^*\diamond\hat{a}) &= \hat{s} + cg^*\diamond\hat{a}, \\
\exp(c\hat{R})\hat{a}^\dagger\diamond g\exp(-c\hat{R}) &= \hat{a}^\dagger\diamond g + cg\diamond\hat{a}, \\
\exp(c\hat{R})\hat{R}^\dagger\exp(-c\hat{R}) &= \hat{R}^\dagger + c\hat{s} + c^2\hat{R}, \\
\exp(c\hat{R})\hat{s}\exp(-c\hat{R}) &= \hat{s} + 2c\hat{R}, \\
\exp(c\hat{s})\hat{a}^\dagger\diamond g\exp(-c\hat{s}) &= \exp(c)\hat{a}^\dagger\diamond g, \\
\exp(c\hat{s})\hat{R}^\dagger\exp(-c\hat{s}) &= \exp(2c)\hat{R}^\dagger,
\end{aligned} \tag{4.156}
$$

where c is an arbitrary complex value.

Exercise 4.8. Ignoring all the operators of the quadrature operator algebra that contain parameter functions, we end up with the set $\{\hat{R},\hat{R}^\dagger,\hat{s}\}$. Show that one can form linear combinations of these operators to produce the $\mathfrak{su}(1,1)$ algebra $\{\hat{K}_x,\hat{K}_y,\hat{K}_z\}$, satisfying the commutation relations

$$
\left[\hat{K}_x,\hat{K}_y\right] = -\mathrm{i}\hat{K}_z, \quad \left[\hat{K}_y,\hat{K}_z\right] = \mathrm{i}\hat{K}_x, \quad \text{and} \quad \left[\hat{K}_z,\hat{K}_x\right] = \mathrm{i}\hat{K}_y. \tag{4.157}
$$

4.5.3.2 Solving the equation

Having discussed the required operator algebra, we can return to the calculation of the inner products. To aid our calculation, we introduce an auxiliary variable t into the exponents in (4.149). The expected form of the normal-ordered product of operators consists of a sequence of exponentiated operators where we introduce unknown functions of t into the exponents. The resulting equation reads

$$
\begin{aligned}
&\exp\left(c_1 t g_1^* \diamond \hat{a}\right) \exp\left(c_2 t \hat{R}\right) \exp\left(c_3 t \hat{a}^\dagger \diamond g_2\right) \exp\left(c_4 t \hat{R}^\dagger\right) \\
&= \exp[h_0(t)] \exp\left[\hat{a}^\dagger \diamond h_1(t)\right] \exp\left[h_2(t)\hat{R}^\dagger\right] \\
&\quad \times \exp\left[h_3(t)\hat{s}\right] \exp\left[h_4(t) \diamond \hat{a}\right] \exp\left[h_5(t)\hat{R}\right],
\end{aligned}
\tag{4.158}
$$

where $h_m(t)$ with $m = 0 \cdots 5$ denotes unknown functions. The unknown functions go to zero for $t = 0$ and the original expression is recovered with $t = 1$. Factors of $\sqrt{2}$ are absorbed into the parameter functions g_1^* and g_2. Note that $h_1(t)$ and $h_4(t)$ are also functions of the wavevector and are thus contracted with the ladder operators.

Next, we apply a derivative with respect to t and then remove as many of the exponentiated operators as possible by operating on both sides of the equation with the respective inverse operators applied on their right-hand sides. The resulting equation now consists of several terms having the form $\exp(\hat{x})\hat{y}\exp(-\hat{x})$. These terms can be simplified with the aid of the generic expressions in (4.156), which are based on the commutation relations in (4.151) and (4.155). After these simplifications, the equation can be separated into six differential equations, based on the forms of the operators in each term. The six differential equations are given by

$$
\begin{aligned}
\partial_t h_0(t) &= \frac{1}{2} c_2 d(t) h_1(t) \diamond h_1(t) + c_1 d(t) g_1^* \diamond h_1(t) + c_2 c_3 t h_1(t) \diamond g_2 \\
&\quad + \frac{1}{2} c_1^2 c_4 t^2 g_1^* \diamond g_1^* + c_1 c_3 t g_1^* \diamond g_2, \\
\partial_t h_1(t) &= c_2 d(t) h_2(t) h_1(t) + c_2 c_4 t h_1(t) + c_1 c_4 t g_1^* \\
&\quad + c_1 d(t) h_2(t) g_1^* + c_2 c_3 t h_2(t) g_2 + c_3 g_2, \\
\partial_t h_2(t) &= c_4 + 2 c_2 c_4 t h_2(t) + c_2 d(t) h_2(t)^2, \\
\partial_t h_3(t) &= c_2 d(t) h_2(t) + c_2 c_4 t, \\
\partial_t h_4(t) &= \left[c_2 d(t) h_1(t) + c_1 d(t) g_1^* + c_2 c_3 t g_2\right] \exp[h_3(t)], \\
\partial_t h_5(t) &= c_2 d(t) \exp[2 h_3(t)],
\end{aligned}
\tag{4.159}
$$

where $d(t) = 1 + c_2 c_4 t^2$. The solutions of these differential equations are

$$
\begin{aligned}
h_0(t) &= \frac{1}{2} \frac{\left(c_1^2 c_4 t g_1^* \diamond g_1^* + c_2 c_3^2 t g_2 \diamond g_2 + 2 c_1 c_3 g_1^* \diamond g_2\right) t^2}{1 - c_2 c_4 t^2}, \\
h_1(t) &= \frac{c_1 c_4 t^2 g_1^* + c_3 t g_2}{1 - c_2 c_4 t^2}, \quad h_2(t) = \frac{c_4 t}{1 - c_2 c_4 t^2}, \quad h_3(t) = -\ln\left(1 - c_2 c_4 t^2\right), \\
h_4(t) &= \frac{c_1 t g_1^* + c_2 c_3 t^2 g_2}{1 - c_2 c_4 t^2}, \quad h_5(t) = \frac{c_2 t}{1 - c_2 c_4 t^2}.
\end{aligned}
\tag{4.160}
$$

These functions can now be substituted back into (4.158). To obtain the expression for the overlap, we apply the vacuum state on both sides. For this purpose, we note that

$$
\langle \mathrm{vac} | \exp(K\hat{s}) | \mathrm{vac} \rangle = \exp\left(\frac{1}{2} K \Omega\right),
\tag{4.161}
$$

based on (4.153). So, the generic overlap becomes

$$\langle g|h\rangle = G_0^* H_0 \exp[h_0(t)] \exp\left[\frac{1}{2}h_3(t)\Omega\right]\bigg|_{t=1} . \tag{4.162}$$

At this point, the cardinal number Ω enters the formalism. Henceforth, it is often found in expressions. Any formalism for an infinite number of degrees of freedom inevitably contains such divergent cardinal numbers, unless some fundamental cut-off scale renders everything finite. It is nevertheless nothing to be concerned about because such divergent numbers always cancel in calculations of observable quantities to produce well-defined finite results.

4.5.3.3 Overlaps

For the case of $\langle q|p\rangle$, we set $c_1 = 1$, $c_2 = -1$, $c_3 = \mathrm{i}$, $c_4 = 1$, $g_1^* = \sqrt{2}q$, $g_2 = \sqrt{2}p$, $G_0^* = Q_0^*$, $H_0 = P_0$ and finally $t = 1$. The resulting inner product reads

$$\langle q|p\rangle = \frac{Q_0^* P_0}{2^{\Omega/2}} \exp\left(\frac{1}{2}q \diamond q + \frac{1}{2}p \diamond p + \mathrm{i}q \diamond p\right). \tag{4.163}$$

For convenience, we assume that

$$Q_0^* P_0 = 2^{\Omega/2} \exp\left(-\frac{1}{2}q \diamond q - \frac{1}{2}p \diamond p\right). \tag{4.164}$$

As a result, the overlap becomes

$$\langle q|p\rangle = \exp(\mathrm{i}q \diamond p) . \tag{4.165}$$

Next, we consider $\langle q|q'\rangle$, for which we substitute $c_1 = 1$, $c_2 = -1$, $c_3 = 1$, $c_4 = -1$, $g_1^* = \sqrt{2}q$, $g_2 = \sqrt{2}q'$, $G_0^* = Q_0^*$, and $H_0 = Q_0$. The two relevant functions become

$$\begin{aligned} h_0(t) &= -\frac{t^2}{1-t^2}\left(tq \diamond q + tq' \diamond q' - 2q \diamond q'\right), \\ h_3(t) &= -\ln\left(1-t^2\right). \end{aligned} \tag{4.166}$$

These functions are both singular at $t = 1$. Therefore, we need to consider the overlap as a limit. For this purpose, we substitute $t = 1 - \epsilon$ and consider the limit where $\epsilon \to 0$.

Based on (4.164), the q- and p-dependencies of the prefactors Q_0^* and P_0 can be

$$Q_0 = \kappa_q \exp\left(-\frac{1}{2}q \diamond q\right) \quad \text{and} \quad P_0 = \kappa_p \exp\left(-\frac{1}{2}p \diamond p\right), \tag{4.167}$$

where κ_q and κ_p are real-valued constants to be determined. The form of Q_0 implies that the overlap becomes

$$\langle q|q'\rangle = \kappa_q^2 \lim_{\epsilon\to 0} \frac{1}{(2\epsilon)^{\Omega/2}} \exp\left[-\frac{1}{2\epsilon}\left(q \diamond q + q' \diamond q' - 2q \diamond q'\right)\right]. \tag{4.168}$$

Note that

$$q \diamond q + q' \diamond q' - 2q \diamond q' \equiv \|q - q'\|^2 \geq 0. \tag{4.169}$$

For $\|q - q'\|^2 > 0$, the limit in (4.168) gives zero and for $\|q - q'\|^2 = 0$, the limit becomes divergent (infinite). Hence, unless $\|q - q'\|^2 = 0$, the inner product is zero. The quantity $\|q-q'\|^2$ represents a *metric* derived from the *Euclidean norm* $\|q\|$. It thus implies that the quadrature field variable q ranges over a normed vector space (of real-valued functions) with the Euclidean norm.

4.5.3.4 Dirac delta functional

It is tempting to conclude that

$$\langle q|q'\rangle = \Lambda_q \delta[q - q'], \tag{4.170}$$

where Λ_q is an unknown orthogonality constant and $\delta[\cdot]$ represents a *Dirac delta functional*. However, a Dirac delta functional would be defined within a *functional integral*. It must then satisfy the requirement

$$\int W[q]\langle q|q'\rangle \mathcal{D}[q] = \Lambda_q W[q'], \tag{4.171}$$

where $W[q]$ is an arbitrary functional of q and $\mathcal{D}[q]$ is the *functional integration measure*, representing integration over the functional space of all real-valued functions with a finite Euclidean norm. (Functional integration is discussed in more detail in Appendix C.) Using (4.168), we have

$$\begin{aligned}
\int W[q]\langle q|q'\rangle \, \mathcal{D}[q] &= \kappa_q^2 \lim_{\epsilon\to 0} \frac{1}{(2\epsilon)^{\Omega/2}} \int W[q] \exp\left(\frac{-\|q - q'\|^2}{2\epsilon}\right) \mathcal{D}[q] \\
&= \kappa_q^2 \lim_{\epsilon\to 0} \int W[\sqrt{2\epsilon} q_0 + q'] \exp\left(-\|q_0\|^2\right) \mathcal{D}[q_0] \\
&= \kappa_q^2 W[q'] \int \exp\left(-\|q_0\|^2\right) \mathcal{D}[q_0] = \kappa_q^2 \pi^{\Omega/2} W[q'].
\end{aligned} \tag{4.172}$$

Here, we first shifted the q-variable and then absorbed the ϵ-factor into q, which then emerged from the measure, causing it to cancel the ϵ-factor in front. At the same time, the argument of the functional becomes independent of the *integration field variable*. In the end, the result has the form required for a Dirac delta functional and it gives the expression for the orthogonality constant

$$\Lambda_q = \kappa_q^2 \pi^{\Omega/2}. \tag{4.173}$$

Selecting $\kappa_q = \pi^{-\Omega/4}$, we get $\Lambda_q = 1$. So, the q-basis obeys an *orthogonality condition*

$$\langle q|q'\rangle = \delta[q-q']. \tag{4.174}$$

Following a similar calculation for the p-basis, one obtains

$$\int W[p]\langle p|p'\rangle\mathcal{D}[p] = \Lambda_p W[p'] = \kappa_p^2\pi^{\Omega/2}W[p']. \tag{4.175}$$

With the choice of κ_q and (4.164), we get $\Lambda_p = (2\pi)^{\Omega}$. The orthogonality condition for the p-basis is then given by

$$\langle p|p'\rangle = (2\pi)^{\Omega}\delta[p-p']. \tag{4.176}$$

Therefore, the prefactors in the definitions of the rendering operators for the quadrature eigenvectors are

$$\begin{aligned} Q_0 &= \pi^{-\Omega/4}\exp\left(-\frac{1}{2}q\diamond q\right), \\ P_0 &= 2^{\Omega/2}\pi^{\Omega/4}\exp\left(-\frac{1}{2}p\diamond p\right). \end{aligned} \tag{4.177}$$

The complete definition of the quadrature eigenvectors in terms of their rendering operators are now given by

$$\begin{aligned} |q\rangle &= \pi^{-\Omega/4}\exp\left(-\frac{1}{2}q\diamond q + \sqrt{2}\hat{a}^{\dagger}\diamond q - \frac{1}{2}\hat{a}^{\dagger}\diamond\hat{a}^{\dagger}\right)|\text{vac}\rangle, \\ |p\rangle &= 2^{\Omega/2}\pi^{\Omega/4}\exp\left(-\frac{1}{2}p\diamond p + \mathrm{i}\sqrt{2}\hat{a}^{\dagger}\diamond p + \frac{1}{2}\hat{a}^{\dagger}\diamond\hat{a}^{\dagger}\right)|\text{vac}\rangle. \end{aligned} \tag{4.178}$$

Since these two sets of eigenvectors each form orthogonal sets, we can use them as *orthogonal quadrature bases* for the expansions of quantum optical states representing all the degrees of freedom. Below, we consider their *completeness.*

4.5.4 Quadrature expansion of coherent states

We often need to compute the overlaps between coherent states and quadrature bases elements. For $\langle q|\alpha\rangle$, we obtain

$$\begin{aligned} \langle q|\alpha\rangle &= \langle\text{vac}|\hat{\Xi}_q^{\dagger}|\alpha\rangle = \pi^{-\Omega/4}\langle\text{vac}|\exp\left(-\frac{1}{2}q\diamond q + \sqrt{2}q\diamond\hat{a} - \frac{1}{2}\hat{a}\diamond\hat{a}\right)|\alpha\rangle \\ &= \pi^{-\Omega/4}\langle\text{vac}|\alpha\rangle\exp\left(-\frac{1}{2}q\diamond q + \sqrt{2}q\diamond\alpha - \frac{1}{2}\alpha\diamond\alpha\right) \\ &= \pi^{-\Omega/4}\exp\left(-\frac{1}{2}\alpha^*\diamond\alpha - \frac{1}{2}q\diamond q + \sqrt{2}q\diamond\alpha - \frac{1}{2}\alpha\diamond\alpha\right). \end{aligned} \tag{4.179}$$

Since the rendering operators in (4.178) are both *holomorphic functionals* of the creation operator, we used the adjoint of (4.113). In a similar way, we get

$$\langle p|\alpha\rangle = (4\pi)^{\Omega/4}\exp\left(-\frac{1}{2}\alpha^*\diamond\alpha-\frac{1}{2}p\diamond p-\mathrm{i}\sqrt{2}p\diamond\alpha+\frac{1}{2}\alpha\diamond\alpha\right). \tag{4.180}$$

In analogy to (3.118), we can relate the quadrature field variables to complex-valued field variables that are equivalent to the parameter functions associated with the coherent states. The relationship is a generalization of (3.118), given by

$$\alpha_s(\mathbf{k}) = \frac{1}{\sqrt{2}}[q_s(\mathbf{k})+\mathrm{i}p_s(\mathbf{k})] \quad \text{and} \quad \alpha_s^*(\mathbf{k}) = \frac{1}{\sqrt{2}}[q_s(\mathbf{k})-\mathrm{i}p_s(\mathbf{k})]. \tag{4.181}$$

When the complex parameter functions of the coherent states are related to real-valued parameter functions q_0 and p_0 of the quadrature bases, generalizing (3.123), the expressions of the inner products become

$$\begin{aligned}
\langle q|\alpha\rangle &= \pi^{-\Omega/4}\exp\left[-\frac{1}{2}(q-q_0)\diamond(q-q_0)+\mathrm{i}\left(q-\frac{1}{2}q_0\right)\diamond p_0\right],\\
\langle \alpha|q\rangle &= \pi^{-\Omega/4}\exp\left[-\frac{1}{2}(q-q_0)\diamond(q-q_0)-\mathrm{i}\left(q-\frac{1}{2}q_0\right)\diamond p_0\right],\\
\langle p|\alpha\rangle &= (4\pi)^{\Omega/4}\exp\left[-\frac{1}{2}(p-p_0)\diamond(p-p_0)-\mathrm{i}q_0\diamond\left(p-\frac{1}{2}p_0\right)\right],\\
\langle \alpha|p\rangle &= (4\pi)^{\Omega/4}\exp\left[-\frac{1}{2}(p-p_0)\diamond(p-p_0)+\mathrm{i}q_0\diamond\left(p-\frac{1}{2}p_0\right)\right].
\end{aligned} \tag{4.182}$$

Here, we also show the adjoint expressions. These expressions are reminiscent of their equivalents without the spatiotemporal degrees of freedom in Section 3.2.7.

4.5.5 Completeness of quadrature bases

Now we investigate the *completeness* of these quadrature bases. For this purpose, we consider the quantity

$$\hat{B} = \int |q\rangle\,\langle q|\ \mathcal{D}[q], \tag{4.183}$$

expressed as a functional integral equivalent to (4.171) with q representing an *integration field variable*. (*Functional integration* is discussed in Appendix C.) Overlapping $\hat{B}$ on both sides by two different coherent states, we obtain

$$\begin{aligned}
\langle\alpha_1|\,\hat{B}\,|\alpha_2\rangle &= \int\langle\alpha_1|q\rangle\langle q|\alpha_2\rangle\ \mathcal{D}[q]\\
&= \pi^{-\Omega/2}\int\exp\left[-\frac{1}{2}(q-q_1)\diamond(q-q_1)-\mathrm{i}p_1\diamond\left(q-\frac{1}{2}q_1\right)\right.\\
&\quad\left.-\frac{1}{2}(q-q_2)\diamond(q-q_2)+\mathrm{i}p_2\diamond\left(q-\frac{1}{2}q_2\right)\right]\ \mathcal{D}[q],
\end{aligned} \tag{4.184}$$

where we used (4.182), and defined

$$\alpha_1 = \frac{1}{\sqrt{2}}(q_1 + \mathrm{i}p_1) \quad \text{and} \quad \alpha_2 = \frac{1}{\sqrt{2}}(q_2 + \mathrm{i}p_2). \tag{4.185}$$

The functional integral over q in (4.184) produces a factor of $\pi^{\Omega/2}$, which removes the factor in front. The result, expressed in terms of α_1 and α_2, reads

$$\langle \alpha_1 | \hat{B} | \alpha_2 \rangle = \exp\left(-\frac{1}{2}\|\alpha_1\|^2 - \frac{1}{2}\|\alpha_2\|^2 + \alpha_1^* \diamond \alpha_2\right) \equiv \langle \alpha_1 | \alpha_2 \rangle, \tag{4.186}$$

which follows from (4.114). Since $|\alpha_1\rangle$ and $|\alpha_2\rangle$ can represent arbitrary coherent states, it then follows that

$$\int |q\rangle \langle q| \; \mathcal{D}[q] = \mathbb{1}. \tag{4.187}$$

A similar analysis for the p basis gives

$$\int |p\rangle \langle p| \; \mathcal{D}^{\circ}[p] = \mathbb{1}, \tag{4.188}$$

where

$$\mathcal{D}^{\circ}[p] \triangleq \mathcal{D}\left[\frac{p}{2\pi}\right] = \frac{\mathcal{D}[p]}{(2\pi)^{\Omega}}. \tag{4.189}$$

Based on the expressions in (4.187) and (4.188), we can express any operator that is defined as a holomorphic functional of one of the quadrature operators in terms of the quadrature basis of that operator. By applying the identity resolve in terms of the basis of that operator, we obtain

$$\begin{aligned} h[\hat{q}] &= \int |q\rangle \, h[q] \, \langle q| \; \mathcal{D}[q], \\ h[\hat{p}] &= \int |p\rangle \, h[p] \, \langle p| \; \mathcal{D}^{\circ}[p], \end{aligned} \tag{4.190}$$

where we used (4.135).

Here, we obtain another significant result. With analogy to Section 4.4.3, where it is shown that the tensor products of discrete Fock states form a complete orthogonal basis, we summarize here our observations concerning the quadrature bases:

(a) These quadrature bases, with all degrees of freedom incorporated, are complete orthogonal bases for the Hilbert space of all quantum optical states.

(b) The elements of these bases are *not* normalizable and, therefore, are not elements of the Hilbert space of all quantum optical states.

(c) These bases each contain an uncountable infinite number of elements, but because they are not elements of the Hilbert space of all quantum optical states, they do not have any implication on the separability of the Hilbert space.

The elements of the quadrature bases belong to a set beyond the Hilbert space of all quantum optical states. Despite the fact that these bases do not satisfy the requirements for a basis in the standard functional analysis [15] context, they are more convenient to use for calculations than the Fock basis considered in Section 4.4.3.

4.5.6 State representation

Equipped with a complete orthogonal basis for the space of all quantum optical states, we can express an arbitrary state as an expansion in terms of this basis. It becomes a generalization of (3.22) in which the coordinate basis (representing the particle-number degrees of freedom only) is replaced by the quadrature basis (incorporating all the degrees of freedom) and the integral over the basis become a functional integral. The resulting expression (for a pure state) reads

$$|\psi\rangle = \int |q\rangle \, \psi[q] \, \mathcal{D}[q], \tag{4.191}$$

in terms of the q-basis, where $\psi[q]$ is a *coefficient functional* or *wave functional.*

4.6 Discussion

In our quest to incorporate the spatiotemporal and spin degrees of freedom with the particle-number degree of freedom, we started with a discussion on the quantization of the electromagnetic field. Since the classical description of electromagnetism in terms of Maxwell's equation involves all the spatiotemporal and spin degrees of freedom, it demonstrates how the incorporation of these degrees of freedom with the particle-number degrees of freedom is accomplished in terms of the definition of the generalization of the ladder operators as functions of the wavevectors and carrying spin indices.

However, for a state-based formulation, we need to find a way to represent arbitrary quantum optical states in terms of all the degrees of freedom. It naturally leads to the quest for a complete orthogonal basis. For this purpose we use the ladder operators obtained in the quantization process to incorporate the spatiotemporal and spin degrees of freedom with the particle-number degree of freedom into a complete orthogonal basis. In this way, we consider the *fixed-spectrum Fock states*, the *fixed-spectrum coherent states*, and finally the *eigenvectors* of the spin- and wavevector-dependent quadrature operators. These eigenvectors are simply referred to as the *quadrature bases* (not to be confused with the *fixed-spectrum quadrature bases*). Although all three these types of bases play significant roles in analyses in quantum optics, the quadrature bases play a special role. They naturally lead to the notion of a *functional phase space* on which states and operators can be represented as *functionals* of quadrature field variables.

The discrete Fock states also form a complete orthogonal basis, but their discreteness makes them unsuitable for technical calculations. The coherent states provide a continuous basis, but they are not orthogonal. Therefore, the quadrature bases, which are continuous bases and complete orthogonal bases, provide the answer for our quest.

Unlike the Fock states and the coherent states, the elements of the quadrature bases are not normalizable and can therefore not individually represent quantum states. However, they can be used to provide expansions of any state. *The elements of a complete basis for an infinite number of degrees of freedom cannot be normalizable, continuous, and orthogonal at the same time.* In each of these three types of bases, one of these three properties is sacrificed: the Fock basis is not continuous; coherent states are not orthogonal; and quadrature bases elements are not normalizable. One needs to decide which property can be done without. We choose to sacrifice normalizability. However, the functional phase space thus defined can also be represented in terms of fixed-spectrum coherent states. As a result, the formalism based on quadrature bases strongly overlaps with a coherent state based formalism.

The three bases that are considered as candidates for the formulation of quantum optical states are not the only possible bases that incorporate all the degrees of freedom without constraint. In Section 6.2.4, we show that there are an infinite number of different bases that can be used to represent any quantum optical state without constraint. However, all these additional bases have similar properties to those of the coherent states. Therefore, we don't need to consider them further here.

The expressions of the quantum optical states in terms of wave functionals, are converted to Wigner functionals, which can also represent mixed states. The details are discussed in the next chapter, where the *Wigner functional formalism* is presented.

Bibliography

[1] M. E. Peskin and D. V. Schroeder. *An Introduction to Quantum Field Theory*. Addison-Wesley Publishing Company, Reading, Massachusetts, USA, 1995.

[2] C. Itzykson and J.-B. Zuber. *Quantum Field Theory*. McGraw-Hill, New York, USA, 1985.

[3] F. Mandl and G. Shaw. *Quantum Field Theory*. John Wiley & Sons, New York, USA, 1984.

[4] S. Weinberg. *The Quantum Theory of Fields, Volume I*. Cambridge University Press, New York, 1995.

[5] S. Weinberg. *The Quantum Theory of Fields, Volume II*. Cambridge University Press, New York, 1996.

[6] S. L. Braunstein and P. Van Loock. Quantum information with continuous variables. *Rev. Mod. Phys.*, 77:513–577, 2005.

[7] A. I. Lvovsky and M. G. Raymer. Continuous-variable optical quantum-state tomography. *Rev. Mod. Phys.*, 81:299–332, 2009.

[8] G. Adesso, S. Ragy, and A. R. Lee. Continuous variable quantum information: Gaussian states and beyond. *Open Syst. Inf. Dyn.*, 21:1440001, 2014.

[9] P. A. Zyla and et al. Particle data group. *Prog. Theor. Exp. Phys.*, 2020:083C01, 2020.

[10] M. A. Nielsen and I. L. Chuang. *Quantum Computation and Quantum Information*. Cambridge University Press, Cambridge, England, 2000.

[11] A. Einstein, B. Podolsky, and N. Rosen. Can quantum-mechanical description of physical reality be considered complete? *Phys. Rev.*, 47:777–780, 1935.

[12] E. Schrödinger. Discussion of probability relations between separated systems. *Math. Proc. Camb. Philos. Soc.*, 31:555–563, 1935.

[13] J. F. Clauser, M. A. Horne, A. Shimony, and R. A. Holt. Proposed experiment to test local hidden-variable theories. *Phys. Rev. Lett.*, 23:880–884, 1969.

[14] A. Aspect, P. Grangier, and G. Roger. Experimental realization of Einstein–Podolsky–Rosen–Bohm Gedankenexperiment. *Phys. Rev. Lett.*, 49:91–94, 1982.

[15] E. Kreyszig. *Introductory Functional Analysis with Applications*. John Wiley & Sons, New York, USA, 1978.

5 Wigner functional theory

The key benefit of the quadrature bases obtained in Chapter 4 is that they provide a powerful, yet convenient, tool for the development of a functional formalism that incorporates all the degrees of freedom that are relevant in quantum optical systems. The development of such a functional formalism as a functional extension of the *Moyal formalism* [1–3] is the topic of this chapter.

The relevant results of Chapter 4 on which we base this development can be summarized as the inner products between quadrature bases elements and the completeness conditions obtained in Sections 4.5.3 and 4.5.5, respectively. The inner products are

$$\begin{aligned}\langle q|q'\rangle &= \delta[q-q'], \quad \langle p|p'\rangle = (2\pi)^{\Omega}\delta[p-p'],\\ \langle q|p\rangle &= \exp\left(\mathrm{i}q\diamond p\right), \quad \langle p|q\rangle = \exp\left(-\mathrm{i}q\diamond p\right),\end{aligned} \tag{5.1}$$

where $\delta[\cdot]$ is the *Dirac delta functional*, and the *completeness conditions* are

$$\int |q\rangle\langle q|\,\mathcal{D}[q] = \mathbb{1} \quad \text{and} \quad \int |p\rangle\langle p|\,\mathcal{D}^{\circ}[p] = \mathbb{1}. \tag{5.2}$$

We are now almost ready to formulate a *Wigner functional theory*, based on these results. A few loose ends need to be tied up first.

5.1 Functional Fourier analysis

The expressions in (5.1) and (5.2) naturally lead to a functional integral approach for any analysis involving the quadrature bases obtained in Chapter 4. For instance, if we apply the identity operator resolved in one quadrature basis to the elements of the other quadrature basis, we obtain

$$\begin{aligned}\mathbb{1}|p\rangle &= \int |q\rangle\langle q|p\rangle\,\mathcal{D}[q] = \int |q\rangle \exp(\mathrm{i}q\diamond p)\,\mathcal{D}[q],\\ \mathbb{1}|q\rangle &= \int |p\rangle\langle p|q\rangle\,\mathcal{D}^{\circ}[p] = \int |p\rangle \exp(-\mathrm{i}q\diamond p)\,\mathcal{D}^{\circ}[p],\end{aligned} \tag{5.3}$$

where $\mathcal{D}^{\circ}[p]$ incorporates a constant $(2\pi)^{-\Omega}$, as defined in (4.189). These functional integrals express each quadrature basis in terms of the other quadrature basis with the quadrature variables q and p acting as the respective *integration field variables*. Apart from the fact that it clearly shows the mutually unbiased nature of these two bases, it also shows that they are related by functional integrals that look like Fourier integrals.

To demonstrate that these integrals really are Fourier integrals, we insert identities resolved in one quadrature basis into the inner product between elements of the other quadrature basis. The resulting integrals

https://doi.org/10.1515/9783111445342-007

$$
\begin{aligned}
\langle q|\mathbb{1}|q'\rangle &= \int \langle q|p\rangle\langle p|q'\rangle\, \mathcal{D}^{\circ}[p] \\
&= \int \exp(\mathrm{i}q \diamond p)\exp(-\mathrm{i}q' \diamond p)\, \mathcal{D}^{\circ}[p] = \delta[q-q'], \\
\langle p|\mathbb{1}|p'\rangle &= \int \langle p|q\rangle\langle q|p'\rangle\, \mathcal{D}[q] \\
&= \int \exp(-\mathrm{i}q \diamond p)\exp(\mathrm{i}q \diamond p')\, \mathcal{D}[q] = (2\pi)^{\Omega}\delta[p-p'],
\end{aligned} \tag{5.4}
$$

show that the exponential functionals with the contracted field variables in their arguments form a functional basis that is both *orthogonal* and complete. So, the functional integrals in (5.3) are indeed functional Fourier integrals.

The *functional Fourier transform* is thus given by

$$
\tilde{\psi}[p] = \int \langle p|q\rangle\langle q|\psi\rangle\, \mathcal{D}[q] = \int \exp(-\mathrm{i}q \diamond p)\psi[q]\, \mathcal{D}[q], \tag{5.5}
$$

where $\psi[q]$ and $\tilde{\psi}[p]$ are the *wave functionals* defined in (4.134). The *functional inverse Fourier transform* is obtained in a similar way as

$$
\psi[q] = \int \langle q|p\rangle\langle p|\psi\rangle\, \mathcal{D}^{\circ}[p] = \int \exp(\mathrm{i}q \diamond p)\tilde{\psi}[p]\, \mathcal{D}^{\circ}[p]. \tag{5.6}
$$

The measures $\mathcal{D}[q]$ and $\mathcal{D}^{\circ}[p]$ run over all finite-energy real-valued continuously differentiable functions.[a] Functional integration is discussed in Appendix C.

5.2 Functional operator identities

The Fourier relationships in (5.5) and (5.6) allow us to represent the quadrature operators in terms of their dual bases. The results are the functional equivalents of the particle-number degrees of freedom expressions obtained in Chapter 3. Considering the respective eigenvalue equations and representing the quadrature basis elements in terms of the opposite basis, we get

$$
\begin{aligned}
\hat{p}_s(\mathbf{k})\,|p\rangle = |p\rangle\, p_s(\mathbf{k}) &= \int |q\rangle\, p_s(\mathbf{k}) \exp(\mathrm{i}q \diamond p)\, \mathcal{D}[q] \\
&= \int |q\rangle \left[-\mathrm{i}\frac{\delta}{\delta q_s(\mathbf{k})}\right] \exp(\mathrm{i}q \diamond p)\, \mathcal{D}[q], \\
\hat{q}_s(\mathbf{k})\,|q\rangle = |q\rangle\, q_s(\mathbf{k}) &= \int |p\rangle\, q_s(\mathbf{k}) \exp(-\mathrm{i}q \diamond p)\, \mathcal{D}^{\circ}[p] \\
&= \int |p\rangle \left[\mathrm{i}\frac{\delta}{\delta p_s(\mathbf{k})}\right] \exp(-\mathrm{i}q \diamond p)\, \mathcal{D}^{\circ}[p].
\end{aligned} \tag{5.7}
$$

---

a In mathematical terms, these field variables are *Schwartz functions*. Formally, $q, p \in \mathcal{S}(\mathbb{R}^3, \mathbb{C})$, which is dense in $L^2(\mathbb{R}^3, \mathbb{C})$.

Here, we introduce the *functional derivative* that produces

$$\frac{\delta f_r(\mathbf{k}')}{\delta f_s(\mathbf{k})} = (2\pi)^3\omega\delta_{r,s}\delta(\mathbf{k}-\mathbf{k}') \equiv \mathbf{1}_{r,s}(\mathbf{k}-\mathbf{k}'). \tag{5.8}$$

It follows that the quadrature operators can be expressed as

$$\begin{aligned}\hat{p}_s(\mathbf{k}) &= \int |p\rangle\, p_s(\mathbf{k})\, \langle p|\; \mathcal{D}^\circ[p] = \int |q\rangle \left[-\mathrm{i}\frac{\delta}{\delta q_s(\mathbf{k})}\right] \langle q|\; \mathcal{D}[q],\\ \hat{q}_s(\mathbf{k}) &= \int |q\rangle\, q_s(\mathbf{k})\, \langle q|\; \mathcal{D}[q] = \int |p\rangle \left[\mathrm{i}\frac{\delta}{\delta p_s(\mathbf{k})}\right] \langle p|\; \mathcal{D}^\circ[p],\end{aligned} \tag{5.9}$$

which are the functional equivalents of (3.36) and (3.37).

One can use the definitions of the quadrature bases in terms of the opposite bases, given in (5.3), to define the functional versions of the shift operators or *Weyl operators*, as obtained for the particle-number degree of freedom only in (3.35). As such, it demonstrates that the quadrature operators serve as *generators of translation* when applied to elements of the opposite basis.

Exercise 5.1. Use the expressions in (5.3) and (4.135) to show that

$$\exp(\mathrm{i}q_0 \diamond \hat{p})\;|q\rangle = |q - q_0\rangle \quad\text{and}\quad \exp(-\mathrm{i}\hat{q} \diamond p_0)\;|p\rangle = |p - p_0\rangle\,. \tag{5.10}$$

5.3 Definition of the Wigner functional

The *Wigner functional* of an operator is defined, in analogy to (3.212) [4], as the partial functional Fourier transform of the operator overlapped by two oppositely shifted quadrature basis elements (the q-basis). It reads

$$W[q,p] \triangleq \int \left\langle q + \frac{1}{2}x \middle| \hat{\rho} \middle| q - \frac{1}{2}x \right\rangle \exp(-\mathrm{i}x \diamond p)\; \mathcal{D}[x], \tag{5.11}$$

where $\hat{\rho}$ is a density operator that incorporates all degrees of freedom and x is the *integration field variable*. In the case of pure states, the overlapped density operator leads to *wave functionals*

$$\left\langle q + \frac{1}{2}x \middle| \psi \right\rangle \left\langle \psi \middle| q - \frac{1}{2}x \right\rangle = \psi\left[q + \frac{1}{2}x\right] \psi^*\left[q - \frac{1}{2}x\right]. \tag{5.12}$$

where $\psi[q] = \langle q|\psi\rangle$, as introduced in (4.134).

Exercise 5.2. Derive the equivalent expression for the Wigner functional in terms of the p-basis instead of the q-basis:

$$W[q,p] = \int \left\langle p + \frac{1}{2}y \middle| \hat{\rho} \middle| p - \frac{1}{2}y \right\rangle \exp\left(iq \diamond y\right) \mathcal{D}^{\circ}[y]. \tag{5.13}$$

Hint: Insert identities resolved in terms of the p-basis on either side of the operator in (5.11).

Exercise 5.3. Show that the Wigner functional of a density operator is real valued.

Hint: Use the fact that a density operator is Hermitian.

Exercise 5.4. Show that the Wigner functionals of the quadrature operators $\hat{q}_s(\mathbf{k})$ and $\hat{p}_s(\mathbf{k})$ are the quadrature field variables $q_s(\mathbf{k})$ and $p_s(\mathbf{k})$, respectively.

Hint: For $\hat{p}$, insert an identity resolved in terms of the p-basis on one side of the operator.

In some discussions, it is more convenient to represent a Wigner functional calculation as a process that maps the operators that are defined on the Hilbert space to functionals on the phase space. The notation we use in such cases is

$$W_{\hat{A}}[q,p] \triangleq \text{Wigner}\{\hat{A}\}, \tag{5.14}$$

where $\hat{A}$ can be any operator of the Hilbert space.

Exercise 5.5. Show that the Wigner functionals of the quadrature eigenvectors $|q_0\rangle$ are given by

$$\text{Wigner}\{|q_0\rangle\langle q_0|\} = \delta[q - q_0]. \tag{5.15}$$

5.3.1 Marginal probability distribution

The Wigner functional is regarded as a *quasi-probability distribution*; it can have negative values and does not qualify as a true probability density. However, one can compute a *marginal probability distribution* from it by integrating either over p, or q, or a linear combination of them. As a demonstration, we consider the density operator of a pure state $\hat{\rho} = |\psi\rangle\langle\psi|$. The Wigner functional is then given in terms of wave functionals by

$$W[q,p] = \int \psi\left[q + \frac{1}{2}x\right] \psi^{*}\left[q - \frac{1}{2}x\right] \exp(-ix \diamond p)\, \mathcal{D}[x]. \tag{5.16}$$

Integrating it over p, we obtain

$$\int W[q,p]\, \mathcal{D}^{\circ}[p] = \int \psi\left[q + \frac{1}{2}x\right] \psi^{*}\left[q - \frac{1}{2}x\right] \delta[x]\, \mathcal{D}[x] = |\psi[q]|^2. \tag{5.17}$$

The resulting squared modulus of the wave functional represents the probability distribution as a function of q. For a general density operator, the result is $\langle q|\hat{\rho}|q\rangle = \rho[q,q]$. Integrating such a probability distribution over the remaining variable, we obtain

$$\int W[q,p]\,\mathcal{D}^{\circ}[p]\,\mathcal{D}[q] = \int |\psi[q]|^2\,\mathcal{D}[q] = 1, \tag{5.18}$$

as required for the *conservation of probability.*

Exercise 5.6. Show that the functional integration over q of the Wigner functional given in (5.11) produces the marginal probability distribution as a function of p.

Hint: Use (5.13).

5.3.2 Operator-based calculation of the Wigner functional

We can convert the calculation of the Wigner functional into a purely operator-based calculation with the aid of the shift operators defined in (5.10)

$$\begin{aligned} W[q,p] &= \int \left\langle q+\frac{1}{2}x\middle|\hat{\rho}\middle|q-\frac{1}{2}x\right\rangle \exp(-\mathrm{i}x\diamond p)\,\mathcal{D}[x] \\ &= \mathcal{N}_0\int \langle q+q'|\,\hat{\rho}\,|q-q'\rangle \exp(-\mathrm{i}2q'\diamond p)\,\mathcal{D}[q'] \\ &= \mathcal{N}_0\int \langle q'|\exp(\mathrm{i}q\diamond\hat{p})\,\hat{\rho}\exp(-\mathrm{i}q\diamond\hat{p})\exp(\mathrm{i}2\hat{q}\diamond p)\,|-q'\rangle\,\mathcal{D}[q'] \\ &= \mathcal{N}_0\,\mathrm{tr}\left\{\exp(\mathrm{i}q\diamond\hat{p})\,\hat{\rho}\exp(-\mathrm{i}q\diamond\hat{p})\exp(\mathrm{i}2\hat{q}\diamond p)\,\hat{\Pi}\right\} \\ &= \mathcal{N}_0\,\mathrm{tr}\left\{\exp(-\mathrm{i}\hat{q}\diamond p)\exp(\mathrm{i}q\diamond\hat{p})\,\hat{\rho}\exp(-\mathrm{i}q\diamond\hat{p})\exp(\mathrm{i}\hat{q}\diamond p)\,\hat{\Pi}\right\} \\ &= \mathcal{N}_0\,\mathrm{tr}\left\{\exp(-\mathrm{i}\hat{q}\diamond p+\mathrm{i}q\diamond\hat{p})\,\hat{\rho}\exp(\mathrm{i}\hat{q}\diamond p-\mathrm{i}q\diamond\hat{p})\,\hat{\Pi}\right\}, \end{aligned} \tag{5.19}$$

where

$$\mathcal{N}_0 \triangleq 2^{\Omega}, \tag{5.20}$$

and

$$\hat{\Pi} \triangleq \int |-q'\rangle\langle q'|\,\mathcal{D}[q'] \equiv \int |-p'\rangle\langle p'|\,\mathcal{D}^{\circ}[p], \tag{5.21}$$

is the *flip operator* or *parity operator*. In the derivation, we redefined the integration variable, which produces $\mathcal{N}_0$. Then we used the shift operators in (5.10) together with (4.135) to convert all the exponential functionals into exponentiated operators, independent of the integration field variable. The expression can then be represented as a trace in which the remaining functional integral produces the parity operator defined in (5.21). Half of the operator that contains $\hat{q}$ is pulled through the parity operator to the other side, causing a change in the sign in the exponent. Finally, the Baker–Campbell–Hausdorff formula in (3.40) is used to combine the exponentiated operators into displacement operators, while the phase factors thus produced cancel each other.

The operator-based expression for the Wigner functional now reads

$$W[q,p] = \mathcal{N}_0 \operatorname{tr}\left\{\hat{D}[q,p]\hat{\Pi}\hat{D}^{\dagger}[q,p]\hat{\rho}\right\}, \tag{5.22}$$

where the displacement operator in terms of quadrature operators is given by

$$\hat{D}[q,p] = \exp\left(\mathrm{i}\hat{q}\diamond p - \mathrm{i}q\diamond\hat{p}\right), \tag{5.23}$$

as a generalization of (3.119). It is obtained from (4.110) with the aid of (4.131) and (4.181). We can also represent the displacement operator in the operator-based expression for the Wigner functional using the complex parameter function α as in (4.110). By implication, the Wigner functional can also be expressed as a functional of the *complex field variable* α related to the quadrature variables via (4.181). Then it reads

$$W[\alpha] = \mathcal{N}_0 \operatorname{tr}\left\{\hat{D}[\alpha]\hat{\Pi}\hat{D}^{\dagger}[\alpha]\hat{\rho}\right\}. \tag{5.24}$$

Exercise 5.7. Show that the Wigner functional of the parity-operator is

$$\text{Wigner}\{\hat{\Pi}\} = \pi^{\Omega}\delta[q]\delta[p]. \tag{5.25}$$

5.4 Characteristic functional

For the particle-number degree of freedom only, a *characteristic function* is obtained as the two-dimensional *symplectic Fourier transform* of the Wigner function. This Fourier transform is called symplectic due to the sign difference for the two one-dimensional Fourier transforms. This relative minus sign between the terms in the exponent appears because one of the two transformations is an inverse Fourier transform.

Generalized in the context of a functional phase space with all the degrees of freedom included, it becomes a *characteristic functional*, given by the *functional symplectic Fourier transform* of the Wigner functional with respect to both quadrature variables

$$\begin{aligned}\chi[\xi,\zeta] &= \int W[q,p]\exp(\mathrm{i}\zeta\diamond p - \mathrm{i}q\diamond\xi)\,\mathcal{D}[q]\,\mathcal{D}^{\circ}[p] \\ &= \int \left\langle q+\frac{1}{2}\zeta\middle|\hat{\rho}\middle|q-\frac{1}{2}\zeta\right\rangle \exp(-\mathrm{i}q\diamond\xi)\,\mathcal{D}[q].\end{aligned} \tag{5.26}$$

In terms of *complex field variables*, we can also express it as

$$\chi[\beta] = \int W[\alpha]\exp(\beta^{*}\diamond\alpha - \alpha^{*}\diamond\beta)\,\mathcal{D}^{\circ}[\alpha], \tag{5.27}$$

where

$$\beta \triangleq \frac{1}{\sqrt{2}}(\zeta + \mathrm{i}\xi), \tag{5.28}$$

and

$$\mathcal{D}^{\circ}[\alpha] \triangleq \mathcal{D}[q]\mathcal{D}^{\circ}[p] = \mathcal{D}\left[q, \frac{p}{2\pi}\right] = \frac{\mathcal{D}[q,p]}{(2\pi)^{\Omega}}. \tag{5.29}$$

The Wigner functional is recovered from the characteristic functional via the inverse symplectic Fourier transforms

$$\begin{aligned} W[q,p] &= \int \chi[\xi,\zeta] \exp(\mathrm{i}q \diamond \xi - \mathrm{i}\zeta \diamond p)\, \mathcal{D}[\zeta]\, \mathcal{D}^{\circ}[\xi], \quad \text{or} \\ W[\alpha] &= \int \chi[\beta] \exp(\alpha^* \diamond \beta - \beta^* \diamond \alpha)\, \mathcal{D}^{\circ}[\beta]. \end{aligned} \tag{5.30}$$

One of the main purposes of the characteristic functional is to serve as a *generating functional for the moments* of the Wigner functional. (Generating functionals are discussed in more detail in Appendix A.) The moments with respect to the real-valued quadrature field variables are computed by applying *functional derivatives* with respect to the real-valued field variables of the characteristic functional

$$\left(-\mathrm{i}\frac{\delta}{\delta\zeta}\right)^m \left(\mathrm{i}\frac{\delta}{\delta\xi}\right)^n \chi[\xi,\zeta]\bigg|_{\zeta=\xi=0} = \int W[q,p] p^m q^n\, \mathcal{D}[q]\, \mathcal{D}^{\circ}[p]. \tag{5.31}$$

A similar expression can be used for moments with respect to the complex-valued field variables. It reads

$$\left(-\frac{\delta}{\delta\beta}\right)^m \left(\frac{\delta}{\delta\beta^*}\right)^n \chi[\beta]\bigg|_{\beta=\beta^*=0} = \int W[\alpha]\alpha^{*m}\alpha^n\, \mathcal{D}^{\circ}[\alpha]. \tag{5.32}$$

The functional derivative with respect to a complex field variable is defined in terms of functional derivatives with respect to the real-valued field variables in terms of which the complex field variable is expressed via the *chain rule for functional derivatives*

$$\begin{aligned} \frac{\delta f[q[\alpha],p[\alpha]]}{\delta\alpha} &= \frac{\delta f[q,p]}{\delta q}\frac{\delta q[\alpha]}{\delta\alpha} + \frac{\delta f[q,p]}{\delta p}\frac{\delta p[\alpha]}{\delta\alpha} \\ &= \frac{1}{\sqrt{2}}\frac{\delta f[q,p]}{\delta q} - \mathrm{i}\frac{1}{\sqrt{2}}\frac{\delta f[q,p]}{\delta p}. \end{aligned} \tag{5.33}$$

A similar rule can be derived for the functional derivative with respect to α^*.

Exercise 5.8. Show that

$$\frac{\delta\alpha^*}{\delta\alpha} = \frac{\delta\alpha}{\delta\alpha^*} = 0. \tag{5.34}$$

5.4.1 Operator-based calculation

The characteristic functional is obtained from a purely operator-based calculation in a way similar to that of Section 5.3.2. Using the final expression in (5.26), we get

$$\begin{aligned}\chi[\xi,\zeta] &= \int \langle q|\exp\left(\mathrm{i}\frac{1}{2}\zeta\diamond\hat{p}\right)\hat{\rho}\exp\left(\mathrm{i}\frac{1}{2}\zeta\diamond\hat{p}\right)\exp(-\mathrm{i}\hat{q}\diamond\xi)|q\rangle\ \mathcal{D}[q] \\ &= \mathrm{tr}\left\{\hat{\rho}\exp\left(\mathrm{i}\frac{1}{2}\zeta\diamond\hat{p}\right)\exp(-\mathrm{i}\hat{q}\diamond\xi)\exp\left(\mathrm{i}\frac{1}{2}\zeta\diamond\hat{p}\right)\right\} \\ &= \mathrm{tr}\left\{\hat{\rho}\exp(\mathrm{i}\zeta\diamond\hat{p}-\mathrm{i}\hat{q}\diamond\xi)\right\}. \end{aligned} \tag{5.35}$$

Here, we used the shift operators defined in (5.10) together with the pull-back process in (4.135) to convert all the exponential functionals into exponentiated operators. The integral over q produces the identity operator via the completeness condition in the q-basis in (5.2). The exponentiated operators are combined with the aid of the Baker–Campbell–Hausdorff formula in (3.40) to produce an adjoint displacement operator.

Exercise 5.9. Use the Baker–Campbell–Hausdorff formula in (3.40) to show that

$$\exp\left(\mathrm{i}\frac{1}{2}\zeta\diamond\hat{p}\right)\exp(-\mathrm{i}\hat{q}\diamond\xi)\exp\left(\mathrm{i}\frac{1}{2}\zeta\diamond\hat{p}\right) = \exp(\mathrm{i}\zeta\diamond\hat{p}-\mathrm{i}\hat{q}\diamond\xi). \tag{5.36}$$

The characteristic functional can thus be obtained with

$$\chi[\beta] \triangleq \mathrm{char}\{\hat{\rho}\} = \mathrm{tr}\left\{\hat{\rho}\hat{D}^{\dagger}[\beta]\right\}, \tag{5.37}$$

where β is defined in (5.28). It thus gives a mapping between operators on the Hilbert space and the characteristic functionals on phase space. In view of the fact that the displacement operators form a complete orthogonal basis for all operators, as shown in Section 6.2.3, the trace represents an inner product between the state and the displacement operator. The characteristic functional thus serves as a coefficient function with which an operator on the Hilbert space can be expressed by an expansion

$$\hat{A} = \int \chi_{\hat{A}}[\beta]\hat{D}[\beta]\ \mathcal{D}^{\circ}[\beta] = \int \mathrm{char}\{\hat{A}\}\hat{D}[\beta]\ \mathcal{D}^{\circ}[\beta], \tag{5.38}$$

in terms of the displacement operators.

5.5 Weyl transformation

If the calculation of the Wigner functional for an operator retains all the information, it should be possible to invert the process. To demonstrate this inversion, we start by computing the inverse Fourier transform of (5.11). The result is

$$\int W[q,p]\exp(ix\diamond p)\,\mathcal{D}^{\circ}[p] = \left\langle q+\frac{1}{2}x\middle|\hat{\rho}\middle|q-\frac{1}{2}x\right\rangle. \tag{5.39}$$

Then we redefine the q and x field variables so that

$$q_1 \triangleq q+\frac{1}{2}x \quad \text{and} \quad q_2 \triangleq q-\frac{1}{2}x. \tag{5.40}$$

The expression then becomes

$$\int W\left[\frac{q_1+q_2}{2},p\right]\exp[ip\diamond(q_1-q_2)]\,\mathcal{D}^{\circ}[p] = \langle q_1|\hat{\rho}|q_2\rangle. \tag{5.41}$$

Now we multiply the expression by $|q_1\rangle$ and $\langle q_2|$ and integrate over q_1 and q_2 to convert these products of quadrature bases elements into identity operators according to (5.2):

$$\begin{aligned}&\int |q_1\rangle\, W\left[\frac{q_1+q_2}{2},p\right]\exp[ip\diamond(q_1-q_2)]\,\langle q_2|\;\mathcal{D}^{\circ}[p]\,\mathcal{D}[q_1,q_2]\\ =&\int |q_1\rangle\,\langle q_1|\hat{\rho}|q_2\rangle\,\langle q_2|\;\mathcal{D}[q_1,q_2] = \mathbb{1}\hat{\rho}\mathbb{1}.\end{aligned} \tag{5.42}$$

In the end, we restore the field variables q and x. The density operator is thus given in terms of its Wigner functional by

$$\hat{\rho} = \int \left|q+\frac{1}{2}x\right\rangle W[q,p]\exp(ix\diamond p)\left\langle q-\frac{1}{2}x\right|\;\mathcal{D}^{\circ}[p]\,\mathcal{D}[q]\,\mathcal{D}[x]. \tag{5.43}$$

This inverse transformation from the Wigner functional back to the operator is called the *Weyl transformation*. It shows that the mapping between the Hilbert space of all quantum optical states and the space of all Wigner functionals on the functional phase space is invertible, as shown in Figure 5.1. The mappings from one to the other are represented by the Wigner functional calculation and the Weyl transformation, respectively.

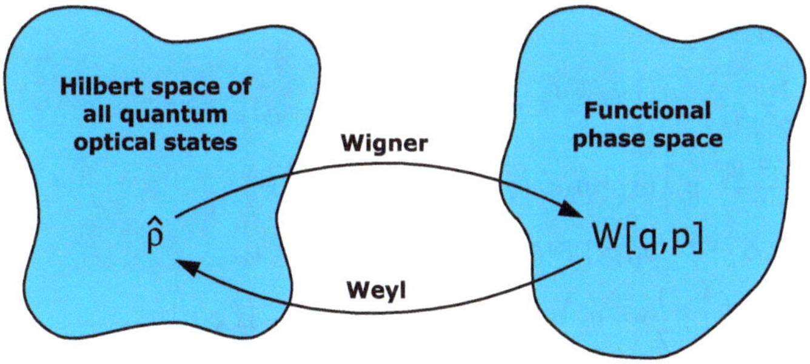

Figure 5.1: The mapping between the Hilbert space of all quantum optical states and the functional phase space is invertible.

Similar to the notation introduced for the Wigner functional calculation in (5.14), we also define the Weyl transformation as a mapping from functionals on the phase space to operators defined on the Hilbert space. The notation in this case is given by

$$\hat{A} = \text{Weyl}\{W_{\hat{A}}[q,p]\}. \tag{5.44}$$

Using the Weyl transformation, we can readily show that the *operator trace* is represented in terms of a functional integration over the functional phase space on which the Wigner functional of that operator is defined:

$$\text{tr}\{\hat{A}\} = \int W_{\hat{A}}[q,p]\, \mathcal{D}[q]\, \mathcal{D}^{\circ}[p] \equiv \int W_{\hat{A}}[\alpha]\, \mathcal{D}^{\circ}[\alpha]. \tag{5.45}$$

The normalization of a density operator thus implies that

$$\text{tr}\{\hat{\rho}\} = \int W_{\hat{\rho}}[q,p]\, \mathcal{D}[q]\, \mathcal{D}^{\circ}[p] \equiv \int W_{\hat{\rho}}[\alpha]\, \mathcal{D}^{\circ}[\alpha] = 1. \tag{5.46}$$

5.6 Star product

The way Wigner functionals are combined to represent the products of operators leads to the definition of the *star product* (also called the *Moyal product*). Here, we derive the integral expression for the star product. For this purpose, we compute the Wigner functional of a product of two operators, both of which are expressed in terms of *Weyl transformations* of their Wigner functionals. Using the notation for these processes given in (5.14) and (5.44), we can express the final Wigner functional as

$$\begin{aligned} W_{\hat{A}\hat{B}} &\triangleq \text{Wigner}\left\{\hat{A}\hat{B}\right\} \\ &= \text{Wigner}\left\{\text{Weyl}\left\{W_{\hat{A}}\right\}\text{Weyl}\left\{W_{\hat{B}}\right\}\right\} \triangleq W_{\hat{A}} \star W_{\hat{B}}. \end{aligned} \tag{5.47}$$

The detail calculation then leads to

$$\begin{aligned} W_{\hat{A}\hat{B}} &= \int \left\langle q + \frac{1}{2}x \middle| q_a \right\rangle W_{\hat{A}}\left[\frac{q_a+q_b}{2}, p_1\right] \exp[\mathrm{i}p_1 \diamond (q_a - q_b)]\, \langle q_b | q_c \rangle \\ &\quad \times W_{\hat{B}}\left[\frac{q_c+q_d}{2}, p_2\right] \exp[\mathrm{i}p_2 \diamond (q_c - q_d)] \left\langle q_d \middle| q - \frac{1}{2}x \right\rangle \\ &\quad \times \exp(-\mathrm{i}x \diamond p)\, \mathcal{D}^{\circ}[p_1,p_2]\, \mathcal{D}[q_a,q_b,q_c,q_d,x] \\ &= \int \exp\left[\mathrm{i}p_1 \diamond \left(q + \frac{1}{2}x - q_b\right) + \mathrm{i}p_2 \diamond \left(q_b - q + \frac{1}{2}x\right) - \mathrm{i}x \diamond p\right] \\ &\quad \times W_{\hat{A}}\left[\frac{q}{2} + \frac{q_b}{2} + \frac{x}{4}, p_1\right] W_{\hat{B}}\left[\frac{q}{2} + \frac{q_b}{2} - \frac{x}{4}, p_2\right] \mathcal{D}^{\circ}[p_1,p_2]\, \mathcal{D}[q_b,x] \\ &= \mathcal{N}_0^2 \int \exp[2\alpha_1^* \diamond (\alpha_2 - \alpha) + 2\alpha_2^* \diamond (\alpha - \alpha_1) \\ &\quad + 2\alpha^* \diamond (\alpha_1 - \alpha_2)] W_{\hat{A}}[\alpha_1]\, W_{\hat{B}}[\alpha_2]\, \mathcal{D}^{\circ}[\alpha_1,\alpha_2], \end{aligned} \tag{5.48}$$

where we converted the p's and q's in the final expression to α's, using (4.181).

Exercise 5.10. Show that the Wigner functional for the product of three operators (the *triple star product*) is given by

$$\begin{aligned}\text{Wigner}\left\{\hat{A}\hat{B}\hat{C}\right\} = \int W_{\hat{A}}\left[\frac{1}{2}(\alpha_a + \alpha + \alpha_b)\right] W_{\hat{B}}\left[\alpha_a\right] W_{\hat{C}}\left[\frac{1}{2}(\alpha_a + \alpha - \alpha_b)\right] \\ \times \exp[(\alpha^* - \alpha_a^*) \diamond \alpha_b - \alpha_b^* \diamond (\alpha - \alpha_a)]\, \mathcal{D}^\circ[\alpha_a, \alpha_b].\end{aligned} \tag{5.49}$$

Hint: Apply (5.48) twice, redefine the field variables and evaluate two functional integrals.

Exercise 5.11. Show that the star product for characteristic functionals is represented by a functional integration over a single field variable, given by

$$\text{char}\left\{\hat{A}\hat{B}\right\} = \int \chi_{\hat{A}}[\beta - \beta_1]\chi_{\hat{B}}[\beta_1] \exp\left(-\frac{1}{2}\beta^* \diamond \beta_1 + \frac{1}{2}\beta_1^* \diamond \beta\right) \mathcal{D}^\circ[\beta_1]. \tag{5.50}$$

Hint: Convert the Wigner functionals in (5.48) into characteristic functionals, and evaluate as many of the resulting functional integrals as possible.

Exercise 5.12. Use the expression in (5.48) to show that the trace of the product of two operators is given by

$$\text{tr}\left\{\hat{A}\hat{B}\right\} = \int W_{\hat{A}}[\alpha] W_{\hat{B}}[\alpha]\, \mathcal{D}^\circ[\alpha]. \tag{5.51}$$

Exercise 5.13. Show that the trace of the product of three operators is given by

$$\begin{aligned}\text{tr}\left\{\hat{A}\hat{B}\hat{C}\right\} = \mathcal{N}_0^2 \int W_{\hat{A}}[\alpha_1] W_{\hat{B}}[\alpha_2] W_{\hat{C}}[\alpha_3] \exp[2\alpha_1^* \diamond (\alpha_2 - \alpha_3) \\ + 2\alpha_2^* \diamond (\alpha_3 - \alpha_1) + 2\alpha_3^* \diamond (\alpha_1 - \alpha_2)]\, \mathcal{D}^\circ[\alpha_1, \alpha_2, \alpha_3].\end{aligned} \tag{5.52}$$

Hint: Use (5.49).

5.7 Covariance kernel

In Section 3.4, we discuss pure Gaussian states in the context of only the particle-number degree of freedom. In the case of mixed states, it is more convenient to represent them in terms of Wigner functions on phase space. The Wigner functions of such Gaussian states are again Gaussian functions. When the other degrees of freedom are incorporated, they become *Gaussian functionals*. The most general Gaussian functional can be represented in terms of a *covariance kernel*, together with an arbitrary displacement in the field variables. For this discussion, we use the information provided in Section 3.3.

The term *covariance* comes from the interpretation of the functions on phase space as statistical distributions where the quadrature variables q and p are treated as random variables. The covariance is then given in terms of expectation values of the form

$$V_{qp} = \langle (q - \langle q \rangle)(p - \langle p \rangle) \rangle = \langle qp \rangle - \langle q \rangle \langle p \rangle, \tag{5.53}$$

where $\langle\cdot\rangle$ represents an *expectation value*. For an arbitrary function of the two random variables $h(q,p)$, such an expectation value is compute with

$$\langle h(q,p)\rangle = \int h(q,p) f_{q,p}(q,p)\, \mathrm{d}q\, \mathrm{d}p, \tag{5.54}$$

where $f_{q,p}(q,p)$ is a *joint probability density function* for two random variables.

5.7.1 Gaussian Wigner functionals

When the phase space becomes a functional phase space and the probability density function is replaced by the Wigner functional, the definition of the expectation value can still be formally expressed in a similar way by

$$\langle h[q,p]\rangle = \int h[q,p] W[q,p]\, \mathcal{D}[q]\, \mathcal{D}^{\circ}[p], \tag{5.55}$$

but the physical interpretation of the Wigner functional as a statistical distribution is not valid anymore because it can have negative values. Note that (5.55) is the functional phase space equivalent of the expectation value defined in (3.18).

An arbitrary *Gaussian Wigner functional* can be expressed as

$$W_{\mathrm{g}}[q,p] = \mathcal{N} \exp\left[-(\mathbf{q}-\mathbf{q}_0)^T \diamond \mathbf{K} \diamond (\mathbf{q}-\mathbf{q}_0)\right], \tag{5.56}$$

where $\mathcal{N}$ is a normalization constant, $\mathbf{q}(\mathbf{k}) = [q(\mathbf{k})\ p(\mathbf{k})]^T$ denotes the quadrature field variables (with the spin indices not shown), $\mathbf{q}_0(\mathbf{k}) = [q_0(\mathbf{k})\ p_0(\mathbf{k})]^T$ represents shifts, and

$$\mathbf{K}(\mathbf{k}_1,\mathbf{k}_2) = \begin{bmatrix} K_{qq}(\mathbf{k}_1,\mathbf{k}_2) & K_{qp}(\mathbf{k}_1,\mathbf{k}_2) \\ K_{pq}(\mathbf{k}_1,\mathbf{k}_2) & K_{pp}(\mathbf{k}_1,\mathbf{k}_2) \end{bmatrix}, \tag{5.57}$$

is a 2×2 matrix of kernels, collectively called the *Gaussian kernel*. The symmetry properties imposed by the contractions in (5.56) imply that K_{qq} and K_{pp} are symmetric, and that $K_{qp}^T = K_{pq}$.

The expectation values can be expressed in terms of the moments of the Wigner functional, which are calculated from the characteristic functional as in (5.31). For the Gaussian functional in (5.56), the characteristic functional is generically given by

$$\chi[\xi,\zeta] = \exp\left(-\frac{1}{4}\mathbf{c}^T \diamond \mathbf{K}^{-1} \diamond \mathbf{c} + \mathrm{i}\zeta \diamond p_0 - \mathrm{i}q_0 \diamond \xi\right), \tag{5.58}$$

where $\mathbf{c}(\mathbf{k}) = [\xi(\mathbf{k})\ \zeta(\mathbf{k})]^T$. The first moments lead to $\langle \mathbf{q}(\mathbf{k})\rangle = \mathbf{q}_0(\mathbf{k})$ and $\langle \mathbf{p}(\mathbf{k})\rangle = \mathbf{p}_0(\mathbf{k})$. The *covariance kernel* is obtained with the aid of the second moment:

$$\mathbf{V}(\mathbf{k}_1,\mathbf{k}_2) = \langle [\mathbf{q}(\mathbf{k}_1)-\mathbf{q}_0(\mathbf{k}_1)]\, [\mathbf{q}(\mathbf{k}_2)-\mathbf{q}_0(\mathbf{k}_2)]\rangle = \langle \mathbf{q}(\mathbf{k}_1)\mathbf{q}(\mathbf{k}_2)\rangle - \mathbf{q}_0(\mathbf{k}_1)\mathbf{q}_0(\mathbf{k}_2). \tag{5.59}$$

Using (5.31) to compute the covariance kernel from (5.58), we obtain

$$\mathbf{V} = \frac{1}{2}\mathbf{K}^{-1} \quad \text{implying that} \quad \mathbf{K} = 2\mathbf{V}^{-1}. \tag{5.60}$$

The Gaussian kernel is thus proportional to the inverse of the covariance kernel. Note that the covariance kernel in (5.59) can be obtained for any state. It does not have to be a Gaussian state.

There are certain conditions for Gaussian functionals of the form (5.56) to represent valid Wigner functionals of quantum states. Here, we consider these conditions in terms of its covariance kernel. This analysis is the functional equivalent of the original analysis to determine the constraints on the covariance matrix [5, 6]. The conditions follow from the basic requirement for minimum uncertainty, which is considered in Section 3.3.

In terms of the separate quadrature field variables, the covariance becomes a 2×2 matrix of kernels obtained from expectation values

$$\mathbf{V}(\mathbf{k}_1, \mathbf{k}_2) = \begin{bmatrix} V_{qq}(\mathbf{k}_1, \mathbf{k}_2) & V_{qp}(\mathbf{k}_1, \mathbf{k}_2) \\ V_{pq}(\mathbf{k}_1, \mathbf{k}_2) & V_{pp}(\mathbf{k}_1, \mathbf{k}_2) \end{bmatrix}, \tag{5.61}$$

where

$$\begin{aligned} V_{qq}(\mathbf{k}_1, \mathbf{k}_2) &\triangleq \langle q(\mathbf{k}_1) q(\mathbf{k}_2)\rangle - q_0(\mathbf{k}_1) q_0(\mathbf{k}_2), \\ V_{qp}(\mathbf{k}_1, \mathbf{k}_2) &\triangleq \langle q(\mathbf{k}_1) p(\mathbf{k}_2)\rangle - q_0(\mathbf{k}_1) p_0(\mathbf{k}_2), \\ V_{pq}(\mathbf{k}_1, \mathbf{k}_2) &\triangleq \langle p(\mathbf{k}_1) q(\mathbf{k}_2)\rangle - p_0(\mathbf{k}_1) q_0(\mathbf{k}_2), \\ V_{pp}(\mathbf{k}_1, \mathbf{k}_2) &\triangleq \langle p(\mathbf{k}_1) p(\mathbf{k}_2)\rangle - p_0(\mathbf{k}_1) p_0(\mathbf{k}_2). \end{aligned} \tag{5.62}$$

The equivalent expression in terms of operators is

$$\mathbf{V} = \begin{bmatrix} \langle \hat{q}^2 \rangle - \langle \hat{q}\rangle^2 & \frac{1}{2}\langle\{\hat{q},\hat{p}\}\rangle - \langle\hat{q}\rangle\langle\hat{p}\rangle \\ \frac{1}{2}\langle\{\hat{q},\hat{p}\}\rangle - \langle\hat{q}\rangle\langle\hat{p}\rangle & \langle \hat{p}^2 \rangle - \langle \hat{p}\rangle^2 \end{bmatrix}, \tag{5.63}$$

where $\{\hat{q},\hat{p}\} = \hat{q}\hat{p} + \hat{p}\hat{q}$. In terms of only the particle-number degree of freedom, the covariance matrix becomes a 2×2 matrix V_2. Its determinant reproduces the terms in the uncertainty relationship in (3.149), so that

$$\begin{aligned} \det\{V_2\} &= \left(\langle \hat{q}^2 \rangle - \langle \hat{q}\rangle^2\right)\left(\langle \hat{p}^2 \rangle - \langle \hat{p}\rangle^2\right) - \left(\frac{1}{2}\langle\{\hat{q},\hat{p}\}\rangle - \langle\hat{q}\rangle\langle\hat{p}\rangle\right)^2 \\ &= \sigma_q^2\sigma_p^2 - \left(\frac{1}{2}\langle\{\hat{q},\hat{p}\}\rangle - \langle\hat{q}\rangle\langle\hat{p}\rangle\right)^2 \geq \frac{1}{4}. \end{aligned} \tag{5.64}$$

5.7.2 Finite-dimensional systems

The relationship in (5.64) can be generalized to a multimode scenario, with an arbitrary finite number of dimensions as represented by the number of spatiotemporal modes [6]. One can reproduce the relationship in (5.64) for each spatiotemporal mode by diagonalizing the multimode covariance matrix with the aid of a suitable symplectic transformation. In this way, the different spatiotemporal degrees of freedom become decoupled, so that every pair of associated q and p variables can be treated independently. The symplectic transformation renders the covariance matrix into a block diagonal form containing 2×2 sub-matrices on the diagonal, each of which satisfies the minimum uncertainty condition of the form (5.64). It then follows that the augmented covariance matrix $V - B$ is a positive semidefinite matrix, where

$$B = \frac{1}{2} \bigoplus_{n} \sigma_y, \tag{5.65}$$

is the *symplectic structure matrix*, with σ_y being the Pauli y-matrix. By implication, the eigenvalues of $V - B$ must be larger than or equal to 0 for V to be a valid covariance matrix. The inverse of such a covariance matrix can parameterize the Wigner function of a valid Gaussian state. However, the conditions derived for the finite-dimensional case is valid for all state, not only for Gaussian states.

5.7.3 Kernel-based formulation

To convert the matrix-based version of this analysis into a kernel-based version, we convert the symplectic transformation applied to the matrices into an equivalent transformation applied to the kernels. The symplectic nature of the transformation is imposed by the condition that the transformed quadrature operators obey the same *commutation relations*. When these operators are replaced by field variables, the allowed transformation kernels retain their symplectic nature. In other words, if a canonical transformation of the quadrature operators is represented by

$$\begin{aligned}
\hat{q} \to \hat{q}' &= G_{qq} \diamond \hat{q} + G_{qp} \diamond \hat{p} = \hat{q} \diamond G_{qq}^T + \hat{p} \diamond G_{qp}^T, \\
\hat{p} \to \hat{p}' &= G_{pq} \diamond \hat{q} + G_{pp} \diamond \hat{p} = \hat{q} \diamond G_{pq}^T + \hat{p} \diamond G_{pp}^T,
\end{aligned} \tag{5.66}$$

where the G's represent the canonical transformation kernels, so that

$$\begin{aligned}
[\hat{q}', \hat{p}'] &= \mathrm{i} G_{qq} \diamond G_{pp}^T - \mathrm{i} G_{qp} \diamond G_{pq}^T = \mathrm{i}\mathbf{1}, \\
[\hat{q}', \hat{q}'] &= \mathrm{i} G_{qq} \diamond G_{qp}^T - \mathrm{i} G_{qp} \diamond G_{qq}^T = 0, \\
[\hat{p}', \hat{p}'] &= \mathrm{i} G_{pq} \diamond G_{pp}^T - \mathrm{i} G_{pp} \diamond G_{pq}^T = 0,
\end{aligned} \tag{5.67}$$

then it implies that the symplectic condition

$$\begin{bmatrix} G_{qq} & G_{qp} \\ G_{pq} & G_{pp} \end{bmatrix} \diamond \begin{bmatrix} 0 & \mathbf{1} \\ -\mathbf{1} & 0 \end{bmatrix} \diamond \begin{bmatrix} G_{qq}^T & G_{pq}^T \\ G_{qp}^T & G_{pp}^T \end{bmatrix} = \begin{bmatrix} 0 & \mathbf{1} \\ -\mathbf{1} & 0 \end{bmatrix}, \tag{5.68}$$

is satisfied. The field variables are also transformed by the same kernels

$$\begin{aligned} q \to q' &= G_{qq} \diamond q + G_{qp} \diamond p = q \diamond G_{qq}^T + p \diamond G_{qp}^T, \\ p \to p' &= G_{pq} \diamond q + G_{pp} \diamond p = q \diamond G_{pq}^T + p \diamond G_{pp}^T. \end{aligned} \tag{5.69}$$

The finite-dimensional analysis in the previous section suggests that we can generalize the situation to covariance kernels. It would imply that the covariance kernel of the Wigner functional of a valid quantum state can be diagonalized by a symplectic transformation represented in terms of such kernels, similar to how it works for matrices. As such, it then follows that a suitable symplectic transformation applied to a valid covariance kernel would produce

$$\mathbf{G}^T \diamond \mathbf{V} \diamond \mathbf{G} = \mathcal{V}_2 \mathbf{1}, \tag{5.70}$$

where $\mathbf{1}$ is the identity kernel and $\mathcal{V}_2$ is a 2×2 matrix of spectral functions

$$\mathcal{V}_2(\mathbf{k}) = \begin{bmatrix} \mathcal{V}_{qq}(\mathbf{k}) & \mathcal{V}_{qp}(\mathbf{k}) \\ \mathcal{V}_{pq}(\mathbf{k}) & \mathcal{V}_{pp}(\mathbf{k}) \end{bmatrix}, \tag{5.71}$$

with the property that

$$\mathcal{V}_{qq}(\mathbf{k})\mathcal{V}_{pp}(\mathbf{k}) - \mathcal{V}_{qp}(\mathbf{k})\mathcal{V}_{pq}(\mathbf{k}) \geq \frac{1}{4} \quad \forall\, \mathbf{k}. \tag{5.72}$$

Based on the existence of such a symplectic transformation, in analogy with the finite-dimensional case, we then arrive at an equivalent conclusion, namely that the covariance kernel $\mathbf{V}$ obtained from the Wigner functional representing a valid quantum state satisfies the condition that $\mathbf{V} - \mathbf{B}$ is a positive semidefinite kernel, where $\mathbf{B}$ represents the *symplectic structure kernel* given by

$$\mathbf{B} = \frac{1}{2}\sigma_y \mathbf{1}. \tag{5.73}$$

5.7.4 Complex field variable formulation

So far, we have considered the condition for a valid covariance kernel in terms of the quadrature field variables q and p, in association with the quadrature operators $\hat{q}$ and $\hat{p}$. If we prefer to work with the *complex field variables* α and α^* instead, which are associated with the ladder operators $\hat{a}$ and $\hat{a}^\dagger$, then this symplectic transformation is converted to an equivalent transformation for the complex field variables. This transformation is given by a Bogoliubov transformation. In terms of the ladder operators the Bogoliubov transformation is represented by

$$\begin{aligned}
\hat{a} \to \hat{a}' &= \frac{1}{\sqrt{2}}(\hat{q}' + i\hat{p}') = u \diamond \hat{a} + v \diamond \hat{a}^{\dagger}, \\
\hat{a}^{\dagger} \to \hat{a}'^{\dagger} &= \frac{1}{\sqrt{2}}(\hat{q}' - i\hat{p}') = v^{*} \diamond \hat{a} + u^{*} \diamond \hat{a}^{\dagger},
\end{aligned} \tag{5.74}$$

where the Bogoliubov kernels are related to the canonical transformation kernels by

$$\begin{aligned}
u &= \frac{1}{2}(G_{qq} - iG_{qp} + iG_{pq} + G_{pp}), \\
v &= \frac{1}{2}(G_{qq} + iG_{qp} + iG_{pq} - G_{pp}), \\
u^{*} &= \frac{1}{2}(G_{qq} + iG_{qp} - iG_{pq} + G_{pp}), \\
v^{*} &= \frac{1}{2}(G_{qq} - iG_{qp} - iG_{pq} - G_{pp}).
\end{aligned} \tag{5.75}$$

The *commutation relations*, together with (5.67), then produce

$$\begin{aligned}
[\hat{a}, \hat{a}^{\dagger}] &\to \left[u \diamond \hat{a} + v \diamond \hat{a}^{\dagger}, \hat{a} \diamond v^{\dagger} + \hat{a}^{\dagger} \diamond u^{\dagger}\right] = \mathbf{1}, \\
[\hat{a}, \hat{a}] &\to \left[u \diamond \hat{a} + v \diamond \hat{a}^{\dagger}, \hat{a} \diamond u^{T} + \hat{a}^{\dagger} \diamond v^{T}\right] = 0, \\
[\hat{a}^{\dagger}, \hat{a}^{\dagger}] &\to \left[v^{*} \diamond \hat{a} + u^{*} \diamond \hat{a}^{\dagger}, \hat{a} \diamond v^{\dagger} + \hat{a}^{\dagger} \diamond u^{\dagger}\right] = 0,
\end{aligned} \tag{5.76}$$

leading to the conditions for a *Bogoliubov transformation* (see Section 3.4.3)

$$\begin{aligned}
u \diamond u^{\dagger} - v \diamond v^{\dagger} &= \mathbf{1}, \\
u \diamond v^{T} - v \diamond u^{T} &= 0, \\
v^{*} \diamond u^{\dagger} - u^{*} \diamond v^{\dagger} &= 0.
\end{aligned} \tag{5.77}$$

It shows that the symplectic condition for quadrature field variables or quadrature operators translates to the Bogoliubov condition for ladder operators or their associated complex field variables. In terms of matrices, the Bogoliubov condition reads

$$\begin{bmatrix} u & v \\ v^{*} & u^{*} \end{bmatrix} \diamond \begin{bmatrix} \mathbf{1} & 0 \\ 0 & -\mathbf{1} \end{bmatrix} \diamond \begin{bmatrix} u^{\dagger} & v^{T} \\ v^{\dagger} & u^{T} \end{bmatrix} = \begin{bmatrix} \mathbf{1} & 0 \\ 0 & -\mathbf{1} \end{bmatrix}, \tag{5.78}$$

which corresponds to the equivalent expression for finite dimensions [6].

The process whereby the covariance matrix is diagonalized with the aid of a symplectic transformation in finite-dimensions [5, 6] is generalized to the diagonalization of the covariance kernel with the aid of a Bogoliubov transformation for the functional formalism. We base the existence of such a diagonalization process on the understanding that physical processes are given by Bogoliubov transformations, together with possible displacements. If a physical state is related to the vacuum state through a physical process, then it is represented by a Bogoliubov transformation and a displacement operation. By undoing the displacement operation and the Bogoliubov transformation, we can return to the vacuum state, which obeys the basic minimum uncertainty relations.

The transformation that converts the quadrature variables q and p into the complex variables α and α^* is unitary

$$\begin{bmatrix} \alpha \\ \alpha^* \end{bmatrix} = \frac{1}{\sqrt{2}} \begin{bmatrix} 1 & \mathrm{i} \\ 1 & -\mathrm{i} \end{bmatrix} \begin{bmatrix} q \\ p \end{bmatrix} = \mathbf{U}\mathbf{q}. \tag{5.79}$$

The same transformation converts quadrature operators into ladder operators

$$\begin{bmatrix} \hat{a} \\ \hat{a}^\dagger \end{bmatrix} = \frac{1}{\sqrt{2}} \begin{bmatrix} 1 & \mathrm{i} \\ 1 & -\mathrm{i} \end{bmatrix} \begin{bmatrix} \hat{q} \\ \hat{p} \end{bmatrix}. \tag{5.80}$$

Therefore, their associated covariance matrices are also related by this unitary transformation $\mathbf{V} \to \mathbf{V}' = \mathbf{U}\mathbf{V}\mathbf{U}^\dagger$. This transformation does not change the eigenvalue spectrum, but the two-dimensional matrix in the symplectic structure kernel $\mathbf{B}$ (i. e., the Pauli y-matrix σ_y) becomes

$$\begin{aligned} \sigma_y \to \mathbf{U}\sigma_y\mathbf{U}^\dagger &= \frac{1}{2} \begin{bmatrix} 1 & \mathrm{i} \\ 1 & -\mathrm{i} \end{bmatrix} \begin{bmatrix} 0 & -\mathrm{i} \\ \mathrm{i} & 0 \end{bmatrix} \begin{bmatrix} 1 & 1 \\ -\mathrm{i} & \mathrm{i} \end{bmatrix} \\ &= \begin{bmatrix} -1 & 0 \\ 0 & 1 \end{bmatrix} \equiv -\sigma_z, \end{aligned} \tag{5.81}$$

where σ_z is the Pauli z-matrix. The equation for the validity of the covariance kernel in terms of α and α^* retains the same form, with the *Bogoliubov structure kernel* $\mathbf{C}$ replacing the symplectic structure kernel. In this case, the condition for validity requires that $\mathbf{V}' + \mathbf{C}$ is a positive semidefinite kernel, where

$$\mathbf{C} = \frac{1}{2}\sigma_z\mathbf{1}, \tag{5.82}$$

is the Bogoliubov structure kernel. The Gaussian kernel of a valid Wigner functional expressed in terms of α and α^* is proportional to the inverse of the covariance kernel $\mathbf{V}'$, satisfying this condition.

5.8 Other functional phase space representations

The Wigner functional is not the only representation of states and operators on functional phase space. Some of the other representations are related to the Wigner functionals by convolutions; either as

$$R[\alpha] = \int G[\alpha - \alpha']W[\alpha']\,\mathcal{D}^\circ[\alpha'] \quad \text{or}$$
$$W[\alpha] = \int G[\alpha - \alpha']R[\alpha']\,\mathcal{D}^\circ[\alpha']. \tag{5.83}$$

Two representations that are often encountered are the *Husimi Q*-functions [9] and the *Glauber–Sudarshan P*-functions [7, 8]. Although we mostly work with the Wigner functionals in the rest of the book, it is instructive to consider these representations in the context of a functional phase space.

5.8.1 Glauber–Sudarshan *P*-functionals

In terms of only the particle-number degree of freedom, the *P*-functions $P(\alpha)$ are related to density operators $\hat{\rho}$ through an integral over all coherent states by

$$\hat{\rho} = \frac{1}{2\pi}\int |\alpha\rangle\, P(\alpha)\, \langle\alpha|\; \mathrm{d}^2\alpha. \tag{5.84}$$

It is generalized for a functional phase space in terms of the functional integral over all fixed-spectrum coherent states, as

$$\hat{\rho} = \int |\alpha\rangle\, P[\alpha]\, \langle\alpha|\; \mathcal{D}^\circ[\alpha], \tag{5.85}$$

where $|\alpha\rangle$ represents the fixed-spectrum coherent states, and $P[\alpha]$ is a functional of the complex-valued field variable α. We call them *Glauber–Sudarshan functionals* or *P*-functionals. Due to the Hermitian property of density matrices, the *P*-functionals are real valued. Normalization requires that

$$\mathrm{tr}\{\hat{\rho}\} = \int P[\alpha]\,\mathcal{D}^\circ[\alpha] = 1. \tag{5.86}$$

As with Wigner functionals, we see that the operator trace becomes a functional integral of the *P*-functional over the functional phase space.

To invert (5.85), we overlap it by two fixed-spectrum coherent states with parameter functions that are equal, apart from a relative minus sign. It leads to

$$\langle\alpha_1|\,\hat{\rho}\,|-\alpha_1\rangle = \int P[\alpha]\exp\left(-\|\alpha\|^2 - \|\alpha_1\|^2 + \alpha_1^* \diamond \alpha - \alpha^* \diamond \alpha_1\right)\mathcal{D}^\circ[\alpha], \tag{5.87}$$

which represents a *symplectic functional Fourier integral*, similar to what we have for the characteristic functional. We can therefore perform the corresponding inverse symplectic functional Fourier transform, leading to

$$P[\alpha] = \exp\left(\|\alpha\|^2\right)\int \exp\left(\|\alpha_1\|^2\right)\langle\alpha_1|\,\hat{\rho}\,|-\alpha_1\rangle \exp\left(\alpha^* \diamond \alpha_1 - \alpha_1^* \diamond \alpha\right)\mathcal{D}^\circ[\alpha_1], \tag{5.88}$$

provided that $\exp\left(\|\alpha_1\|^2\right)\langle\alpha_1|\hat{\rho}|-\alpha_1\rangle$ is a finite energy functional. Since the latter condition cannot always be satisfied, we find that the P-functionals can be severely singular (as, for instance, for the Fock states).

Wigner functionals are given in terms of the P-functionals by applying (5.11) to (5.85)

$$W[q,p] = \int\left\langle q+\frac{1}{2}x\middle|\alpha_0\right\rangle P[\alpha_0]\left\langle\alpha_0\middle|q-\frac{1}{2}x\right\rangle \exp(-ix\diamond p)\,\mathcal{D}[x]\,\mathcal{D}^{\circ}[\alpha_0], \tag{5.89}$$

and substituting the overlaps from (4.182) with the required shifts. After evaluating the functional integral over x, we obtain

$$W[q,p] = \mathcal{N}_0\int P[\alpha_0]\exp\left(-2\|\alpha-\alpha_0\|^2\right)\,\mathcal{D}^{\circ}[\alpha_0]. \tag{5.90}$$

The result is a functional convolution, as in (5.83), of the P-functional by a minimum uncertainty width Gaussian functional.

To express the P-functional in terms of the Wigner functional, we need to perform the opposite process. Alternatively, we can perform a Fourier transform. The Fourier transform of a convolution gives a product of the respective Fourier transformed functions. The transfer functional is a Gaussian, with which one can divide both sides. Since a Gaussian is nowhere zero, its inverse does not blow up, except at infinity. In this way, one would remove the Gaussian transfer functional and recover the Wigner functional after an inverse Fourier transform. However, the inverse Gaussian is not a finite energy functional. So, its inverse Fourier transform is not well-defined. It thus requires that the Wigner functional is broad enough that its Fourier transform would be narrow enough to render the combination of it with the inverse Gaussian into a finite energy functional. If the Wigner functional is that of a coherent state, the P-functional becomes a Dirac delta functional.

5.8.2 Husimi Q-functionals

For the particle-number degree of freedom only, the *Husimi* Q-functions are obtained from the density operators by

$$Q(\alpha) = \langle\alpha|\hat{\rho}|\alpha\rangle\,. \tag{5.91}$$

It readily generalizes to functionals when the density operators incorporate all degrees of freedom and are overlapped by fixed-spectrum coherent states:

$$Q[\alpha] = \langle\alpha|\hat{\rho}|\alpha\rangle\,. \tag{5.92}$$

Due to the positivity of the density operators, Q-functionals are positive for all α. The completeness relation for the fixed-spectrum coherent states (see Section 6.1) ensures that

$$\int Q[\alpha]\,\mathcal{D}^{\circ}[\alpha] = 1. \tag{5.93}$$

The Q-functionals are complementary to the *P*-functionals. When the expectation value of an *observable* $\hat{A}$ is computed for a given state, the result is computed with the aid of a trace over the product of the density operator and the observable

$$\langle\hat{A}\rangle = \mathrm{tr}\{\hat{\rho}\hat{A}\}. \tag{5.94}$$

If one of the two operators is represented in terms of a *P*-functional, it leads to

$$\begin{aligned}\langle\hat{A}\rangle &= \int \mathrm{tr}\left\{|\alpha\rangle\, P_{\hat{\rho}}[\alpha]\,\langle\alpha|\,\hat{A}\right\}\,\mathcal{D}^{\circ}[\alpha] \\ &= \int P_{\hat{\rho}}[\alpha]\,\langle\alpha|\,\hat{A}\,|\alpha\rangle\;\mathcal{D}^{\circ}[\alpha] = \int P_{\hat{\rho}}[\alpha]Q_{\hat{A}}[\alpha]\,\mathcal{D}^{\circ}[\alpha].\end{aligned} \tag{5.95}$$

So, the other operator is converted into its *Q*-functional.

To consider the relationship of the Husimi *Q*-functionals to the Wigner functionals, we replace the density operator by its *Weyl transformation* (5.43) in terms of the Wigner functional into the definition of the Husimi *Q*-functional in (5.92)

$$Q[\alpha_0] = \int \left\langle \alpha_0 \middle| q + \frac{1}{2}x \right\rangle W[q,p]\exp(\mathrm{i}x \diamond p)\left\langle q - \frac{1}{2}x \middle| \alpha_0 \right\rangle\,\mathcal{D}[q]\,\mathcal{D}^{\circ}[p]\,\mathcal{D}[x]. \tag{5.96}$$

Using (4.182), we get

$$Q[\alpha_0] = \mathcal{N}_0 \int W[\alpha]\exp\left(-2\|\alpha-\alpha_0\|^2\right)\,\mathcal{D}^{\circ}[\alpha], \tag{5.97}$$

where we evaluated the functional integral over x and combined the field variables q and p into the complex field variable α. The result is a functional convolution of the Wigner functional, as in (5.83), with a minimum uncertainty width Gaussian functional. It is called the *Weierstrass transformation.*

We can invert the expression to obtain the Wigner functional from the *Q*-functional using the same process to invert (5.90). It again produces an inverse Gaussian, which requires that the *Q*-functional is broad enough so that its Fourier transform would be narrow enough to render the combination of it with the inverse Gaussian into a finite energy functional. This requirement places a restriction on the nature of Husimi *Q*-functionals.

5.8.3 Characteristic functionals for other representations

The characteristic functionals for the *P*- and *Q*-functionals are defined similarly to the way they are defined for the Wigner functionals in (5.27). They are given in terms of symplectic Fourier transforms by

$$\chi_P[\beta] = \int P[\alpha] \exp(\beta^* \diamond \alpha - \alpha^* \diamond \beta)\, \mathcal{D}^{\circ}[\alpha],$$
$$\chi_Q[\beta] = \int Q[\alpha] \exp(\beta^* \diamond \alpha - \alpha^* \diamond \beta)\, \mathcal{D}^{\circ}[\alpha], \tag{5.98}$$

where β is defined in (5.28). The inverse transformations are

$$P[\alpha] = \int \chi_P[\beta] \exp(\alpha^* \diamond \beta - \beta^* \diamond \alpha)\, \mathcal{D}^{\circ}[\beta],$$
$$Q[\alpha] = \int \chi_Q[\beta] \exp(\alpha^* \diamond \beta - \beta^* \diamond \alpha)\, \mathcal{D}^{\circ}[\beta]. \tag{5.99}$$

Exercise 5.14. Show that the characteristic functional for the Wigner functional $\chi[\beta]$ is related to those of the P- and Q-functionals by

$$\chi[\beta] = \chi_P[\beta] \exp\left(-\frac{1}{2}\|\beta\|^2\right),$$
$$\chi_Q[\beta] = \chi[\beta] \exp\left(-\frac{1}{2}\|\beta\|^2\right). \tag{5.100}$$

5.8.4 Operator ordering

There exists an association between the type of operator ordering and the different phase space representations. Here, we derive this association in the functional context. We start with the Baker–Campbell–Hausdorff formula in (3.41) to define different versions of the displacement operator based on their operator ordering:

$$\hat{D}_s[\beta] \triangleq \exp\left(\frac{1}{2}s\|\beta\|^2\right)\hat{D}[\beta]$$
$$= \begin{cases} \exp\left(\hat{a}^\dagger \diamond \beta\right) \exp\left(-\beta^* \diamond \hat{a}\right) & \text{for } s = 1, \\ \exp\left(\hat{a}^\dagger \diamond \beta - \beta^* \diamond \hat{a}\right) & \text{for } s = 0, \\ \exp\left(-\beta^* \diamond \hat{a}\right) \exp\left(\hat{a}^\dagger \diamond \beta\right) & \text{for } s = -1. \end{cases} \tag{5.101}$$

For $s = 1$, the operator is *normal ordered*, and for $s = -1$, it is *antinormal ordered*. For $s = 0$, we get the usual displacement operator, which is *symmetrically* or *Weyl ordered*.

Next, we generalize the expression for the characteristic functional in (5.37) for these three orderings by defining

$$\chi_s[\xi, \zeta] \triangleq \exp\left(\frac{1}{2}s\|\beta\|^2\right) \operatorname{tr}\left\{\hat{\rho}\hat{D}^\dagger[\beta]\right\} = \operatorname{tr}\left\{\hat{\rho}\hat{D}_s^\dagger[\beta]\right\}. \tag{5.102}$$

The corresponding functional phase space representations then follow from the symplectic Fourier transform as in (5.30), leading to

$$W_s[\alpha] = \operatorname{tr}\left\{\hat{\rho}\hat{T}_s[\alpha]\right\}, \tag{5.103}$$

where

$$\hat{T}_s[\alpha] \triangleq \int \hat{D}_s^{\dagger}[\beta] \exp(\alpha^* \diamond \beta - \beta^* \diamond \alpha)\, \mathcal{D}^{\circ}[\beta], \tag{5.104}$$

is called the *Stratonovich–Weyl* operator [10].

Exercise 5.15. Show that for $s = 1$ and $s = -1$, we get the Glauber–Sudarshan P-functional $W_1[\alpha] = P[\alpha]$ and the Husimi Q-functional $W_{-1}[\alpha] = Q[\alpha]$, respectively.

In Section 6.2.3, it is shown that displacement operators obey an orthogonality condition. For the three different operator orderings, one can generalize the orthogonality condition to read

$$\operatorname{tr}\left\{\hat{D}_{-s}^{\dagger}[\zeta]\hat{D}_s[\beta]\right\} = (2\pi)^{\Omega}\delta[\zeta - \beta]. \tag{5.105}$$

This orthogonality condition can be used to express an arbitrary operator in term of a specific ordering

$$\hat{O}_s = \int F_s[\alpha]\hat{D}_s[\alpha]\, \mathcal{D}^{\circ}[\alpha], \tag{5.106}$$

where

$$F_s[\alpha] = \operatorname{tr}\left\{\hat{D}_{-s}^{\dagger}[\alpha]\hat{O}\right\}. \tag{5.107}$$

Computing the functional phase space representation associated with the opposite ordering for the operator, we obtain

$$\begin{aligned} W_{-s}[\alpha] &= \operatorname{tr}\left\{\hat{O}_s\hat{T}_{-s}[\alpha]\right\} \\ &= \int F_s[\beta] \operatorname{tr}\left\{\hat{D}_s[\beta]\hat{D}_{-s}^{\dagger}[\eta]\right\} \exp(\alpha^* \diamond \eta - \eta^* \diamond \alpha)\, \mathcal{D}^{\circ}[\beta, \eta] \\ &= \int F_s[\eta] \exp(\alpha^* \diamond \eta - \eta^* \diamond \alpha)\, \mathcal{D}^{\circ}[\eta], \end{aligned} \tag{5.108}$$

where we used (5.105). It thus follows that the coefficient function is the characteristic functional for $\hat{O}$ in the opposite ordering. The Wigner functionals are their own opposites in this sense. Therefore, the Wigner functional formalism does not need to incorporate any other type of functional phase space representation.

Bibliography

[1] H. J. Groenewold. On the principles of elementary quantum mechanics. *Physica*, 12:405–460, 1946.
[2] J. E. Moyal. Quantum mechanics as a statistical theory. *Math. Proc. Camb. Philos. Soc.*, 45:99–124, 1949.

[3] T. L. Curtright and C. K. Zachos. Quantum mechanics in phase space. *Asia Pac. Phys. Newsl.*, 1:37–46, 2012.
[4] E. Wigner. On the quantum correction for thermodynamic equilibrium. *Phys. Rev.*, 40:749–759, 1932.
[5] R. Simon, E. C. G. Sudarshan, and N. Mukunda. Gaussian-wigner distributions in quantum mechanics and optics. *Phys. Rev. A*, 36:3868–3880, 1987.
[6] R. Simon, N. Mukunda, and B. Dutta. Quantum-noise matrix for multimode systems: U(n) invariance, squeezing, and normal forms. *Phys. Rev. A*, 49:1567–1583, 1994.
[7] E. C. G. Sudarshan. Equivalence of semiclassical and quantum mechanical descriptions of statistical light beams. *Phys. Rev. Lett.*, 10:277–279, 1963.
[8] R. J. Glauber. Coherent and incoherent states of the radiation field. *Phys. Rev.*, 131:2766–2788, 1963.
[9] K. Husimi. Some formal properties of the density matrix. *Nippon Sugaku-Buturigakkwai Kizi Dai 3 Ki*, 22:264–314, 1940.
[10] R. L. Stratonovich. On distributions in representation space. *Soviet Physics JETP*, 4:891–898, 1957.

6 Wigner functionals of states and operators

The basic development of the Wigner functional formalism is provided in Chapter 5. There are still various techniques that can be developed to assist calculations within this formalism. Here, we proceed with the development of this formalism by using it to compute the Wigner functionals of the most common states and operators found in quantum optics, and to discuss various related aspects of these states and operators. Various techniques are demonstrated in the process. The calculations often involve functional integrations, which are discussed in Appendix C, and demonstrate helpful methodology, often involving generating functions as presented in Appendix A.

The results we obtain here are the functional equivalents of the Wigner functions that are often found in literature on quantum optics [1, 2]. While some continuous variable phase space formulations in quantum optics still incorporate ladder operators or quadrature operators in their descriptions [3, 4], we do not follow that approach. The Wigner functionals encountered here are functionals of the field variables that parameterize the functional phase space.

6.1 Coherent states

The *fixed-spectrum coherent states* are defined and discussed in Section 4.4. Here, we compute their Wigner functionals and consider some of their pertinent properties. We also present a technique to compute the Wigner functionals of other states and operators using coherent states.

6.1.1 Wigner functional

To calculate the Wigner functional of a fixed-spectrum coherent state, we substitute $\hat{\rho} \rightarrow |\alpha_0\rangle\langle\alpha_0|$ into (5.11). Its Wigner functional is then given by

$$W_{\text{coh}}[q,p] = \int \left\langle q + \frac{1}{2}x \middle| \alpha_0 \right\rangle \left\langle \alpha_0 \middle| q - \frac{1}{2}x \right\rangle \exp(-\mathrm{i}p \diamond x)\, \mathcal{D}[x]. \tag{6.1}$$

The overlaps between the quadrature basis elements and the fixed-spectrum coherent states are provided in (4.182). Substituting these expressions with the required shifts into (6.1), we obtain

$$W_{\text{coh}}[q,p] = \frac{1}{\pi^{\Omega/2}} \int \exp\left[-(q-q_0)\diamond(q-q_0) - \frac{1}{4}x \diamond x - \mathrm{i}x \diamond (p-p_0)\right] \mathcal{D}[x], \tag{6.2}$$

where we represent the complex parameter function of the coherent state in terms of its real and imaginary parts

https://doi.org/10.1515/9783111445342-008

$$\alpha_0 \rightarrow \frac{1}{\sqrt{2}} [q_0 + \mathrm{i} p_0] . \tag{6.3}$$

After evaluating the functional integral over x, with (C.15), we obtain

$$W_{\mathrm{coh}}[q, p] = \mathcal{N}_0 \exp\left(-\|q - q_0\|^2 - \|p - p_0\|^2\right) , \tag{6.4}$$

where $\mathcal{N}_0$ is the normalization constant. Reverting back to the complex field variable and parameter function, we finally get

$$W_{\mathrm{coh}}[\alpha] = \mathcal{N}_0 \exp\left(-2\|\alpha - \alpha_0\|^2\right) . \tag{6.5}$$

The Wigner functional for the fixed-spectrum coherent state is a Gaussian functional, centred at the *parameter function* $\alpha_{0,s}(\mathbf{k})$, a complex-valued spectral function as defined in (4.105). For $\alpha_0 = 0$, we obtain the Wigner functional for the *vacuum state*.

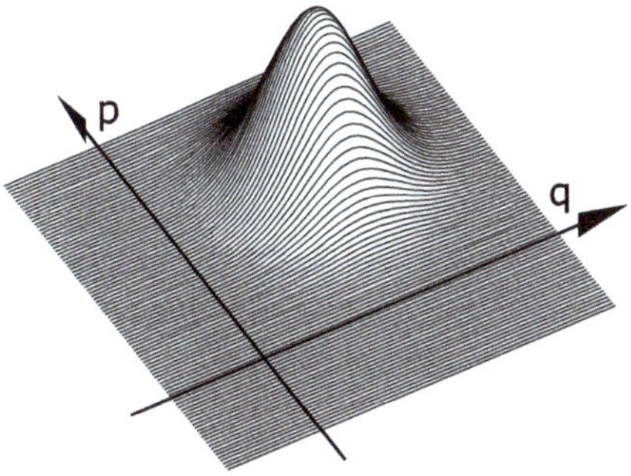

Figure 6.1: Wigner function of a coherent state on a two-dimensional phase space.

The coherent state Wigner functional is an infinite-dimensional generalization of the two-dimensional Wigner function of a coherent state represented in terms of the particle-number degree of freedom only. To visualize the fixed-spectrum coherent state Wigner functional, we show its two-dimensional analogue in Figure 6.1.

6.1.1.1 Normalization

The normalization of the coherent state is imposed by the normalization condition in (5.46). After shifting the function to the origin (which does not affect its normalization), we have the normalization requirement

$$\mathcal{N}_0 \int \exp\left(-2\|\alpha\|^2\right) \mathcal{D}^{\circ}[\alpha] = 1. \tag{6.6}$$

Remember that the functional integration measure $\mathcal{D}^{\circ}[\alpha]$, as defined in (5.29), contains a constant factor of $(2\pi)^{-1}$ per spatiotemporal degree of freedom. The number of spatiotemporal degrees of freedom (or the number of spatiotemporal dimensions) for the functional phase space is equivalent to the cardinality of the Fourier space, which we represent by Ω, as defined in (4.154). Hence, the integral is multiplied by a factor $(2\pi)^{-\Omega}$.

Evaluating the functional integration in (6.6), using (C.22), we get

$$\mathcal{N}_0 \frac{\pi^{\Omega}}{(2\pi)^{\Omega}} = 1. \tag{6.7}$$

So, the normalization constant for the Wigner functional of the fixed-spectrum coherent state is the same constant we obtained in (5.20). The maximum value of the Wigner functional of a coherent state is the maximum that any Wigner functional can have. Since this value is divergent, it may not seem very helpful to know what it is. However, keeping careful track of this quantity by labelling it with its specific symbol, we can confirm its cancellation when quantities that must come out to be finite are calculated.

6.1.1.2 Purity

Another property of coherent states that affects the normalization constant is their purity. It is given by the trace of the square of the density operator. According to (5.51), the trace of the product of two operators is equivalent to the integral over the product of their Wigner functionals. Therefore, we just integrate the square of the Wigner functional of a state to determine its purity. For the coherent state, it becomes

$$\text{purity} = \mathcal{N}_0^2 \int \exp\left(-4\|\alpha - \alpha_0\|^2\right) \mathcal{D}^{\circ}[\alpha]. \tag{6.8}$$

Again we can remove the shift without affecting the purity of the state. The result can be converted back to the trace of the state, as defined for its normalization, by scaling the integration variable according to

$$\alpha \to \frac{1}{\sqrt{2}}\alpha \quad \text{and} \quad \alpha^* \to \frac{1}{\sqrt{2}}\alpha^*. \tag{6.9}$$

The only effect is a constant (the Jacobian) emerging from the measure

$$\mathcal{D}^{\circ}[\alpha] \to 2^{-\Omega}\mathcal{D}^{\circ}[\alpha]. \tag{6.10}$$

Hence,

$$\text{purity} = \mathcal{N}_0^2 2^{-\Omega} \int \exp\left(-2\|\alpha\|^2\right) \mathcal{D}^{\circ}[\alpha] = \mathcal{N}_0 2^{-\Omega} = 1, \tag{6.11}$$

where we applied the normalization of the coherent state as given in (6.6), and substituted (5.20). It follows that the coherent state with a normalization constant given by (5.20) is a pure state, as expected.

The same argument can be made for all *pure Gaussian states* (pure states with Wigner functionals given by Gaussian functionals). It follows that they all have the same normalization constant given by (5.20).

6.1.2 Completeness of fixed-spectrum coherent states

We postponed the discussion about the completeness of the coherent states until we have formulated Wigner functional theory because it simplifies the calculation. Here, we consider the completeness of the fixed-spectrum coherent states, using their Wigner functionals given in (6.5) with (5.20). In terms of the Weyl representation (5.43), the density operator for the fixed-spectrum coherent state is given by

$$\hat{\rho}_{\text{coh}} = |\alpha_0\rangle\langle\alpha_0| = \mathcal{N}_0 \int \left|q + \frac{1}{2}x\right\rangle \exp\left(-2\|\alpha - \alpha_0\|^2\right) \exp(ip \diamond x) \times \left\langle q - \frac{1}{2}x\right| \mathcal{D}[x,q]\,\mathcal{D}^{\circ}[p], \tag{6.12}$$

where α is related to q and p as shown in (4.181).

For the completeness condition, we consider the operator defined by

$$\hat{L} = \int |\alpha_0\rangle\langle\alpha_0|\,\mathcal{D}[\alpha_0] = \mathcal{N}_0 \int \left|q + \frac{1}{2}x\right\rangle \exp\left(-2\|\alpha - \alpha_0\|^2\right) \exp(ip \diamond x) \times \left\langle q - \frac{1}{2}x\right| \mathcal{D}[x,q]\,\mathcal{D}^{\circ}[p]\,\mathcal{D}[\alpha_0], \tag{6.13}$$

where the parameter function of the coherent state becomes a field variable that we integrate over. Since the original field variable α and the new field variable α_0 belong to the same set of functions, we can change the integration field variable from one to the other in (6.6) without changing the result. Therefore, we obtain

$$\int \exp\left(-2\|\alpha - \alpha_0\|^2\right) \mathcal{D}[\alpha_0] = \pi^{\Omega}. \tag{6.14}$$

The operator $\hat{L}$ then becomes

$$\hat{L} = (2\pi)^{\Omega} \int \left|q + \frac{1}{2}x\right\rangle \exp(ip \diamond x) \left\langle q - \frac{1}{2}x\right| \mathcal{D}[x,q]\,\mathcal{D}^{\circ}[p]. \tag{6.15}$$

The integral over p gives a Dirac delta functional, which enforces $x = 0$ when we evaluate the integral over x. Hence,

$$\hat{L} = (2\pi)^{\Omega} \int \left|q + \frac{1}{2}x\right\rangle \delta[x] \left\langle q - \frac{1}{2}x\right| \mathcal{D}[x,q] = (2\pi)^{\Omega} \int |q\rangle\langle q|\,\mathcal{D}[q] = (2\pi)^{\Omega}\,\mathbb{1}. \tag{6.16}$$

where we used the completeness condition for the q-basis (3.21). The *completeness condition* for the fixed-spectrum coherent states thus reads

$$\frac{1}{(2\pi)^{\Omega}}\int |\alpha\rangle\langle\alpha|\ \mathcal{D}[\alpha] \equiv \int |\alpha\rangle\langle\alpha|\ \mathcal{D}^{\circ}[\alpha] = \mathbb{1}, \tag{6.17}$$

which is the functional equivalent of (3.114).

6.1.3 Coherent-state-assisted approach

It is often convenient to use identity operators resolved in terms of coherent states, as in (6.17), to aid the calculation of Wigner functionals for states or operators. Consider the general case for an operator $\hat{A}$, leading to

$$\begin{aligned} W_{\hat{A}}[q,p] &= \int \left\langle q+\frac{1}{2}x\middle|\hat{A}\middle|q-\frac{1}{2}x\right\rangle \exp(-\mathrm{i}p\diamond x)\ \mathcal{D}[x] \\ &= \int \left\langle q+\frac{1}{2}x\middle|\alpha_1\right\rangle \langle\alpha_1|\hat{A}|\alpha_2\rangle \left\langle\alpha_2\middle|q-\frac{1}{2}x\right\rangle \\ &\quad\times \exp(-\mathrm{i}p\diamond x)\ \mathcal{D}[x]\ \mathcal{D}^{\circ}[\alpha_1,\alpha_2], \end{aligned} \tag{6.18}$$

where α_1 and α_2 are the integration field variables, serving as parameter functions of the fixed-spectrum coherent states, respectively resolving the two identity operators that we inserted on either side of $\hat{A}$. Replacing the overlaps using (4.182) with the appropriate shifts and field variables, and evaluating the functional integral over x, we get

$$\begin{aligned} W_{\hat{A}}[q,p] = 2^{\Omega}\int \exp\Big[&-q^2-p^2+\sqrt{2}q\diamond(\alpha_1+\alpha_2^*)-\mathrm{i}\sqrt{2}p\diamond(\alpha_1-\alpha_2^*)-\frac{1}{2}\alpha_1^*\diamond\alpha_1 \\ &-\frac{1}{2}\alpha_2^*\diamond\alpha_2^*-\alpha_2^*\diamond\alpha_1\Big]\ \langle\alpha_1|\hat{A}|\alpha_2\rangle\ \mathcal{D}^{\circ}[\alpha_1,\alpha_2]. \end{aligned} \tag{6.19}$$

Finally, we express the result in terms of α, instead of q and p:

$$\begin{aligned} W_{\hat{A}}[\alpha] = \mathcal{N}_0\int \exp\Big(&-2\alpha^*\diamond\alpha+2\alpha^*\diamond\alpha_1+2\alpha_2^*\diamond\alpha-\frac{1}{2}\alpha_1^*\diamond\alpha_1 \\ &-\frac{1}{2}\alpha_2^*\diamond\alpha_2-\alpha_2^*\diamond\alpha_1\Big)\ \langle\alpha_1|\hat{A}|\alpha_2\rangle\ \mathcal{D}^{\circ}[\alpha_1,\alpha_2]. \end{aligned} \tag{6.20}$$

The expression provides a method to compute the Wigner functional for an operator $\hat{A}$ from the functional produced by the overlap of this operator by two fixed-spectrum coherent states $\langle\alpha_1|\hat{A}|\alpha_2\rangle$. This calculation involves functional integrations that can usually be evaluated without problem, sometimes with the aid of generating function(al)s.

6.1.3.1 Polynomial operators

If the operator is a polynomial of ladder operators, the overlap by the two coherent states would produce a bivariate polynomial in field variables α_1^* and α_2, times an exponential function, leading to an expression of the form

$$\langle \alpha_1 | \hat{A} | \alpha_2 \rangle = C[\alpha_1^*, \alpha_2] \exp\left(-\frac{1}{2}\alpha_1^* \diamond \alpha_1 - \frac{1}{2}\alpha_2^* \diamond \alpha_2 + \alpha_1^* \diamond \alpha_2 \right), \tag{6.21}$$

where C represents the bivariate polynomial functional in α_1^* and α_2. In such cases, we can use source terms for the two field variables to remove the bivariate polynomial from the functional integral, by replacing the field variables in the normal-ordered version of C by *functional derivatives* with respect to the auxiliary field variables. The overlapped operator is then given by

$$\langle \alpha_1 | \hat{A} | \alpha_2 \rangle = C_{\hat{A}}[\delta_\nu, \delta_{\mu^*}] \exp\left(\alpha_1^* \diamond \nu + \mu^* \diamond \alpha_2 - \frac{1}{2}\alpha_1^* \diamond \alpha_1 \right.$$
$$\left. - \frac{1}{2}\alpha_2^* \diamond \alpha_2 + \alpha_1^* \diamond \alpha_2 \right)\Big|_{\mu^*=\nu=0} . \tag{6.22}$$

The functional integral in (6.20) with the exponential part in this expression can now be evaluated separately. The result can be used to express the Wigner functional of the polynomial operator $\hat{A}$ as

$$W_{\hat{A}}[\alpha] = C_{\hat{A}}[\delta_\nu, \delta_{\mu^*}] \exp\left(\mu^* \diamond \alpha + \alpha^* \diamond \nu - \frac{1}{2}\mu^* \diamond \nu \right)\Big|_{\mu^*=\nu=0} . \tag{6.23}$$

Here, the exponential factor acts as a *generating functional* and $C_{\hat{A}}[\delta_\nu, \delta_{\mu^*}]$ represents a *construction operation* composed of functional derivatives.

Exercise 6.1. Show that the exponential factor in (6.23) is produced when the exponential factor in (6.22) is inserted into (6.20) and the functional integrals are evaluated.

The procedure to obtain the Wigner functional for a polynomial operator can thus be performed without the need to perform a functional integration. It is done as follows:

1. First, ensure the operator is in normal order. If not, use the commutation relations in (4.15) to convert it to normal order.
2. Replace the ladder operators by functional derivatives

$$\hat{a}_s^\dagger(\mathbf{k}) \to \frac{\delta}{\delta \nu_s(\mathbf{k})} \quad \text{and} \quad \hat{a}_s(\mathbf{k}) \to \frac{\delta}{\delta \mu_s^*(\mathbf{k})}, \tag{6.24}$$

   to define the functional derivative operation $C[\delta_\nu, \delta_{\mu^*}]$.
3. Apply the functional derivative operation on the generating functional shown in (6.23) and set the auxiliary field variables to zero to obtain the Wigner functional.

Exercise 6.2. Show that the star product of polynomial operators can be computed by

$$W_{\hat{A}} \star W_{\hat{B}} = C_{\hat{A}}[\delta_{\nu}, \delta_{\mu^*}]\, C_{\hat{B}}[\delta_{\xi}, \delta_{\zeta^*}] \exp\left[(\mu^* + \zeta^*) \diamond \alpha + \alpha^* \diamond (\nu + \xi) - \frac{1}{2}(\mu^* + \zeta^*) \diamond (\nu + \xi) + \mu^* \diamond \xi\right]\Bigg|_{\mu^*=\nu=\zeta^*=\xi=0}, \tag{6.25}$$

where $C_{\hat{A}}$ and $C_{\hat{B}}$ are the construction operations associated with the respective polynomial operators.

6.1.4 Schrödinger cat state

A *Schrödinger cat state* is a superposition of two or more quantum states that each consist of multiple particles (photons). Here, we consider a Schrödinger cat state that is composed of two fixed-spectrum coherent states $|\zeta\rangle$ and $|-\zeta\rangle$. Its density operator reads

$$\hat{\rho}_{\text{cat}} = \frac{1}{2\mathcal{N}[\zeta]} (|\zeta\rangle + |-\zeta\rangle)(\langle\zeta| + \langle-\zeta|), \tag{6.26}$$

where ζ and $-\zeta$ are the parameter functions for the two coherent states, respectively, and the normalization constant is

$$\mathcal{N}[\zeta] = 1 + \langle\zeta|-\zeta\rangle = 1 + \exp\left(-2\|\zeta\|^2\right). \tag{6.27}$$

The Wigner functional of the Schrödinger cat state can be computed in the same way that the Wigner functional of a fixed-spectrum coherent states is computed. It is represented by the sum of four terms,

$$\text{Wigner}\{\hat{\rho}_{\text{cat}}\} = \frac{1}{2\mathcal{N}[\zeta]} (\text{Wigner}\{|\zeta\rangle\langle\zeta|\} + \text{Wigner}\{|\zeta\rangle\langle-\zeta|\} + \text{Wigner}\{|-\zeta\rangle\langle\zeta|\} + \text{Wigner}\{|-\zeta\rangle\langle-\zeta|\}). \tag{6.28}$$

Two of these terms are *cross-Wigner functionals*, having kets and bras for different pure states Wigner$\{|\zeta\rangle\langle-\zeta|\}$ and Wigner$\{|-\zeta\rangle\langle\zeta|\}$. They combine to form an interference term. The result is

$$W_{\text{cat}}[\alpha] = \frac{\mathcal{N}_0}{2\mathcal{N}[\zeta]} \left[\exp\left(-2\|\alpha - \zeta\|^2\right) + \exp\left(-2\|\alpha + \zeta\|^2\right) + 2\exp\left(-2\|\alpha\|^2\right)\cos\left(4\,\text{Im}\{\alpha^* \diamond \zeta\}\right)\right]. \tag{6.29}$$

The Wigner function of a Schrödinger cat state on a two-dimensional phase space (for the particle-number degree of freedom only) is shown in Figure 6.2 for illustration. It consists of three parts, representing the three terms in the square brackets in (6.29). Two of these parts are proportional to the Wigner functionals of the original two fixed spectrum coherent states that were combined in the superposition. Due to the choice of their parameter functions (one being the negative of the other), they are located at geometrically opposite locations with respect to the origin. The third part is a modulated version of the Wigner functional for the vacuum state located at the origin. The modulation produces a sinusoidal interference pattern, the frequency of which depends on the magnitude of the parameter functions of the original fixed-spectrum coherent states. As a result, the central part of the Wigner functional contains negative values, which are often associated with the "quantumness" of a quantum state.

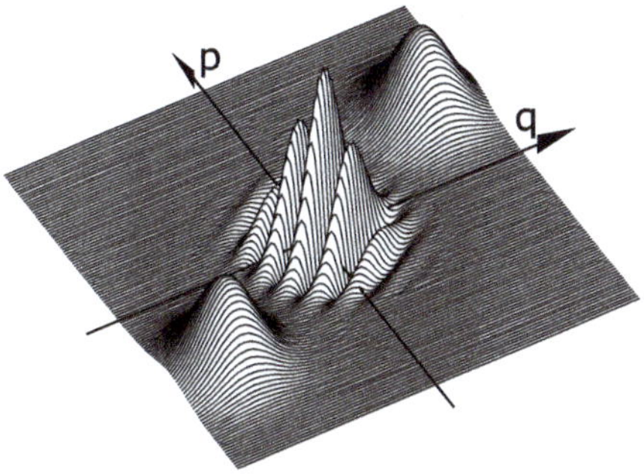

Figure 6.2: Wigner function of a Schrödinger cat state on a two-dimensional phase space.

The time evolution of such a Schrödinger cat state is given by a rotation of the functional around the origin. One can compute the *marginal probability distribution* by integrating the Wigner functional along one of the quadrature field variables. The resulting marginal probability distribution produces oscillations as a function of time of the two displaced Gaussian functionals, periodically producing interference between them. In terms of the two-dimensional version shown in Figure 6.2, the oscillating time evolution of its marginal probability distribution is equivalent to the situation shown in Figure 2.7, where the oscillating field in a GRIN lens is shown. It thus follows that the oscillating time evolution of the marginal probability distribution of a Schrödinger cat state on a two-dimensional phase space is exactly analogous to the GRIN lens scenario in one dimension.

6.2 Displacement operator

6.2.1 Wigner functional

The displacement operator is closely associated with the coherent states, in that coherent states are displaced vacuum states. Here, we first compute its Wigner functional, using the Baker–Campbell–Hausdorff formula in (3.41) to separate the displacement operator into a normal-ordered product of exponential operators:

$$\hat{D}[\alpha_0] = \exp\left(\hat{a}^\dagger \diamond \alpha_0 - \alpha_0^* \diamond \hat{a}\right) = \exp\left(-\frac{1}{2}\|\alpha_0\|^2\right) \exp\left(\hat{a}^\dagger \diamond \alpha_0\right) \exp(-\alpha_0^* \diamond \hat{a}). \quad (6.30)$$

For the coherent-state-assisted approach, we compute the overlap

$$\begin{aligned}\langle\alpha_1|\,\hat{D}[\alpha_0]\,|\alpha_2\rangle &= \exp\left(-\frac{1}{2}\|\alpha_0\|^2\right)\langle\alpha_1|\exp\left(\hat{a}^\dagger \diamond \alpha_0\right)\exp\left(-\alpha_0^* \diamond \hat{a}\right)|\alpha_2\rangle \\ &= \exp\left(-\frac{1}{2}\|\alpha_0\|^2\right)\exp\left(\alpha_1^* \diamond \alpha_0\right)\langle\alpha_1|\alpha_2\rangle\exp\left(-\alpha_0^* \diamond \alpha_2\right) \\ &= \exp\left(-\frac{1}{2}\|\alpha_0\|^2 + \alpha_1^* \diamond \alpha_0 - \frac{1}{2}|\alpha_1|^2 - \frac{1}{2}|\alpha_2|^2 + \alpha_1^* \diamond \alpha_2 - \alpha_0^* \diamond \alpha_2\right). \end{aligned} \quad (6.31)$$

Substituting this result into (6.20), we get

$$\begin{aligned} W_{\hat{D}} &\triangleq \text{Wigner}\left\{\hat{D}[\alpha_0]\right\} \\ &= \mathcal{N}_0 \int \exp\Big(-2\alpha^* \diamond \alpha + 2\alpha^* \diamond \alpha_1 + 2\alpha_2^* \diamond \alpha - \alpha_1^* \diamond \alpha_1 - \alpha_2^* \diamond \alpha_2 \\ &\quad - \alpha_2^* \diamond \alpha_1 + \alpha_1^* \diamond \alpha_0 + \alpha_1^* \diamond \alpha_2 - \alpha_0^* \diamond \alpha_2 - \frac{1}{2}\alpha_0^* \diamond \alpha_0\Big)\,\mathcal{D}^\circ[\alpha_1, \alpha_2]. \end{aligned} \quad (6.32)$$

The two functional integrals are evaluated to obtain the expression for the Wigner functional of the displacement operator. As a demonstration, we show the two isotropic Gaussian functional integrations in detail, using (C.22). The integration over α_1 gives

$$\begin{aligned} W_{\hat{D}} &= \mathcal{N}_0 \int \exp\Big[-2\alpha^* \diamond \alpha - \alpha_2^* \diamond \alpha_2 + 2\alpha_2^* \diamond \alpha - \alpha_0^* \diamond \alpha_2 - \frac{1}{2}\alpha_0^* \diamond \alpha_0 \\ &\quad - \alpha_1^* \diamond \alpha_1 + (2\alpha^* - \alpha_2^*) \diamond \alpha_1 + \alpha_1^* \diamond (\alpha_0 + \alpha_2)\Big]\,\mathcal{D}^\circ[\alpha_1, \alpha_2] \\ &= \mathcal{N}_0 \int \exp\Big(-2\alpha^* \diamond \alpha - \alpha_2^* \diamond \alpha_2 + 2\alpha_2^* \diamond \alpha - \alpha_0^* \diamond \alpha_2 - \frac{1}{2}\alpha_0^* \diamond \alpha_0 \\ &\quad - \alpha_2^* \diamond \alpha_2 - \alpha_2^* \diamond \alpha_0 + 2\alpha^* \diamond \alpha_2 + 2\alpha^* \diamond \alpha_0\Big)\,\mathcal{D}^\circ[\alpha_2]. \end{aligned} \quad (6.33)$$

Then we perform the integration over α_2 to get

$$\begin{aligned} W_{\hat{D}} &= \mathcal{N}_0 \int \exp\Big[-2\alpha^* \diamond \alpha + 2\alpha^* \diamond \alpha_0 - \frac{1}{2}\alpha_0^* \diamond \alpha_0 \\ &\quad - 2\alpha_2^* \diamond \alpha_2 + (2\alpha^* - \alpha_0^*) \diamond \alpha_2 + \alpha_2^* \diamond (2\alpha - \alpha_0)\Big]\,\mathcal{D}^\circ[\alpha_2] \\ &= \exp\left(\alpha^* \diamond \alpha_0 - \alpha_0^* \diamond \alpha\right). \end{aligned} \quad (6.34)$$

In conclusion, the Wigner functional for the displacement operator, produced by evaluating the functional integrals over α_1 and α_2 in (6.32), reads

$$W_{\hat{D}}[\alpha;\alpha_0] = \exp\left(\alpha^* \diamond \alpha_0 - \alpha_0^* \diamond \alpha\right) = \exp\left(\mathrm{i}q \diamond p_0 - \mathrm{i}q_0^* \diamond p\right). \tag{6.35}$$

It has a form similar to the operator itself.

6.2.2 Displacement of an arbitrary state

Displacement operators are not only used to produce coherent states from the vacuum. They can be applied to arbitrary states. When a state is represented by its density operator, the displacement is performed by

$$\hat{\rho} \to \hat{D}[\alpha_0]\hat{\rho}\hat{D}^\dagger[\alpha_0]. \tag{6.36}$$

Using the triple star product in (5.49), we calculate the expression for the Wigner functional of the product of these three operators. In terms of (6.35), the expression reads

$$\begin{aligned}
\mathrm{Wigner}\{\hat{D}\hat{\rho}\hat{D}^\dagger\} = & \int \exp\left[\frac{1}{2}\left(\alpha_a^* + \alpha^* + \alpha_b^*\right) \diamond \alpha_0 - \frac{1}{2}\alpha_0^* \diamond (\alpha_a + \alpha + \alpha_b)\right] \\
& \times \exp\left[-\frac{1}{2}\left(\alpha_a^* + \alpha^* - \alpha_b^*\right) \diamond \alpha_0 + \frac{1}{2}\alpha_0^* \diamond (\alpha_a + \alpha - \alpha_b)\right] \\
& \times W_{\hat{\rho}}[\alpha_a] \exp[(\alpha^* - \alpha_a^*) \diamond \alpha_b - \alpha_b^* \diamond (\alpha - \alpha_a)]\, \mathcal{D}^\circ[\alpha_a, \alpha_b] \\
= & \int W_{\hat{\rho}}[\alpha_a] \exp\left[\alpha_b^* \diamond (\alpha_a - \alpha + \alpha_0)\right. \\
& \left. - \left(\alpha_a^* - \alpha^* + \alpha_0^*\right) \diamond \alpha_b\right]\, \mathcal{D}^\circ[\alpha_a, \alpha_b].
\end{aligned} \tag{6.37}$$

Since there are no quadratic terms for α_b in the exponent, the functional integration over α_b produces a Dirac delta functional, which we can evaluate with the aid of (C.1). The result is

$$\mathrm{Wigner}\{\hat{D}\hat{\rho}\hat{D}^\dagger\} = (2\pi)^\Omega \int W_{\hat{\rho}}[\alpha_a]\, \delta[\alpha_a - \alpha + \alpha_0]\, \mathcal{D}^\circ[\alpha_a] = W_{\hat{\rho}}[\alpha - \alpha_0]. \tag{6.38}$$

It thus follows that the effect of the displacement operation on an arbitrary Wigner functional is a shift in its argument.

6.2.3 Orthogonality and completeness

Displacement operators form a complete orthonormal set of operators in terms of which all the linear operators on a Hilbert space can be expanded. To justify this statement, we define the inner product for the space of all these linear operators by the *Hilbert–Schmidt inner product* [5], which is given by the trace of the product of one operator with the Hermitian adjoint of the other operator

$$\left\langle \hat{D}(\alpha), \hat{D}(\beta) \right\rangle \triangleq \mathrm{tr}\{\hat{D}^{\dagger}(\alpha)\hat{D}(\beta)\}. \tag{6.39}$$

We compute this trace over the product of operators, using their representations in terms of Wigner functionals. The *orthogonality of displacement operators* is therefore given in terms of a functional integral of the product of the Wigner functionals of two displacement operators. The integral is readily evaluated to produce Dirac delta functionals with the aid of (5.4) or (C.1). It reads

$$\begin{aligned}
\mathrm{tr}\{\hat{D}^{\dagger}(\zeta)\hat{D}(\beta)\} &= \int W_{\hat{D}}^{*}[\alpha;\zeta]W_{\hat{D}}[\alpha;\beta]\ \mathcal{D}^{\circ}[\alpha] \\
&= \int \exp(-\mathrm{i}p_0 \diamond q + \mathrm{i}q_0 \diamond p)\exp(\mathrm{i}p_1 \diamond q - \mathrm{i}q_1 \diamond p)\ \mathcal{D}[q]\ \mathcal{D}^{\circ}[p]. \\
&= (2\pi)^{\Omega}\delta[p_0 - p_1]\delta[q_0 - q_1],
\end{aligned} \tag{6.40}$$

where the parameter functions of the two displacement operators are represented in terms of real-valued parameter functions by

$$\zeta = \frac{1}{\sqrt{2}}(q_0 + \mathrm{i}p_0) \quad \text{and} \quad \beta = \frac{1}{\sqrt{2}}(q_1 + \mathrm{i}p_1). \tag{6.41}$$

We can also show that the displacement operators obey a *completeness condition.* In this case, we express the tensor product of two displacement operators with the aid of Weyl representations (5.43) in terms of their Wigner functionals

$$\begin{aligned}
&\int \hat{D}_A^{\dagger}(\zeta) \otimes \hat{D}_B(\zeta)\ \mathcal{D}^{\circ}[\zeta] \\
&= \int \left|q + \frac{1}{2}x\right\rangle_A \left|q' + \frac{1}{2}x'\right\rangle_B \exp(\mathrm{i}p \diamond x + \mathrm{i}p' \diamond x') \\
&\quad \times \left\{\int \exp(-\mathrm{i}p_0 \diamond q + \mathrm{i}q_0 \diamond p)\exp(\mathrm{i}p_0 \diamond q' - \mathrm{i}q_0 \diamond p')\ \mathcal{D}^{\circ}[p_0]\ \mathcal{D}[q_0]\right\} \\
&\quad \times \left\langle q - \frac{1}{2}x\right|_A \left\langle q' - \frac{1}{2}x'\right|_B\ \mathcal{D}^{\circ}[p,p']\ \mathcal{D}[q,x,q',x'].
\end{aligned} \tag{6.42}$$

Again, we use (5.4) or (C.1) to perform the calculation, leading to

$$\begin{aligned}
&\int \hat{D}_A^{\dagger}(\zeta) \otimes \hat{D}_B(\zeta)\ \mathcal{D}^{\circ}D[\zeta] \\
&= \int \left|q + \frac{1}{2}x\right\rangle_A \left|q' + \frac{1}{2}x'\right\rangle_B \exp(\mathrm{i}p \diamond x + \mathrm{i}p' \diamond x')(2\pi)^{\Omega}\delta[p - p']\delta[q - q'] \\
&\quad \times \left\langle q - \frac{1}{2}x\right|_A \left\langle q' - \frac{1}{2}x'\right|_B\ \mathcal{D}^{\circ}[p,p']\ \mathcal{D}[q,x,q',x'] \\
&= \int \left|q + \frac{1}{2}x\right\rangle_A \left|q - \frac{1}{2}x\right\rangle_B \left\langle q - \frac{1}{2}x\right|_A \left\langle q + \frac{1}{2}x\right|_B\ \mathcal{D}[q,x] \\
&= \int |q_1\rangle_A |q_2\rangle_B \langle q_2|_A \langle q_1|_B\ \mathcal{D}[q_1,q_2] = \mathbb{1}_{AB} \otimes \mathbb{1}_{BA},
\end{aligned} \tag{6.43}$$

where we performed a change of integration variables to replace q, x by q_1, q_2. It eventually turns the expression into the resolution of two identity operators. The subscripts A and B demonstrate how the resulting tensor product of identity operators produces a crossover relative to the operation of the displacement operators. They connect the output of one displacement operator with the input of the other one.

The orthogonality and completeness relations follow from the fact that the Wigner functional of a displacement operator is formally equivalent to the functional generalization of a two-dimensional plane wave, which is known to form a complete orthogonal basis for functions on a two-dimensional plane. The one-to-one correspondence between these Wigner functionals and the displacement operators bestow on the latter the same properties. Therefore, all linear operators that can be expressed purely in terms of the quadrature operators can be expressed as linear combinations of displacement operators. Such expansions are given in terms of the characteristic functionals, discussed in Section 5.4, as coefficient functionals.

6.2.4 Completeness of arbitrary displaced states

The result obtained in (6.43) has a remarkable implication. It means that we can produce a set of states to serve as an over complete basis by taking any state and performing all possible displacements on it. Consider, for example, an arbitrary pure state represented by $\hat{\rho} = |\psi\rangle\langle\psi|$. Applying displacement operators on this state and integrating over all possible displacements, we get from (6.43) that

$$\int \hat{D}_A^\dagger(\zeta)\,|\psi\rangle\langle\psi|\,\hat{D}_B(\zeta)\,\mathcal{D}^\circ[\zeta] = \mathbb{1}\,\mathrm{tr}\{|\psi\rangle\langle\psi|\,\mathbb{1}\} = \mathbb{1}, \tag{6.44}$$

which represents a *completeness condition*. The identity is resolved in terms of all the displaced versions of this arbitrary initial pure state. This type of completeness is therefore not a unique feature of the coherent states, which are obtained as all the possible displacements of the vacuum state.

6.3 Fixed-spectrum Fock states

The *fixed-spectrum Fock states* are discussed in Section 4.2 and defined in (4.64). Here, we use the coherent-state-assisted approach of Section 6.1.3 to compute the Wigner functionals of the fixed-spectrum Fock states. The overlap between such a Fock state and a fixed-spectrum coherent state is

$$\langle n_F|\alpha_0\rangle = \exp\left(-\frac{1}{2}\|\alpha_0\|^2\right)\frac{(F^*\diamond\alpha_0)^n}{\sqrt{n!}}, \tag{6.45}$$

where we used (4.65) and (4.108). Hence,

$$\langle\alpha_1|n_F\rangle\langle n_F|\alpha_2\rangle = \exp\left(-\frac{1}{2}\|\alpha_1\|^2 - \frac{1}{2}\|\alpha_2\|^2\right)\frac{1}{n!}\left(\alpha_1^* \diamond FF^* \diamond \alpha_2\right)^n . \tag{6.46}$$

When we substitute it into (6.20) the integrand is not in the form of a Gaussian functional. So, we cannot evaluate the functional integrals. Fortunately, we can overcome this obstacle with the aid of generating functions, which are discussed in Appendix A. There are different options in applying generating functions to simplify the calculations from this point onwards.

6.3.1 One-parameter generating function

The first approach is to simplify the computation by representing the overlap in terms of a one-parameter generating function

$$\mathcal{K} \triangleq \sum_n \eta^n \langle\alpha_1|n_F\rangle\langle n_F|\alpha_2\rangle = \exp\left(-\frac{1}{2}\|\alpha_1\|^2 - \frac{1}{2}\|\alpha_2\|^2 + \eta\alpha_1^* \diamond FF^* \diamond \alpha_2\right), \tag{6.47}$$

where η is the generating parameter. The overlap is recovered by

$$\langle\alpha_1|n_F\rangle\langle n_F|\alpha_2\rangle = \frac{1}{n!}\partial_\eta^n \mathcal{K}\Big|_{\eta=0} . \tag{6.48}$$

Substituting $\langle\alpha_1|\hat{A}|\alpha_2\rangle \rightarrow \mathcal{K}$ into (6.20), we obtain

$$\begin{aligned} \mathcal{W}_{|n\rangle}(\eta) = \mathcal{N}_0 \int \exp\,(&-2\alpha^* \diamond \alpha + 2\alpha^* \diamond \alpha_1 + 2\alpha_2^* \diamond \alpha - \alpha_1^* \diamond \alpha_1 \\ &- \alpha_2^* \diamond \alpha_2 - \alpha_2^* \diamond \alpha_1 + \eta\alpha_1^* \diamond FF^* \diamond \alpha_2)\ \mathcal{D}^\circ[\alpha_1, \alpha_2]. \end{aligned} \tag{6.49}$$

The notation $\mathcal{W}$ is used instead of W to remind us that it is a *generating function* for Wigner functionals and not a Wigner functional itself. After performing the functional integrations over α_1 and α_2, we get

$$\mathcal{W}_{|n\rangle}(\eta) = \frac{\mathcal{N}_0 \exp\left(-2\|\alpha\|^2\right)}{\det\{\mathbf{1} + \eta FF^*\}} \exp\left[4\eta\alpha^* \diamond FF^* \diamond (\mathbf{1} + \eta FF^*)^{-1} \diamond \alpha\right], \tag{6.50}$$

where $\det\{\mathbf{1} + \eta FF^*\}$ represents a *functional determinant*, as discussed in Appendix C.3. The combination FF^* forms a single-mode projection kernel. Since F is normalized, such a kernel is idempotent $FF^* \diamond FF^* = FF^*$. So, the inverse and determinant in (6.50) are equivalent to those discussed in Appendix D.1, which helps us to obtain

$$\begin{aligned} \det\{\mathbf{1} + \eta FF^*\} &= 1 + \eta, \\ (\mathbf{1} + \eta FF^*)^{-1} &= \mathbf{1} - \frac{\eta}{1+\eta} FF^* . \end{aligned} \tag{6.51}$$

With these replacements, we can simplify the expression of the one-parameter generating function for the fixed-spectrum Fock states. It reads

$$\mathcal{W}_{|n\rangle}(\eta) = \frac{\mathcal{N}_0}{1+\eta} \exp\left(-2\|\alpha\|^2 + \frac{4\eta}{1+\eta}\alpha^* \diamond FF^* \diamond \alpha\right). \tag{6.52}$$

Comparing it to the generating function for Laguerre polynomials [6], which is given by (2.130) for $\ell = 0$,

$$\mathcal{L}(v) = \frac{1}{1-v} \exp\left(-\frac{xv}{1-v}\right) = \sum_n v^n L_n(x), \tag{6.53}$$

we see that the Wigner functional for a Fock state is given by

$$W_{|n\rangle}[\alpha] = \mathcal{N}_0(-1)^n L_n\left(4\alpha^* \diamond FF^* \diamond \alpha\right) \exp\left(-2\|\alpha\|^2\right), \tag{6.54}$$

where n is the *occupation number*. The Wigner functional in (6.54) represents a *polynomial Gaussian functional* that becomes negative in some regions on phase space for $n > 0$. For $\eta = 0$, (6.52) produces the expression for the Wigner functional of the vacuum state, having occupation number $n = 0$.

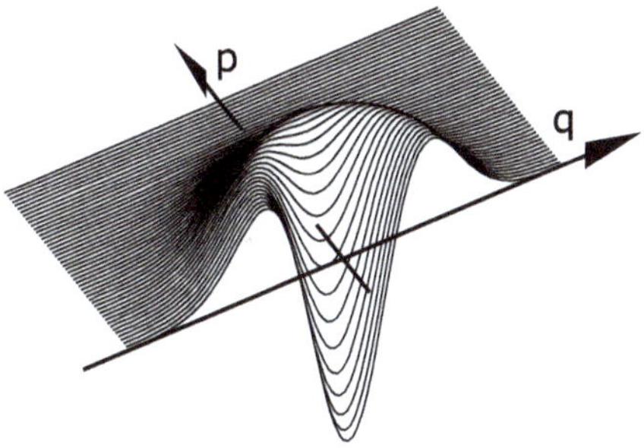

Figure 6.3: Wigner function of a single-photon Fock state on a two-dimensional phase space.

The single-photon Fock state is given by

$$W_{|1\rangle}[\alpha] = \partial_\eta \mathcal{W}_{|n\rangle}(\eta)\big|_{\eta=0} = \mathcal{N}_0 \exp\left(-2\|\alpha\|^2\right)\left(4\alpha^* \diamond FF^* \diamond \alpha - 1\right). \tag{6.55}$$

The parameter function F effectively reduces the Wigner functional to a Wigner function on a two-dimensional plane. This Wigner function is shown in Figure 6.3. It shows that the Wigner function is negative in a region centred at the origin.

Exercise 6.3. Convert the Wigner functional of the single-photon Fock state in (6.55) to quadrature variables and show that the marginal probability distributions (discussed in Section 5.3.1) are positive semidefinite.

Exercise 6.4. Show that all the Wigner functionals obtained from (6.52) are normalized.

Hint: The functional integral of the generating function becomes a generating function for the inverse normalization constants of the individual functionals.

Exercise 6.5. Use (A.9) and the generating function in (6.52) to show that the different fixed-spectrum Fock states associated with the same parameter function are all mutually orthogonal.

Hint: Use two generating functions with different generating parameters.

Exercise 6.6. Use the generating function in (6.52) to compute the purities of the fixed-spectrum Fock states.

Hint: Use the result of the previous exercise.

6.3.2 Two-parameter generating function

The generating function in (6.52) cannot be used for cases where the state is a superposition of Fock states. For such situations, we need different generating parameters for the kets and bras. Here, we provide an alternative expression for the generating function of the overlap by separating the parameter function dependent term in the exponent of the generating function into two terms. So, instead of (6.47), we define

$$\mathcal{K}' = \exp\left(-\frac{1}{2}\|\alpha_1\|^2 - \frac{1}{2}\|\alpha_2\|^2 + \eta_1\alpha_1^* \diamond F + \eta_2 F^* \diamond \alpha_2\right), \tag{6.56}$$

which implies that

$$\langle\alpha_1|n_F\rangle\langle n_F|\alpha_2\rangle = \frac{1}{n!}\frac{\partial^n}{\partial\eta_1^n}\frac{\partial^n}{\partial\eta_2^n}\mathcal{K}'\bigg|_{\eta_1=\eta_2=0}. \tag{6.57}$$

After substituting $\langle\alpha_1|\hat{A}|\alpha_2\rangle \to \mathcal{K}'$ into (6.20) and evaluating the functional integrals over α_1 and α_2, we obtain the following two-parameter generating function for the Wigner functionals of the fixed-spectrum Fock states

$$\mathcal{W}_{|n\rangle}(\eta_1, \eta_2) = \mathcal{N}_0 \exp\left(-2\|\alpha\|^2 + 2\eta_1\alpha^* \diamond F + 2\eta_2 F^* \diamond \alpha - \eta_1\eta_2\right). \tag{6.58}$$

Exercise 6.7. Show that when the same number of derivatives are applied on both generating parameters in (6.58), the resulting Wigner functionals are the same as those in (6.54).

6.3.3 Multiparameter generating function

The generating function in (6.58) still assumes a single parameter function. Therefore, the Fock states that it generates are restricted to a subspace associated with this parameter function. To produce states consisting of combinations of different Fock states with different parameter functions, we need a generating function that contains all those parameter functions. Such a situation can be achieved by replacing

$$\begin{aligned}\eta_1 F &\rightarrow \sum_m \mu_m F_m \triangleq \mathcal{F}(\{\mu\}),\\ \eta_2 F^* &\rightarrow \sum_m \nu_m F_m^* \triangleq \mathcal{F}^*(\{\nu\}),\\ \eta_1\eta_2 &\rightarrow \sum_m \nu_m \mu_m = \mathcal{F}^*(\{\nu\}) \diamond \mathcal{F}(\{\mu\}),\end{aligned} \tag{6.59}$$

in (6.58), where all the parameter functions F_m are mutually orthogonal and μ_m and ν_m are the associated generating parameters. The generating function becomes

$$\begin{aligned}\mathcal{W}_{|n\rangle}(\{\mu\},\{\nu\}) = \mathcal{N}_0 \exp\big[&-2\alpha^* \diamond \alpha + 2\alpha^* \diamond \mathcal{F}(\{\mu\}) + 2\mathcal{F}^*(\{\nu\}) \diamond \alpha\\ &- \mathcal{F}^*(\{\nu\}) \diamond \mathcal{F}(\{\mu\})\big].\end{aligned} \tag{6.60}$$

It can for instance be used to generate Bell states, as defined in (4.98), with arbitrary different parameter functions.

6.3.4 Number operator

An example of an operator whose Wigner functional can be obtained with the functional derivative procedure in Section 6.1.3 is the number operator in (4.51). Being in normal order, it is readily converted into a functional derivative operation:

$$\hat{n} \rightarrow C_{\hat{n}}\left[\delta_\nu, \delta_{\mu^*}\right] = \delta_\nu \diamond \delta_{\mu^*}. \tag{6.61}$$

When applied to the generating functional in (6.23), it produces

$$W_{\hat{n}}[\alpha] = \delta_\nu \diamond \delta_{\mu^*} \exp\left(\mu^* \diamond \alpha + \alpha^* \diamond \nu - \frac{1}{2}\mu^* \diamond \nu\right)\Big|_{\mu^*=\nu=0} = \alpha^* \diamond \alpha - \frac{1}{2}\Omega, \tag{6.62}$$

where $\Omega = \text{tr}\{\mathbf{1}\}$ appears due to the third term in the exponent of the generating functional when the functional derivative operation is applied to it.

For comparison, we also compute the Wigner functional of the number operator using the general coherent-state-assisted approach. First, we compute the overlap of the number operator by two coherent states

$$\begin{aligned}\langle\alpha_1|\,\hat{n}\,|\alpha_2\rangle &= \langle\alpha_1|\,\hat{a}^\dagger \diamond \hat{a}\,|\alpha_2\rangle = \alpha_1^* \diamond \alpha_2 \langle\alpha_1|\alpha_2\rangle\\ &= \alpha_1^* \diamond \alpha_2 \exp\left(-\frac{1}{2}\|\alpha_1\|^2 - \frac{1}{2}\|\alpha_2\|^2 + \alpha_1^* \diamond \alpha_2\right),\end{aligned} \tag{6.63}$$

where we used (4.114). We convert the expression to Gaussian form by representing it as a generating function

$$\mathcal{G} = \exp\left(-\frac{1}{2}\|\alpha_1\|^2 - \frac{1}{2}\|\alpha_2\|^2 + J\alpha_1^* \diamond \alpha_2\right), \tag{6.64}$$

so that

$$\langle \alpha_1 | \hat{n} | \alpha_2 \rangle = \partial_J \mathcal{G}\big|_{J=1}. \tag{6.65}$$

Then we substitute $\langle \alpha_1 | \hat{A} | \alpha_2 \rangle \rightarrow \mathcal{G}$ into (6.20) to obtain a functional integral for the generating function of the number operator's Wigner functional. It reads

$$\begin{aligned} \mathcal{W}_{\mathcal{G}} = \mathcal{N}_0 \int \exp(&-2\alpha^* \diamond \alpha + 2\alpha^* \diamond \alpha_1 + 2\alpha_2^* \diamond \alpha \\ &- \alpha_1^* \diamond \alpha_1 - \alpha_2^* \diamond \alpha_2 - \alpha_2^* \diamond \alpha_1 + J\alpha_1^* \diamond \alpha_2)\ \mathcal{D}^\circ[\alpha_1, \alpha_2]. \end{aligned} \tag{6.66}$$

As before, we use $\mathcal{W}$ instead of W to remind us that it is a *generating function* for a Wigner functional and not the Wigner functional itself. The expression represents two isotropic Gaussian functional integrations, which can be evaluated in sequence with the aid of (C.22), leading to

$$\mathcal{W}_{\mathcal{G}} = \frac{\mathcal{N}_0}{(1+J)^{\Omega}} \exp\left(-2\frac{1-J}{1+J}\alpha^* \diamond \alpha\right). \tag{6.67}$$

Then we evaluate the derivative and set $J = 1$, to get

$$W_{\hat{n}}[\alpha] = \partial_J \mathcal{W}_{\mathcal{G}}\big|_{J=1} = \alpha^* \diamond \alpha - \frac{1}{2}\Omega, \tag{6.68}$$

in agreement with (6.62).

Exercise 6.8. Compute the Wigner functional of $\hat{s}$ given in (4.153).

Exercise 6.9. Compute the average number of photons in a fixed-spectrum coherent state, using the Wigner functionals of the number operator and the fixed-spectrum coherent state.
Hint: Use (5.51).

Exercise 6.10. Use the generating function in (6.52) to compute a generating function for the occupation numbers of all fixed-spectrum Fock states $n = \langle n_F | \hat{n} | n_F \rangle$.

While the Wigner functional for the number operator gives a means to compute the expectation value of the photon number, the *variance* in the photon number needs the square of the number operator, which requires the star product of two such Wigner functionals. Alternatively, we can use the procedure in Section 6.1.3 to compute the Wigner functional for the square of the number operator.

The square of the number operator is not in normal order. When the ladder operators are commuted to express it in normal order, we end up with two terms

$$\hat{n}^2 = \int \hat{a}_s^\dagger(\mathbf{k}_1)\hat{a}_s^\dagger(\mathbf{k}_2)\hat{a}_s(\mathbf{k}_1)\hat{a}_s(\mathbf{k}_2)\,\mathrm{d}_\omega k_1\,\mathrm{d}_\omega k_2 + \int \hat{a}_s^\dagger(\mathbf{k})\hat{a}_s(\mathbf{k})\,\mathrm{d}_\omega k. \tag{6.69}$$

Applying the procedure in Section 6.1.3, we then obtain (see exercise below)

$$\text{Wigner}\left\{\hat{n}^2\right\} = \left(\alpha^* \diamond \alpha - \frac{1}{2}\Omega\right)^2 - \frac{1}{4}\Omega. \tag{6.70}$$

When the two number operators represent two measurements applied on a state, the combined operator is directly expressed in normal order without commutations. Then we only have the first term. So, the squared number operator in normal order is

$$:\hat{n}^2: = \int \hat{a}_s^\dagger(\mathbf{k}_1)\hat{a}_s^\dagger(\mathbf{k}_2)\hat{a}_s(\mathbf{k}_1)\hat{a}_s(\mathbf{k}_2)\,\mathrm{d}_\omega k_1\,\mathrm{d}_\omega k_2. \tag{6.71}$$

The notation $:\hat{O}:$ represents an operator $\hat{O}$ in normal order (without commutations).

Exercise 6.11. Use the procedure in Section 6.1.3 to show that the Wigner functional of the squared number operator in normal order is given by

$$\text{Wigner}\left\{:\hat{n}^2:\right\} = \left(\alpha^* \diamond \alpha - \frac{1}{2}\Omega\right)^2 - \left(\alpha^* \diamond \alpha - \frac{1}{2}\Omega\right) - \frac{1}{4}\Omega, \tag{6.72}$$

whereas $\text{Wigner}\left\{\hat{n}^2\right\}$ is given by (6.70).

6.4 Linear lossless optics

The purpose of the Wigner functional formalism is to model physical quantum optical systems in terms of all the degrees of freedom. So far, we have only computed the Wigner functionals of a few states and operators. These operators are not specifically associated with physical systems. Here, we introduce the modelling of physical systems by discussing a fairly simple scenario: linear lossless optical systems. In this discussion, we gradually develop an understanding of the modelling process of such physical systems in terms of the formalism.

All operations on quantum states are unitary processes maintaining their normalization. However, when the process includes loss, the situation becomes more complicated. In general, such systems are said to be *open* in that they are involved in interactions with other systems that are not included in the system under consideration. Such interactions generally lead to *loss*, which affects the normalization of the system under investigation. The effect of loss is considered in Section 6.7. Here, we consider *closed systems* that do not interact with any other systems, and thus retain their normalization.

In its simplest form, the effect of such unitary processes on certain quantum states is equivalent to the linear process applied to classical fields. For such cases, the effect of linear optical processes (or unitary processes) on quantum states is understood in the context of linear systems theory, as discussed in Section 2.1. Such a process is represented by a direct unitary operation applied to a state. It can be seen as a two-port system with one input and one output. For the quantum system, it is represented by the process

$$|\psi_{\text{in}}\rangle \to |\psi_{\text{out}}\rangle = \hat{U}\,|\psi_{\text{in}}\rangle\,, \tag{6.73}$$

where the unitary operator $\hat{U}$ represents the linear process. For classical fields, it is represented by the superposition integral in (2.37), expressed as

$$f_{\text{in}}(\mathbf{k}) \to g_{\text{out}}(\mathbf{k}) = \int K(\mathbf{k},\mathbf{k}')f_{\text{in}}(\mathbf{k}')\,\mathrm{d}_{\omega}k', \tag{6.74}$$

with $\mathrm{d}_{\omega}k$ defined in (4.49), or as $g_{\text{out}} = K \diamond f_{\text{in}}$ in compact notation.

Although the superposition integral can be defined for *configuration space*, we prefer the description in the Fourier domain in terms of the three-dimensional *wavevector* **k**. Often, we also consider representations in terms of *optical beam variables*, with the two-dimensional *transverse wavevector* **K** and the *angular frequency* ω.

6.4.1 Single-photon processes

The processes represented in (6.73) and (6.74) are equivalent for certain kinds of quantum states. One example is the single-photon states, as defined in (4.61). Since they are linear in the sense that they do not incorporate nonlinear effects (interactions), such processes, as applied to these quantum states and classical fields, respectively, represent the same physical process. Our understanding of classical systems and fields can assist in our understanding of quantum systems and how they are modelled in terms of the functional formalism. Therefore, we start with scenarios where they represent the same physical process.

In terms of (4.61), we define the input and output single-photon states as

$$|\psi_{\text{in}}\rangle \triangleq \int |\mathbf{k}\rangle\, F(\mathbf{k})\,\mathrm{d}_{\omega}k \quad \text{and} \quad |\psi_{\text{out}}\rangle \triangleq \int |\mathbf{k}\rangle\, G(\mathbf{k})\,\mathrm{d}_{\omega}k, \tag{6.75}$$

ignoring the spin indices. Here, $F(\mathbf{k})$ and $G(\mathbf{k})$ are the *parameter functions* of the single-photon (fixed-spectrum Fock) states, represented as normalized *angular spectra*. They are normalized so that $\|F(\mathbf{k})\|^2 = \|G(\mathbf{k})\|^2 = 1$ to ensure the normalization of the states. These parameter functions also represent the Fourier domain wave functions, because $\langle\mathbf{k}|\psi_{\text{in}}\rangle = F(\mathbf{k})$ and $\langle\mathbf{k}|\psi_{\text{out}}\rangle = G(\mathbf{k})$.

The unitary operators for the single-photon states can be modelled as generalizations of the single-photon projection operator, given by

$$\hat{P}_1 = \int |\mathbf{k}\rangle \langle \mathbf{k}| \, \mathrm{d}_\omega k. \tag{6.76}$$

The generalization is done by inserting a kernel function to define a unitary operator:

$$\hat{U}_1 = \int |\mathbf{k}\rangle \, T(\mathbf{k}, \mathbf{k}') \, \langle \mathbf{k}'| \, \mathrm{d}_\omega k \, \mathrm{d}_\omega k'. \tag{6.77}$$

When applied to the single-photon input state, as in (6.73), we get

$$\begin{aligned} \hat{U}_1 |\psi_{\text{in}}\rangle &= \int |\mathbf{k}\rangle \, T(\mathbf{k}, \mathbf{k}_1)\langle \mathbf{k}_1|\mathbf{k}_2\rangle F(\mathbf{k}_2) \, \mathrm{d}_\omega k_2 \, \mathrm{d}_\omega k_1 \, \mathrm{d}_\omega k \\ &= \int |\mathbf{k}\rangle \, T(\mathbf{k}, \mathbf{k}_1) F(\mathbf{k}_1) \, \mathrm{d}_\omega k_1 \, \mathrm{d}_\omega k, \end{aligned} \tag{6.78}$$

by using (4.59). Since it produces the output state defined in (6.75), we get

$$G(\mathbf{k}) = \int T(\mathbf{k}, \mathbf{k}_1) F(\mathbf{k}_1) \, \mathrm{d}_\omega k', \tag{6.79}$$

which is formally equivalent to (6.74) for the classical process. The only difference is that the angular spectra $f_{\text{in}}(\mathbf{k})$ and $g_{\text{out}}(\mathbf{k})$ are not normalized in the classical case. However, the linearity of the process implies that the magnitude of the function is unaffected, provided that the process is unitary. We know that the quantum process must be unitary to maintain the normalization of the states. Hence,

$$\begin{aligned} \hat{U}_1^\dagger \hat{U}_1 &= \int |\mathbf{k}\rangle \, T^\dagger(\mathbf{k}, \mathbf{k}_1) T(\mathbf{k}_1, \mathbf{k}_2) \, \langle \mathbf{k}_2| \, \mathrm{d}_\omega k_2 \, \mathrm{d}_\omega k_1 \, \mathrm{d}_\omega k \\ &= \int |\mathbf{k}\rangle \, \mathbf{1}(\mathbf{k}, \mathbf{k}_1) \, \langle \mathbf{k}_1| \, \mathrm{d}_\omega k_1 \, \mathrm{d}_\omega k \equiv \mathbb{1}. \end{aligned} \tag{6.80}$$

Since linear processes in classical systems are also unitary, as shown in Section 2.1.5, it follows that both $T(\mathbf{k}, \mathbf{k}')$ and $K(\mathbf{k}, \mathbf{k}')$ are unitary kernels. If they describe the same physical system, then $T(\mathbf{k}, \mathbf{k}') \equiv K(\mathbf{k}, \mathbf{k}')$. It thus follows that the quantum process for single photons is formally equivalent to the classical process, apart from the magnitudes of the functions involved.

The equivalence between classical and quantum processes can also be demonstrated for other cases, such as coherent states. However, due to the multiple photons in coherent states, we first need to discuss how such multiphoton processes are modelled.

6.4.2 Multiphoton processes

The operators that we defined above only work for single-photon states. For multiphoton states, we need tensor products of these single-photon operators. Representing such operators in terms of the single-photon wavevector basis as generalizations of the single-photon projection operator, we end up with complicated expressions in terms of superpositions of tensor products of single-photon wavevector basis elements. Moreover, it is difficult to ensure that the different terms in the expansion of such operators only operate on the specific terms in the expansion of the states for which they are intended.

Instead, such multiphoton operators are usually expressed as exponential operators with polynomials in terms of ladder operators in their exponents. Due to the use of ladder operators, such representations of physical systems are often considered to be a second quantization approach.

Historically, *second quantization* was developed as part of quantum field theory [7, 8], which is employed with great success in particle physics. The unitary evolution operator, which contains the *Hamiltonian* in the argument of an exponential function, can be expanded as a *Dyson expansion* [7] in a perturbative approach. The progressive orders in such an expansion represents an increasing number of interactions with progressively decreasing *probability amplitudes*. In other words, exponentiated operators of this form describe physical processes where particles (photons) can undergo zero, one, two, three, or more interactions with progressively smaller probabilities.

The nature of the dynamics that is studied in particle physics differs in an essential way from what one usually finds in quantum optical systems. For example, compare the dynamics of fundamental particles with the scenario of a simple transmission function, such as an aperture or a lens, which are often found in setups for quantum optical experiments. Contrary to the particle physics scenario, *every photon needs to be processed exactly once* by the transmission function. There is no progressively varying probability amplitude for interactions relevant in such a scenario.

As an illustration, we consider an arbitrary exponential operator with a polynomial of ladder operators in its exponent, represented as a unitary operator

$$\hat{U} = \exp\left[\mathrm{i}P\left(\hat{a}, \hat{a}^{\dagger}\right)\right], \tag{6.81}$$

where $P(\cdot)$ is a Hermitian multivariate polynomial. Expanding the operator in such a way that all the terms are in normal order, we can express it as

$$\hat{U} = \sum_{m=0}^{\infty} \hat{C}_m\left(\hat{a}^{\dagger}\right)\hat{a}^m, \tag{6.82}$$

where

$$\hat{C}_m\left(\hat{a}^{\dagger}\right) = \sum_{n=0}^{\infty} C_{mn}\hat{a}^{\dagger n}, \tag{6.83}$$

is a coefficient operator in the form of a polynomial of the creation operator.[a] Applying this operator on an n-photon Fock state, we obtain

---

a Ignore for the moment the fact that, if the coefficient operator does not contain the same number of creation operators as the number of annihilation operators in the term, the process does not respect the *conservation of particle number*, which is expected to hold for linear lossless operations in optics.

$$\hat{U}|n\rangle = \sum_{m=0}^{\infty} \hat{C}_m \hat{a}^m |n\rangle = \sum_{m=0}^{n} \hat{C}_m |n-m\rangle \sqrt{\frac{n!}{(n-m)!}}. \tag{6.84}$$

Those terms with $m > n$ are zero, because they lead to the annihilation of the vacuum state. The term with $m = n$ turns the Fock state into a vacuum state and the creation operators in the coefficient operator $\hat{C}_{m=n}$ can then produce the appropriate output state. In the case of a transmission function, these creation operations are responsible for producing the output photons with the appropriate modulation after the input photons have been removed by the annihilation operation. The problem comes in with those terms for which $m < n$. Such terms are not reduced to the vacuum state by the product of annihilation operators. Their coefficient operators then produce additional photons in association with that term in the expansion. In other words, there would be photons that can pass through the operation without being processed. In the case of a transmission function, for instance, these unaffected photons would not be modulated with the appropriate transmission function. The result is that the operator does not seem to reconstruct the physical process correctly.

This illustration is based on the idea that the only way to process photons formally is to remove them from the input state with an annihilation process and then reproduce them again with a creation process in association with the necessary kernels or functions representing the effect of the process. However, there is another way to use the ladder operators to model such a process. The ladder operators can be used to *count* the number of photons in each term in the expansion of the state and then use that information to provide the correct number of kernels or functions to implement the process.

To illustrate such a process, we start by considering a special kind of process that is diagonal in a particular basis (e. g., filters and transmission functions). Such cases allow us to use ladder operators, combined into number operators to obtain the required physical result. If the number operator can count the number of photons and then produce the correct number of transfer functions or transmission functions, the result would model the physical process correctly.

The proposed operator for a transmission function t, for example, is

$$\hat{T} = \exp\left[\ln(t)\hat{n}\right]. \tag{6.85}$$

Operating on a Fock state, it produces

$$\hat{T}|n\rangle = |n\rangle \exp\left[\ln(t)n\right] = |n\rangle\, t^n. \tag{6.86}$$

Now each of the n photons has a factor of the transmission function t.

There is an issue with this expression that needs to be addressed. The transmission function is a function of the spatial degrees of freedom $t(\mathbf{x})$. After the operation, these spatial degrees of freedom must be tied to the spatial degrees of freedom of the respective photons in the Fock state. In other words, if the parameter function associated with a Fock state is defined in configuration space as $f(\mathbf{x})$, then the operation must produce

$f(\mathbf{x}) \to t(\mathbf{x})f(\mathbf{x})$. Such a process necessitates the formulation of the number operator in configuration space.

6.4.2.1 Shift invariant linear lossless processes

For the moment, we avoid issues related to the transformation between the configuration space and the Fourier domain by considering a *phase-only transfer function* that is diagonal in the Fourier domain, as found with *shift invariant* systems (see Section 2.1.3). The *phase-only* property is required to maintain unitarity. We define the operator as

$$\hat{T} = \exp\left[\int L(\mathbf{k})\hat{a}^{\dagger}(\mathbf{k})\hat{a}(\mathbf{k})\, \mathrm{d}_{\omega}k\right], \tag{6.87}$$

in analogy with the transmission function. When applied on a single-photon state, the operator in the exponent produces

$$\begin{aligned}\int L(\mathbf{k})\hat{a}^{\dagger}(\mathbf{k})\hat{a}(\mathbf{k})\, \mathrm{d}_{\omega}k\, |1_F\rangle &= \int L(\mathbf{k})\hat{a}^{\dagger}(\mathbf{k})\hat{a}(\mathbf{k})\, |\mathbf{k}'\rangle\, F(\mathbf{k}')\, \mathrm{d}_{\omega}k\, \mathrm{d}_{\omega}k' \\ &= \int |\mathbf{k}\rangle\, L(\mathbf{k})F(\mathbf{k})\, \mathrm{d}_{\omega}k.\end{aligned} \tag{6.88}$$

The terms in the expansion of the operator would therefore produce

$$\frac{1}{m!}\left[\int L(\mathbf{k})\hat{a}^{\dagger}(\mathbf{k})\hat{a}(\mathbf{k})\, \mathrm{d}_{\omega}k\right]^{m} |1_F\rangle = \int |\mathbf{k}\rangle\, \frac{1}{m!}L^{m}(\mathbf{k})F(\mathbf{k})\, \mathrm{d}_{\omega}k. \tag{6.89}$$

Hence,

$$\hat{T}\, |1_F\rangle = \int |\mathbf{k}\rangle\, \exp[L(\mathbf{k})]F(\mathbf{k})\, \mathrm{d}_{\omega}k. \tag{6.90}$$

Comparing this result with those in Section 2.1.3, we identify $\exp[L(\mathbf{k})]$ as the *transfer function* [9] of the linear *spectral filtering* process. For a unitary process, the magnitude of the transfer function must be equal to 1. Therefore, it is a *phase-only transfer function* that can be expressed as $\exp[L(\mathbf{k})] = \exp[\mathrm{i}\theta(\mathbf{k})]$, so that $L(\mathbf{k}) = \mathrm{i}\theta(\mathbf{k})$, where $\theta(\mathbf{k})$ is a real-valued phase function. The transformation of the parameter function is then given by $F(\mathbf{k}) \to \exp[\mathrm{i}\theta(\mathbf{k})]F(\mathbf{k})$.

We can now consider Fock states with more photons. The analysis becomes significantly more complicated. To simplify the notation, we define

$$\hat{L} \triangleq \int L(\mathbf{k})\hat{a}^{\dagger}(\mathbf{k})\hat{a}(\mathbf{k})\, \mathrm{d}_{\omega}k, \tag{6.91}$$

so that $\hat{T} = \exp(\hat{L})$. It then follows that $\hat{T}^{-1} = \exp(-\hat{L})$. To handle the case of an arbitrary Fock state, we define a generating function

$$\sum_{n} \frac{1}{n!}J^{n}\hat{a}_F^{\dagger n} = \exp\left(J\hat{a}_F^{\dagger}\right), \tag{6.92}$$

for a parameter function F. Then a fixed-spectrum Fock state with n photons is

$$|n_F\rangle = \frac{1}{\sqrt{n!}} \partial_J^n \exp\left(J\hat{a}_F^\dagger\right)\Big|_{J=0} |\text{vac}\rangle . \tag{6.93}$$

Applying the transformation on the generating function, we get

$$\begin{aligned}\hat{T} \exp\left(J\hat{a}_F^\dagger\right) |\text{vac}\rangle &= \hat{T} \exp\left(J\hat{a}_F^\dagger\right) \hat{T}^{-1}\hat{T} |\text{vac}\rangle \\ &= \hat{T} \exp\left(J\hat{a}_F^\dagger\right) \hat{T}^{-1} |\text{vac}\rangle , \end{aligned} \tag{6.94}$$

where we used the fact that $\exp(\hat{L}) |\text{vac}\rangle = |\text{vac}\rangle$. Using (3.103) to evaluate the product of operators, we have

$$\begin{aligned}\hat{T} \exp\left(J\hat{a}_F^\dagger\right) \hat{T}^{-1} &= \exp(\hat{L}) \exp\left(J\hat{a}_F^\dagger\right) \exp(-\hat{L}) \\ &= \exp\left(J\hat{a}_F^\dagger\right) + \left[\hat{L}, \exp\left(J\hat{a}_F^\dagger\right)\right] + \frac{1}{2!}\left[\hat{L}, \left[\hat{L}, \exp\left(J\hat{a}_F^\dagger\right)\right]\right] \\ &\quad + \frac{1}{3!}\left[\hat{L}, \left[\hat{L}, \left[\hat{L}, \exp\left(J\hat{a}_F^\dagger\right)\right]\right]\right] + \cdots . \end{aligned} \tag{6.95}$$

To evaluate the commutation $[\hat{L}, \exp(J\hat{a}_F^\dagger)]$, which can also be done with (3.103), we first need to calculate the commutation

$$\left[\hat{L}, \hat{a}_F^\dagger\right] = \int \hat{a}^\dagger(\mathbf{k}) L(\mathbf{k}) F(\mathbf{k}) \, \mathrm{d}_\omega k \triangleq \hat{a}_{LF}^\dagger . \tag{6.96}$$

From (3.103), we then get

$$\exp\left(J\hat{a}_F^\dagger\right) \hat{L} \exp\left(-J\hat{a}_F^\dagger\right) = \hat{L} + \left[J\hat{a}_F^\dagger, \hat{L}\right] + \frac{1}{2!}\left[J\hat{a}_F^\dagger, \left[J\hat{a}_F^\dagger, \hat{L}\right]\right] + \cdots = \hat{L} - J\hat{a}_{LF}^\dagger , \tag{6.97}$$

so that

$$\left[\hat{L}, \exp\left(J\hat{a}_F^\dagger\right)\right] = J\hat{a}_{LF}^\dagger \exp\left(J\hat{a}_F^\dagger\right). \tag{6.98}$$

We can now evaluate the commutations in (6.95). The sequence does not end, but with the necessary care it can be summed, leading to

$$\begin{aligned}\exp(\hat{L}) \exp\left(J\hat{a}_F^\dagger\right) \exp(-\hat{L}) &= \exp\left\{J \int \hat{a}^\dagger(\mathbf{k}) \exp[L(\mathbf{k})] F(\mathbf{k}) \, \mathrm{d}_\omega k\right\} \\ &\triangleq \exp\left[J\hat{a}_{\exp(L)F}^\dagger\right] . \end{aligned} \tag{6.99}$$

Using this generating function to produce the result of the transformation of the Fock state, we eventually obtain

$$\hat{T} |n_F\rangle = \frac{1}{\sqrt{n!}} \partial_J^n \exp\left[J\hat{a}_{\exp(L)F}^\dagger\right]\Big|_{J=0} |\text{vac}\rangle = \frac{1}{\sqrt{n!}} \left[\hat{a}_{\exp(L)F}^\dagger\right]^n |\text{vac}\rangle = \left|n_{\exp(L)F}\right\rangle . \tag{6.100}$$

As a result, we see that the transformation represented in terms of the exponential operator in (6.87) performs a transformation of the parameter function of the Fock state $F \rightarrow \exp(L)F$. For the unitary case, we already know that $L = \mathrm{i}\theta$, so that the transformation implements a *phase modulation* of the parameter function in the Fourier domain. Since all the photons are parameterized by the same parameter function, the operator thus processes all the photons in exactly the same way.

6.4.2.2 General linear lossless processes

The diagonal form of the operator in (6.87) represents shift invariant systems and is not as general as it can be. We selected the diagonal form of shift invariant systems to alleviate the complexity of the analysis to some extent. The general form for an operator modelling such a system is

$$\hat{T} = \exp\left[\int \hat{a}^{\dagger}(\mathbf{k})K(\mathbf{k},\mathbf{k}')\hat{a}(\mathbf{k}')\,\mathrm{d}_{\omega}k\,\mathrm{d}_{\omega}k'\right] = \exp\left(\hat{a}^{\dagger}\diamond K\diamond\hat{a}\right), \tag{6.101}$$

where $K(\mathbf{k},\mathbf{k}')$ is an arbitrary *anti-Hermitian* kernel, that is, $K^{\dagger} = -K$. It ensures that $\hat{T}$ is unitary. Such an operator is referred to as a *phase operator* due to the implied generalized phase modulation process as imposed by the requirement for unitarity. We can now follow the same procedure to see what happens when this phase operator is applied to a Fock state. The result is the same, apart from some differences in the required notation.

Here, we only demonstrate the case where the operator is applied to a single-photon state. The exponent produces

$$\begin{aligned}\hat{L}\,|1_F\rangle &= \int \hat{a}^{\dagger}(\mathbf{k})K(\mathbf{k},\mathbf{k}')\hat{a}(\mathbf{k}')\,\mathrm{d}_{\omega}k\,\mathrm{d}_{\omega}k'\,|1_F\rangle \\ &= \int |\mathbf{k}\rangle\,K(\mathbf{k},\mathbf{k}')F(\mathbf{k}')\,\mathrm{d}_{\omega}k\,\mathrm{d}_{\omega}k' \equiv |1_{K\diamond F}\rangle\end{aligned} \tag{6.102}$$

The terms in the expansion of the exponential operator thus produce

$$\begin{aligned}\frac{1}{m!}\hat{L}^{m}\,|1_F\rangle &= \frac{1}{m!}\left[\int \hat{a}^{\dagger}(\mathbf{k})K(\mathbf{k},\mathbf{k}')\hat{a}(\mathbf{k}')\,\mathrm{d}_{\omega}k\,\mathrm{d}_{\omega}k'\right]^{m}|1_F\rangle \\ &= \int |\mathbf{k}\rangle\,\frac{1}{m!}K^{\diamond m}(\mathbf{k},\mathbf{k}')F(\mathbf{k}')\,\mathrm{d}_{\omega}k\,\mathrm{d}_{\omega}k',\end{aligned} \tag{6.103}$$

where $K^{\diamond m}$ represents m $\diamond$-contractions of the kernel. The combined operator applied to a single-photon state gives

$$\hat{T}\,|1_F\rangle = \int |\mathbf{k}\rangle\,\exp_{\diamond}[K(\mathbf{k},\mathbf{k}')]F(\mathbf{k}')\,\mathrm{d}_{\omega}k\,\mathrm{d}_{\omega}k', \tag{6.104}$$

where $\exp_\diamond(\cdot)$ is defined in (C.17), representing an exponentiated kernel function whose expansion contains products represented by $\diamond$-contractions and the leading-order term is the identity kernel **1** defined in (4.54).

Using the same procedure as for shift invariant processes, we can apply this general process to multiphoton states. It produces all possible combinations of multiple operations on individual photons. However, the resulting expansion again factorizes, so that the transformation is directly applied on the parameter function. Hence, each photon undergoes the same transformation. The exponentiated operator in (6.101) thus models the physical process correctly by implementing an operation where all photons are processed as they should be.

6.4.3 Wigner functional of the phase operator

The calculation to demonstrate the phase modulation process performed by a phase operator when applied to an arbitrary state is easier to perform in terms of Wigner functionals, provided that the Wigner functional of the phase operator is known. It is challenging to calculate the Wigner functional of a generic phase operator. Fortunately, we only need to do it once. Therefore, we now turn to the calculation of the Wigner functional of the phase operator, using the coherent-state-assisted approach of Section 6.1.3.

The exponentiated operator in (6.101) is an example of a *Gaussian operator*. Such Gaussian operators are exponentials with up to second-order polynomials of ladder operators in their arguments. They include displacement operators, as discussed in Section 6.2. Below, more examples, such as the beamsplitter (Section 6.6) and the squeezing operator (Section 6.8), are encountered.

6.4.3.1 Normal ordering

For the coherent-state-assisted approach, the ladder operators need to be in normal order so that the terms in the Taylor expansion can be written as $(\hat{a}^\dagger)^m\hat{a}^m$ instead of $(\hat{a}^\dagger\hat{a})^n$. For this purpose, we relate the current form of the phase operator to the required form:

$$\begin{aligned}\exp\left(t\hat{a}^\dagger\diamond K\diamond\hat{a}\right) &= \sum_{n=0}^{\infty}\frac{t^n}{n!}\left(\hat{a}^\dagger\diamond K\diamond\hat{a}\right)^n\\ &= \sum_{m=0}^{\infty}\int B_m(\mathbf{k}_1..\mathbf{k}_m,\mathbf{q}_1..\mathbf{q}_m,t)\left[\prod_{r=1}^{m}\hat{a}^\dagger(\mathbf{k}_r)\right]\\ &\quad\times\left[\prod_{r=1}^{m}\hat{a}(\mathbf{q}_r)\right]\prod_{r=1}^{m}\mathrm{d}_\omega k_r\,\mathrm{d}_\omega q_r,\end{aligned}\tag{6.105}$$

where we insert an auxiliary variable t in the exponent of the phase operator. We then equate the expansion of the phase operator in powers of t to a normal-ordered expansion with unknown coefficient functions represented by $B_m(\cdot,t)$. The original form of the operator is recovered for $t = 1$ and for $t = 0$ it becomes the identity operator. It thus follows that $B_0(t) = 1$ and $B_m(\cdot, 0) = 0$ for all $m > 0$. The $\mathbf{k}_r$'s and $\mathbf{q}_r$'s are wavevectors associated with the creation and annihilation operators, respectively. For the sake of simplifying the notation, we ignore the spin indices in this derivation.

Taking the derivative of (6.105) with respect to t, we obtain

$$\sum_{n=0}^{\infty} \frac{t^n}{n!} \left(\hat{a}^\dagger \diamond K \diamond \hat{a}\right)^{n+1} = \sum_{m=1}^{\infty} \int \partial_t B_m(\mathbf{k}_1..\mathbf{k}_m, \mathbf{q}_1..\mathbf{q}_m, t) \left[\prod_{r=1}^{m} \hat{a}^\dagger(\mathbf{k}_r)\right] \times \left[\prod_{r=1}^{m} \hat{a}(\mathbf{q}_r)\right] \prod_{r=1}^{m} \mathrm{d}_\omega k_r \mathrm{d}_\omega q_r, \tag{6.106}$$

where we redefined $n \to n+1$ on the left-hand side. The summation on the right-hand side starts at $m = 1$ because $\partial_t B_0(t) = 0$. Multiplying (6.105) on the right by $\hat{a}^\dagger \diamond K \diamond \hat{a}$ and subtracting the result from (6.106), we obtain

$$\begin{aligned} 0 = & \sum_{m=1}^{\infty} \int \partial_t B_m(\mathbf{k}_1..\mathbf{k}_m, \mathbf{q}_1..\mathbf{q}_m, t) \left[\prod_{r=1}^{m} \hat{a}^\dagger(\mathbf{k}_r)\right] \left[\prod_{r=1}^{m} \hat{a}(\mathbf{q}_r)\right] \prod_{r=1}^{m} \mathrm{d}_\omega k_r \mathrm{d}_\omega q_r \\ & - \sum_{m=0}^{\infty} \int B_m(\mathbf{k}_1..\mathbf{k}_m, \mathbf{q}_1..\mathbf{q}_m, t) \left[\prod_{r=1}^{m} \hat{a}^\dagger(\mathbf{k}_r)\right] \left[\prod_{r=1}^{m} \hat{a}(\mathbf{q}_r)\right] \\ & \times \hat{a}^\dagger(\mathbf{k}_0) K(\mathbf{k}_0, \mathbf{q}_0) \hat{a}(\mathbf{q}_0) \prod_{r=0}^{m} \mathrm{d}_\omega k_r \mathrm{d}_\omega q_r. \end{aligned} \tag{6.107}$$

One can use the identity $[ab, c] = [a, c]b + a[b, c]$ to show that

$$\begin{aligned} & \left[\prod_{r=1}^{m} \hat{a}^\dagger(\mathbf{k}_r)\right] \left[\prod_{r=1}^{m} \hat{a}(\mathbf{q}_r)\right] \hat{a}^\dagger(\mathbf{k}_0) \hat{a}(\mathbf{q}_0) \\ = & \left[\prod_{r=0}^{m} \hat{a}^\dagger(\mathbf{k}_r)\right] \left[\prod_{r=0}^{m} \hat{a}(\mathbf{q}_r)\right] + \sum_{s=1}^{m} \mathbf{1}(\mathbf{q}_s, \mathbf{k}_0) \left[\prod_{r=1}^{m} \hat{a}^\dagger(\mathbf{k}_r)\right] \left[\prod_{r=0, r\neq s}^{m} \hat{a}(\mathbf{q}_r)\right]. \end{aligned} \tag{6.108}$$

Hence,

$$\begin{aligned} 0 = \int \Bigg\{ & \sum_{m=1}^{\infty} \partial_t B_m(\mathbf{k}_1..\mathbf{k}_m, \mathbf{q}_1..\mathbf{q}_m, t) \left[\prod_{r=1}^{m} \hat{a}^\dagger(\mathbf{k}_r)\right] \left[\prod_{r=1}^{m} \hat{a}(\mathbf{q}_r)\right] \\ & - \sum_{m=1}^{\infty} K(\mathbf{k}_1, \mathbf{q}_1) B_{m-1}(\mathbf{k}_2..\mathbf{k}_m, \mathbf{q}_2..\mathbf{q}_m, t) \left[\prod_{r=1}^{m} \hat{a}^\dagger(\mathbf{k}_r)\right] \left[\prod_{r=1}^{m} \hat{a}(\mathbf{q}_r)\right] \\ & - \sum_{m=1}^{\infty} B_m(\mathbf{k}_1..\mathbf{k}_m, \mathbf{q}_1..\mathbf{q}_m, t) \int \sum_{s=1}^{m} K(\mathbf{q}_s, \mathbf{q}_0) \left[\prod_{r=1}^{m} \hat{a}^\dagger(\mathbf{k}_r)\right] \\ & \times \left[\prod_{r=0, r\neq s}^{m} \hat{a}(\mathbf{q}_r)\right] \mathrm{d}_\omega q_0 \Bigg\} \prod_{r=1}^{m} \mathrm{d}_\omega k_r \mathrm{d}_\omega q_r, \end{aligned} \tag{6.109}$$

where we redefined the indices in the second term to match those in the other terms. It then follows that the summations can be combined, which implies that each of the terms in the summation over m must equal zero, respectively. Note, however, that the integration variables in the second term can be redefined so that K can depend on any of them. It leads to the summation over all such cases divided by the number of cases.

As a result, we have the following recursive differential equations:

$$\begin{aligned}
&\partial_t B_m(\mathbf{k}_1..\mathbf{k}_m, \mathbf{q}_1..\mathbf{q}_m, t) \\
&= \frac{1}{m}\sum_{r=1}^{m} K(\mathbf{k}_r, \mathbf{q}_r) B_{m-1}(\mathbf{k}_1..\mathbf{k}_{r-1}, \mathbf{k}_{r+1}..\mathbf{k}_m, \mathbf{q}_1..\mathbf{q}_{r-1}, \mathbf{q}_{r+1}..\mathbf{q}_m, t) \\
&+ \int \sum_{r=1}^{m} B_m(\mathbf{k}_1..\mathbf{k}_m, \mathbf{q}_1..\mathbf{q}_{r-1}, \mathbf{q}_0, \mathbf{q}_{r+1}..\mathbf{q}_m, t) K(\mathbf{q}_0, \mathbf{q}_r)\, \mathrm{d}_\omega q_0 .
\end{aligned} \tag{6.110}$$

Using $B_0(t) = 1$ and the boundary conditions $B_m(..., 0) = 0$ for $m > 0$, we solve these equations consecutively to obtain the general solution

$$B_m(\{\mathbf{k}_r\}, \{\mathbf{q}_r\}, t) = \frac{1}{m!}\prod_{r=1}^{m}\{\exp_\diamond[tK(\mathbf{k}_r, \mathbf{q}_r)] - \mathbf{1}\} . \tag{6.111}$$

Substituting this result into (6.105) and setting $t = 1$, we obtain an expression for the phase operator in normal order, given by

$$\begin{aligned}
\exp\left(\hat{a}^\dagger \diamond K \diamond \hat{a}\right) = \sum_{m=0}^{\infty} \frac{1}{m!} \int &\left[\prod_{r=1}^{m} \hat{a}^\dagger(\mathbf{k}_r)\right] \left[\prod_{r=1}^{m} \{\exp_\diamond[K(\mathbf{k}_r, \mathbf{q}_r)] - \mathbf{1}\}\right] \\
&\times \left[\prod_{r=1}^{m} \hat{a}(\mathbf{q}_r)\right] \prod_{r=1}^{m} \mathrm{d}_\omega k_r \mathrm{d}_\omega q_r .
\end{aligned} \tag{6.112}$$

6.4.3.2 Wigner functional

The coherent-state-assisted approach of Section 6.1.3 is now used to calculate the Wigner functional for the phase operator. First, the expression in (6.112) is overlapped by two different coherent states

$$\begin{aligned}
&\langle \alpha_1 | \exp\left(\hat{a}^\dagger \diamond K \diamond \hat{a}\right) | \alpha_2 \rangle \\
&= \sum_{m=0}^{\infty} \frac{1}{m!} \int \langle \alpha_1 | \left[\prod_{r=1}^{m} \hat{a}^\dagger(\mathbf{k}_r)\right] \left[\prod_{r=1}^{m} \{\exp_\diamond[K(\mathbf{k}_r, \mathbf{q}_r)] - \mathbf{1}\}\right] \\
&\quad \times \left[\prod_{r=1}^{m} \hat{a}(\mathbf{q}_r)\right] | \alpha_2 \rangle \prod_{r=1}^{m} \mathrm{d}_\omega k_r \mathrm{d}_\omega q_r \\
&= \exp\left(-\frac{1}{2}\|\alpha_1\|^2 - \frac{1}{2}\|\alpha_2\|^2 + \alpha_1^* \diamond \alpha_2\right) \sum_{m=0}^{\infty} \frac{1}{m!} \{\alpha_1^* \diamond [\exp_\diamond(K) - \mathbf{1}] \diamond \alpha_2\}^m \\
&= \exp\left[-\frac{1}{2}\|\alpha_1\|^2 - \frac{1}{2}\|\alpha_2\|^2 + \alpha_1^* \diamond \exp_\diamond(K) \diamond \alpha_2\right] .
\end{aligned} \tag{6.113}$$

Then, substituting (6.113) into (6.20) and evaluating the isotropic functional integrals over α_1 and α_2, we obtain

$$\begin{aligned}\text{Wigner}\{\hat{T}\} &= \mathcal{N}_0 \int \exp\left[-2\alpha^* \diamond \alpha + 2\alpha^* \diamond \alpha_1 + 2\alpha_2^* \diamond \alpha - \alpha_1^* \diamond \alpha_1\right.\\ &\quad \left.- \alpha_2^* \diamond \alpha_2 - \alpha_2^* \diamond \alpha_1 + \alpha_1^* \diamond \exp_\diamond(K) \diamond \alpha_2\right] \mathcal{D}^\circ[\alpha_1, \alpha_2]\\ &= \frac{\mathcal{N}_0 \exp\{2\alpha^* \diamond [\exp_\diamond(K) - \mathbf{1}] \diamond [\exp_\diamond(K) + \mathbf{1}]^{-1} \diamond \alpha\}}{\det\{\mathbf{1} + \exp_\diamond(K)\}}.\end{aligned} \tag{6.114}$$

The resulting Wigner functional for the phase operator can be expressed as

$$W_{\hat{T}}[\alpha] = \text{Wigner}\left\{\exp\left(\hat{a}^\dagger \diamond K \diamond \hat{a}\right)\right\} = \mathcal{N}_K \exp\left(2\alpha^* \diamond \tau \diamond \alpha\right), \tag{6.115}$$

where

$$\begin{aligned}\mathcal{N}_K &\triangleq \frac{\mathcal{N}_0}{\det\{\mathbf{1} + \exp_\diamond(K)\}},\\ \tau &\triangleq \tanh_\diamond\left(\frac{1}{2}K\right) = [\exp_\diamond(K) - \mathbf{1}] \diamond [\exp_\diamond(K) + \mathbf{1}]^{-1}.\end{aligned} \tag{6.116}$$

For the process to be unitary, K must be anti-Hermitian. Then $\exp_\diamond(K)$ is unitary, which in turn implies that τ is anti-Hermitian, so that the Wigner functional is unitary, as expected.[b] The result obtained here is useful in various situations.

What if $K(\mathbf{k}_1, \mathbf{k}_2) = i\pi\mathbf{1}(\mathbf{k}_1, \mathbf{k}_2)$? It would mean that $\exp_\diamond(K) = -\mathbf{1}$, so that we would get $\exp_\diamond(K) + \mathbf{1} = 0$, leading to a divergence in the definition of τ in (6.116). In this case, $\hat{T} = \exp(i\pi\hat{n}) \equiv (-1)^{\hat{n}}$. For the Wigner functional in this case, we return to (6.114):

$$\begin{aligned}\text{Wigner}\{\exp(i\pi\hat{n})\} &= \mathcal{N}_0 \int \exp\left(-2\alpha^* \diamond \alpha + 2\alpha^* \diamond \alpha_1 + 2\alpha_2^* \diamond \alpha - \alpha_1^* \diamond \alpha_1\right.\\ &\quad \left.- \alpha_2^* \diamond \alpha_2 - \alpha_2^* \diamond \alpha_1 - \alpha_1^* \diamond \alpha_2\right) \mathcal{D}^\circ[\alpha_1, \alpha_2]\\ &= \mathcal{N}_0 \int \exp\left[-2\alpha^* \diamond \alpha + 2\alpha^* \diamond \alpha_1 - \alpha_1^* \diamond \alpha_1 - \alpha_1^* \diamond (2\alpha - \alpha_1)\right] \mathcal{D}^\circ[\alpha_1]\\ &= \mathcal{N}_0 \int \exp\left(-2\alpha^* \diamond \alpha + 2\alpha^* \diamond \alpha_1 - 2\alpha_1^* \diamond \alpha\right) \mathcal{D}^\circ[\alpha_1]\\ &= \frac{1}{\mathcal{N}_0} \int \exp\left(-2\alpha^* \diamond \alpha + \alpha^* \diamond \alpha_3 - \alpha_3^* \diamond \alpha\right) \mathcal{D}^\circ[\alpha_3]\\ &= \pi^\Omega \delta[\alpha] \exp(-2\alpha^* \diamond \alpha) \equiv \pi^\Omega \delta[\alpha].\end{aligned} \tag{6.117}$$

Comparing this Wigner functional with that of the *parity operator* in (5.25), we conclude that an alternative expression for the parity operator is

---

b If $\hat{T}$ is not unitary, the process it represents would not be *trace preserving*. A better way to consider such non-unitary transformations is to model them in terms of beamsplitters (see Section 6.6.4).

$$\hat{\Pi} = \exp(i\pi\hat{n}) \equiv (-1)^{\hat{n}}. \tag{6.118}$$

Exercise 6.12. Show that, by representing the anti-Hermitian kernel $K(\mathbf{k}_1, \mathbf{k}_2)$ as

$$\hat{a}^\dagger \diamond K \diamond \hat{a} = i\hat{a}^\dagger \diamond \theta \diamond \hat{a}, \tag{6.119}$$

in terms of a Hermitian kernel $\theta(\mathbf{k}_1, \mathbf{k}_2)$, we obtained an expressions for the Wigner functional of the phase operator given by

$$W_{\hat{T}}[\alpha] = \mathcal{N}_\theta \exp\left[i2\alpha^* \diamond \tan_\diamond\left(\frac{1}{2}\theta\right) \diamond \alpha\right], \tag{6.120}$$

where

$$\mathcal{N}_\theta = \frac{\exp(-i\frac{1}{2}\operatorname{tr}\{\theta\})}{\det\{\cos_\diamond(\frac{1}{2}\theta)\}}. \tag{6.121}$$

(The global phase can be discarded.)

6.4.4 Phase transformation of an arbitrary state

Having obtained the Wigner functional for the phase operator in (6.115), we can use it to consider the effect of a phase transformation on the Wigner functional of an arbitrary state. The result proves to be more general than what we could achieve in terms of operators. So, we apply $\hat{\rho} \to \hat{T}\hat{\rho}\hat{T}^\dagger$ in terms of Wigner functionals, where $\hat{T}$ is the unitary phase operator given in (6.101). The Wigner functional for the transformed state is then obtained with the aid of the *triple star product* given in (5.49). It becomes

$$\begin{aligned}
\text{Wigner}\{\hat{T}\hat{\rho}\hat{T}^\dagger\} = {} & \mathcal{N}_K\mathcal{N}_K^\dagger \int \exp\left[\frac{1}{2}(\alpha_a^* + \alpha^* + \alpha_b^*) \diamond \tau \diamond (\alpha_a + \alpha + \alpha_b)\right. \\
& \left. + \frac{1}{2}(\alpha_a^* + \alpha^* - \alpha_b^*) \diamond \tau^\dagger \diamond (\alpha_a + \alpha - \alpha_b)\right] W_{\hat{\rho}}[\alpha_a] \\
& \times \exp[(\alpha^* - \alpha_a^*) \diamond \alpha_b - \alpha_b^* \diamond (\alpha - \alpha_a)]\, \mathcal{D}^\circ[\alpha_a, \alpha_b] \\
= {} & \mathcal{N}_K\mathcal{N}_K^\dagger \int W_{\hat{\rho}}[\alpha_a] \exp[(\alpha^* + \alpha^* \diamond \tau - \alpha_a^* + \alpha_a^* \diamond \tau) \diamond \alpha_b \\
& - \alpha_b^* \diamond (\alpha - \tau \diamond \alpha - \alpha_a - \tau \diamond \alpha_a)]\, \mathcal{D}^\circ[\alpha_a, \alpha_b],
\end{aligned} \tag{6.122}$$

because $\tau^\dagger = -\tau$. Before we evaluate the integration over α_b, we perform a chance of the integration field variable: $\alpha_b \to (\mathbf{1} - \tau)^{-1} \diamond \alpha_c$ and $\alpha_b^* \to \alpha_c^* \diamond (\mathbf{1} + \tau)^{-1}$. The measure becomes (see Appendix C.2)

$$\mathcal{D}[\alpha_b] \to \frac{1}{\det\{\mathbf{1} - \tau \diamond \tau\}} \mathcal{D}[\alpha_c]. \tag{6.123}$$

Therefore, the functional integral now reads

$$\begin{aligned}\text{Wigner}\{\hat{T}\hat{\rho}\hat{T}^{\dagger}\} = & \frac{\mathcal{N}_K\mathcal{N}_K^{\dagger}}{\det\{\mathbf{1}-\tau\diamond\tau\}}\int W_{\hat{\rho}}[\alpha_a]\exp\left[(\alpha^*\diamond E-\alpha_a^*)\diamond\alpha_c\right.\\ & \left.-\alpha_c^*\diamond\left(E^{\dagger}\diamond\alpha-\alpha_a\right)\right]\mathcal{D}^{\circ}[\alpha_a,\alpha_b]\\ = & \frac{\mathcal{N}_K\mathcal{N}_K^{\dagger}}{\det\{\mathbf{1}-\tau\diamond\tau\}}W_{\hat{\rho}}\left[\alpha^*\diamond E, E^{\dagger}\diamond\alpha\right],\end{aligned} \tag{6.124}$$

where

$$E \triangleq (\mathbf{1}+\tau)\diamond(\mathbf{1}-\tau)^{-1}. \tag{6.125}$$

The functional integration over α_c produces a Dirac delta functional, which then allows us to evaluate the functional integration over α_a as well. Using (6.116), we can show that

$$E \equiv \exp_{\diamond}(K) \quad \text{and} \quad \frac{\mathcal{N}_K\mathcal{N}_K^{\dagger}}{\det\{\mathbf{1}-\tau\diamond\tau\}} \equiv 1, \tag{6.126}$$

leading to the final result

$$\text{Wigner}\{\hat{T}\hat{\rho}\hat{T}^{\dagger}\} = W_{\hat{\rho}}\left[\alpha^*\diamond\exp_{\diamond}(K), \exp_{\diamond}(-K)\diamond\alpha\right]. \tag{6.127}$$

We emphasize the role of the transformation on both the field variable and its complex conjugate by expressing the Wigner functional as $W[\alpha^*,\alpha]$ instead of $W[\alpha]$. The prefactors all cancel when we use (6.116). The result shows that the effect of the *phase operator* on the Wigner functional is a phase transformation of the field variables given by

$$\alpha^* \to \alpha^*\diamond\exp_{\diamond}(K) \quad \text{and} \quad \alpha \to \exp_{\diamond}(-K)\diamond\alpha. \tag{6.128}$$

For the shift-invariant phase operators considered in Section 6.4.2, the kernel functions are diagonal, leading to $K(\mathbf{k},\mathbf{k}') = i\theta(\mathbf{k})\mathbf{1}(\mathbf{k},\mathbf{k}')$. As a result, the kernel becomes $E = \exp(i\theta)\mathbf{1}$, where $\exp(i\theta)$ is the *phase-only transfer function* of the spectral filter. The transformation imposed by such a shift-invariant phase operators is

$$\text{Wigner}\{\hat{T}\hat{\rho}\hat{T}^{\dagger}\} = W_{\hat{\rho}}\left[\alpha^*\exp(i\theta), \exp(-i\theta)\alpha\right]. \tag{6.129}$$

The field variable is multiplied by the phase factor representing the phase-only transfer function, as opposed to being contracted with it.

The result in (6.127) is much more general than what is accomplished in Section 6.4.2 with the operator-based derivation. While it is shown in Section 6.4.2 that a shift-invariant phase operator applied to a Fock states causes its parameter function to be transformed as $F \to \exp(K)F$, we show here that the field variables for an *arbitrary* state are transformed by a *general* phase operator as given in (6.128). The equivalent transformation of the Fock state parameter function (for a general phase operator) can be readily obtained by applying (6.128) to the generating function in (6.52). It leads to $F \to \exp_\diamond(K) \diamond F$, which also explains the opposite sign in the argument of the exponentiated kernel: field variables always transform with the Hermitian adjoint of the kernel with which the parameter functions are transformed. We can now also determine how the parameter function of a coherent state transforms by using (6.128).

Exercise 6.13. Show that when we apply the phase transformation in (6.128) on a coherent state, it implies a transformation of the parameter function α_0 given by

$$\alpha_0 \to \exp_\diamond(K) \diamond \alpha_0. \tag{6.130}$$

6.4.5 Phase-only spatial modulation

Another special type of lossless operation is when an optical element performs a phase modulation directly on the parameter function in *configuration space*. It is an operation that is implemented by the *transmission function* of a *phase-only optical element* (such as a *thin lens*). The general form of the kernel in the Fourier domain for this case, as found in (2.41), is a *convolution kernel*. Since transmission functions are two-dimensional functions of the transverse coordinates $t(\mathbf{X})$, the convolution kernel is naturally represented in terms of optical beam variables by

$$E(\mathbf{k},\mathbf{k}') = 2\pi\delta(\omega-\omega')\int t(\mathbf{X})\exp[\mathrm{i}\mathbf{X}\cdot(\mathbf{K}-\mathbf{K}')]\,\mathrm{d}^2X. \tag{6.131}$$

For the lossless case, the *transmission function* is a two-dimensional phase factor $t(\mathbf{X}) = \exp[\mathrm{i}\phi(\mathbf{X})]$ where $\phi(\mathbf{X})$ is a real-valued phase function. It implies that the complex conjugate is $t^*(\mathbf{X}) = 1/t(\mathbf{X})$ and that the kernel is unitary $E^\dagger = E^{-1}$. It follows that the kernel in the Wigner functional for such a process is given by

$$\begin{aligned}\tau(\mathbf{k}-\mathbf{k}') &= (\mathbf{1}-E)\diamond(\mathbf{1}+E)^{-1}\\ &= 2\pi\delta(\omega-\omega')\int\frac{1-t(\mathbf{X})}{1+t(\mathbf{X})}\exp[\mathrm{i}\mathbf{X}\cdot(\mathbf{K}-\mathbf{K}')]\,\mathrm{d}^2X\\ &= -\mathrm{i}2\pi\delta(\omega-\omega')\int\tan\left[\frac{1}{2}\phi(\mathbf{X})\right]\exp[\mathrm{i}\mathbf{X}\cdot(\mathbf{K}-\mathbf{K}')]\,\mathrm{d}^2X,\end{aligned} \tag{6.132}$$

The transformation of the field variables for *phase-only spatial modulation* is

$$\alpha^* \to \alpha^*\diamond E \quad \text{and} \quad \alpha \to E^{-1}\diamond\alpha, \tag{6.133}$$

with E defined in (6.131).

6.4.6 Unitary temporal evolution

In Section 3.6.5, we considered the unitary temporal evolution of states and operators in terms of the particle-number degree of freedom only. Here, we extend the analysis to include all the other degrees of freedom and we perform the analysis on functional phase space in terms of Wigner functionals. To determine how Wigner functionals transform as a result of unitary evolution, we need to compute the Wigner functional for the unitary evolution operator given in (3.232). First, we consider the interaction-free Hamiltonian and compute its Wigner functional.

6.4.6.1 Interaction-free Hamiltonian

When we include all the other degrees of freedom, the *interaction-free Hamiltonian* is given by (4.28). For convenience, we represent this Hamiltonian as

$$\hat{H} = \frac{1}{2}\hat{a}^{\dagger} \diamond \mathcal{E} \diamond \hat{a} + \frac{1}{2}\hat{a} \diamond \mathcal{E} \diamond \hat{a}^{\dagger}, \tag{6.134}$$

where $\mathcal{E}_{r,s}(\mathbf{k}_1, \mathbf{k}_2) = \hbar\omega_1 \mathbf{1}_{r,s}(\mathbf{k}_1, \mathbf{k}_2)$ with $\mathbf{1}$ defined in (4.54). It may seem confusing, because the definition in (4.54) already contains ω. However, the one in (4.54) is required to remove ω in the denominator of the $\diamond$-contraction.

To compute the Wigner functional of the Hamiltonian in (6.134), we use the procedure in Section 6.1.3 for polynomial operators. According to the procedure, we first convert the Hamiltonian to normal order and then into a functional derivative operation, leading to

$$\hat{H} \rightarrow C_{\hat{H}}\left[\delta_{\nu}, \delta_{\mu^*}\right] = \delta_{\nu} \diamond \mathcal{E} \diamond \delta_{\mu^*} + \frac{1}{2}\Omega_{\mathcal{E}}, \tag{6.135}$$

where

$$\Omega_{\mathcal{E}} \triangleq \operatorname{tr}\{\mathcal{E}\} = 2\hbar \int \omega\delta(0)\, \mathrm{d}^3k, \tag{6.136}$$

which is a *divergent constant*, representing the *zero-point energy*. The factor of 2 comes from the two spin states. Then we apply the functional derivative operation on the generating functional and set the auxiliary field variables to zero to obtain

$$\begin{aligned} W_{\hat{H}}[\alpha] &\triangleq \text{Wigner}\{\hat{H}\} \\ &= C_{\hat{H}}\left[\delta_{\nu}, \delta_{\mu^*}\right] \exp\left(\mu^* \diamond \alpha + \alpha^* \diamond \nu - \frac{1}{2}\mu^* \diamond \nu\right)\Big|_{\mu^*=\nu=0} \\ &= \left(\alpha^* \diamond \mathcal{E} \diamond \alpha - \frac{1}{2}\Omega_{\mathcal{E}}\right) + \frac{1}{2}\Omega_{\mathcal{E}} = \alpha^* \diamond \mathcal{E} \diamond \alpha. \end{aligned} \tag{6.137}$$

We see that the divergent constants cancel. If instead, we use the *normal-ordered Hamiltonian* $:\hat{H}:$ (without the zero-point energy term), we would get

$$\text{Wigner}\{:\hat{H}:\} = \text{Wigner}\{\hat{a}^{\dagger} \diamond \mathcal{E} \diamond \hat{a}\} = \alpha^{*} \diamond \mathcal{E} \diamond \alpha - \frac{1}{2}\Omega_{\mathcal{E}}. \tag{6.138}$$

6.4.6.2 Unitary evolution operator

For temporal evolution, we need to compute the Wigner functional of the unitary evolution operator, which contains the Hamiltonian in its exponent. Here, we discard the zero-point energy term (which only contributes a global phase factor that can be discarded) and express it in normal order $:\hat{H}:$ as in (6.138). When this Hamiltonian is placed in the exponent of the unitary evolution operator in (3.232), Plank's constant cancels so that we end up with (3.240) with its argument in normal order. The result is a phase operator in the form of (6.101)

$$\hat{U} = \exp\left(\frac{-\mathrm{i}}{\hbar}\int_0^t \hat{a}^{\dagger} \diamond \mathcal{E} \diamond \hat{a}\, \mathrm{d}t'\right) = \exp\left[\hat{a}^{\dagger} \diamond (-\mathrm{i}\omega t\mathbf{1}) \diamond \hat{a}\right]. \tag{6.139}$$

However, even with its argument in normal order, this unitary evolution operator is not in normal order. Therefore, we use the procedure in Section 6.4.3 to obtain the normal-ordered expression given in (6.112) with $K = -\mathrm{i}\omega t\mathbf{1}$. The Wigner functional for the unitary operator then follows from (6.115). For the current case, it reads

$$\text{Wigner}\{\hat{U}(t)\} = \frac{\mathcal{N}_0}{\det\{\mathbf{1} + E(t)\}} \exp\left[2\alpha^{*} \diamond \tau(t) \diamond \alpha\right]. \tag{6.140}$$

where

$$\begin{aligned} E(t) &= \exp(-\mathrm{i}\omega t)\mathbf{1}, \\ \tau(t) &= -(\mathbf{1} - E) \diamond (\mathbf{1} + E)^{-1} = -\mathrm{i}\tan\left(\frac{1}{2}\omega t\right)\mathbf{1}. \end{aligned} \tag{6.141}$$

The unitary evolution of an arbitrary state, as represented in terms of its Wigner functional, is given by (6.128). Here, we have

$$\alpha^{*}(\mathbf{k}) \to \alpha^{*}(\mathbf{k})\exp(-\mathrm{i}\omega_{\mathbf{k}}t) \quad \text{and} \quad \alpha(\mathbf{k}) \to \exp(\mathrm{i}\omega_{\mathbf{k}}t)\alpha(\mathbf{k}), \tag{6.142}$$

where $\omega_{\mathbf{k}} = c|\mathbf{k}|$. The transformation of the state becomes

$$\text{Wigner}\left\{\hat{U}(t)\hat{\rho}\hat{U}^{\dagger}(t)\right\} = W_{\hat{\rho}}\left[\exp(-\mathrm{i}\omega_{\mathbf{k}}t)\alpha^{*}, \exp(\mathrm{i}\omega_{\mathbf{k}}t)\alpha\right]. \tag{6.143}$$

The interaction-free unitary evolution of a state on functional phase space leads to rotations on concentric circular orbits around the origin. The rotation frequency depends on the *angular frequency* $\omega_{\mathbf{k}}$ as determined by the wavevector $\mathbf{k}$ on which the field variable depends.

6.4.7 Phase modulation as a continuous process

So far, apart from the temporal evolution, we have considered the linear optical operation or phase transformation as a once-off process, leading to the transformation of the state given in (6.128). In many scenarios, the phase transformation is a continuous process during propagation of the state through some structure or medium along a spatial direction. Here, we describe such a continuous process with the aid of an *evolution equation*. We do this by applying a z-derivative to the state after applying (6.128).

The effect of a continuous phase modulation process on a Wigner functional at a point z along the propagation direction is given by

$$\text{Wigner}\left\{\hat{U}(z)\hat{\rho}\hat{U}^{\dagger}(z)\right\} = W_{\hat{\rho}}\left[\alpha^{*}\diamond E(z), E^{\dagger}(z)\diamond\alpha\right], \tag{6.144}$$

where $E(z)$ is a unitary kernel, so that $E^{\dagger} = E^{-1}$. The evolving state becomes a function of z, as given by the original state in which the field variables are transformed with z-dependent kernels. The z-derivative produces

$$\frac{\mathrm{d}}{\mathrm{d}z}W_{\hat{\rho}}\left[\bar{\alpha}^{*},\bar{\alpha}\right] = \alpha^{*}\diamond\partial_{z}E(z)\diamond\frac{\delta W_{\hat{\rho}}\left[\bar{\alpha}^{*},\bar{\alpha}\right]}{\delta\bar{\alpha}^{*}} + \frac{\delta W_{\hat{\rho}}\left[\bar{\alpha}^{*},\bar{\alpha}\right]}{\delta\bar{\alpha}}\diamond\partial_{z}E^{\dagger}(z)\diamond\alpha, \tag{6.145}$$

where $\bar{\alpha}^{*} \triangleq \alpha^{*}\diamond E(z)$ and $\bar{\alpha} \triangleq E^{\dagger}(z)\diamond\alpha$. Let us define

$$\partial_{z}E(z) \triangleq -\mathrm{i}\Phi(z), \tag{6.146}$$

where $\Phi(\mathbf{k},\mathbf{k}',z)$ is a *Hermitian phase kernel*. Then we get

$$\frac{\mathrm{d}}{\mathrm{d}z}W_{\hat{\rho}}\left[\bar{\alpha}^{*},\bar{\alpha}\right] = \mathrm{i}\frac{\delta W_{\hat{\rho}}\left[\bar{\alpha}^{*},\bar{\alpha}\right]}{\delta\bar{\alpha}}\diamond\Phi(z)\diamond\bar{\alpha} - \mathrm{i}\bar{\alpha}^{*}\diamond\Phi(z)\diamond\frac{\delta W_{\hat{\rho}}\left[\bar{\alpha}^{*},\bar{\alpha}\right]}{\delta\bar{\alpha}^{*}}, \tag{6.147}$$

in which all the field variables are barred. However, we note that the transformation $\alpha \to \bar{\alpha} = E^{\dagger}(z)\diamond\alpha$ maps the field variables from the phase space back onto the same phase space. Therefore, we can revert back to the original unbarred notation, so that the evolution equation becomes

$$\frac{\mathrm{d}W_{\hat{\rho}}[\alpha]}{\mathrm{d}z} = \mathrm{i}\frac{\delta W_{\hat{\rho}}[\alpha]}{\delta\alpha}\diamond\Phi(z)\diamond\alpha - \mathrm{i}\alpha^{*}\diamond\Phi(z)\diamond\frac{\delta W_{\hat{\rho}}[\alpha]}{\delta\alpha^{*}}. \tag{6.148}$$

The resulting evolution equation represents an arbitrary continuous phase modulation process as determined by the *phase kernel* $\Phi(z)$. Note that the right-hand side of the equation is Hermitian, as required for consistency with the Hermitian Wigner functional on the left-hand side. For a given phase kernel, one can solve the equation to obtain the evolving state for that physical scenario.

6.4.8 Free space propagation

A ubiquitous example of a continuous phase modulation process is *free space propagation* [9], discussed in Section 2.6. Here, we consider it as a unitary process applied to photonic quantum states. Any continuous process experienced by a photonic state must include some form of propagation. Inside a bounded structure (such as an optical fibre) where the optical field separates into a superposition of bounded modes, the propagation is determined by a propagation phase factor $\exp(\mathrm{i}\beta_n z)$, where β_n is a propagation constant for the n-th mode. A more complicated scenario is *free space propagation*, where the optical field undergoes *diffraction*. Here, we focus on *paraxial propagation* or *Fresnel propagation*. (See Section 2.6.1.)

This analysis gives us the opportunity to formulate a standard approach to derive an *evolution equation* for quantum states from a given classical equation of motion. In other words, if we want to determine the effect of a classical process on a quantum state, we can follow the steps outlined below.

The first task is to find the expression for the *generator of the evolution*, which is analogous to the Hamiltonian as the generator of temporal evolution. For this purpose, we formulate the *classical equation of motion* as an equation for the evolution of the *classical angular spectrum* of the field. In the case of (paraxial) Fresnel propagation, the starting point is the *paraxial wave equation* of (2.90), in which the optical field is represented by the slow-varying monochromatic electric phasor field $\mathbf{E}(\mathbf{x})$. We now express this field in terms of a z-dependent transverse monochromatic angular spectrum $\mathbf{G}(\mathbf{K}, z)$, by performing a transverse Fourier transform. It is converted to a scalar equation by selecting a specific polarization. The resulting equation reads

$$\partial_z G(\mathbf{K}, z) = -\mathrm{i}\frac{|\mathbf{K}|^2}{2k} G(\mathbf{K}, z), \tag{6.149}$$

in terms of optical beam variables (without ω due to the monochromatic assumption), as can be seen from (2.145). The equation shows that the paraxial free space propagation process leads to a quadratic modulation in the spatial frequency domain.

The next step is to convert the classical equation of motion for the angular spectrum into an equation for the quantum field by replacing the z-dependent transverse angular spectrum by a Fourier domain *field operator*. The latter is obtained by performing a fixed-z Fourier transform, as defined in (2.107), on the annihilation or creation part of the *quantized electric field*, as given in (4.46) in optical beam variables. The ladder (annihilation and creation) operators in terms of which the quantized electric field is expressed, have a *generic commutation relation* as provided in (4.42). Here, we represent it as

$$[\hat{a}(\mathbf{K}, \omega), \hat{a}^{\dagger}(\mathbf{K}', \omega')] = (2\pi)^3 \mathcal{C}\delta(\mathbf{K} - \mathbf{K}')\delta(\omega - \omega') \triangleq \mathbf{1}, \tag{6.150}$$

where $\mathcal{C}$ (instead of $\mathcal{C}'$) is the factor that specifies the transformation properties of the commutation relation, and we ignore the spin indices. For commutation relations that are consistent with the Lorentz covariant formulation, we can set $C = k_z$. The quantized electric field operator in terms of optical beam variables on which the Fourier transform is to be applied is given by the first term of (4.46), expressed here as

$$\hat{\mathbf{E}}^{(+)}(\mathbf{x}, t) = \mathrm{i} \int \sqrt{\frac{\mu_0 k_z \hbar}{2C}} \vec{\eta}_s \hat{a}_s(\mathbf{K}, \omega) \exp(-\mathrm{i}\omega t + \mathrm{i}\mathbf{k} \cdot \mathbf{x}) \frac{\omega \, \mathrm{d}\omega \, \mathrm{d}^2 k}{(2\pi)^3 k_z}. \tag{6.151}$$

Then the Fourier domain scalar field operators are

$$\begin{aligned} \hat{G}(\mathbf{K}, \omega, z) &= \int \vec{\eta}^* \cdot \hat{\mathbf{E}}^{(+)}(\mathbf{x}, t) \exp(\mathrm{i}\omega t - \mathrm{i}\mathbf{K} \cdot \mathbf{X}) \, \mathrm{d}t \, \mathrm{d}^2 x \\ &= \mathrm{i}\sqrt{\frac{k^2 \hbar}{2\epsilon_0 k_z C}} \hat{a}(\mathbf{K}, \omega, z), \end{aligned} \tag{6.152}$$

and its Hermitian adjoint. The ladder operators are expressed in the (spatial) Heisenberg picture as $\hat{a}(\mathbf{K}, \omega, z) = \hat{a}(\mathbf{K}, \omega) \exp(\mathrm{i}k_z z)$ and the reference plane is defined at $z = 0$. The equal-z *commutation relation* for these Fourier domain scalar field operators is

$$[\hat{G}(\mathbf{K}, \omega, z), \hat{G}^\dagger(\mathbf{K}', \omega', z)] = (2\pi)^3 \frac{\hbar k^2}{2\epsilon_0 k_z} \delta(\mathbf{K} - \mathbf{K}')\delta(\omega - \omega'). \tag{6.153}$$

We note that the expression of this commutation relation does not depend on the definition of C. The quantized *evolution equation* for paraxial free space propagation is then obtained by replacing the angular spectrum in (6.149) by the Fourier domain scalar field operator in (6.152). It reads

$$\partial_z \hat{G}(\mathbf{K}, \omega, z) = -\mathrm{i}\frac{|\mathbf{K}|^2}{2k} \hat{G}(\mathbf{K}, \omega, z). \tag{6.154}$$

Next, we find the expression for the propagation operator (i. e., the generator for evolution along a spatial direction) that we need to substitute into the quantum propagation equation for operators in (3.237). It is done by comparing the quantized version of the classical evolution equation in (6.154) with the quantum propagation equation for an ansatz of the propagation operator. The ansatz is given generically by

$$\hat{P}_\Delta = \hat{G}^\dagger \diamond P \diamond \hat{G} = \int \hat{G}^\dagger(\mathbf{K}, \omega, z) P(\mathbf{K}, \omega, \mathbf{K}', \omega') \hat{G}(\mathbf{K}', \omega', z) \, \mathrm{d}_C k \, \mathrm{d}_C k', \tag{6.155}$$

where

$$\mathrm{d}_C k \triangleq \frac{\mathrm{d}^2 k \, \mathrm{d}\omega}{(2\pi)^3 C}. \tag{6.156}$$

Although the $\diamond$-contractions provide a compact notation, we need to remember that a $\diamond$-contraction is always defined in terms of the commutation relation for the ladder operators so that the associated identity removes a $\diamond$-contraction: $\mathbf{1} \diamond \alpha = \alpha$; hence, the C

in the denominator. Based on the form of (6.154), we know that $P(\mathbf{K},\omega,\mathbf{K}',\omega')$ is diagonal. Therefore, we can represent it as

$$P(\mathbf{K},\omega,\mathbf{K}',\omega') = P_0(\mathbf{K})(2\pi)^3 C\delta(\mathbf{K}-\mathbf{K}')\delta(\omega-\omega') \equiv P_0\mathbf{1}. \tag{6.157}$$

The quantum propagation equation for operators (3.237) then gives

$$\begin{aligned} -\mathrm{i}\hbar\partial_z\hat{G}(\mathbf{K},\omega,z) &= [\hat{G}(\mathbf{K},\omega,z),\hat{P}_\Delta] \\ &= \frac{\hbar k^2}{2\epsilon_0 k_z C}\int P(\mathbf{K},\omega,\mathbf{K}',\omega')\hat{G}(\mathbf{K}',\omega',z)\,\mathrm{d}_C k' \\ &= \frac{\hbar k^2}{2\epsilon_0 k_z C}P_0(\mathbf{K})\hat{G}(\mathbf{K},\omega,z). \end{aligned} \tag{6.158}$$

The comparison with (6.154) leads to

$$P_0(\mathbf{K}) = -\frac{\epsilon_0 k_z C|\mathbf{K}|^2}{k^3}, \tag{6.159}$$

so that the kernel of the propagation operator is

$$P(\mathbf{K},\omega,\mathbf{K}',\omega') = -(2\pi)^3\frac{\epsilon_0 C^2|\mathbf{K}|^2}{k^2}\delta(\mathbf{K}-\mathbf{K}')\delta(\omega-\omega'), \tag{6.160}$$

where we substitute $k_z = k$ under the *paraxial approximation.*

Having determined the expression for the propagation operator, we can now move to the Schrödinger picture to consider the evolution of states as determined by this propagation operator. We substitute the propagation operator into the evolution equation in (3.229) and then express it in terms of Wigner functionals. The density operator of the state becomes the Wigner functional of the state $W_{\hat{\rho}}[\alpha]$ and the propagation operator is replaced by its Wigner functional. While the former is an arbitrary functional, the latter can be computed using the coherent-state-assisted approach for polynomial operators provided at the end of Section 6.1.3. However, here we have field operators instead of the ladder operators. Therefore, it is better to proceed with the general coherent-state-assisted approach of Section 6.1.3.

The first step in the coherent-state-assisted approach is to overlap the propagation operator by coherent states on both sides. When the field operator is applied to a coherent state, then according to (6.152) (in the Schrödinger picture) it produces

$$\hat{G}(\mathbf{K},\omega)\,|\alpha\rangle = \mathrm{i}\,|\alpha\rangle\sqrt{\frac{\hbar k^2}{2\epsilon_0 k_z C}}\alpha(\mathbf{K},\omega), \tag{6.161}$$

and its Hermitian adjoint, where $\alpha(\mathbf{K},\omega)$ is the complex-valued parameter function of the coherent state in optical beam variables. The overlap of the propagation operator by coherent states thus leads to

$$\begin{aligned}\langle\alpha_1|\hat{P}_\Delta|\alpha_2\rangle &= \langle\alpha_1|\hat{G}^\dagger\diamond P\diamond\hat{G}|\alpha_2\rangle\\ &= \hbar\exp\left(-\frac{1}{2}\|\alpha_1\|^2-\frac{1}{2}\|\alpha_2\|^2+\alpha_1^*\diamond\alpha_2\right)\alpha_1^*\diamond P'\diamond\alpha_2,\end{aligned} \tag{6.162}$$

where we define a new kernel P' given by

$$P'(\mathbf{K},\omega,\mathbf{K}',\omega') = -\frac{|\mathbf{K}|^2}{2k}(2\pi)^3 C\delta(\mathbf{K}-\mathbf{K}')\delta(\omega-\omega'), \tag{6.163}$$

after setting $k_z = k$. We pulled Planck's constant out of the definition of P'. It will eventually cancel because free space propagation is a linear process that does not involve interactions requiring the presence of Planck's constant.

The next step in the coherent-state-assisted approach would normally be to substitute (6.162) into the integral expression in (6.20). Instead, we use a *generating functional* with a *construction operation* to represent the overlap $\langle\alpha_1|\hat{P}_\Delta|\alpha_2\rangle$. As such, it is similar to the procedure provided at the end of Section 6.1.3. The present case is not very complicated and does not really justify the use of the generating functional, but it helps to produce the evolution equation and serves as a simple example to demonstrate the process, which is again used in Chapter 8. The generating functional is formed by adding source terms for the field variables into the exponent of the overlap between the two coherent states. It reads

$$\mathcal{G} = \exp\left(-\frac{1}{2}\|\alpha_1\|^2-\frac{1}{2}\|\alpha_2\|^2+\alpha_1^*\diamond\alpha_2+\alpha_1^*\diamond\nu+\mu^*\diamond\alpha_2\right), \tag{6.164}$$

where ν and μ^* are auxiliary field variables. The construction process is obtained by replacing the field variables in the polynomial factor in (6.162) by functional derivatives with respect to these auxiliary field variables to reproduce the field variables when applied to the generating functional. It is thus given by

$$C_\delta = \frac{\delta}{\delta\nu}\diamond P'\diamond\frac{\delta}{\delta\mu^*}, \tag{6.165}$$

so that

$$\langle\alpha_1|\hat{P}_\Delta|\alpha_2\rangle = \hbar\, C_\delta\{\mathcal{G}\}|_{\nu=\mu^*=0}\,. \tag{6.166}$$

We substitute the generating functional, instead of the overlapped operator, into (6.20) and evaluate the functional integrals for the coherent-state-assisted approach. The result is a generating functional for the Wigner functional of the propagation operator

$$\mathcal{W} = \exp\left(\mu^*\diamond\alpha+\alpha^*\diamond\nu-\frac{1}{2}\mu^*\diamond\nu\right), \tag{6.167}$$

which is equivalent to the generating function obtained in the procedure at the end of Section 6.1.3.[c]

The *evolution equation* is now obtained from (3.229) by replacing the density operator by its Wigner functional. The product of the density operator with the propagation operator becomes the star product of the state's Wigner functional and that of the propagation operator, which is given in terms of the generating functional and the construction operation. The result represents a *Moyal (Poisson) bracket* [10–12], given by

$$-\mathrm{i}\partial_z W_{\hat{\rho}} = \mathcal{C}_\delta \left\{\mathcal{W} \star W_{\hat{\rho}} - W_{\hat{\rho}} \star \mathcal{W}\right\}\Big|_{\nu=\mu^*=0}, \tag{6.168}$$

in terms of the generating functional and the construction process, where $\star$ is the *star product*, discussed in Section 5.6. A factor of $\hbar$ is cancelled from both sides. The calculation of the two star-product terms in the Moyal bracket can be evaluated even when the Wigner functional of the state is not known, because the functional integrations involve Dirac delta functionals that only affect the arguments of the unknown Wigner functional of the state. They are given by

$$\begin{aligned}
\mathcal{W} \star W_{\hat{\rho}} &= \exp\left(\mu^* \diamond \alpha + \alpha^* \diamond \nu - \frac{1}{2}\mu^* \diamond \nu\right) W_{\hat{\rho}}\left[\alpha^* + \frac{1}{2}\mu^*, \alpha - \frac{1}{2}\nu\right], \\
W_{\hat{\rho}} \star \mathcal{W} &= \exp\left(\mu^* \diamond \alpha + \alpha^* \diamond \nu - \frac{1}{2}\mu^* \diamond \nu\right) W_{\hat{\rho}}\left[\alpha^* - \frac{1}{2}\mu^*, \alpha + \frac{1}{2}\nu\right].
\end{aligned} \tag{6.169}$$

Exercise 6.14. Use the expressions in (6.20) and (6.167) to compute those in (6.169).

The final step is to apply the construction process of (6.165) on the two star-product terms, according to the process in (6.168), thus producing the evolution equation. After some simplifications, the evolution equation reads

$$-\mathrm{i}\partial_z W_{\hat{\rho}} = -\frac{\delta W_{\hat{\rho}}}{\delta \alpha} \diamond P' \diamond \alpha + \alpha^* \diamond P' \diamond \frac{\delta W_{\hat{\rho}}}{\delta \alpha^*}. \tag{6.170}$$

In terms of the expression for the kernel in (6.163), it becomes

$$-\mathrm{i}\partial_z W_{\hat{\rho}} = \int \frac{\delta W_{\hat{\rho}}}{\delta \alpha(\mathbf{K},\omega)} \frac{|\mathbf{K}|^2}{2k} \alpha(\mathbf{K},\omega) - \alpha^*(\mathbf{K},\omega) \frac{|\mathbf{K}|^2}{2k} \frac{\delta W_{\hat{\rho}}}{\delta \alpha^*(\mathbf{K},\omega)}\, \mathrm{d}_C k. \tag{6.171}$$

---

c In effect, we could have followed a procedure similar to the one in Section 6.1.3 in which the field operators are replaced by functional derivatives according to

$$\hat{G}(\mathbf{K},\omega) \to \mathrm{i}\sqrt{\frac{\hbar k^2}{2\epsilon_0 k_z C}} \frac{\delta}{\delta \mu^*(\mathbf{K},\omega)} \quad \text{and} \quad \hat{G}^\dagger(\mathbf{K},\omega) \to -\mathrm{i}\sqrt{\frac{\hbar k^2}{2\epsilon_0 k_z C}} \frac{\delta}{\delta \nu(\mathbf{K},\omega)}.$$

Since the kernel is diagonal, it leads to a single integral over the wavevectors. One can represent such diagonal kernels generically by combining them with the identity kernel defined in (6.150). It allows us to write the evolution equation as

$$\partial_z W_{\hat{\rho}} = \mathrm{i}\frac{\delta W_{\hat{\rho}}}{\delta \alpha} \diamond T\mathbf{1} \diamond \alpha - \mathrm{i}\alpha^* \diamond T\mathbf{1} \diamond \frac{\delta W_{\hat{\rho}}}{\delta \alpha^*}, \tag{6.172}$$

where we multiplied by i, and

$$T(\mathbf{K}) = \frac{|\mathbf{K}|^2}{2k}, \tag{6.173}$$

is the function on the diagonal. The resulting evolution equation corresponds to the one in (6.148) for continuous phase modulation.

The solution of the evolution equation in (6.172) has the form of a continuous phase modulation, as given in (6.144), with

$$E(z) = \exp_\diamond\left(-\mathrm{i}\frac{z|\mathbf{K}|^2}{2k}\right). \tag{6.174}$$

It can be compared with classical *Fresnel propagation* in Section 2.6.1. The opposite sign in the argument is due to the fact that parameter functions and field variables transform with opposite signs.

6.5 Thermal states

The vacuum state belongs to both the Fock states (for $n = 0$) and to the coherent states (for $\alpha = 0$). Another set of states that the vacuum state belongs to is the *thermal states*. However, the thermal states differ from the Fock states and the coherent states in that they are not pure states. They are *mixed states* except for the vacuum state.

Here, we first discuss the thermal states for *blackbody radiation* and compute its Wigner functional. Its properties are considered in more detail. Finally, we generalize the expression to define general thermal states.

6.5.1 Density operator of the blackbody thermal state

The thermal state for blackbody radiation is often defined as a density operator

$$\hat{\rho}_{\mathrm{th}} = \frac{\exp\left(-\hat{H}/k_B T\right)}{\mathrm{tr}\left\{\exp\left(-\hat{H}/k_B T\right)\right\}} = \frac{\exp\left(-\hat{\omega}/\omega_T\right)}{\mathrm{tr}\left\{\exp\left(-\hat{\omega}/\omega_T\right)\right\}}, \tag{6.175}$$

where $\hat{H}$ is the *interaction-free Hamiltonian* for the quantized electromagnetic field in (4.28), T is the temperature, and k_B is the Boltzmann's constant.[d] The *frequency operator* $\hat{\omega}$ is the generalized version of (3.241) with all degrees of freedom included

$$\hat{\omega} = \sum_s \int a_s^\dagger(\mathbf{k}) a_s(\mathbf{k}) \frac{\mathrm{d}^3 k}{(2\pi)^3} = \sum_s \int \omega a_s^\dagger(\mathbf{k}) a_s(\mathbf{k})\, \mathrm{d}_\omega k, \tag{6.176}$$

and ω_T is the *thermal frequency*

$$\omega_T \triangleq \frac{k_B T}{\hbar}. \tag{6.177}$$

The Hamiltonian is given by (6.134), and can also be expressed as

$$\hat{H} = \hbar\hat{\omega} + \Omega_{\mathcal{E}}, \tag{6.178}$$

in terms of the frequency operator. The *divergent constant* $\Omega_{\mathcal{E}}$ is defined in (6.136). It produces a phase factor in the last expression in (6.175) that cancels out. The trace in the denominator of (6.175) normalizes the operator. This normalization is addressed after we have computed the Wigner functional.

6.5.2 Wigner functional for the blackbody thermal state

The Wigner functional of the unitary evolution operator is obtained in (6.140). The Wigner functional for the blackbody thermal state is similar. The density operator in (6.175) has the form of a phase operator

$$\hat{\rho}_{\mathrm{th}} = \mathcal{N}_{\mathrm{th}} \exp(-\hat{a}^\dagger \diamond \beta \diamond \hat{a}), \tag{6.179}$$

where $\mathcal{N}_{\mathrm{th}}$ is a normalization constant, and

$$\beta_{r,s}(\mathbf{k},\mathbf{k}') \triangleq \frac{\omega_{\mathbf{k}}}{\omega_T} \mathbf{1}_{r,s}(\mathbf{k},\mathbf{k}'), \tag{6.180}$$

with $\omega_{\mathbf{k}} = c|\mathbf{k}|$. Unlike the phase operator, which is unitary, the density operator is Hermitian. Nevertheless, we can treat them formally the same when calculating their Wigner functionals. Therefore, the Wigner functional of the blackbody thermal state in (6.179) can be obtained from the calculation in Section 6.4.3. Substituting $K \to -\beta$ into (6.116) in (6.115) and normalizing the result, we obtain

$$W_{\mathrm{th}}[\alpha] = \mathcal{N}_0 \det\{\theta\} \exp\left(-2\alpha^* \diamond \theta \diamond \alpha\right), \tag{6.181}$$

---

d The Boltzmann constant is $k_B = 1.38 \times 10^{-23}$ J/K (joule per kelvin) for temperature T in kelvin [K].

where

$$\theta \triangleq \tanh_{\diamond}\left(\frac{1}{2}\beta\right) = \tanh\left(\frac{\omega}{2\omega_T}\right)\mathbf{1}. \tag{6.182}$$

The normalization constant $\mathcal{N}_0 \det\{\theta\}$ is computed by requiring that the trace (i. e., the integral over the Wigner functional) equals 1. Since the density operator of the blackbody thermal state has the form of a phase operator, its Wigner functional in (6.181) resembles (6.115). However, while the Wigner functional of the thermal state is normalized, that of the phase operator in (6.115) is not normalizable, and the kernel of the thermal state θ is Hermitian, while τ in (6.115) is anti-Hermitian.

The thermal state is a mixed state, and its Wigner functional is centred at the origin. For $T \to 0$, we have $\theta \to \mathbf{1}$, leading to the Wigner functional of the vacuum state.

6.5.2.1 Determinant

The *functional determinant* in the normalization constant of the Wigner functional for the blackbody thermal state is expressed by

$$\det\{\theta\} = \det\{(\mathbf{1} - E) \diamond (\mathbf{1} + E)^{-1}\} = \frac{\det\{\mathbf{1} - E\}}{\det\{\mathbf{1} + E\}}, \tag{6.183}$$

where

$$E = \exp\left(-\frac{\omega}{\omega_T}\right)\mathbf{1} \triangleq X(\omega)\mathbf{1}. \tag{6.184}$$

The two determinants can be represented in terms of the *kernel trace*, as defined in (C.18), so that

$$\begin{aligned}\det\{\mathbf{1} \pm X(\omega)\mathbf{1}\} &= \exp\left(\operatorname{tr}\{\ln_{\diamond}[\mathbf{1} \pm X(\omega)\mathbf{1}]\}\right)\\ &= \exp\left[\sum_{n=1}^{\infty}\frac{(\pm 1)^n}{n}\int \exp\left(-\frac{nc|\mathbf{k}|}{\omega_T}\right)2\delta(0)\,\mathrm{d}^3k\right]\\ &= \exp\left[\frac{16\pi\omega_T^3\delta(0)}{c^3}\sum_{n=1}^{\infty}\frac{(\pm 1)^n}{n^4}\right],\end{aligned} \tag{6.185}$$

where $\delta(0)$ comes from $\operatorname{tr}\{X(\omega)\mathbf{1}\}$, which causes the argument of $\delta(\mathbf{k})$ to be set equal to zero. As a result, $\delta(0)$ carries the units of [distance^3]. The full determinant thus becomes

$$\det\{\theta\} = \exp\left[\frac{16\pi\omega_T^3\delta(0)}{c^3}\sum_{n=1}^{\infty}\frac{(-1)^n - 1}{n^4}\right] = \exp\left[-\frac{\pi^5\omega_T^3}{3c^3}\delta(0)\right], \tag{6.186}$$

which is effectively zero.

6.5.2.2 Energy in the blackbody thermal state

The reason for the vanishing determinant is that the trace in the exponent diverges. To find the cause of this divergence, we compute the energy in the blackbody thermal state in terms of the density operator in (6.175) or the Wigner functional in (6.181). In terms of Wigner functionals, the energy in a state is given by the functional integral over the product of the Wigner functionals of the state and the energy operator. The *normal-ordered Hamiltonian* serves as the energy operator. For the blackbody thermal state, the energy in the state is

$$\langle E_{\mathrm{th}} \rangle = \int W_{:\hat{H}:}[\alpha] W_{\mathrm{th}}[\alpha] \, \mathcal{D}^{\circ}[\alpha], \tag{6.187}$$

where the Wigner functionals $W_{:\hat{H}:}[\alpha]$ and $W_{\mathrm{th}}[\alpha]$ are given in (6.138) and (6.181), respectively. Using a generating function to place the Wigner functional of the Hamiltonian into the exponent, we get

$$\begin{aligned} \langle E_{\mathrm{th}} \rangle &= \mathcal{N}_0 \det\{\theta\} \int \left(\alpha^* \diamond \mathcal{E} \diamond \alpha - \frac{1}{2}\Omega_{\mathcal{E}} \right) \exp\left(-2\alpha^* \diamond \theta \diamond \alpha\right) \, \mathcal{D}^{\circ}[\alpha] \\ &= \mathcal{N}_0 \det\{\theta\} \, \partial_J \int \exp\left[-2\alpha^* \diamond \left(\theta - \frac{1}{2} J \mathcal{E} \right) \diamond \alpha - \frac{1}{2} J \Omega_{\mathcal{E}} \right] \, \mathcal{D}^{\circ}[\alpha] \bigg|_{J=0} \\ &= \partial_J \frac{\det\{\theta\} \exp(-\frac{1}{2} J \Omega_{\mathcal{E}})}{\det\left\{\theta - \frac{1}{2} J \mathcal{E}\right\}} \bigg|_{J=0} . \end{aligned} \tag{6.188}$$

To compute the derivative, we use the identity for the derivative of a determinant given in (D.34). As a result, the expectation value of the energy is

$$\langle E_{\mathrm{th}} \rangle = \frac{1}{2} \operatorname{tr}\left\{\mathcal{E} \diamond \theta^{-1}\right\} - \frac{1}{2}\Omega_{\mathcal{E}}. \tag{6.189}$$

Using the definition in (6.182), we simplify the inverse so that

$$\theta^{-1} = (\mathbf{1} + E) \diamond (\mathbf{1} - E)^{-1} = \mathbf{1} + \frac{2X}{1 - X}\mathbf{1}, \tag{6.190}$$

where X is given in (6.184). As a result, we obtain

$$\begin{aligned} \langle E_{\mathrm{th}} \rangle &= \frac{1}{2} \operatorname{tr}\{\mathcal{E}\} + \operatorname{tr}\left\{\mathcal{E}\frac{X}{1-X}\right\} - \frac{1}{2}\Omega_{\mathcal{E}} = \sum_{n=1}^{\infty} \operatorname{tr}\{\mathcal{E}X^n\} \\ &= \sum_{n=1}^{\infty} 2\delta(0)\hbar \int \omega \exp\left(-\frac{n\omega}{\omega_T}\right) \, \mathrm{d}^3k \\ &= \sum_{n=1}^{\infty} \frac{48\pi\hbar\omega_T^4 \delta(0)}{n^4 c^3} = \frac{8\pi^5 \hbar \omega_T^4}{15 c^3} \delta(0). \end{aligned} \tag{6.191}$$

Since we use the normal-ordered Hamiltonian, the divergent constant zero-point energy $\frac{1}{2}\operatorname{tr}\{\mathcal{E}\} = \frac{1}{2}\Omega_{\mathcal{E}}$ cancels. The result is still divergent due to $\delta(0)$. It implies that the energy in the state is infinite. We conclude that the state given in (6.175) is not a physical state that can be prepared in a laboratory.

6.5.3 Physically realistic thermal states

The question is, if the state in (6.175) is not a physically realistic thermal state due to its infinite energy, how can we model a finite energy thermal state that can be prepared in a laboratory? The modelling of such a finite-energy thermal state provides us with another opportunity to work through an exercise in modelling.

For the expression of a *finite-energy thermal state*, we use a more restrictive expression for θ. The trace in the expectation value of the energy can be expressed as

$$\frac{1}{2}\operatorname{tr}\left\{\mathcal{E}\diamond\theta^{-1}\right\} = \frac{\hbar}{2}\int \theta^{-1}(\mathbf{k},\mathbf{k})\,\frac{\mathrm{d}^3k}{(2\pi)^3}. \tag{6.192}$$

Hence, any thermal state kernel function with an inverse that contains a finite-energy function on its diagonal leads to a finite expectation value for the energy of the state. It implies that the kernel of a finite-energy thermal state cannot be diagonal, because a kernel that is finite on the diagonal and zero everywhere else is of measure zero.

Based on these considerations, we introduce a modification of (6.184) where we replace the identity kernel by an *idempotent kernel* P having the same dimensions as the identity kernel. Then E becomes

$$E = \exp\left(-\frac{\omega}{\omega_T}\right)P(\mathbf{k},\mathbf{k}') = XP. \tag{6.193}$$

Thanks to the assumed idempotency of P, we can simplify the inverse of θ, so that

$$\theta^{-1} = (\mathbf{1}+XP)\diamond(\mathbf{1}-XP)^{-1} = \mathbf{1}+\frac{2X}{1-X}P. \tag{6.194}$$

As a result, the expectation value for the energy becomes

$$\langle E_{\text{th}}\rangle = \hbar\sum_{n=1}^{\infty}\int \exp\left(-\frac{n\omega}{\omega_T}\right)P(\mathbf{k},\mathbf{k})\,\frac{\mathrm{d}^3k}{(2\pi)^3}. \tag{6.195}$$

The result is now finite, provided that P is finite on the diagonal.

For a physically realistic scenario, the idempotent kernel P represents a finite aperture (or detector area) and a finite duration of the measurement. Such a scenario can be modelled by a *convolution kernel* (see Section 2.1.7). It is given by

$$\begin{aligned} P(\mathbf{k}_1,\mathbf{k}_2) = T(\mathbf{K}_1-\mathbf{K}_2,\omega_1,\omega_2) &= c\sqrt{\omega_1\omega_2}\int p(\mathbf{X})\exp[-\mathrm{i}\mathbf{X}\cdot(\mathbf{K}_1-\mathbf{K}_2)]\,\mathrm{d}^2x \\ &\quad\times\int d(t)\exp[\mathrm{i}(\omega_1-\omega_2)t]\,\mathrm{d}t, \end{aligned} \tag{6.196}$$

in optical beam variables, where $p(\mathbf{X})$ is a dimensionless binary transmission function, representing an aperture, and $d(t)$ is another dimensionless binary function, representing the duration of the process. The factor of $\sqrt{\omega_1\omega_2}$ helps to maintain Lorentz covariance. Such binary functions are equal to 1 inside a region (an area A or a period Δt) and zero outside. The factor of c serves to provide the correct units. It appears because the right-hand side is expressed in optical beam variables, while the idempotent kernel on the left-hand side is given as a function of three-dimensional wavevectors.

6.5.3.1 Properties of convolution kernels

We have encountered convolution kernels before, in Section 2.1.7. Here, we expand on that discussion by providing some additional information about their properties that become pertinent in the current context. These properties are demonstrated by considering contractions between two such kernels. To simplify the expressions, we simply represent the convolution kernel (for the Lorentz covariant case) as

$$T(\mathbf{k}_1 - \mathbf{k}_2) = \sqrt{\omega_1\omega_2} \int h(\mathbf{x}) \exp[-i\mathbf{x} \cdot (\mathbf{k}_1 - \mathbf{k}_2)]\, d^3x. \tag{6.197}$$

Contractions of these convolution kernels can be written in a simplified way as the Fourier transform of the product of their transmission functions:[e]

$$\begin{aligned} T_1 \diamond T_2 &= \int T_1(\mathbf{k}_1 - \mathbf{k}')T_2(\mathbf{k}' - \mathbf{k}_2)\, d_\omega k' \\ &= \int \sqrt{\omega_1\omega'} \int h_1(\mathbf{x}_1) \exp[-i\mathbf{x}_1 \cdot (\mathbf{k}_1 - \mathbf{k}')]\, d^3x_1 \\ &\quad \times \sqrt{\omega'\omega_2} \int h_2(\mathbf{x}_2) \exp[-i\mathbf{x}_2 \cdot (\mathbf{k}' - \mathbf{k}_2)]\, d^3x_2\, d_\omega k' \\ &= \sqrt{\omega_1\omega_2} \int h_1(\mathbf{x})h_2(\mathbf{x}) \exp[-i\mathbf{x} \cdot (\mathbf{k}_1 - \mathbf{k}_2)]\, d^3x. \end{aligned} \tag{6.198}$$

As a result, convolution kernels commute $T_1 \diamond T_2 = T_2 \diamond T_1$. It also follows from (6.198) that a kernel $K_\diamond\{T\}$, defined in terms of an expansion of $\diamond$-contractions of the convolution kernel, can be written as the Fourier transform of the function $K_h(\mathbf{x})$ with an equivalent expansion of the transmission function

$$K_\diamond\{T\} = \sqrt{\omega_1\omega_2} \int K_h(\mathbf{x}) \exp[-i\mathbf{x} \cdot (\mathbf{k}_1 - \mathbf{k}_2)]\, d^2x. \tag{6.199}$$

Here, $K_\diamond\{T\}$ and $K_h(\mathbf{x})$ are *holomorphic kernels or functions*

$$K_\diamond\{T\} = \sum_n c_n T^{\diamond n} \quad \text{and} \quad K_h(\mathbf{x}) = \sum_n c_n h^n(\mathbf{x}), \tag{6.200}$$

---

e We retain the term "transmission function" and we express it as a function in three dimensions for this discussion even though a transmission function is strictly speaking a two-dimensional function.

where the coefficients c_n are the same in both expansions.

When the two convolution kernels in (6.198) are the same $T_1 = T_2 = T$, the contraction simplifies to

$$T \diamond T = \sqrt{\omega_1\omega_2} \int h^2(\mathbf{x}) \exp[-i\mathbf{x} \cdot (\mathbf{k}_1 - \mathbf{k}_2)] \, d^3x. \tag{6.201}$$

If $h(\mathbf{x})$ is a binary function, then $h^2(\mathbf{x}) = h(\mathbf{x})$.[f] It then follows that $T \diamond T = T$. Hence, with a binary transmission function, the convolution kernel is *idempotent*. Such binary transmission functions are often encountered in optical systems where they can, for example, represent the transmission functions of apertures.

6.5.3.2 Finite energy thermal state

Returning to our model of a physically realistic thermal state, we substitute (6.196) into (6.195) to obtain

$$\langle E_{\text{th}} \rangle = c\hbar A \Delta t \sum_{n=1}^{\infty} \int \exp\left(-\frac{n\omega}{\omega_T}\right) \frac{d^3k}{(2\pi)^3} = \frac{\pi^2 \hbar A \Delta t \omega_T^4}{30c^2}. \tag{6.202}$$

The resulting expectation value of the energy of the thermal state is now finite, thanks to the incorporation of the idempotent convolution kernel. It represents the energy that is observed from a thermal source through an aperture with an area of A during a time interval Δt. Dividing the expression throughout by Δt, we obtain the *thermal power* P_{th} radiated from this aperture:

$$P_{\text{th}} = \frac{\pi^2 \hbar A \omega_T^4}{30c^2}. \tag{6.203}$$

6.5.4 Single-mode thermal states

Another way to model a physically realizable thermal state is as a *single-mode thermal state*. In this case, we define the inverse kernel by

$$\theta^{-1} = \mathbf{1} + 2N\Theta\Theta^*, \tag{6.204}$$

where $\Theta_s(\mathbf{k})$ is a normalized parameter function and N is the average number of photons in the thermal state. Such a thermal state represents a situation where thermal radiation is spatiotemporally filtered so that it is parameterized by a specific parameter function,

f The equation $a^2 = a$ has only two solutions: $a = 0$ and $a = 1$. Hence, the requirement for idempotency is satisfied if $h : \mathbb{R}^3 \rightarrow \{0, 1\}$.

represented by Θ. A filtering process applied to a general thermal state can be modelled with the aid of an inhomogeneous beamsplitter, as discussed in Section 6.6.4.

6.5.5 General thermal states

In the preceding sections, we have extensively discussed the *blackbody thermal state*. However, thermal states can exist in more general forms. Henceforth, we represent a *general thermal state* as in (6.181), but where the thermal state kernel θ represents any Hermitian kernel with $0 < \det\{\theta\} \leq 1$.

Exercise 6.15. Show that the purity of the general thermal state given in (6.181) is given by $\det\{\theta\}$.

Exercise 6.16. Show that the average number of photons in the general thermal state in (6.181) is given by

$$\langle n \rangle_{\text{th}} = \frac{1}{2}\operatorname{tr}\{\theta^{-1}\} - \frac{1}{2}\Omega, \tag{6.205}$$

and the average number of photons in the single-mode thermal state, with its kernel given in (6.204), is N.

6.6 Beamsplitter

The beamsplitter is an optical component that serves various purposes and is often encountered in optical setups. It is a four-port device with two input ports and two output ports. Light beams (or photons) that enter the two respective input ports are divided between the two output ports according to some splitting process. Being a lossless device, the beamsplitter is modelled by a unitary process. While it is an essential part of many optical systems, it also plays a significant role in a conceptual sense. Its representation as a unitary process provides the basic mechanism to transform a state that is originally defined on one phase space (or Hilbert space) into a state defined on two phase spaces (or two Hilbert spaces).

We can distinguish between the normal situation of a *homogeneous beamsplitter* where the splitting process is independent of the spatiotemporal degrees of freedom, and an *inhomogeneous beamsplitter* where the splitting process depends on the spatiotemporal degrees of freedom of the photons in the beam. A typical example of the latter is an *aperture*. In the following sections, we calculate the Wigner functionals of these beamsplitters and consider some applications.

The formulation of a beamsplitter plays a significant role in the modelling of systems with *loss*. In the example of an aperture, an inhomogeneous beamsplitter can be used to direct the light that is blocked by the aperture to a loss channel. Tracing out the *loss channel*, we obtain the state of the light that can pass through the aperture at the other output port. Loss is discussed in more detail in Section 6.7.

6.6.1 Homogeneous beamsplitter

The unitary operation of an ideal homogeneous beamsplitter is often represented with a 2×2 unitary matrix U. Similar to unitary kernels, discussed in Section 2.1.5, a matrix U is unitary when its Hermitian adjoint is equal to its inverse $U^{\dagger} = U^{-1}$. When the determinant of a unitary matrix is $\det\{U\} = 1$, it belongs to a Lie group [13] called the *special unitary group* SU(n). (See the discussion of Lie groups in Section 2.8.1.) The relevant Lie group for the ideal homogeneous beamsplitter is SU(2). The matrices in the SU(2) group in terms of which we model the ideal homogeneous beamsplitter can be parameterized by three (angle of phase) parameters. For convenience, we write it as

$$U_{\text{bs}} = \begin{bmatrix} \Phi_{\text{i}} & 0 \\ 0 & \Phi_{\text{i}}^{*} \end{bmatrix} \begin{bmatrix} C & -S \\ S & C \end{bmatrix} \begin{bmatrix} \Phi_{\text{o}} & 0 \\ 0 & \Phi_{\text{o}}^{*} \end{bmatrix}, \tag{6.206}$$

where

$$\Phi_{\text{i}} = \exp(\text{i}\phi_{\text{i}}), \quad \Phi_{\text{o}} = \exp(\text{i}\phi_{\text{o}}), \quad C = \cos(\varphi), \quad \text{and} \quad S = \sin(\varphi), \tag{6.207}$$

with ϕ_{i} and ϕ_{o} being the input and output phases, respectively, and φ is an angle determining the reflectivity or *splitting ratio* of the beamsplitter.

For a *balanced beamsplitter* with a 50:50 splitting ratio, we have $\varphi = \frac{1}{4}\pi$. The input and output phases are not very important because they are easily changed by small changes in the distances before and after the beamsplitter. Therefore, one often finds different representations of the unitary matrix for a beamsplitter, using different assumptions about these phases. The one that we use here is symmetric in form:

$$U_{\text{bs}}(\varphi) = \begin{bmatrix} C & \text{i}S \\ \text{i}S & C \end{bmatrix} = \begin{bmatrix} \cos(\varphi) & \text{i}\sin(\varphi) \\ \text{i}\sin(\varphi) & \cos(\varphi) \end{bmatrix}. \tag{6.208}$$

In Figure 6.4, we show a diagram of the operation of an unbalanced homogeneous beamsplitter. It shows how two optical beams are divided and recombined. The arrowed lines represent the photons from the two beams. Thin lines represent a smaller probability amplitude than the thicker lines.

For the purpose of calculations in quantum optics, the homogeneous beamsplitter is represented by a unitary operator (see Section 3.2.2) expressed in terms of ladder operators. The unitary operator for the homogeneous beamsplitter, incorporating all the spatiotemporal degrees of freedom, is given by

$$\hat{U}_{\text{bs}}(\varphi) = \exp\left[\text{i}\left(\hat{a}_A^{\dagger} \diamond \hat{a}_B + \hat{a}_B^{\dagger} \diamond \hat{a}_A\right)\varphi\right], \tag{6.209}$$

where A and B represent the two input ports and also the two output ports.

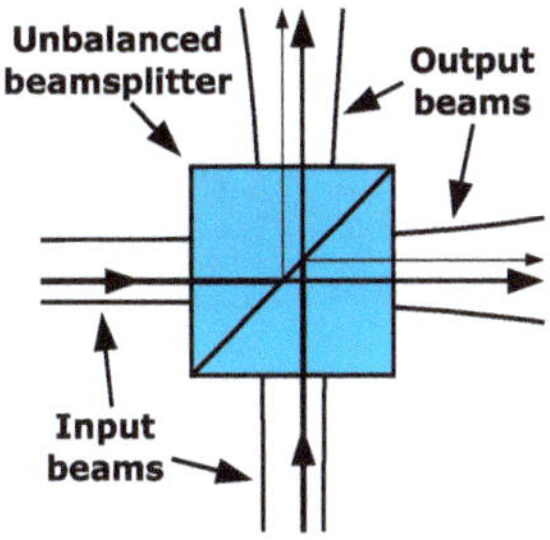

Figure 6.4: Diagram of an unbalanced homogeneous beamsplitter showing how it divides and recombines two beams passing through it.

6.6.2 Wigner functional of the homogeneous beamsplitter

To calculate the Wigner functional of the homogeneous beamsplitter, we use the coherent-state-assisted approach of Section 6.1.3. Therefore, we need to compute the overlap of the unitary operator of the homogeneous beamsplitter by two coherent states. However, instead of working with (6.209), we use the definition of the operation of the beamsplitter on the ladder operators to transform the displacement operators of the overlapping coherent states. The operation of the beamsplitter is defined in terms of the transformation

$$\begin{aligned} \hat{a}_A &\to C\hat{a}_A + \mathrm{i}S\hat{a}_B, \quad \hat{a}_A^\dagger \to C\hat{a}_A^\dagger - \mathrm{i}S\hat{a}_B^\dagger, \\ \hat{a}_B &\to C\hat{a}_B + \mathrm{i}S\hat{a}_A, \quad \hat{a}_B^\dagger \to C\hat{a}_B^\dagger - \mathrm{i}S\hat{a}_A^\dagger, \end{aligned} \tag{6.210}$$

where C and S are defined in (6.207). When the beamsplitter operator is overlapped by the coherent states, we can apply this transformation on the displacement operators of the coherent states on the right-hand side. It leads to

$$\begin{aligned} \hat{U}_{\text{bs}} |\alpha_2\rangle_A |\beta_2\rangle_B &= \hat{U}_{\text{bs}} \exp\left(\hat{a}_A^\dagger \diamond \alpha_2 - \alpha_2^* \diamond \hat{a}_A + \hat{a}_B^\dagger \diamond \beta_2 - \beta_2^* \diamond \hat{a}_B\right) |\text{vac}\rangle \\ &\to \exp\left[\left(C\hat{a}_A^\dagger - \mathrm{i}S\hat{a}_B^\dagger\right) \diamond \alpha_2 - \alpha_2^* \diamond (C\hat{a}_A + \mathrm{i}S\hat{a}_B)\right. \\ &\quad \left. + \left(C\hat{a}_B^\dagger - \mathrm{i}S\hat{a}_A^\dagger\right) \diamond \beta_2 - \beta_2^* \diamond (C\hat{a}_B + \mathrm{i}S\hat{a}_A)\right] |\text{vac}\rangle \\ &= \exp\left[\hat{a}_A^\dagger \diamond (C\alpha_2 - \mathrm{i}S\beta_2) - (C\alpha_2^* + \mathrm{i}S\beta_2^*) \diamond \hat{a}_A\right] \\ &\quad \times \exp\left[\hat{a}_B^\dagger \diamond (C\beta_2 - \mathrm{i}S\alpha_2) - (C\beta_2^* + \mathrm{i}S\alpha_2^*) \diamond \hat{a}_B\right] |\text{vac}\rangle \\ &= |C\alpha_2 - \mathrm{i}S\beta_2\rangle_A |C\beta_2 - \mathrm{i}S\alpha_2\rangle_B, \end{aligned} \tag{6.211}$$

where we first combined the two displacement operators, then performed the transformation, before we separated the result into two separate displacement operators for the output A and B channels again. The resulting *transformation of the parameter functions* comes out to be

$$\begin{aligned}\alpha_2 &\to C\alpha_2 - iS\beta_2, \quad \alpha_2^* \to C\alpha_2^* + iS\beta_2^*, \\ \beta_2 &\to C\beta_2 - iS\alpha_2, \quad \beta_2^* \to C\beta_2^* + iS\alpha_2^*.\end{aligned} \tag{6.212}$$

The overlap can then be evaluated with the aid of (4.114)

$$\begin{aligned}&\langle\beta_1|_B \langle\alpha_1|_A \hat{U}_{\text{bs}} |\alpha_2\rangle_A |\beta_2\rangle_B \\ &= \langle\alpha_1|C\alpha_2 - iS\beta_2\rangle_A \langle\beta_1|C\beta_2 - iS\alpha_2\rangle_B \\ &= \exp\Big(-\frac{1}{2}\alpha_1^* \diamond \alpha_1 - \frac{1}{2}\alpha_2^* \diamond \alpha_2 - \frac{1}{2}\beta_1^* \diamond \beta_1 - \frac{1}{2}\beta_2^* \diamond \beta_2 \\ &\quad + C\alpha_1^* \diamond \alpha_2 + C\beta_1^* \diamond \beta_2 - iS\alpha_1^* \diamond \beta_2 - iS\beta_1^* \diamond \alpha_2\Big).\end{aligned} \tag{6.213}$$

This expression is now substituted into the integral expression for the coherent-state-assisted approach given in (6.20). After evaluating all the functional integrals, we obtain the Wigner functional of the beamsplitter operator

$$W_{\hat{A}}[\alpha,\beta] = \frac{\mathcal{N}_0}{(1+C)^{\Omega}} \exp\left[-i\frac{2S}{1+C}(\beta^* \diamond \alpha + \alpha^* \diamond \beta)\right]. \tag{6.214}$$

6.6.3 Applying the homogeneous beamsplitter

To test the expression for the Wigner functional of the homogeneous beamsplitter, we apply it to an arbitrary state entering the two ports. The result requires the evaluation of a *double* triple star product. Therefore, we need to apply (5.49) twice to obtain

$$\begin{aligned}&\text{Wigner}\left\{\hat{U}_{\text{bs}}\hat{\rho}\hat{U}_{\text{bs}}^{\dagger}\right\} \\ &= \mathcal{N}_{\text{bs}}^2 \int \exp\Big[-i\frac{\eta}{4}(\alpha_a^* + \alpha^* + \alpha_b^*) \diamond (\beta_a + \beta + \beta_b) \\ &\quad + i\frac{\eta}{4}(\alpha_a^* + \alpha^* - \alpha_b^*) \diamond (\beta_a + \beta - \beta_b) - i\frac{\eta}{4}(\beta_a^* + \beta^* + \beta_b^*) \diamond (\alpha_a + \alpha + \alpha_b) \\ &\quad + i\frac{\eta}{4}(\beta_a^* + \beta^* - \beta_b^*) \diamond (\alpha_a + \alpha - \alpha_b)\Big] \exp[(\alpha^* - \alpha_a^*) \diamond \alpha_b - \alpha_b^* \diamond (\alpha - \alpha_a) \\ &\quad + (\beta^* - \beta_a^*) \diamond \beta_b - \beta_b^* \diamond (\beta - \beta_a)] W_{\hat{\rho}}[\alpha_a^*, \alpha_a, \beta_a^*, \beta_a]\, \mathcal{D}^{\circ}[\alpha_a, \alpha_b, \beta_a, \beta_b] \\ &= \mathcal{N}_{\text{bs}}^2 \int \exp\Big[\Big(\beta^* - \beta_a^* - i\frac{\eta}{2}\alpha_a^* - i\frac{\eta}{2}\alpha^*\Big) \diamond \beta_b - \beta_b^* \diamond \Big(\beta - \beta_a + i\frac{\eta}{2}\alpha_a + i\frac{\eta}{2}\alpha\Big) \\ &\quad + \Big(\alpha^* - \alpha_a^* - i\frac{\eta}{2}\beta_a^* - i\frac{\eta}{2}\beta^*\Big) \diamond \alpha_b - \alpha_b^* \diamond \Big(\alpha - \alpha_a + i\frac{\eta}{2}\beta_a + i\frac{\eta}{2}\beta\Big)\Big] \\ &\quad \times W_{\hat{\rho}}[\alpha_a^*, \alpha_a, \beta_a^*, \beta_a]\, \mathcal{D}^{\circ}[\alpha_a, \alpha_b, \beta_a, \beta_b],\end{aligned} \tag{6.215}$$

where we defined

$$\eta \triangleq \frac{2S}{1+C} \quad \text{and} \quad \mathcal{N}_{\text{bs}} \triangleq \frac{\mathcal{N}_0}{(1+C)^{\Omega}}. \tag{6.216}$$

The functional integrations over α_b and β_b produce Dirac delta functionals, leading to replacements of the arguments of the input Wigner functional when the functional integrations over α_a and β_a are performed. We start with the functional integration over β_b, and then over β_a. With an appropriate redefinition of $\alpha_b \to \frac{1}{2}[1+\cos(\varphi)]\alpha_c$, which conveniently removes the normalization factors $\mathcal{N}_{\text{bs}}$, we can then evaluate the functional integration over α_c, and finally over α_a. The result is

$$\begin{aligned} W_{\hat{\rho}}[\alpha,\beta] &\to \text{Wigner}\left\{\hat{U}_{\text{bs}}\hat{\rho}\hat{U}_{\text{bs}}^{\dagger}\right\} \\ &= W_{\hat{\rho}}[C\alpha^* - iS\beta^*, C\alpha + iS\beta, C\beta^* - iS\alpha^*, C\beta + iS\alpha]. \end{aligned} \tag{6.217}$$

So, the homogenous beamsplitter performs a *transformation of the field variables* in the arguments of the Wigner functional of the initial states, given by

$$\begin{aligned} \alpha &\to C\alpha + iS\beta, \quad \alpha^* \to C\alpha^* - iS\beta^*, \\ \beta &\to C\beta + iS\alpha, \quad \beta^* \to C\beta^* - iS\alpha^*. \end{aligned} \tag{6.218}$$

Note that the transformation of the field variables has the opposite sign for S compared to the transformation of the parameter functions given in (6.212). Although it is informative to have the expression for the Wigner functional of the beamsplitter, practical calculations involving beamsplitters in general simply employ the transformations in (6.218) on the field variables of the input state to obtain the output state directly.

Exercise 6.17. Compute the states obtained at the output ports of a 50:50 beamsplitter when a thermal state and a vacuum state enter the respective input ports of the beamsplitter.

6.6.4 Inhomogeneous beamsplitter

An *inhomogeneous beamsplitter* is a device that redirects photons in an optical beam according to the spatiotemporal degrees of freedom of those photons. As an example, consider a *D-shaped mirror*. Such D-shaped mirrors have sharp straight edges that are often used in optical setups to cut optical beams in half, sending one half in a different direction. The probability for a photon to be reflected by the mirror is determined by the integrated *probability distribution* of the photon over the area of the mirror. More sophisticated scenarios can be envisaged where the deflection of a photon is determined by the contraction between the photon's parameter function and some kernel associated with the beamsplitter.

The operation of an inhomogeneous beamsplitter is demonstrated by the diagram in Figure 6.5. It models a hole in a perfectly reflective surface. Depending on the position of an arrow, it either passes through the hole or is reflected by the surface around the hole. The arrowed lines in the two beams coming from the respective input ports are distinguished by showing them either as solid lines or as dashed lines.

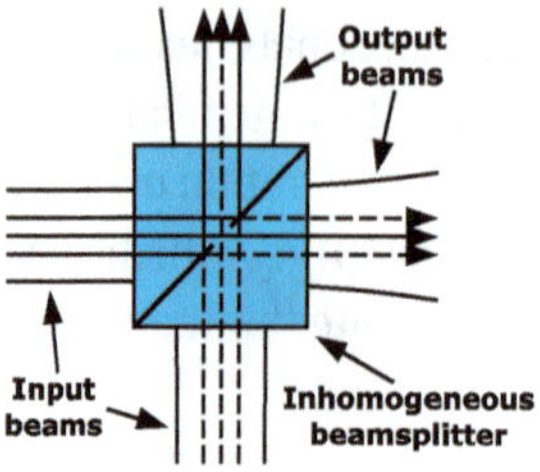

Figure 6.5: Diagram of an inhomogeneous beamsplitter showing how it divides and recombines two beams passing through it.

Polarizing beamsplitters can also be regarded as inhomogeneous beamsplitters. In this case, the probability for a photon to be directed toward a given output port is determined by its spin degree of freedom (state of polarization).

To model the inhomogeneous beamsplitter, we directly express it as a Wigner functional that is obtained from a generalization of the Wigner functional for the homogeneous beamsplitter. So, based on the form of the Wigner functional for a homogeneous beamsplitter in (6.214), the Wigner functional for an inhomogeneous beamsplitter reads

$$W_{\text{ibs}}[\alpha,\beta] = \mathcal{N}_R \exp\left(-\mathrm{i}2\alpha^* \diamond R \diamond \beta - \mathrm{i}2\beta^* \diamond R^\dagger \diamond \alpha\right), \tag{6.219}$$

where R is an arbitrary kernel and $\mathcal{N}_R$ is a prefactor that maintains unitarity. The expression for the Wigner functional of the inhomogeneous beamsplitter has the same form as the Wigner functional of the homogeneous beamsplitter. The unitary nature of the Wigner functional (which mean that $W_{\text{ibs}}^* \star W_{\text{ibs}} = 1$) is apparent because the argument of the exponential is anti-Hermitian.

Exercise 6.18. Show that the Wigner functional in (6.219) represents a unitary operator with

$$\mathcal{N}_R = \det\{\mathbf{1} + R^\dagger \diamond R\} = \det\{\mathbf{1} + R \diamond R^\dagger\}. \tag{6.220}$$

6.6.4.1 Applying the generic inhomogeneous beamsplitter

We now apply the generic inhomogeneous beamsplitter to an arbitrary *bipartite state* entering the two ports. It again requires the evaluation of a double triple star product. After some simplifications, the expression becomes

$$\begin{aligned}
&\text{Wigner}\left\{\hat{U}_{\text{ibs}}\hat{\rho}\hat{U}_{\text{ibs}}^\dagger\right\} \\
&= \mathcal{N}_R^{-2} \int \exp\left[\left(\alpha^* - \alpha_a^* - \mathrm{i}\beta_a^* \diamond R^\dagger - \mathrm{i}\beta^* \diamond R^\dagger\right) \diamond \alpha_b\right. \\
&\quad - \alpha_b^* \diamond (\alpha - \alpha_a + \mathrm{i}R \diamond \beta_a + \mathrm{i}R \diamond \beta) + (\beta^* - \beta_a^* - \mathrm{i}\alpha_a^* \diamond R - \mathrm{i}\alpha^* \diamond R) \diamond \beta_b \\
&\quad \left. - \beta_b^* \diamond \left(\beta - \beta_a + \mathrm{i}R^\dagger \diamond \alpha_a + \mathrm{i}R^\dagger \diamond \alpha\right)\right] W_{\hat{\rho}}[\alpha_a^*, \alpha_a, \beta_a^*, \beta_a]\, \mathcal{D}^\circ[\alpha_a, \alpha_b, \beta_a, \beta_b].
\end{aligned} \tag{6.221}$$

As before, the functional integrations over α_b and β_b produce Dirac delta functionals, leading to replacements in the arguments of the Wigner functional upon the evaluation of the functional integrations over α_a and β_a. Here, we start with the integration over α_b, followed by the integration over α_a,

$$\begin{aligned}
&\text{Wigner}\left\{\hat{U}_{\text{ibs}}\hat{\rho}\hat{U}_{\text{ibs}}^{\dagger}\right\}\\
&= \mathcal{N}_R^2 \int \exp\left[\left(\beta^* - \beta^* \diamond R^{\dagger} \diamond R - \beta_a^* - \beta_a^* \diamond R^{\dagger} \diamond R - \mathrm{i}2\alpha^* \diamond R\right) \diamond \beta_b\right.\\
&\quad \left. - \beta_b^* \diamond \left(\beta - R^{\dagger} \diamond R \diamond \beta - \beta_a - R^{\dagger} \diamond R \diamond \beta_a + \mathrm{i}2R^{\dagger} \diamond \alpha\right)\right]\\
&\quad \times W_{\hat{\rho}}[\alpha^* - \mathrm{i}\beta_a^* \diamond R^{\dagger} - \mathrm{i}\beta^* \diamond R^{\dagger}, \alpha + \mathrm{i}R \diamond \beta_a + \mathrm{i}R \diamond \beta, \beta_a^*, \beta_a]\, \mathcal{D}^{\circ}[\beta_a, \beta_b]\\
&= \mathcal{N}_R^2 \int W_{\hat{\rho}}[\alpha^* - \mathrm{i}\beta_a^* \diamond R^{\dagger} - \mathrm{i}\beta^* \diamond R^{\dagger}, \alpha + \mathrm{i}R \diamond \beta_a + \mathrm{i}R \diamond \beta, \beta_a^*, \beta_a]\\
&\quad \times \exp\left[(\beta^* \diamond \mathcal{K}_- - \beta_a^* \diamond \mathcal{K}_+ - \mathrm{i}2\alpha^* \diamond R) \diamond \beta_b\right.\\
&\quad \left. - \beta_b^* \diamond \left(\mathcal{K}_- \diamond \beta - \mathcal{K}_+ \diamond \beta_a + \mathrm{i}2R^{\dagger} \diamond \alpha\right)\right] \mathcal{D}^{\circ}[\beta_a, \beta_b],
\end{aligned} \tag{6.222}$$

where

$$\mathcal{K}_- \triangleq \mathbf{1} - R^{\dagger} \diamond R \quad \text{and} \quad \mathcal{K}_+ \triangleq \mathbf{1} + R^{\dagger} \diamond R. \tag{6.223}$$

Now, we redefine

$$\beta_b \to (\mathbf{1} + R^{\dagger} \diamond R)^{-1} \diamond \beta_c = \mathcal{K}_+^{-1} \diamond \beta_c. \tag{6.224}$$

The resulting Jacobian removes the normalization factors. Thus, we obtain

$$\begin{aligned}
\text{Wigner}\left\{\hat{U}_{\text{ibs}}\hat{\rho}\hat{U}_{\text{ibs}}^{\dagger}\right\} &= \int W_{\hat{\rho}}[\alpha^* - \mathrm{i}\beta_a^* \diamond R^{\dagger} - \mathrm{i}\beta^* \diamond R^{\dagger}, \alpha + \mathrm{i}R \diamond \beta_a + \mathrm{i}R \diamond \beta, \beta_a^*, \beta_a]\\
&\quad \times \exp\left[\left(\beta^* \diamond \mathcal{K}_- \diamond \mathcal{K}_+^{-1} - \mathrm{i}2\alpha^* \diamond R \diamond \mathcal{K}_+^{-1} - \beta_a^*\right) \diamond \beta_c\right.\\
&\quad \left. - \beta_c^* \diamond \left(\mathcal{K}_+^{-1} \diamond \mathcal{K}_- \diamond \beta + \mathrm{i}2\mathcal{K}_+^{-1} \diamond R^{\dagger} \diamond \alpha - \beta_a\right)\right] \mathcal{D}^{\circ}[\beta_a, \beta_c]\\
&= W_{\hat{\rho}}[\alpha^* \diamond \mathcal{C} - \mathrm{i}\beta^* \diamond \mathcal{S}^{\dagger}, \mathcal{C} \diamond \alpha + \mathrm{i}\mathcal{S} \diamond \beta,\\
&\quad \beta^* \diamond \mathcal{C}^{\dagger} - \mathrm{i}2\alpha^* \diamond \mathcal{S}, \mathcal{C}^{\dagger} \diamond \beta + \mathrm{i}\mathcal{S}^{\dagger} \diamond \alpha].
\end{aligned} \tag{6.225}$$

where we used

$$R \diamond (\mathbf{1} + R^{\dagger} \diamond R)^{-1} = (\mathbf{1} + R \diamond R^{\dagger})^{-1} \diamond R, \tag{6.226}$$

and defined

$$\begin{aligned}
\mathcal{C} &\triangleq (\mathbf{1} - R \diamond R^{\dagger}) \diamond (\mathbf{1} + R \diamond R^{\dagger})^{-1} = (\mathbf{1} + R \diamond R^{\dagger})^{-1} \diamond (\mathbf{1} - R \diamond R^{\dagger}),\\
\mathcal{C}^{\dagger} &\triangleq (\mathbf{1} - R^{\dagger} \diamond R) \diamond (\mathbf{1} + R^{\dagger} \diamond R)^{-1} = (\mathbf{1} + R^{\dagger} \diamond R)^{-1} \diamond (\mathbf{1} - R^{\dagger} \diamond R),\\
\mathcal{S} &\triangleq 2(\mathbf{1} + R \diamond R^{\dagger})^{-1} \diamond R = 2R \diamond (\mathbf{1} + R^{\dagger} \diamond R)^{-1},
\end{aligned}$$

$$\mathcal{S}^{\dagger} \triangleq 2(\mathbf{1}+R^{\dagger}\diamond R)^{-1}\diamond R^{\dagger} = 2R^{\dagger}\diamond(\mathbf{1}+R\diamond R^{\dagger})^{-1}. \tag{6.227}$$

So, the transformations for a generic inhomogeneous beamsplitter are

$$\begin{aligned} \alpha &\to \mathcal{C}\diamond\alpha + \mathrm{i}\mathcal{S}\diamond\beta, \quad \alpha^* \to \alpha^*\diamond\mathcal{C} - \mathrm{i}\beta^*\diamond\mathcal{S}^{\dagger}, \\ \beta &\to \mathcal{C}^{\dagger}\diamond\beta + \mathrm{i}\mathcal{S}^{\dagger}\diamond\alpha, \quad \beta^* \to \beta^*\diamond\mathcal{C}^{\dagger} - \mathrm{i}\alpha^*\diamond\mathcal{S}. \end{aligned} \tag{6.228}$$

6.6.4.2 Projective beamsplitter

In many (if not most) applications of inhomogeneous beamsplitters, the kernel can be modelled as an *idempotent* Hermitian projection operation. Such an inhomogeneous beamsplitter is called a *projective beamsplitter*. The photons have varying probabilities to be reflected, depending on their spatiotemporal degrees of freedom. These probabilities are determined by the application of the projection kernels on the parameter functions of the photons.

So, we replace the kernel R by a kernel Q that is idempotent ($Q\diamond Q = Q$) and Hermitian ($Q^{\dagger} = Q$). The inverse $\mathcal{K}_+^{-1}$ can then be simplified

$$\mathcal{K}_+^{-1} \equiv (\mathbf{1}+Q\diamond Q)^{-1} = (\mathbf{1}+Q)^{-1} = \mathbf{1} - \frac{1}{2}Q. \tag{6.229}$$

Moreover,

$$\begin{aligned} (\mathbf{1}-Q\diamond Q)\diamond(\mathbf{1}+Q\diamond Q)^{-1} &= (\mathbf{1}-Q)\diamond\left(\mathbf{1}-\frac{1}{2}Q\right) = \mathbf{1}-Q \triangleq P, \\ 2Q\diamond(\mathbf{1}+Q\diamond Q)^{-1} &= 2Q\diamond\left(\mathbf{1}-\frac{1}{2}Q\right) = Q. \end{aligned} \tag{6.230}$$

As a result,

$$\mathcal{C} = \mathcal{C}^{\dagger} \equiv P \quad \text{and} \quad \mathcal{S} = \mathcal{S}^{\dagger} \equiv Q. \tag{6.231}$$

The transformations for the projective beamsplitter thus simplify to

$$\begin{aligned} \alpha &\to P\diamond\alpha + \mathrm{i}Q\diamond\beta, \quad \alpha^* \to \alpha^*\diamond P - \mathrm{i}\beta^*\diamond Q, \\ \beta &\to P\diamond\beta + \mathrm{i}Q\diamond\alpha, \quad \beta^* \to \beta^*\diamond P - \mathrm{i}\alpha^*\diamond Q. \end{aligned} \tag{6.232}$$

Henceforth, when we refer to the “inhomogeneous beamsplitter,” it implies the projective beamsplitter, unless stated otherwise.

6.7 Loss

In photonic quantum systems, a loss involves the uncontrolled absorption of photons by the environment. Loss generally has a detrimental effect on quantum resources: it can turn a pure state into a mixed state, leading to a loss in quantum coherence; it reduces the amount of entanglement in entangled states; and it reduces the squeezing in a squeezed state. Although coherent states only suffer reductions in their average numbers of photons as a result of loss, they are usually not regarded as useful resources in photonic quantum information systems. The more valuable a quantum state is in such photonic quantum information systems, the more detrimental the effect of loss is on such a quantum state.

Loss can either be caused by the medium through which a photon propagates, or by the structure of an optical component through which it passes. In the former case, all the photons in a state have the same probability to be lost, regardless of their (spatiotemporal or spin) properties. We refer to it as a *homogeneous loss*. In the latter case, the structure of the optical component affects the probability for a photon to be lost, based on its spatiotemporal or spin properties. We refer to it as an *inhomogeneous loss*.

An optical component usually produces the loss as a once-off process. For a lossy medium, on the other hand, the loss is generally a continuous process. As a result, the quantum state *evolves* as it propagates through such a medium. A once-off process neglects the effect of free space propagation.

6.7.1 Homogeneous loss

First, we consider a *homogeneous loss* where all the photons in a state have the same probability to be lost, regardless of their degrees of freedom. The standard approach to model such a loss is with the aid of a homogeneous beamsplitter.

Say we want to compute the effect of photon loss on a state $|\psi\rangle$. Together with the vacuum state, it is represented as a combined state $|\psi\rangle_A |\text{vac}\rangle_B$, acting as a *bipartite input state* entering the two input ports of the beamsplitter. First, we apply the unitary transformation that represents the beamsplitter on the bipartite input state. The state after the beamsplitter is then given by

$$\hat{\rho}_{\text{bs}} = \hat{U}_{\text{bs}} \left(|\psi\rangle_A |\text{vac}\rangle_B \langle\psi|_A \langle\text{vac}|_B\right) \hat{U}_{\text{bs}}^{\dagger}, \tag{6.233}$$

where $\hat{U}_{\text{bs}}$ represents the unitary operator for the beamsplitter. The output port of the beamsplitter labelled by B, which receives photons from the state $|\psi\rangle$ via a reflection, is now associated with the loss. The amount of loss (probability that a photon is lost) is governed by the splitting ratio of the beamsplitter. To implement the loss, we perform a partial trace over the lost degrees of freedom. The state that remains after the loss is obtained at the other output port of the beamsplitter. It reads

$$\hat{\rho}_{\text{out}} = \text{tr}_B \left\{\hat{U}_{\text{bs}} \left(|\psi\rangle_A |\text{vac}\rangle_B \langle\psi|_A \langle\text{vac}|_B\right) \hat{U}_{\text{bs}}^{\dagger}\right\}. \tag{6.234}$$

In terms of Wigner functionals the process is performed as follows. The Wigner functional of the state is first multiplied by the vacuum state's Wigner functional with different field variables representing the different input ports. The action of the beamsplitter on the combined Wigner functional is implemented by applying the transformation in (6.218) to its arguments. The angle φ that specifies the splitting ratio, determines the amount of loss. The lost degrees of freedom are removed by integrating out the field variables of the lost channel, which are the same as those of the input vacuum state.

6.7.1.1 Effect of homogeneous loss on coherent states

As an example, we consider the effect of homogeneous loss on a coherent state. The combined Wigner functional for the coherent state and the vacuum state is

$$W[\alpha,\beta] = \mathcal{N}_0^2 \exp\left(-2\|\alpha-\alpha_0\|^2 - 2\|\beta\|^2\right). \tag{6.235}$$

After the beamsplitter, it becomes

$$W_{\text{bs}}[\alpha,\beta] = \mathcal{N}_0^2 \exp\left(-2\|\alpha C + \mathrm{i}\beta S - \alpha_0\|^2 - 2\|\beta C + \mathrm{i}\alpha S\|^2\right). \tag{6.236}$$

While C and S are given in terms of the angle φ as in (6.207), they can also be expressed as $C = \sqrt{1-L}$ and $S = \sqrt{L}$, where L represents the probability of a photon to be lost. When we trace out the loss channel, the result is

$$\begin{aligned} W_{\text{loss}}[\alpha] &= \mathcal{N}_0^2 \int \exp\left(-2\|\alpha C + \mathrm{i}\beta S - \alpha_0\|^2 - 2\|\beta C + \mathrm{i}\alpha S\|^2\right) \mathcal{D}^\circ[\beta] \\ &= \mathcal{N}_0 \exp\left(-2\|\alpha - \alpha_0 C\|^2\right). \end{aligned} \tag{6.237}$$

The resulting state is still a (pure) coherent state, but the amplitude of the parameter function is reduced by a factor $C = \sqrt{1-L}$.

6.7.1.2 Effect of homogeneous loss on a Fock state

Unlike its effect on coherent states, the effect of a homogeneous loss on Fock states is that it produces a mixed state. As a worked example, we consider the case where the loss is applied to a fixed-spectrum Fock state. Before using the Wigner functional formalism, we first perform the calculation for this problem in terms of ladder operators.

We expect to see that the loss reduces the *occupation number* of the Fock state, causing it to become mixed. The mixture is produces because different numbers of photons can be lost leaving behind different Fock states with varying probabilities. The loss mechanism can be introduced in the normal way with a homogenous beamsplitter, using (6.210). It leads to

$$\begin{aligned} |n_F\rangle = \frac{\hat{a}_F^{\dagger n}}{\sqrt{n!}}|\text{vac}\rangle &\to \frac{1}{\sqrt{n!}}\left(C\hat{a}_F^\dagger - \mathrm{i}S\hat{a}_L^\dagger\right)^n|\text{vac}\rangle \\ &= \frac{1}{\sqrt{n!}}\sum_{m=0}^{n}\frac{n!}{m!(n-m)!}\left(C\hat{a}_F^\dagger\right)^m\left(-\mathrm{i}S\hat{a}_L^\dagger\right)^{n-m}|\text{vac}\rangle \end{aligned}$$

$$= \sum_{m=0}^{n} \sqrt{\frac{n!}{m!(n-m)!}} C^m (-iS)^{n-m} |m_F\rangle |(n-m)_L\rangle , \tag{6.238}$$

where $\hat{a}_L^\dagger$ is the creation operator for the lost photon. One can then trace out the lost photons. The result is a mixture of different Fock states

$$\text{tr}_L\{\hat{U}_{\text{bs}}^\dagger |n_F\rangle \langle n_F| \hat{U}_{\text{bs}}\} = \sum_{m=0}^{n} \frac{n!(1-L)^m L^{n-m}}{m!(n-m)!} |m_F\rangle \langle m_F| , \tag{6.239}$$

where we expressed C and S in terms of the loss probability L.

Exercise 6.19. Show that the state in (6.239) is normalised.

Below, we perform the same calculation in terms of Wigner functionals. To facilitate a comparison, we convert the result in (6.239) into a generating function for Wigner functionals. Since the linear combination of pure states is represented by the same linear combination of their Wigner functionals, we can replace $|m_F\rangle \langle m_F| \rightarrow W_{m,F}$. To convert the result to a generating function, we multiply the expression by η^n, where η is the generating parameter, and sum over n,

$$\begin{aligned} \mathcal{W}(\eta) &= \sum_{n=0}^{\infty} \eta^n \sum_{m=0}^{n} \frac{n!(1-L)^m L^{n-m}}{m!(n-m)!} W_{m,F} \\ &= \sum_{m=0}^{\infty} \sum_{p=0}^{\infty} \eta^{p+m} \frac{(p+m)!(1-L)^m L^p}{m!p!} W_{m,F}. \end{aligned} \tag{6.240}$$

For the final expression, we changed the order of the summation and redefined the index $n \rightarrow p + m$. First, we evaluate the summation over p, leading to

$$\mathcal{W}(\eta) = \sum_{m=0}^{\infty} \frac{\eta^m (1-L)^m}{(1-\eta L)^{1+m}} W_{m,F}. \tag{6.241}$$

Using the expression in (6.54), we write this result as

$$\mathcal{W}(\eta) = \frac{\mathcal{N}_0}{(1-\eta L)} \sum_{m=0}^{\infty} \left[\frac{-\eta(1-L)}{(1-\eta L)} \right]^m L_m (4\alpha^* \diamond FF^* \diamond \alpha) \exp\left(-2\|\alpha\|^2\right). \tag{6.242}$$

Based on (6.53), we see that the result of the summation gives a similar expression, with the replacements

$$\nu \rightarrow \frac{-\eta(1-L)}{(1-\eta L)} \quad \text{and} \quad x \rightarrow 4\alpha^* \diamond FF^* \diamond \alpha, \tag{6.243}$$

and a multiplication with $\exp(-2\|\alpha\|^2)/(1-\eta L)$. So, the generating function for Fock states that have suffered a loss is

$$\mathcal{W}(\eta) = \frac{\mathcal{N}_0}{1+\eta-2\eta L} \exp\left[-2\|\alpha\|^2 + \frac{4\eta(1-L)\alpha^* \diamond FF^* \diamond \alpha}{1+\eta-2\eta L}\right]. \tag{6.244}$$

6.7.1.3 Using Wigner functionals

Now we perform the same calculation with the aid of Wigner functionals. For this purpose, the transformation in (6.218) is applied to the product of the generating function (for the Wigner functionals of fixed-spectrum Fock states) in (6.52) and the Wigner functional of the vacuum state. The result is

$$\begin{aligned} W[\alpha,\beta] &= W_{\hat{\rho}}[\alpha] W_{\text{vac}}[\beta] = \frac{\mathcal{N}_0^2}{1+\eta} \exp\left(-2\|\alpha\|^2 + \frac{4\eta}{1+\eta}\alpha^* \diamond FF^* \diamond \alpha - 2\|\beta\|^2\right) \\ &\to \text{Wigner}\left\{\hat{U}_{\text{bs}}(\hat{\rho} \otimes |\text{vac}\rangle\langle\text{vac}|)\hat{U}_{\text{bs}}^{\dagger}\right\} \\ &= \frac{\mathcal{N}_0^2}{1+\eta} \exp\left[-2\alpha^* \diamond \left(\mathbf{1} - \frac{2C^2\eta}{1+\eta}FF^*\right) \diamond \alpha + \mathrm{i}\frac{4SC\eta}{1+\eta}\alpha^* \diamond FF^* \diamond \beta \right. \\ &\quad \left. - \mathrm{i}\frac{4SC\eta}{1+\eta}\beta^* \diamond FF^* \diamond \alpha - 2\beta^* \diamond \left(\mathbf{1} - \frac{2S^2\eta}{1+\eta}FF^*\right) \diamond \beta\right]. \end{aligned} \tag{6.245}$$

Then we trace (integrate) over β, leading to

$$\begin{aligned} &\int \text{Wigner}\left\{\hat{U}_{\text{bs}}(\hat{\rho} \otimes |\text{vac}\rangle\langle\text{vac}|)\hat{U}_{\text{bs}}^{\dagger}\right\} \mathcal{D}^{\circ}[\beta] \\ &= \frac{\mathcal{N}_0}{(1+\eta)\det\left\{\mathbf{1} - \frac{2S^2\eta}{1+\eta}FF^*\right\}} \exp\left[-2\alpha^* \diamond \left(\mathbf{1} - \frac{2C^2\eta}{1+\eta}FF^*\right) \diamond \alpha \right. \\ &\quad \left. + 2\left(\frac{4SC\eta}{1+\eta}\right)^2 \alpha^* \diamond FF^* \diamond \left(\mathbf{1} - \frac{2S^2\eta}{1+\eta}FF^*\right)^{-1} \diamond FF^* \diamond \alpha\right]. \end{aligned} \tag{6.246}$$

The inverse can we expressed as (see Appendix D)

$$\left(\mathbf{1} - \frac{2S^2\eta}{1+\eta}FF^*\right)^{-1} = \mathbf{1} + \frac{2S^2\eta}{1+\eta-2S^2\eta}FF^*, \tag{6.247}$$

and the determinant can be simplified so that the prefactor becomes

$$\frac{\mathcal{N}_0}{(1+\eta)\det\left\{\mathbf{1} - \frac{2S^2\eta}{1+\eta}FF^*\right\}} = \frac{\mathcal{N}_0}{1+\eta-2S^2\eta}. \tag{6.248}$$

The resulting expression reads

$$\int \text{Wigner}\left\{\hat{U}_{\text{bs}}(\hat{\rho} \otimes |\text{vac}\rangle\langle\text{vac}|)\hat{U}_{\text{bs}}^{\dagger}\right\} \mathcal{D}^{\circ}[\beta]$$

$$= \frac{\mathcal{N}_0}{1+\eta-2S^2\eta}\exp\left(-2\|\alpha\|^2+\frac{4C^2\eta\alpha^*\diamond FF^*\diamond\alpha}{1+\eta-2S^2\eta}\right). \tag{6.249}$$

We can now substitute $S^2 = L$ and $C^2 = 1 - L$, leading to the same expression obtained in (6.244). The result is a generating function for the mixed state that is produced when a fixed-spectrum Fock state suffers a homogeneous loss.

6.7.2 Inhomogeneous loss

A situation that is often found in practical optical systems is where photons are lost depending on their spatiotemporal or spin degrees of freedom. A typical example is when photons must pass through an aperture, as shown in Figure 6.6. The probability for a photon to pass through such an aperture is determined by the integrated probability distribution of the photon over the area of the aperture, which in turn depends on the spatial properties of that photon. We compute this probability with an overlap between the parameter function of the photon and the transmission function of the aperture. Such an overlap implies that the inhomogeneous loss process is governed by a projection process. Therefore, we refer to it as a *projective loss*.

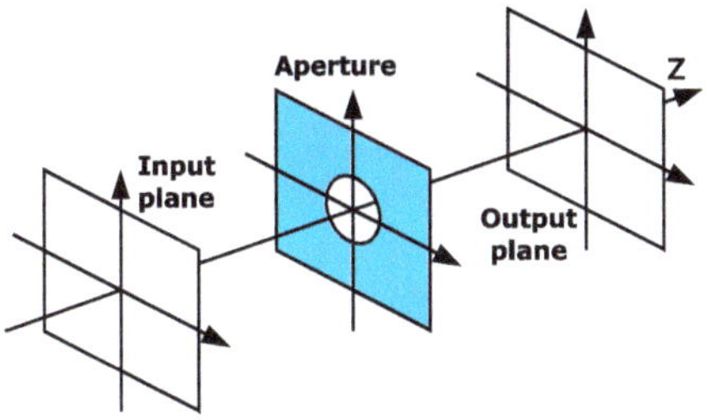

Figure 6.6: Free space propagation system with an aperture.

While apertures are ubiquitous in optical systems, they are not the only devices that produce losses based on the structure of the optical components. In general, the loss of a photon in such scenarios is determined by a projection applied on the photon's parameter function, giving the probability for it to be directed into the loss channel.

6.7.2.1 Projective loss suffered by a Fock state

As an example, we apply the projective beamsplitter to consider loss suffered by a Fock state passing through an aperture. The input state to the projective beamsplitter is given by the tensor product of a Fock state, represented by the one-parameter generating function for the Wigner functionals of Fock states (6.52), and a vacuum state. So, the Wigner functional for the combined state is

$$\mathcal{W}(\eta) = \frac{\mathcal{N}_0}{1+\eta} \exp\left(-2\|\alpha\|^2 + \frac{4\eta\alpha^* \diamond FF^* \diamond \alpha}{1+\eta} - 2\|\beta\|^2\right). \tag{6.250}$$

Next, we apply the transformation given in (6.232)

$$\begin{aligned}\mathcal{W}(\eta) \to \frac{\mathcal{N}_0}{1+\eta} \exp\Big(&-2\alpha^* \diamond \alpha + \frac{4\eta}{1+\eta}\alpha^* \diamond P \diamond FF^* \diamond P \diamond \alpha \\ &+ \mathrm{i}\frac{4\eta}{1+\eta}\alpha^* \diamond P \diamond FF^* \diamond Q \diamond \beta - \mathrm{i}\frac{4\eta}{1+\eta}\beta^* \diamond Q \diamond FF^* \diamond P \diamond \alpha \\ &- 2\beta^* \diamond \beta + \frac{4\eta}{1+\eta}\beta^* \diamond Q \diamond FF^* \diamond Q \diamond \beta\Big).\end{aligned} \tag{6.251}$$

The integral over β performs the trace over the lost degrees of freedom. For this purpose, we need the inverse (see Appendix D)

$$\left(\mathbf{1} - \frac{2\eta}{1+\eta}Q \diamond FF^* \diamond Q\right)^{-1} = \mathbf{1} + \frac{2\eta Q \diamond FF^* \diamond Q}{1+\eta-2\eta L}, \tag{6.252}$$

where the loss is now given by

$$L \triangleq F^* \diamond Q \diamond F = 1 - F^* \diamond P \diamond F, \tag{6.253}$$

representing the probability for a single photon to be blocked by the aperture. We also need to evaluate the determinant

$$\det\left\{\mathbf{1} - \frac{2\eta}{1+\eta}Q \diamond FF^* \diamond Q\right\} = 1 - \frac{2\eta L}{1+\eta}. \tag{6.254}$$

So, the integral over β becomes

$$\begin{aligned}\mathcal{W}'(\eta) = {} & \frac{\mathcal{N}_0}{1+\eta} \exp\left(-2\alpha^* \diamond \alpha + \frac{4\eta}{1+\eta}\alpha^* \diamond P \diamond FF^* \diamond P \diamond \alpha\right) \\ & \times \int \exp\Big[-2\beta^* \diamond \left(\mathbf{1} - \frac{2\eta}{1+\eta}Q \diamond FF^* \diamond Q\right) \diamond \beta \\ & + \mathrm{i}\frac{4\eta}{1+\eta}\left(\alpha^* \diamond P \diamond FF^* \diamond Q \diamond \beta - \beta^* \diamond Q \diamond FF^* \diamond P \diamond \alpha\right)\Big]\, \mathcal{D}^\circ[\beta] \\ = {} & \frac{\mathcal{N}_0 \exp(-2\alpha^* \diamond \alpha)}{1+\eta-2\eta L} \exp\left(\frac{4\eta\alpha^* \diamond P \diamond FF^* \diamond P \diamond \alpha}{1+\eta-2\eta L}\right).\end{aligned} \tag{6.255}$$

Now we just need to recognize that the projected parameter function $P \diamond F$ needs to be normalized. We define the normalized function as

$$G = \frac{P \diamond F}{\sqrt{F^* \diamond P \diamond F}} = \frac{P \diamond F}{\sqrt{1-L}}. \tag{6.256}$$

Then the expression becomes

$$\mathcal{W}'(\eta) = \frac{\mathcal{N}_0}{1+\eta-2\eta L}\exp\left[-2\|\alpha\|^2 + \frac{4\eta(1-L)\|G^* \diamond \alpha\|^2}{1+\eta-2\eta L}\right], \tag{6.257}$$

which is similar to what we obtained for homogeneous loss in (6.244). The difference is that the parameter function of the Fock states is affected by the shape of the aperture.

6.7.3 Loss as a continuous process

There are not many scenarios where loss occurs as a continuous process that becomes so complicated that it justifies the use of a Wigner functional calculation. In the analysis to follow, we provide a simple demonstration of the computation process. As such, it informs similar analyses that we encounter later. However, the *spatiotemporal degrees of freedom* do not play any role in this loss process. It therefore only involves the *particle-number degree of freedom.*

A potential scenario where the spatiotemporal degrees of freedom may play a role when loss acts as a continuous process is in *structured media*, such as *photonic crystals* [14]. However, the free propagation of photons (without loss) through such a structure is not described by the free space propagation considered in Section 6.4.8. It needs a separate careful analysis before the effects imposed by a continuous inhomogeneous loss can be analysed. Such a comprehensive analysis is not easily done generically. It needs the details of the physical scenario under investigation. Nevertheless, the procedure provided here can assist in performing such an analysis.

Often loss is caused by a medium through which a photonic state propagates. The medium may introduce scattering or absorption that removes photons from the state. In such a case, *the probability for the loss of a photon is proportional to the propagation distance through the medium.* The loss is modelled as a hypothetical beamsplitter with a power reflectivity *per unit length*, so that

$$S(z) = \sqrt{\zeta z} \quad \text{and} \quad C(z) = \sqrt{1-\zeta z}. \tag{6.258}$$

However, by modelling it in terms of the square root of z, we inadvertently assume that z is positive. It presents a problem; for negative z the model does not make sense. To avoid the issue, we expand the equation to the second order in the angle φ that determines the splitting ratio. Over a short propagation distance, this angle is small so that

$$S(\varphi) = \sin(\varphi) \approx \varphi \quad \text{and} \quad C(\varphi) = \cos(\varphi) \approx 1 - \frac{1}{2}\varphi^2. \tag{6.259}$$

The transformation for the beamsplitter in (6.218) can then be expressed by expanding S and C in terms of φ as

$$\alpha \to \left(1-\frac{1}{2}\varphi^2\right)\alpha + \mathrm{i}\varphi\beta, \quad \alpha^* \to \left(1-\frac{1}{2}\varphi^2\right)\alpha^* - \mathrm{i}\varphi\beta^*,$$

$$\beta \to \left(1 - \frac{1}{2}\varphi^2\right)\beta + i\varphi\alpha, \quad \beta^* \to \left(1 - \frac{1}{2}\varphi^2\right)\beta^* - i\varphi\alpha^*. \tag{6.260}$$

The transformation of the Wigner functional for the combined state reads

$$W_{\hat{\rho}}[\alpha]W_{\text{vac}}[\beta] \to W_1(\varphi) = W_{\hat{\rho}}\left[\left(1 - \frac{1}{2}\varphi^2\right)\alpha + i\varphi\beta\right] W_{\text{vac}}\left[\left(1 - \frac{1}{2}\varphi^2\right)\beta + i\varphi\alpha\right]. \tag{6.261}$$

We can now compute the Taylor series expansion of the transformed state to second order in φ around $\varphi = 0$. The reason for expanding it to second order is because the first-order terms vanish when we trace out the photons that are lost. To see why it happens, we first consider only the first-order terms. The Taylor series expansion of the transformed state around $\varphi = 0$ is then represented by

$$\begin{aligned} W_1(\varphi) = W_1(0) - i\varphi \Bigg[&\beta^* \diamond \frac{\delta W_1(0)}{\delta\alpha^*} - \frac{\delta W_1(0)}{\delta\alpha} \diamond \beta \\ &+ \alpha^* \diamond \frac{\delta W_1(0)}{\delta\beta^*} - \frac{\delta W_1(0)}{\delta\beta} \diamond \alpha \Bigg] + \text{higher-order terms}, \end{aligned} \tag{6.262}$$

where $W_1(0) \equiv W_{\hat{\rho}}[\alpha]W_{\text{vac}}[\beta]$. Now we trace over β to remove the photons that are lost. For the first-order terms on the right hand-side, we have

$$\int \beta W_{\text{vac}}[\beta^*, \beta]\, \mathcal{D}^\circ[\beta] = \int \beta^* W_{\text{vac}}[\beta^*, \beta]\, \mathcal{D}^\circ[\beta] = 0, \tag{6.263}$$

because the vacuum state is centred at the origin. Moreover, when the integrand is a total derivative, the result is 0 because the normalized Wigner functional tends to 0 at infinity. Then we have

$$\int \frac{\delta W[\beta]}{\delta\beta^*}\, \mathcal{D}^\circ[\beta] = \int \frac{\delta W[\beta]}{\delta\beta}\, \mathcal{D}^\circ[\beta] = 0. \tag{6.264}$$

So, without the second-order terms, the trace over β leads to the trivial equation

$$\int \frac{dW_1(\varphi)}{d\varphi}\, \mathcal{D}^\circ[\beta] = \frac{dW_{\hat{\rho}}[\alpha](\varphi)}{d\varphi} = 0. \tag{6.265}$$

As a result, we find that all the first-order terms vanish when a functional integration over the field variable β is performed to remove the lost degrees of freedom. Therefore, we need the second-order terms to obtain a non-trivial equation.

The expansion of the transformed state with the second-order terms in φ reads

$$\begin{aligned} W_1(\varphi) = W_1(0) + \text{first-order terms} - \frac{1}{2}\varphi^2 \Bigg[&\alpha \diamond \frac{\delta W_1(0)}{\delta\beta\delta\beta} \diamond \alpha + \alpha^* \diamond \frac{\delta W_1(0)}{\delta\beta^*\delta\beta^*} \diamond \alpha^* \\ &+ \beta \diamond \frac{\delta W_1(0)}{\delta\alpha\delta\alpha} \diamond \beta + \beta^* \diamond \frac{\delta W_1(0)}{\delta\alpha^*\delta\alpha^*} \diamond \beta^* + 2\beta \diamond \frac{\delta W_1(0)}{\delta\alpha\delta\beta} \diamond \alpha \\ &+ 2\beta^* \diamond \frac{\delta W_1(0)}{\delta\alpha^*\delta\beta^*} \diamond \alpha^* - 2\alpha^* \diamond \frac{\delta W_1(0)}{\delta\beta^*\delta\beta} \diamond \alpha - 2\alpha^* \diamond \frac{\delta W_1(0)}{\delta\beta^*\delta\alpha} \diamond \beta \end{aligned}$$

$$
\begin{aligned}
&- 2\beta^* \diamond \frac{\delta W_1(0)}{\delta \alpha^* \delta \beta} \diamond \alpha - 2\beta^* \diamond \frac{\delta W_1(0)}{\delta \alpha^* \delta \alpha} \diamond \beta + \frac{\delta W_1(0)}{\delta \alpha} \diamond \alpha + \alpha^* \diamond \frac{\delta W_1(0)}{\delta \alpha^*} \\
&+ \frac{\delta W_1(0)}{\delta \beta} \diamond \beta + \beta^* \diamond \frac{\delta W_1(0)}{\delta \beta^*} \Bigg] + \text{higher-order terms,}
\end{aligned} \tag{6.266}
$$

where the "first-order terms" are those that we obtained in (6.262).

When we apply the trace over β to remove the photons that are lost, in addition to removing the first-order terms, as we already saw, many of the second-order terms also become zero. For example, the second derivatives with respect to β are zero

$$
\int \frac{\delta^2 W[\beta]}{\delta \beta^* \delta \beta^*} \mathcal{D}^\circ[\beta] = \int \frac{\delta^2 W[\beta]}{\delta \beta \delta \beta^*} \mathcal{D}^\circ[\beta] = \int \frac{\delta^2 W[\beta]}{\delta \beta \delta \beta} \mathcal{D}^\circ[\beta] = 0, \tag{6.267}
$$

for the same reason that the first derivatives are zero, as shown in (6.264).

For the higher-order moments, we can use the *characteristic functional* of the vacuum state as a *generating functional for the moments*

$$
\chi_{\text{vac}}[\mu] = \int W_{\text{vac}}[\beta] \exp\left(\mu^* \diamond \beta - \beta^* \diamond \mu\right) \mathcal{D}^\circ[\beta] = \exp\left(-\frac{1}{2}\mu^* \diamond \mu\right). \tag{6.268}
$$

It then follows that

$$
\int \beta(\mathbf{k}_1)\beta(\mathbf{k}_2) W_{\text{vac}}[\beta] \mathcal{D}^\circ[\beta] = \int \beta^*(\mathbf{k}_1)\beta^*(\mathbf{k}_2) W_{\text{vac}}[\beta] \mathcal{D}^\circ[\beta] = 0. \tag{6.269}
$$

However,

$$
\int \beta^*(\mathbf{k}_1)\beta(\mathbf{k}_2) W_{\text{vac}}[\beta] \mathcal{D}^\circ[\beta] = -\left.\frac{\delta^2 \chi_{\text{vac}}[\mu]}{\delta \mu(\mathbf{k}_1)\mu^*(\mathbf{k}_2)}\right|_{\mu^*=\mu=0} = \frac{1}{2}\mathbf{1}(\mathbf{k}_1, \mathbf{k}_2). \tag{6.270}
$$

Some of the terms contain a contraction to the same field variable with respect to which a *functional derivative* is applied on the Wigner functional. In such cases, we can use *functional integration by parts*. Generically, they become

$$
\begin{aligned}
\int \frac{\delta W[\beta]}{\delta \beta(\mathbf{k}_1)} \beta(\mathbf{k}_2) \mathcal{D}^\circ[\beta] &= -\delta(\mathbf{k}_1 - \mathbf{k}_2), \\
\int \beta^*(\mathbf{k}_1) \frac{\delta W[\beta]}{\delta \beta^*(\mathbf{k}_2)} \mathcal{D}^\circ[\beta] &= -\delta(\mathbf{k}_1 - \mathbf{k}_2),
\end{aligned} \tag{6.271}
$$

assuming that $W[\beta]$ is normalized. These identities lead to

$$
\begin{aligned}
\int \beta \diamond \frac{\delta W_1}{\delta \alpha \delta \beta} \diamond \alpha \mathcal{D}^\circ[\beta] &= -\frac{\delta W_{\hat{\rho}}}{\delta \alpha} \diamond \alpha, \\
\int \beta^* \diamond \frac{\delta W_1}{\delta \alpha^* \delta \beta^*} \diamond \alpha^* \mathcal{D}^\circ[\beta] &= -\alpha^* \diamond \frac{\delta W_{\hat{\rho}}}{\delta \alpha^*}, \\
\int \alpha^* \diamond \frac{\delta W_1}{\delta \beta^* \delta \alpha} \diamond \beta \mathcal{D}^\circ[\beta] = \int \beta^* \diamond \frac{\delta W_1}{\delta \alpha^* \delta \beta} &\diamond \alpha \mathcal{D}^\circ[\beta] = 0,
\end{aligned}
$$

$$\int \frac{\delta W_1}{\delta \beta} \diamond \beta \mathcal{D}^\circ[\beta] = \int \beta^* \diamond \frac{\delta W_1}{\delta \beta^*} \mathcal{D}^\circ[\beta] = -\Omega W_{\hat{\rho}}. \tag{6.272}$$

With the aid of these results, we evaluate the functional integration of (6.266) over β. Most of the terms are removed, leaving us with

$$\begin{aligned} W_{\hat{\rho}}(\varphi) = W_{\hat{\rho}}(0) + \frac{1}{2}\varphi^2 \bigg[& \frac{\delta W_{\hat{\rho}}(0)}{\delta \alpha} \diamond \alpha + \alpha^* \diamond \frac{\delta W_{\hat{\rho}}(0)}{\delta \alpha^*} \\ & + 2\Omega W_{\hat{\rho}}(0) + \frac{\delta}{\delta \alpha^*} \diamond \frac{\delta}{\delta \alpha} W_{\hat{\rho}}(0) \bigg] + \text{higher-order terms.} \end{aligned} \tag{6.273}$$

Converting the φ-dependence into a z-dependence via

$$\varphi^2 \approx \sin^2(\varphi) \to \zeta \Delta z, \tag{6.274}$$

we get

$$W_{\hat{\rho}}(0) \to W_{\hat{\rho}}(z) \quad \text{and} \quad W_{\hat{\rho}}(\varphi) \to W_{\hat{\rho}}(z + \Delta z). \tag{6.275}$$

The quantity ζ, which formally represents the beamsplitter's reflectivity per unit distance, is now interpreted as an *absorption coefficient* (loss per unit distance). Dividing the expression by Δz and taking the limit $\Delta z \to 0$ (which also removes the higher-order terms), we obtain a differential equation

$$\partial_z W_{\hat{\rho}}(z) = \frac{1}{2}\zeta \left[\frac{\delta W_{\hat{\rho}}(z)}{\delta \alpha} \diamond \alpha + \alpha^* \diamond \frac{\delta W_{\hat{\rho}}(z)}{\delta \alpha^*} + 2\Omega W_{\hat{\rho}}(z) + \frac{\delta}{\delta \alpha^*} \diamond \frac{\delta}{\delta \alpha} W_{\hat{\rho}}(z) \right], \tag{6.276}$$

where Ω is the divergent constant defined in (4.154). The result in (6.276) is equivalent to a particle-number degree of freedom only analysis, because the homogenous loss completely decouples from any spatiotemporal effects.

The *evolution equation* in (6.276) does not include the effect of free space propagation. For the inclusion of free space propagation, the two terms on the right-hand side of (6.172) need to be added to the right-hand side of (6.276).

Exercise 6.20. Show that (6.276) is *trace preserving*, by showing that the right-hand side becomes zero when it is integrated over α.

Exercise 6.21. Use (6.276) to compute an evolution equation for the *average number of photons*, given by $N(z) = \text{tr}\{\hat{\rho}(z)\hat{n}\}$, by using the Wigner functional for the number operator in (6.62). Show that it leads to

$$\partial_z N(z) = -\zeta N(z), \tag{6.277}$$

with the expected solution

$$N(z) = N(0) \exp(-\zeta z). \tag{6.278}$$

6.8 Squeezing

The squeezing process is discussed in Section 3.4.2 in the context of the particle-number degree of freedom. When we consider squeezing operators with all the degrees of freedom included, the squeezing parameter becomes a *squeezing kernel*. Most of the properties of the squeezing process remain the same, including its association with the *Bogoliubov transformation*.

Here, we compute the Wigner functional for the *squeezing operator*. It is applied to an arbitrary state, showing that it comes down to a Bogoliubov transformation of the Wigner functional's arguments. When applied to the Wigner functional of a vacuum state, such a Bogoliubov transformation produces the *squeezed vacuum state*, which is relevant for spontaneous parametric down-conversion, as discussed in Chapter 8.

In the absence of spatiotemporal and spin degrees of freedom, the most general form of the squeezing operator is given in (3.167), in terms of the complex-valued *squeezing parameter* ξ. When we incorporate the spatiotemporal degrees of freedom, the squeezing operator is generalized to

$$\hat{S} = \exp\left(\frac{1}{2}\hat{a} \diamond \xi^{\dagger} \diamond \hat{a} - \frac{1}{2}\hat{a}^{\dagger} \diamond \xi \diamond \hat{a}^{\dagger}\right), \tag{6.279}$$

where $\xi(\mathbf{k}_1, \mathbf{k}_2)$ now represents a *squeezing kernel*. Due to the symmetry of the expression, the squeezing kernel is symmetric, so that $\xi^T = \xi$ and $\xi^{\dagger} = \xi^*$.

6.8.1 Wigner functional of the squeezing operator

We use the coherent-state-assisted approach to compute the Wigner functional of the squeezing operator. For this purpose, the squeezing operator in (6.279) must be converted into a normal-ordered product of operators. It requires two complicated steps. The first step is done with a process similar to what we used in Section 4.5.3. To simplify the subsequent expressions, we define the notation

$$\begin{aligned}\hat{N}[g] &\triangleq \hat{a}^{\dagger} \diamond g \diamond \hat{a},\\ \hat{M}[h^*] &\triangleq \hat{a} \diamond h^* \diamond \hat{a},\\ \hat{M}^{\dagger}[h] &\triangleq \hat{a}^{\dagger} \diamond h \diamond \hat{a}^{\dagger},\end{aligned} \tag{6.280}$$

where g, h, and h^* are arbitrary kernels, but h and h^* are symmetric. The first separation of the squeezing operator into a product of exponentiated operators is expressed as

$$\begin{aligned}&\exp\left(\frac{1}{2}t\hat{M}[\xi^*] - \frac{1}{2}t\hat{M}^{\dagger}[\xi]\right)\\ &= \exp\left[h_0(t)\right] \exp\left(\hat{M}^{\dagger}[h_1(t)]\right) \exp\left(\hat{N}[h_2(t)]\right) \exp\left(\hat{M}[h_3(t)]\right),\end{aligned} \tag{6.281}$$

where t is an auxiliary variable. The unknown function $h_0(t)$ and the unknown kernel functions $h_1(t)$, $h_2(t)$, and $h_3(t)$ all vanish at $t = 0$. Since a term of the form $\hat{a} \diamond K \diamond \hat{a}^\dagger$ can be converted into one with the form $\hat{N}[K^T]$ via the commutation relations, it is not represented in the exponents on the right-hand side. From the form of the expression, we know that ξ, h_1, and h_3 are symmetric.

Taking a derivative with respect to t and removing as many of the exponentiated operators as possible, we get

$$\begin{aligned}\left[\partial_t \hat{S}(t)\right] \hat{S}^\dagger(t) = {} & \frac{1}{2}\hat{M}[\xi^*] - \frac{1}{2}\hat{M}^\dagger[\xi] \\ = {} & \partial_t h_0(t) + \hat{M}^\dagger[\partial_t h_1(t)] + \exp\left(\hat{M}^\dagger[h_1(t)]\right)\hat{N}[\partial_t h_2(t)]\exp\left(-\hat{M}^\dagger[h_1(t)]\right) \\ & + \exp\left(\hat{M}^\dagger[h_1(t)]\right)\exp\left(\hat{N}[h_2(t)]\right)\hat{M}[\partial_t h_3(t)] \\ & \times \exp\left(-\hat{N}[h_2(t)]\right)\exp\left(-\hat{M}^\dagger[h_1(t)]\right). \end{aligned} \tag{6.282}$$

The products of exponentiated operators can be simplified with the aid of (3.103). The *commutations* we need for this purpose are those among the operators defined in (6.280). They are readily shown to produce

$$\begin{aligned} [\hat{N}[g], \hat{M}^\dagger[h]] &= 2\hat{M}^\dagger[g \diamond h], \\ [\hat{M}[h^*], \hat{N}[g]] &= 2\hat{M}[h^* \diamond g], \\ [\hat{M}[h^*], \hat{M}^\dagger[h]] &= 4\hat{N}[h \diamond h^*] + 2\,\mathrm{tr}\{h \diamond h^*\}. \end{aligned} \tag{6.283}$$

We use these commutation relations, together with the identity in (3.103), to compute

$$\begin{aligned} \exp(\hat{M}^\dagger[h])\hat{N}[g]\exp(-\hat{M}^\dagger[h]) &= \hat{N}[g] - 2\hat{M}^\dagger[g \diamond h], \\ \exp(\hat{N}[g])\hat{M}[h^*]\exp(-\hat{N}[g]) &= \hat{M}[h^* \diamond \exp_\diamond(-2g)], \\ \exp(\hat{M}^\dagger[h])\hat{M}[h^*]\exp(-\hat{M}^\dagger[h]) &= \hat{M}[h^*] - 4\hat{N}[h \diamond h^*] - 2\,\mathrm{tr}\{h \diamond h^*\} \\ &\quad + 4\hat{M}^\dagger[h \diamond h^* \diamond h]. \end{aligned} \tag{6.284}$$

They are used to simplify the equation in (6.282), leading to

$$\begin{aligned} \frac{1}{2}\hat{M}[\xi^*] - \frac{1}{2}\hat{M}^\dagger[\xi] = {} & \partial_t h_0 + \hat{M}^\dagger[\partial_t h_1] + \hat{N}[\partial_t h_2] - 2\hat{M}^\dagger[(\partial_t h_2) \diamond h_1] \\ & + \hat{M}[(\partial_t h_3) \diamond \exp_\diamond(-2h_2)] - 4\hat{N}[h_1 \diamond (\partial_t h_3) \diamond \exp_\diamond(-2h_2)] \\ & - 2\,\mathrm{tr}\{h_1 \diamond (\partial_t h_3) \diamond \exp_\diamond(-2h_2)\} \\ & + 4\hat{M}^\dagger[h_1 \diamond (\partial_t h_3) \diamond \exp_\diamond(-2h_2) \diamond h_1]. \end{aligned} \tag{6.285}$$

Based on the type of operator or whether we have a constant, this equation can be separated into four equations

$$
\begin{aligned}
0 &= \partial_t h_0 - 2\,\mathrm{tr}\{h_1 \diamond (\partial_t h_3) \diamond \exp_\diamond(-2h_2)\}, \\
0 &= \partial_t h_2 - 4h_1 \diamond (\partial_t h_3) \diamond \exp_\diamond(-2h_2), \\
\frac{1}{2}\xi^* &= (\partial_t h_3) \diamond \exp_\diamond(-2h_2), \\
-\frac{1}{2}\xi &= \partial_t h_1 - 2(\partial_t h_2) \diamond h_1 + 4h_1 \diamond (\partial_t h_3) \diamond \exp_\diamond(-2h_2) \diamond h_1.
\end{aligned} \tag{6.286}
$$

With some simplification, they lead to separate differential equations for the four unknown functions

$$
\begin{aligned}
\partial_t h_0 &= \mathrm{tr}\{h_1 \diamond \xi^*\}, \quad \partial_t h_1 = 2h_1 \diamond \xi^* \diamond h_1 - \frac{1}{2}\xi, \\
\partial_t h_2 &= 2h_1 \diamond \xi^*, \quad \partial_t h_3 = \frac{1}{2}\xi^* \diamond \exp_\diamond(2h_2).
\end{aligned} \tag{6.287}
$$

As a guide for the solutions of these differential equations, we consider the case when the unknown functions only depend on t. The simpler one-dimensional versions of these differential equations are

$$
\begin{aligned}
\partial_t H_1(t) &= 2\xi_0^* H_1^2(t) - \frac{1}{2}\xi_0, \\
\partial_t H_2(t) = 2\partial_t H_0(t) &= 2\xi_0^* H_1(t), \\
\partial_t H_3(t) &= \frac{1}{2}\xi_0^* \exp\left[2H_2(t)\right],
\end{aligned} \tag{6.288}
$$

where $\xi_0 = |\xi_0| \exp(i\theta)$ is a complex constant, representing the squeezing parameter. The solutions of these simpler differential equations are

$$
\begin{aligned}
H_1(t) &= -\frac{1}{2}\exp(i\theta)\tanh(|\xi_0|t), \\
H_2(t) &= -\ln\left[\cosh(|\xi_0|t)\right], \\
H_3(t) &= \frac{1}{2}\exp(-i\theta)\tanh(|\xi_0|t).
\end{aligned} \tag{6.289}
$$

To solve the differential equations in (6.287), we start by integrating them to remove the derivatives. Since all their initial conditions are zero, the integral equations are

$$
\begin{aligned}
h_0(t) &= \int_0^t \mathrm{tr}\{h_1(t') \diamond \xi^*\}\,\mathrm{d}t', \quad h_1(t) = \int_0^t 2h_1(t') \diamond \xi^* \diamond h_1(t')\,\mathrm{d}t' - \frac{1}{2}t\xi, \\
h_2(t) &= \int_0^t 2h_1(t') \diamond \xi^*\,\mathrm{d}t', \quad h_3(t) = \int_0^t \frac{1}{2}\xi^* \diamond \exp_\diamond[2h_2(t')]\,\mathrm{d}t'.
\end{aligned} \tag{6.290}
$$

These equations can now be solved in sequence by repeated back substitution, starting with the second equation which only contains $h_1(t)$ as an unknown.

After the integral for $h_1(t)$ is solved through repeated back substitution, the integrals over the t's are evaluated. Up to the 9th order, the resulting expression for h_1 is

$$\begin{aligned} h_1(t) = & -\frac{1}{2}t\xi + \frac{1}{6}t^3\xi \diamond \xi^* \diamond \xi - \frac{1}{15}t^5\xi \diamond \xi^* \diamond \xi \diamond \xi^* \diamond \xi \\ & + \frac{17}{630}t^7\xi \diamond \xi^* \diamond \xi \diamond \xi^* \diamond \xi \diamond \xi^* \diamond \xi \\ & - \frac{31}{2835}t^9\xi \diamond \xi^* \diamond \xi \diamond \xi^* \diamond \xi \diamond \xi^* \diamond \xi \diamond \xi^* \diamond \xi + \cdots . \end{aligned} \tag{6.291}$$

Hence, $h_1(t)$ is symmetric because ξ is symmetric. This expansion corresponds to the expansion of H_1 in (6.289). Therefore, we formally express the full wavevector-dependent $h_1(t)$ in analogy to (6.289) as

$$h_1(t) = -\frac{1}{2}\tanh_\diamond(t\xi), \tag{6.292}$$

with the understanding that the expansion of $\tanh_\diamond(K)$ with an arbitrary kernel K is given by −2 times the expansion in (6.291) with $\xi \to K$ and $t = 1$.

Next, we consider the third equation in (6.290), which is the differential equation for h_2. Its solution follows directly from the solution of $h_1(t)$. Substituting (6.291) into the differential equation for h_2 and integrating it over t, we get

$$\begin{aligned} h_2(t) = & -\frac{1}{2}t^2\xi \diamond \xi^* + \frac{1}{12}t^4\xi \diamond \xi^* \diamond \xi \diamond \xi^* - \frac{1}{45}t^6\xi \diamond \xi^* \diamond \xi \diamond \xi^* \diamond \xi \diamond \xi^* \\ & + \frac{17}{2520}t^8\xi \diamond \xi^* \diamond \xi \diamond \xi^* \diamond \xi \diamond \xi^* \diamond \xi \diamond \xi^* \\ & - \frac{31}{141750}t^{10}\xi \diamond \xi^* \diamond \xi \diamond \xi^* \diamond \xi \diamond \xi^* \diamond \xi \diamond \xi^* \diamond \xi \diamond \xi^* + \cdots . \end{aligned} \tag{6.293}$$

It follows that $h_2(t)$ is Hermitian $h_2^\dagger(t) = h_2(t)$.[g] Again, the expansion agrees with that which is obtained from the one-dimensional solution for $H_2(t)$. Formally, we express the wavevector-dependent kernel function $h_2(t)$ as

$$h_2(t) = -\ln_\diamond\left[\cosh_\diamond(t\xi)\right]. \tag{6.294}$$

It then follows directly from the expression for $h_2(t)$ that

$$h_0(t) = -\frac{1}{2}\operatorname{tr}\left\{\ln_\diamond\left[\cosh_\diamond(t\xi)\right]\right\} \tag{6.295}$$

is the solution for $h_0(t)$ from the first equation in (6.290).

g We can interchange $\xi \leftrightarrow \xi^*$ leading to an expression that is not in general equal to the one in (6.293).

The solution of the last differential equations in (6.290) for $h_3(t)$ is obtained by expanding its right-hand side after substituting (6.293) into it and integrating the result over t. It reads

$$\begin{aligned} h_3(t) = {} & \frac{1}{2}t\xi^* - \frac{1}{6}t^3\xi^* \diamond \xi \diamond \xi^* + \frac{1}{15}t^5\xi^* \diamond \xi \diamond \xi^* \diamond \xi \diamond \xi^* \\ & - \frac{17}{630}t^7\xi^* \diamond \xi \diamond \xi^* \diamond \xi \diamond \xi^* \diamond \xi \diamond \xi^* \\ & + \frac{31}{2835}t^9\xi^* \diamond \xi \diamond \xi^* \diamond \xi \diamond \xi^* \diamond \xi \diamond \xi^* \diamond \xi \diamond \xi^* + \cdots , \end{aligned} \quad (6.296)$$

which turns out to be minus the complex conjugate (or Hermitian adjoint) of $h_1(t)$. Therefore, we formally express it as

$$h_3(t) = \frac{1}{2}\tanh_\diamond(\xi^* t) \equiv -h_1^*(t) = -h_1^\dagger(t). \quad (6.297)$$

Although we represent these solutions formally by the expressions in terms of the hyperbolic trigonometric kernel functions, their proper definitions are provided by the expansions. The formal expressions are used in the subsequent derivation simply because they lead to simpler expressions.

If ξ can be expressed in terms of a *polar decomposition* [5], given by

$$\xi = \exp_\diamond(\mathrm{i}\theta) \diamond |\xi| \quad \text{and} \quad \xi^* = \exp_\diamond(-\mathrm{i}\theta) \diamond |\xi|, \quad (6.298)$$

the expressions for the kernels and the function would become

$$\begin{aligned} h_0(t) &= -\frac{1}{2}\operatorname{tr}\{\ln_\diamond[\cosh_\diamond(t|\xi|)]\}, \\ h_1(t) &= -\frac{1}{2}\exp_\diamond(\mathrm{i}\theta) \diamond \tanh_\diamond(t|\xi|), \\ h_2(t) &= -\ln_\diamond[\cosh_\diamond(t|\xi|)], \\ h_3(t) &= \frac{1}{2}\exp_\diamond(-\mathrm{i}\theta) \diamond \tanh_\diamond(t|\xi|). \end{aligned} \quad (6.299)$$

The expressions for $h_0(t)$ and $h_2(t)$ would then be real valued. However, such a polar decomposition may not exist if ξ is not *normal* (i. e., $\xi \diamond \xi^* \neq \xi^* \diamond \xi$).

The result of the first step in the normal-ordering process is obtained by substituting the solutions for these h-functions into (6.281) and setting $t = 1$. The result reads

$$\begin{aligned} \hat{S} &= \exp(h_0)\exp\left(\hat{a}^\dagger \diamond h_1 \diamond \hat{a}^\dagger\right)\exp\left(\hat{a}^\dagger \diamond h_2 \diamond \hat{a}\right)\exp(\hat{a} \diamond h_3 \diamond \hat{a}) \\ &= [\det\{\cosh_\diamond(\xi)\}]^{-1/2}\exp\left[-\frac{1}{2}\hat{a}^\dagger \diamond \tanh_\diamond(\xi) \diamond \hat{a}^\dagger\right] \\ &\quad \times \exp\left\{-\hat{a}^\dagger \diamond \ln_\diamond[\cosh_\diamond(\xi)] \diamond \hat{a}\right\}\exp\left[\frac{1}{2}\hat{a} \diamond \tanh_\diamond(\xi^*) \diamond \hat{a}\right]. \end{aligned} \quad (6.300)$$

The second step is to convert the operator with h_2 into normal order. This step is already done in Section 6.4.3. It gives

$$\exp\left(\hat{a}^{\dagger}\diamond h_2\diamond\hat{a}\right)=\sum_{m=0}^{\infty}\frac{1}{m!}\int\left[\prod_{r=1}^{m}\hat{a}^{\dagger}(\mathbf{k}_r)\right]\left[\prod_{r=1}^{m}(h_4-\mathbf{1})\right]\times\left[\prod_{r=1}^{m}\hat{a}(\mathbf{q}_r)\right]\prod_{r=1}^{m}\mathrm{d}_{\omega}k_r\mathrm{d}_{\omega}q_r, \tag{6.301}$$

where

$$h_4\triangleq\exp_{\diamond}(h_2)\equiv\left[\cosh_{\diamond}(\xi)\right]^{-1}. \tag{6.302}$$

Proceeding with the calculation of the Wigner functional using the coherent-state-assisted approach, we evaluate the overlap of the squeezing operator by two coherent states. It gives

$$\begin{aligned}\langle\alpha_1|\,\hat{S}\,|\alpha_2\rangle=&\,\mathcal{N}_c\,\langle\alpha_1|\exp\left(\hat{a}^{\dagger}\diamond h_1\diamond\hat{a}^{\dagger}\right)\left\{\sum_{m=0}^{\infty}\frac{1}{m!}\int\left[\prod_{r=1}^{m}\hat{a}^{\dagger}(\mathbf{k}_r)\right]\left[\prod_{r=1}^{m}(h_4-\mathbf{1})\right]\right.\\&\times\left.\left[\prod_{r=1}^{m}\hat{a}(\mathbf{q}_r)\right]\prod_{r=1}^{m}\mathrm{d}_{\omega}k_r\mathrm{d}_{\omega}q_r\right\}\exp\left(\hat{a}\diamond h_3\diamond\hat{a}\right)|\alpha_2\rangle\\=&\,\mathcal{N}_c\exp\left(-\frac{1}{2}\|\alpha_1\|^2-\frac{1}{2}\|\alpha_2\|^2+\alpha_1^*\diamond h_4\diamond\alpha_2\right.\\&\left.+\alpha_1^*\diamond h_1\diamond\alpha_1^*+\alpha_2\diamond h_3\diamond\alpha_2\right),\end{aligned} \tag{6.303}$$

where we defined

$$\mathcal{N}_c\triangleq\exp\left(h_0\right)\equiv\frac{1}{\sqrt{\det\left\{\cosh_{\diamond}(\xi)\right\}}}. \tag{6.304}$$

Next, we substitute (6.303) into (6.20), leading to

$$\begin{aligned}\text{Wigner}\{\hat{S}\}=\mathcal{N}_0\mathcal{N}_c\int&\exp\left(-2\alpha^*\diamond\alpha+2\alpha^*\diamond\alpha_1+2\alpha_2^*\diamond\alpha\right.\\&-\alpha_1^*\diamond\alpha_1-\alpha_2^*\diamond\alpha_2-\alpha_2^*\diamond\alpha_1+\alpha_1^*\diamond h_4\diamond\alpha_2\\&\left.+\alpha_1^*\diamond h_1\diamond\alpha_1^*+\alpha_2\diamond h_3\diamond\alpha_2\right)\,\mathcal{D}^{\circ}[\alpha_1,\alpha_2].\end{aligned} \tag{6.305}$$

Both functional integrals are in the form of anisotropic Gaussian functional integrals, as considered in Appendix C.4. First, we evaluate the integral over α_2, for which we substitute $A\rightarrow\frac{1}{2}\mathbf{1}$, $B\rightarrow 0$, $B^*\rightarrow -h_3$, $F\rightarrow\alpha_1-2\alpha$, and $F^*\rightarrow-\alpha_1^*\diamond h_4$ into (C.30). The result of this functional integration is

$$
\begin{aligned}
\text{Wigner}\{\hat{S}\} &= \mathcal{N}_0\mathcal{N}_c \int \exp\left(-2\alpha^* \diamond \alpha - \alpha_1^* \diamond \alpha_1 + 2\alpha^* \diamond \alpha_1 + \alpha_1^* \diamond h_1 \diamond \alpha_1^*\right) \\
&\quad \times \exp\left[\alpha_1^* \diamond h_4 \diamond (2\alpha - \alpha_1) + (2\alpha - \alpha_1) \diamond h_3 \diamond (2\alpha - \alpha_1)\right] \mathcal{D}^\circ[\alpha_1] \\
&= \mathcal{N}_0\mathcal{N}_c \exp\left(-2\alpha^* \diamond \alpha + 4\alpha \diamond h_3 \diamond \alpha\right) \\
&\quad \times \int \exp\left[-\alpha_1^* \diamond (\mathbf{1} + h_4) \diamond \alpha_1 + \alpha_1 \diamond h_3 \diamond \alpha_1 + \alpha_1^* \diamond h_1 \diamond \alpha_1^*\right. \\
&\quad \left. + 2\left(\alpha^* - 2\alpha \diamond h_3\right) \diamond \alpha_1 + 2\alpha_1^* \diamond h_4 \diamond \alpha\right] \mathcal{D}^\circ[\alpha_1],
\end{aligned}
\tag{6.306}
$$

where we used the fact that $h_3^T = h_3$. For the integration over α_1, we then substitute $A \to \frac{1}{2}(\mathbf{1} + h_4)$, $B \to -h_1$, $B^* \to -h_3$, $F \to -2h_4 \diamond \alpha$, and $F^* \to -2(\alpha^* - 2\alpha \diamond h_3)$ into (C.30) to get the result

$$
\begin{aligned}
\text{Wigner}\{\hat{S}\} &= \frac{\mathcal{N}_0\mathcal{N}_c \exp\left(-2\alpha^* \diamond \alpha + 4\alpha \diamond h_3 \diamond \alpha\right)}{\sqrt{\det\{\mathbf{1} + h_4\} \det\{\mathbf{1} + h_4^* - 4h_3 \diamond (\mathbf{1} + h_4)^{-1} \diamond h_1\}}} \\
&\quad \times \exp\left\{2\left(\alpha^* - 2\alpha \diamond h_3\right) \diamond (\mathbf{1} + h_4)^{-1} \diamond h_4 \diamond \alpha\right. \\
&\quad + 2\left[\alpha \diamond h_4^* + 2\left(\alpha^* - 2\alpha \diamond h_3\right) \diamond (\mathbf{1} + h_4)^{-1} \diamond h_1\right] \\
&\quad \diamond \left[\mathbf{1} + h_4^* - 4h_3 \diamond (\mathbf{1} + h_4)^{-1} \diamond h_1\right]^{-1} \\
&\quad \left. \diamond \left[\alpha^* - 2h_3 \diamond \alpha + 2h_3 \diamond (\mathbf{1} + h_4)^{-1} \diamond h_4 \diamond \alpha\right]\right\},
\end{aligned}
\tag{6.307}
$$

using $h_4^T = h_4^*$. We use the properties of the h-kernels to simplify this expression. Due to the way h_1, h_3, and h_4 are constructed in terms of ξ and ξ^*, it follows that

$$
\begin{aligned}
h_1 \diamond \xi^* &= \xi \diamond h_1^* = -\xi \diamond h_3, \\
\xi^* \diamond h_1 &= h_1^* \diamond \xi = -h_3 \diamond \xi, \\
h_4 \diamond \xi &= \xi \diamond h_4^*, \\
(\mathbf{1} + h_4)^{-1} \diamond h_1 &= h_1 \diamond \left(\mathbf{1} + h_4^*\right)^{-1}, \\
h_3 \diamond (\mathbf{1} + h_4)^{-1} &= \left(\mathbf{1} + h_4^*\right)^{-1} \diamond h_3.
\end{aligned}
\tag{6.308}
$$

Moreover, the differential equations for $h_1(t)$ and $h_3(t)$ in (6.287), together with the definition of $h_4(t)$ in (6.302) and the fact that $h_1^*(t) = -h_3(t)$, lead to the following equations

$$
\begin{aligned}
\partial_t h_1(t) &= 2h_1(t) \diamond \xi^* \diamond h_1(t) - \frac{1}{2}\xi = -2\xi \diamond h_3(t) \diamond h_1(t) - \frac{1}{2}\xi \\
&= -2h_1(t) \diamond h_3(t) \diamond \xi - \frac{1}{2}\xi, \\
\partial_t h_1(t) &= -\partial_t h_3^*(t) = -\frac{1}{2}\xi \diamond h_4^*(t) \diamond h_4^*(t) = -\frac{1}{2}h_4(t) \diamond h_4(t) \diamond \xi.
\end{aligned}
\tag{6.309}
$$

From these two equations for $h_1(t)$, it then follows that

$$
4h_3(t) \diamond h_1(t) = h_4^*(t) \diamond h_4^*(t) - \mathbf{1},
$$

$$4h_1(t) \diamond h_3(t) = h_4(t) \diamond h_4(t) - \mathbf{1}. \tag{6.310}$$

The argument of the inverse in (6.307) thus becomes

$$\begin{aligned}
&\mathbf{1} + h_4^* - 4h_3 \diamond (\mathbf{1} + h_4)^{-1} \diamond h_1 \\
&= \mathbf{1} + h_4^* - 4h_3 \diamond h_1 \diamond (\mathbf{1} + h_4^*)^{-1} \\
&= (\mathbf{1} + h_4^*)^{-1} \diamond [(\mathbf{1} + h_4^*) \diamond (\mathbf{1} + h_4^*) - h_4^* \diamond h_4^* + \mathbf{1}] = 2\mathbf{1}.
\end{aligned} \tag{6.311}$$

Using these relationships, together with

$$(\mathbf{1} + h_4)^{-1} \diamond h_4 = \mathbf{1} - (\mathbf{1} + h_4)^{-1}, \tag{6.312}$$

we simplify the expression in (6.307) to get

$$\begin{aligned}
\text{Wigner}\{\hat{S}\} = \frac{\mathcal{N}_c \mathcal{N}_0}{\sqrt{\det\{\mathbf{1} + h_4\}\mathcal{N}_0}} & \exp\left[2\alpha^* \diamond (\mathbf{1} + h_4)^{-1} \diamond h_1 \diamond \alpha^* \right. \\
& \left. - 2\alpha \diamond (\mathbf{1} + h_4^*)^{-1} \diamond h_1^* \diamond \alpha\right].
\end{aligned} \tag{6.313}$$

The Wigner functional of the squeezing operator can then be expressed as

$$\text{Wigner}\{\hat{S}\} = \mathcal{N}_\tau \exp\left(\alpha \diamond \tau^* \diamond \alpha - \alpha^* \diamond \tau \diamond \alpha^*\right), \tag{6.314}$$

where

$$\begin{aligned}
\mathcal{N}_\tau &= \sqrt{\frac{\mathcal{N}_0 \det\{h_4\}}{\det\{\mathbf{1} + h_4\}}} = \frac{1}{\det\{\cosh_\diamond\left(\frac{1}{2}\xi\right)\}}, \\
\tau &= -2\,(\mathbf{1} + h_4)^{-1} \diamond h_1 = \tanh_\diamond\left(\frac{1}{2}\xi\right), \\
\tau^* &= -2\,(\mathbf{1} + h_4^*)^{-1} \diamond h_1^* = \tanh_\diamond\left(\frac{1}{2}\xi^*\right).
\end{aligned} \tag{6.315}$$

For the benefit of subsequent discussions, we also note that

$$\tau \diamond \tau^* = -(h_4 - \mathbf{1}) \diamond (\mathbf{1} + h_4)^{-1} = 2\,(\mathbf{1} + h_4)^{-1} - \mathbf{1}. \tag{6.316}$$

Hence

$$\mathbf{1} - \tau \diamond \tau^* = 2h_4 \diamond (\mathbf{1} + h_4)^{-1} \quad \text{and} \quad \mathbf{1} + \tau \diamond \tau^* = 2\,(\mathbf{1} + h_4)^{-1}. \tag{6.317}$$

Unlike the expressions in terms of the functional hyperbolic trigonometric functions, those in terms of the h-kernels are generally valid, because they follow from the symmetry properties of these kernels and their differential equations.

6.8.2 Squeezing of an arbitrary state

The Wigner functional for the squeezing operator in (6.314) is used to apply a squeezing process on an arbitrary state. Using the expression for the triple star product in (5.49), with some simplifications, we obtain

$$\begin{aligned}\text{Wigner}\{\hat{S}\hat{\rho}\hat{S}^{\dagger}\} = \mathcal{N}_{\tau}^{2} \int W_{\hat{\rho}}\,[\alpha_a] \exp\,[(\alpha_a \diamond \tau^* - \alpha_a^* + \alpha \diamond \tau^* + \alpha^*) \diamond \alpha_b \\ - \alpha_b^* \diamond (\tau \diamond \alpha_a^* - \alpha_a + \tau \diamond \alpha^* + \alpha)]\ \mathcal{D}[\alpha_a, \alpha_b].\end{aligned} \tag{6.318}$$

There are no quadratic terms for α_b. As a result, the functional integral over α_b produces a Dirac delta functional. However, its argument is a combination of α_a and α_a^*. To get a better picture of how such a Dirac delta functional operates, we redefine the integration field variables in terms of α_b by

$$\begin{aligned}\alpha_c &\triangleq \alpha_b + \tau \diamond \alpha_b^*, \\ \alpha_c^* &\triangleq \alpha_b^* + \alpha_b \diamond \tau^*.\end{aligned} \tag{6.319}$$

The inverse transformation is

$$\begin{aligned}\alpha_b &= \chi \diamond \alpha_c - \chi \diamond \tau \diamond \alpha_c^*, \\ \alpha_b^* &= \chi^* \diamond \alpha_c^* - \chi^* \diamond \tau^* \diamond \alpha_c,\end{aligned} \tag{6.320}$$

where

$$\chi \triangleq (\mathbf{1} - \tau \diamond \tau^*)^{-1}. \tag{6.321}$$

To compute the Jacobian for the change of integration variables from α_b to α_c, we use the expression (C.12). It leads to

$$\begin{aligned}\text{Jacobian} &= \left|\det\left\{\chi \diamond \chi^* - \chi \diamond \tau \diamond (\chi^*)^{-1} \diamond \chi^* \diamond \tau^* \diamond \chi^*\right\}\right| \\ &= \left|\det\left\{\chi^*\right\}\right| = \det\left\{\cosh_{\diamond}^2\left(\frac{1}{2}\xi^*\right)\right\} \equiv \mathcal{N}_{\tau}^{-2}.\end{aligned} \tag{6.322}$$

Therefore, it cancels the normalization constant.

The expression for the Wigner functional of the transformed state becomes

$$\begin{aligned}\text{Wigner}\{\hat{S}\hat{\rho}\hat{S}^{\dagger}\} = \int \exp\{[\alpha^* \diamond \chi \diamond (1 + \tau \diamond \tau^*) + 2\alpha \diamond \tau^* \diamond \chi - \alpha_a^*] \diamond \alpha_c \\ - \alpha_c^* \diamond [2\chi \diamond \tau \diamond \alpha^* + \chi \diamond (1 + \tau \diamond \tau^*) \diamond \alpha - \alpha_a]\} \\ \times W_{\hat{\rho}}\,[\alpha_a]\ \mathcal{D}[\alpha_a, \alpha_c].\end{aligned} \tag{6.323}$$

Using (6.317), we express the kernels as

$$
\begin{aligned}
(\mathbf{1} + \tau \diamond \tau^*) \diamond \chi = h_4^{-1} &= \cosh_\diamond(\xi) \triangleq \mathcal{C}, \\
2\chi \diamond \tau = -2h_4^{-1} \diamond h_1 &= \sinh_\diamond(\xi) \triangleq \mathcal{S}, \\
2\tau^* \diamond \chi = -2h_1^* \diamond h_4^{-1} &= \sinh_\diamond(\xi^*) \triangleq \mathcal{S}^*.
\end{aligned} \tag{6.324}
$$

Hence,

$$
\begin{aligned}
\text{Wigner}\{\hat{S}\hat{\rho}\hat{S}^\dagger\} = &\int \exp\{\alpha_c^* \diamond [\alpha_a - \mathcal{C} \diamond \alpha - \mathcal{S} \diamond \alpha^*] \\
&- [\alpha_a^* - \alpha \diamond \mathcal{S}^* - \alpha^* \diamond \mathcal{C}] \diamond \alpha_c\} W_{\hat{\rho}}[\alpha_a]\, \mathcal{D}[\alpha_a, \alpha_c] \\
= &\, W_{\hat{\rho}}[\alpha^* \diamond \mathcal{C} + \alpha \diamond \mathcal{S}^*, \mathcal{C} \diamond \alpha + \mathcal{S} \diamond \alpha^*].
\end{aligned} \tag{6.325}
$$

The arguments of the result represent a *Bogoliubov transformation*. While those discussed in Section 3.4.3 involve ladder operators, the Bogoliubov transformation here is a transformation of field variables. We represent this Bogoliubov transformation as

$$
\begin{aligned}
\alpha &\to U \diamond \alpha + V \diamond \alpha^*, \\
\alpha^* &\to \alpha^* \diamond U + \alpha \diamond V^*,
\end{aligned} \tag{6.326}
$$

where $U = \mathcal{C}$ and $V = \mathcal{S}$ represent kernels, as defined in (6.324) for the case above. These *Bogoliubov kernels* replace the Bogoliubov coefficients of Section 3.4.3. The kernels provide more information and they have properties not relevant for the coefficients.

The transformation of the argument of the Wigner functional is required to be unitary. It implies that the change in integration field variables given by a general Bogoliubov transformation

$$
\begin{aligned}
\alpha &= U \diamond \bar{\alpha} + V \diamond \bar{\alpha}^*, \\
\alpha^* &= \bar{\alpha}^* \diamond U^\dagger + \bar{\alpha} \diamond V^\dagger,
\end{aligned} \tag{6.327}
$$

must not change the normalization of the state. So, the Jacobian produced by the change in integration field variables must give

$$
\left| \det\left\{ \frac{\delta\alpha}{\delta\bar{\alpha}} \diamond \frac{\delta\alpha^*}{\delta\bar{\alpha}^*} - \frac{\delta\alpha}{\delta\bar{\alpha}^*} \diamond \frac{\delta\alpha^*}{\delta\bar{\alpha}} \right\} \right| = \left| \det\left\{ U \diamond U^\dagger - V \diamond V^\dagger \right\} \right| = 1. \tag{6.328}
$$

As shown in (5.77), this requirement is satisfied provided that

$$
U \diamond U^\dagger - V \diamond V^\dagger = \mathbf{1}. \tag{6.329}
$$

Exercise 6.22. Show that the Bogoliubov kernels defined in (6.324) satisfy the condition in (6.329).

Hint: Use the expressions in term of the h-kernels, which are valid regardless of the validity of the polar decomposition of the squeezing kernel ξ.

Exercise 6.23. Use (C.12) to show that the Jacobian for a Bogoliubov transformation is given by (6.328). Hint: use (5.77).

6.8.3 Bogoliubov transformation

A review of the Bogoliubov transformation with all the degrees of freedom included is provided to augment the discussion in Section 3.4.3. A general Bogoliubov transformation applied to ladder operators is expressed by

$$\begin{aligned}\hat{a} &\to U\diamond\hat{a}+V\diamond\hat{a}^{\dagger},\\ \hat{a}^{\dagger} &\to \hat{a}^{\dagger}\diamond U^{\dagger}+\hat{a}\diamond V^{\dagger},\end{aligned} \tag{6.330}$$

in terms of the *Bogoliubov kernels* U and V. It is a generalization of the transformation of field variables in (6.326) given in terms of ladder operators. According to the definition of a Bogoliubov transformation, the resulting operators must obey the same *commutation relations* as the ladder operators on which the transformation is applied. Hence,

$$\begin{aligned}\left[U\diamond\hat{a}+V\diamond\hat{a}^{\dagger},\hat{a}^{\dagger}\diamond U^{\dagger}+\hat{a}\diamond V^{\dagger}\right] &= U\diamond U^{\dagger}-V\diamond V^{\dagger}=\mathbf{1},\\ \left[U\diamond\hat{a}+V\diamond\hat{a}^{\dagger},U\diamond\hat{a}+V\diamond\hat{a}^{\dagger}\right] &= U\diamond V^{T}-V\diamond U^{T}=0,\\ \left[\hat{a}^{\dagger}\diamond U^{\dagger}+\hat{a}\diamond V^{\dagger},\hat{a}^{\dagger}\diamond U^{\dagger}+\hat{a}\diamond V^{\dagger}\right] &= V^{*}\diamond U^{\dagger}-U^{*}\diamond V^{\dagger}=0,\end{aligned} \tag{6.331}$$

in agreement with (5.77). While we already found that the first requirement is satisfied in (6.329), another requirement is that $U\diamond V^{T}=V\diamond U^{T}$.

The inverse Bogoliubov transformation should recover the original operators. Assume that the inverse Bogoliubov transformation is represented by

$$\begin{aligned}\hat{a} &\to U_i\diamond\hat{a}+V_i\diamond\hat{a}^{\dagger}\\ \hat{a}^{\dagger} &\to \hat{a}^{\dagger}\diamond U_i^{\dagger}+\hat{a}\diamond V_i^{\dagger}.\end{aligned} \tag{6.332}$$

When we apply it to the right-hand side of (6.330), the latter becomes

$$\begin{aligned}\hat{a} &\to U\diamond(U_i\diamond\hat{a}+V_i\diamond\hat{a}^{\dagger})+V\diamond(\hat{a}^{\dagger}\diamond U_i^{\dagger}+\hat{a}\diamond V_i^{\dagger})^{T}\\ &=(U\diamond U_i+V\diamond V_i^{*})\diamond\hat{a}+(V\diamond U_i^{*}+U\diamond V_i)\diamond\hat{a}^{\dagger},\\ \hat{a}^{\dagger} &\to (\hat{a}^{\dagger}\diamond U_i^{\dagger}+\hat{a}\diamond V_i^{\dagger})\diamond U^{\dagger}+(U_i\diamond\hat{a}+V_i\diamond\hat{a}^{\dagger})^{T}\diamond V^{\dagger}\\ &=\hat{a}^{\dagger}\diamond\left(U_i^{\dagger}\diamond U^{\dagger}+V_i^{T}\diamond V^{\dagger}\right)+\hat{a}\diamond\left(U_i^{T}\diamond V^{\dagger}+V_i^{\dagger}\diamond U^{\dagger}\right).\end{aligned} \tag{6.333}$$

It requires that

$$\begin{aligned}U\diamond U_i+V\diamond V_i^{*} &= U_i^{\dagger}\diamond U^{\dagger}+V_i^{T}\diamond V^{\dagger}=\mathbf{1},\\ V\diamond U_i^{*}+U\diamond V_i &= U_i^{T}\diamond V^{\dagger}+V_i^{\dagger}\diamond U^{\dagger}=0.\end{aligned} \tag{6.334}$$

From the comparison of these equations with those in (6.331), it follows that $U_i = U^\dagger$ and $V_i = -V^T$. As a result, the inverse Bogoliubov transformation is represented by

$$\begin{aligned} \hat{a} &\to U^\dagger \diamond \hat{a} - V^T \diamond \hat{a}^\dagger, \\ \hat{a}^\dagger &\to \hat{a}^\dagger \diamond U - \hat{a} \diamond V^*. \end{aligned} \tag{6.335}$$

Since the Bogoliubov transformation is often associated with a squeezing operation, as shown in (6.325), the inverse Bogoliubov transformation is likewise associated with the inverse squeezing operation. Moreover, being unitary, the inverse squeezing operation is obtained simply by changing the sign of the squeezing kernel $\zeta \to -\zeta$, which in turn implies that $\tau \to -\tau$. Hence, when associated with a squeezing operation, U is Hermitian and V is symmetric:

$$U^\dagger = U \quad \text{and} \quad V^T = V, \tag{6.336}$$

which corresponds to the fact that C and S in (6.324) are respectively Hermitian and symmetric, as a consequence of the way these kernels are constructed from the kernel of the squeezing operator. However, these properties are not required for a general Bogoliubov transformation, as indicated in (6.331).

6.8.4 Squeezed vacuum state

A state that we often encounter in the context of parametric down-conversion (see Chapter 8) is the *squeezed vacuum state*. The expression of its Wigner functional is readily obtained by applying the general Bogoliubov transformation of (6.330) to the Wigner functional of the vacuum state. It leads to

$$\begin{aligned} W_{\text{vac}}[\alpha] &= \mathcal{N}_0 \exp\left(-2\alpha^* \diamond \alpha\right) \\ &\to \mathcal{N}_0 \exp\left[-2\left(\alpha^* \diamond U^\dagger + \alpha \diamond V^\dagger\right) \diamond \left(U \diamond \alpha + V \diamond \alpha^*\right)\right] \\ &= \mathcal{N}_0 \exp\left(-2\alpha^* \diamond A \diamond \alpha - \alpha^* \diamond B \diamond \alpha^* - \alpha \diamond B^* \diamond \alpha\right), \end{aligned} \tag{6.337}$$

where

$$\begin{aligned} A &\triangleq U^\dagger \diamond U + V^T \diamond V^*, \\ B &\triangleq U^\dagger \diamond V + V^T \diamond U^*, \\ B^* &\triangleq V^\dagger \diamond U + U^T \diamond V^*. \end{aligned} \tag{6.338}$$

Here, we used the symmetries imposed on B and B^* by their contractions to the field variables in the last two expressions. The relationships in (6.338) are more general than what a squeezing process produces, because they do not assume (6.336). The final result in (6.337) is the generic form for the Wigner functional of a *squeezed vacuum state*, regardless of the properties of the Bogoliubov kernels.

Although the squeezing process applied to the vacuum state with the aid of a squeezing operator leads to this squeezed vacuum state, the converse is not always true. In Chapter 8, the parametric down-conversion process is shown to produce a squeezed vacuum state as in (6.337), but for which we cannot readily provide a corresponding *squeezing operator* in the form of (6.314).

Exercise 6.24. Compute the normalization factor for the squeezed vacuum state in (6.337). Under what conditions would $\mathcal{N}_0$ represent the correct normalization factor?

Exercise 6.25. Compute the purity of a squeezed vacuum state. Show that, for the squeezed vacuum state in (6.337) to be pure, it must have a normalization factor given by $\mathcal{N}_0$.

Exercise 6.26. Use (6.338), the properties in (6.331), and a new assumed property of the Bogoliubov kernels,

$$U^{\dagger} \diamond V = V^T \diamond U^*, \tag{6.339}$$

to show that

$$A \diamond B = B \diamond A^* \quad \text{and} \quad A \diamond A - B \diamond B^* = \mathbf{1}. \tag{6.340}$$

Using the definitions of the kernels in terms of the squeezing kernel in (6.324), so that $U = \cosh_\diamond(\xi)$ and $V = \sinh_\diamond(\xi)$, we can show that the Wigner functional in (6.337) is indeed a squeezed state. However, to do that, we need to assume that we can express the squeezing kernel in terms of a polar decomposition. Then the phase factors $\exp_\diamond(-i\theta)$ in these kernels produce rotations on the complex planes (two-dimensional phase spaces), associated with the different dimensions. Such rotations can be removed with a redefinition of the complex field variables α (when it is allowed). Then the expression can be simplified significantly by expressing it in terms of the quadrature field variables. We can also incorporate the rotations into the definitions of these quadrature field variables. The kernels then become exponential functions, leading to

$$W_{\text{squ}}[\alpha] = \mathcal{N}_0 \exp\left[-q_\theta \diamond \exp_\diamond(-2|\xi|) \diamond q_\theta - p_\theta \diamond \exp_\diamond(2|\xi|) \diamond p_\theta\right], \tag{6.341}$$

where q_θ and p_θ are rotated quadrature field variables.

The expression in (6.341) demonstrates the squeezing effect on the Wigner functional. Along one direction (the q_θ-direction), the Wigner functional is expanded by the kernel $\exp_\diamond(|\xi|)$. Along the orthogonal direction, the p_θ-direction, the Wigner functional is shrunk by the kernel $\exp_\diamond(-|\xi|)$. These two kernels are each other's inverse, so that their contraction produces **1**, representing the minimum uncertainty area on each of the two-dimensional phase spaces associated with the different spatiotemporal degrees of freedom. The orientations of the squeezing directions are controlled by the phase factor, which determines the definitions of q_θ and p_θ. Note that θ and $|\xi|$ depend on the spatiotemporal degrees of freedom. Different orientations and amounts of squeezing can exist for different spatiotemporal degrees of freedom.

Exercise 6.27. Show that, if

$$a = \frac{1}{\sqrt{2}}(q + \mathrm{i}p), \tag{6.342}$$

then

$$\exp_\diamond(-\mathrm{i}\theta) \diamond a = \frac{1}{\sqrt{2}}(q_\theta + \mathrm{i}p_\theta), \tag{6.343}$$

where

$$q_\theta = \cos_\diamond(\theta) \diamond q + \sin_\diamond(\theta) \diamond p \quad \text{and} \quad p_\theta = \cos_\diamond(\theta) \diamond p - \sin_\diamond(\theta) \diamond q. \tag{6.344}$$

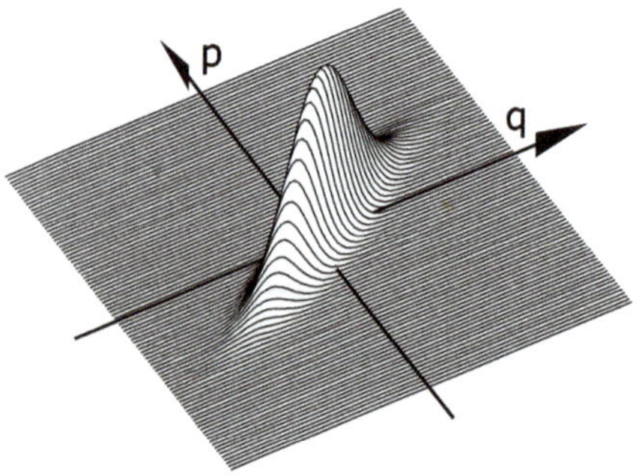

Figure 6.7: Wigner function of a squeezed vacuum state on a two-dimensional phase space.

In Figure 6.7, the rotated Wigner function of a squeezed vacuum state is shown on a two-dimensional phase space. It as an analogy of the Wigner functional of a squeezed vacuum state on infinite-dimensional functional phase space.

6.8.5 Identities for pure squeezed state kernels

6.8.5.1 Purity based

Based on the result in (C.29) and the fact that the Wigner functional of a state is normalised, we see that the determinants in (C.29) for the *pure* squeezed vacuum state in (6.337) must cancel. This condition can be satisfied if

$$A^* - B^* \diamond A^{-1} \diamond B = A^{*-1}. \tag{6.345}$$

This relationship is additional to the requirements that A is Hermitian and invertible, and that B is symmetric, but not in general invertible. It is a stronger statement than the requirement that the determinants cancel. The additional relationship is Hermitian, but not symmetric. So, the transpose (or complex conjugate) gives

$$A - B \diamond A^{*-1} \diamond B^{*} = A^{-1}. \tag{6.346}$$

We can now consider

$$B \diamond A^{*-1} \diamond B^{*} \diamond A^{-1} \diamond B = [A - A^{-1}] \diamond A^{-1} \diamond B = B \diamond A^{*-1} \diamond [A^{*} - A^{*-1}]. \tag{6.347}$$

Hence,

$$A^{-1} \diamond A^{-1} \diamond B = B \diamond A^{*-1} \diamond A^{*-1}. \tag{6.348}$$

Applying A on the left and A^{*} on the right, we get

$$A^{-1} \diamond B \diamond A^{*} = A \diamond B \diamond A^{*-1}. \tag{6.349}$$

Doing it again, we obtain

$$B \diamond A^{*} \diamond A^{*} = A \diamond A \diamond B. \tag{6.350}$$

The same can be done for the complex conjugates, leading to

$$\begin{aligned} A^{*-1} \diamond A^{*-1} \diamond B^{*} &= B^{*} \diamond A^{-1} \diamond A^{-1}, \\ A^{*-1} \diamond B^{*} \diamond A &= A^{*} \diamond B^{*} \diamond A^{-1}, \\ B^{*} \diamond A \diamond A &= A^{*} \diamond A^{*} \diamond B^{*}. \end{aligned} \tag{6.351}$$

It follows that $B \diamond B^{*}$ and $A \diamond A$ commute, and so do $B^{*} \diamond B$ and $A^{*} \diamond A^{*}$.

With these identities, one can show that

$$\begin{aligned} A^{-1} \diamond B \diamond A^{*} \diamond B^{*} \diamond A^{-1} &= A^{-1} \diamond B \diamond A^{*-1} \diamond B^{*} \diamond A \\ &= A^{-1} \diamond \left(A - A^{-1}\right) \diamond A = A - A^{-1}. \end{aligned} \tag{6.352}$$

The complex conjugate gives

$$A^{*-1} \diamond B^{*} \diamond A \diamond B \diamond A^{*-1} = A^{*} - A^{*-1}. \tag{6.353}$$

Exercise 6.28. Show that the average number of photons in the squeezed vacuum state in (6.337) reads

$$\langle \hat{n} \rangle = \frac{1}{2} \mathrm{tr}\{A - \mathbf{1}\}. \tag{6.354}$$

6.8.5.2 Based on additional property

With the aid of (6.339) and (6.340), the above relations can be simplified further. Hence,

$$A^{-1} \diamond B \diamond A^{*} = A \diamond B \diamond A^{*-1} = B. \tag{6.355}$$

6.8.6 Bogoliubov eigenstates and their completeness

In Section 3.4.4, we computed the eigenstates for the Bogoliubov operators in terms of the particle-number degree of freedom only. Repeating the computation of the Bogoliubov eigenstates, now with all the other degrees of freedom incorporated, by following the same steps as in Section 3.4.4, we end up with the eigenstates formally expressed by

$$|\alpha_0; \xi\rangle \triangleq \hat{D}[\alpha_0]\hat{S}[\xi]\,|\text{vac}\rangle\,, \tag{6.356}$$

but now α_0 represents a *parameter function* for the displacement and ξ represents a *squeezing kernel* for the squeezing process.

It follows from Section 6.2.4 that the set of all such eigenstates with the same squeezing kernel ξ and all possible displacements α_0 resolves the identity

$$\int |\alpha_0; \xi\rangle\,\langle\alpha_0; \xi|\;\mathcal{D}^{\circ}[\alpha_0] = \mathbb{1}. \tag{6.357}$$

The set of all possible displacements α_0 is exactly equivalent to the functional phase space in which the parameter function α_0 acts as the field variable. Such a set of Bogoliubov eigenstates can therefore be used for representations on functional phase space in the same way that the set of all fixed-spectrum coherent states serves to define the *P*- and *Q*-functionals discussed in Section 5.8. Hence, we can define functional representations in terms of the Bogoliubov eigenstates that are analogous to the Husimi Q-functionals and Glauber–Sudarshan P-functionals. They are readily defined as

$$Q_{\xi}[\alpha] = \langle\alpha; \xi|\,\hat{\rho}\,|\alpha; \xi\rangle\,, \tag{6.358}$$

and

$$\hat{\rho} = \int |\alpha; \xi\rangle\, P_{\xi}[\alpha]\,\langle\alpha; \xi|\;\mathcal{D}^{\circ}[\alpha]. \tag{6.359}$$

The corresponding Wigner functional is given by a rescaling of the q and p field variables, which can be readily removed by a redefinition of these field variables. Therefore, the Wigner functional is essentially equivalent for all such cases. Since there are an infinite number of different squeezing kernels, there are also an infinite number of different such sets of Bogoliubov eigenstates resolving the identity. Therefore, an infinite different number of generalized *P*- and *Q*-functionals can be defined.

6.8.7 Effect of loss on squeezing

In Section 6.7.1, it is shown that a coherent state remains pure after it has suffered a loss. The same is not true for a squeezed state. Here, we apply a homogeneous beamsplitter to the combination of a squeezed vacuum state and a vacuum state. To simplify the analysis, we assume that the kernels are real-valued and symmetric. (For general complex kernels, the squeezing directions are different for the different spatiotemporal degrees of freedom, making it difficult to see the effect of the loss.) Then we can express them as $A = \cosh_\diamond(2\xi)$ and $B = \sinh_\diamond(2\xi)$ in terms of a real-valued squeezing kernel ξ. We can then convert the expression to real-valued quadrature field variables. The combined Wigner functional prior to the beamsplitter then becomes

$$\begin{aligned} W_{\text{loss}}[q,p] = {} & \mathcal{N}_0^2 \exp\left[-q \diamond \exp_\diamond(2\xi) \diamond q - p \diamond \exp_\diamond(-2\xi) \diamond p\right] \\ & \times \exp\left(-q' \diamond q' - p' \diamond p'\right), \end{aligned} \tag{6.360}$$

where

$$\alpha \to \frac{1}{\sqrt{2}}(q + ip) \quad \text{and} \quad \beta \to \frac{1}{\sqrt{2}}(q' + ip'). \tag{6.361}$$

For quadrature field variables, the transformations for the beamsplitter are

$$\begin{aligned} q &\to qC - p'S, \quad & p &\to pC + q'S, \\ q' &\to q'C - pS, \quad & p' &\to p'C + qS, \end{aligned} \tag{6.362}$$

where $S = \sqrt{L}$ and $C = \sqrt{1-L}$ for a loss L. It is equivalent to (6.218). After applying these transformations, we perform the functional integrations over q' and p', using (C.15). Then we make some simplifications, leading to

$$W_{\text{loss}}[q,p] = \mathcal{N}_{\text{ms}} \exp\left(-q \diamond K_q \diamond q - p \diamond K_p \diamond p\right), \tag{6.363}$$

where

$$\begin{aligned} \mathcal{N}_{\text{ms}} &= \frac{\mathcal{N}_0}{\sqrt{\det\left\{(1-L)^2\mathbf{1} + 2(1-L)L\cosh_\diamond(2|\xi|) + L^2\mathbf{1}\right\}}}, \\ K_q &= \left[(1-L)\exp_\diamond(-2|\xi|) + L\mathbf{1}\right]^{-1}, \\ K_p &= \left[(1-L)\exp_\diamond(2|\xi|) + L\mathbf{1}\right]^{-1}. \end{aligned} \tag{6.364}$$

Comparing this result to (6.341), we see that the loss affects the squeezing. In fact, the contraction of the two kernels is not equal to **1** anymore.

One can also readily show that the state is not pure anymore. The purity is given by the integral of the square of the Wigner functional. Through a simple redefinition of the integration variable $\alpha \to \alpha/\sqrt{2}$, one can bring the expression back into the form of an integral over a single Wigner functional if it is pure. However, since the normalization is maintained by the determinants, the redefinition does not reproduce the normalization integral. Hence, the state is not pure.

6.8.8 Twin-beam squeezed vacuum state

A squeezed vacuum state is often represented as a *bipartite state* where the pairs of correlated photons are divided between two beams so that all photons in the one beam are correlated to photons in the other beam. Here, we call such a state a *twin-beam squeezed vacuum state*. In a less ideal situation, some of the photons would also be correlated to other photons in the same beam.

This state is also often called a *two-mode* squeezed vacuum state under the assumption that the two beams can be distinguished as being associated with two distinct spatial modes.[h] In the general case, the photons in the state can carry multiple modes. To avoid any confusion here, the term *twin-beam* squeezed vacuum state is used instead.

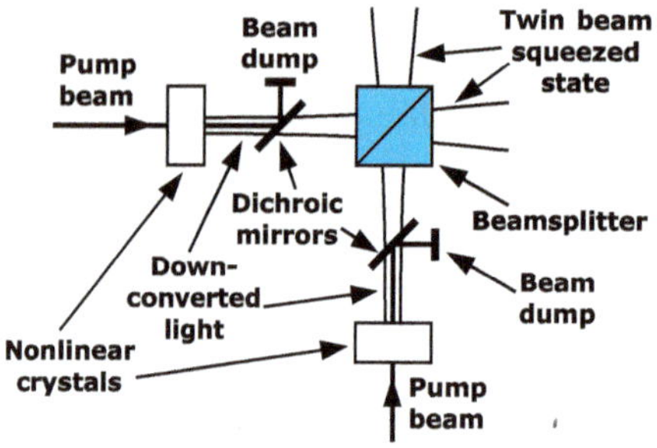

Figure 6.8: Optical system for the preparation of a twin-beam squeezed vacuum state.

There are different ways to produce a twin-beam squeezed vacuum state. An ideal way to obtain such a state [15, 16], as shown in Figure 6.8, is to start with two specially prepared single-beam squeezed vacuum states that are then combined via a beamsplitter. Using (6.337), we express the input state for the beamsplitter as

$$\begin{aligned} W_{\text{in}}[\alpha,\beta] = \mathcal{N}_0^2 \exp\left(-2\alpha^* \diamond A \diamond \alpha - \alpha^* \diamond B \diamond \alpha^* - \alpha \diamond B^* \diamond \alpha\right) \\ \times \exp\left(-2\beta^* \diamond \bar{A} \diamond \beta - \beta^* \diamond \bar{B} \diamond \beta^* - \beta \diamond \bar{B}^* \diamond \beta\right), \end{aligned} \tag{6.365}$$

where the barred kernels distinguish them from the unbarred ones. After the 50:50 beamsplitter, the state becomes

h The term "mode" may also sometimes refer to distinct output paths (from a beamsplitter for example), which is here represented by different phase spaces.

$$
\begin{aligned}
W_{\text{out}}[\alpha,\beta] = \mathcal{N}_0^2 \exp\Big[& \mathrm{i}\beta^* \diamond (A-\bar{A}) \diamond \alpha - \alpha^* \diamond (A+\bar{A}) \diamond \alpha - \mathrm{i}\alpha^* \diamond (A-\bar{A}) \diamond \beta \\
& - \beta^* \diamond (A+\bar{A}) \diamond \beta - \frac{1}{2}\alpha^* \diamond (B-\bar{B}) \diamond \alpha^* + \mathrm{i}\alpha^* \diamond (B+\bar{B}) \diamond \beta^* \\
& + \frac{1}{2}\beta^* \diamond (B-\bar{B}) \diamond \beta^* - \frac{1}{2}\alpha \diamond (B^* - \bar{B}^*) \diamond \alpha - \mathrm{i}\alpha \diamond (B^* + \bar{B}^*) \diamond \beta \\
& + \frac{1}{2}\beta \diamond (B^* - \bar{B}^*) \diamond \beta \Big]
\end{aligned} \tag{6.366}
$$

For identical single-beam squeezed vacuum states, the barred kernels are equal to the unbarred kernels. The state after the beamsplitter then becomes

$$
W_{\text{tbsqv}}[\alpha,\beta] = \mathcal{N}_0^2 \exp\left(-2\alpha^* \diamond A \diamond \alpha - 2\beta^* \diamond A \diamond \beta - 2\alpha^* \diamond B \diamond \beta^* - 2\alpha \diamond B^* \diamond \beta\right), \tag{6.367}
$$

where we absorbed factors of i into the uneven kernels $-\mathrm{i}B \to B$ and $\mathrm{i}B^* \to B^*$.

The resulting expression for the twin-beam squeezed vacuum state in (6.367) is often used in scenarios where correlation measurements are made on *photon pairs*, as produced in parametric down-conversion, which is discussed in Chapters 8 and 9. However, such a twin-beam squeezed vacuum state is often an idealization, unless it is produced by the physical process depicted in Figure 6.8. A more realistic scenario involves *post processing*, as discussed in the next section.

Exercise 6.29. Show that by tracing out one of the beams of the twin-beam squeezed vacuum state in (6.367), the result has the form of a thermal state, as in (6.181), with $\theta = A^{-1}$.

Exercise 6.30. Compute the Schmidt number (4.95) of the squeezed vacuum state in (6.367).

6.9 Post processing

The Wigner functional of the squeezed vacuum state in (6.337), as produced by parametric down-conversion discussed in Chapter 8, is a functional of only one field variable. So, it is defined on a single functional phase space and is thus referred to as a *single-beam squeezed vacuum state*. The term "beam" is used in the context of quantum optics.

In many applications of such squeezed vacuum states produced by parametric down-conversion, one needs to work with two or more field variables, for example, when multiple measurements are to be made on the state. With two field variables, the state is treated as a *twin-beam squeezed vacuum state*. A Wigner functional with just one field variable can sometimes be used (depending on the application) when two measurements are applied to it. The condition is that the two detector kernels do not overlap so that they commute. Such a treatment may lead to simpler expressions. For a more comprehensive discussion of such measurements, see Section 7.6.

To produce a twin-beam state, starting from a single-beam squeezed state, the latter is subjected to a *post-processing operation* that separates the state into two beams represented by two disjoint functional phase spaces, each having its own field variable. The process whereby the output functional phase space is converted into two disjoint functional phase spaces can be identified as a physical process that is performed on the single-beam state. Such a post-processing operation is often implicitly done on the state prior to a measurement. The Wigner functional then becomes a functional of both field variables of the respective phase spaces.

There exist various forms of post processing that convert the output functional phase space into two disjoint phase spaces. However, most of them can be modelled as some form of beamsplitter. Here, we are specifically interested in post processing performed with the aid of an *inhomogeneous beamsplitter*, as discussed in the Section 6.6.4. In most cases, it is implemented as a projection operation, represented by *idempotent kernel functions*. The operation of such a beamsplitter on the Wigner functional of a state (combined with a vacuum state) transforms its arguments according to the transformations given in (6.232).

6.9.1 One aperture or half-space

As an example, we consider the case where a single mirror is used to separate a part of a beam representing the state from the rest. The mirror is defined by an aperture. For a *D-shaped mirror* the aperture can be modelled as the half-plane, shown in Figure 6.9. The rest of the field that is reflected by the mirror may still be used and is not traced out. A vacuum state is combined in a tensor product for the state at the other input port of the inhomogeneous beamsplitter. The resulting state is a functional of two field variables α and β, representing the reflected and unreflected degrees of freedom, given by

$$W_{\text{mir}}[\alpha,\beta] = W_{\text{in}}[P \diamond \alpha - \mathrm{i}Q \diamond \beta] W_{\text{vac}}[P \diamond \beta - \mathrm{i}Q \diamond \alpha]. \tag{6.368}$$

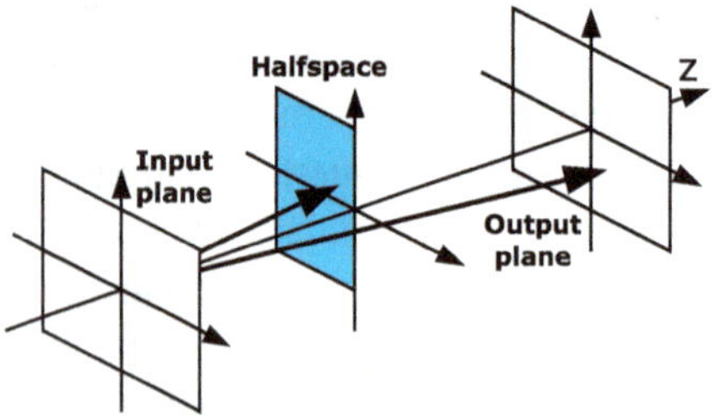

Figure 6.9: Free space propagation system with an aperture blocking off half of the transverse plane.

When the input state is the squeezed vacuum state in (6.337), the transformation of the field variables produces

$$\begin{aligned}W_{\text{mir}}[\alpha,\beta] = \mathcal{N}_0^2 \exp\big(&-2\alpha^* \diamond A_{pp} \diamond \alpha - \alpha \diamond B_{pp}^* \diamond \alpha - \alpha^* \diamond B_{pp} \diamond \alpha^* \\ &- 2\beta^* \diamond A_{qq} \diamond \beta + \beta \diamond B_{qq}^* \diamond \beta + \beta^* \diamond B_{qq} \diamond \beta^* \\ &- \mathrm{i}2\beta^* \diamond A_{qp} \diamond \alpha - \mathrm{i}\beta \diamond B_{qp}^* \diamond \alpha + \mathrm{i}\beta^* \diamond B_{qp} \diamond \alpha^* \\ &- \mathrm{i}2\alpha^* \diamond A_{pq} \diamond \beta - \mathrm{i}\alpha \diamond B_{pq}^* \diamond \beta + \mathrm{i}\alpha^* \diamond B_{pq} \diamond \beta^* \\ &- 2\beta^* \diamond P \diamond \beta - 2\alpha^* \diamond Q \diamond \alpha\big),\end{aligned} \tag{6.369}$$

where $B_{qp} \triangleq Q \diamond B \diamond P$, and so forth. In such scenarios, it is often reasonable to assume that the *conservation of momentum* causes $B_{pp} = 0$ and $A_{qp} = A_{pq} = 0$. For a D-shaped mirror, we would also have $B_{qq} = 0$. So, the expressions can be simplified to read

$$\begin{aligned}W_{\text{mir}}[\alpha,\beta] &= \mathcal{N}_0^2 \exp\big(-2\alpha^* \diamond A_{pp} \diamond \alpha - 2\alpha^* \diamond Q \diamond \alpha - 2\beta^* \diamond A_{qq} \diamond \beta \\ &\quad - 2\beta^* \diamond P \diamond \beta + \mathrm{i}2\alpha^* \diamond B_{pq} \diamond \beta^* - \mathrm{i}2\beta \diamond B_{qp}^* \diamond \alpha\big) \\ &= \mathcal{N}_0^2 \exp\big[-2\alpha^* \diamond (\mathbf{1} + E_{pp}) \diamond \alpha - 2\beta^* \diamond (\mathbf{1} + E_{qq}) \diamond \beta \\ &\quad + \mathrm{i}2\alpha^* \diamond B_{pq} \diamond \beta^* - \mathrm{i}2\beta \diamond B_{qp}^* \diamond \alpha\big],\end{aligned} \tag{6.370}$$

where $E_{pp} \triangleq P \diamond E \diamond P = P \diamond (A - \mathbf{1}) \diamond P$, and similar for E_{qq}. The result is formally a pure state, because the apertures are introduced with a unitary process. However, the terms that were dropped are not exactly zero in a physical scenario. Therefore, the physical state is only approximately pure.

Exercise 6.31. Apply this single aperture post processing on the single-mode thermal state whose kernel is given in (6.204). Assume that the mode of the thermal state is an eigenstate of the post processing projection kernel and show that the single-mode thermal state remains the same.

6.9.2 Two apertures

Another way to separate an output parametric down-converted state into two beams is by selecting two regions in the far-field configuration space with the aid of two inhomogeneous (projective) beamsplitters. These regions are often determined by the receiving apertures of optical systems processing or measuring the light that propagates in that direction. We represent the two receiving apertures by *idempotent kernels* representing the projection operations associated with the two beamsplitters.

Provided that the two apertures do not overlap, so that the contraction of the two different projection kernels gives zero, the order of the beamsplitters is not important; they commute. As a result, it is equivalent to having one screen with two apertures, as shown in Figure 6.10. We can therefore apply the transformations in (6.232) twice. For convenience, we redefine the projection operators to interchange P and Q and define P_1 and P_2 to represent the two apertures. The resulting transformations of the three field variables are (not showing the Hermitian adjoints)

$$\begin{aligned} \zeta &\to (\mathbf{1} - P_1 - P_2) \diamond \zeta - \mathrm{i}P_1 \diamond \alpha - \mathrm{i}P_2 \diamond \beta, \\ \alpha &\to (\mathbf{1} - P_1) \diamond \alpha - \mathrm{i}P_1 \diamond \zeta, \\ \beta &\to (\mathbf{1} - P_2) \diamond \beta - \mathrm{i}P_2 \diamond \zeta, \end{aligned} \tag{6.371}$$

representing a *three-port beamsplitter*. The original parametric down-converted state has ζ as its field variable. It is multiplied by two vacuum states, respectively with α and β as field variables. After the substitutions, the field variable ζ is integrated out, to trace out the lost photons that did not pass through either of the two apertures. The resulting state is a functional of both α and β, representing the degrees of freedom passing through the respective apertures. Formally, the process can be represented as

$$\begin{aligned} W_{\mathrm{pp}}[\alpha, \beta] = \int & W_{\mathrm{PDC}}[(\mathbf{1} - P_1 - P_2) \diamond \zeta - \mathrm{i}P_1 \diamond \alpha - \mathrm{i}P_2 \diamond \beta] \\ & \times W_{\mathrm{vac}}[(\mathbf{1} - P_1) \diamond \alpha - \mathrm{i}P_1 \diamond \zeta] W_{\mathrm{vac}}[(\mathbf{1} - P_2) \diamond \beta - \mathrm{i}P_2 \diamond \zeta] \, \mathcal{D}^{\circ}[\zeta]. \end{aligned} \tag{6.372}$$

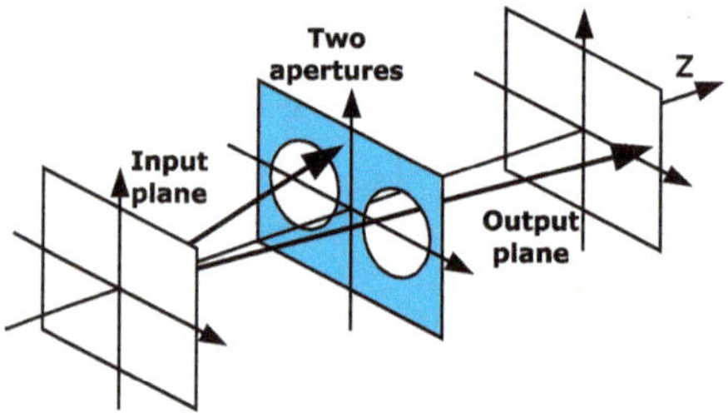

Figure 6.10: Free space propagation system with two apertures in a screen on the transverse plane.

6.9.2.1 Twin-beam squeezed vacuum

When the two-aperture post processing is applied on the squeezed vacuum state, it is converted to a twin-beam state. The input state is a squeezed vacuum state with a Wigner functional given by (6.337). The input to the three-port inhomogeneous beamsplitter is given by $W_{\mathrm{in}}[\alpha, \beta, \zeta] = W_{\mathrm{vac}}[\alpha] W_{\mathrm{vac}}[\beta] W_{\mathrm{sqv}}[\zeta]$, together with the two vacuum states' Wigner functionals. The transformation in (6.371) for the two-aperture post processing converts it into a complicated Gaussian Wigner functional with many terms in its exponent. It is simplified in a way similar to the case for one aperture, by assuming that the two apertures are in correlated locations, so that those kernels that do not have corresponding projections on either side can be set to zero. As a result, all the shift terms for ζ fall away. The functional integral over ζ can then be evaluated with the aid of the expressions in Appendix C. Due to the lost photons, the result is not normalized. The normalized result is

$$
\begin{aligned}
W_{\text{pp}}[\alpha,\beta] &= \int W_{\text{out}}[\alpha,\beta,\zeta]\,\mathcal{D}^{\circ}[\zeta] \\
&= \mathcal{N}_{\text{tbsqv}} \exp\left[-2\alpha^* \diamond (\mathbf{1}+E_{11}) \diamond \alpha - 2\beta^* \diamond (\mathbf{1}+E_{22}) \diamond \beta \right. \\
&\quad \left. + 2\alpha \diamond B_{12}^* \diamond \beta + 2\alpha^* \diamond B_{12} \diamond \beta^* \right],
\end{aligned} \tag{6.373}
$$

where $\mathcal{N}_{\text{tbsqv}}$ is the normalization constant, $E_{11} = P_1 \diamond (A - \mathbf{1}) \diamond P_1$, and so forth. The last two terms in the exponents picked up different signs due to the imaginary factors in the definition of the transformations in (6.371). Apart from these sign changes, the expression has the same form for the twin-beam squeezed state as obtained in (6.367).

6.9.2.2 Twin-beam Bogoliubov

To simplify calculations, we can introduce a *twin-beam Bogoliubov transformation*. It is done by applying the inhomogeneous beamsplitter transformations of (6.232) to the Bogoliubov transformation in (6.330). The result is

$$
\begin{aligned}
\alpha &\rightarrow U \diamond P \diamond \alpha + \mathrm{i}U \diamond Q \diamond \beta + V \diamond P^* \diamond \alpha^* - \mathrm{i}V \diamond Q^* \diamond \beta^*, \\
\alpha^* &\rightarrow \alpha^* \diamond P \diamond U^\dagger - \mathrm{i}\beta^* \diamond Q \diamond U^\dagger + \alpha \diamond P^* \diamond V^\dagger + \mathrm{i}\beta \diamond Q^* \diamond V^\dagger .
\end{aligned} \tag{6.374}
$$

Note that the projection kernels are complex conjugated on the right-hand side of V and on the left-hand side of $V^\dagger$.

The two projection operations P and Q are associated with two disjoint subspaces of the functional space, represented by $\mathcal{P}$ and $\mathcal{Q}$, respectively. These two spaces are associated with the operations of the Bogoliubov kernels. They transform the input parameter function to either one of these two subspaces. It is also assumed that the input parameter function belongs to only one of these two subspaces. The effect of U is to map a parameter function back to the same subspaces. On the other hand, V always maps a parameter function from one subspace to the other subspace.

When the transformation in (6.374) is performed on the field variables of an input Wigner functional, it often leads to terms in which opposite subspaces are contracted. Such terms become zero. To make it easier to identify such cases, we can interchange the order of the projection operations and the Bogoliubov kernels according to the rules

$$
U \diamond P = P \diamond U, \quad U \diamond Q = Q \diamond U, \quad V \diamond P^* = Q \diamond V, \quad V \diamond Q^* = P \diamond V. \tag{6.375}
$$

After performing these interchanges on (6.374), they become

$$
\begin{aligned}
\alpha &= P \diamond U \diamond \alpha + \mathrm{i}Q \diamond U \diamond \beta + Q \diamond V \diamond \alpha^* - \mathrm{i}P \diamond V \diamond \beta^*, \\
\alpha^* &= \alpha^* \diamond U^\dagger \diamond P - \mathrm{i}\beta^* \diamond U^\dagger \diamond Q + \alpha \diamond V^\dagger \diamond Q + \mathrm{i}\beta \diamond V^\dagger \diamond P.
\end{aligned} \tag{6.376}
$$

To see the effect of these transformations, we apply them on the field variables of the Wigner functional of a vacuum state. After discarding terms with incompatible projection kernels contracted on each other, and employing the idempotency of the projection kernels, we get

$$
\begin{aligned}
W_{\text{vac}}[\alpha] &= \mathcal{N}_0 \exp\left(-2\alpha^* \diamond \alpha\right) \\
&\to \mathcal{N}_0 \exp\left[-2\alpha^* \diamond \left(U^\dagger \diamond P \diamond U + V^T \diamond Q^* \diamond V^*\right) \diamond \alpha\right. \\
&\quad - 2\beta^* \diamond \left(U^\dagger \diamond Q \diamond U + V^T \diamond P^* \diamond V^*\right) \diamond \beta \\
&\quad + \mathrm{i}2\alpha^* \diamond \left(U^\dagger \diamond P \diamond V + V^T \diamond Q^* \diamond U^*\right) \diamond \beta^* \\
&\quad \left. - \mathrm{i}2\beta \diamond \left(V^\dagger \diamond P \diamond U + U^T \diamond Q^* \diamond V^*\right) \diamond \alpha\right].
\end{aligned} \tag{6.377}
$$

Having removed all the terms with incompatible subspace overlaps, we can interchange the projection operations and the Bogoliubov kernels back so that the Bogoliubov kernels are contracted on each other. Then we can combine them into the kernels for the squeezed vacuum state, as defined in (6.338). The transformed vacuum state thus reads

$$
\begin{aligned}
W_{\text{sqvac}} = \mathcal{N}_0^2 \exp\left(-2\alpha^* \diamond P \diamond A \diamond P \diamond \alpha - 2\beta^* \diamond Q \diamond A \diamond Q \diamond \beta\right. \\
\left. - 2\alpha^* \diamond P \diamond B \diamond Q^* \diamond \beta^* - 2\beta \diamond Q^* \diamond B^* \diamond P \diamond \alpha\right),
\end{aligned} \tag{6.378}
$$

where the additional factors of $\pm\mathrm{i}$ are absorbed into B and B^*.

We also need to consider the vacuum state that was not Bogoliubov transformed and that entered the other input port of the beamsplitter. It becomes

$$
\begin{aligned}
W_{\text{vac}}[\beta] &= \mathcal{N}_0 \exp\left(-2\beta^* \diamond \beta\right) \\
&\to \mathcal{N}_0 \exp\left(-2\beta^* \diamond P \diamond \beta - 2\alpha^* \diamond Q \diamond \alpha\right).
\end{aligned} \tag{6.379}
$$

In this way, both field variables cover the entire functional space involving both subspaces $\mathcal{P}$ and $\mathcal{Q}$. The part of these field variables lying in the opposite subspaces compared to those in (6.378) are provided by the other vacuum state entering the other port of the inhomogeneous beamsplitter without being Bogoliubov transformed. However, when measurements are applied to the state, it is assumed that α is restricted to $\mathcal{P}$ and β is restricted to $\mathcal{Q}$. Therefore, the second vacuum state and the projection kernels in (6.378) become superfluous, so that the Wigner functional of the twin-beam squeezed state can be expressed succinctly as presented in (6.367).

Formally, these restrictions are done by dressing the measurement kernels for the respective field variables by the appropriate projection kernels:

$$
D_\alpha \to P \diamond D_\alpha \diamond P, \quad D_\beta \to Q \diamond D_\beta \diamond Q. \tag{6.380}
$$

Unless we are concerned about the unitarity of the process, we can completely discard the part of the state produced by the vacuum state which is not Bogoliubov transformed. The only relevant transformations are those associated with α as the input field variable.

Since it is associated with the subspace $\mathcal{P}$, only the terms associated with P in (6.376) are relevant for the transformations of α in the input states. The transformations reduce to

$$\begin{aligned} \alpha &= U \diamond \alpha + V \diamond \beta^*, \\ \alpha^* &= \alpha^* \diamond U^\dagger + \beta \diamond V^\dagger, \end{aligned} \tag{6.381}$$

where we absorb the factor of i into the field variable β. The resulting transformation now represents the *twin-beam Bogoliubov transformation*. It is directly applied to the single-beam input state, which only depends on α.

Although the twin-beam Bogoliubov transformation makes derivations simpler, it can affect the invertibility of kernels. Since it effectively discards parts of the phase space, a kernel defined on the full phase space would strictly speaking not be invertible. However, if such a kernel was invertible on the original full phase space, we can assume that we can carry this property over to the restricted phase space, provided that mappings produced by the inverted kernel would not produce elements in the discarded part of the space for input elements that are not from the discarded part of the space. As an example, consider the kernel V. It maps elements in $\mathcal{P}$ to elements in $\mathcal{Q}$. If V can be inverted, then the inverse would map elements in $\mathcal{Q}$ to elements in $\mathcal{P}$. However, V^{-1} would only be allowed to receive input elements from $\mathcal{Q}$ and not from $\mathcal{P}$. In general V is not considered to be invertible.

6.10 Discussion

The topics that are covered in this chapter do not represent an exhaustive discussion of all the aspects that are relevant to quantum optics in the context of photonic quantum information systems. Even if we include those topics that are covered in the next chapter on measurements, we still excluded many scenarios that are often associated with quantum optics.

It is not our intention to be exhaustive. Instead the main goals of this chapter are: to provide Wigner functional representations of the most common quantum optical states and operators and to show how they are calculated; to demonstrate the use of phase space techniques in analyses using Wigner functional theory; and to discuss the process of modelling physical systems.

The most common quantum optical states are the coherent states, Fock states, thermal states and squeezed vacuum states. These states are all considered in terms of their representations incorporating all the spatiotemporal and spin degrees of freedom. As a result, they are often specified as *fixed-spectrum* states, when they can be parameterized in terms of a single (spectral) parameter function. In the case of the squeezed vacuum states, they are paremeterized in terms of kernels.

The most common operators (apart from the ladder operators) are the number operator, the displacement operator, the phase operator, the squeezing operators, and the unitary operators for the beamsplitter. Although the Wigner functionals for these operators are calculated in detail (often requiring complicated analyses), it is shown that their operations on states can be easily represented by transformations of the field variables in terms of which the Wigner functionals of the states are expressed. These four operators are all unitary, maintaining the normalization of the states on which they operate.

Displacement operators introduce an additional property that follows from their completeness and orthogonality. They serve as an operator basis in terms of which other operators can be expressed. They also lead to the definitions of bases for states in terms of which the identity can be resolved.

The interaction-free Hamiltonian is another operator that plays a significant role in quantum optics, often making an appearance in the analyses. However, since it is readily modelled in terms of the exponent of a phase operator, it is not discussed separately.

The squeezing operator leads to the use of Bogoliubov transformations applied to the field variables of Wigner functionals. However, we find that the most general Bogoliubov transformation is more general than those that are related to squeezing operations. As a result, we have the tantalizing situation where we arrive at a set of transformations that are not related to a known unitary transformation in their most general form. The consequences of this observation show up again in the analyses of subsequent chapters.

Modelling is one of the crucial skills of anybody involved with the analysis or design of physical systems. We discuss the modelling of such physical systems in the context of various scenarios, including linear lossless optics leading to generalized phase modulation using phase operators; the modelling of physical thermal states; beamsplitters that are used to model loss; and in systems that implement post-processing. These discussions provide opportunities to address various aspects that are associated with quantum systems. More discussions on modelling are provided in the next chapter in the context of measurement systems.

Among the phase space techniques that are demonstrated in this chapter, there are the development of evolution equations and the use of generating functions. Moreover, there are numerous techniques that are introduced during the course of the discussions to solve minor challenges in analyses.

Generating functions and functionals play a vital role in almost all analyses that we perform for various reasons. By using a generating function in a calculation, we obtain the result for an infinite number of functions from one calculation. All that is necessary is to preform the extraction process (usually with some derivatives) to obtain a specific result. Without such generating functions, we would not be able to evaluate many of the functional integrals because their integrands are not in Gaussian form. In this sense, generating function(al)s provide a double benefit. While they give a single result that is applicable for an infinite set of functions, those that we use are always in Gaussian form allowing us to evaluate the functional integrals in which they appear.

A powerful technique to study the dynamics of physical systems and their effect on quantum states is with the use of evolution equations. Such systems are those where a state evolves while propagating through them. In quantum mechanics, evolution equations usually take the form of Schrödinger equations or Heisenberg equations. In this chapter, we develop an approach to formulate evolution equations when the evolving state is represented in terms of Wigner functionals that include all the degrees of freedom. Such evolution equations can be more complicated. The derivation of such an evolution equation is presented here for free space propagation (as a simple example), using a generating functional together with a construction process for the basic dynamics. It is also used to derive an evolution equation for loss as a continuous process. The same approach is used in Chapter 8 to study parametric down-conversion.

One of the topics that we do not discuss in this chapter (or in the rest of the book) is the interaction between light and matter. There are a number of reasons for this omission. Firstly, such interactions often involve atoms. However, such atoms are usually modelled in terms of discrete variables, for which the Wigner functional formalism is not suitable. In those cases where Wigner functional theory would be useful, the material that we do discuss should provide sufficient information on how to model and analyse such scenarios. Moreover, the interactions between light and matter represent quantum interactions. We consider such a quantum interaction extensively in the context of parametric down-conversion in Chapter 8.

Bibliography

[1] D. F. Walls and G. J. Milburn. *Quantum Optics*. Springer, Berlin, 1995.

[2] C. C. Gerry and P. L. Knight. *Optical Coherence and Quantum Optics*. Cambridge University Press, New York, 2005.

[3] S. L. Braunstein and P. Van Loock. Quantum information with continuous variables. *Rev. Mod. Phys.*, 77:513–577, 2005.

[4] G. Adesso, S. Ragy, and A. R. Lee. Continuous variable quantum information: Gaussian states and beyond. *Open Syst. Inf. Dyn.*, 21:1440001, 2014.

[5] M. A. Nielsen and I. L. Chuang. *Quantum Computation and Quantum Information*. Cambridge University Press, Cambridge, England, 2000.

[6] M. Abramowitz and I. A. Stegun. *Handbook of Mathematical Functions*. Dover, Toronto, 1972.

[7] M. E. Peskin and D. V. Schroeder. *An Introduction to Quantum Field Theory*. Addison-Wesley Publishing Company, Reading, Massachusetts, USA, 1995.

[8] S. Weinberg. *The Quantum Theory of Fields, Volume I*. Cambridge University Press, New York, 1995.

[9] J. W. Goodman. *Introduction to Fourier Optics, 2nd ed.* McGraw-Hill, New York, USA, 1996.

[10] H. J. Groenewold. On the principles of elementary quantum mechanics. *Physica*, 12:405–460, 1946.

[11] J. E. Moyal. Quantum mechanics as a statistical theory. *Math. Proc. Camb. Philos. Soc.*, 45:99–124, 1949.

[12] T. L. Curtright and C. K. Zachos. Quantum mechanics in phase space. *Asia Pac. Phys. Newsl.*, 1:37–46, 2012.

[13] W. -K. Tung. *Group Theory in Physics*. World Scientific, Singapore, 1985.

[14] K. Sakoda. *Optical Properties of Photonic Crystals*. Springer, Berlin, 2001.
[15] A. Furusawa, J. L. Sorensen, S. L. Braunstein, C. A. Fuchs, H. J. Kimble, and E. S. Polzik. Unconditional quantum teleportation. *Science*, 282:706–709, 1998.
[16] N. Takei, N. Lee, D. Moriyama, J. S. Neergaard-Nielsen, and A. Furusawa. Time-gated Einstein–Podolsky–Rosen correlation. *Phys. Rev. A*, 74:060101(R), 2006.

7 Measurements

Numerous different experimental techniques have been developed over time to make observations of quantum phenomena. These techniques range from photographic plates, phosphor screens, and bubble chambers in the early years, up to modern day CCD arrays, avalanche photodiodes, and the highly sophisticated detector systems used in supercolliders such as the Large Hadron Collider.

In all these devices, a tiny individual interaction is converted into a large enough effect so that it can be recorded. For example, an avalanche photodiode triggers the detection of a single photon by releasing the energy stored in a whole bunch of electrons, so that it is registered as a pulse of electric current.

Regardless of the mechanism that is used to magnify the initial interaction, all these detection processes start with a single fundamental interaction, which is then magnified to give enough energy to be registered. As a result, all such measurements inherit the properties that are associated with interactions. In Chapter 1, we explained that these fundamental interactions are quantized and localized. It thus follows that these measurements also reveal quantized and localized results. The observation of a localized quantization effect does not mean that a single particle (a "lump of matter" as understood in the classical sense) has been observed. It just means that a quantum of the field is observed by the magnification of a single interaction that is quantized and localized.

A general theory of measurements applicable to quantum physics usually needs to accommodate the measuring system together with the object (system or state) that is being measured. Moreover, it also needs to take into account the fact that the object being measured ends up in a specific state after the measurement. Fortunately, the situation in quantum optics is usually simpler because photons are destroyed during the interaction that initiates the measurement. Even when the state on which the measurement is applied may contain multiple photons that survive after the measurement, it is often discarded (traced out) once the measurement is done. Sometimes the measurement is used as an operation applied to the state so that the remainder of the state after the measurement takes on a specific form suitable for subsequent processing. We discuss such scenarios in the context of *heralding* below.

Modern theories of measurements also allow for the most general type of measurements in terms of formulation of *positive operator-valued measurements* (POVM) [1]. Such formulations represent more versatile descriptions of the most general measurements. Nevertheless, practical implementations of such general measurement protocols in quantum optics are usually done in terms of basic projective measurements. The more sophisticated measurement processes are then obtained from calculations made on the raw projective measurement results. For this reason, such practical measurements are here mostly modelled in terms of projective measurements. An exception is *homodyne detection*, which we discuss separately below.

A specific measurement is generally associated with an *observable* defined by a Hermitian operator. Such Hermitian operators often have multiple eigenstates associated

https://doi.org/10.1515/9783111445342-009

with nonzero real eigenvalues. A single measurement for such an observable produces a result represented by any one of these eigenstates. By making repeated measurements on identical copies of the state, one can build up a statistical distribution over the different eigenstates for the state being measured. On the other hand, the observables associated with *projective measurements* have only one eigenstate with a nonzero eigenvalue. To obtain the same statistical distribution for the state being measured, one can use different projective measurements associated with all the different basis elements in terms of which the state is defined.

It is not our intention to discuss an abstract theory of measurements for quantum physics. There are many adequate treatments of such theories [1]. Instead, we analyse various practical measurement systems typically encountered in quantum optics experiments. As such, the discussions in this chapter provide various demonstrations of how to model physical measurement systems in quantum optics and photonic quantum information technology.

Measurements in quantum optics can take on various forms, depending on the degrees of freedom that are involved and the nature of the information being extracted. To analyse the effects of the physical experimental parameters, we require a formalism that can incorporate all the degrees of freedom that are affected by these parameters. The Wigner functional formalism developed in Chapter 5 is well suited for this purpose.

7.1 Wigner functionals in measurement calculations

First, we provide some general formal definitions associated with measurements and provide a few general approaches to express the associated calculations in terms of Wigner functionals. As mentioned above, a measurement can usually be associated with a Hermitian operator $\hat{A}$ that we call the *observable*. It can be represented as

$$\hat{A} = \sum_m |\phi_m\rangle \lambda_m \langle\phi_m| , \tag{7.1}$$

where $|\phi_m\rangle$ is the m-th eigenstate and λ_m denotes its associated eigenvalue. When we apply this observable on an arbitrary state $|\psi\rangle$, it produces

$$\hat{A}|\psi\rangle = \sum_m |\phi_m\rangle \lambda_m \langle\phi_m|\psi\rangle . \tag{7.2}$$

The result has the form of a superposition of all the possible outcomes that can be produced by a single measurement. Here, λ_m represents the measurement value obtained from the single measurement and $\langle\phi_m|\psi\rangle$ is the *complex probability amplitude* for that measurement outcome. *Born's rule* states that the *probability* for the outcome is given by $|\langle\phi_m|\psi\rangle|^2$.

Quantum theory does not provide us with the ability to predict which of the different outcomes of an observable is obtained for a single measurement. In fact, it is argued that this inability to predict the specific outcome is a fundamental limitation. It is therefore often more typical to measure the *expectation value* of an observable. For this purpose, the observable is applied repeatedly to identical copies of the state and the results are averaged to obtain the expectation value.

Given a general state in terms of its density operator $\hat{\rho}$, we compute the *expectation value of an observable* with the trace

$$\langle\hat{A}\rangle = \text{tr}\{\hat{\rho}\hat{A}\}. \tag{7.3}$$

For a pure state $\hat{\rho} = |\psi\rangle\langle\psi|$, the observable then produces

$$\langle\hat{A}\rangle = \langle\psi|\hat{A}|\psi\rangle = \sum_m \lambda_m |\langle\phi_m|\psi\rangle|^2. \tag{7.4}$$

The *observable* associated with a *projective measurement* is special in that it has only one nonzero eigenvalue. It is thus represented by

$$\hat{P} = |\phi\rangle\langle\phi|, \tag{7.5}$$

where $|\phi\rangle$ is the single eigenstate of the projection operator. In this case, the expectation value,

$$\langle\hat{P}\rangle = \text{tr}\{\hat{\rho}\hat{P}\} = \langle\phi|\hat{\rho}|\phi\rangle, \tag{7.6}$$

coincides with the probability for a single successful measurement. One can use a set of such projection operators with different eigenstates to build up a statistical distribution from multiple measurements by recording the number for the successful measurements of each of the projection operators and dividing it by the total number of successful measurements for all the projection operators.

In terms of Wigner functions (for the particle-number degree of freedom only), the trace of the product of operators can be expressed as

$$\langle\hat{A}\rangle = \text{tr}\{\hat{\rho}\hat{A}\} = \int W_{\hat{\rho}}(q,p)W_{\hat{A}}(q,p)\,\mathrm{d}q\,\mathrm{d}p, \tag{7.7}$$

where $W_{\hat{\rho}}(q,p)$ and $W_{\hat{A}}(q,p)$ are the Wigner functions for the density operator and the observable, respectively. The equivalent expression in terms of *Wigner functionals* reads

$$\langle\hat{A}\rangle = \int W_{\hat{\rho}}[\alpha]W_{\hat{A}}[\alpha]\,\mathcal{D}^{\circ}[\alpha]. \tag{7.8}$$

The nature of $W_{\hat{A}}[\alpha]$ is determined by the detail of the measurement. For example, the Hermitian operator often used as the *observable* for intensity is the *number operator*, defined in (4.48), with its Wigner functional given in (6.62). However, the measurable quantity associated with the number operator is the *occupation number* (the number of photons in a state). The result of the measurement gives us the average number of photons in a state $\langle n\rangle$, expressed in terms of the Wigner functional by

$$\langle n\rangle = \int W_{\hat{\rho}}[\alpha] \left(\alpha^* \diamond \alpha - \frac{1}{2}\Omega\right) \mathcal{D}^\circ[\alpha]. \tag{7.9}$$

If the experimental conditions for the measurement is such that it gives the number of photons per unit area and per unit time, then the intensity is obtained from the measured distribution by multiplying it by the energy per photon: $I = \hbar\omega\langle n\rangle$. However, it is shown in Section 7.3 below that the situation is more complicated when we consider practical intensity measurements.

7.1.1 Generating functions in measurement calculations

The Wigner functional of a state is often given by an exponential function

$$W_{\hat{\rho}}[\alpha] = \mathcal{N} \exp(-S[\alpha]), \tag{7.10}$$

where $\mathcal{N}$ is a normalization constant, and $S[\alpha]$ is a second-order polynomial functional in the field variables α and α^*. In such a case, the state is a *Gaussian state*. Sometimes the functional is in fact a generating function(al) for the Wigner functionals of different states, in which case $S[\alpha]$ also contains special terms with auxiliary parameters or fields.

When the Wigner functional of the observable $W_{\hat{A}}[\alpha]$ is itself a second-order polynomial functional in α and α^* (as with the number operator, for example), then we can add it to $S[\alpha]$ as a source term, multiplied by an auxiliary parameter, in the exponent of the Wigner functional for the state

$$W_{\hat{\rho}}[\alpha] \to \mathcal{W}(J) = \mathcal{N} \exp\left(-S[\alpha] + J W_{\hat{A}}[\alpha]\right), \tag{7.11}$$

where J is the auxiliary (generating) parameter. The result serves as a *generating function for the measurements*. The expectation value is then produced by

$$\langle\hat{A}\rangle = \left\langle \frac{\partial}{\partial J}\mathcal{W}(J)\right\rangle\bigg|_{J=0} = \mathcal{N} \int W_{\hat{A}}[\alpha] \exp\left(-S[\alpha]\right) \mathcal{D}^\circ[\alpha]. \tag{7.12}$$

The angular brackets representing the expectation value implies a trace process that is implemented with a functional integral over α. For instance, to check that the state is properly normalized, one can show that

$$\langle\mathcal{W}(0)\rangle \triangleq \mathcal{N} \int \exp\left(-S[\alpha]\right) \mathcal{D}^\circ[\alpha] = 1. \tag{7.13}$$

Various other ways exist to define the generating functions, depending on the particular applications. See Appendix A.

7.1.2 Construction operation

For more complicated measurements represented by higher than second-order polynomials in α and α^*, it is still possible to construct the measurement with the aid of source terms. The generating function(al) is augmented by a *construction operation* consisting of derivatives with respect to the generating parameters, thus *constructing* the *observable* from the generating function(al).

For example, consider a generating functional given by

$$\mathcal{W}[J] = \mathcal{N} \exp\left(-S + \mu \diamond W_{\hat{A}}\right), \tag{7.14}$$

where μ is an auxiliary source field and $W_{\hat{A}}$ is expressed in terms of the field variables α and α^*. (Usually $W_{\hat{A}}$ is just the sum of linear source terms for these field variables.) Such a process allows us to use *functional derivatives* to compute more complicated quantities. A simple construction operation could, for instance, be $C\{\mathcal{W}\} = F \diamond \delta_\mu \mathcal{W}|_{\mu=0}$, where $F(\mathbf{k})$ is a given spectral function. Applied to the generating functional, it produces

$$\left\langle \int F(\mathbf{k}) \frac{\delta}{\delta\mu(\mathbf{k})} \mathcal{W}[\mu]\, \mathrm{d}k \right\rangle\Bigg|_{\mu=0} = \int F \diamond W_{\hat{A}} \exp\left(-S\right)\, \mathcal{D}^{\circ}[\alpha]. \tag{7.15}$$

The construction method is useful for complicated measurable quantities.

7.2 Detectors

The physical devices with which most optical measurements are performed are *detectors*. These devices have specific properties that affect the efficacy of the measurements. Before we can discuss measurements in quantum optics, we first need to look at how these detectors are modelled. The different kinds of detectors are broadly grouped into *classical detectors* and *quantum detectors*, based on their sensitivity and ability to detect single photons.

Classical detectors, which include PIN diodes and CCD arrays, usually register *photo-currents* as a function of the intensity (or photon flux) of the light falling on the detector. In general, the magnitude of the photo-current is a nonlinear function of the incident intensity. However, over a certain range, the function is approximately linear so that the photo-current within this range is proportional to the intensity with some off-set. The photo-current is represented as a *signal*, which is a function of time. It represents a temporally filtered version of the light signal. The filtering is represented by a convolution of the input light signal by the impulse response of the detector. Moreover, the light is integrated over the *area of the detector* and it also has a varying sensitivity (detection efficiency) as a function of the wavelength.

Quantum detectors, such as *avalanche photon diodes*, share the same properties of classical detectors, but they also incorporate mechanisms to enhance the response of the device to the extent that a single photon can be registered as an electronic pulse (a "click"). Usually, such a device can nevertheless not distinguish between one or more photons that arrive simultaneously to cause such a click. Therefore, the device is generally not *photon-number resolving*. The temporal resolution is determined by a *gating* process. Although the gating process is similar to the convolution of the input with an impulse response, the gating period can be set fixed in time with a fixed duration.

Recently, new mechanisms have been invented that allow the device to provide a reasonable estimate of the number of photons arriving at the same time, provided that there are not too many of them. It thus allows for the possibility of *photon-number-resolving detection*. Although we often assume photon-number-resolving detectors in our calculations, we can also model cases that are not photon-number resolving.

In our calculations involving optical measurements, we model the detectors with the aid of *detector kernel functions*. Focusing on the number operator, we incorporate a detector kernel D into it to define a *generalized* (or *localized*) number operator

$$\hat{n}_D \triangleq \hat{a}^\dagger \diamond D \diamond \hat{a}. \tag{7.16}$$

In general, such a detector kernel can be modelled by an *idempotent kernel* times a *detection efficiency* $0 < \eta < 1$. With perfect detection efficiency, it follows that $D \diamond D = D$. The detector kernel is in general Hermitian, but not invertible. Since the number operator is composed of ladder operators in the Fourier domain, the detector kernel function is represented in the Fourier domain. The Wigner functional of the generalized number operator can be readily computed as in (6.62). It is given by

$$W_{\hat{n}}[\alpha] = \alpha^* \diamond D \diamond \alpha - \frac{1}{2}\operatorname{tr}\{D\}, \tag{7.17}$$

where $\operatorname{tr}\{D\}$ counts the average number of modes that can be observed by (can "pass through") the detector kernel.

To compute the intensity (or photon-number) function of the image that a camera observes, one performs a trace of the state times a *localized number operator* that represents the detection process of a pixel on the camera. The generalized number operator represents a *localized* number operator when its kernel implements some form of localization, as typically associated with a physical detector having a finite area and a finite integration time.

Alternative methods to model such localized measurements involve the use of field operators, which are related to the ladder operator by Fourier transforms. Since the detector kernel approach is general enough to include those cases that are modelled with field operators, we do not use field operators to model quantum optics measurements.

7.2.1 Single-mode detector kernel

Often the detector is set up to detect a single spatiotemporal mode. For example, in coincidence detection setups, the light is often coupled into a *single-mode fibre*, which is a very effective spatial filter that only allows the single propagating mode of the fibre to be detected. In other scenarios, the spatial properties of the detectors is not of any importance, apart from the size of the detector. It allows one to model such a detector as a single-mode detector, which is easier to use in calculations.

For example, in a CCD array, which is often used in quantum optics experiments, the size of the pixel determines the resolution.[a] In such a case, the detector for the single pixel is modelled in terms of a number operator by

$$\hat{n}_D = \eta\hat{a}^{\dagger} \diamond MM^* \diamond \hat{a}, \tag{7.18}$$

where η is the *detection efficiency*. The *single-mode detector kernel*, which is given by $D(\mathbf{k}_1, \mathbf{k}_2) = M(\mathbf{k}_1)M^*(\mathbf{k}_2)$, is associated with one such pixel. It serves as a *projection kernel* that projects out the specific spatiotemporal degrees of freedom associated with the detector mode. The angular spectrum of the detector mode $M(\mathbf{k})$ is normalized so that $\|M\|^2 = M^* \diamond M = 1$, thus ensuring *idempotency* of the kernels (up to the detection efficiency). Henceforth, we usually ignore the detection efficiency by assuming that we can set $\eta = 1$, unless stated otherwise.

If the shape of the mode is not important, such as when only the size of the mode is needed to provide information about the resolution, the mode is assumed to be a Gaussian function with a size given by the resolution in the detector plane. Although the measured temporal spectrum is usually determined by a *line filter* (an optical element that only allows a relatively small bandwidth of wavelengths to pass through) in the experimental setup, we incorporate it for convenience into the definition of the detector. In such a case, we define

$$M(\mathbf{k}) = \sqrt{2\pi}w_0 h(\omega - \omega_0, \delta_0) \exp\left(-\frac{1}{4}w_0^2|\mathbf{K}|^2\right), \tag{7.19}$$

where w_0 is the spatial mode size in the detector plane, determining the output resolution, $h(\omega - \omega_0, \delta_0)$ is a temporal spectrum centred at ω_0 with a bandwidth δ_0.

When the single-mode detector represents a pixel in a CCD array, the mode function needs additional parameters for the location of the pixel in the array. Without any additional optics, a shifted location produces a linear phase term in the Fourier domain. The detector mode can then be represented by

$$M_{mn}(\mathbf{k}) = \sqrt{2\pi}w_0 h(\omega - \omega_0, \delta_0) \exp\left(-\frac{1}{4}w_0^2|\mathbf{K}|^2 - \mathrm{i}\mathbf{K}\cdot\mathbf{X}_{mn}\right), \tag{7.20}$$

---

a Strictly speaking, it is the *grid spacing* that determines the resolution and the pixel size relative to this grid spacing affects the *detection efficiency*.

where $\mathbf{X}_{mn} = m\Delta x\vec{x} + n\Delta y\vec{y}$ is the two-dimensional position vector for the location of the detector (pixel) in the CCD array.

7.2.2 Convolution detector kernels

Regardless of whether the detector is a pixel in an CCD array or not, it can in general be modelled with a convolution kernel in optical beam variables, as discussed in Sections 2.1.7 and 6.5.3. Here, we consider the non-Lorentz covariant formulation. This detector kernel is given by

$$D(\mathbf{K}_1, \omega_1, \mathbf{K}_2, \omega_2) = \int v(t) \exp[\mathrm{i}(\omega_1 - \omega_2)t]\, \mathrm{d}t \times \int p(\mathbf{X} - \mathbf{X}_0) \exp[-\mathrm{i}\mathbf{X} \cdot (\mathbf{K}_1 - \mathbf{K}_2)]\, \mathrm{d}^2x, \tag{7.21}$$

where the *gating function* $v(t)$ represents a binary time pulse for the integration time of the detector, and the *pixel function* $p(\mathbf{X} - \mathbf{X}_0)$ is a binary area function representing the active area of the detector, shifted to the location $\mathbf{X}_0$ (as we had for the pixels in the CCD array). It does not incorporate any loss. (A homogeneous loss can be incorporated into the *detection efficiency*.) Hence, it is idempotent $D \diamond D = D$, which implies that $p(\mathbf{X})$ and $v(t)$ are binary functions. As usual, the dimensions of the kernel are given by the inverse of those for the $\diamond$-contraction, as dictated by its integration measure. For (7.21), the measure of the relevant $\diamond$-contraction is given by (2.105).

7.2.3 Detector array

Instead of considering the single pixels in a CCD array, one can model the entire CCD array in terms of a set of suitable detector modes. Although this model is not used again, it is instructive to see how to do it. First, we define a *hat function*

$$\Pi(s) = \begin{cases} 1 & \text{for } |s| \leq \frac{1}{2}, \\ 0 & \text{otherwise,} \end{cases} \tag{7.22}$$

so that

$$\sum_{m=-\infty}^{\infty} \Pi(s - m) = 1. \tag{7.23}$$

The two-dimensional pixel functions are then defined as

$$d_{mn}(x,y) = \Pi\left(\frac{x}{\Delta x} - m\right) \Pi\left(\frac{y}{\Delta y} - n\right), \tag{7.24}$$

where the indices m and n distinguish the separate pixel functions along the two orthogonal directions in the array, and Δx and Δy are the widths of the pixels along the two orthogonal directions (often $\Delta x = \Delta y$). These pixel functions are thus *binary functions* (having function values equal to either 0 or 1), representing the areas of the individual pixel in the CCD array. It then follows that these two-dimensional pixel functions can be used to tile the entire two-dimensional detector plane, so that

$$\sum_{mn} d_{mn}(x,y) = 1. \tag{7.25}$$

The detector kernels are then obtained from these pixel functions with the aid of two-dimensional Fourier transforms. The resulting detector kernels are idempotent *convolution kernels*, given by

$$D_{mn}(\mathbf{K}_1 - \mathbf{K}_2) = \int d_{mn}(\mathbf{X}) \exp[-\mathrm{i}\mathbf{X} \cdot (\mathbf{K}_1 - \mathbf{K}_2)] \, \mathrm{d}^2 x, \tag{7.26}$$

where $\mathbf{K}$ is the two-dimensional transverse part of the wavevector, and $\mathbf{X}$ is a two-dimensional position vector on the plane of the CCD array. For the moment, we assumed monochromatic conditions. The binary pixel functions thus act like transmission functions or aperture functions. For a binary function, it follows that $d_{mn}^2(\mathbf{X}) = d_{mn}(\mathbf{X})$. Therefore, the associated convolution kernel is idempotent $D_{mn} \diamond D_{mn} = D_{mn}$.

7.2.4 Detection systems

In addition to the physical detector, a detection system may also incorporate an optical system that directs the light onto the detector. Such optical systems often incorporate lenses in the form of $2f$ or $4f$ systems (see Section 2.7), performing Fourier transforms or imaging on the photonic quantum state at the input to the detection system. For the purpose of calculations, such systems can be modelled as phase modulation systems, as discussed in Section 6.4. They may also incorporate imperfections such as apertures that can be modelled with the inhomogeneous beamsplitter discussed in Section 6.6.4. Such apertures introduce finite resolutions as discussed in Section 2.7.3. All the optics are combined with the basic model of the detector to define an idempotent kernel.

When the detector is located in the output plane of a $2f$ system, we can incorporate the lens system into the detector mode function. In such a case, the location of the detector, represented by a shift in the detector mode function on the output plane of a $2f$ system, is directly related to a shift in the angular spectrum in the input plane of the $2f$ system. The effective detector mode in the input plane of the $2f$ system is then given by

$$M(\mathbf{k}) = \sqrt{2\pi} w_0 h(\omega - \omega_0, \delta_0) \exp\left(-\frac{1}{4} w_0^2 |\mathbf{K} - \mathbf{K}_0|^2\right), \tag{7.27}$$

where w_0 is the mode size in the input place, and $\mathbf{K}_0$ represents the shifts. The mode size is related to the detector size (of a pixel) in the output plane w_{det} by

$$w_0 = \frac{\lambda f}{\pi w_{\text{det}}}, \tag{7.28}$$

as shown in (2.165). Here, λ and f are the free space wavelength and the focal length of the lens, respectively. Similarly, the shift $\mathbf{K}_0$ is related to the detector position in the output plane $\mathbf{X}_0$ by

$$\mathbf{K}_0 = \frac{k\mathbf{X}_0}{f} = \frac{2\pi\mathbf{X}_0}{\lambda f}, \tag{7.29}$$

as in (2.164), with k being the wavenumber in free space.

7.2.5 Infinite resolution detector kernel

If we are not interested in the effect of the finite resolution in the detector plane, as imposed by the finite size of the detector, we can assume that the size of the detector is infinitely small. However, we need to consider such a situation carefully. Starting with the single-mode detector kernel modelled as a Gaussian function, we apply a limit where $w_{\text{det}} \to 0$ to convert the detector mode into a Dirac delta function to get the *infinite resolution detector kernel.*

Usually, when we represent a Dirac delta function with a limit process

$$\delta(x) = \lim_{\epsilon\to 0} f(x;\epsilon), \tag{7.30}$$

as shown in (2.17), we select a function $f(x;\epsilon)$ that is normalized in the sense that it integrates to 1,

$$\int f(x;\epsilon)\, \mathrm{d}x = 1. \tag{7.31}$$

On the other hand, a spatial mode, such as the detector mode, is normalized in such a way that its squared modulus integrates to 1,

$$\int |M(x;\epsilon)|^2\, \mathrm{d}x = 1. \tag{7.32}$$

Such a normalized mode would not work in the limit process for the Dirac delta function, because the integral of the square of a Dirac delta function is divergent. Therefore, we need to modify the prefactor of the mode so that we can treat it in the same way. This modification can be readily done for the Gaussian function, which is used in (2.17). Considering a one-dimensional Gaussian function

$$g(x; w) = N \exp\left(-\frac{x^2}{w^2}\right), \tag{7.33}$$

where N is the normalization constant. The *Euclidean norm* requires that

$$\int |g(x; w)|^2 \, dx = \frac{\sqrt{\pi} N^2 w}{\sqrt{2}} = 1, \tag{7.34}$$

so that

$$N = \frac{2^{1/4}}{\pi^{1/4} \sqrt{w}}. \tag{7.35}$$

Using this normalization constant when integrating the Gaussian mode, we get

$$\int g(x; w) \, dx = (2\pi)^{1/4} \sqrt{w}. \tag{7.36}$$

So, for the limit process, we need to separate the normalization constant into a part that contributes in the limit process in which we can replace $w \to \epsilon$ and a part that remains outside. For the one-dimensional case, it thus becomes

$$(2\pi)^{1/4} \sqrt{w} \lim_{\epsilon \to 0} \frac{1}{\sqrt{\pi}\epsilon} \exp\left(-\frac{x^2}{\epsilon^2}\right) = (2\pi)^{1/4} \sqrt{w} \delta(x). \tag{7.37}$$

The small value of $\sqrt{w}$ in the remaining prefactor now introduces the expected reduction in the *detection probability* per pixel due to the small size of a pixel detector compared to the overall size of the beam in the output plane.

For the two-dimensional detector mode, given in (7.27), the calculation is done in the input (front-focal) plane of the $2f$ system with integration over the Fourier domain. It requires that we define ϵ inversely proportional to the mode size w_0 in the input plane. The latter is defined in terms of the mode size w_{det} for the detector (pixel) in the output (back-focal) plane in (7.28). So, when w_{det} becomes small, w_0 becomes large. We also need to remember the 2π-factors that come with Fourier domain integrations. Ignoring the angular frequency part, we express the limit process with the detector mode as

$$\frac{\sqrt{2\pi} w_{\text{det}}}{\lambda_d f} \lim_{\epsilon \to 0} \frac{4\pi}{\epsilon^2} \exp\left(-\frac{|\mathbf{K} - \mathbf{K}_0|^2}{\epsilon^2}\right) = \frac{\sqrt{2\pi} w_{\text{det}}}{\lambda_d f} \delta\left(\mathbf{K} - \frac{k_d}{f} \mathbf{X}_0\right), \tag{7.38}$$

where we replaced $\mathbf{K}_0$ according to (7.29), with $k \to k_d$, in the argument of the two-dimensional Dirac delta function. Here, $\mathbf{X}_0$ is the location of the pixel in the output plane. In this way, $\mathbf{X}_0$ becomes the coordinates in terms of which the configuration space function in the output plane is defined. The single-mode detector kernel then becomes

$$D(\mathbf{k}_1, \mathbf{k}_2) = M(\mathbf{k}_1) M^*(\mathbf{k}_2) = \frac{k_d^2 w_{\text{det}}^2}{2\pi f^2} S(\omega_1, \omega_2) \delta\left(\mathbf{K}_1 - \frac{k_d}{f} \mathbf{X}_0\right) \delta\left(\mathbf{K}_2 - \frac{k_d}{f} \mathbf{X}_0\right), \tag{7.39}$$

where $S(\omega_1, \omega_2)$ represents the angular frequency kernel function. We need to remember that the kernel in (7.39) does not have a well-defined trace. It is, however, convenient for a quick calculation, because it allows a direct replacement of the front-focal plane Fourier domain variables with the output plane coordinates.

7.3 Intensity

Classically, the intensity of an optical field is given by the squared modulus of its complex amplitude that represents the electric field (under a *scalar approximation*). The photocurrent that is measured in a detector is (approximately) proportional to the intensity integrated over the area of the detector.

In the context of quantum physics, the measurement of intensity is understood differently. When we say that we measure the intensity, what we mean in the quantum context is that we are measuring the average number of photons in the field. Being a dimensionless quantity, the average number of photons needs to be multiplied by the appropriate parameters to convert it into an intensity. Formally, the average number of photons is usually obtained as the expectation value of the number operator, which can be seen as the overlap of an unnormalized state with itself after one photon has been removed by the application of the annihilation operator. The number of photons in that state follows from the *bosonic enhancement* of the probability for the observation of a single photon. However, the actual physical process in the detection of the intensity involves the detection of all photons allowed by the quantum efficiency of the detector. The detection process that gives the intensity can therefore also be represented by a different operation. The average number of photons can be expressed in terms of the probabilities to observe a certain number of photons. Here, we consider both these viewpoints.

7.3.1 Particle-number degree of freedom only

First, we consider the more traditional number operator approach. Crudely (ignoring all the other degrees of freedom), we express such a measurement as

$$\langle n \rangle = \langle \psi | \hat{n} | \psi \rangle = \langle \psi | \hat{a}^{\dagger} \hat{a} | \psi \rangle . \tag{7.40}$$

Here, we assumed the state is pure. More generally, it is given in terms of a density operator by $\langle n \rangle = \mathrm{tr}\{\hat{\rho}\hat{n}\}$. In that case, we can write the state as an ensemble average of pure states. For an arbitrary pure state, expressed as an expansion of Fock states, as in (3.63), the complex coefficients C_n in the expansion satisfy

$$\sum_n |C_n|^2 = 1, \tag{7.41}$$

where $|C_n|^2$ gives the probability for n photons. Then the expectation value reads

$$\langle\psi|\,\hat{n}\,|\psi\rangle = \sum_{mn} C_m^*\langle m|n\rangle n C_n = \sum_n n|C_n|^2 \equiv \langle n\rangle. \tag{7.42}$$

Next, we consider the alternative approach, where the detections of all the photons are considered as projective measurements associated with projection operators for the number of photons. The probability to measure n photons P_n is given by

$$P_n = \mathrm{tr}\{\hat{P}_n\hat{\rho}\}, \tag{7.43}$$

where $\hat{P}_n = |n\rangle\,\langle n|$ is the projection operator for n-photons. The expectation value of the number of photons is then expressed as

$$\langle n\rangle = \sum_n nP_n. \tag{7.44}$$

Applying this process to a pure state expanded in terms of Fock states, we get

$$\begin{aligned}\langle n\rangle &= \sum_n n\ \mathrm{tr}\left\{\hat{P}_n \sum_{m,r} |m\rangle\, C_m C_r^*\, \langle r|\right\} = \sum_n n \sum_{m,n} \langle n|m\rangle C_m C_r^* \langle r|n\rangle \\ &= \sum_n n|C_n|^2 = \langle n\rangle.\end{aligned} \tag{7.45}$$

So, we see that the two approaches give the same result in the case where we ignore all the other degrees of freedom. The number operator is thus represented in terms of projection operators by

$$\hat{n} = \sum_n n\hat{P}_n. \tag{7.46}$$

When we introduce spatiotemporal degrees of freedom into the state, the equivalence between the two ways to compute the average number of photons is maintained provided that the number operator and the projection operators involve all the degrees of freedom without bias. However, such a situation does not describe physical measurement systems. In a physical measurement of the intensity, the detection system invariably introduces restrictions on the spatiotemporal degrees of freedom. It is therefore necessary to model these restrictions in terms of the number operator and the projection operators. In the next section, we discuss the modelling of such physical detectors.

7.3.2 Localized measurements

The calculation in Section 7.3.1 does not represent the actual physical process of the measurement very well. While the physical process does not depend on the spectral function of the state being measured, it does impose constraints in terms of *spatiotemporal degrees of freedom*. In Section 7.2, we see that the physical detectors have finite areas. Moreover, physical measurements have finite durations or finite *integration periods*.

Consider, for example, a *localized number operator* as modelled by the single-mode detector kernel in (7.18) with $\eta = 1$, used for measurements on a fixed-spectrum Fock states as defined in Section 4.2.5 in terms of fixed-spectrum ladder operators. Based on (4.66), the effect of the localized number operator on such a fixed-spectrum Fock state is

$$\begin{aligned} \hat{n}_D |n_F\rangle &= \hat{a}^\dagger \diamond MM^* \diamond \hat{a} |n_F\rangle \\ &= \hat{a}^\dagger \diamond MM^* \diamond F |(n-1)_F\rangle \sqrt{n} \\ &= |(n-1)_F\rangle |1_M\rangle (M^* \diamond F) \sqrt{n}, \end{aligned} \tag{7.47}$$

where M is the detector mode and F is the parameter function of the Fock state. For the case where $M = F$, the final result is $|n_F\rangle n$. However, it is in general unlikely that the parameter function of the fixed-spectrum Fock states exactly matches the mode of the detector. For the case where they are not the same, we represent the overlap as $M^* \diamond F = \mu$ with $|\mu| < 1$. It then follows that[b]

$$\langle n_F| \hat{n}_D |n_F\rangle = \langle n_F|(n-1)_F, 1_M\rangle \mu \sqrt{n} = |\mu|^2 n. \tag{7.48}$$

The factor of $|\mu|^2$ represents a loss. However, this loss is independent of the occupation number. Therefore, any state that is expanded in terms of these fixed-spectrum Fock states, all with the same spectrum, produces an overall factor of $|\mu|^2$.

Next, we consider the same calculation in terms of the projection operators. We first need to modify the definition of the projection operators to introduce the detector kernel. There are different ways in which the detector kernel can be incorporated into the projection operators. Here, we first consider an intuitive generalization, *which turns out to be wrong*. It is obtained by generalizing the *single-photon projection operator* for all degrees of freedom, by introducing a detector kernel, so that

$$\hat{P}_1 = \sum_s \int |\mathbf{k}, s\rangle \langle \mathbf{k}, s| \, \mathrm{d}_\omega k = \sum_s \int \hat{a}_s^\dagger(\mathbf{k}) |\text{vac}\rangle \langle \text{vac}| \hat{a}_s(\mathbf{k}) \, \mathrm{d}_\omega k \tag{7.49}$$

$$\begin{aligned} \to \hat{P}_1^D &= \sum_{r,s} \int |\mathbf{k}, r\rangle D_{r,s}(\mathbf{k}, \mathbf{k}') \langle \mathbf{k}', s| \, \mathrm{d}_\omega k \, \mathrm{d}_\omega k' \\ &= \sum_{r,s} \int \hat{a}_r^\dagger(\mathbf{k}) |\text{vac}\rangle D_{r,s}(\mathbf{k}, \mathbf{k}') \langle \text{vac}| \hat{a}_s(\mathbf{k}) \, \mathrm{d}_\omega k. \end{aligned} \tag{7.50}$$

For the equivalent two-photon projection operator, two detector kernels are included that are pairwise contracted to the ladder operators on opposite sides, and so forth. The notation becomes significantly simpler for single-mode detector kernels. The projection operators are then simply expressed as

---

b Note that $|(n-1)_F\rangle |1_M\rangle = |(n-1)_F, 1_M\rangle$ is not a normalized state.

$$\hat{P}_n^M = |n_M\rangle \langle n_M| \,. \tag{7.51}$$

Overlapping these projection operators by two fixed-spectrum Fock states, we get

$$\langle n_F| \hat{P}_n^M |n_F\rangle = \langle n_F|n_M\rangle\langle n_M|n_F\rangle = |\mu|^{2n}, \tag{7.52}$$

based on (4.65). If the localized number operator can be expressed in terms of these projection operators, by

$$\hat{n}'_D = \sum_n n\hat{P}_n^M, \tag{7.53}$$

then we would get

$$\langle n_F| \hat{n}'_D |n_F\rangle = n|\mu|^{2n}, \tag{7.54}$$

which is different from what we had for $\hat{n}_D$. So, something is wrong.

Exercise 7.1. Show that projection operators with idempotent detector kernels are idempotent.

7.3.2.1 Physical understanding

To understand what the difference is between the two approaches, we consider a simplified scenario. We divide the plane in which the detector lies into discrete points denoted by an index m, so that we can write an arbitrary *single-photon state* as

$$|\psi\rangle = \sum_m |\mathbf{x}_m\rangle \, \psi(\mathbf{x}_m), \tag{7.55}$$

where $\psi(\mathbf{x})$ is the single-photon wavefunction. A multiphoton state with N photons where all photons have the same wavefunction is then given by[c]

$$|N_\psi\rangle = \sum_{m..p} |\mathbf{x}_m\rangle \cdots |\mathbf{x}_p\rangle \mathcal{N}_{m..p}\psi(\mathbf{x}_m)\cdots\psi(\mathbf{x}_p), \tag{7.56}$$

where $\mathcal{N}_{m..p}$ is the appropriate combinatorial factor to ensure normalization. Say the detector is located as $\mathbf{x} = \mathbf{x}_0$, then we see that the multiphoton state contains terms with a varying number of photons located at $\mathbf{x}_0$, ranging from 0 to N. Each term has a complementary number of photons located at all the other possible locations.

Assume that the detector has a perfect quantum efficiency so that $\eta = 1$. Then a term with a certain number of photons located at the detector leads to the successfully detection of all those photons, giving a photon-current proportional to that number of

---

c Note that the order of the kets has no meaning because the state is symmetrized.

photons. All the other photons located at other points are not detected. They are traced out, leading to a mixed state given by

$$\begin{aligned}\hat{\rho}_N(\mathbf{x}_0) &= \text{tr}_{\mathbf{x}\neq\mathbf{x}_0}\{|N_\psi\rangle\langle N_\psi|\} \\ &= \sum_{n=0}^{N} |n(\mathbf{x}_0)\rangle \frac{N!}{n!(N-n)!}|\psi(\mathbf{x}_0)|^{2n}\left[1-|\psi(\mathbf{x}_0)|^2\right]^{N-n}\langle n(\mathbf{x}_0)|,\end{aligned} \tag{7.57}$$

where $|n(\mathbf{x}_0)\rangle$ is a Fock state with n photons located at $\mathbf{x}_0$. The normalization of the wavefunction implies that

$$\sum_{\mathbf{x}\neq\mathbf{x}_0} |\psi(\mathbf{x})|^2 = 1 - |\psi(\mathbf{x}_0)|^2. \tag{7.58}$$

Exercise 7.2. Show that the mixed state in (7.57) is properly normalized.

We see that, even if the number of photons is fixed for the entire state, the number of photons detected at the location of the detector can vary. This situation is not obtained from the projection operators that we defined above, unless all the photons are located at the detector. Therefore, the projection operator approach that we considered above cannot give the correct result.

7.3.2.2 Localized projection operator

Based on this understanding, we can now propose a *localized projection operator* that gives the correct result. Instead of requiring all photons to be at the same location, the projection operator can be separated into two parts, one part operates on those photons that are found at the detector location and the other part operates on those that are everywhere else. Each part is associated with a specific number of photons. One can represent such a projection operator as

$$\hat{R}_{N,n} \triangleq \hat{P}_{n,D}\hat{Q}_{N-n,D}. \tag{7.59}$$

where $\hat{P}_{n,D}$ is the same projection operator that we had before, and $\hat{Q}_{m,D}$ differs from $\hat{P}_{n,D}$ only in that the detector kernel D is replaced by $\mathbf{1} - D$, where $\mathbf{1}$ is the identity operator for all single-photon states. Their general expressions (using the summation convention for the spin indices) are

$$\begin{aligned}\hat{P}_{n,D} = \frac{1}{n!}\int & \hat{a}_r^\dagger(\mathbf{k}_1)\cdots\hat{a}_u^\dagger(\mathbf{k}_n)\,|\text{vac}\rangle\, D_{r,s}(\mathbf{k}_1,\mathbf{k}_1')\cdots D_{u,v}(\mathbf{k}_n,\mathbf{k}_n') \\ &\times \langle\text{vac}|\,\hat{a}_s(\mathbf{k}_1')\cdots\hat{a}_v(\mathbf{k}_n')\, \mathrm{d}_\omega k_1 \ldots \mathrm{d}_\omega k_n\, \mathrm{d}_\omega k_1' \cdots \mathrm{d}_\omega k_n',\end{aligned} \tag{7.60}$$

and

$$\hat{Q}_{m,D} = \frac{1}{n!} \int \hat{a}_r^\dagger(\mathbf{k}_1) \cdots \hat{a}_u^\dagger(\mathbf{k}_m) |\text{vac}\rangle$$
$$\times [\mathbf{1}(\mathbf{k}_1, \mathbf{k}_1') - D_{r,s}(\mathbf{k}_1, \mathbf{k}_1')] \cdots [\mathbf{1}(\mathbf{k}_m, \mathbf{k}_m') - D_{u,v}(\mathbf{k}_m, \mathbf{k}_m')]$$
$$\times \langle\text{vac}| \hat{a}_s(\mathbf{k}_1') \cdots \hat{a}_v(\mathbf{k}_m')\, \mathrm{d}_\omega k_1 \cdots \mathrm{d}_\omega k_m\, \mathrm{d}_\omega k_1' \cdots \mathrm{d}_\omega k_m'. \tag{7.61}$$

The modified projection operator is thus given by

$$\hat{R}_{N,n} = \frac{1}{n!(N-n)!} \int \hat{a}_r^\dagger(\mathbf{k}_1) \cdots \hat{a}_u^\dagger(\mathbf{k}_N) |\text{vac}\rangle D_{r,s}(\mathbf{k}_1, \mathbf{k}_1') \cdots D_{u,v}(\mathbf{k}_n, \mathbf{k}_n')$$
$$\times [\mathbf{1}(\mathbf{k}_{n+1}, \mathbf{k}_{n+1}') - D_{r,s}(\mathbf{k}_{n+1}, \mathbf{k}_{n+1}')] \cdots [\mathbf{1}(\mathbf{k}_N, \mathbf{k}_N') - D_{u,v}(\mathbf{k}_N, \mathbf{k}_N')]$$
$$\times \langle\text{vac}| \hat{a}_s(\mathbf{k}_1') \cdots \hat{a}_v(\mathbf{k}_N')\, \mathrm{d}_\omega k_1 \cdots \mathrm{d}_\omega k_N\, \mathrm{d}_\omega k_1' \cdots \mathrm{d}_\omega k_N'. \tag{7.62}$$

The localized projection operator $\hat{R}_{N,n}$ selects out all those states with a total of N photons of which only n of them are located at the detector, and the rest are anywhere else but at the detector. It gives the probability $P_{N,n}$ for a state with N photons:

$$P_{N,n} = \langle N_F| \hat{R}_{N,n} |N_F\rangle = \frac{N!}{n!(N-n)!} |\mu|^{2n} \left(1 - |\mu|^2\right)^{N-n}. \tag{7.63}$$

The expectation value for the number of photons at the detector for a fixed-spectrum pure state expressed in terms of fixed-spectrum Fock state by

$$|\psi_F\rangle = \sum_{N=0}^{\infty} |N_F\rangle\, C_N, \tag{7.64}$$

is then given by

$$\sum_{N=0}^{\infty} |C_N|^2 \sum_{n=0}^{N} n \langle N_F| \hat{R}_{N,n} |N_F\rangle = \sum_{N=0}^{\infty} |C_N|^2 \sum_{n=0}^{N} n \frac{N!}{n!(N-n)!} |\mu|^{2n} \left(1 - |\mu|^2\right)^{N-n}$$
$$= \sum_{N=0}^{\infty} |C_N|^2 N |\mu|^2 = |\mu|^2 \langle N\rangle\, . \tag{7.65}$$

It gives the same result as obtained for the number operator. We have now resolved the discrepancy between the two approaches and modified the projection operators to give the same result as the number operator approach.

It may seem that we could have just stayed with the number operators, which give the correct result based on the *bosonic enhancement* at the location of the detector. However, these projection operators are used again below. Therefore, it helps to understand how they need to be modelled correctly.

7.3.3 Squeezed state intensity

As an example, we consider the measurement of the intensity of a squeezed vacuum state, as represented by a measurement of the photon-number density per unit time. The Wigner functional expression for such a measurement is given by

$$\begin{aligned}\langle n\rangle_{\text{sq}} = & \int \mathcal{N}_0 \exp\left(-2\alpha^* \diamond A \diamond \alpha - \alpha^* \diamond B \diamond \alpha^* - \alpha \diamond B^* \diamond \alpha\right) \\ & \times \left(\alpha^* \diamond D \diamond \alpha - \frac{1}{2}\operatorname{tr}\{D\}\right) \mathcal{D}^\circ[\alpha]. \end{aligned} \tag{7.66}$$

The integrand is converted to Gaussian form by representing the number operator's Wigner functional as a generating function. The functional integration then leads to

$$\begin{aligned}\langle n\rangle_{\text{sq}} = & \partial_J \int \mathcal{N}_0 \exp\Big(- 2\alpha^* \diamond A \diamond \alpha - \alpha^* \diamond B \diamond \alpha^* - \alpha \diamond B^* \diamond \alpha \\ & + J\alpha^* \diamond D \diamond \alpha - \frac{1}{2}J\operatorname{tr}\{D\}\Big) \mathcal{D}^\circ[\alpha]\Big|_{J=0} \\ = & \; \partial_J \frac{\exp\left(-\frac{1}{2}J\operatorname{tr}\{D\}\right)}{\sqrt{\det\{A_J\}}\sqrt{\det\{A_J^* - B^* \diamond A_J^{-1} \diamond B\}}}\Bigg|_{J=0}, \end{aligned} \tag{7.67}$$

where

$$A_J \triangleq A - \frac{1}{2}JD. \tag{7.68}$$

To evaluate the derivative, we compute derivatives of the *functional determinant* and the inverse inside it, using the identities provided in Appendix D.5. After applying the derivative with respect to J with the aid of the two identities and setting $J = 0$, the expression becomes

$$\begin{aligned}\langle n\rangle_{\text{sq}} = & \Big[-\frac{1}{2}\operatorname{tr}\{D\} + \frac{1}{4}\operatorname{tr}\left\{A^{-1} \diamond D\right\} \\ & + \frac{1}{4}\operatorname{tr}\left\{\left(A^* - B^* \diamond A^{-1} \diamond B\right)^{-1} \diamond \left(D^* + B^* \diamond A^{-1} \diamond D \diamond A^{-1} \diamond B\right)\right\}\Big] \\ & \times \left[\det\{A\}\det\{A^* - B^* \diamond A^{-1} \diamond B\}\right]^{-1/2}. \end{aligned} \tag{7.69}$$

Since the squeezed vacuum state is a pure state, we can apply the identities in Section 6.8.5 to simplify this result. Then it reads

$$\langle n\rangle_{\text{sq}} = \frac{1}{2}\operatorname{tr}\{A \diamond D\} - \frac{1}{2}\operatorname{tr}\{D\} = \frac{1}{2}\operatorname{tr}\{(A - \mathbf{1}) \diamond D\}. \tag{7.70}$$

We see that only A is involved in the measurement of the intensity.

7.4 Photon statistics

What is the *probability* to detect a given number of photons? For the *particle-number degree of freedom* only, this probability is obtained with

$$P_n = \text{tr}\{\hat{\rho}\hat{P}_n\} = \langle n|\,\hat{\rho}\,|n\rangle\,, \tag{7.71}$$

where $\hat{P}_n = |n\rangle\langle n|$ is the *photon-number projection operator* for n photons, represented in terms of the n-photon Fock state. We encountered these projection operators in the context of intensity measurements in Section 7.3. Instead of using the heuristic discrete points that we used there, we perform a more thorough analysis in terms of wavevectors. Here, we consider the application of these projection operators in the context of photon statistics, while incorporating the spatiotemporal degrees of freedom, giving us an opportunity to consider them in terms of Wigner functional theory.

7.4.1 Ideal photon-number projection operators

When we include all the degrees of freedom in the definition of the *ideal photon-number projection operators*, we end up with an expression

$$\hat{P}_n = \frac{1}{n!}\hat{P}_1^{\otimes n} = \frac{1}{n!}\left(\sum_s \int |\mathbf{k},s\rangle\langle\mathbf{k},s|\;\mathrm{d}_\omega k\right)^{\otimes n}, \tag{7.72}$$

where $\hat{P}_1$ is the single-photon projection operator defined in (7.49) and the superscript $\otimes n$ represents a tensor product of n copies of this operator. Such an ideal photon-number projection operator does not take the spatiotemporal restrictions imposed by a physical detector system into account. (Such an ideal projection operator can be used to model a *bucket detector* when the light falling on the detector is already constrained, making any additional constraints superfluous.)

To compute the Wigner functional for these ideal projection operators, we use the coherent-state-assisted approach. Overlapping the projection operators with two coherent states, we have

$$\begin{aligned}
\langle\alpha_1|\,\hat{P}_n\,|\alpha_2\rangle &= \frac{1}{n!}\langle\alpha_1|\left(\int |\mathbf{k}\rangle\langle\mathbf{k}|\;\mathrm{d}_\omega k\right)^{\otimes n}|\alpha_2\rangle \\
&= \frac{1}{n!}\int \langle\alpha_1|\left[\prod_{u=1}^{n}\hat{a}^\dagger(\mathbf{k}_u)\right]|\text{vac}\rangle\langle\text{vac}|\left[\prod_{v=1}^{n}\hat{a}(\mathbf{k}_v)\right]|\alpha_2\rangle\prod_{t=1}^{n}\mathrm{d}_\omega k_t \\
&= \frac{1}{n!}\int\left[\prod_{u=1}^{n}\alpha_1^*(\mathbf{k}_u)\right]\langle\alpha_1|\text{vac}\rangle\langle\text{vac}|\alpha_2\rangle\left[\prod_{v=1}^{n}\alpha_2(\mathbf{k}_v)\right]\prod_{t=1}^{n}\mathrm{d}_\omega k_t \\
&= \frac{1}{n!}\exp\left(-\frac{1}{2}\|\alpha_1\|^2 - \frac{1}{2}\|\alpha_2\|^2\right)(\alpha_1^*\diamond\alpha_2)^n\,.
\end{aligned} \tag{7.73}$$

Next, we define a generating function

$$\mathcal{G} = \exp\left(-\frac{1}{2}\|\alpha_1\|^2 - \frac{1}{2}\|\alpha_2\|^2 + J\alpha_1^* \diamond \alpha_2\right). \tag{7.74}$$

The original overlap is reproduced by

$$\langle\alpha_1|\,\hat{P}_n\,|\alpha_2\rangle = \frac{1}{n!}\,\partial_J^n \mathcal{G}\Big|_{J=0}. \tag{7.75}$$

The equivalent generating function for the projection operators is

$$\hat{\mathcal{P}}(J) \triangleq \sum_{n=0}^{\infty} J^n \hat{P}_n, \tag{7.76}$$

where $\hat{P}_0 = |\text{vac}\rangle\,\langle\text{vac}|$. The equivalence between this generating function and the overlap for the Wigner functional is given by

$$\frac{1}{n!}\,\partial_J^n\,\langle\alpha_1|\,\hat{\mathcal{P}}(J)\,|\alpha_2\rangle\Big|_{J=0} = \frac{1}{n!}\,\partial_J^n \mathcal{G}\Big|_{J=0}. \tag{7.77}$$

When we substitute $\mathcal{G}$ into (6.20) and evaluate the functional integrations, we get

$$\begin{aligned} \mathcal{W}(J) &= \mathcal{N}_0 \int \exp\left(-2\alpha^* \diamond \alpha + 2\alpha^* \diamond \alpha_1 + 2\alpha_2^* \diamond \alpha - \alpha_1^* \diamond \alpha_1\right. \\ &\quad \left. - \alpha_2^* \diamond \alpha_2 - \alpha_2^* \diamond \alpha_1 + J\alpha_1^* \diamond \alpha_2\right) \mathcal{D}^\circ[\alpha_1, \alpha_2] \\ &= \frac{\mathcal{N}_0}{(1+J)^\Omega} \exp\left(-2\frac{1-J}{1+J}\alpha^* \diamond \alpha\right). \end{aligned} \tag{7.78}$$

It is a generating function for the Wigner functional of the projection operator for a fixed number of photons. Following from (7.75), the probability for n photons in a state $\hat{\rho}$ is now obtained by

$$\text{tr}\{\hat{\rho}\hat{P}_n\} = \frac{1}{n!}\,\partial_J^n \int \mathcal{W}(J) W_{\hat{\rho}}[\alpha]\,\mathcal{D}^\circ[\alpha]\Big|_{J=0}, \tag{7.79}$$

where $W_{\hat{\rho}}[\alpha]$ is the Wigner functional of the state.

7.4.2 Localized photon-number projection operators

In practice, one cannot always implement an ideal photon-number projection operation. The physical detector system invariably involves a filtering process on the *spatiotemporal degrees of freedom* as well. To model this effect, we incorporate a kernel into the photon-number projection operator. Such a localized photon-number projection operator is developed in Section 7.3.2 in terms of detector kernels. It is shown that the projection operator defined in (7.62) produces results that are consistent with those obtained from a localized number operator.

In the case where the full projection operator consists of the tensor product of two different projection operators, the generating function for the Wigner functional of the full projection operator is the product of their respective generating functions. However, the calculation of the combinatorics is rather complicated. Instead, we follow the previous derivation, incorporating both projection operators. Here, we replace the detector kernel by the idempotent projection kernel $P(\mathbf{k},\mathbf{k}')$, and define its complement by $Q = \mathbf{1} - P$. In this case, the overlap produces

$$\begin{aligned}
\langle\alpha_1|\,\hat{R}_{N,n}\,|\alpha_2\rangle &= \frac{1}{n!(N-n)!}\,\langle\alpha_1|\left(\int |\mathbf{k}\rangle\, P(\mathbf{k},\mathbf{k}')\,\langle\mathbf{k}'|\ \mathrm{d}_\omega k\ \mathrm{d}_\omega k'\right)^{\otimes n} \\
&\quad\times\left(\int |\mathbf{k}\rangle\, Q(\mathbf{k},\mathbf{k}')\,\langle\mathbf{k}'|\ \mathrm{d}_\omega k\ \mathrm{d}_\omega k'\right)^{\otimes(N-n)}|\alpha_2\rangle \\
&= \frac{1}{n!(N-n)!}\int \langle\alpha_1|\left[\prod_{u=1}^{N}\hat{a}^\dagger(\mathbf{k}_u)\right]|\mathrm{vac}\rangle\left[\prod_{u=1}^{n}P(\mathbf{k}_u,\mathbf{k}_u')\right] \\
&\quad\times\left[\prod_{v=n+1}^{N}Q(\mathbf{k}_v,\mathbf{k}_v')\right]\langle\mathrm{vac}|\left[\prod_{v=1}^{N}\hat{a}(\mathbf{k}_v')\right]|\alpha_2\rangle\prod_{t=1}^{N}\mathrm{d}_\omega k_t\ \mathrm{d}_\omega k_t' \\
&= \frac{1}{n!(N-n)!}\exp\left(-\frac{1}{2}\|\alpha_1\|^2-\frac{1}{2}\|\alpha_2\|^2\right) \\
&\quad\times\left(\alpha_1^*\diamond P\diamond\alpha_2\right)^n\left(\alpha_1^*\diamond Q\diamond\alpha_2\right)^{N-n}.
\end{aligned}\tag{7.80}$$

The required generating function in this case is

$$\mathcal{G}_{\hat{R}}(J,K) = \exp\left(-\frac{1}{2}\|\alpha_1\|^2-\frac{1}{2}\|\alpha_2\|^2+J\alpha_1^*\diamond P\diamond\alpha_2+K\alpha_1^*\diamond Q\diamond\alpha_2\right),\tag{7.81}$$

from which any overlap can be obtained by

$$\langle\alpha_1|\,\hat{R}_{N,n}\,|\alpha_2\rangle = \frac{1}{n!(N-n)!}\,\partial_J^n\partial_K^{N-n}\mathcal{G}_{\hat{R}}(J,K)\Big|_{J=K=0}.\tag{7.82}$$

From (6.20), we now get

$$\mathcal{W}(J,K) = \frac{\mathcal{N}_0\exp\left[-2\alpha^*\diamond(1+JP+KQ)^{-1}\diamond(1-JP+KQ)\diamond\alpha\right]}{\det\{1+JP+KQ\}}.\tag{7.83}$$

Using the identities in Appendix D and the fact that P and Q are idempotent kernels, respectively, associated with orthogonal complements, we compute the determinant:

$$\det\{1+JP+KQ\} = (1+J)^{\mathrm{tr}\{P\}}(1+K)^{\mathrm{tr}\{Q\}},\tag{7.84}$$

and the inverse:

$$(1+JP+KQ)^{-1}=1-\frac{JP}{1+J}-\frac{KQ}{1+K}. \tag{7.85}$$

Then the expression for the generating function simplifies to

$$\mathcal{W}(J,K)=\frac{\mathcal{N}_0\exp(-2\alpha^*\diamond\alpha)}{(1+J)^{\mathrm{tr}\{P\}}(1+K)^{\mathrm{tr}\{Q\}}}\exp\left(\frac{4J\alpha^*\diamond P\diamond\alpha}{1+J}+\frac{4K\alpha^*\diamond Q\diamond\alpha}{1+K}\right). \tag{7.86}$$

Usually, the total number of photons in the state is of no concern; only the number of photons that can pass through P is of interest. Therefore, we sum over (or trace over) all the other photons, those that can pass through Q. It is easily done by setting $K=1$. Noting that $Q=\mathbf{1}-P$ and $\Omega=\mathrm{tr}\{\mathbf{1}\}$, we get

$$\mathcal{P}(J)\triangleq\mathcal{W}(J,1)=\mathcal{N}_J^{\mathrm{tr}\{D\}}\exp\left(-2\mathcal{J}\alpha^*\diamond D\diamond\alpha\right), \tag{7.87}$$

where we replace P by D to represent an idempotent detector kernel, and defined

$$\mathcal{J}\triangleq\frac{1-J}{1+J}\quad\text{and}\quad\mathcal{N}_J\triangleq\frac{2}{1+J}. \tag{7.88}$$

The expression in (7.87) is extremely versatile. It is the most useful generating function for the purpose of measurements, because it gives the photon statistics from which most other quantities can be computed. We denote it by the special notation $\mathcal{P}$.

Exercise 7.3. Use (7.87) to compute a generating function for the photon statistics of a coherent state with parameter function ζ. Show that it gives the generating function for the *Poisson distribution* obtained in (3.107) with $\langle n\rangle=\zeta^*\diamond D\diamond\zeta$.

Exercise 7.4. Show that the generating function for the photon statistics of a squeezed vacuum state, computed with (7.87) using a single-mode detector kernel, is given by

$$\mathcal{G}_{\mathrm{stat}}^{(\mathrm{sq})}=\frac{\mathcal{N}_J}{\sqrt{(1+\mathcal{J}\mu)^2-\mathcal{J}^2|\beta|^2}}=\frac{2}{\sqrt{[1+J+(1-J)\mu]^2-(1-J)^2|\beta|^2}}, \tag{7.89}$$

where

$$\mu=M^*\diamond A\diamond M\quad\text{and}\quad\beta=M\diamond A^*\diamond B^*\diamond A^{-1}\diamond M. \tag{7.90}$$

Hint: Use the identities for pure squeezed vacuum state kernels in Section 6.8.5 and those for determinants and inverses in Appendix D to simplify the result of the functional integration.

Exercise 7.5. Use the generating function for the photon statistics of a squeezed vacuum state in (7.89) to show that the average number of photons and the variance in the number of photons in a squeezed state are, respectively,

$$\langle n\rangle=\frac{1}{2}(\mu-1)\quad\text{and}\quad\sigma^2=\frac{1}{4}\left(\mu^2+|\beta|^2-1\right). \tag{7.91}$$

Part of the significance of $\mathcal{P}(J)$ is the fact that it readily reproduces results from other operators associated with measurements. The identity operator is obtained for $J = 1$; it is easy to show that $\mathcal{P}(1) = 1$. The number operator is given by the sum over all projection operators with coefficients equal to the integer n, as shown in (7.46). It thus follows that the Wigner functional of the localized number operator is obtained from $\mathcal{P}(J)$ by

$$W_{\hat{n}} = \partial_J \mathcal{P}(J)\big|_{J=1} . \tag{7.92}$$

So, it is not necessary to use of the localized number operator in calculations, because $\mathcal{P}(J)$ is already in Gaussian form (as required for functional integrations) and it can reproduce the results obtained for the localized number operator.

7.4.3 Thermal state photon statistics

To demonstrate the typical calculations associated with the measurement of photon statistics, we consider the interesting case of a thermal state. For this purpose, the generating function for the Wigner functionals of the ideal n-photon projection operators in (7.78) can be used, or its equivalent with a more realistic detector kernel given in (7.87). The thermal state is represented generically in (6.181). Here, we consider a thermal state that is physically realizable, having a kernel of the form given in (6.194). An example is the single-mode thermal state with a kernel defined in (6.204).

First, we use the general projection operator applied to a thermal state with finite energy. The generating function for the Wigner functionals of these general n-photon projection operators is given in (7.87). The measurement is computed from the trace

$$\begin{aligned}
\mathrm{tr}\{\hat{\mathcal{P}}(J)\hat{\rho}_{\mathrm{th}}\} &= \int \mathcal{P}(J) W_{\mathrm{th}}[\alpha]\ \mathcal{D}^{\circ}[\alpha] \\
&= \mathcal{N}_0 \det\{\theta\} \mathcal{N}_J^{\mathrm{tr}\{D\}} \int \exp\left(-2\mathcal{J}\alpha^* \diamond D \diamond \alpha - 2\alpha^* \diamond \theta \diamond \alpha\right)\ \mathcal{D}^{\circ}[\alpha] \\
&= \frac{\mathcal{N}_J^{\mathrm{tr}\{D\}}}{\det\left\{\mathbf{1} + \mathcal{J}D \diamond \theta^{-1}\right\}} = \frac{2^{\mathrm{tr}\{D\}}}{\det\left\{(1+J)\mathbf{1} + (1-J)D \diamond \theta^{-1}\right\}},
\end{aligned} \tag{7.93}$$

where $\hat{\mathcal{P}}(J)$ is defined in (7.76) with an assumed detector kernel incorporated, and $\mathcal{J}$ and $\mathcal{N}_J$ are defined in (7.88).

7.4.3.1 Ideal projections

In the case where the projection operators are ideal n-photon projection operators, we can replace the detector kernel by the identity. We also take the thermal state kernel as given by (6.194). Then the trace becomes

$$\mathrm{tr}\{\hat{\mathcal{P}}(J)\hat{\rho}_{\mathrm{th}}\} = \frac{\mathcal{N}_0}{\det\left\{(1-J)\theta^{-1} + (1+J)\right\}} = \frac{\det\{\mathbf{1} - XP\}}{\det\{\mathbf{1} - JXP\}}, \tag{7.94}$$

where

$$X = \exp\left(-\frac{\omega}{\omega_T}\right), \tag{7.95}$$

and P is an idempotent kernel that ensures a physical thermal state with a finite energy or finite power. To compute the *functional determinant*, we expand

$$\ln(\mathbf{1} - JXP) = -\sum_{m=1}^{\infty} \frac{J^m X^m}{m} P. \tag{7.96}$$

The m-dependent traces can be computed, under the assumption that P has a given function on its diagonal. Here, we consider the case where P is given by the convolution kernel in (6.196). The traces are then expressed as

$$\begin{aligned}\operatorname{tr}\{X^m P\} &= \int \exp\left(-\frac{m\omega}{\omega_T}\right) P(\mathbf{k},\mathbf{k})\, \mathrm{d}_\omega k \\ &= \int \exp\left(-\frac{m\omega}{\omega_T}\right) c\omega \int p(\mathbf{X})\, \mathrm{d}^2x \int d(t)\, \mathrm{d}t\, \mathrm{d}_\omega k \\ &= cA\Delta t \int \exp\left(-\frac{m\omega}{\omega_T}\right) \frac{\mathrm{d}^3 k}{(2\pi)^3} = \frac{\eta_T}{m^3},\end{aligned} \tag{7.97}$$

where

$$\eta_T \triangleq \frac{A\Delta t \omega_T^3}{\pi^2 c^2}, \tag{7.98}$$

is a dimensionless quantity. The average number of photons in the state is $N = \eta_T \zeta(3)$, where $\zeta(\cdot)$ is the Riemann zeta-function [2]. Therefore,

$$\det\{\mathbf{1} - JXP\} = \exp\left(-\sum_{m=1}^{\infty} \frac{\eta_T J^m}{m^4}\right) = \exp\left[-\eta_T \operatorname{polylog}(4, J)\right], \tag{7.99}$$

where $\operatorname{polylog}(n, x)$ is the general *poly-logarithm function* defined as

$$\operatorname{polylog}(n, x) = \sum_{m=1}^{\infty} \frac{x^m}{m^n} \quad \text{for } |x| < 1, \tag{7.100}$$

or

$$\operatorname{polylog}(n, x) = \frac{x}{\Gamma(n)} \int_0^{\infty} \frac{z^{n-1}}{\exp(z) - x}\, \mathrm{d}z \quad \text{for } \operatorname{Re}\{n\} > 0. \tag{7.101}$$

Setting $J = 1$, we get

$$\det\{\mathbf{1} - XP\} = \exp\left(-\frac{\pi^4 \eta_T}{90}\right). \tag{7.102}$$

So, the generating function becomes

$$\mathrm{tr}\{\hat{\mathcal{P}}(J)\hat{\rho}_{\mathrm{th}}\} = \exp\left[\eta_T \,\mathrm{polylog}(4,J) - \frac{\pi^4\eta_T}{90}\right]. \tag{7.103}$$

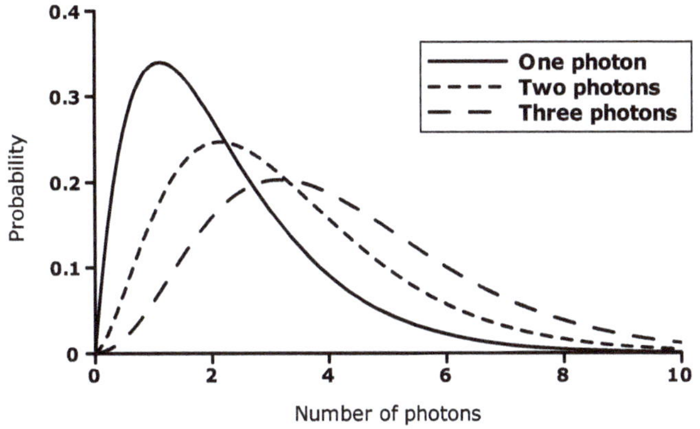

Figure 7.1: Probability curves for thermal states with an average of one, two or three photons.

Plotting the probability curves for thermal states with different average numbers of photons as functions of the number of photons in the state in Figure 7.1, we find that the probabilities for larger numbers of photons are larger than those for smaller numbers of photons when the average number of photons in the state is large enough. Without the other degrees of freedom, a thermal state in terms of only the particle-number degree of freedom produces probabilities for larger numbers of photons that are always smaller than those for smaller numbers of photons. The reason why the inclusion of the other degrees of freedom changes this situation is that there are far more degrees of freedom associated with larger numbers of photons than smaller numbers of photons. If we trace over these extra degrees of freedom, the probabilities are enhanced.

7.4.3.2 Single-mode projections

We can reproduce the scenario where the probabilities for larger numbers of photons are always smaller than those for smaller numbers of photons by considering single-mode projective measurements. Such measurements effectively remove the contributions of the other degrees of freedom. For this purpose, we replace the detector kernel in (7.93) by a single-mode detector kernel $D = MM^*$. We also use a single-mode kernel for the thermal state as given in (6.204).

The trace now becomes

$$\mathrm{tr}\{\hat{\mathcal{P}}(J)\hat{\rho}_{\mathrm{th}}\} = \frac{2}{\det\{(1+J)\mathbf{1} + (1-J)MM^* \diamond (\mathbf{1} + 2N\Theta\Theta^*)\}} = \frac{1}{[1+(1-J)N|\tau|^2]}, \tag{7.104}$$

where M and Θ are normalized modes associated with the detector and the thermal state, respectively, and $\tau \triangleq M^* \diamond \Theta$. It can be regarded as an *overlap efficiency* affecting the detection efficiency. The determinant is evaluated with the aid of Appendix D. The photon statistics (probabilities for the different numbers of photons) is now obtained as a power series expansion in terms of the generating parameter J. For $\tau = 1$ (when $M = \Theta$), these probabilities are

$$P_n = \frac{N^n}{(1+N)^{n+1}}. \tag{7.105}$$

For a given average number of photons N in the state, the probabilities for larger numbers of photons now always exceed those for smaller numbers of photons.

7.5 First-order correlations

One of the most characteristic aspects of quantum physics is the effect of *quantum correlations*. Historically, these correlations started within the classical context as *optical interference* measurements, which led to the notion of *coherence*, as discussed in Section 2.10. Optical interference measures the *first-order correlations* between different optical fields (or different parts of the same optical field), by performing measurements with one detector where the different optical fields overlap. It is represented by the *first-order correlation function* in (2.232).

The quantum version of the *first-order correlation function* presented in Section 2.10 is obtained by replacing the classical fields in Section 2.10.1 by *field operators*, as discussed in Section 4.1. It then reads

$$\Gamma^{(1)}(\mathbf{x}_1, t_1, \mathbf{x}_2, t_2) = \left\langle \hat{E}^{+}(\mathbf{x}_1, t_1)\hat{E}^{-}(\mathbf{x}_2, t_2)\right\rangle. \tag{7.106}$$

The intensity is proportional to the photon-number distribution. It is given by the first-order correlation function evaluated at the same point $I(\mathbf{x}, t) \propto \langle n(\mathbf{x}, t)\rangle = \Gamma^{(1)}(\mathbf{x}, t, \mathbf{x}, t)$. The quantum version of the *complex degree of coherence* is given by

$$g^{(1)}(\mathbf{x}_1, t_1, \mathbf{x}_2, t_2) = \frac{\Gamma^{(1)}(\mathbf{x}_1, t_1, \mathbf{x}_2, t_2)}{\sqrt{\Gamma^{(1)}(\mathbf{x}_1, t_1, \mathbf{x}_1, t_1)\Gamma^{(1)}(\mathbf{x}_2, t_2, \mathbf{x}_2, t_2)}}, \tag{7.107}$$

corresponding to the classical version in (2.235). The properties of the complex degree of coherence in quantum scenarios are the same as those in the classical case: the maximum value of $g^{(1)} = 1$ occurs when the two points coincide.

Although optical interferometry, discussed in Section 2.10.1, is not directly associated with quantum optics, interferometric systems, such as those shown in Figure 2.11, are often used as part of quantum optical systems. Therefore, we consider the operation of such a measurement on a quantum state.

If the input state is a pure state, the unitary transformation performed by the interferometer can be expressed as

$$|\psi_{\text{out}}\rangle = \hat{U}_{\text{bs}}^{\dagger}\hat{U}_{\phi}\hat{U}_{\text{bs}}\left(|\psi_{\text{in}}\rangle_A \otimes |\text{vac}\rangle_B\right), \tag{7.108}$$

where $\hat{U}_{\text{bs}}$ is the unitary operator for the beamsplitter and $\hat{U}_{\phi}$ is the unitary operator for the relative phase modulation. The second beamsplitter is implemented with the adjoint unitary operator so that it would cancel the first one if there is no phase modulation. In the end, we trace over the unobserved output port. The entire process is expressed in terms of the density operator for the input state as

$$\hat{\rho}_{\text{out},A} = \text{tr}_B\left\{\hat{U}_{\text{bs}}^{\dagger}\hat{U}_{\phi}\hat{U}_{\text{bs}}\left(\hat{\rho}_{\text{in},A} \otimes \hat{\rho}_{\text{vac},B}\right)\hat{U}_{\text{bs}}^{\dagger}\hat{U}_{\phi}^{\dagger}\hat{U}_{\text{bs}}\right\}. \tag{7.109}$$

For convenience, we can combine the three unitary operators into one operator representing the entire unitary operator for the interferometer $\hat{U}_{\text{intf}} = \hat{U}_{\text{bs}}\hat{U}_{\phi}\hat{U}_{\text{bs}}^{\dagger}$.

Formulating the measurement process of an interferometer in terms of Wigner functionals, we obtain the combined process with $W_{\text{intf}} = W_{\text{bs}} \star W_{\phi} \star W_{\text{bs}}^{*}$. However, the transformations of the field variables in (6.218) representing the homogeneous beamsplitter and those in (6.128) representing the phase modulation provide an easier way to implement the process. The phase modulation in (6.128) is applied to one field variable α, while the other field variable β is modulated by the adjoint unitary kernels. The combined transformation from the input to the output is given by

$$\begin{aligned} \alpha &\to \cos_{\diamond}(\Phi) \diamond \alpha - \sin_{\diamond}(\Phi) \diamond \beta, \\ \beta &\to \cos_{\diamond}(\Phi) \diamond \beta + \sin_{\diamond}(\Phi) \diamond \alpha, \end{aligned} \tag{7.110}$$

with the replacement $K(\mathbf{k}_1, \mathbf{k}_2) \to i\Phi(\mathbf{k}_1, \mathbf{k}_2)$, where $\Phi(\mathbf{k}_1, \mathbf{k}_2)$ is a Hermitian phase kernel in the unitary kernel in (6.128) that implements the relative phase modulation in the interferometer. When applied to a coherent state, the phase modulation can be transferred from the field variable to the parameter function, thus reproducing the expected result obtained for classical fields.

Exercise 7.6. Apply the transformations in (7.110) to the Wigner functional of a coherent state, where the phase kernel is a constant phase factor $\Phi = i\phi\mathbf{1}$, and show that the parameter functions at the two output ports respectively become

$$\alpha_0 \to \sin(\phi)\alpha_0 \quad \text{and} \quad \alpha_0 \to \cos(\phi)\alpha_0. \tag{7.111}$$

We can use (7.87) with a single-mode detector kernel to implement measurements on the output from one of the two output ports. Instead of applying the unitary transformations for the beamsplitters and the phase modulation to the state, we can apply them to the generating function for the measurement process. The resulting generating function is

$$\mathcal{P}_{\text{tra}}(J) = \mathcal{N}_J \exp\left[-\frac{1}{2}\mathcal{J}\left|\alpha^* \diamond (M_1 + M_2) + \mathrm{i}\beta^* \diamond (M_1 - M_2)\right|^2\right], \tag{7.112}$$

where

$$M_1 = \exp_{\diamond}(\mathrm{i}\Phi) \diamond M \quad \text{and} \quad M_2 = \exp_{\diamond}(-\mathrm{i}\Phi) \diamond M, \tag{7.113}$$

represent the oppositely transformed detector modes, respectively. These phase modulations can, for example, implement relative shifts in the locations of these detector modes. For such a case, we have

$$\exp_{\diamond}[\mathrm{i}\Phi(\mathbf{k}_1, \mathbf{k}_2)] \to \exp(\mathrm{i}\mathbf{x}_0 \cdot \mathbf{k}_1)\mathbf{1}(\mathbf{k}_1, \mathbf{k}_2), \tag{7.114}$$

where $\mathbf{x}_0$ represents the shift. In this way, the correlation between different parts in an optical field can be measured.

One input port of an interferometer usually receives a vacuum state. The integral over the vacuum state's field variable can be evaluated, leading to

$$\mathcal{P}_{\text{inf}} = \frac{2}{1 + J + (1 - J)\tau} \exp\left[-\frac{2(1 - J)\left|\alpha^* \diamond \cos_{\diamond}(\Phi) \diamond M\right|^2}{1 + J + (1 - J)\tau}\right], \tag{7.115}$$

after some simplification, where

$$\tau \triangleq M^* \diamond \sin_{\diamond}(\Phi) \diamond \sin_{\diamond}(\Phi) \diamond M. \tag{7.116}$$

This generating function is multiplied with the input state's Wigner functional and the field variable is integrated (traced out), providing a generating function for the photon statistics at the output port as a function of the phase modulation parameter (which can be the relative shift as explained above). The average photon number is obtained from this generating function by applying one derivative with respect to J and setting $J = 1$.

Exercise 7.7. Use the generating function in (7.115) to compute the photon-number distribution at the output port for a coherent state with a parameter function given by ζ. Show that the result reproduces (3.107) by interpreting $\langle n\rangle = |\zeta^* \diamond M_c|^2$.

Exercise 7.8. Apply the generating function in (7.115) to a general thermal state, as given in (6.181). Show that the average number of photons in the output state is

$$\langle n\rangle_{\text{th-inf}} = \frac{1}{2} M^* \diamond \cos_{\diamond}(\Phi) \diamond (\theta^{-1} - \mathbf{1}) \diamond \cos_{\diamond}(\Phi) \diamond M, \tag{7.117}$$

and that the output photon-number distribution is given by

$$P_n^{(\text{th-inf})} = \frac{\langle n\rangle_{\text{th-inf}}^n}{(1 + \langle n\rangle_{\text{th-inf}})^{n+1}}. \tag{7.118}$$

Replace θ by the single-mode thermal state kernel in (6.204) to obtain a simpler expression.

7.6 Second-order correlations

In Section 7.5, we consider first-order correlations, which are equivalent to classical correlations as observed in classical interferometry. Here, we proceed to *second-order correlations*, which include quantum interference effects. Hanbury Brown and Twiss made observations of correlations between the intensities measured by two detectors [3]. Although the Hanbury Brown-Twiss correlations can be explained in terms of classical stochastic fields, these observations start to introduce the notion of quantum interference, which only reveals itself with two detectors. Quantum correlations, which are distinguished from classical correlations, are found in phenomena such as *entanglement*, *antibunching*, and the *Hong–Ou–Mandel effect*.

7.6.1 Hanbury Brown-Twiss

The generic example of a second-order correlation in the classical context is the Hanbury Brown-Twiss effect. It shows up in the correlation between intensity measurements at two different spacetime points

$$\Gamma^{(2)}(\mathbf{x}_1, t_1, \mathbf{x}_2, t_2) \triangleq \langle I(\mathbf{x}_1, t_1) I(\mathbf{x}_2, t_2) \rangle . \tag{7.119}$$

In terms of the scalar optical field, the *second-order correlation function* becomes

$$\Gamma^{(2)}(x_1, x_2) = \langle f^*(x_1) f(x_1) f^*(x_2) f(x_2) \rangle . \tag{7.120}$$

Here, and in what follows, we simplify the notation by representing the spacetime point $(\mathbf{x}_n, t_n)$ simply as (x_n). To see how the classical field can give rise to the correlations observed in the Hanbury Brown-Twiss experiment, one can model the field as a stochastic field with a normally distributed far field.

In analogy with the normalized first-order correlation function, defined in (2.235), the *normalized second-order correlation function* is formally defined as

$$g^{(2)}(x_1, x_2) \triangleq \frac{\langle I(x_1) I(x_2) \rangle}{\langle I(x_1) \rangle \langle I(x_2) \rangle}, \tag{7.121}$$

in terms of intensities, which can also be expressed in terms of the scalar optical field. The intensity can be modelled as an average intensity plus some intensity fluctuations $I(x) = I_0 + \Delta I$, where $\langle \Delta I \rangle = 0$ so that $\langle I(x) \rangle = I_0$, but $\left\langle \Delta I^2 \right\rangle \geq 0$. It then follows that when the spacetime points coincide we have $\left\langle I^2(x) \right\rangle = I_0^2 + \left\langle \Delta I^2 \right\rangle$. Therefore, $\left\langle I^2(x) \right\rangle \geq \langle I(x) \rangle^2$, which implies that

$$g^{(2)}(x, x) = \frac{\left\langle I^2(x) \right\rangle}{\langle I(x) \rangle^2} \geq 1, \tag{7.122}$$

in contrast to what we found for the normalized first-order correlation function. As a result, the "normalization" of the second-order correlation does not implies that its maximum value is 1.

When the two points do not coincide, but are still close enough to each other so that the intensities are almost equal, we can represent the ensemble elements in terms of a mean value with some difference, as

$$I(x_1) = I + \Delta I \quad \text{and} \quad I(x_2) = I - \Delta I, \tag{7.123}$$

where ΔI is the difference due to the variation in the field. For a small enough separation, the difference is small compared to I and can be positive or negative. The normalized second-order correlation then becomes

$$g^{(2)}(x_1, x_2) = \frac{\langle I^2 \rangle - \langle \Delta I^2 \rangle}{\langle I \rangle^2 - \langle \Delta I \rangle^2}. \tag{7.124}$$

For a homogeneous stationary field, the variations add up to zero, so that $\langle \Delta I \rangle \approx 0$. Then

$$g^{(2)}(x_1, x_2) \approx \frac{\langle I^2 \rangle - \langle \Delta I^2 \rangle}{\langle I \rangle^2} \leq \frac{\langle I^2 \rangle}{\langle I \rangle^2}. \tag{7.125}$$

As a result,

$$g^{(2)}(x_1, x_2) \leq g^{(2)}(x, x). \tag{7.126}$$

The probability to see two photons decreases as the two spacetime points move further away from each other. It implies that light tends to "bunch" together due to its bosonic nature, leading to the phenomenon called *bunching*.

7.6.2 Quantum correlations

In the context of quantum physics, we consider the intensities (or the average numbers of photons) in terms of the *localized number operator* given in (7.16). First we need to address a subtlety. Replacing the two intensities in (7.119) by localized number operators, we obtain a product of four ladder operators. Although the order of the classical fields in the expression has no significance, we need to pay attention to the order of the ladder operators because these operators do not always commute. Modelling the measurement in terms of electric field operators, as defined in Section 4.1.3, we find that the commutation produces a divergent term when the locations for the two measurements coincide, as can be seen from (4.32). To avoid this potential divergent term, these operators are rearranged so that they are in *normal order* so that all the creation operators are placed to the left of the annihilation operators. It differs from the situation where an operator is converted to normal order *with the aid of the commutation relations*. Here, the operator is directly written in normal order *without commutation operations*, as explained at the end of Section 6.3.4.

We do not need to use electric field operators to model the measurement. Instead, the measurement is modelled in terms of the localized number operators, which does not produce the same divergences. Nevertheless, one should maintain the use of the normal order. For an arbitrary quantum state given by a density operator $\hat{\rho}$, the second-order correlation is then expressed as

$$\Gamma^{(2)}(x_1, x_2) \triangleq \mathrm{tr}\{\hat{\rho}{:}\hat{n}(x_1)\hat{n}(x_2){:}\}, \tag{7.127}$$

where ${:}\hat{n}(x_1)\hat{n}(x_2){:}$ represents the product of localized number operators *in normal order*. We can write it explicitly as

$$\begin{aligned} {:}\hat{n}(x_1)\hat{n}(x_2){:} \triangleq \int & \hat{a}^{\dagger}(\mathbf{k}_1)\hat{a}^{\dagger}(\mathbf{k}_3)D(\mathbf{k}_1, \mathbf{k}_2; x_1)D(\mathbf{k}_3, \mathbf{k}_4; x_2) \\ & \times \hat{a}(\mathbf{k}_2)\hat{a}(\mathbf{k}_4)\, \mathrm{d}_{\omega}k_1\, \mathrm{d}_{\omega}k_2\, \mathrm{d}_{\omega}k_3\, \mathrm{d}_{\omega}k_4, \end{aligned} \tag{7.128}$$

where $D(\cdot)$ represents the (localized) detector kernel. The normalized version of the second-order correlation function is

$$g^{(2)}(x_1, x_2) \triangleq \frac{\mathrm{tr}\{\hat{\rho}{:}\hat{n}(x_1)\hat{n}(x_2){:}\}}{\mathrm{tr}\{\hat{\rho}\hat{n}(x_1)\}\,\mathrm{tr}\{\hat{\rho}\hat{n}(x_2)\}}. \tag{7.129}$$

It is often referred to as the *second-order degree of coherence*. The quantities in the denominator are identified as the average numbers of photons in the state at the two locations, which are proportional to the average intensities at those locations

$$\mathrm{tr}\{\hat{\rho}\hat{n}(x)\} \triangleq \langle n(x)\rangle \propto \langle I(x)\rangle\,. \tag{7.130}$$

We also have

$$\mathrm{tr}\{\hat{\rho}{:}\hat{n}(x_1)\hat{n}(x_2){:}\} \equiv \left\langle n^2(x)\right\rangle \propto \left\langle I^2(x)\right\rangle, \tag{7.131}$$

in correspondence to what we get for classical states. However, the situation can become more complicated for quantum states.

Considering the (not normal-ordered) product $\hat{n}(x_1)\hat{n}(x_2)$ and performing a commutation of the central two ladder operators, we end up with two terms

$$\begin{aligned} \hat{n}(x_1)\hat{n}(x_2) = \int & \hat{a}^{\dagger}(\mathbf{k}_1)\hat{a}^{\dagger}(\mathbf{k}_3)D(\mathbf{k}_1, \mathbf{k}_2; x_1)D(\mathbf{k}_3, \mathbf{k}_4; x_2)\hat{a}(\mathbf{k}_2)\hat{a}(\mathbf{k}_4) \\ & + \hat{a}^{\dagger}(\mathbf{k}_1)D(\mathbf{k}_1, \mathbf{k}_2; x_1)D(\mathbf{k}_2, \mathbf{k}_4; x_2) \\ & \times \hat{a}(\mathbf{k}_4)\, \mathrm{d}_{\omega}k_1\, \mathrm{d}_{\omega}k_2\, \mathrm{d}_{\omega}k_3\, \mathrm{d}_{\omega}k_4 \\ = {} & {:}\hat{n}(x_1)\hat{n}(x_2){:} + \hat{a}^{\dagger} \diamond D(x_1) \diamond D(x_2) \diamond \hat{a}. \end{aligned} \tag{7.132}$$

The first term contains the four ladder operators in normal order. The second term has two ladder operators in normal order contracted with the contraction of the two detector kernels. Usually, the two kernels represent two separate detectors with non-overlapping locations; their contraction gives zero so that the term with the two ladder operators drops away. In such a case, the two localized number operators commute.

What happens when the detectors overlap? Strictly speaking, it is not physically possible for two different detectors to be located at the same point in space and time. Nevertheless, one may argue that it is possible to use a beamsplitter in such a way that the two detectors are effectively located at the same point in space. In such a case, the two sets of ladder operators associated with different output ports of a beamsplitter commute with one another. Formally, we can consider a situation for which the two localized number operators do not commute. Such ladder operators need to be placed in normal order, as in (7.128).

7.6.2.1 Second-order coherence

Generalizing first-order coherence discussed in Section 2.10 to higher orders, one may propose that *second-order coherence* is attained when $g^{(2)} = 1$ (and similar for all higher-order coherence) [4]. However, if second-order coherence is the *ability of a (quantum) field to produce interference* and the second-order correlation peak is associated with the manifestation of such second-order interference, for which $g^{(2)} \geq 1$, then the condition where $g^{(2)} = 1$ would not maximize interference and thus not be a reasonable condition for second-order coherence.[d]

Another proposal is that *higher-order coherence* is revealed by the presence of off-diagonal elements in the density matrix of a quantum state. However, any definition of higher-order coherence is expected to be invariant to local unitary transformations. It is often possible to transform such off-diagonal elements in the density matrix away with a local unitary transformation.

Although the presence of such off-diagonal elements does not provide a unitary invariant notion of coherence, it reminds us of the concept of *purity*, because the lack of such off-diagonal elements generally implies a mixed state. Mixed states tend to be more diagonal with smaller off-diagonal elements. It is indeed true that the ability of a quantum state to produce interference is destroyed when such a state becomes mixed. There are situations where a pure state does not produce (first-order) interference, such as when there is entanglement (or classical non-separability) among different degrees of freedom in the state. In such a case, one can nevertheless observe higher-order coherence. Therefore, the coherence of a state may be generally quantified by the purity of that state. It is easily calculated, as shown in (3.17), and is invariant with respect to local unitary transformations.

---

d See the discussion in Section 12.6 of Mandel and Wolf [5].

We do not use coherence as a quantum property, apart from its first-order form in the context of optical interferometry. The purity of a quantum state captures to a large extent the properties that can be associated with the generalization of first-order coherence. The notion of second-order (or higher order) coherence is therefore not a concept that we find useful in our discussions here.

7.6.3 Fock states

To illustrate second-order quantum correlations, we consider the case of a correlation measurement on a fixed-spectrum Fock state with an *occupation* number $n > 2$. First, we apply two annihilation operators to such a Fock state. Since the spin is not relevant in this discussion, we ignored the spin indices. Using (4.66), we obtain

$$\hat{a}(\mathbf{k}_1)\hat{a}(\mathbf{k}_2)\,|n_F\rangle = |(n-2)_F\rangle\, F(\mathbf{k}_1)F(\mathbf{k}_2)\sqrt{n-1}\sqrt{n}. \tag{7.133}$$

It then follows that

$$\langle n_F|:\hat{n}(x_1)\hat{n}(x_2):|n_F\rangle = (n-1)n[F^* \diamond D(x_1) \diamond F][F^* \diamond D(x_2) \diamond F], \tag{7.134}$$

where $F^* \diamond D \diamond F$ represents the overlap of the detector kernel by the parameter function of the Fock state. The individual detectors produce

$$\langle n_F|\,\hat{n}(x_m)\,|n_F\rangle = nF^* \diamond D(x_m) \diamond F, \tag{7.135}$$

for $m = 1, 2$. So, provided that $F^* \diamond D(x_m) \diamond F \neq 0$ for either of the two detectors, the normalized second-order correlation for the Fock state is given by

$$g^{(2)}_{\text{Fock}}(x_1, x_2) = 1 - \frac{1}{n} < 1, \tag{7.136}$$

regardless of the locations of the detectors. Without normal ordering, we get

$$\langle n_F|\,\hat{n}(x_1) = \langle n_F|\,\hat{a}^\dagger \diamond D(x_1) \diamond \hat{a} = \sqrt{n}\,\langle (n-1)_F|\,\langle 1_{F^* \diamond D(x_1)}|\,. \tag{7.137}$$

and

$$\hat{n}(x_2)\,|n_F\rangle = \hat{a}^\dagger \diamond D(x_2) \diamond \hat{a}\,|n_F\rangle = |(n-1)_F\rangle\,|1_{D(x_2)\diamond F}\rangle\,\sqrt{n}. \tag{7.138}$$

Overlapping these two states, we need to consider all possible ways in which the single-photon parts of the states can be contracted. Hence,

$$\begin{aligned}\langle n_F|\,\hat{n}(x_1)\hat{n}(x_2)\,|n_F\rangle &= (n-1)n[F^* \diamond D(x_1) \diamond F][F^* \diamond D(x_2) \diamond F]\\ &\quad + nF^* \diamond D(x_1) \diamond D(x_2) \diamond F.\end{aligned} \tag{7.139}$$

It corresponds to the two terms in (7.132). The last term becomes zero when the two detector kernels do not overlap. The first term is equal to the result for the normal-ordered expression. Therefore, when we consider only those scenarios where the two measurements are made at different spacetime points, the measurement comes out the same regardless of the normal ordering.

7.6.4 Antibunching

While *bunching* is the phenomenon where photons tend to bunch together in the same spacetime location, the opposite, called *antibunching*, is the phenomenon where photons tend to avoid having the same spatiotemporal properties. Based on (7.126), we see that light has a tendency to exhibit bunching in classical scenarios.

Above, we saw that $g^{(2)}$ is smaller than 1 for certain quantum states such as the Fock states. This result is in contrast to (7.122). So, something is different in the quantum case. The result for Fock states shows that the inverse of the occupation number is subtracted from 1 to give a value of the normalized second-order correlation. It tells us that the photons tend to be uncorrelated, especially when the number of photons in the state is small. However, it does not give any information of the spatiotemporal function of these correlations, because the result is independent of the spacetime coordinates.

With antibunching, the second-order correlation peak for measurements at the same spacetime point is lower than for separate spacetime points. To understand the implication, we model the average intensity as a summation over an ensemble of fluctuating intensities, each multiplied by a probability P_n,

$$\langle I(x)\rangle = \sum_n P_n I_n(x). \tag{7.140}$$

The second-order correlation then becomes

$$\left\langle I^2(x)\right\rangle = \sum_n P_n I_n^2(x). \tag{7.141}$$

We can express the normalized second-order correlation for measurements at the same spacetime point as

$$g^{(2)}(x,x) = \frac{\sum_n^N P_n I_n^2(x)}{\left[\sum_n^N P_n I_n(x)\right]^2} = 1 + \frac{\sum_n^N P_n\left[I_n(x) - \langle I(x)\rangle\right]^2}{\langle I(x)\rangle^2}. \tag{7.142}$$

The only way to get $g^{(2)}(x,x) < 1$ is to have $P_n < 0$ for some of the P_n's, which is not allowed if these P_n's are true probabilities. The same argument applies when we consider the average number of photons instead of the average intensity. By implication, the probabilistic model of average intensities or average numbers of photons in terms of ensembles is not valid anymore. It demonstrates the difference between what is considered “classical” and what is considered “quantum” in this context.

7.6.5 Wigner functionals

In terms of Wigner functionals, the calculation of second-order correlations involves the Wigner functionals of three operators. The trace is represented by the functional integral, but since there are three operators, we cannot in general just multiply their Wigner functionals. One can combine the Wigner functionals of the two localized number operators by a star product and then multiply it by the Wigner functional of the state. However, for two non-overlapping detectors, when $D(x_1) \diamond D(x_2) = D(x_2) \diamond D(x_1) = 0$, the star product reduces to a multiplication.

Exercise 7.9. Use (6.25) to compute the star product of two localized number operators and compare it with (6.70) when their detector kernels coincide $D_1 = D_2$. Show that, if their detector kernels are non-overlapping so that $D_1 \diamond D_2 = D_2 \diamond D_1 = 0$, the star product of two localized number operators produces

$$W_{\hat{n}1} \star W_{\hat{n}2} = W_{\hat{n}1} W_{\hat{n}2}. \tag{7.143}$$

Therefore, we can represent the second-order correlation for non-overlapping measurements in terms of Wigner functionals by

$$\Gamma^{(2)}(x_1, x_2) \triangleq \int W_{\hat{\rho}}[\alpha] W_{\hat{n}}[\alpha](x_1) W_{\hat{n}}[\alpha](x_2)\, \mathcal{D}^{\circ}[\alpha], \tag{7.144}$$

where x_1 and x_2 represent the locations of the detectors. Instead of using the Wigner functionals of the localized number operators, we can use the generating functions for the Wigner functionals of the localized projection operators given in (7.87). In that case, the second-order correlation is obtained as

$$\Gamma^{(2)}(x_1, x_2) = \partial_{J,K} \int W_{\hat{\rho}} \mathcal{P}(x_1; J) \mathcal{P}(x_2; K)\, \mathcal{D}^{\circ}[\alpha] \bigg|_{J=K=1}. \tag{7.145}$$

To compute the second-order degree of coherence in (7.129), we also need the intensities (or photon-numbers) at the two locations. Such quantities are all obtained from the generating function that is produced by the functional integral

$$\mathcal{C}(J, K) \triangleq \int W_{\hat{\rho}} \mathcal{P}(x_1; J) \mathcal{P}(x_2; K)\, \mathcal{D}^{\circ}[\alpha]. \tag{7.146}$$

The second-order correlation is

$$\Gamma^{(2)}(x_1, x_2) \triangleq \partial_{J,K} \mathcal{C}(J, K)\big|_{J=K=1}. \tag{7.147}$$

For the intensities, we remove one of the detectors, which is readily done by setting its generating parameter equal to 1 (without the derivative), which converts it to the identity.[e] Thus, we get

$$\begin{aligned} I(x_1) &\triangleq \partial_J \mathcal{C}(J,K)\big|_{J=K=1}, \\ I(x_2) &\triangleq \partial_K \mathcal{C}(J,K)\big|_{J=K=1}. \end{aligned} \tag{7.148}$$

An alternative situation is where the state on which the measurements are made is already in the form of a twin-beam state and is thus represented by a Wigner functional of two field variables $W_{\hat{\rho}}[\alpha,\beta]$. In that case, the two measurements are done independently on the two field variables. It can be represented by two generating functions for photon statistics, as given in (7.87), but as functionals of the two different field variables. The corresponding generating functions for the measurements is then given by

$$\mathcal{C}(J,K) = \int W_{\hat{\rho}}[\alpha,\beta]\mathcal{P}[\alpha](x_1;J)\mathcal{P}[\beta](x_2;K)\,\mathcal{D}^{\circ}[\alpha,\beta]. \tag{7.149}$$

The subsequent calculations remain the same.

Regardless of whether the state is a single-beam state with two non-overlapping measurements applied to it or whether the state is a twin-beam state with measurements made on each beam, the resulting generating function is used in the same way. It allows one to compute correlations between arbitrary numbers of detected photons. We can represent this generating function as

$$\mathcal{C}(J,K) = \sum_{mn} J^m K^n C_{mn}, \tag{7.150}$$

where

$$C_{mn}(x_1,x_2) \triangleq \operatorname{tr}\{\hat{\rho}\hat{P}_m(x_1)\hat{P}_n(x_2)\}, \tag{7.151}$$

in terms of operators, with $\hat{P}_m(x)$ representing the projection operator for m photons at a location x.

All second-order correlations can be computed in this way. The details are determined by the nature of the state on which the measurements are applied and by the nature of the measurements as specified by their detector kernels. For example, temporal correlations require that the states are parameterized by kernels or parameter functions that contain temporal information, likely in the form of temporal frequency spectra, and the detector kernels also contain temporal information, typically specified by temporal convolution kernels.

e The effect of setting a generating parameter equal to 1 is equivalent to tracing over the degrees of freedom that it represents.

The generating function for the correlated detection of photons provides us with a versatile tool to calculate photon statistics. Such calculations can take on different forms. First, we note that $\mathcal{C}(1,1) = 1$, ensuring proper normalization. As shown above, we can use it to obtain *coincidence counts*, which is given by the *probability* for detecting one photon on both sides:

$$P(1,1) = \partial_J \partial_K \mathcal{C}(J,K)\big|_{J=K=0} . \tag{7.152}$$

It also allows us to compute the *singles counts*, representing the probabilities for detecting one photon on one side while tracing over the other side. Substituting $K = 1$ (implying that $\mathcal{K} = 0$), we trace over the one side, leading to

$$P(1,-) = \sum_n P(1,n) = \partial_J \mathcal{C}(J,1)\big|_{J=0} . \tag{7.153}$$

The opposite is obtained by setting $J = 1$ (which means that $\mathcal{J} = 0$) to produce

$$P(-,1) = \sum_m P(m,1) = \partial_K \mathcal{C}(1,K)|_{K=0} . \tag{7.154}$$

The probability for detecting one photon on one side and nothing on the other sides is

$$P(1,0) = \partial_J \mathcal{C}(J,0)\big|_{J=0} . \tag{7.155}$$

Higher-order statistics (detecting more than one photon on a side) is given by higher derivatives. For example, the probability for detecting one photon on one side and two photons on the other side is given by

$$P(1,2) = \partial_J \partial_K^2 \mathcal{C}(J,K)\big|_{J=K=0} . \tag{7.156}$$

7.6.5.1 Antibunching from Wigner functionals

Returning to the topic of *antibunching* discussed in Section 7.6.4, we reconsider the analysis in terms of the Wigner functional formalism, where states are represented by Wigner functionals. The correlation measurements with two detectors is computed with the aid of (7.144) and, for one detector, we get the distribution of the average number of photons, given by

$$\langle n(x)\rangle = \int W_{\hat{\rho}}[\alpha] W_{\hat{n}}[\alpha](x)\, \mathcal{D}^{\circ}[\alpha]. \tag{7.157}$$

The second-order correlation can be expressed as

$$\begin{aligned}\Gamma^{(2)}(x_1,x_2) = \langle n(x_1)\rangle \langle n(x_2)\rangle + \int & W_{\hat{\rho}}[\alpha][W_{\hat{n}}[\alpha](x_1) - \langle n(x_1)\rangle] \\ & \times [W_{\hat{n}}[\alpha](x_2) - \langle n(x_2)\rangle]\, \mathcal{D}^{\circ}[\alpha]. \end{aligned} \tag{7.158}$$

The normalized second-order correlation then becomes

$$g^{(2)}(x_1,x_2) = 1 + \frac{1}{\langle n(x_1)\rangle\,\langle n(x_2)\rangle}\int W_{\hat{\rho}}[\alpha][W_{\hat{n}}[\alpha](x_1) - \langle n(x_1)\rangle] \\ \times [W_{\hat{n}}[\alpha](x_2) - \langle n(x_2)\rangle]\,\mathcal{D}^{\circ}[\alpha]. \tag{7.159}$$

For the Fock states, we found that this normalized second-order correlation is smaller than 1 regardless of the location of the detectors. It implies that the second term in (7.159) must be negative. This expression is only valid provided that $x_1 \neq x_2$. In case $x_1 = x_2$, the product of the two Wigner functionals for the measurements must be replaced by their star product, as shown in the exercise to derive (7.143).

To investigate this phenomenon further, we analyse the physical setups with which such correlation measurements are made. As we stated above, the notion of a second-order correlation obtained from measurements *at the same spacetime point* needs to be considered carefully. While theoretical discussions about such a notion are often found, our focus here is on the practical implementation of such measurements.

There are two general approaches to achieve such measurements. One is to perform two measurements at two different spacetime points for multiple cases, revealing the trend for the values of the correlation as the two points are brought closer together. The other approach is to separate the state into two parts with a beamsplitter and perform measurements on the respective parts associated with points that would have coincided without the separation. We consider both scenarios, starting with the beamsplitter.

7.6.5.2 Second-order correlation using a beamsplitter

Consider an arbitrary state represented by $W_{\hat{\rho}}[\alpha]$ being sent through a beamsplitter with a vacuum state entering the other port. At each of the two output ports of the beamsplitter, we apply the generating function in (7.87) to measure the photon statistics. Since we are interested in the measurements at locations that represent the same spacetime point in the state, the detector kernels in the two generating functions for the photon statistics are the same idempotent kernel D, but the generating parameters are different: J and K. The beamsplitter is implemented as a transformation of the Wigner functionals' arguments given by (6.218). The resulting expression for the generating function of the correlation photon statistics in terms of Wigner functionals, is

$$\mathcal{C}(J,K) = \int W_{\hat{\rho}}[C\alpha + iS\beta]W_{\text{vac}}[C\beta + iS\alpha]\mathcal{P}[\alpha](J)\mathcal{P}[\beta](K)\,\mathcal{D}^{\circ}[\alpha,\beta]. \tag{7.160}$$

where $\mathcal{P}$ is the generating function provided in (7.87) for the projection operators measuring the photon statistics. For a 50:50 beamsplitter, we set $C = S = \frac{1}{\sqrt{2}}$.

Since there is only one unknown state, we can perform a change of integration field variables and evaluate one of the functional integrations. We shift β into α by replacing

$$\alpha \to \sqrt{2}\xi - i\beta \quad \text{and} \quad \alpha^* \to \sqrt{2}\xi^* + i\beta^*, \tag{7.161}$$

where ξ is the new integration field variable. The argument of $W_{\hat{\rho}}$ becomes independent of β. It leads to

$$\begin{aligned}
\mathcal{C}(J,K) &= \mathcal{N}_0^2\mathcal{N}_J^{\text{tr}\{D\}}\mathcal{N}_K^{\text{tr}\{D\}} \int W_{\hat{\rho}}[\xi^*,\xi] \exp\left[-2\left(\sqrt{2}\beta^* - \mathrm{i}\xi^*\right)\diamond\left(\sqrt{2}\beta + \mathrm{i}\xi\right)\right. \\
&\quad \left. - 2\mathcal{J}(\sqrt{2}\xi^* + \mathrm{i}\beta^*)\diamond D \diamond(\sqrt{2}\xi - \mathrm{i}\beta) - 2\mathcal{K}\beta^*\diamond D\diamond\beta\right] \mathcal{D}^\circ[\beta,\xi] \\
&= \frac{\mathcal{N}_J^{\text{tr}\{D\}}\mathcal{N}_K^{\text{tr}\{D\}}}{\det\{\mathbf{1} + \frac{1}{2}\mathcal{J}D + \frac{1}{2}\mathcal{K}D\}} \int W_{\hat{\rho}}[\xi^*,\xi] \\
&\quad \times \exp\left[2\xi^*\diamond(\mathbf{1}+\mathcal{J}D)\diamond\left(\mathbf{1}+\frac{1}{2}\mathcal{J}D+\frac{1}{2}\mathcal{K}D\right)^{-1}\diamond(\mathbf{1}+\mathcal{J}D)\diamond\xi\right. \\
&\quad \left. - 2\xi^*\diamond(\mathbf{1}+2\mathcal{J}D)\diamond\xi\right] \mathcal{D}^\circ[\xi],
\end{aligned} \tag{7.162}$$

where

$$\begin{aligned}
\mathcal{J} &= \frac{1-J}{1+J}, \quad \mathcal{N}_J = \left(\frac{2}{1+J}\right)^{\text{tr}\{D\}}, \\
\mathcal{K} &= \frac{1-K}{1+K}, \quad \mathcal{N}_K = \left(\frac{2}{1+K}\right)^{\text{tr}\{D\}}.
\end{aligned} \tag{7.163}$$

We evaluate the inverse and determinant (see Appendix D) as

$$\begin{aligned}
\left(\mathbf{1}+\frac{1}{2}\mathcal{J}D+\frac{1}{2}\mathcal{K}D\right)^{-1} &= \mathbf{1} - \frac{\frac{1}{2}\mathcal{J}+\frac{1}{2}\mathcal{K}}{1+\frac{1}{2}\mathcal{J}+\frac{1}{2}\mathcal{K}}D, \\
\det\left\{\mathbf{1}+\frac{1}{2}\mathcal{J}D+\frac{1}{2}\mathcal{K}D\right\} &= \left(1+\frac{1}{2}\mathcal{J}+\frac{1}{2}\mathcal{K}\right)^{\text{tr}\{D\}}.
\end{aligned} \tag{7.164}$$

Thus, we can simplify the expression to obtain

$$\begin{aligned}
\mathcal{C}(J,K) &= \frac{\mathcal{N}_J^{\text{tr}\{D\}}\mathcal{N}_K^{\text{tr}\{D\}}}{\left(1+\frac{1}{2}\mathcal{J}+\frac{1}{2}\mathcal{K}\right)^{\text{tr}\{D\}}} \int W_{\hat{\rho}}[\xi^*,\xi] \\
&\quad \times \exp\left(-\frac{\mathcal{J}+\mathcal{K}+2\mathcal{J}\mathcal{K}}{1+\frac{1}{2}\mathcal{J}+\frac{1}{2}\mathcal{K}}\xi^*\diamond D\diamond\xi\right) \mathcal{D}^\circ[\xi].
\end{aligned} \tag{7.165}$$

To extract the two number operators from the generating function, we apply one derivative with respect to each of the two generating parameters and then set them equal to 1. The result reads

$$\begin{aligned}
\Gamma^{(2)}(x,x) &= \partial_K\partial_J\mathcal{C}(J,K)\big|_{J=K=1} \\
&= \frac{1}{4}\int W_{\hat{\rho}}[\xi^*,\xi]\left[\left(\xi^*\diamond D\diamond\xi - \frac{1}{2}\,\text{tr}\{D\}\right)^2\right. \\
&\quad \left. - \left(\xi^*\diamond D\diamond\xi - \frac{1}{2}\,\text{tr}\{D\}\right) - \frac{1}{4}\,\text{tr}\{D\}\right] \mathcal{D}^\circ[\xi].
\end{aligned} \tag{7.166}$$

Comparing the polynomial expression in the square brackets with the Wigner functional in (6.72) by replacing $D \rightarrow \mathbf{1}$ and $\text{tr}\{D\} \rightarrow \Omega$, we find that the Wigner functional for the combined operator, representing the two measurements after the respective output ports of the beamsplitter, is equivalent to the Wigner functional for the two number operators in normal order. (They would be identical for the same detector kernel.) The practical implementation of correlation measurements at the coinciding spacetime points with the aid of a beamsplitter therefore corresponds exactly to the formulation of such correlation measurements with number operators in normal order.

The expression in (7.166) can be used to determine whether a state represented by its Wigner functional displays antibunching. For that purpose, one can allow the two detectors to be located at different points.

7.6.5.3 Second-order temporal correlations

Many applications in quantum information technology are based on the use of single photons, thus requiring a single-photon source. While weak coherent states have a higher probability to produce single photons than multiple photons, it is still preferable to have a source that produces only single photons at a time. It is necessary to test such sources to assess their quality as single-photon sources. The usual way to perform such a test is to see whether one can detect two photons from such a source at the same time. It thus involves two measurements, which implies a second-order correlation measurement. Instead of separating these two measurements with the aid of a beamsplitter, as done in the previous section, we can do it by separating them in time. It thus leads to *second-order temporal correlations*. The idea is then to show that the correlation approaches zero as the separation in time becomes smaller.

Ironically, if such a correlation produces a nonzero result for $\tau > 0$, then obviously the state produced by such a "single-photon" source is not a *single-photon state*. It is indeed a *multiphoton state* with special correlation properties ensuring that the second-order correlation becomes zero for $\tau = 0$.

In terms of separate instances in time t_1 and t_2, we represent the second-order temporal correlations as

$$\begin{aligned} \Gamma^{(2)}(t_1, t_2) &= \text{tr}\{\hat{\rho}{:}\hat{n}(t_1)\hat{n}(t_2){:}\} \\ &= \int \text{tr}\{\hat{\rho}\hat{a}^{\dagger}(\mathbf{k}_1)\hat{a}^{\dagger}(\mathbf{k}_3)D(\mathbf{k}_1, \mathbf{k}_2; t_1)D(\mathbf{k}_3, \mathbf{k}_4; t_2) \\ &\quad \times \hat{a}(\mathbf{k}_2)\hat{a}(\mathbf{k}_4)\}\, \text{d}_{\omega}k_1\, \text{d}_{\omega}k_2\, \text{d}_{\omega}k_3\, \text{d}_{\omega}k_4. \end{aligned} \tag{7.167}$$

It is assumed that the source produces a continuous sequence of single photons. Therefore, the system is shift invariant, so that the correlation is only a function of the time interval $\Gamma^{(2)}(t_1, t_2) \rightarrow \Gamma^{(2)}(t_2 - t_1) = \Gamma^{(2)}(\tau)$. We replace $t_2 = t_1 + \tau$ and compute a time average over a period T. In this case, the measurements are never made with $\tau = 0$.

Therefore, the kernels $D(t)$ and $D(t+\tau)$ do not overlap, which implies that the two number operators commute. It is therefore not necessary to represent the product of number operators in normal order. Hence,

$$\Gamma^{(2)}(\tau) = \frac{1}{T}\int_0^T \mathrm{tr}\{\hat{\rho}\hat{n}(t)\hat{n}(t+\tau)\}\,\mathrm{d}t \quad \text{for } \tau > \Delta t, \tag{7.168}$$

where Δt represents the *gating time* of the detectors.

In terms of Wigner functionals, the expressions for the number operators are the same as before, apart from the fact that they are parameterized by a time t. It thus reads

$$W_{\hat{n}}[\alpha](t) \triangleq \text{Wigner}\{\hat{n}(t)\} = \alpha^* \diamond D(t) \diamond \alpha - \frac{1}{2}\,\mathrm{tr}\{D(t)\}. \tag{7.169}$$

The expression for the second-order temporal correlation thus becomes

$$\Gamma^{(2)}(\tau) = \frac{1}{T}\int_0^T \int W_{\hat{\rho}}[\alpha] W_{\hat{n}}[\alpha](t) W_{\hat{n}}[\alpha](t+\tau)\,\mathcal{D}^\circ[\alpha]\,\mathrm{d}t. \tag{7.170}$$

It is similar to the more general case in (7.166), but here the separation is maintained so that $\tau > \Delta t$. It can be used to assess the quality of the state produced by a single-photon source. Instead of using the Wigner functionals for the number operators, one can also use the generating function in (7.87) for the photon statistics. In addition to the number operators, the generating function for the photon statistics allows one to compute various other quantities, as shown in Section 7.6.5.

7.6.6 Second-order correlations in squeezed vacuum states

Having discussed various general aspects of second-order correlation measurements, we now turn to a specific example. Here, we consider the case of the second-order correlation measurements applied to a generic pure *squeezed vacuum state*. (In Chapter 9, detailed calculations are made of second-order correlation measurements applied to parametric down-converted states.) The squeezed vacuum state is represented by the Wigner functional in (6.337) with unspecified squeezed state kernels A and B.

The setup for such second-order correlation measurements on a squeezed vacuum state is shown in Figure 7.2. The *pump beam* illuminates the nonlinear crystal to produce down-converted light under non-collinear phase-matching conditions so that the *signal* and *idler* photons propagate with a nonzero down-conversion angle. Such non-collinear phase-matching conditions are discussed in Section 8.6.2.

Substituting (6.337) and (7.87) into (7.146), we get an anisotropic functional integral that can be evaluated with (C.29) to produce

$$\begin{aligned} \mathcal{C}(\mathcal{J},\mathcal{K}) &= \mathcal{N}_0 \mathcal{N}_{\mathcal{J}}^{\text{tr}\{D\text{s}\}} \mathcal{N}_{\mathcal{K}}^{\text{tr}\{D\text{i}\}} \int \exp\left[-2\alpha^* \diamond (A + \mathcal{J}D_{\text{s}} + \mathcal{K}D_{\text{i}}) \diamond \alpha \right. \\ &\quad \left. - \alpha^* \diamond B \diamond \alpha^* - \alpha \diamond B^* \diamond \alpha\right] \mathcal{D}^{\circ}[\alpha] \\ &= \frac{\mathcal{N}_{\mathcal{J}}^{\text{tr}\{D\text{s}\}} \mathcal{N}_{\mathcal{K}}^{\text{tr}\{D\text{i}\}}}{\sqrt{\det\{A_0\}}\sqrt{\det\{A_0^* - B^* \diamond A_0^{-1} \diamond B\}}}. \end{aligned} \tag{7.171}$$

Here, the two detectors associated with the signal and idler photons are denoted by subscripts s and i, respectively, and

$$A_0 \triangleq A + \mathcal{J}D_{\text{s}} + \mathcal{K}D_{\text{i}}. \tag{7.172}$$

Although the detector kernels are idempotent and non-overlapping, the *functional determinants* and inverses cannot in general be simplified further in their current form. We can obtain a formal expression for the second-order correlation by applying the calculation in (7.147) to (7.171), as shown below, but calculations based on this result tend to be tedious. Fortunately, there are experimental conditions that can be exploited to impose simplifying assumptions. For example, under the appropriate experimental conditions, the detectors can be modelled as *single-mode detectors*, allowing the expression to be simplified. Alternatively, under *weak squeezing*, one can make an expansion in terms of the squeezing parameter. We consider these options below.

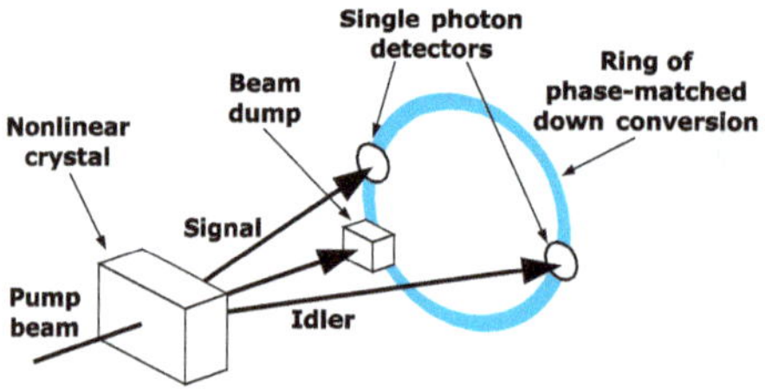

Figure 7.2: Optical setup for correlation measurements on a squeezed vacuum state produced by spontaneous parametric down-conversion.

The results that we obtain in the following sections, reveal that expressions relating to the intensity (or photon number) contain the Hermitian kernel A, while those relating to the correlation information contain the symmetric kernel B. For this reason, one may refer to A as the *intensity kernel* and to B as the *correlation kernel*.

7.6.6.1 Generic correlation

Here, we use the generating function in (7.171) to compute the second-order correlation. For this purpose, we use the identities in (D.34) and (D.32) for the derivatives of the determinants and the inverses.

After applying the derivatives to the generating function and setting the generating parameters equal to 1, we simplify the expression using the identities in Appendix D. We also employ the fact that a trace (or a determinant) is equal to its transpose. Since A and the detector kernels D_s and D_i are Hermitian, we get $(A \diamond D_s)^T = D_s^* \diamond A^*$. We use these properties to simplify the final expression, which then reads

$$\begin{aligned} \Gamma^{(2)} &= \partial_{J,K} C(J,K)\big|_{J=K=1} \\ &= \frac{1}{4}\operatorname{tr}\{(A-\mathbf{1})\diamond D_s\}\operatorname{tr}\{(A-\mathbf{1})\diamond D_i\} + \frac{1}{4}\operatorname{tr}\{A\diamond D_s\diamond A\diamond D_i\} \\ &\quad + \frac{1}{4}\operatorname{tr}\left\{A^*\diamond B^*\diamond A^{-1}\diamond D_s\diamond A^{-1}\diamond B\diamond A^*\diamond D_i^*\right\}. \end{aligned} \tag{7.173}$$

We can use (6.340) to simplify this expression further, so that

$$\begin{aligned} \Gamma^{(2)} &= \frac{1}{4}\operatorname{tr}\{(A-\mathbf{1})\diamond D_s\}\operatorname{tr}\{(A-\mathbf{1})\diamond D_i\} + \frac{1}{4}\operatorname{tr}\{A\diamond D_s\diamond A\diamond D_i\} \\ &\quad + \frac{1}{4}\operatorname{tr}\{B^*\diamond D_s\diamond B\diamond D_i^*\}. \end{aligned} \tag{7.174}$$

The photon-number distributions for the two detectors are obtained as

$$\begin{aligned} \langle n\rangle_s &= \partial_J C(J,K)\big|_{J=K=1} = \frac{1}{2}\operatorname{tr}\{(A-\mathbf{1})\diamond D_s\}, \\ \langle n\rangle_i &= \partial_K C(J,K)\big|_{J=K=1} = \frac{1}{2}\operatorname{tr}\{(A-\mathbf{1})\diamond D_i\}. \end{aligned} \tag{7.175}$$

So, the second-order degree of coherence is given by

$$g^{(2)} = 1 + \frac{\operatorname{tr}\{A\diamond D_s\diamond A\diamond D_i\} + \operatorname{tr}\{B^*\diamond D_s\diamond B\diamond D_i^*\}}{\operatorname{tr}\{(A-\mathbf{1})\diamond D_s\}\operatorname{tr}\{(A-\mathbf{1})\diamond D_i\}}. \tag{7.176}$$

If the two detectors are far enough apart, we get $\operatorname{tr}\{A\diamond D_s\diamond A\diamond D_i\} \approx 0$, but the last term represents the correlations that come from the entangled photons due to the B-kernels and can be quite large.

The only simplifying assumption that is used to obtain (7.176) is (6.340). However, every quantity needs to be computed separately and the resulting calculations can be rather tedious. If the generating function in (7.171) can be simplified, such calculations would become simpler, even if the initial simplification process is still tedious. Next, we consider such simplifying assumptions.

7.6.6.2 Single-mode detectors

The first simplifying assumption is introduced by modelling the detectors with single-mode detector kernels as in (7.171). The complexity of an arbitrary detector kernel is seldom of interest. The general purpose of the detector kernels is to fix the location of detection and perhaps to introduce a scale for the size of the detector. Other details are usually of little concern. Therefore, we model the detector simply as a spatiotemporal mode, using the Gaussian function for ease of calculations. Here, we do not specify the modal functions because we do not use the expressions of the squeezed vacuum state kernels. Such detailed calculations are postponed until Chapter 9.

For single-mode detectors, we assume that

$$D_s(\mathbf{k},\mathbf{k}') = M_s(\mathbf{k})M_s^*(\mathbf{k}') \quad \text{and} \quad D_i(\mathbf{k},\mathbf{k}') = M_i(\mathbf{k})M_i^*(\mathbf{k}'), \tag{7.177}$$

where the normalized detector modes are distinguished by their locations

$$M_s(\mathbf{k}) \triangleq M(\mathbf{k};x_1) \quad \text{and} \quad M_i(\mathbf{k}) \triangleq M(\mathbf{k};x_2). \tag{7.178}$$

It implies that $\mathrm{tr}\{D_s\} = \mathrm{tr}\{D_i\} = 1$. In what follows, these modes are contracted with the squeezed vacuum state kernels. To simplify notation, we define such contractions by

$$\begin{aligned}
&F_x \triangleq A \diamond M_x, \quad F_x^* \triangleq M_x^* \diamond A, \quad G_x \triangleq A^{-1} \diamond M_x, \quad G_x^* \triangleq M_x^* \diamond A^{-1},\\
&H_x^* \triangleq B^* \diamond A^{-1} \diamond M_x, \quad H_x \triangleq M_x^* \diamond A^{-1} \diamond B,\\
&N_x^* \triangleq A^* \diamond B^* \diamond A^{-1} \diamond M_x \equiv B^* \diamond M_x,\\
&N_x \triangleq M_x^* \diamond A^{-1} \diamond B \diamond A^* \equiv M_x^* \diamond B,
\end{aligned} \tag{7.179}$$

where the subscript x can represent either s or i, and we used (6.340) to simplify the last two definitions. Note that the B-kernels change the complex conjugation of the modes. Superpositions of modes are denoted by T's and are defined as they show up. We also encounter overlaps that produce complex constants. They are defined by

$$\begin{aligned}
&\mu_{xy} \triangleq M_x^* \diamond A \diamond M_y \equiv \mu_{yx}^*, \quad \nu_{xy} \triangleq M_x^* \diamond A^{-1} \diamond M_y \equiv \nu_{yx}^*,\\
&\beta_{xy} \triangleq M_x^* \diamond A^{-1} \diamond B \diamond A^* \diamond M_y^* \equiv M_x^* \diamond B \diamond M_y^* \equiv \beta_{yx},\\
&\beta_{xy}^* \triangleq M_x \diamond A^* \diamond B^* \diamond A^{-1} \diamond M_y \equiv M_x \diamond B^* \diamond M_y \equiv \beta_{yx}^*,
\end{aligned} \tag{7.180}$$

where the subscripts x and y represent any combination of s and i, and we again used (6.340) to simplify the definitions of β_{xy} and β_{xy}^*.

For the single-mode detector kernels, the inverse and determinants can be simplified with the aid of expressions in Appendix D. First, we consider

$$\begin{aligned}
A_0^{-1} &= (A + \mathcal{J}M_sM_s^* + \mathcal{K}M_iM_i^*)^{-1}\\
&= A^{-1} - \frac{\mathcal{J}(1+\mathcal{K}\nu_{ii})G_sG_s^* - \mathcal{J}\mathcal{K}\nu_{is}^*G_sG_i^* - \mathcal{J}\mathcal{K}\nu_{is}G_iG_s^* + \mathcal{K}(1+\mathcal{J}\nu_{ss})G_iG_i^*}{(1+\mathcal{J}\nu_{ss})(1+\mathcal{K}\nu_{ii}) - \mathcal{J}\mathcal{K}\nu_{is}\nu_{is}^*}\\
&= A^{-1} - \frac{T_aG_s^* + T_bG_i^*}{d_0},
\end{aligned} \tag{7.181}$$

where

$$T_a \triangleq \mathcal{J}(1+\mathcal{K}\nu_{ii})G_s - \mathcal{J}\mathcal{K}\nu_{is}G_i,$$
$$T_b \triangleq \mathcal{K}(1+\mathcal{J}\nu_{ss})G_i - \mathcal{J}\mathcal{K}\nu_{is}^*G_s,$$
$$d_0 \triangleq (1+\mathcal{J}\nu_{ss})(1+\mathcal{K}\nu_{ii}) - \mathcal{J}\mathcal{K}\nu_{is}\nu_{is}^*. \quad (7.182)$$

Substituting this inverse into the long determinant, we simplify it further to get

$$\begin{aligned}
&\det\{A_0^* - B^* \diamond A_0^{-1} \diamond B\} \\
&= \det\left\{A^* + \mathcal{J}M_s^*M_s + \mathcal{K}M_i^*M_i - B^* \diamond A^{-1} \diamond B + B^* \diamond \frac{T_aG_s^* + T_bG_i^*}{d_0} \diamond B\right\} \\
&= \det\left\{A^{*-1} + \mathcal{J}M_s^*M_s + \mathcal{K}M_i^*M_i + \frac{T_c^*H_s + T_d^*H_i}{d_0}\right\}, \quad (7.183)
\end{aligned}$$

where we used (6.345), and defined

$$T_c^* \triangleq \mathcal{J}(1+\mathcal{K}\nu_{ii})H_s^* - \mathcal{J}\mathcal{K}\nu_{is}H_i^*,$$
$$T_d^* \triangleq \mathcal{K}(1+\mathcal{J}\nu_{ss})H_i^* - \mathcal{J}\mathcal{K}\nu_{is}^*H_s^*. \quad (7.184)$$

The first three terms in the argument of the determinant are combined into a new kernel, which we define as an inverse that can be expressed as

$$\begin{aligned}
A_1 &\triangleq \left(A^{-1} + \mathcal{J}M_sM_s^* + \mathcal{K}M_iM_i^*\right)^{-1} \\
&= A - \frac{\mathcal{J}(1+\mathcal{K}\mu_{ii})F_sF_s^* - \mathcal{J}\mathcal{K}\mu_{is}^*F_sF_i^* - \mathcal{J}\mathcal{K}\mu_{is}F_iF_s^* + \mathcal{K}(1+\mathcal{J}\mu_{ss})F_iF_i^*}{(1+\mathcal{J}\mu_{ss})(1+\mathcal{K}\mu_{ii}) - \mathcal{J}\mathcal{K}\mu_{is}\mu_{is}^*}. \quad (7.185)
\end{aligned}$$

This kernel is pulled out of the determinant, leading to

$$\begin{aligned}
\det\{A_0^* - B^* \diamond A_0^{-1} \diamond B\} &= \frac{1}{\det\{A_1\}}\det\left\{\mathbf{1} + \frac{A_1^* \diamond T_c^*H_s + A_1^* \diamond T_d^*H_i}{d_0}\right\} \\
&= \frac{1}{\det\{A_1\}}\left(1 + \frac{H_s \diamond A_1^* \diamond T_c^*}{d_0}\right)\left(1 + \frac{H_i \diamond A_1^* \diamond T_d^*}{d_0}\right) \\
&\quad - \frac{(H_s \diamond A_1^* \diamond T_d^*)(H_i \diamond A_1^* \diamond T_c^*)}{\det\{A_1\}d_0^2}, \quad (7.186)
\end{aligned}$$

where

$$H_x \diamond A_1^* \diamond T_c^* = \mathcal{J}(1+\mathcal{K}\nu_{ii})H_x \diamond A_1^* \diamond H_s^* - \mathcal{J}\mathcal{K}\nu_{is}H_x \diamond A_1^* \diamond H_i^*,$$
$$H_x \diamond A_1^* \diamond T_d^* = \mathcal{K}(1+\mathcal{J}\nu_{ss})H_x \diamond A_1^* \diamond H_i^* - \mathcal{J}\mathcal{K}\nu_{is}^*H_x \diamond A_1^* \diamond H_s^*, \quad (7.187)$$

with

$$\begin{aligned}
H_x \diamond A_1^* \diamond H_y^* &= \mu_{xy} - \nu_{xy} - \left[\mathcal{J}(1+\mathcal{K}\mu_{ii})\beta_{xs}\beta_{sy}^* - \mathcal{J}\mathcal{K}\mu_{is}\beta_{xs}\beta_{iy}^* - \mathcal{J}\mathcal{K}\mu_{is}^*\beta_{xi}\beta_{sy}^*\right. \\
&\quad \left. + \mathcal{K}(1+\mathcal{J}\mu_{ss})\beta_{xi}\beta_{iy}^*\right]\left[(1+\mathcal{J}\mu_{ss})(1+\mathcal{K}\mu_{ii}) - \mathcal{J}\mathcal{K}\mu_{is}\mu_{is}^*\right]^{-1}. \quad (7.188)
\end{aligned}$$

The smaller determinants are simplified as

$$\begin{aligned}\det\{A_0\} &= \det\{A + \mathcal{J} M_s M_s^* + \mathcal{K} M_i M_i^*\}\\ &= \det\{A\}[(1 + \mathcal{J} \nu_{ss})(1 + \mathcal{K}\nu_{ii}) - \mathcal{J}\mathcal{K}\nu_{is}\nu_{is}^*],\\ \det\{A_1\} &= \frac{1}{\det\{A^{-1} + \mathcal{J} M_s M_s^* + \mathcal{K} M_i M_i^*\}}\\ &= \frac{\det\{A\}}{(1 + \mathcal{J}\mu_{ss})(1 + \mathcal{K}\mu_{ii}) - \mathcal{J}\mathcal{K}\mu_{is}\mu_{is}^*}.\end{aligned} \tag{7.189}$$

All that remains is to substitute all these results into the full expression in (7.171) and perform some final simplifications. The resulting expression of the generating function for the correlations reads

$$\begin{aligned}\mathcal{C} = \mathcal{N}_J\mathcal{N}_K \Big[&(1 + \mathcal{J}\mu_{ss})^2(1 + \mathcal{K}\mu_{ii})^2 - \mathcal{J}^2(1 + \mathcal{K}\mu_{ii})^2|\beta_{ss}|^2\\ &- \mathcal{K}^2(1 + \mathcal{J}\mu_{ss})^2|\beta_{ii}|^2 - 2\mathcal{J}\mathcal{K}(1 + \mathcal{J}\mu_{ss})(1 + \mathcal{K}\mu_{ii})\left(|\mu_{is}|^2 + |\beta_{si}|^2\right)\\ &+ \mathcal{J}^2\mathcal{K}^2\left(\left|\beta_{si}^2 - \beta_{ss}\beta_{ii}\right|^2 + \left|\mu_{si}^2 - \beta_{ss}\beta_{ii}^*\right|^2 - 2|\mu_{si}|^2|\beta_{si}|^2 - |\beta_{ss}|^2|\beta_{ii}|^2\right)\\ &+ 2\mathcal{J}\mathcal{K}^2(1 + \mathcal{J}\mu_{ss})\left(\mu_{si}\beta_{ii}\beta_{si}^* + \mu_{si}^*\beta_{ii}^*\beta_{si}\right)\\ &+ 2\mathcal{J}^2\mathcal{K}(1 + \mathcal{K}\mu_{ii})\left(\mu_{si}\beta_{ss}^*\beta_{si} + \mu_{si}^*\beta_{ss}\beta_{si}^*\right)\Big]^{-1/2}.\end{aligned} \tag{7.190}$$

Although it is still a fairly complicated expression, those computed from it are often much simpler. For example, the second-order correlation is

$$\Gamma^{(2)} = \partial_{J,K}\mathcal{C}\big|_{J=K=1} = \frac{1}{4}(\mu_{ss} - 1)(\mu_{ii} - 1) + \frac{1}{4}|\mu_{is}|^2 + \frac{1}{4}|\beta_{si}|^2. \tag{7.191}$$

It can be readily confirmed that this result is the same one that would be obtained by substituting the single-mode detector kernel into (7.173).

We can replace $A \to \mathbf{1} + E$ in the definition of μ_{xy} in (7.180) and define

$$\eta_{xy} \triangleq M_x^* \diamond E \diamond M_y. \tag{7.192}$$

Then we can replace $\mu_{ss} \to 1 + \eta_{ss}$, $\mu_{ii} \to 1 + \eta_{ii}$ and $\mu_{is} \to \eta_{is}$ (because $M_s^* \diamond M_i = 0$ for non-overlapping detectors) into the second-order correlation so that it becomes

$$\Gamma^{(2)} = \frac{1}{4}\eta_{ss}\eta_{ii} + \frac{1}{4}|\eta_{is}|^2 + \frac{1}{4}|\beta_{si}|^2. \tag{7.193}$$

The generating function in (7.190) can be used to compute all the measurable quantities for the photon statistics associated with single-mode correlation measurements on a squeezed vacuum state. It is correctly normalized, as can be seen when computing the trace over both measurements by setting $J = K = 1$. The result is $\mathcal{C}(1,1) = 1$. The singles counts are computed as in (7.153) and (7.154), to obtain

$$\partial_J C(J,1)\big|_{J=0} = \frac{2\left(\mu_{ss}^2 - 1 - |\beta_{ss}|^2\right)}{\left[(1+\mu_{ss})^2 - |\beta_{ss}|^2\right]^{3/2}},$$
$$\partial_K C(1,K)\big|_{K=0} = \frac{2\left(\mu_{ii}^2 - 1 - |\beta_{ii}|^2\right)}{\left[(1+\mu_{ii})^2 - |\beta_{ii}|^2\right]^{3/2}}. \qquad (7.194)$$

The other probabilities can have rather complicated expressions.

7.6.6.3 Non-collinear

The expression for the single-mode correlation generating function obtained in (7.190) is still fairly complex. Other experimental conditions that are usually valid for such correlation measurements (anticipating results obtained in Chapter 8) can be used to make further simplifications. To measure quantum correlations among photons in a squeezed vacuum state with two detectors, they need to be spatially separated. The phase-matching conditions for the parametric process with which the squeezed vacuum state is produced also needs to be *non-collinear* to optimize the *measurement probability*. Due to the properties of the A- and B-kernels, their contributions tend to depend on the modes with which they are contracted. While A can only match modes on the same side, B can only match modes on opposite sides. As a result, we can assume that $\mu_{si} = \mu_{si} = \beta_{ss} = \beta_{ii} = 0$. With these quantities set to zero, (7.190) simplifies significantly:

$$\mathcal{C}_{nc}(J,K) = \frac{\mathcal{N}_J\mathcal{N}_K}{(1+\mathcal{J}\mu_{ss})(1+\mathcal{K}\mu_{ii}) - \mathcal{J}\mathcal{K}|\beta_{si}|^2}. \qquad (7.195)$$

The corresponding second-order correlation becomes

$$\Gamma^{(2)} = \partial_{J,K}\mathcal{C}_{nc}\big|_{J=K=1} = \frac{1}{4}(\mu_{ss}-1)(\mu_{ii}-1) + \frac{1}{4}|\beta_{si}|^2 = \frac{1}{4}\eta_{ss}\eta_{ii} + \frac{1}{4}|\beta_{si}|^2, \qquad (7.196)$$

where we replaced $\mu_{ss} \to 1+\eta_{ss}$ and $\mu_{ii} \to 1+\eta_{ii}$ in the last expression. The singles counts simplify to

$$\partial_J C(J,1)\big|_{J=0} = \frac{2\eta_{ss}}{(2+\eta_{ss})^2} \approx \frac{1}{2}\eta_{ss},$$
$$\partial_K C(1,K)\big|_{K=0} = \frac{2\eta_{ii}}{(2+\eta_{ii})^2} \approx \frac{1}{2}\eta_{ii}. \qquad (7.197)$$

The approximations are valid for $\eta_{xx} \ll 1$.

Considering the expression in (7.196), we see that the first term is given by the product of the two singles-counts provided in (7.197). This term represents the *accidental coincidence counts*. Such accidental coincidences happen when two uncorrelated photons are detected at the two detectors, respectively. In practical experiments, the counts are represented as *count rates* for the number of counts per unit time. To compute the *accidental coincidence count rate*, one multiplies the product of the two *singles count rates* by a *gating time* that is the time period within which the detection of two photons at the respective detectors is considered as a coincidence detection.

The coincidence count, as computed in (7.152), is given by

$$\begin{aligned} P(1,1) &= \partial_J \partial_K C(J,K)\big|_{J=K=0} \\ &= \frac{4(\eta_{ss}\eta_{ii} - |\beta_{si}|^2)}{[(2+\eta_{ss})(2+\eta_{ii}) - |\beta_{si}|^2]^2} + \frac{32|\beta_{si}|^2}{[(2+\eta_{ss})(2+\eta_{ii}) - |\beta_{si}|^2]^3}. \end{aligned} \tag{7.198}$$

For $\eta_{ss}, \eta_{ii} \ll \beta_{si}$ it becomes

$$\partial_J \partial_K C(J,K)\big|_{J=K=0} \approx \frac{1}{4}|\beta_{si}|^2. \tag{7.199}$$

7.6.6.4 Non-photon-number-resolving detectors

The correlations provided above all assume that the detectors are able to determine how many photons have been detected. Such detectors are called *photon-number-resolving detectors*. Such detectors are difficult to produce with the current level of technology. Therefore, most detectors in these applications are not photon-number resolving. We can use the generating function to compute the correlation result that is obtained with *non-photon-number-resolving detectors*. For such a calculation, we subtract the vacuum from the trace on both sides, respectively. The resulting *coincidence count* is

$$\begin{aligned} C_{cc}^{(npnr)} &= C(1,1) - C(0,1) - C(1,0) + C(0,0) \\ &= 1 - \frac{2}{2+\eta_{ss}} - \frac{2}{2+\eta_{ii}} + \frac{4}{(2+\eta_{ss})(2+\eta_{ii}) - |\beta_{si}|^2}. \end{aligned} \tag{7.200}$$

The approximate value that we would obtain from (7.200) for non-photon-number-resolving detection when $\eta_{ss}, \eta_{ii} \ll \beta_{si}$ is the same as in (7.199).

We can see that the coincidence count obtained with non-photon-number-resolving detectors also contains a contribution from accidental coincidence counts. In this case, the singles counts are given by

$$C_{sc}^{(npnr)} = C(1,1) - C(0,1) = 1 - \frac{2}{2+\eta_{ss}}, \tag{7.201}$$

and similar for the idler. The *accidental coincidence count* therefore becomes

$$\begin{aligned} C_{acc}^{(npnr)} &= [C(1,1) - C(0,1)]\,[C(1,1) + C(1,0)] \\ &= 1 - \frac{2}{2+\eta_{ss}} - \frac{2}{2+\eta_{ii}} + \frac{4}{(2+\eta_{ss})(2+\eta_{ii})}. \end{aligned} \tag{7.202}$$

The presence of $|\beta_{si}|^2$ in the denominator in (7.200) give rise to an enhancement in the overall correlations, representing the contribution of the true coincidences.

7.6.6.5 Weak squeezing

Another approximation that can be imposed under suitable experimental conditions is the *weak squeezing approximation*. Many experiments with squeezed vacuum states are performed with the aim of preparing entangled *biphotons*. These single pairs of entangled photons can be used in correlation experiments. To suppress the production of more than one such pair at a time, the chosen experimental parameters reduce the efficiency of the down-conversion process. As a result, the squeezing parameter is small and can be used as an expansion parameter. For this purpose, we express the kernels of the squeezed vacuum state as

$$A(\mathbf{k},\mathbf{k}') = \mathbf{1}(\mathbf{k},\mathbf{k}') + \xi^2 E(\mathbf{k},\mathbf{k}') \quad \text{and} \quad B(\mathbf{k},\mathbf{k}') = \xi B(\mathbf{k},\mathbf{k}'), \tag{7.203}$$

where we show the squeezing parameter ξ explicitly. In terms of these expressions, the two determinants in (7.171) become

$$\begin{aligned} \det\{A_0\} &= \det\{\mathbf{1} + \xi^2 E + \mathcal{J}D_{\mathrm{s}} + \mathcal{K}D_{\mathrm{i}}\}, \\ \det\{A_1\} &= \det\{A_0^* - B^* \diamond A_0^{-1} \diamond B\} \\ &= \det\{\mathbf{1} + \xi^2 E^* + \mathcal{J}D_{\mathrm{s}}^* + \mathcal{K}D_{\mathrm{i}}^* \\ &\quad - \xi^2 B^* \diamond (\mathbf{1} + \xi^2 E + \mathcal{J}D_{\mathrm{s}} + \mathcal{K}D_{\mathrm{i}})^{-1} \diamond B\}. \end{aligned} \tag{7.204}$$

For the single biphoton case, we expand the generating function to second-order in ξ. When $D_{\mathrm{s}} \diamond D_{\mathrm{i}} = D_{\mathrm{i}} \diamond D_{\mathrm{s}} = 0$, we simplify the result so that

$$\mathbf{1} + \mathcal{J}D_{\mathrm{s}} + \mathcal{K}D_{\mathrm{i}} = (\mathbf{1} + \mathcal{J}D_{\mathrm{s}}) \diamond (\mathbf{1} + \mathcal{K}D_{\mathrm{i}}), \tag{7.205}$$

which allows us to evaluate the inverses and determinants with the aid of the identities in Appendix D. Substituting the expressions for the kernels in (7.203) into (6.346) for a pure squeezed vacuum state, we find that

$$2E - B \diamond B^* \approx 0. \tag{7.206}$$

With the aid of these simplifications, the expansion of the generating function to second order in the squeezing parameter becomes

$$\begin{aligned} \mathcal{C} \approx{}& 1 + \frac{1}{8}\xi^2 \operatorname{tr}\{B \diamond B^*\} + \frac{1}{4}\xi^2(1-J)(1-K)\operatorname{tr}\{D_{\mathrm{s}} \diamond B \diamond D_{\mathrm{i}}^* \diamond B^*\} \\ & - \frac{1}{4}\xi^2(1-J)\operatorname{tr}\{D_{\mathrm{s}} \diamond B \diamond B^*\} - \frac{1}{4}\xi^2(1-K)\operatorname{tr}\{D_{\mathrm{i}} \diamond B \diamond B^*\} \\ & + \frac{1}{8}\xi^2(1-J)^2\operatorname{tr}\{D_{\mathrm{s}} \diamond B \diamond D_{\mathrm{s}}^* \diamond B^*\} \\ & + \frac{1}{8}\xi^2(1-K)^2\operatorname{tr}\{D_{\mathrm{i}} \diamond B \diamond D_{\mathrm{i}}^* \diamond B^*\}. \end{aligned} \tag{7.207}$$

The second-order correlation for weak squeezing is then given by

$$\Gamma^{(2)} = \partial_{J,K} C(J,K)\big|_{J=K=0} \approx \frac{1}{4}\xi^2 \operatorname{tr}\{D_s \diamond B \diamond D_i^* \diamond B^*\}. \tag{7.208}$$

The same expression can be obtained by substituting (7.203) into (7.174) and retaining the leading-order terms. To obtain the probability for larger numbers of photons, the expansion needs to be computed to larger orders in ξ. For the single-mode detectors, this result is equivalent to (7.199).

7.6.7 Hong–Ou–Mandel effect

The *Hong–Ou–Mandel effect* is a truly quantum interference effect because it involves more than one photon. However, it is purely associated with discrete numbers of particles and thus not really relevant in the context of functional phase space. Here, we discuss the effect simply for the sake of providing background for Section 9.7.

When two indistinguishable photons are sent into the two input ports of a 50:50 beamsplitter, as depicted in Figure 7.3, they produce a superposition of the two possible paths from their input ports to the two output ports. Here, we represent the unitary transformation of the beamsplitter with a relative minus sign in one of the two transformations. When the product of the superpositions for the two photons is multiplied out, the two cross-terms with a photon exiting each of the two output ports (representing the same state for indistinguishable photons) have opposite signs. So, they cancel and the only remaining possibilities are those where both photons exit the same output port.

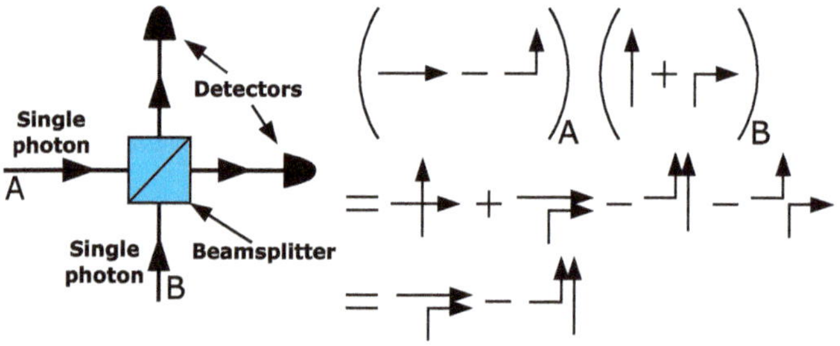

Figure 7.3: Diagrammatic representation of the Hong–Ou–Mandel effect.

The cancellation happens between terms in a quantum superposition each representing two photons. It is therefore a truly quantum interference effect and cannot be reproduced in terms of a classical description.

If the photons are distinguishable, as would be the case if they have different states of polarization or different spatial modes, the two terms with one photon exiting each of the two output ports produces a Ψ_- Bell state, as defined in (4.98). A coincidence observed at both output ports thus represents the measurement of a Ψ_- Bell state.

7.7 Noise

A ubiquitous aspect of measurements is *noise*. The concept of noise begs the question of a *signal* that carries information. In the context of measurement, the signal represents the information that is obtained from the measurement process. The noise is any additional undesirable signal that obscures the information in the desirable signal. In this sense, noise produces unwanted contributions to the measurement results. It is ubiquitous in both classical and quantum systems.

There are other undesirable effects that can affect the measurement of a signal. These undesirable effects include *distortions* and *loss*. Distortions are undesirable effects intrinsic to the system such a nonlinearities that distort the signal. Loss, which is discussed in Section 6.7, produces an *attenuation* of the signal caused by the loss of power (loss of photons) in the signal due to absorption in the system. Here, noise is considered as being distinct from these other inimical effects.

In the classical context, noise is considered to be *additive*, in the form of unwanted additional radiation coming from other sources and passing through the system together with the signal. In the quantum context, noise can also be produced by the discreteness of the detection of individual quanta. The random nature of these detections produce a random fluctuation in the measured signal. It is referred to as *shot noise* or just *quantum noise* and is an intrinsic property of quantum measurements.

The pertinent quantity to consider when dealing with noise is the *signal-to-noise ratio* (SNR), given by the ratio of the power in the signal P_S to the power in the noise P_N,

$$\mathrm{SNR} \triangleq \frac{P_S}{P_N}. \tag{7.209}$$

Since the power in the signal can vary, it is not the absolute power in the noise that is important, but rather their ratio as given by the signal-to-noise ratio. It is usually measured in *decibel*, which is defined as

$$\mathrm{SNR}_{\mathrm{dB}} \triangleq 10 \log_{10}\left(\frac{P_S}{P_N}\right). \tag{7.210}$$

7.7.1 Quantum noise

Let us first discuss quantum noise produced by the statistical properties of quantum states. In Section 7.4, we discussed methods to calculate the photon statistics of quantum states, as revealed by photon-number-resolving measurements. The tool for this purpose is the generating function in (7.87). In Section 7.4.3, we used it to obtain the photon statistics for a thermal state, as shown in Figure 7.1.

For a coherent state, the photon statistics takes on the form of a *Poisson distribution*, as discussed among the properties of coherent states in Section 3.2 and shown in Figure 3.2. It is also shown there that the variance of the Poisson distribution is equal to its average. It follows that its standard deviation is equal to the square root of the average. The power of the radiation represented by a quantum state is proportional to the average number of photons (or photon rate) in that state. On the other hand, the power of the quantum noise in this radiation is proportional to the standard deviation of the photon statistics of the number of photons (or photon rate). For a coherent state, the signal-to-noise ratio is then given by

$$\mathrm{SNR} = \sqrt{\langle n \rangle}. \tag{7.211}$$

It is a reflection of the uncertainty in the measurement result due to the intrinsic statistical properties of the coherent state and is referred to as the *standard quantum limit*. Since coherent states represent the best scenario that we can have in the classical case, this intrinsic signal-to-noise ratio represents the best case. However, since the average number of photons $\langle n \rangle$ can be arbitrarily large in classical scenarios, the "best classical scenario" is not much of a limitation.

The standard quantum limit becomes an issue in scenarios where the average number of photons is limited for some reason that may be imposed by the physical system. In such scenarios, one would prefer to use other quantum states that produce better signal-to-noise ratios as a function of the number of photons in the state. Such an approach is the topic of *quantum metrology* [6, 7].

Squeezed vacuum states are such quantum states. In Section 3.4.2, it is shown that squeezing causes the width of a state to be reduced along a specific direction in phase space (quadrature direction). A consequence is that the width of this state is then increased along the orthogonal quadrature direction. The effect of this squeezing process on a vacuum state is shown in Figure 6.7.

If we measure quadrature q of such a squeezed state instead of the occupation number n, we can obtain signal-to-noise ratios much better than the standard quantum limit in (7.211). The improvement in the intrinsic signal-to-noise ratio for such a scenario is governed by $\exp(\xi)$, where ξ is the squeezing parameter. To make such a measurement, we need to perform the measurement along a specific direction in phase space associated with the direction along which the state is squeezed. The Wigner functional of a state rotates as a function of time around the origin at a rate given by the frequency. So, the measurement process must be synchronized with the rotation of the state. The *homodyne measurement system*, discuss in Section 7.9, facilitates such a synchronized measurement of quadrature. Other systems that exploit this property of squeezing include *nonlinear interferometers* (also called *SU11 interferometers*) [8, 9].

7.7.2 General state with noise

Our interest in the effect of noise in quantum optical systems is not restricted to quantum noise. Additive classical noise can be enough of a nuisance in quantum systems to justify its consideration. To study the effect of noise in quantum scenarios, we combine it with the quantum states using a beamsplitter and then compute the quantum statistics. The total process can be represented by

$$P_n = \operatorname{tr}\left\{\hat{U}_{\text{bs}}\left(\hat{\rho}_{\text{signal}} \otimes \hat{\rho}_{\text{noise}}\right)\hat{U}_{\text{bs}}^{\dagger}\hat{P}_n\right\}, \tag{7.212}$$

where $\hat{\rho}_{\text{signal}}$ and $\hat{\rho}_{\text{noise}}$ are the quantum states of the signal and the noise, respectively, $\hat{U}_{\text{bs}}$ is the unitary operator for the beamsplitter, and $\hat{P}_n$ is the projection operator for n photons representing the photon statistics measurement.

The Wigner functional calculation can be done with the aid of (7.87) to produce a generating function for the photon statistics, and the beamsplitter is implemented with the transformation of the field variables given in (6.218). The resulting expression reads

$$\mathcal{G}_P(J) = \int W_{\text{signal}}[C\beta + iS\alpha]W_{\text{noise}}[C\alpha + iS\beta]\mathcal{P}[\alpha](J)\,\mathcal{D}^{\circ}[\alpha,\beta], \tag{7.213}$$

where $\mathcal{P}[\alpha](J)$ is the generating function in (7.87).

The balanced introduction of noise to the signal requires a 50:50 beamsplitter, for which $C = S = \frac{1}{\sqrt{2}}$. However, by retaining a variable beamsplitter reflectivity, we can investigate the separate contributions to the photon statistics. The statistics is measured at one of the output ports of the beamsplitter (the one represented by α) while the other is traced out by integrating over β to remove the unwanted degrees of freedom.

The quantum state of additive noise is modelled as a thermal state with its Wigner functional given by (6.181). Here, we use it with an unspecified thermal state kernel θ. The measurement is performed with a detector kernel D inserted into (7.87). The expression then becomes

$$\begin{aligned}\mathcal{G}_P(J) = \int & \mathcal{N}_J^{\operatorname{tr}\{D\}} \exp\left[-2\mathcal{J}\alpha^* \diamond D \diamond \alpha\right] \\ & \times \mathcal{N}_0 \det\{\theta\} \exp\left[-2\left(C\alpha^* - iS\beta^*\right) \diamond \theta \diamond \left(C\alpha + iS\beta\right)\right] \\ & \times W_{\hat{\rho}}[C\beta^* - iS\alpha^*, C\beta + iS\alpha]\,\mathcal{D}^{\circ}[\alpha,\beta],\end{aligned} \tag{7.214}$$

where $\mathcal{J}$ and $\mathcal{N}_J$ are defined in (7.88).

Before considering specific states, we can evaluate one of the two functional integrations. For that purpose, we shift one of the integration field variables out of the argument of the Wigner functional of the unknown state by redefining β. It is done by replacing

$$\beta \to \frac{1}{C}\eta - i\frac{S}{C}\alpha \quad \text{and} \quad \beta^* \to \frac{1}{C}\eta^* + i\frac{S}{C}\alpha^*, \tag{7.215}$$

where η is the new field variable, leading to

$$\begin{aligned}\mathcal{G}_P(J) &= \int \mathcal{N}_0 \det\{\theta\} \exp\left[-2(\alpha^* - \mathrm{i}S\eta^*) \diamond \theta \diamond (\alpha + \mathrm{i}S\eta) C^{-2}\right] \\ &\quad \times \mathcal{N}_J^{\mathrm{tr}\{D\}} \exp\left[-2\mathcal{J}\alpha^* \diamond D \diamond \alpha\right] W_{\hat{\rho}}[\eta^*, \eta] C^{-2\Omega} \mathcal{D}^\circ[\alpha, \eta] \\ &= \frac{\mathcal{N}_0 \det\{\theta\} \mathcal{N}_J^{\mathrm{tr}\{D\}}}{C^{2\Omega}} \int \exp\left[-2\alpha^* \diamond \left(C^{-2}\theta + \mathcal{J}D\right) \diamond \alpha - \mathrm{i}\frac{2S}{C^2}\alpha^* \diamond \theta \diamond \eta\right. \\ &\quad \left. + \mathrm{i}\frac{2S}{C^2}\eta^* \diamond \theta \diamond \alpha - \frac{2S^2}{C^2}\eta^* \diamond \theta \diamond \eta\right] W_{\hat{\rho}}[\eta^*, \eta]\, \mathcal{D}^\circ[\alpha, \eta]. \end{aligned} \tag{7.216}$$

The transformation of the integration field variables produces a constant factor of $C^{-2\Omega}$. Now we evaluate the functional integral over α to get

$$\begin{aligned}\mathcal{G}_P(J) &= \frac{\det\{\theta\} \mathcal{N}_J^{\mathrm{tr}\{D\}} C^{-2\Omega}}{\det\{C^{-2}\theta + \mathcal{J}D\}} \int \exp\left[2\frac{S^2}{C^2}\eta^* \diamond \theta \diamond \left(\theta + C^2\mathcal{J}D\right)^{-1} \diamond \theta \diamond \eta\right. \\ &\quad \left. - 2\frac{S^2}{C^2}\eta^* \diamond \theta \diamond \eta\right] W_{\hat{\rho}}[\eta^*, \eta]\, \mathcal{D}^\circ[\eta] \\ &= \frac{\mathcal{N}_J^{\mathrm{tr}\{D\}}}{\det\{\mathbf{1} + C^2\mathcal{J}D \diamond \theta^{-1}\}} \int W_{\hat{\rho}}[\eta^*, \eta] \\ &\quad \times \exp\left[-2S^2\mathcal{J}\eta^* \diamond \left(\mathbf{1} + C^2\mathcal{J}D \diamond \theta^{-1}\right)^{-1} \diamond D \diamond \eta\right] \mathcal{D}^\circ[\eta]. \end{aligned} \tag{7.217}$$

Based on the discussion in Section 6.5 in the context of blackbody radiation, we know that the kernel of a physical thermal state cannot be diagonalized by the plane wave basis, to produce a kernel that is proportional to the identity operator $\theta = T(\mathbf{k})\mathbf{1}$, where $T(\mathbf{k})$ is a nonzero spectral function. Such a diagonal thermal state kernel would be unphysical in the sense that it represents an infinite energy state. Moreover, one does not have control over the properties of the noise. So, we prefer to leave the kernel of the thermal state representing the noise unspecified as an arbitrary physical thermal state.

Instead, we do have control over the detector with which we perform the measurements of the photon statistic. Therefore, we assume that the detector kernel can be modelled as a single-mode detector kernel $D = MM^*$, so that $\mathrm{tr}\{D\} = 1$. In that case, the inverse and determinant can be simplified as

$$\begin{aligned}\left(\mathbf{1} + C^2\mathcal{J}MM^* \diamond \theta^{-1}\right)^{-1} &= \mathbf{1} - \frac{C^2\mathcal{J}MM^* \diamond \theta^{-1}}{1 + C^2\mathcal{J}\tau}, \\ \det\left\{\mathbf{1} + C^2\mathcal{J}MM^* \diamond \theta^{-1}\right\} &= 1 + C^2\mathcal{J}\tau, \end{aligned} \tag{7.218}$$

where

$$\tau \triangleq M^* \diamond \theta^{-1} \diamond M \tag{7.219}$$

denotes the thermal radiation detected by the detector. With these simplifications, the expression in (7.217) becomes

$$\mathcal{G}_P(J) = \mathcal{M} \int W_{\hat{\rho}}[\eta^*, \eta] \exp\left(-2\mathcal{K}\eta^* \diamond MM^* \diamond \eta\right) \mathcal{D}^\circ[\eta], \tag{7.220}$$

where

$$\begin{aligned} \mathcal{M} &= \frac{2}{1 + J + (1-J)C^2\tau}, \\ \mathcal{K} &= \frac{(1-J)S^2}{1 + J + (1-J)C^2\tau}. \end{aligned} \tag{7.221}$$

The generating function in (7.220) can be used to produce all the photon statistics that can be obtained from the signal with the additive noise. As an exercise, consider the two extremes:

Exercise 7.10. Show that, for $S = 1$ and $C = 0$, the expression in (7.220) reproduces the generating function for the photon statistics of the signal only, given by

$$\mathcal{G}_P(J) = \mathcal{N}_J \int W_{\hat{\rho}}[\eta^*, \eta] \exp\left(-2\mathcal{J}\eta^* \diamond MM^* \diamond \eta\right) \mathcal{D}^\circ[\eta]. \tag{7.222}$$

Exercise 7.11. Show that, for $S = 0$ and $C = 1$, the expression in (7.220) reproduces the generating function for the photon statistics of the thermal state only

$$\mathcal{G}_P(J) = \frac{1}{1 + (1-J)\frac{1}{2}(\tau - 1)}, \tag{7.223}$$

which is equivalent to (7.105).

7.7.2.1 Coherent state with noise

As an example, we first consider the case where the signal is represented in terms of a coherent state with a parameter function ζ. Inserting the Wigner functional for such a coherent state into (7.220), we get

$$\begin{aligned} \mathcal{G}_P(J) &= \mathcal{M} \int \mathcal{N}_0 \exp\left[-2(\eta^* - \zeta^*) \diamond (\eta - \zeta) - 2\mathcal{K}\eta^* \diamond MM^* \diamond \eta\right] \mathcal{D}^\circ[\eta] \\ &= \frac{\mathcal{M}}{\det\{\mathbf{1} + \mathcal{K}MM^*\}} \exp\left[2\zeta^* \diamond (\mathbf{1} + \mathcal{K}MM^*)^{-1} \diamond \zeta - 2\zeta^* \diamond \zeta\right]. \end{aligned} \tag{7.224}$$

The inverse and determinant are

$$\begin{aligned} \det\{\mathbf{1} + \mathcal{K}MM^*\} &= 1 + \mathcal{K} = \frac{2 + (1-J)C^2(\tau - 1)}{1 + J + (1-J)C^2\tau}, \\ (\mathbf{1} + \mathcal{K}MM^*)^{-1} &= \mathbf{1} - \frac{\mathcal{K}}{1 + \mathcal{K}} MM^* = \mathbf{1} - \frac{(1-J)S^2 MM^*}{2 + (1-J)C^2(\tau - 1)}. \end{aligned} \tag{7.225}$$

So, the generating function (with a variable reflectivity) becomes

$$\mathcal{G}_P(J) = \frac{1}{1 + (1-J)C^2\frac{1}{2}(\tau - 1)} \exp\left[-\frac{(1-J)S^2|\nu|^2}{1 + (1-J)C^2\frac{1}{2}(\tau - 1)}\right], \tag{7.226}$$

where $\nu = \zeta^* \diamond M$ represents the overlap between the parameter function of the coherent state and the detector mode.

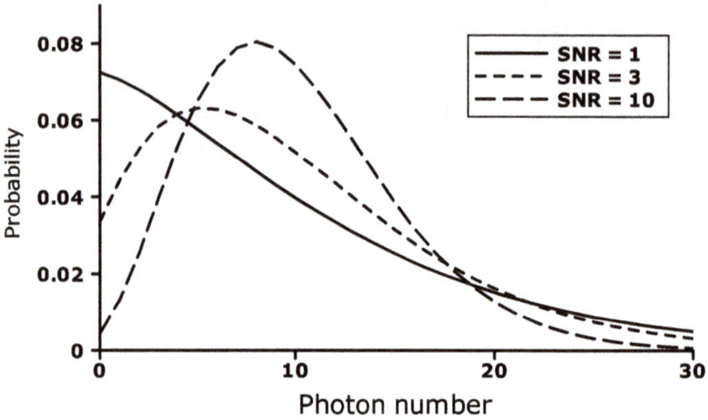

Figure 7.4: Probability curves for photons in a coherent state with thermal noise for signal-to-noise ratios of 1, 3, and 10. The average number of photons in the combined state is 10.

The two extremes then allow us to compute the average number of photons in the signal and noise, respectively, using

$$\langle n \rangle \triangleq \partial_J \mathcal{G}_P(J)\big|_{J=1} . \tag{7.227}$$

For $S = 1$ and $C = 0$ (signal), the expression for the photon statistics becomes the generating function for the Poisson distribution

$$\mathcal{G}_P(J) = \exp\left[-(1-J)|\nu|^2\right], \tag{7.228}$$

as found in (3.107). It gives the average number of photons as $\langle n \rangle = |\nu|^2 \triangleq N_s$, which is proportional to the signal power. The generating function for the noise only (obtained with $S = 0$ and $C = 1$), is still given by (7.223), with the average number of photons being $\langle n \rangle = \frac{1}{2}(\tau - 1) \triangleq N_n$, which is proportional to the noise power. So, the signal-to-noise ratio, observed by the single-mode detector, is given by

$$\mathrm{SNR}_{\mathrm{coh}} = \frac{N_s}{N_n} = \frac{2|\zeta^* \diamond M|^2}{M^* \diamond (\theta^{-1} - \mathbf{1}) \diamond M}. \tag{7.229}$$

The generating function for the photon statistics can now be expressed as

$$\mathcal{G}_P(J) = \frac{1}{1+(1-J)\frac{1}{2}N_n} \exp\left[-\frac{(1-J)\frac{1}{2}N_s}{1+(1-J)\frac{1}{2}N_n}\right], \tag{7.230}$$

where we set $C = S = \frac{1}{\sqrt{2}}$.

In Figure 7.4, the photon-number distributions are shown for signal-to-noise ratios of 1, 3, and 10, with a total average number of photons equal to 10. For SNR = 1, the distribution looks more like that of a thermal state, whereas for SNR = 10, it starts to resemble the Poisson distribution of a coherent state.

7.7.2.2 Squeezed vacuum state with noise

Our next example is where a squeezed vacuum state represents the signal. We substitute the Wigner functional for a squeezed vacuum state into (7.220). It produces

$$\begin{aligned}\mathcal{G}_P(J) &= \mathcal{M}\int \mathcal{N}_0 \exp\left(-2\eta^* \diamond A \diamond \eta - \eta^* \diamond B \diamond \eta^* - \eta \diamond B^* \diamond \eta\right.\\ &\quad \left. - 2\mathcal{K}\eta^* \diamond MM^* \diamond \eta\right) \mathcal{D}^\circ[\eta]\\ &= \frac{\mathcal{M}}{\sqrt{\det\{A + \mathcal{K}MM^*\}}} \frac{1}{\sqrt{\det\{A + \mathcal{K}MM^* - B \diamond (A^* + \mathcal{K}M^*M)^{-1} \diamond B^*\}}}.\end{aligned} \tag{7.231}$$

where $\mathcal{M}$ and $\mathcal{K}$ are given in (7.221). To simplify this expression, we first use the following expressions:

$$\begin{aligned}\det\{A + \mathcal{K}MM^*\} &= \det\{A\}(1 + \mathcal{K}\nu),\\ (A + \mathcal{K}MM^*)^{-1} &= A^{-1} - \frac{\mathcal{K}}{1+\mathcal{K}\nu} A^{-1} \diamond MM^* \diamond A^{-1},\end{aligned} \tag{7.232}$$

where

$$\nu \triangleq M^* \diamond A^{-1} \diamond M. \tag{7.233}$$

The long determinant in (7.231) then becomes

$$\begin{aligned}&\det\{A + \mathcal{K}MM^* - B \diamond (A^* + \mathcal{K}M^*M)^{-1} \diamond B^*\}\\ &= \det\{A^{-1} + \mathcal{K}MM^* + \mathcal{K}'HH^*\}\\ &= \frac{1}{\det\{A\}}\left[(1+\mathcal{K}\mu)(1+\mathcal{K}'\mu - \mathcal{K}'\nu) - \mathcal{K}'\mathcal{K}|\beta|^2\right] = \frac{(1+\mathcal{K}\mu)^2 - \mathcal{K}^2|\beta|^2}{\det\{A\}(1+\mathcal{K}\nu)},\end{aligned} \tag{7.234}$$

where we used the identities in Section 6.8.5 and Appendix D, and defined

$$\begin{aligned}\mathcal{K}' &= \frac{\mathcal{K}}{1+\mathcal{K}\nu}, \quad \mu = M^* \diamond A \diamond M, \quad H = M^* \diamond A^{-1} \diamond B,\\ \beta &= M^* \diamond A^{-1} \diamond B \diamond A^* \diamond M^* \equiv M^* \diamond B \diamond M^*.\end{aligned} \tag{7.235}$$

The full expression then becomes

$$
\begin{aligned}
\mathcal{G}_P(J) &= \frac{\mathcal{M}}{\sqrt{\det\{A\}(1+\mathcal{K}\nu)\det\{A^{-1}+\mathcal{K}MM^*+\mathcal{K}'HH^*\}}} \\
&= \frac{\mathcal{M}}{\sqrt{(1+\mathcal{K}\mu)^2-\mathcal{K}^2|\beta|^2}} \\
&= \frac{2}{\sqrt{[1+J+(1-J)C^2\tau+(1-J)S^2\mu]^2-(1-J)^2S^4|\beta|^2}}.
\end{aligned}
\tag{7.236}
$$

Note that we can obtain this expression from (7.89) by replacing $\mathcal{N}_J$ and $\mathcal{J}$ by $\mathcal{M}$ and $\mathcal{K}$, respectively. When substituting $S = 0$ and $C = 1$, we recover the generating function for the photon statistics of the thermal noise only. For $S = 1$ and $C = 0$, we get (7.89).

Exercise 7.12. Compute the signal-to-noise ratio for the squeezed vacuum state from (7.236) in terms of μ and τ.

7.8 Heralding

Although the discussions in this chapter are focused on measurements in quantum optics, there are scenarios where measurements are used as a means for state preparation. We can argue that such forms of state preparation can also be viewed as part of a measurement protocol.

In this section, we consider the processes whereby states are prepared with the aid of such measurements, which generally fall under the term *heralding*. The idea is that the required state is only successfully produced when a specific measurement is successfully performed. In that case, the successful measurement *heralds* the existence of the required state. The latter is then used in subsequent processing followed by some final measurements. The final measurements must be performed in coincidence with the heralding measurement to ensure that those final measurements are only recorded when the required state exists. The requirement of a coincidence condition implies that the heralding, together with the final measurements, can be combined into a measurement protocol performed on the initial state, which is often a Gaussian state.

Heralding is a post-selection process. It means that the process selects specific parts of a state satisfying certain conditions. However, the validity of these conditions for the creation of a post-selected state generally comes with a reduced probability, which means that it is not normalized. To obtain a valid state from such a process, the result needs to be normalized by dividing it by its trace.

Our goal is to model the physical measurement process to see how any constraints in the spatiotemporal properties of the measurement system affects the result. For this purpose, the Wigner functional formalism is used with the generating function for the

Wigner functionals of generalized projection operators, as developed in Sections 7.3 and 7.4, and given in (7.87).

There are various examples of states produced by such heralding processes. *Single-photon Fock states* [10] are obtained from weak twin-beam squeezed vacuum states by measuring a photon on one side to herald a single photon on the other side. Similarly, *multiphoton Fock states* [11, 12] are obtained from squeezed vacuum states by measuring a certain number of photons on one side using a *photon-number-resolving detector* to herald the creation of a Fock state with the same number of photons on the other side. *Photon-subtracted states* [13] are produced from squeezed vacuum states by using an unbalanced beamsplitter, sending a weak beam to a single-photon detector. *Photon-added states* [14] are obtained from Gaussian states (squeezed or coherent states) by using an *optical parametric amplifier* (part of *stimulated parametric down-conversion*) instead of an unbalanced beamsplitter, sending a weak beam (the idler) to a single-photon detector. *Schrödinger cat states* [15] are approximated by photon-subtracted squeezed vacuum states. There are various generalization of these heralded state preparation protocols. *Quantum teleportation* [16] is also a heralded process, but instead of state preparation, its purpose is to transfer an existing unknown state to a remote system. In what follows, we consider single-photon Fock states preparation, photon subtraction and photon addition. Teleportation is discussed in Section 9.7 in the context of *upconversion*.

The significance of photon-subtracted and photon-added states is that their Wigner functionals could be negative in some regions of phase space. Such *Wigner negativity* is considered to be an indication of quantum properties that can be used as a *resource* for quantum information processing. Here, we are not specifically interested in the quantum information aspects of these states, but we discuss the extent to which such states can be produced in practical heralded state preparation systems.

7.8.1 Generic process

In terms of the quantum operator notation, a heralded state is obtained from an initial state via the process

$$\hat{\rho}_{\text{her}} = \mathcal{N}\, \text{tr}_B\{\hat{P}_D \hat{U}_{\text{bs}} (\hat{\rho}_{\text{in}} \otimes \hat{\rho}_{\text{vac}})\, \hat{U}_{\text{bs}}^{\dagger}\}. \tag{7.237}$$

Here, $\hat{P}_D$ is a projection operator representing the heralding detector, $\hat{U}_{\text{bs}}$ is the unitary operator for a generic beamsplitter (homogeneous, inhomogeneous, or nonlinear, depending on the application), $\hat{\rho}_{\text{in}}$ is the initial state, and $\hat{\rho}_{\text{vac}}$ is the vacuum state. The trace represents a partial trace that is only performed on the part of the state on which the heralding detection is performed. The heralding process is not *trace preserving*. So, the result after the trace needs to be normalized; hence, the normalization constant $\mathcal{N}$. A diagram for the generic heralding process is shown in Figure 7.5.

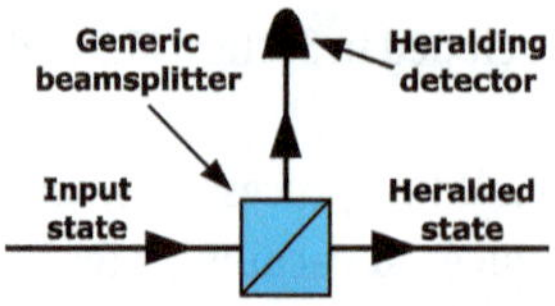

Figure 7.5: Generic heralding process for quantum state preparation.

In terms of Wigner functionals, the two unitary operations for the beamsplitters operating on the tensor product of the initial state and the vacuum state can be represented by two triple star products. However, in the actual calculation, the beamsplitter is usually implemented as a transformation of the arguments of the Wigner functionals. The trace with the heralding detector is then obtained by a final integration of the product. The full expression is

$$W_{\text{her}}[\alpha] = \int \mathcal{P}[\beta]\,[W_{\text{bs}} \star (W_{\text{in}} W_{\text{vac}}) \star W_{\text{bs}}^{*}]\,[\alpha, \beta]\,\mathcal{D}^{\circ}[\beta], \tag{7.238}$$

where $\mathcal{P}$ is the generating function for photon statistics given in (7.87).

7.8.2 Single-photon Fock state preparation

To prepare Fock states, one generally needs a photon-number-resolving detector. Here, we only consider single-photon Fock states under the assumption that the probability of more photons is suppressed by the weakness of the squeezing. For a weakly squeezed vacuum state, the *biphoton* term dominates over terms with larger numbers of photon pairs. In this case, we assume that the initial state is a *twin-beam squeezed vacuum state*, as discussed in Sections 6.8.8 and 6.9.2, which removes the need for the beamsplitter. The calculation for single-photon Fock state preparation thus involves a single trace over the product of the twin-beam squeezed state in (6.367) and the photon-statistics generating function in (7.87). A single-mode detector kernel is used for this purpose. The calculation then produces a generating function expressed as

$$\begin{aligned}
\mathcal{W}_{\text{ph}}(J) &= \int W_{\text{tbsqv}}[\alpha, \beta]\mathcal{P}[\beta]\,\mathcal{D}^{\circ}[\beta] \\
&= \frac{2\mathcal{N}_0^2}{1+J} \int \exp\left(-2\beta^{*} \diamond A \diamond \beta - 2\alpha^{*} \diamond A \diamond \alpha\right. \\
&\quad \left. - 2\beta^{*} \diamond B \diamond \alpha^{*} - 2\alpha \diamond B^{*} \diamond \beta - 2\mathcal{J}\beta^{*} \diamond MM^{*} \diamond \beta\right)\,\mathcal{D}^{\circ}[\beta] \\
&= \exp\left[2\alpha \diamond B^{*} \diamond (A + \mathcal{J}MM^{*})^{-1} \diamond B \diamond \alpha^{*}\right] \\
&\quad \times \frac{2\mathcal{N}_0 \exp(-2\alpha^{*} \diamond A \diamond \alpha)}{(1+J)\det\{A + \mathcal{J}MM^{*}\}}.
\end{aligned} \tag{7.239}$$

The inverse and determinant can be simplified, using (D.23) and (D.28), so that

$$\det\{A + \mathcal{J}MM^*\} = \det\{A\}(1 + \mathcal{J}M^* \diamond A^{-1} \diamond M),$$
$$(A + \mathcal{J}MM^*)^{-1} = A^{-1} - \mathcal{J}\frac{A^{-1} \diamond MM^* \diamond A^{-1}}{1 + \mathcal{J}M^* \diamond A^{-1} \diamond M}. \tag{7.240}$$

The generating function thus becomes

$$\begin{aligned}\mathcal{W}_{\text{ph}} &= \frac{2\mathcal{N}_0 \exp(-2\alpha^* \diamond A \diamond \alpha + 2\alpha \diamond B^* \diamond A^{-1} \diamond B \diamond \alpha^*)}{\det\{A\}(1 + J + \nu - J\nu)} \\ &\quad \times \exp\left(-2\mathcal{J}\frac{\alpha^* \diamond HH^* \diamond \alpha}{1 + \mathcal{J}\nu}\right) \\ &= \frac{2\mathcal{N}_0 \exp(-2\alpha^* \diamond A^{-1} \diamond \alpha)}{\det\{A\}(1 + J + \nu - J\nu)} \exp\left[-2\frac{(1 - J)\alpha^* \diamond HH^* \diamond \alpha}{1 + J + \nu - J\nu}\right],\end{aligned} \tag{7.241}$$

where we used the assumed purity of the twin-beam squeezed state to simplify the expression with the aid of Section 6.8.5. We also used the definitions

$$\begin{aligned}\mu &= M^* \diamond A \diamond M, \quad \nu = M^* \diamond A^{-1} \diamond M, \\ H^* &= B^* \diamond A^{-1} \diamond M = M \diamond A^{*-1} \diamond B^*, \\ H &= M^* \diamond A^{-1} \diamond B = B \diamond A^{*-1} \diamond M^*,\end{aligned} \tag{7.242}$$

from (7.180). With the aid of Section 6.8.5, it follows that

$$\begin{aligned}H^* \diamond A \diamond H &= M \diamond A^{*-1} \diamond B^* \diamond A \diamond B \diamond A^{*-1} \diamond M^* \\ &= M^* \diamond A \diamond M - M^* \diamond A^{-1} \diamond M = \mu - \nu.\end{aligned} \tag{7.243}$$

Since the Wigner functionals that are produced from the resulting generating function with the aid of derivatives with respect the J are not normalized, we need to compute the appropriate normalization constants for each of them. The generating function can be used to compute a generating function for the inverse normalization constants. By integrating the generating function over α, we obtain

$$\begin{aligned}\mathcal{W}_{\text{norm}} &= \frac{2}{\det\{A\}(1 + J + \nu - J\nu)}\left[\det\left\{A^{-1} + \frac{(1 - J)HH^*}{1 + J + \nu - J\nu}\right\}\right]^{-1} \\ &= \frac{2}{1 + J + \nu - J\nu}\left[1 + \frac{(1 - J)(\mu - \nu)}{1 + J + \nu - J\nu}\right]^{-1} = \frac{2}{1 + J + (1 - J)\mu},\end{aligned} \tag{7.244}$$

where we again used the identities in Section 6.8.5 to simplify the contractions of the squeezed vacuum state kernels. The result is then simplified with the aid of the identities in Appendix D and (7.243). It become a relatively simple generating function in terms of μ instead of ν.

The normalized heralded single-photon Wigner functional is calculated from the two generating functions as follows

$$W_1[\alpha] = \frac{\partial_J \mathcal{W}_{\text{ph}}|_{J=0}}{\partial_J \mathcal{W}_{\text{norm}}|_{J=0}}$$
$$= \frac{\mathcal{N}_0(1+\mu)^2 \exp\left(-2\alpha^* \diamond A^{-1} \diamond \alpha\right)}{\det\{A\}(1+\nu)^3(\mu-1)}$$
$$\times \exp\left(-\frac{2}{1+\nu}\alpha^* \diamond HH^* \diamond \alpha\right)\left(4\alpha^* \diamond HH^* \diamond \alpha - 1 + \nu^2\right). \tag{7.245}$$

We see that the result is a thermal state, multiplied by a polynomial functional. In the limit of *weak squeezing*, we can assume that $A^{-1} \approx \mathbf{1} - E$ where $E = A - \mathbf{1}$ is the second-order term in the expansion of the kernels in terms of the squeezing parameter. Therefore, $\mu = 1 + \eta$ and $\nu \approx 1 - \eta$, where

$$\eta \triangleq M^* \diamond E \diamond M. \tag{7.246}$$

Furthermore, B and B^* are first order in the squeezing parameter, so that we can approximate $A^{-1} \diamond B \approx B$ and $B^* \diamond A^{-1} \approx B^*$. Using these approximations, we expand the argument of the exponent and the polynomial factor, respectively, to leading order in the squeezing parameter. Then the expression simplifies to

$$W_1[\alpha] \approx \mathcal{N}_0 \exp\left(-2\alpha^* \diamond \alpha\right)\left(\frac{4\alpha \diamond B^* \diamond MM^* \diamond B \diamond \alpha^*}{2\eta} - 1\right). \tag{7.247}$$

The envelope becomes that of a pure vacuum state in the weak squeezing limit, but we still have some distortion of the mode due to the B-kernels in the polynomial. In effect, these kernels perform a transformation of the normalized detector mode to some unnormalized distorted mode. We can compute the normalization factor by performing the overlap $M^* \diamond B \diamond B^* \diamond M \approx 2M^* \diamond E \diamond M = 2\eta$, using (7.206). Therefore, we can define normalized modes

$$M_b \triangleq \frac{M^* \diamond B}{\sqrt{2\eta}}, \quad \text{and} \quad M_b^* \triangleq \frac{M \diamond B^*}{\sqrt{2\eta}}. \tag{7.248}$$

The Wigner functional of the single-photon Fock state then becomes

$$W_1[\alpha] \approx \mathcal{N}_0 \exp\left(-2\alpha^* \diamond \alpha\right)\left(4\alpha^* \diamond M_b M_b^* \diamond \alpha - 1\right). \tag{7.249}$$

Since M_b represents a specific mode produced by a transformation of the detector mode, the resulting Wigner functional has the expected shape of the Wigner function for a single-photon Fock state only on a two-dimensional subspace of the infinite dimensional functional phase space. In this two-dimensional subspace, the field variable α is proportional to the normalized mode M_b. Therefore, we can replace $\alpha \to \alpha_0 M_b$, where α_0 is a complex-valued variable, into the Wigner functional of the single-photon Fock state and trace over the orthogonal part of the phase space to obtain the two-dimensional (radially symmetric) *Wigner function*

$$W_{\text{spfs}}(r) = 2\exp\left(-r^2\right)\left(2r^2 - 1\right), \tag{7.250}$$

where $r^2 = 2|\alpha_0|^2$. The Wigner function of an ideal single-photon Fock state on a two-dimensional phase space is shown in Figure 6.3. In Section 7.9.3, we return to the heralded single-photon Fock state for the homodyne measurement of its Wigner function.

7.8.3 Photon-subtracted states

One way to produce a Wigner functional with a negative region is to subtract a photon from a squeezed vacuum state, whose Wigner functional is positive everywhere. Formally, one may think of a photon subtraction process as an annihilation operator applied to the state. In a practical experiment, photon subtraction is performed with the aid of heralding. A small portion of the state is separated off by an *unbalanced homogeneous beamsplitter* having a small reflectivity and sent to a photon-number-resolving detector. The detection of n photons then heralds the formation of an n-photon subtracted state.

For a known input state, one can follow the obvious process of calculating a generating function for the heralded subtracted state with the aid of the photon statistics generating function in (7.87) and the homogeneous beamsplitter transformation in (6.232). However, we want to derive a procedure to compute the single-photon subtracted state for an arbitrary unknown input state.

The procedure that we follow here is similar to the one that was followed when we developed an evolution equation for continuous homogeneous loss in Section 6.7.3. We implement the beamsplitter by the transformations of the field variables given in (6.218). For the unbalanced beamsplitters with a small reflectivity, we represent $S = \zeta$ and $C = \sqrt{1-\zeta^2}$ where ζ is the *amplitude reflectivity* (the intensity reflectivity is ζ^2). Then the beamsplitter transformations become

$$\begin{aligned} \alpha &\to \sqrt{1-\zeta^2}\alpha + i\zeta\beta, \quad \alpha^* \to \sqrt{1-\zeta^2}\alpha^* - i\zeta\beta^*, \\ \beta &\to \sqrt{1-\zeta^2}\beta + i\zeta\alpha, \quad \beta^* \to \sqrt{1-\zeta^2}\beta^* - i\zeta\alpha^*. \end{aligned} \tag{7.251}$$

These transformations are applied to the product of the Wigner functionals of the unknown input state and a vacuum state. The transformed state reads

$$\begin{aligned} W_{\text{tr}}[\alpha,\beta] = {}& \mathcal{N}_0 \exp\left[-2\left(\sqrt{1+\zeta^2}\beta^* - i\zeta\alpha^*\right)\diamond\left(\sqrt{1+\zeta^2}\beta + i\zeta\alpha\right)\right] \\ & \times W_{\text{in}}\left[\sqrt{1+\zeta^2}\alpha^* - i\zeta\beta^*, \sqrt{1+\zeta^2}\alpha + i\zeta\beta\right]. \end{aligned} \tag{7.252}$$

Under the assumption that the reflectivity is very small, we can expand $W_{\text{tr}}[\alpha,\beta]$ as a power series in ζ. Then we can multiply the terms by (7.87) as a function of β and integrate over β. It produces a generating function for the detection of a specific number of photons. The number of photons that we can detect depends on the order to which we make the expansion. Here, we only consider single-photon subtraction for which we need to make the expansion up to ζ^2. For a two-photon subtracted state, we would need an expansion up to ζ^4. All the uneven terms become zero when we trace over β and the zeroth-order term does not produce any result for the measurement because it represents the vacuum. So, we focus on the second-order term.

To compute the second-order term in the expansion, we take two derivatives with respect to ζ and set $\zeta = 0$. The result is then multiplied by $\frac{1}{2}\zeta^2$. The differentiation process produces *functional derivatives* of the Wigner functional W_{in}. The result for the second-order term is

$$\begin{aligned}
&\frac{1}{2}\zeta^2\,\partial_\zeta^2 W_{\text{tr}}[\alpha,\beta]\Big|_{\zeta=0} \\
&= \mathcal{N}_0\left[2\beta^*\diamond\left(\frac{\delta^2 W_{\text{in}}}{\delta\alpha^*\delta\alpha}+2\alpha\frac{\delta W_{\text{in}}}{\delta\alpha}+2\frac{\delta W_{\text{in}}}{\delta\alpha^*}\alpha^*+2\mathbf{1}W_{\text{in}}+4\alpha W_{\text{in}}\alpha^*\right)\diamond\beta\right. \\
&\quad -\beta^*\diamond\left(\frac{\delta^2 W_{\text{in}}}{\delta\alpha^*\delta\alpha^*}+2\frac{\delta W_{\text{in}}}{\delta\alpha^*}\alpha+2\alpha\frac{\delta W_{\text{in}}}{\delta\alpha^*}+4\alpha W_{\text{in}}\alpha\right)\diamond\beta^* \\
&\quad -\beta\diamond\left(\frac{\delta^2 W_{\text{in}}}{\delta\alpha\delta\alpha}+2\frac{\delta W_{\text{in}}}{\delta\alpha}\alpha^*+2\alpha^*\frac{\delta W_{\text{in}}}{\delta\alpha}+4\alpha^* W_{\text{in}}\alpha^*\right)\diamond\beta \\
&\quad \left.-\frac{\delta W_{\text{in}}}{\delta\alpha}\diamond\alpha-\alpha^*\diamond\frac{\delta W_{\text{in}}}{\delta\alpha^*}-4\alpha^*\diamond\alpha W_{\text{in}}\right]\frac{1}{2}\zeta^2\exp\left(-2\beta^*\diamond\beta\right).
\end{aligned} \tag{7.253}$$

Now we perform photon measurements on the part of the state that depends on β. For this purpose, we multiply the second-order term by (7.87), expressed as a functional of β and evaluate a functional integration over β. The β-dependent part of the integrand consists of an exponential factor times factors of β^* and β representing moments of the functional of β. To compute these moments, we can use the *characteristic functional* as a *generating functional for the moments*, similar to what we did in Section 6.7.3. Here, we incorporate the vacuum state into the generating function

$$\mathcal{P}_{\text{vac}}[\beta](J) \triangleq W_{\text{vac}}[\beta]\mathcal{P}[\beta](J) = \mathcal{N}_J^{\text{tr}\{D\}}\exp\left(-2\beta^*\diamond\beta-2\mathcal{J}\beta^*\diamond D\diamond\beta\right). \tag{7.254}$$

Then we compute the characteristic functional of $\mathcal{P}_{\text{vac}}[\beta](J)$:

$$\begin{aligned}
\chi_{\mathcal{P}}[\mu] &= \mathcal{N}_J^{\text{tr}\{D\}}\mathcal{N}_0\int\exp\left(-2\beta^*\diamond\beta-2\mathcal{J}\beta^*\diamond D\diamond\beta+\mu^*\diamond\beta-\beta^*\diamond\mu\right)\,\mathcal{D}^\circ[\beta] \\
&= \exp\left[-\frac{1}{2}\mu^*\diamond\mu+\frac{1}{4}(1-J)\mu^*\diamond D\diamond\mu\right],
\end{aligned} \tag{7.255}$$

where we use the idempotency of D, together with expressions from Appendix D. As shown in (5.32), the different moments are produced by functional derivatives with respect to the auxiliary field variables μ^* and μ for factors of β and β^*, respectively (with the appropriate sign change for β^*), after which the auxiliary field variables are set equal to zero. The relevant moments in this case are

$$\begin{aligned}\int \beta^*(\mathbf{k})\beta(\mathbf{k}')\mathcal{P}_{\text{vac}}[\beta]\,\mathcal{D}^\circ[\beta] &= -\left.\frac{\delta^2\chi_{\mathcal{P}}[\mu]}{\delta\mu(\mathbf{k})\delta\mu^*(\mathbf{k}')}\right|_{\mu^*=\mu=0}\\ &= \frac{1}{2}\mathbf{1}(\mathbf{k}',\mathbf{k}) - \frac{1}{4}(1-J)D(\mathbf{k}',\mathbf{k}),\\ \int \mathcal{P}_{\text{vac}}[\beta]\,\mathcal{D}^\circ[\beta] &= \chi_{\mathcal{P}}[\mu]|_{\mu^*=\mu=0} = 1.\end{aligned} \tag{7.256}$$

The other moments are zero, as in (6.263). With the aid of these moments, we obtain an expression for the generating function of the heralded single-photon subtracted state that reads

$$\begin{aligned}\mathcal{W}_{1\text{ps}}[\alpha] &= \int W_{\text{tr}}[\alpha,\beta]\mathcal{P}[\beta](J)\,\mathcal{D}^\circ[\beta]\\ &= \frac{1}{2}\zeta^2\left[\frac{\delta}{\delta\alpha}\diamond\frac{\delta}{\delta\alpha^*}W_{\text{in}} + \frac{\delta W_{\text{in}}}{\delta\alpha}\diamond\alpha + \alpha^*\diamond\frac{\delta W_{\text{in}}}{\delta\alpha^*} + 2\Omega W_{\text{in}}\right.\\ &\quad - \frac{1}{2}(1-J)\left(\frac{\delta}{\delta\alpha}\diamond D\diamond\frac{\delta}{\delta\alpha^*}W_{\text{in}} + 2\,\text{tr}\{D\}W_{\text{in}} + 2\frac{\delta W_{\text{in}}}{\delta\alpha}\diamond D\diamond\alpha\right.\\ &\quad\left.\left. + 2\alpha^*\diamond D\diamond\frac{\delta W_{\text{in}}}{\delta\alpha^*} + 4\alpha^*\diamond D\diamond\alpha W_{\text{in}}\right)\right],\end{aligned} \tag{7.257}$$

where $\text{tr}\{\mathbf{1}\} = \Omega$. The resulting generating function is linear in the generating parameter J. It can therefore only produce the single-photon subtracted state.

While the initial Wigner functional is normalized, the one obtained from (7.257) is not. The required normalization constant is computed with the aid of a generating function that is obtained by integrating (7.257) over α. An integrand that is the total derivative of a normalized Wigner functional produces 0 because such a Wigner functional tends to 0 at infinity. Therefore,

$$\int \frac{\delta^2 W_{\text{in}}}{\delta\alpha^*\delta\alpha}\,\mathcal{D}^\circ[\alpha] = 0. \tag{7.258}$$

Using *partial functional integration*, as given in (6.271), we get

$$\int \alpha(\mathbf{k})\frac{\delta W_{\text{in}}}{\delta\alpha(\mathbf{k}')}\,\mathcal{D}^\circ[\alpha] = \int \alpha^*(\mathbf{k})\frac{\delta W_{\text{in}}}{\delta\alpha^*(\mathbf{k}')}\,\mathcal{D}^\circ[\alpha] = -\delta(\mathbf{k}-\mathbf{k}'). \tag{7.259}$$

With the aid of these identities, we obtain the inverse normalization constant

$$\begin{aligned}\mathcal{N}^{-1} &= \partial_J \int \mathcal{W}_{1\text{ps}}\,\mathcal{D}^\circ[\alpha]\Big|_{J=0}\\ &= \frac{1}{4}\zeta^2\int\left(\frac{\delta}{\delta\alpha}\diamond D\diamond\frac{\delta}{\delta\alpha^*}W_{\text{in}} + 2\,\text{tr}\{D\}W_{\text{in}} + 2\frac{\delta W_{\text{in}}}{\delta\alpha}\diamond D\diamond\alpha\right.\\ &\quad\left. + 2\alpha^*\diamond D\diamond\frac{\delta W_{\text{in}}}{\delta\alpha^*} + 4\alpha^*\diamond D\diamond\alpha W_{\text{in}}\right)\mathcal{D}^\circ[\alpha]\\ &= \zeta^2\left[\int(\alpha^*\diamond D\diamond\alpha)W_{\text{in}}[\alpha]\,\mathcal{D}^\circ[\alpha] - \frac{1}{2}\,\text{tr}\{D\}\right].\end{aligned} \tag{7.260}$$

One integral remains to be evaluated, for which the initial Wigner functional needs to be known. An example is considered below.

The single-photon subtracted state for an arbitrary initial state is

$$\begin{aligned} W_{1\text{ps}}[\alpha] &= N\left\{\partial_J \mathcal{W}_{1\text{ps}}\big|_{J=0}\right\} \\ &= \mathcal{N}\left(\frac{\delta}{\delta\alpha}\diamond D\diamond\frac{\delta}{\delta\alpha^*}W_{\text{in}} + 2\,\text{tr}\{D\}W_{\text{in}} + 2\frac{\delta W_{\text{in}}}{\delta\alpha}\diamond D\diamond\alpha \right. \\ &\quad \left. + 2\alpha^*\diamond D\diamond\frac{\delta W_{\text{in}}}{\delta\alpha^*} + 4\alpha^*\diamond D\diamond\alpha W_{\text{in}}\right). \end{aligned} \tag{7.261}$$

Here, $N\{\cdot\}$ is a normalization process and $\mathcal{N}$ is the normalization constant.

We can now consider different input states. A coherent state remains unaffected by a photon subtraction process. A thermal state is affected, but its Wigner functional remains positive everywhere.

Exercise 7.13. Show that photon subtraction does not affect a coherent state.
Hint: Remember to normalize the result.

Exercise 7.14. Show that photon subtraction does not produce negative regions in the resulting Wigner functional when the initial state is a thermal state.
Hint: Remember that the kernel of a thermal state is positive definite.

7.8.3.1 Photon-subtracted squeezed vacuum

When we substitute the Wigner functional of a squeezed vacuum state

$$W_{\text{in}} \to W_{\text{sq}} = \mathcal{N}_0 \exp\left(-2\alpha^*\diamond A\diamond\alpha - \alpha^*\diamond B\diamond\alpha^* - \alpha\diamond B^*\diamond\alpha\right), \tag{7.262}$$

into (7.261), we obtain a Wigner functional in the form of a polynomial Gaussian state that becomes negative at the origin. It reads

$$W_{1\text{ps}}^{(\text{sq})} = 4\mathcal{N}\left[(\alpha^*\diamond E + \alpha\diamond B^*)\diamond D\diamond(E\diamond\alpha + B\diamond\alpha^*) - \frac{1}{2}\,\text{tr}\{E\diamond D\}\right]W_{\text{sq}}, \tag{7.263}$$

where $E = A - \mathbf{1}$. At the origin, where $\alpha = 0$, we get

$$W_{1\text{ps}}^{(\text{sq})}[\alpha = 0] = -2\mathcal{N}\,\text{tr}\{E\diamond D\} < 0. \tag{7.264}$$

For larger numbers of subtracted photons, the number of regions where the Wigner functional is negative increases. For even numbers of subtracted photons none of the negative regions is located at the origin.

When we consider the single-mode detector kernel $D = MM^*$, the squeezed vacuum state kernels perform transformations on the detector mode M. The resulting expression for the normalized heralded state becomes

$$W_{1ps}^{(sq)} = 4\mathcal{N}\left(\left|\alpha^* \diamond E \diamond M + \alpha \diamond B^* \diamond M\right|^2 - \frac{1}{2}\eta\right) W_{sq}, \tag{7.265}$$

where η is given in (7.246). Since photon subtraction does not in general assume a weak squeezed vacuum state as input (as assumed in the case of single-photon state preparation, discussed in Section 7.8.2), we do not use the notation for the normalized modes defined in (7.248) here. Therefore, $E \diamond M$ and $B^* \diamond M$ represent *unnormalized* transformed detector modes, restricting the Wigner function to a four-dimensional subspace of the infinite-dimensional functional phase space. Detailed calculations of the transformation of the detector mode by these kernels are provided in Section 9.4.

7.8.4 Photon-added states

Another way to produce a Wigner functional with negative regions is to add photons to the state. It is a more powerful way to obtain states with negative regions than photon subtraction (considered in Section 7.8.3), because photon addition can produce negative regions with input states for which photon subtraction does not produce negative regions. Formally, photon addition is achieved by applying a creation operator to the state. In a practical experiment, photon addition is performed by using the state as seed to stimulate a parametric down-conversion process and then herald the photon addition by detecting a single photon in the *difference frequency beam*.

The process of heralded photon addition is expressed by

$$\hat{\rho}_{hpa} = \text{tr}_B\left\{\hat{P}_B\hat{U}_{ibs}\left[\left(\hat{S}\hat{\rho}_{in}\hat{S}^\dagger\right) \otimes \hat{\rho}_{vac}\right]\hat{U}_{ibs}^\dagger\right\}, \tag{7.266}$$

in terms of operators, where the input state is squeezed, and then together with a vacuum state, sent through an inhomogeneous beamsplitter to separate the squeezed vacuum state into two beams, after which one of the beams is measured. The measurement is done with a *photon-number-resolving detector* $\hat{P}_B$. Here, we again consider the process for an arbitrary unknown state instead of performing the calculations in terms of Wigner functionals for a given input state. Therefore, we focus on single-photon addition. When we convert (7.266) into a Wigner functional expression, in which we apply the necessary transformations of the arguments of the unknown state and the vacuum state to implement the Bogoliubov transformation and the inhomogeneous beamsplitter, we end up with an expression that we cannot evaluate further unless we know the unknown state. To perform this analysis while keeping the unknown state arbitrary, we introduce an approximation.

To proceed, we use our knowledge of the *squeezing* process gained in Section 6.8.2 to consider the practical implementation of the process of photon addition. The latter is a heralded process that requires the separation of the state into two phase spaces. It is assumed that the squeezing parameter for the stimulated process is weak to ensure that the contribution of more photons is suppressed. As a result, we can treat the squeezing parameter as an expansion parameter. In terms of Wigner functionals, the state after the inhomogeneous beamsplitter is produced directly by the application of the *twin-beam Bogoliubov transformations* provided in (6.381) to the input state. It combines the nonlinear process with the beamsplitter into a nonlinear beamsplitter. The vacuum state is combined with the unknown state to represent the input state. (The vacuum state is now also affected by the nonlinear process, which is expected in view of spontaneous parametric down-conversion.) For weak squeezing, we display the squeezing parameter ξ explicitly and replace the Bogoliubov kernels by

$$U \to \mathbf{1} + \xi^2 Y \quad \text{and} \quad V \to \xi V, \tag{7.267}$$

where Y represents the sub-leading part of U. Therefore, the twin-beam Bogoliubov transformations that we use here is given by

$$\begin{aligned} \alpha &\to (\mathbf{1} + \xi^2 Y) \diamond \alpha + \xi V \diamond \beta^*, \\ \beta &\to (\mathbf{1} + \xi^2 Y) \diamond \beta + \xi V \diamond \alpha^*. \end{aligned} \tag{7.268}$$

For the current analysis, we'll also assume that U is Hermitian and V is symmetric as imposed by the squeezing operation discussed in Section 6.8.2. It implies that $Y^\dagger = Y$ (so that $Y^T = Y^*$) and $V^T = V$ (which also means that $V^\dagger = V^*$). With these transformations, the Wigner functional of the input state becomes the Bogoliubov transformed state

$$\begin{aligned} W_{\text{bt}}[\alpha, \beta] = {} & W_{\text{in}}\left[\alpha^* + \xi^2 \alpha^* \diamond Y + \xi \beta \diamond V^*, \alpha + \xi^2 Y \diamond \alpha + \xi V \diamond \beta^*\right] \\ & \times \mathcal{N}_0 \exp\left[-2\left(\beta^* + \xi^2 \beta^* \diamond Y + \xi \alpha \diamond V^*\right)\right. \\ & \left. \diamond \left(\beta + \xi^2 Y \diamond \beta + \xi V \diamond \alpha^*\right)\right]. \end{aligned} \tag{7.269}$$

The calculation now proceeds along the same steps that we used for photon subtraction in Section 7.8.3. We expand the Bogoliubov transformed input Wigner functional to second order in the squeezing parameter. It leads to

$$\begin{aligned} & \frac{1}{2}\xi^2\, \partial_\xi^2 W_{\text{bt}}[\alpha, \beta]\Big|_{\zeta=0} \\ & = \xi^2 \left[\beta^* \diamond V \diamond \frac{\delta^2 W_{\text{in}}}{\delta\alpha\delta\alpha^*} \diamond V^* \diamond \beta + \frac{1}{2}\beta^* \diamond V \diamond \frac{\delta^2 W_{\text{in}}}{\delta\alpha\delta\alpha} \diamond V \diamond \beta^*\right. \\ & + \frac{1}{2}\beta \diamond V^* \diamond \frac{\delta^2 W_{\text{in}}}{\delta\alpha^*\delta\alpha^*} \diamond V^* \diamond \beta + \frac{\delta W_{\text{in}}}{\delta\alpha} \diamond Y \diamond \alpha + \alpha^* \diamond Y \diamond \frac{\delta W_{\text{in}}}{\delta\alpha^*} \\ & - 2\left(\alpha \diamond V^* \diamond \beta + \beta^* \diamond V \diamond \alpha^*\right)\left(\frac{\delta W_{\text{in}}}{\delta\alpha^*} \diamond V^* \diamond \beta + \beta^* \diamond V \diamond \frac{\delta W_{\text{in}}}{\delta\alpha}\right) \\ & - 4\left(\beta^* \diamond Y \diamond \beta + \alpha \diamond Y \diamond \alpha^*\right) W_{\text{in}} \\ & \left. + 2\left(\alpha \diamond V^* \diamond \beta + \beta^* \diamond V \diamond \alpha^*\right)^2 W_{\text{in}}\right] \mathcal{N}_0 \exp\left(-2\beta^* \diamond \beta\right). \end{aligned} \tag{7.270}$$

With the aid of the generating function of the photon statistics given in (7.87), we calculate the state produced by a single-photon detection. The functional integral over β is evaluated with the aid of the characteristic functional in (7.255). Here, the same expressions are obtained for the moments as provided in (7.256). In analogy to (7.206), the first Bogoliubov identity in (6.331) leads to $V \diamond V^* \approx 2Y$ when we apply (7.267). This simplification and the moments then lead to

$$\begin{aligned}
\mathcal{W}_{1\text{pa}} &= \int W_{\text{bt}}[\alpha,\beta]\mathcal{P}[\beta](J)\,\mathcal{D}^{\circ}[\beta] \\
&= \frac{\delta}{\delta\alpha} \diamond Y \diamond \frac{\delta}{\delta\alpha^*} W_{\text{in}} - \frac{\delta W_{\text{in}}}{\delta\alpha} \diamond Y \diamond \alpha - \alpha^* \diamond Y \diamond \frac{\delta W_{\text{in}}}{\delta\alpha^*} - 2\,\text{tr}\{Y\}\, W_{\text{in}} \\
&\quad - \frac{1}{4}(1-J)\bigg(\frac{\delta}{\delta\alpha^*} \diamond D_V \diamond \frac{\delta}{\delta\alpha} W_{\text{in}} + 4\alpha \diamond D_V \diamond \alpha^* W_{\text{in}} \\
&\quad - 2\,\text{tr}\{D_V\} W_{\text{in}} - 2\alpha \diamond D_V \diamond \frac{\delta W_{\text{in}}}{\delta\alpha} - 2\frac{\delta W_{\text{in}}}{\delta\alpha^*} \diamond D_V \diamond \alpha^* \bigg),
\end{aligned} \tag{7.271}$$

where $D_V \triangleq V^* \diamond D \diamond V$, so that $\text{tr}\{D \diamond Y\} \approx \frac{1}{2}\text{tr}\{D_V\}$. We discard the factor of ξ^2 since it is removed by the normalization anyway. The resulting generating function is linear in the generating parameter J, allowing only a single-photon addition. A higher-order expansion is required to consider the addition of more photons. The single-photon added state for an arbitrary initial state is given by

$$\begin{aligned}
W_{1\text{pa}} &= N\left\{\partial_J \mathcal{W}_{1\text{pa}}\big|_{J=0}\right\} \\
&= \mathcal{N}\bigg[\frac{1}{4}\frac{\delta}{\delta\alpha^*} \diamond D_V \diamond \frac{\delta}{\delta\alpha} W_{\text{in}} + \left(\alpha \diamond D_V \diamond \alpha^* - \frac{1}{2}\text{tr}\{D_V\}\right) W_{\text{in}} \\
&\quad - \frac{1}{2}\alpha \diamond D_V \diamond \frac{\delta W_{\text{in}}}{\delta\alpha} - \frac{1}{2}\frac{\delta W_{\text{in}}}{\delta\alpha^*} \diamond D_V \diamond \alpha^*\bigg],
\end{aligned} \tag{7.272}$$

where $N\{\cdot\}$ is the normalization process, and $\mathcal{N}$ is the normalization constant.

Exercise 7.15. Compute the normalization constant for (7.272) to show that it is given by

$$\mathcal{N}^{-1} = \int (\alpha \diamond D_V \diamond \alpha^*) W_{\text{in}}[\alpha]\,\mathcal{D}^{\circ}[\alpha] + \frac{1}{2}\text{tr}\{D_V\}. \tag{7.273}$$

7.8.4.1 Photon-added coherent states

As an example, we consider the preparation of a photon-added coherent state. In contrast to the case with photon subtraction, the Wigner functional of a photon-added coherent state has negative regions. Substituting the coherent state Wigner functional with parameter function ζ into (7.272), and using a single-mode detector kernel, we obtain

$$\begin{aligned}
W_{1\text{pa}}^{(\text{coh})} &= \frac{1}{2}\mathcal{N}\,[2M^* \diamond V \diamond (2\alpha^* - \zeta^*)(2\alpha - \zeta) \diamond V^* \diamond M \\
&\quad - M^* \diamond V \diamond V^* \diamond M]\, W_{\text{coh}},
\end{aligned} \tag{7.274}$$

which is negative at $\alpha = \frac{1}{2}\zeta$ for $M^* \diamond V \diamond V^* \diamond M > 0$. However, for $\|\zeta\| \gg 1$, the negative amplitude of the Wigner functional at $\alpha = \frac{1}{2}\zeta$ is severely suppressed.

To compute the normalization factor, we use the *characteristic functional* as a *generating functional for the moments*, in the same way we did in (7.255). For the coherent state, this characteristic functional is given by

$$\begin{aligned}\chi_{\mathcal{N}}[\mu] &= \int \mathcal{N}_0 \exp\left[-2(\alpha^* - \zeta^*) \diamond (\alpha - \zeta) + \mu^* \diamond \alpha - \alpha^* \diamond \mu\right] \mathcal{D}^\circ[\alpha] \\ &= \exp\left(\mu^* \diamond \zeta - \zeta^* \diamond \mu - \frac{1}{2}\mu^* \diamond \mu\right). \end{aligned} \tag{7.275}$$

The moment integral thus produces

$$\begin{aligned}\int \alpha^*(\mathbf{k})\alpha(\mathbf{k}')W_{\text{in}}[\alpha]\,\mathcal{D}^\circ[\alpha] &= -\left.\frac{\delta^2\chi_{\mathcal{N}}[\mu]}{\delta\mu(\mathbf{k})\delta\mu^*(\mathbf{k}')}\right|_{\mu^*=\mu=0} \\ &= \zeta^*(\mathbf{k})\zeta(\mathbf{k}') + \frac{1}{2}\mathbf{1}(\mathbf{k}',\mathbf{k}). \end{aligned} \tag{7.276}$$

Therefore, we get the normalization constant

$$\begin{aligned}\mathcal{N}^{-1} &= \partial_J \int \mathcal{W}_{1\text{pa}}\,\mathcal{D}^\circ[\alpha]\Big|_{J=0} \\ &= M^* \diamond V \diamond \zeta^*\zeta \diamond V^* \diamond M + M^* \diamond V \diamond V^* \diamond M. \end{aligned} \tag{7.277}$$

The normalized expression for the Wigner functional of the photon-added coherent state is

$$W_{1\text{pa}}^{(\text{coh})} = \frac{1}{2}\frac{2|M_V^* \diamond (2\alpha - \zeta)|^2 - 1}{|M_V^* \diamond \zeta|^2 + 1} W_{\text{coh}}, \tag{7.278}$$

where we represent the transformed detector mode as a normalized mode

$$M_V \triangleq \frac{V \diamond M^*}{\sqrt{M^* \diamond V \diamond V^* \diamond M}}. \tag{7.279}$$

The normalization cancels out in the polynomial.

In a similar way to what we found for the heralded single-photon Fock state, the polynomial part of the Wigner functional is effectively restricted to a two-dimensional subspace associated with the transformed mode M_V. The displacement of the coherent state on this plane is determined by the overlap between the transformed mode M_V and the parameter function ζ. We perform the replacement $\alpha \to \alpha_0 M_V$, define $\zeta_0 = M_V^* \diamond \zeta$, and trace over the part of the functional phase space that is orthogonal to M_V to obtain the two-dimensional Wigner function on this reduced phase space. The result reads

$$W_{1\text{pa}}^{(\text{coh})}(\alpha_0) = \frac{2|2\alpha_0 - \zeta_0|^2 - 1}{|\zeta_0|^2 + 1} \exp\left(-2|\alpha_0 - \zeta_0|^2\right). \tag{7.280}$$

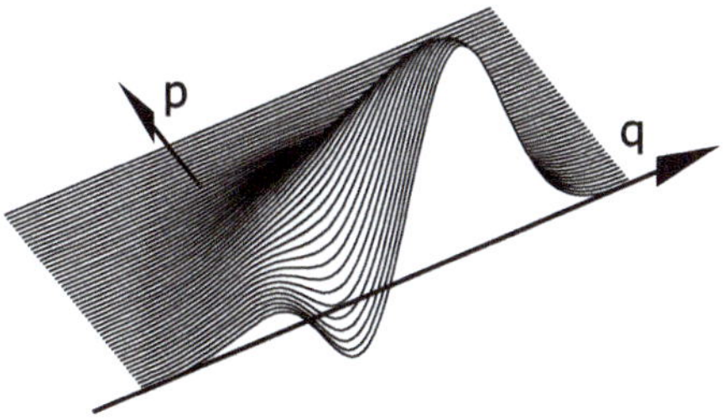

Figure 7.6: Wigner function of a photon-added coherent state.

In Figure 7.6, the two-dimensional Wigner function of the photon-added coherent state is shown for the case where $|\zeta_0| \approx 1$. It is displayed for positive p only to reveal the negative region in the shape of the Wigner function.

7.9 Homodyne measurement

Being Hermitian, quadrature operators can be regarded as *observables*. It is indeed possible to measure the quadratures of a state. The practical way to implement such a measurement is to use *homodyning* [17].

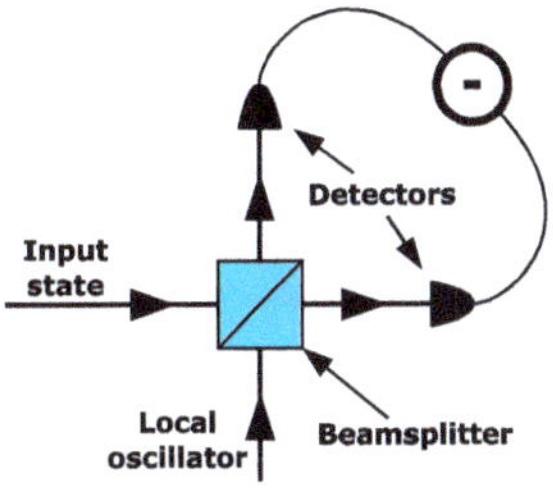

Figure 7.7: Homodyning system.

The generic measurement setup for optical homodyning is shown in Figure 7.7. It consists of a 50:50 beamsplitter and two detectors. One input port of the beamsplitter receives the state to be measured and the other receives a bright laser source, referred to as the *local oscillator*. After the intensities are measured at the two output ports, they are subtracted from each other. The result then represents the quadrature measurement.

Homodyne measurements can be used to perform *quantum state tomography* in terms of the particle-number degree of freedom. It requires the measurement of the photon number as a statistical distribution. When the spatiotemporal degrees of freedom are included, the situation becomes more complicated.

7.9.1 Generic homodyne process

To understand how the homodyne process works and why it can measure the quadratures, we start with the basic operator expression for the system in Figure 7.7:

$$\langle h\rangle = \mathrm{tr}\left\{\hat{U}_{\mathrm{bs}}(\hat{\rho}_{\mathrm{in}} \otimes \hat{\rho}_{\mathrm{lo}})\hat{U}_{\mathrm{bs}}^{\dagger}(\hat{n}_A - \hat{n}_B)\right\}. \tag{7.281}$$

Here, $\hat{U}_{\mathrm{bs}}$ is the unitary operator for the beamsplitter, $\hat{\rho}_{\mathrm{in}}$ is the density operator for the input state, $\hat{\rho}_{\mathrm{lo}}$ is the density operator for the local oscillator, and $\hat{n}_A$ and $\hat{n}_B$ are the two number operators associated with the measurements at the two detectors. Initially, we assume ideal detectors, so that

$$\hat{n}_A - \hat{n}_B = \hat{a}_A^{\dagger} \diamond \hat{a}_A - \hat{a}_B^{\dagger} \diamond \hat{a}_B, \tag{7.282}$$

We can apply the unitary operators for the beamsplitter on these number operators, by using the inverse (with the opposite sign) of the transformations in (6.210). For the 50:50 beamsplitter, we set $C = S = \frac{1}{\sqrt{2}}$ The result is

$$\hat{U}_{\mathrm{bs}}^{\dagger}(\hat{n}_A - \hat{n}_B)\hat{U}_{\mathrm{bs}} = \mathrm{i}\hat{a}_B^{\dagger} \diamond \hat{a}_A - \mathrm{i}\hat{a}_A^{\dagger} \diamond \hat{a}_B \triangleq \hat{h}'. \tag{7.283}$$

The measurement is then given by

$$\langle h'\rangle = \mathrm{tr}\left\{(\hat{\rho}_{\mathrm{in}} \otimes \hat{\rho}_{\mathrm{lo}})\hat{h}'\right\}, \tag{7.284}$$

The local oscillator is assumed to be a coherent state given by $|\gamma\rangle$, with parameter function denoted by γ. After performing a partial trace over the local oscillator degrees of freedom, denoted by the label B, we get

$$\mathrm{tr}_B\left\{\hat{\rho}_{\mathrm{lo}}\hat{h}'\right\} = \langle\gamma|\,\hat{h}'\,|\gamma\rangle = \mathrm{i}\gamma^* \diamond \hat{a}_A - \mathrm{i}\hat{a}_A^{\dagger} \diamond \gamma \triangleq \hat{h}[\gamma]. \tag{7.285}$$

The final expression is then given by

$$\langle h\rangle = \mathrm{tr}_A\left\{\hat{\rho}_{\mathrm{in}}\hat{h}[\gamma]\right\}. \tag{7.286}$$

We represent the parameter function of the local oscillator as

$$\gamma(\mathbf{k}) = \exp(\mathrm{i}\theta)\gamma_0\Gamma(\mathbf{k}), \tag{7.287}$$

where θ is a constant phase, $\gamma_0 = \|\gamma\|$ is the magnitude of the parameter function, and $\Gamma(\mathbf{k})$ is a normalized spectral function, representing the shape of the local oscillator parameter function (mode). For simplicity, we assume that $\Gamma(\mathbf{k})$ is a real-valued function. The result of the partial trace thus produces

$$\hat{h}[\gamma] = \sqrt{2}\gamma_0[\sin(\theta)\hat{q} - \cos(\theta)\hat{p}] \diamond \Gamma = -\sqrt{2}\gamma_0\hat{p}_{\theta} \diamond \Gamma, \tag{7.288}$$

where $\hat{p}_\theta$ is a rotated version of the quadrature operator, equivalent to the definitions of the rotated quadrature variables in (6.344). The result in (7.288) implies that the spatiotemporal properties of the local oscillator impose restrictions on the observed photons. The overlap between the shape function of the local oscillator and the rotated quadrature operator represents a spectral filtering process in terms of the quadrature degree of freedom. Moreover, the orientation of the rotated quadrature operator is determined by the phase of the spectral function of the local oscillator. As a result, the spectral function of the local oscillator can be used in the ideal case to control the spatiotemporal properties of the homodyne detection process.

7.9.1.1 Detector

The idealized number operator that we used in (7.282) does not represent the physical scenario very well. For a more realistic scenario, we insert a detector kernel D modelling the physical measurement system $\hat{a}^\dagger \diamond \hat{a} \to \hat{a}^\dagger \diamond D \diamond \hat{a}$. As a result, the homodyne operator in (7.285) becomes

$$\hat{h}[\gamma] = \mathrm{i}\gamma^* \diamond D \diamond \hat{a}_A - \mathrm{i}\hat{a}_A^\dagger \diamond D \diamond \gamma. \tag{7.289}$$

It is necessary that the two detectors are identical, otherwise an additional unwanted noise term is produced. The mode onto which the homodyne process performs the projection is a transformed version of the local oscillator mode, given by $D \diamond \gamma$.

If we represent the local oscillator in terms of (7.287) and model the detector kernel by a single-mode detector kernel, it leads to

$$\hat{h} = \mathrm{i}\gamma_0 \exp(-\mathrm{i}\theta)\Gamma^* \diamond MM^* \diamond \hat{a}_A - \mathrm{i}\gamma_0 \exp(\mathrm{i}\theta)\hat{a}_A^\dagger \diamond MM^* \diamond \Gamma. \tag{7.290}$$

For $\Gamma \neq M$, their overlap produces a reduced detector efficiency, which we can absorb into the magnitude and phase represented by $\gamma_0 \exp(-\mathrm{i}\theta)$. As a result, we have

$$\hat{h} = \mathrm{i}\gamma_0 \exp(-\mathrm{i}\theta)M^* \diamond \hat{a}_A - \mathrm{i}\gamma_0 \exp(\mathrm{i}\theta)\hat{a}_A^\dagger \diamond M. \tag{7.291}$$

It thus follows that the mode that is selected by the homodyning process is not always determined by the local oscillator when the physical detector kernel is taken into account. Instead, the projection is done by the mode of the detector for the case of a single-mode detector. The only requirement for the local oscillator is that its mode must have a nonzero overlap with that of the detector system, preferably matching it exactly. The role of the local oscillator parameter function is to determine the phase and the magnitude, which is reduced by any modal mismatch compared to the detector mode.

For a more general detector kernel, the detector system may allow multiple modes to be detected. In such a case, the mode of the local oscillator must be able to *pass through* the detector kernel. (Formally, “pass through” means that $\Gamma \diamond D = \Gamma$.) In that case, the mode that is selected by the homodyning process is determined by the local oscillator (or by the part of the local oscillator that can pass through the detector).

7.9.1.2 Wigner functional

For a Wigner functional calculation, we compute the functional integral of the product of the Wigner functionals of the input state and the homodyne operator. Since the latter is a polynomial of ladder operators, we can use the procedure in Section 6.1.3 to compute its Wigner functionals. For the ideal case of (7.285) it reads

$$W_{\hat{h}}[\alpha] = \mathrm{i}\gamma^* \diamond \alpha - \mathrm{i}\alpha^* \diamond \gamma. \tag{7.292}$$

The ideal detectors that we assumed at first represent the homodyne process as a projection onto the mode of the local oscillator. However, when we consider arbitrary detector kernels the Wigner functional of the homodyne operator becomes

$$W_{\hat{h}}[\alpha] = \mathrm{i}\gamma^* \diamond D \diamond \alpha - \mathrm{i}\alpha^* \diamond D \diamond \gamma. \tag{7.293}$$

For a single-mode detector kernel, it reads

$$W_{\hat{h}}[\alpha] = \mathrm{i}\gamma^* \diamond MM^* \diamond \alpha - \mathrm{i}\alpha^* \diamond MM^* \diamond \gamma. \tag{7.294}$$

7.9.1.3 Squeezed vacuum state

Homodyne detection can be used to measure the amount of *squeezing* in a squeezed vacuum state. Here, we consider such a situation as an example of the calculation. The Wigner functional for the homodyne operator in (7.293) is converted to a generating function and multiplied by the Wigner functional for a squeezed vacuum state given in (6.337). The result is then integrated over the field variable, leading to a generating function for homodyne measurements

$$\begin{aligned}
\mathcal{X}(J) = \mathcal{N}_0 \int & \exp\left[-2\alpha^* \diamond A \diamond \alpha - \alpha^* \diamond B \diamond \alpha^* - \alpha \diamond B^* \diamond \alpha\right.\\
& \left. + \mathrm{i}J\left(\gamma^* \diamond D \diamond \alpha - \alpha^* \diamond D \diamond \gamma\right)\right] \mathcal{D}^\circ[\alpha] \\
= & \exp\left[\frac{1}{4}J^2\gamma^* \diamond D \diamond A^{-1} \diamond D \diamond \gamma + \frac{1}{4}J^2\left(\gamma \diamond D^* + \gamma^* \diamond D \diamond A^{-1} \diamond B\right)\right.\\
& \left. \diamond A^* \diamond \left(D^* \diamond \gamma + B^* \diamond A^{-1} \diamond D \diamond \gamma\right)\right] \\
= & \exp\left[\frac{1}{4}J^2\left(2\gamma^* \diamond D \diamond A \diamond D \diamond \gamma + \gamma \diamond D^* \diamond B^* \diamond D \diamond \gamma\right.\right.\\
& \left.\left. + \gamma^* \diamond D \diamond B \diamond D^* \diamond \gamma^*\right)\right],
\end{aligned} \tag{7.295}$$

where we used (C.30) and the identities for a pure squeezed vacuum state provided in Section 6.8.5 to simplify the result.

Assuming that the local oscillator mode can pass through the detector kernel, we can simplify the result to

$$\mathcal{X}(J) = \exp\left[\frac{1}{4}J^2\left(2\gamma^* \diamond A \diamond \gamma + \gamma \diamond B^* \diamond \gamma + \gamma^* \diamond B \diamond \gamma^*\right)\right]. \tag{7.296}$$

The first derivative with respect to J gives the expectation value of $\langle h\rangle$, which is proportional to the expectation value of the rotated quadrature variable. Since the squeezed vacuum state is centred at the origin, this expectation value is zero. The second derivative gives the second moment of the generalized quadrature variable, corresponding to the width of the state on phase space along the direction of the rotated quadrature variable. In this way, we can measure the squeezing of a state. The second derivative is

$$\left\langle h^2\right\rangle = \partial_J^2 \mathcal{X}(J)\big|_{J=0} = \gamma^* \diamond A \diamond \gamma + \frac{1}{2}\gamma \diamond B^* \diamond \gamma + \frac{1}{2}\gamma^* \diamond B \diamond \gamma^*. \tag{7.297}$$

In terms of (7.287), it becomes

$$\left\langle h^2\right\rangle = \gamma_0^2\Gamma^* \diamond A \diamond \Gamma + \frac{1}{2}\gamma_0^2 \exp(\mathrm{i}2\theta)\Gamma \diamond B^* \diamond \Gamma + \frac{1}{2}\gamma_0^2 \exp(-\mathrm{i}2\theta)\Gamma^* \diamond B \diamond \Gamma^*. \tag{7.298}$$

By observing the value as a function of the phase of the local oscillator θ, we obtain the width of the state as it varies along the different directions. The ratio of the maximum width to the minimum width reveals the amount of squeezing in the state.

In the current example, we consider a state whose Wigner functional is centred at the origin of the phase space. For that reason the observed value of the quadrature is zero. In a more general situation, the observed value is not always zero. Continuous variable teleportation is an application that uses the observation of the quadratures of a state to perform a teleportation of the particle-number degrees of freedom of a state.

7.9.2 Homodyne tomography

Homodyne measurements are often applied to quantum states to perform *quantum state tomography* [18, 19] with respect to their particle-number degrees of freedom. For this purpose, the photon-number statistics of the difference in intensity is measured. Instead of using the number operator, which measures the average intensity, we need to use the projection operators for the different numbers of photons.

7.9.2.1 Cross-correlation operator

The subtraction between the measurements is not done for the same number of photons at both detectors. The probabilities for the detection of a certain number of photons at the respective detectors are statistically independent. So, the difference corresponds to the convolution of the statistical distribution at one detector with the mirror image of the one at the other detector. The convolution between a distribution and the mirror image of another distribution produces the *cross-correlation function* of these distributions.

Consider statistical distributions $P_1(n)$ and $P_2(n)$, representing the probabilities of detecting n photons at the two respective detectors distinguished by their subscripts. The cross-correlation between these distributions is

$$R(r) \triangleq \sum_{n=0}^{\infty} P_1(n+r)P_2(n). \tag{7.299}$$

Here, r is any integer. The respective distributions are zero when their arguments become negative. To aid the subsequent calculation, we define a generating function for the cross-correlated values, given by

$$\mathcal{R}(\nu) = \sum_{r=-\infty}^{\infty} \nu^r R(r) = \sum_{n=0}^{\infty} \sum_{r=-\infty}^{\infty} \nu^r P_1(n+r)P_2(n), \tag{7.300}$$

where ν is the generating parameter. Redefining the index as $r \to m - n$, we then get

$$\mathcal{R}(\nu) = \sum_{n=0}^{\infty} \sum_{m=0}^{\infty} \nu^{m-n} P_1(m)P_2(n) = \mathcal{P}_1(\nu)\mathcal{P}_2\left(\frac{1}{\nu}\right), \tag{7.301}$$

where

$$\mathcal{P}_{1,2}(J) \triangleq \sum_{n=0}^{\infty} J^n P_{1,2}(n). \tag{7.302}$$

So, the generating function for the cross-correlation can be obtained directly from the generating functions of the respective distributions.

In the homodyne experiment, the two distributions $P_1(n)$ and $P_2(n)$ are measured simultaneously on different parts of the state. Here, we use the generating function for the cross-correlation to represent it as a combined operator that is traced with the state to determine the cross-correlation distribution. The generating function for such an operator, based on (7.301), has the form

$$\hat{\mathcal{R}}(\nu) = \hat{\mathcal{P}}_1(\nu)\hat{\mathcal{P}}_2\left(\frac{1}{\nu}\right), \tag{7.303}$$

where $\hat{\mathcal{P}}_1(J)$ and $\hat{\mathcal{P}}_2(J)$ are both the generating function in (7.76) for the n-photon projection operators associated with the respective detectors. In terms of Wigner functionals, this generating function is given in (7.87). Since the two detectors are different their kernels commute. Therefore, we do not need to use the star product to combine them.

When we substitute $J \to 1/\nu$ into the expression in (7.87), we get

$$\frac{1-J}{1+J} \to -\frac{1-\nu}{1+\nu},$$

$$\frac{2^{\text{tr}\{D\}}}{(1+J)^{\text{tr}\{D\}}} \to \frac{2^{\text{tr}\{D\}}\nu^{\text{tr}\{D\}}}{(1+\nu)^{\text{tr}\{D\}}}. \tag{7.304}$$

The combined generating function for the cross-correlation is thus given by

$$\mathcal{W}_{\hat{R}} = \frac{4^{\text{tr}\{D\}}\nu^{\text{tr}\{D\}}}{(1+\nu)^{2\,\text{tr}\{D\}}} \exp\left[-2\frac{1-\nu}{1+\nu}\left(\alpha^* \diamond D \diamond \alpha - \beta^* \diamond D \diamond \beta\right)\right], \tag{7.305}$$

We apply the beamsplitter on this generating function by applying the inverse of the field variable transformation in (6.218) for a 50:50 beamsplitter ($S = C = \frac{1}{\sqrt{2}}$) to (7.305). The result reads

$$\mathcal{W}'_{\hat{R}} = \mathcal{N} \exp\left[-i2\mathcal{V}\left(\beta^* \diamond D \diamond \alpha - \alpha^* \diamond D \diamond \beta\right)\right], \tag{7.306}$$

where we define

$$\mathcal{N}(\nu) = \left[\frac{4\nu}{(1+\nu)^2}\right]^{\text{tr}\{D\}} \quad \text{and} \quad \mathcal{V}(\nu) = \frac{1-\nu}{1+\nu}. \tag{7.307}$$

The generating function in (7.306) is multiplied by the Wigner functional for the local oscillator (a coherent state with a parameter function γ). The functional integration over the local oscillator field variable β produces

$$\begin{aligned} \mathcal{W}_{\hat{h}} &= \int W_{\text{lo}}[\beta]\mathcal{W}'_{\hat{R}}[\alpha,\beta]\,\mathcal{D}^\circ[\beta] \\ &= \int \mathcal{N}_0\mathcal{N} \exp\left[-2\|\beta-\gamma\|^2 - i2\mathcal{V}\left(\beta^* \diamond D \diamond \alpha - \alpha^* \diamond D \diamond \beta\right)\right] \mathcal{D}^\circ[\beta] \\ &= \mathcal{N} \exp\left[-i2\mathcal{V}\left(\gamma^* \diamond D \diamond \alpha - \alpha^* \diamond D \diamond \gamma\right) + 2\mathcal{V}^2\alpha^* \diamond D \diamond \alpha\right], \end{aligned} \tag{7.308}$$

where we used the idempotency of the detector kernel to set $D \diamond D = D$.

7.9.2.2 Cross-correlation distribution

The generating function for the cross-correlation distribution produced by the homodyne detection of a state is now obtained from the trace of the Wigner functional of the state times $\mathcal{W}_{\hat{h}}$. Its functional integral expression reads

$$\begin{aligned} \mathcal{W}_{\text{C}} &= \int W_{\text{in}}[\alpha]\mathcal{W}_{\hat{h}}\,\mathcal{D}^\circ[\alpha] \\ &= \mathcal{N}\int W_{\text{in}}[\alpha] \exp\left[-i2\mathcal{V}\left(\gamma^* \diamond D \diamond \alpha - \alpha^* \diamond D \diamond \gamma\right)\right. \\ &\quad \left. + 2\mathcal{V}^2\alpha^* \diamond D \diamond \alpha\right] \mathcal{D}^\circ[\alpha], \end{aligned} \tag{7.309}$$

where $W_{\text{in}}[\alpha]$ is the Wigner functional for the input state. The generating function for the photon statistics, given in (7.309), can be represented by

$$\mathcal{W}_C(\nu) = \sum_{m=-\infty}^{\infty} \nu^m C(m). \tag{7.310}$$

where ν is the generating parameter and $C(m)$ represents the probabilities in the cross-correlation distribution.

Since the index r in (7.300) also runs over negative values, we cannot extract the different cross-correlation probabilities with the aid of derivatives as is usually done with generating functions. To extract the individual cross-correlation probabilities, we use the identity in (3.113). It implies that we can extract the cross-correlation probabilities by replacing $\nu \to \exp(\mathrm{i}\phi)$ in the generating function and integrating over ϕ:

$$\begin{aligned} C(n) &= \sum_{m=-\infty}^{\infty} \frac{1}{2\pi} \int_0^{2\pi} \exp[\mathrm{i}(m-n)\phi]\,\mathrm{d}\phi\; C(m) \\ &= \frac{1}{2\pi} \int_0^{2\pi} \exp(-in\phi)\mathcal{W}_C(e^{\mathrm{i}\phi})\,\mathrm{d}\phi. \end{aligned} \tag{7.311}$$

The effect of the replacement $\nu \to \exp(\mathrm{i}\phi)$ on $\mathcal{N}$ and $\mathcal{V}$ defined in (7.307) is

$$\begin{aligned} \mathcal{N}(e^{\mathrm{i}\phi}) &= \left\{ \frac{4\exp(\mathrm{i}\phi)}{[1+\exp(\mathrm{i}\phi)]^2} \right\}^{\mathrm{tr}\{D\}} = \frac{1}{\cos^{2\,\mathrm{tr}\{D\}}(\frac{1}{2}\phi)}, \\ \mathcal{V}(e^{\mathrm{i}\phi}) &= \frac{1-\exp(\mathrm{i}\phi)}{1+\exp(\mathrm{i}\phi)} = -\mathrm{i}\tan\left(\frac{1}{2}\phi\right). \end{aligned} \tag{7.312}$$

The expression for the cross-correlation probabilities becomes

$$\begin{aligned} C(n) = \frac{1}{2\pi} \int_{-\pi}^{\pi} \frac{\exp(-in\phi)}{\cos^{2\,\mathrm{tr}\{D\}}(\frac{1}{2}\phi)} \int W_{\mathrm{in}}[\alpha] \exp\left[-2\tan^2\left(\frac{1}{2}\phi\right)\alpha^* \diamond D \diamond \alpha \right. \\ \left. - 2\tan\left(\frac{1}{2}\phi\right)(\gamma^* \diamond D \diamond \alpha - \alpha^* \diamond D \diamond \gamma)\right] \mathcal{D}^\circ[\alpha]\,\mathrm{d}\phi, \end{aligned} \tag{7.313}$$

where we changed the original integration boundaries of ϕ to $-\pi \le \phi < \pi$ to accommodate the continuous region for the tangent function.

In homodyne tomography, the distribution of cross-correlation probabilities is treated as a function. It has a finite resolution due to the discreteness of the photon statistics. The magnitude of the local oscillator parameter function γ_0 determines this resolution. We define a continuous variable to replace the discrete index n with the aid of γ_0 by

$$n \to x = \frac{n}{\gamma_0}. \tag{7.314}$$

At the same time, we change the integration variable ϕ, by replacing

$$2\gamma_0 \tan\left(\frac{1}{2}\phi\right) \rightarrow y, \tag{7.315}$$

so that

$$\begin{aligned} \tan\left(\frac{1}{2}\phi\right) &\rightarrow \frac{y}{2\gamma_0}, \quad \phi \rightarrow 2\tan^{-1}\left(\frac{y}{2\gamma_0}\right), \\ \cos^2\left(\frac{1}{2}\phi\right) &\rightarrow \frac{4\gamma_0^2}{y^2+4\gamma_0^2}, \quad \mathrm{d}\phi \rightarrow \frac{4\gamma_0 \mathrm{d}y}{y^2+4\gamma_0^2}. \end{aligned} \tag{7.316}$$

The integration over y extends from $-\infty$ to ∞. However, the Wigner functional for a physical state always introduces a scale that leads to a maximum value Y beyond which the contribution to the integral over y is negligible. For $\gamma_0 \gg Y$, we can assume that[f] $y^2 + 4\gamma_0^2 \approx 4\gamma_0^2$. It then follows that

$$\exp(-in\phi) \rightarrow \exp\left[-\mathrm{i}2x\gamma_0 \tan^{-1}\left(\frac{y}{2\gamma_0}\right)\right] \approx \exp(-\mathrm{i}xy). \tag{7.317}$$

The local oscillator parameter function is now expressed in terms of (7.287) to reveal the role of its magnitude γ_0. It leads to

$$\begin{aligned} \gamma^* \diamond D \diamond \alpha - \alpha^* \diamond D \diamond \gamma &\rightarrow [\exp(-\mathrm{i}\theta)\Gamma^* \diamond D \diamond \alpha - \exp(\mathrm{i}\theta)\alpha^* \diamond D \diamond \Gamma]\gamma_0 \\ &\triangleq -\mathrm{i}\gamma_0\kappa[\alpha](\theta), \end{aligned} \tag{7.318}$$

where $\kappa[\alpha](\theta)$ is a real-valued quantity representing the contractions of α as a function of the phase θ. The cross-correlation probability distribution then becomes

$$C(x,\theta) = \frac{1}{2\pi\gamma_0} \int_{-\infty}^{\infty} \int \exp\left(\mathrm{i}\kappa y - \mathrm{i}xy - \frac{y^2}{2\gamma_0^2}\alpha^* \diamond D \diamond \alpha\right) W_{\text{in}}[\alpha]\, \mathcal{D}^\circ[\alpha]\, \mathrm{d}y. \tag{7.319}$$

Since γ_0 is very large, one may consider discarding the quadratic term in the exponent. The resulting expression would be

$$C(x,\theta) = \frac{1}{2\pi\gamma_0} \int_{-\infty}^{\infty} \int \exp(\mathrm{i}\kappa y - \mathrm{i}xy)\, W_{\text{in}}[\alpha]\, \mathcal{D}^\circ[\alpha]\, \mathrm{d}y, \tag{7.320}$$

reminiscent of an integral for a Dirac delta function. While the integral over y is just a one-dimensional integral that produces a Dirac delta function, the replacement of κ by x inside $W_{\text{in}}[\alpha]$ is done by a functional integral. The process whereby this replacement is accomplished needs to be considered carefully.

---

f However, if $1+\epsilon$ is raised to the power of infinity (Ω), then even for a very small value of ϵ, the result would diverge. We can nevertheless argue that $\mathrm{tr}\{D\}$ is finite so that the approximation is valid.

7.9.2.3 Output Wigner function

The result in (7.319) is a function and not a functional. It can be used to obtain an *observed Wigner function* for the part of the Wigner functional corresponding to the mode selected by the local oscillator and the detector. For this process, we use a *Radon transform* [20]. In terms of x, the function in (7.319) is interpreted as a *marginal probability distribution* obtained from the partial integration of the Wigner functional. This marginal distribution is converted into a corresponding slice of the characteristic function via a Fourier transform

$$\chi(\rho,\theta) = \int C(x,\theta)\exp(\mathrm{i}x\rho)\,\mathrm{d}x. \tag{7.321}$$

The variables ρ and θ are interpreted as the radial and azimuthal coordinates for a two-dimensional function. We convert them into Cartesian coordinates $\{\xi,\zeta\}$, so that $\chi(\rho,\theta) \to \chi'(\xi,\zeta)$. Then we apply a two-dimensional symplectic Fourier transform on this characteristic function

$$W_{\text{obs}}(q,p) = \int \chi'(\xi,\zeta)\exp\left(\mathrm{i}q\xi - \mathrm{i}p\zeta\right)\frac{\mathrm{d}\zeta\,\mathrm{d}\xi}{2\pi}, \tag{7.322}$$

to produce the *observed Wigner function* $W_{\text{obs}}(q,p)$ as a function of the real-valued variables q and p. The details of the conversion between radial coordinates and Cartesian coordinates are discussed below. When we substitute (7.319) into this Fourier transform and evaluate the integrals over x and y, we obtain

$$\begin{aligned} W_{\text{H}}(q,p) &= \int C(x,\theta)\exp(\mathrm{i}px)\exp\left(\mathrm{i}q\xi - \mathrm{i}p\zeta\right)\,\mathrm{d}x\,\frac{\mathrm{d}\zeta\,\mathrm{d}\xi}{2\pi} \\ &= \frac{1}{2\pi\gamma_0}\int \exp\left\{\mathrm{i}\kappa[\alpha](\theta)\rho + \mathrm{i}q\xi - \mathrm{i}p\zeta - \frac{\rho^2}{2\gamma_0^2}\alpha^*\diamond D\diamond\alpha\right\} \\ &\quad\times W_{\text{in}}[\alpha]\,\mathcal{D}^\circ[\alpha]\,\frac{\mathrm{d}\zeta\,\mathrm{d}\xi}{2\pi}. \end{aligned} \tag{7.323}$$

7.9.2.4 Detector system

In general, the detector kernel D is not invertible because it acts like a projection operator restricting the functional phase space to those fields that can pass through D. It means that the functional integration over α in (7.323) cannot in general be evaluated.[g] Even if we discard the quadratic term, the remaining part of the argument in the exponent does not represent the entire functional phase space. The projection induced by Γ is in general even more restrictive than the projection associated with D (unless D is a single-mode kernel).

---

g Strictly speaking, the input Wigner functional should always be well-defined on the entire functional phase space, which means that if we combine it with the detector kernel, the combination should be invertible. Here, we provide a discussion without invoking knowledge about the input Wigner functional.

To evaluate the functional integration, we separate the phase space into the subspaces defined by the projection. For this purpose, we denote the total functional phase space by $\mathcal{A}$, and the subspace onto which D projects by $\mathcal{M}$. It implies that, for $\alpha \in \mathcal{M}$, we have $\alpha^* \diamond D \diamond \alpha \neq 0$.

In the absurd case when $D \diamond \Gamma = \Gamma^* \diamond D = 0$ (the detector cannot measure the mode of the local oscillator), the Γ-dependent linear term in the functional integral in (7.318) is zero, leaving us with

$$W_0(q,p) = \frac{1}{2\pi\gamma_0} \int \exp\left(iq\xi - ip\zeta - \frac{\rho^2}{2\gamma_0^2} \alpha^* \diamond D \diamond \alpha \right) W_{\text{in}}[\alpha]\, \mathcal{D}^\circ[\alpha]\, \frac{d\zeta\, d\xi}{2\pi}. \tag{7.324}$$

The remaining α-dependent part is a rotationally symmetric function of ρ (given in terms of ξ and ζ) centred at the origin (both in the functional phase space and in the space defined by ξ and ζ). Evaluating the Fourier integral over ξ and ζ, we obtain a rotationally symmetric function at the origin of the observed phase space defined by q and p with a width inversely proportional to γ_0.

For a practical homodyning tomography system, one would never have the situation where $\Gamma^* \diamond D = D \diamond \Gamma = 0$. However, for a multimode detector, there would be modes $M_i \in \mathcal{M}$ that are orthogonal to Γ. For these modes, the detector in the homodyning process can register photons that come from the state but not from the local oscillator. Such photons produce a background, leading to a contribution to the observed Wigner functional represented by (7.324). To avoid such unwanted contributions, the detector in a homodyne tomography system must be a *single-mode detector*, so that $D = \eta M M^*$. Since it plays a significant role, we explicitly incorporate the *detection efficiency* η here. It does not mean that Γ can pass through a single-mode detector unaffected. One can in general assume that

$$\Gamma = \mu M + \Gamma_\perp, \tag{7.325}$$

where M is the mode of the detectors (i. e., the part of Γ that can pass through D unaffected), while $\Gamma_\perp^* \diamond M = 0$, and $\mu = \Gamma^* \diamond M$, representing an *overlap efficiency*. In general, such an overlap produces a complex factor. Here we absorb the phase into θ and assume that μ is a real value. It then follows that $D \diamond \Gamma = \eta\mu M$ and $\Gamma^* \diamond D = \eta\mu M^*$. The expression for the observed Wigner function becomes

$$\begin{aligned} W_{\text{H}}(q,p) = \frac{1}{2\pi\gamma_0} \int & \exp\left[-\eta\mu\rho \exp(-i\theta) M^* \diamond \alpha + \eta\mu\rho \exp(i\theta) \alpha^* \diamond M\right] \\ & \times \exp\left(iq\xi - ip\zeta - \frac{\rho^2\eta}{2\gamma_0^2} \alpha^* \diamond M M^* \diamond \alpha \right) W_{\text{in}}[\alpha]\, \mathcal{D}^\circ[\alpha]\, \frac{d\zeta\, d\xi}{2\pi}. \end{aligned} \tag{7.326}$$

In the exponent, α is always contracted with M. In effect, the functional integral separates into an ordinary integral over the complex two-dimensional plane defined by the contraction $M^* \diamond \alpha$, and a functional integral over the remaining functional phase space orthogonal to M. The only part of the integrand that still depends on the latter is the input Wigner functional $W_{\text{in}}[\alpha]$. The systematic way to perform such a separation of a functional integral is to use an *inhomogeneous beamsplitter*. In effect, the integration field variable is replaced by

$$\alpha \to MM^* \diamond \alpha + Q \diamond \beta = \alpha_0 M + Q \diamond \beta, \tag{7.327}$$

where $Q = \mathbf{1} - MM^*$ is the projection kernel for the orthogonal space, and $\alpha_0 = M^* \diamond \alpha$ is a complex variable. The functional integral over β then produces the trace of the input Wigner functional over β,

$$\text{tr}\{W_{\text{in}}\}(\alpha_0) = \int W_{\text{in}}[\alpha_0 M + Q \diamond \beta]\, \mathcal{D}^\circ[\beta] \triangleq W_{\text{tr}}(\alpha_0). \tag{7.328}$$

Thus, we get

$$\begin{aligned} W_{\text{obs}}(q,p) = {} & \frac{1}{2\pi\gamma_0} \int W_{\text{tr}}(\alpha_0) \exp\left[-\eta\mu\rho\exp(-\text{i}\theta)\alpha_0 + \eta\mu\rho\exp(\text{i}\theta)\alpha_0^*\right] \\ & \times \exp\left(\text{i}q\xi - \text{i}p\zeta - \frac{\rho^2\eta}{2\gamma_0^2}|\alpha_0|^2\right) \frac{\text{d}^2\alpha_0}{2\pi} \frac{\text{d}\zeta\, \text{d}\xi}{2\pi}. \end{aligned} \tag{7.329}$$

Now, we convert α_0 into two real-valued quadrature variables

$$\alpha_0 = \frac{1}{\sqrt{2}}(q_0 + \text{i}p_0), \tag{7.330}$$

and the radial coordinates ρ and θ into Cartesian coordinates ζ and ξ, according to

$$\zeta = \sqrt{2}\rho\cos(\theta) \quad \text{and} \quad \xi = \sqrt{2}\rho\sin(\theta), \tag{7.331}$$

so that

$$\rho^2 = \frac{1}{2}(\zeta^2 + \xi^2). \tag{7.332}$$

With these replacements, we can express the observed Wigner function as

$$\begin{aligned} W_{\text{obs}}(q,p) = {} & \int \exp\left[\text{i}\eta\mu q_0\xi - \text{i}\eta\mu p_0\zeta + \text{i}q\xi - \text{i}p\zeta - \frac{\eta}{8\gamma_0^2}\left(\zeta^2 + \xi^2\right)\left(q_0^2 + p_0^2\right)\right] \\ & \times \frac{1}{2\pi\gamma_0} W_{\text{tr}}(q_0, p_0) \frac{\text{d}q_0\, \text{d}p_0}{2\pi} \frac{\text{d}\zeta\, \text{d}\xi}{2\pi}. \end{aligned} \tag{7.333}$$

The integration over ζ and ξ is evaluated to produce

$$W_{\text{obs}}(q,p) = \frac{\gamma_0}{\pi^2\eta} \int \frac{W_{\text{tr}}(q_0,p_0)}{q_0^2 + p_0^2} \exp\left[-\frac{2\gamma_0^2}{\eta} \frac{(q_0\eta\mu + q)^2 + (p_0\eta\mu + p)^2}{q_0^2 + p_0^2}\right] \text{d}q_0\, \text{d}p_0. \tag{7.334}$$

The observed Wigner function is thus obtained from the traced Wigner functional through a linear integral operation (superposition integral) with a kernel (point-spread function) given by

$$K(q_0, p_0, q, p) = \frac{\gamma_0}{(q_0^2 + p_0^2)\pi^2\eta} \exp\left[-\frac{2\gamma_0^2}{\eta} \frac{(q_0\eta\mu + q)^2 + (p_0\eta\mu + p)^2}{q_0^2 + p_0^2}\right]. \tag{7.335}$$

Studying this point-spread function, we identify the distortions that the observed Wigner function incurs. The first effect is a scaling caused by the product of the two efficiencies $\eta\mu$. To see this effect, we redefine the integration variables by $q_0 \to q_1/\eta\mu$ and $p_0 \to p_1/\eta\mu$. The result is

$$\begin{aligned} W_{\mathrm{obs}}(q, p) = {} & \frac{\gamma_0}{\pi^2\eta} \int W_{\mathrm{tr}}\left(\frac{q_1}{\eta\mu}, \frac{p_1}{\eta\mu}\right) \frac{1}{q_1^2 + p_1^2} \\ & \times \exp\left[-2\gamma_0^2\eta\mu^2 \frac{(q_1 + q)^2 + (p_1 + p)^2}{q_1^2 + p_1^2}\right] \mathrm{d}q_1\, \mathrm{d}p_1. \end{aligned} \tag{7.336}$$

The arguments of W_{tr} demonstrate the effect of the scaling. We also see that γ_0 is effectively reduced to $\gamma_0\eta^{1/2}\mu$. The scaling is readily corrected, provided that the scaling factor is known. Therefore, the detection and overlap efficiencies need to be measured as a calibration of the process.

Another effect is the varying resolution with which the observed Wigner function is rendered by the point-spread function. Without the factors of $q_0^2 + p_0^2$ in the denominators, the point-spread function becomes a Dirac delta function in the limit $\gamma_0 \to \infty$, representing the ideal tomography process if we ignore the scaling. The factors of $q_0^2 + p_0^2$ in the denominators cause the kernel to become broader, depending on the distance from the origin. For each point given by q and p, the function value of the observed Wigner function is produced by a weighted integration over an area around that point. For example, at the origin where $q = p = 0$, the point-spread function becomes

$$K(q_0, p_0, 0, 0) = \frac{\gamma_0 \exp\left(-2\gamma_0^2\eta\mu^2\right)}{(q_0^2 + p_0^2)\pi^2\eta}, \tag{7.337}$$

which is singular and produces a divergent result when the observed Wigner function is nonzero at the origin. For other values of $\{q, p\}$ away from the origin, the point-spread function forms a peak located at $\{-q, -p\}$ (if we ignore the scaling) with a width governed by the value of γ_0. It becomes broader as the point $\{q, p\}$ moves away from the origin. In fact, the kernel has a scale invariance: one can scale all the variables by the same factor and it will cancel out apart from an overall change in the amplitude of the kernel. It implies that the width of the peak scales linearly with the distance of the peak from the origin. The width of the peak becomes comparable to the minimum uncertainty area when the average number of photons in the state becomes comparable with the average number of photons in the local oscillator. The resolution of the observed Wigner function is determined by both the distance from the origin and the ratio of the average number of photons in the local oscillator to the average number of photons in the state.

7.9.3 Example: Single-mode Fock state

In Section 7.8.2, we discuss the heralded preparation of a Fock state. There it is shown that the resulting Fock state is parameterized by the transformed mode of the heralding detector. Here, we consider the homodyne tomography of such a *fixed-spectrum Fock state* with its normalized parameter function given by F. The generating function for its Wigner functional is given in (6.52). After applying the inhomogeneous beamsplitter transformation in (6.232) to separate the integration domains, we obtain

$$\begin{aligned}\mathcal{W}[\alpha,\beta](J) = \frac{\mathcal{N}_0^2}{1+J}\exp\bigg[&-2\alpha^*\diamond\left(\mathbf{1}-\frac{2J}{1+J}P\diamond FF^*\diamond P\right)\diamond\alpha \\ &+\mathrm{i}\frac{4J}{1+J}(\beta^*\diamond Q\diamond FF^*\diamond P\diamond\alpha-\alpha^*\diamond P\diamond FF^*\diamond Q\diamond\beta) \\ &-2\beta^*\diamond\left(\mathbf{1}-\frac{2J}{1+J}Q\diamond FF^*\diamond Q\right)\diamond\beta\bigg],\end{aligned} \tag{7.338}$$

where $P = MM^*$ and $Q = \mathbf{1} - MM^*$, with M being the detector mode in the homodyne system. Since we are interested in the case when the parameter function does not exactly match the mode of the detector, we assume that

$$F(\mathbf{k}) = \nu M(\mathbf{k}) + M_{\perp}(\mathbf{k}), \tag{7.339}$$

where $\nu = M^*\diamond F$ and $M^*\diamond M_{\perp} = 0$. Then $P\diamond F = \nu M$ and $Q\diamond F = M_{\perp}$. With these simplifications, the expression becomes

$$\begin{aligned}\mathcal{W}[\alpha,\beta](J) = \frac{\mathcal{N}_0^2}{1+J}\exp\bigg[&-2\alpha^*\diamond\left(\mathbf{1}-\frac{2J|\nu|^2MM^*}{1+J}\right)\diamond\alpha-2\beta^*\diamond\left(\mathbf{1}-\frac{2JM_{\perp}M_{\perp}^*}{1+J}\right)\diamond\beta \\ &+\mathrm{i}\frac{4J\nu^*}{1+J}\beta^*\diamond M_{\perp}M^*\diamond\alpha-\mathrm{i}\frac{4J\nu}{1+J}\alpha^*\diamond MM_{\perp}^*\diamond\beta\bigg].\end{aligned} \tag{7.340}$$

After integrating the part of the α-dependent functional orthogonal to M, we obtain

$$\begin{aligned}\mathcal{W}[\beta](\alpha_0,J) = \frac{2\mathcal{N}_0}{1+J}\exp\bigg[&-2\frac{1+J-2J|\nu|^2}{1+J}|\alpha_0|^2+\mathrm{i}\frac{4J\nu^*\alpha_0}{1+J}\beta^*\diamond M_{\perp} \\ &-\mathrm{i}\frac{4J\nu\alpha_0^*}{1+J}M_{\perp}^*\diamond\beta-2\beta^*\diamond\left(\mathbf{1}-\frac{2J}{1+J}M_{\perp}M_{\perp}^*\right)\diamond\beta\bigg],\end{aligned} \tag{7.341}$$

where $\alpha_0 = M^*\diamond\alpha$ is a complex variable. Next, we evaluate the functional integration over β, leading to

$$
\begin{aligned}
\mathcal{W}_{\text{tr}} &\triangleq \int \mathcal{W}[\beta](\alpha_0,J)\,\mathcal{D}^{\circ}[\beta] \\
&= \frac{2}{1+J}\exp\left(-2\frac{1+J-2J|\nu|^2}{1+J}|\alpha_0|^2\right)\left(\det\left\{\mathbf{1}-\frac{2JM_{\perp}M_{\perp}^{*}}{1+J}\right\}\right)^{-1} \\
&\quad\times\exp\left[8\frac{J^2|\nu|^2|\alpha_0|^2}{(1+J)^2}M_{\perp}^{*}\diamond\left(\mathbf{1}-\frac{2JM_{\perp}M_{\perp}^{*}}{1+J}\right)^{-1}\diamond M_{\perp}\right].
\end{aligned}
\tag{7.342}
$$

The determinant and inverse can be simplified as

$$
\begin{aligned}
\det\left\{\mathbf{1}-\frac{2JM_{\perp}M_{\perp}^{*}}{1+J}\right\} &= 1-\frac{2J}{1+J}\left(1-|\nu|^2\right), \\
\left(\mathbf{1}-\frac{2JM_{\perp}M_{\perp}^{*}}{1+J}\right)^{-1} &= \mathbf{1}+\frac{2JM_{\perp}M_{\perp}^{*}}{1-J+2J|\nu|^2},
\end{aligned}
\tag{7.343}
$$

where $M_{\perp}^{*}\diamond M_{\perp}=1-|\nu|^2$, which follows from (7.339) since $F^{*}\diamond F=1$. Therefore, the expression becomes

$$
\mathcal{W}_{\text{tr}}(\alpha_0,J)=\frac{2}{1+(2|\nu|^2-1)J}\exp\left(-2|\alpha_0|^2+\frac{4J|\nu|^2|\alpha_0|^2}{1+(2|\nu|^2-1)J}\right).
\tag{7.344}
$$

Together with the rest of (7.334), the expression for the generating function is

$$
\begin{aligned}
\mathcal{W}_{\text{obs}}(q,p,J) &= \frac{2\gamma_0}{[1+(2|\nu|^2-1)J]\pi^2\eta}\int\frac{1}{(q_0^2+p_0^2)\pi}\exp\left[-\frac{(1-J)(q_0^2+p_0^2)}{1+(2|\nu|^2-1)J}\right] \\
&\quad\times\exp\left[-\frac{2\gamma_0^2}{\eta}\frac{(q_0\eta\mu+q)^2+(p_0\eta\mu+p)^2}{q_0^2+p_0^2}\right]\,dq_0\,dp_0.
\end{aligned}
\tag{7.345}
$$

Assuming γ_0 is suitably large, and $q\neq 0$ or $p\neq 0$, the kernel becomes zero unless q_0 and p_0 are approximately equal to $-q/\eta\mu$ and $-p/\eta\mu$, respectively. Therefore, we can substitute $q_0^2+p_0^2\to(q^2+p^2)/\eta^2\mu^2$ in the denominator, and evaluate the remaining integrals. The result reads

$$
\begin{aligned}
\mathcal{W}_{\text{obs}} &\approx \frac{2\gamma_0\eta\mu^2}{\pi(1-J)(q^2+p^2)+2\pi(1-J+2J|\nu|^2)\gamma_0^2\eta^3\mu^4} \\
&\quad\times\exp\left[-\frac{2(1-J)(q^2+p^2)\gamma_0^2\eta\mu^2}{(1-J)(q^2+p^2)+2(1-J+2J|\nu|^2)\gamma_0^2\eta^3\mu^4}\right].
\end{aligned}
\tag{7.346}
$$

For large enough γ_0^2, it becomes

$$
\mathcal{W}_{\text{obs}}\approx\frac{1}{\pi(1-J+2J|\nu|^2)\gamma_0\eta^2\mu^2}\exp\left[-\frac{(1-J)(q^2+p^2)}{(1-J+2J|\nu|^2)\eta^2\mu^2}\right].
\tag{7.347}
$$

The Wigner function for the observed single-photon Fock state is now readily produced with a single derivative with respect to J, after which J is set to zero and the function is normalized, which produces

$$W_{\text{spfs}}^{(\text{obs})}(q,p) = \frac{2}{\eta^4\mu^4}\left[2\left(q^2+p^2\right)|\nu|^2 + \left(1-2|\nu|^2\right)\eta^2\mu^2\right]\exp\left(-\frac{q^2+p^2}{\eta^2\mu^2}\right). \tag{7.348}$$

It still contains the product of efficiencies $\eta\mu$ and the overlap ν that affects the shape of the Wigner function. The effect of the product of efficiencies $\eta\mu$ is a simple scaling of the coordinates, as explained above. It is removed by replacing $q \to q\eta\mu$ and $p \to p\eta\mu$. The resulting rescaled normalized Wigner function is

$$W_{\text{spfs}}^{(\text{obs})}(r) = 2\left(2r^2|\nu|^2 + 1 - 2|\nu|^2\right)\exp\left(-r^2\right), \tag{7.349}$$

where $r^2 = q^2 + p^2$. It is negative at the origin ($r = 0$) provided that $|\nu|^2 > \frac{1}{2}$.

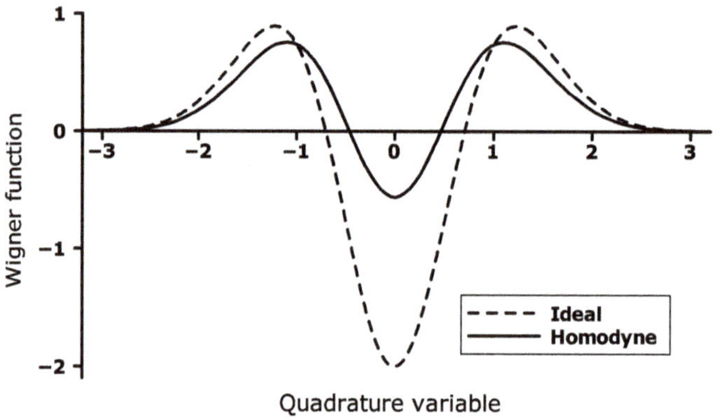

Figure 7.8: Cross-sections of the two-dimensional Wigner functions of an ideal single-photon Fock state and one obtained from the homodyne tomography with a non-ideal overlap between the Fock state parameter function and the homodyne detector mode.

In Figure 7.8, the cross-section of the Wigner function of a single-photon Fock state obtained by homodyne tomography when $|\nu| = 0.8$ is compared to the cross-sections of the ideal single-photon Fock state's Wigner function. It shows how the Wigner function obtained by homodyne tomography is affected by the non-ideal overlap between the Fock state parameter function and the homodyne detector mode.

In experimental scenarios, it may be difficult to match the different modes. There are three modes that need to be matched: the transformed heralding detector mode that parameterizes the heralded Fock state, the local oscillator of the homodyne system, and the detector mode of the homodyne system. The overlap between the latter two produces a factor that combines with the detector efficiency, which is readily removed provided that it is known. The overlap between the homodyne detector mode and the Fock state parameter function produces a complex parameter ν that is not so easily removed. However, if one can measure it, one can try to improve the experimental setup to maximize the magnitude of ν. Using the observed curve of the Wigner function, we compute the ratio of the maximum to the minimum of the curve. It gives an expression

$$\frac{W_{\text{max}}}{W_{\text{min}}} = \frac{2|\nu|^2}{1-2|\nu|^2} \exp\left(\frac{1-4|\nu|^2}{2|\nu|^2}\right), \tag{7.350}$$

assuming $|\nu|^2 > \frac{1}{2}$, which only depends on $|\nu|^2$. It allows us to determine the magnitude of this overlap $|\nu|$. Next, we use the value of r where the Wigner function becomes zero to determine the value of the combined efficiency, with which we can correct the scaling. From (7.348), the null is obtained at

$$r_{\text{null}} = \eta\mu\sqrt{\frac{2|\nu|^2-1}{2|\nu|^2}}, \tag{7.351}$$

provided that $|\nu|^2 > \frac{1}{2}$, leading to a value for $\eta\mu$. In this way, the values of the overlap and the efficiency can be used as feedback to improve the experimental conditions.

7.9.4 Naive homodyning

A simple way to *guess* the observed Wigner function that would be produced by a homodyning system from a given Wigner functional is to replace the field variable by a complex variable times the parameter function. However, such a simple calculation can produce unreliable results. To demonstrate this problem, we compute the marginals of the resulting Wigner function for the example of the Fock that is considered above.

The generating function in (7.346) is used to compute the marginals of the Wigner function that it produces. We integrate over p and divide by 2π. For convenience, we set $\eta = \mu = 1$ assuming that the rescaling has already been done to remove the effect of the efficiencies. The resulting generating function for the marginal distributions reads

$$\begin{aligned} \mathcal{W}_{\text{mar}}(q,J) &= \int \mathcal{W}_{\text{obs}}(q,p,J)\,\frac{dp}{2\pi} \\ &= \frac{1}{\sqrt{\pi(1-J)(1-J+2J|\nu|^2)}} \exp\left[-\frac{(1-J)q^2}{1-J+2J|\nu|^2}\right]. \end{aligned} \tag{7.352}$$

The marginal distribution for the single-photon Fock state is then given by

$$\partial_J \mathcal{W}_{\text{mar}}(q,J)\big|_{J=0} = \frac{1}{\sqrt{\pi}}\left(2|\nu|^2q^2 + 1 - |\nu|^2\right)\exp(-q^2). \tag{7.353}$$

It is positive semidefinite for all the allowed values of $|\nu|$ (i. e., $0 \leq |\nu| \leq 1$). For $|\nu| = 1$, the distribution is zero at the origin. It is positive at the origin for smaller values of $|\nu|$.

We compare this result with what is obtained from *naive homodyning* where we simply substitute $\alpha \to \alpha_0 M$ into the generating function for Wigner functionals of the Fock states, to get

$$\mathcal{W}'(\alpha_0,J) = \frac{2}{1+J}\exp\left(-2|\alpha_0|^2 + \frac{4J|\nu|^2}{1+J}|\alpha_0|^2\right). \tag{7.354}$$

Here ν emerges due to the non-ideal overlap. It can be compared to (7.347) for $\eta = 1$. Performing the integration over p to produce the generating function for the marginal distributions, we obtain

$$\begin{aligned}\mathcal{W}'(q,J) &= \int \mathcal{W}'(q,p,J)\,\frac{\mathrm{d}p}{2\pi} \\ &= \frac{1}{\sqrt{\pi(1+J)(1+J-2J|\nu|^2)}}\exp\left(-q^2 + \frac{2J|\nu|^2q^2}{1+J}\right).\end{aligned} \tag{7.355}$$

In this case, the marginal distribution for the single-photon Fock state is

$$\partial_J \mathcal{W}'(q,J)\big|_{J=0} = \frac{2}{\sqrt{\pi}}\left(|\nu|^2q^2 + |\nu|^2 - \frac{3}{4}\right)\exp(-q^2). \tag{7.356}$$

It is smaller than 0 for $|\nu|^2 < \frac{3}{4}$, which is not valid for a probability distribution. Therefore, the naive approach does not always give a valid Wigner function.

7.10 Discussion

All physical quantum optical systems perform measurements on the photons that pass through them. (If not, one can argue that such a system is in fact just a subsystem that must be used within a larger system that incorporates measurements.) Therefore, measurements play a vital role in all physical quantum optical systems.

Abstract formulations of the concept of a measurement can provide powerful tools to investigate quantum information protocols. However, we do not address such topics here. Instead, our interest is to provide tools to model physical measurement systems that incorporate all the degrees of freedom.

While measurements are ubiquitous, they can also be quite diverse. It is therefore not reasonable to attempt an exhaustive discussion of measurement systems. In this chapter, we address various measurement systems and different aspects of such measurement systems. However, the aim is to provide techniques that can be used to model and analyse arbitrary measurement systems. To that end, we discuss some of the basic aspects of measurement systems, such as the basic detectors and how to model them.

The operators that play significant roles in this context are the number operator and the projection operators. They are discussed extensively with the intension to clarify their roles in the modelling of measurement systems. While the number operator provides a direct way to determine the average number of photons in a state by exploiting the effect of bosonic enhancements, the projection operators extract the probability for a given occupation number from the state. The former gives a direct estimate of the photon number, which is proportional to the intensity. The latter provide the photon statistics. Other operators that also play an increasingly significant role in quantum optical systems are the quadrature operators, which are associated with homodyne detection.

The incorporation of the spatiotemporal degrees of freedom into the models for measurement systems introduce some subtleties that need careful consideration. We provide an in-depth discussion to clarify how detector systems based on the number operator and the projection operators must be used to model the physical system correctly.

The use of a generating function for the Wigner functionals of the projection operators alleviates calculations of photon statistics. Moreover, one can use this generating function to obtain the result that would be produced by the number operator. As a result, this generating function for the photon statistics is among the most versatile tools in the analyses of quantum optical systems.

Using these tools, we considered various measurement scenarios. While the measurements of the output photon-number distribution (output intensity) and the output statistics, are common scenarios, the measurements of correlations are particularly relevant in the context of quantum physics. For this purpose, one needs to distinguish between scenarios where correlations can be identified as classical phenomena and those that as specifically associated with quantum phenomena. Optical interference is a well-studied classical correlation phenomenon, which is also found in quantum optical systems. Quantum correlations are associated with measurements involving two detections, often made in coincidence. They include such phenomena as entanglement, antibunching and the Hong–Ou–Mandel effect. The squeezed vacuum state is a prime example of a state that exhibits quantum correlations. However, a thorough analysis of correlation measurements made on this state in terms of all the degrees of freedom leads to some challenges. To overcome these challenges one can use the typical experimental conditions to introduce controlled approximations. We provide a comprehensive discussion of such correlation measurements, demonstrating the use of these approximations to simplify the calculations. These results provide the background for discussions of measurements applied to parametric down-converted states in Chapter 9.

Noise can be modelled in various ways. Here we are interested in the effect of noise in measurements when all the spatiotemporal degrees of freedom are included. We demonstrate how the noise affects the photon statistics of certain quantum states.

The preparation of special quantum optical states for quantum resources play a significant role in the implementation of photonic quantum information systems. Such state preparation procedures are often based on heralding, which is a post-selection process that produces the required state conditioned on a successful heralding detection. Here we discuss the heralded preparation of Fock states, photon-subtracted states, and photon-added states. Our interest is to see how the incorporation of the spatiotemporal degrees of freedom in the modelling of such heralding systems affects the properties of the heralded states. The performance of photonic quantum information systems depends on the quality of such heralded states used as quantum resources.

Finally, we discuss homodyning, which differs from the other measurement scenarios in that it provides a way to measure quadrature as opposed to occupation number. Here, we provide a derivation of the homodyning system that incorporates the spatiotemporal degrees of freedom. It gives a way to model the distortions that are introduced by the physical system parameters, thus allowing one to improve the performance of such a measurement system.

Bibliography

[1] M. A. Nielsen and I. L. Chuang. *Quantum Computation and Quantum Information*. Cambridge University Press, Cambridge, England, 2000.

[2] M. Abramowitz and I. A. Stegun. *Handbook of Mathematical Functions*. Dover, Toronto, 1972.

[3] R. Hanbury Brown and R. Q. Twiss. Correlation between photons in two coherent beams of light. *Nature*, 177:27–29, 1956.

[4] R. J. Glauber. Coherent and incoherent states of the radiation field. *Phys. Rev.*, 131:2766–2788, 1963.

[5] L. Mandel and E. Wolf. *Introductory Quantum Optics*. Cambridge University Press, New York, 1995.

[6] V. Giovannetti, S. Lloyd, and L. Maccone. Advances in quantum metrology. *Nat. Photonics*, 5:222–229, 2011.

[7] E. Polino, M. Valeri, N. Spagnolo, and F. Sciarrino. Photonic quantum metrology. *AVS Quantum Sci.*, 2:024703, 2020.

[8] B. Yurke, S. L. McCall, and J. R. Klauder. SU(2) and SU(1, 1) interferometers. *Phys. Rev. A*, 33:4033–4054, 1986.

[9] W. N. Plick, J. P. Dowling, and G. S. Agarwal. Coherent-light-boosted, sub-shot noise, quantum interferometry. *New J. Phys.*, 12:083014, 2010.

[10] A. I. Lvovsky, H. Hansen, T. Aichele, O. Benson, J. Mlynek, and S. Schiller. Quantum state reconstruction of the single-photon fock state. *Phys. Rev. Lett.*, 87:050402, 2001.

[11] A. Ourjoumtsev, R. Tualle-Brouri, and P. Grangier. Quantum homodyne tomography of a two-photon fock state. *Phys. Rev. Lett.*, 96:213601, 2006.

[12] J. S. Neergaard-Nielsen, B. M. Nielsen, C. Hettich, K. Mølmer, and E. S. Polzik. Generation of a superposition of odd photon number states for quantum information networks. *Phys. Rev. Lett.*, 97:083604, 2006.

[13] A. Biswas and G. S. Agarwal. Nonclassicality and decoherence of photon-subtracted squeezed states. *Phys. Rev. A*, 75:032104, 2007.

[14] G. S. Agarwal and K. Tara. Nonclassical properties of states generated by the excitations of a coherent state. *Phys. Rev. A*, 43:492–497, 1991.

[15] M. Dakna, T. Anhut, T. Opatrný, L. Knöll, and D.-G. Welsch. Generating Schrödinger-cat-like states by means of conditional measurements on a beam splitter. *Phys. Rev. A*, 55:3184–3194, 1997.

[16] D. Bouwmeester, J.-W. Pan, K. Mattle, M. Eibl, H. Weinfurter, and A. Zeilinger. Experimental quantum teleportation. *Nature*, 390:575–579, 1997.

[17] D. T. Smithey, M. Beck, J. Cooper, and M. G. Raymer. Measurement of number-phase uncertainty relations of optical fields. *Phys. Rev. A*, 48:3159–3167, 1993.
[18] A. I. Lvovsky and M. G. Raymer. Continuous-variable optical quantum-state tomography. *Rev. Mod. Phys.*, 81:299–332, 2009.
[19] F. S. Roux. Distortions produced in optical homodyne tomography. *Phys. Rev. A*, 106:023713, 2022.
[20] G. T. Herman. *Image Reconstruction from Projections: The Fundamentals of Computerized Tomography*. Academic Press, New York, 1980.

Part III: **Parametric down-conversion**

8 Modelling parametric down-conversion

8.1 Introduction

Parametric down-conversion is a nonlinear optical process where photons in an incoming *pump beam* are converted into pairs of outgoing photons, called the *signal* and *idler* photons, while conserving energy and momentum. The *conservation of energy* implies that the angular frequencies of the signal and idler photons add up to that of the pump photon. Likewise the *conservation of momentum* means that the vector sum of the signal and idler wavevectors produces the pump wavevector. The restrictions imposed by energy and momentum conservation do not fully constrain the wavevectors and angular frequencies of the outgoing photons. Therefore, the superposition of all possible combinations that satisfy these constraints is produced, leading to *entanglement* in the *spatiotemporal degrees of freedom*. Considered as a *squeezing process*, parametric down-conversion also leads to *entanglement* in the *particle-number degree of freedom*. In addition, there can also be entanglement in the polarization (*spin* degrees of freedom).

While its ability to produce entangled photonic states is one of the main reasons for the significance of parametric down-conversion in quantum optics and photonic quantum information technology, it features as a key process in various applications due to its versatility. Yet, parametric down-conversion is also one of the most complex processes, which is the reason for its long history of study, as presented by the vast literature on the topic. The reason why it is so complex is partly due to the fact that it is a nonlinear process, but also because it involves all the degrees of freedom of light.

The entanglement in the spatiotemporal degrees of freedom produced in parametric down-conversion [1–3] has been used in many quantum information applications, ranging from quantum key distribution [4–8], quantum teleportation [9, 10], quantum ghost imaging [11, 12], and quantum synchronization protocols [13, 14]. Entanglement in the particle-number degree of freedom, becomes manifest in the production of squeezed states [15–18]. Applications of squeezed states include continuous variable teleportation [19, 20], quantum imaging [21–23], quantum state engineering [24], and quantum metrology [25]. In view of the role of these degrees of freedom in parametric down-conversion, a comprehensive analysis of the process in terms of both spatiotemporal degrees of freedom and particle-number degree of freedom can provide a clearer picture of the process and the states it produces [26].

The involvement of all the degrees of freedom in parametric down-conversion inevitably calls for a functional formalism. The Wigner functional formalism [27–29] that is developed in the preceding chapters is well suited for this purpose. It has been applied to investigate complex scenarios in quantum optics [30, 31].

In this chapter, the Wigner functional formalism is used to derive an *evolution equation* for the down-converted state that is produced by parametric down-conversion. We follow the approach introduced in Section 6.4.8. The idea is to derive an evolution equation that is equivalent to the Schrödinger equation for evolution along a spatial propaga-

https://doi.org/10.1515/9783111445342-011

tion direction. The first order of business is to identify the correct *propagation operator*, which is the generator of evolution along a spatial direction, analogues to the Hamiltonian for evolution in time. This propagation operator is determined by comparing the quantized version of the classical equation of motion to a formal expression of the evolution equation in terms of an ansatz for the propagation operator. It requires that we first derive the classical equation of motion for the process, the classical nonlinear wave equation. Equipped with the propagation operator, we then follow the procedure in Section 6.4.8 to derive the evolution equation for parametric down-conversion.

8.1.1 Nonlinear processes

The calculations that have been provided so far in the preceding chapters in general only dealt with linear non-interacting scenarios. Here, we introduce interactions in the form of the nonlinear parametric process. It provides an opportunity to demonstrate the use of functional methods to deal with the challenges.

Without interactions, the calculations of predictions are relatively easy and generally doable. When interactions are introduced, such calculations become more challenging and may even become completely intractable. In general, calculations in the presence of interactions can be done provided that such interactions are weak enough to allow a perturbative approach.

In particle physics, quantum field theory calculations are often performed with the aid of *perturbation theory* [32]. Such calculations of predictions are done with remarkable success when compared with experimental results [33]. It is assumed that the coupling constant associated with the interaction term is small. In theoretical particle physics, such a perturbative approach is used to expand the calculation in progressively more complicated *Feynman diagrams* representing higher-order contributions to the process under investigation. However, it does not mean that such calculations are simple or easy. Often the perturbative approach leads to higher-order terms with logarithmic divergences that need special treatment in terms of renormalization procedures [32] to obtain finite results.

The success of quantum field theory for applications in particle physics is to some extent a consequence of its approach for quantizing an interaction theory: *quantize as a free field theory* and *add interactions perturbatively*. What it means is that the interaction terms in the Lagrangian density or Hamiltonian density are modelled in terms of the *free field operators*. Such a formulation represents the *interaction picture* [32]. Here, we follow the same philosophy. Although other approaches have been proposed, we exclusively adhere to the successful recipe of perturbative quantum field theory.

The scenarios in quantum optics that are considered here, do not present the same challenges as found in particle physics. When we perform calculations in the presence of interactions, as found in parametric processes, the basic interaction is extremely weak. It thus allows a perturbative approach. Moreover, the higher-order contributions do not produce the same logarithmic divergences found in particle physics for reasons explained below. However, the preparation of high fidelity squeezed states requires intensive pumping of the nonlinear medium to increase the efficiency of the down-conversion process through *bosonic enhancement*. In such cases, the effective coupling constant can become large enough to invalidate the conditions for a perturbative expansion.

Fortunately, there is another way to introduce perturbation theory that is valid for highly efficient parametric down-conversion [34]. Even when such interactions are enhanced with strong pump fields, we can still use a perturbative approach, namely the *strong-field perturbative approach*. Assuming that the pump is represented by an intense coherent state, the width of its Wigner functional on phase space represents a minimum uncertainty width that is much smaller than its distance from the origin. The relative width of the pump's Wigner functional serves as the expansion parameter in this alternative perturbative analysis. Such a perturbative expansion becomes more accurate as the intensity of the pump increases, so much so that anything beyond the leading order is so severely suppressed that it can be safely discarded. This leading order is equivalent to the *semiclassical approximation* (also called the *parametric approximation*) that is often used in analyses of parametric down-conversion.

Under the semiclassical approximation, the interactions become bilinear (a vertex with just two lines like a mass-term) and are therefore unable to form loops. Hence, the calculated results do not lead to logarithmic divergences as in particle physics because the strong-field perturbative approach leads to the semiclassical approximation, which removes all loops in higher-order terms.[a]

There are scenarios where the semiclassical approximation is not suitable, for example when the pump field is depleted. In such cases, a strong-field perturbative approach beyond the leading order is required. Without the semiclassical approximation, the interaction terms contain products of three ladder operators. Commutations among such terms produce terms with products of even more ladder operators. It thus leads to an endless cascade of products of ladder operators. Fortunately, such higher-order terms are still suppressed due to the weakness of the unenhanced nonlinear process. Therefore, the series can be truncated to produce reasonably accurate results.

Analyses of the parametric down-conversion process often impose a variety of assumptions and approximations, in addition to the semiclassical approximation, to make them tractable [16, 17, 35, 36]. Among the other approximations are the *thin-crystal approximation* and the *plane wave approximation*, exploiting the typical experimental conditions for parametric down-conversion applications. Another approximation that we already introduce in Chapter 2 is the *paraxial approximation*. All these approximations are *controlled* in that the extent of their validity can be assessed in terms of the experimental conditions.

---

a Sometimes such loops can appear when different types of fields (represented by different field variables) are considered, but in those cases, the vertices are unenhanced and can be discarded based on being higher-order weak field perturbations.

8.2 Classical nonlinear wave equation

The starting point of our development of the parametric process is to identify the propagation operator that represents the nonlinear process. For this purpose, we start with the classical theory, deriving the classical *nonlinear wave equation* under the *paraxial approximation.* There are different approaches that can be followed to derive the nonlinear wave equation. The approach we follow here is to expand the electric flux density **D** in terms of the electric field **E** for a nonlinear medium, as given in (2.55), and then substitute it into Maxwell's equations.

The expression of the flux density in (2.55) implies a certain definition of the *susceptibility tensors.* In a derivation that starts from an assumed form of the Lagrangian density to which the *Euler–Lagrange equation* is applied, the terms in the equivalent expression for the flux density contain *symmetry factors*, thus implying redefinitions of the susceptibility tensors. For example, the second-order nonlinear term becomes $\chi^{(2)}_{abc} \rightarrow 3\chi^{(2)}_{abc}$. Here, we use the definitions without these symmetry factors.

8.2.1 Nonlinear Maxwell equations

Up to the second-order nonlinearity, the flux density is represented according to (2.55) in vector notation and tensor notation as

$$\mathbf{D} = \vec{e}_a D_a = \epsilon \cdot \mathbf{E} + \epsilon_0 \mathbf{E} \cdot \chi \cdot \mathbf{E} = \vec{e}_a \epsilon_{ab} E_b + \epsilon_0 \vec{e}_a \chi^{(2)}_{abc} E_b E_c. \tag{8.1}$$

Here, $\vec{e}_a$ denotes the unit vectors for the coordinate system.[b] The dot-product in the first term of the vector notation expression indicates that the medium is *anisotropic* and produces polarization effects of the medium such as *retardance* and *walk-off*, discussed in Section 2.9.4. For a birefringent crystal, the term separates into two terms for the *ordinary* and *extraordinary* waves, respectively. The state of polarization of the incident field in such anisotropic nonlinear media plays a significant role in the *phase-matching conditions.* For critical phase-matching conditions, such as type I or type II phase-matching conditions, the incident field is an extraordinary wave, which typically experiences some walk-off during propagation through the anisotropic crystal. These phase-matching conditions are discussed below in Section 8.6.2.

Due to the complexity of the nonlinear term, it is more appropriate to express the Maxwell equations in tensor notation. The source-free version of the second Maxwell equation (2.45) then becomes

$$\varepsilon_{abc} \partial_b H_c = \epsilon_{ab} \partial_t E_b + 2\epsilon_0 \chi_{abc} E_b \partial_t E_c, \tag{8.2}$$

b In this expression, these unit vectors are required to make the expression formally equivalent to the expression in vector notation. However, these unit vectors are removed from any subsequent expressions that only appear in tensor notation.

where ε_{abc} is the *totally antisymmetric tensor*, discussed in Section 2.8.1, and

$$\partial_b \triangleq \frac{\partial}{\partial x^b}. \tag{8.3}$$

The source-free version of the third Maxwell equation (2.46) is

$$\partial_a \epsilon_{ab} E_b + 2\epsilon_0 \chi_{abc} E_b \partial_a E_c = 0. \tag{8.4}$$

In Section 2.9.4, it is shown how the linear versions of these two equations lead to the separation of the field into *ordinary* and *extraordinary* waves. For the nonlinear case, these two waves are coupled so that one generates the other via the nonlinearity.

8.2.2 Nonlinear Helmholtz equation

8.2.2.1 Phasor fields

The electric field in the four Maxwell equations is converted to a *phasor field* by expressing it in terms of an inverse temporal Fourier transform, as shown in Section 2.3.2. Then the equations are Fourier transformed with respect to time. The resulting equations contain integrals over frequency for the nonlinear terms.

When the medium is *linear* (and source free), each frequency component of an electromagnetic field is treated separately, since the different frequency components do not couple to one another. For each frequency component, the electric field is expressed as a complex-valued vector field that only depends on the spatial coordinates. The time dependence is given by a factor of $\exp(-i\omega t)$, with ω being the *angular frequency* of the monochromatic field. For convenience, we remove this time-dependent exponential factor by representing the electric field in terms of a *phasor* field. The physical electric field is the real part of the phasor field times the time-dependent phase factor, given in (2.71), where $\tilde{\mathbf{E}}(\mathbf{x})$ represents the electric phasor field.

In a nonlinear medium, we can represent the electric field in terms of phasor fields, but we need to integrate over all the frequencies, because the nonlinear coupling causes terms with different frequencies. It can be expressed without the $\mathrm{Re}\{\cdot\}$, provided that $\tilde{\mathbf{E}}(\mathbf{x}, -\omega) = \tilde{\mathbf{E}}^*(\mathbf{x}, \omega)$. Hence,

$$\mathbf{E}(\mathbf{x}, t) = \int \tilde{\mathbf{E}}(\mathbf{x}, \omega) \exp(-i\omega t) \frac{d\omega}{2\pi}. \tag{8.5}$$

The same is done for the magnetic field $\tilde{\mathbf{H}}(\mathbf{x}, \omega)$.

We express the fields in Maxwell's equations in terms of their phasor fields and then perform Fourier transforms with respect to time to remove the time dependent exponential factors. It also removes one integration over frequency. The resulting equations in tensor notation read

$$\varepsilon_{abc}\partial_b\widetilde{E}_c(\mathbf{x},\omega) = \mathrm{i}\omega\mu_0\widetilde{H}_a(\mathbf{x},\omega), \tag{8.6}$$

$$\begin{aligned}\varepsilon_{abc}\partial_b\widetilde{H}_c(\mathbf{x},\omega) = & -\mathrm{i}\omega\epsilon_{ab}\widetilde{E}_b(\mathbf{x},\omega)\\ & -\mathrm{i}2\epsilon_0\chi_{abc}\int\omega'\widetilde{E}_b(\mathbf{x},\omega-\omega')\widetilde{E}_c(\mathbf{x},\omega')\,\frac{\mathrm{d}\omega'}{2\pi},\end{aligned} \tag{8.7}$$

$$\partial_a\epsilon_{ab}E_b(\mathbf{x},\omega) = -2\epsilon_0\chi_{abc}\int\widetilde{E}_b(\mathbf{x},\omega')\partial_a\widetilde{E}_c(\mathbf{x},\omega-\omega')\,\frac{\mathrm{d}\omega'}{2\pi}, \tag{8.8}$$

$$\partial_a\widetilde{H}_a(\mathbf{x},\omega) = 0. \tag{8.9}$$

8.2.2.2 Nonlinear Helmholtz equation

The nonlinear Helmholtz equation is derived from the nonlinear Maxwell equations in phasor form following almost the same approach used in Section 2.3.3. First, we compute the curl of (8.6) and substitute (8.7) into it, to get

$$\begin{aligned}\varepsilon_{abc}\partial_b\varepsilon_{cde}\partial_d\widetilde{E}_e(\mathbf{x},\omega) &= \mathrm{i}\omega\mu_0\varepsilon_{abc}\partial_b\widetilde{H}_c(\mathbf{x},\omega)\\ &= \omega^2\mu_0\epsilon_{ab}\widetilde{E}_b(\mathbf{x},\omega)\\ &\quad + \frac{2\omega}{c^2}\chi_{abc}\int\omega'\widetilde{E}_b(\mathbf{x},\omega-\omega')\widetilde{E}_c(\mathbf{x},\omega')\,\frac{\mathrm{d}\omega'}{2\pi},\end{aligned} \tag{8.10}$$

where we used $c^2\mu_0\epsilon_0 = 1$. From the identity,

$$\varepsilon_{abc}\varepsilon_{cde} = \delta_{ad}\delta_{be} - \delta_{ae}\delta_{bd}, \tag{8.11}$$

it follows that

$$\begin{aligned}&\partial_a\partial_b\widetilde{E}_b(\mathbf{x},\omega) - \partial_b\partial_b\widetilde{E}_a(\mathbf{x},\omega) - \omega^2\mu_0\epsilon_{ab}\widetilde{E}_b(\mathbf{x},\omega)\\ &= \frac{2\omega}{c^2}\chi_{abc}\int\omega'\widetilde{E}_b(\mathbf{x},\omega-\omega')\widetilde{E}_c(\mathbf{x},\omega')\,\frac{\mathrm{d}\omega'}{2\pi}.\end{aligned} \tag{8.12}$$

The equation is converted to a more familiar form where the left-hand side is expressed in terms of vector notation

$$\nabla^2\widetilde{\mathbf{E}}(\mathbf{x},\omega) - \nabla\left[\nabla\cdot\widetilde{\mathbf{E}}(\mathbf{x},\omega)\right] + \frac{k_0^2}{\epsilon_0}\widetilde{\mathbf{D}}(\mathbf{x},\omega) = \text{nonlinear term}, \tag{8.13}$$

with $k_0 = \omega/c$ being the vacuum wavenumber, and the nonlinear term is expressed in terms of tensor notation as

$$\text{nonlinear term} \triangleq -2k_0^2\vec{e}_a\chi_{abc}\int\frac{\omega'}{\omega}\widetilde{E}_b(\mathbf{x},\omega-\omega')\widetilde{E}_c(\mathbf{x},\omega')\,\frac{\mathrm{d}\omega'}{2\pi}, \tag{8.14}$$

with the unit vectors for the coordinate system $\vec{e}_a$ included to match the vector notation of the left-hand side.

We now consider the ordinary and extraordinary polarized fields, as discussed in Section 2.9.4. Each field is expressed as a plane wave expansion in terms of three-dimensional integrals over the wavevectors. Here, and for the duration of this derivation, the wavevectors denoted by $\mathbf{k}$ represent those that are associated with a specific dielectric constant or refractive index of the dielectric medium, and are related to the vacuum wavevectors $\mathbf{k}_0$ by $\mathbf{k} = n\mathbf{k}_0$, where n is the pertinent refractive index. For the ordinary polarized fields, we have

$$\widetilde{\mathbf{E}}_o(\mathbf{x},\omega) \triangleq \int F_o(\mathbf{k},\omega)\vec{s}(\mathbf{k}) \exp(\mathrm{i}\mathbf{k}\cdot\mathbf{x}) \frac{\mathrm{d}^3k}{(2\pi)^2}, \tag{8.15}$$

where $\vec{s}(\mathbf{k})$ is defined in (2.229). The flux density phasor is related by

$$\widetilde{\mathbf{D}}_o \triangleq \epsilon_0 n_o^2 \widetilde{\mathbf{E}}_o, \tag{8.16}$$

where n_o is the ordinary refractive index. The left-hand side in (8.13) becomes

$$\begin{aligned}&\int F_o(\mathbf{k},\omega)\left[-|\mathbf{k}|^2\vec{s} - \mathrm{i}\mathbf{k}\,(\mathrm{i}\mathbf{k}\cdot\vec{s}) + k_0^2 n_o^2\vec{s}\right] \exp(\mathrm{i}\mathbf{k}\cdot\mathbf{x}) \frac{\mathrm{d}^3k}{(2\pi)^2}\\ &= \nabla^2\widetilde{\mathbf{E}}_o + k_0^2 n_o^2 \widetilde{\mathbf{E}}_o,\end{aligned} \tag{8.17}$$

because $\mathbf{k}\cdot\vec{s} = 0$. For the extraordinary polarized fields, we have

$$\widetilde{\mathbf{E}}_e(\mathbf{x},\omega) \triangleq \int F_e(\mathbf{k},\omega)\left[\vec{p}(\mathbf{k}) + \Delta_X(\mathbf{k})\vec{k}(\mathbf{k})\right] \exp(\mathrm{i}\mathbf{k}\cdot\mathbf{x}) \frac{\mathrm{d}^3k}{(2\pi)^2}, \tag{8.18}$$

from the definitions in (2.229) and (2.230). In this case, D_e is related to the component of the extraordinary electric field perpendicular to $\mathbf{k}$. Hence,

$$\widetilde{\mathbf{D}}_e = \epsilon_0 n_{\text{neff}}^2 \int F_e(\mathbf{k},\omega)\vec{p}(\mathbf{k}) \exp(\mathrm{i}\mathbf{k}\cdot\mathbf{x}) \frac{\mathrm{d}^3k}{(2\pi)^2} \triangleq \epsilon_0 n_{\text{neff}}^2 \widetilde{\mathbf{E}}_{e\perp}, \tag{8.19}$$

where n_{neff} is the effective refractive index defined in (B.1) and $\widetilde{\mathbf{E}}_{e\perp}$ is the part of $\widetilde{\mathbf{E}}_e$ that is orthogonal to $\vec{k}(\mathbf{k})$. Substituted into the left-hand side of (8.13), it then leads to

$$\begin{aligned}&\int F_e(\mathbf{k},\omega)\left[-|\mathbf{k}|^2\vec{p}(\mathbf{k}) + k_0^2 n_{\text{neff}}^2\vec{p}(\mathbf{k})\right] \exp(\mathrm{i}\mathbf{k}\cdot\mathbf{x}) \frac{\mathrm{d}^3k}{(2\pi)^2}\\ &= \nabla^2\widetilde{\mathbf{E}}_{e\perp} + k_0^2 n_{\text{eff}}^2 \widetilde{\mathbf{E}}_{e\perp},\end{aligned} \tag{8.20}$$

because the component along $\vec{k}$ is removed.

The differential operators remove the components in the direction of the wavevector inside the integral. So, in both cases the left-hand side takes on the usual form of the Helmholtz equation. The differences are that the refractive indices are different for the two fields, and in the case of the extraordinary polarized fields, the equation only involves the part of the field that consists of plane waves with components orthogonal to the wavevector.

For the nonlinear term, we combine the two field components to become

$$\begin{aligned}\tilde{E}_a(\mathbf{x},\omega) &= \int \Big[F_{\mathrm{e}}(\mathbf{k},\omega)\vec{p}(\mathbf{k}) + \Delta_X(\mathbf{k})F_{\mathrm{e}}(\mathbf{k},\omega)\vec{k}(\mathbf{k}) \\ &\quad + F_{\mathrm{o}}(\mathbf{k},\omega)\vec{s}(\mathbf{k})\Big]_a \exp(\mathrm{i}\mathbf{k}\cdot\mathbf{x})\,\frac{\mathrm{d}^3k}{(2\pi)^2} \\ &\approx \int [F_{\mathrm{e}}(\mathbf{k},\omega)\vec{p}(\mathbf{k}) + F_{\mathrm{o}}(\mathbf{k},\omega)\vec{s}(\mathbf{k})]_a \exp(\mathrm{i}\mathbf{k}\cdot\mathbf{x})\,\frac{\mathrm{d}^3k}{(2\pi)^2} \\ &\triangleq \tilde{E}_a^{(\mathrm{e})}(\mathbf{x},\omega) + \tilde{E}_a^{(\mathrm{o})}(\mathbf{x},\omega),\end{aligned} \tag{8.21}$$

where the index a represents the different components extracted from the vector expression in the square brackets. Since $\Delta_X(\mathbf{k})$ is small compared to the other components, we discard it. It gives rise to the walk-off effect that we saw in Section 2.9.4, but its contribution to the nonlinear process is small. So, we express the nonlinear term as

$$\begin{aligned}\text{nonlinear term} = &-2k_0^2\vec{e}_a\chi_{abc}\int \frac{\omega'}{\omega}\left[\tilde{E}_b^{(\mathrm{o})}(\mathbf{x},\omega-\omega') + \tilde{E}_b^{(\mathrm{e})}(\mathbf{x},\omega-\omega')\right] \\ &\times \left[\tilde{E}_c^{(\mathrm{o})}(\mathbf{x},\omega) + \tilde{E}_c^{(\mathrm{e})}(\mathbf{x},\omega)\right]\frac{\mathrm{d}\omega'}{2\pi}.\end{aligned} \tag{8.22}$$

Expanded, it produces four terms. However, not all the terms can produce physical fields because they do not (or cannot) satisfy the phase-matching conditions. Different phase-matching conditions based on the polarization of the different fields that take part in the process are discussed in Section 8.6.2.

8.2.2.3 Scalar fields

For convenience, we assume that a given phase-matching condition is imposed by using fields with the necessary polarizations. The equation is then converted into a scalar equation by selecting the specific states of polarization that satisfy a given phase-matching condition. In other words, we assume that all fields can be represented as

$$\vec{e}_a\tilde{E}_a(\mathbf{x},\omega) = \tilde{\mathbf{E}}(\mathbf{x},\omega) = \vec{\eta}\tilde{E}(\mathbf{x},\omega), \tag{8.23}$$

in terms of scalar fields $\tilde{E}(\mathbf{x},\omega)$ times the appropriate polarization vectors $\vec{\eta}$. Then we compute the inner product of the equation with the complex conjugate of the polarization vector of the input field. The result is the *nonlinear scalar Helmholtz equation*:

$$\nabla^2\tilde{E} + k_0^2n^2\tilde{E} = -k_0^2\frac{d_{\mathrm{pmc}}}{\epsilon_0}\int \frac{\omega'}{\omega}\tilde{E}(\mathbf{x},\omega-\omega')\tilde{E}(\mathbf{x},\omega')\,\frac{\mathrm{d}\omega'}{2\pi}, \tag{8.24}$$

where the scalar field $\tilde{E}$ can either be the ordinary or extraordinary wave (those inside the integral are not necessarily the same), n is the appropriate refractive index, and

$$d_{\text{pmc}} \triangleq 2\epsilon_0 \chi_{abc}^{(2)} \eta_a^{(\text{s})} \eta_b^{(\text{i})} \eta_c^{(\text{p})*}, \tag{8.25}$$

is the *nonlinear coefficient* for a given phase-matching condition. It comprises the contraction of the polarization vectors of the pump, signal and idler beams on the susceptibility tensor of the nonlinear medium. The identity of the different fields as being either ordinary or extraordinary waves becomes clear later within the context of a given phase-matching condition. Until then, we treat these fields generically.

8.2.3 Nonlinear paraxial wave equation

A paraxial expansion, discussed in Section 2.3.5, is applied to the nonlinear Helmholtz equation, retaining only terms up to sub-leading order, keeping in mind that the susceptibility tensor χ introduces a significant suppression of its own. Therefore, only the leading order is retained in the nonlinear term. The result is the nonlinear paraxial wave equation. Under the paraxial approximation, the scalar phasor fields are represented as in (2.83). When substituted into the left-hand side of (8.24), discarding the second derivative in z, it produces the expression on the left-hand side of (2.91).

The paraxial approximation produces a z-dependent phase factor in the nonlinear term. The complete nonlinear equation then becomes

$$\begin{aligned} &\nabla_{xy}^2 \bar{E}(\mathbf{x}, \omega) + \text{i}2k\partial_z \bar{E}(\mathbf{x}, \omega) \\ &= -k_0^2 \frac{d_{\text{pmc}}}{\epsilon_0} \int \frac{\omega_1}{\omega} \exp(-\text{i}\Delta kz) \bar{E}(\mathbf{x}, \omega - \omega_1) \bar{E}(\mathbf{x}, \omega_1) \frac{\text{d}\omega_1}{2\pi}, \end{aligned} \tag{8.26}$$

where $k = nk_0$, and all the z-dependent phase factors are moved to the right-hand side to produce a combined phase factor with

$$\Delta k = k - k_1 - k_2, \tag{8.27}$$

where the wavenumbers k_1 and k_2 are associated with the two fields in the nonlinear term, respectively. The result is the *nonlinear scalar paraxial wave equation*. At this point, the z-dependent phase factor is expressed in terms of the wavenumbers, as expected for the paraxial approximation. This phase factor tends to reduce the efficiency of the down-conversion process. For *critical phase-matching*, the wavenumber mismatch in (8.27) becomes zero, so that the z-dependent phase factor is removed from the nonlinear term, leading to efficient down-conversion. In other phase-matching conditions (quasi phase-matching) more sophisticated techniques are used (see Section 8.6.2).

8.2.4 Classical evolution equation

8.2.4.1 Transverse Fourier transform

The paraxial phasor fields in the nonlinear paraxial wave equation are converted to z-dependent angular spectra by expressing the paraxial phasor fields in terms of transverse spatial inverse Fourier transforms. They are given by

$$\bar{E}(\mathbf{x},\omega) = \int F(\mathbf{K},\omega,z)\exp(\mathrm{i}\mathbf{K}\cdot\mathbf{X})\,\frac{\mathrm{d}^2k}{(2\pi)^2}, \tag{8.28}$$

where $F(\mathbf{K},\omega,z)$ is a z-dependent angular spectrum for the phasor field, $\mathbf{K}$ is the two-dimensional transverse part of the wavevector, and $\mathbf{X}$ is the two-dimensional position vector on the transverse plane. The derivation naturally represents these spectra as functions of the optical beam variables $(\mathbf{K},\omega)$, discussed in Sections 2.4.2 and 4.1.5. The left-hand side of the nonlinear paraxial wave equation becomes

$$\begin{aligned}
&\nabla_{xy}^2\bar{E}(\mathbf{x},\omega) + \mathrm{i}2k\partial_z\bar{E}(\mathbf{x},\omega)\\
&= \int\left[-|\mathbf{K}|^2F(\mathbf{K},\omega,z) + \mathrm{i}2k\partial_zF(\mathbf{K},\omega,z)\right]\exp(\mathrm{i}\mathbf{K}\cdot\mathbf{X})\,\frac{\mathrm{d}^2k}{(2\pi)^2},
\end{aligned} \tag{8.29}$$

and the right-hand side becomes

$$\begin{aligned}
&-k_0^2\frac{d_{\mathrm{pmc}}}{\epsilon_0}\int\frac{\omega_1}{\omega}\exp(-\mathrm{i}\Delta kz)\bar{E}(\mathbf{x},\omega-\omega_1)\bar{E}(\mathbf{x},\omega_1)\,\frac{\mathrm{d}\omega_1}{2\pi}\\
&= -k_0^2\frac{d_{\mathrm{pmc}}}{\epsilon_0}\int\frac{\omega_1}{\omega}\exp(-\mathrm{i}\Delta kz)F(\mathbf{K}_2,\omega-\omega_1,z)F(\mathbf{K}_1,\omega_1,z)\\
&\quad\times\exp[\mathrm{i}(\mathbf{K}_1+\mathbf{K}_2)\cdot\mathbf{X}]\,\frac{\mathrm{d}\omega_1}{2\pi}\,\frac{\mathrm{d}^2k_1}{(2\pi)^2}\,\frac{\mathrm{d}^2k_2}{(2\pi)^2}.
\end{aligned} \tag{8.30}$$

Next, we perform a two-dimensional Fourier transform with respect to $\mathbf{X}$ on both sides. The resulting equation reads

$$\begin{aligned}
-\mathrm{i}2k\partial_zF(\mathbf{K},\omega,z) = &-|\mathbf{K}|^2F(\mathbf{K},\omega,z) + k_0^2\frac{d_{\mathrm{pmc}}}{\epsilon_0}\int\frac{\omega'}{\omega}\exp(-\mathrm{i}\Delta kz)\\
&\times F(\mathbf{K}-\mathbf{K}',\omega-\omega',z)F(\mathbf{K}',\omega',z)\,\mathrm{d_b}k',
\end{aligned} \tag{8.31}$$

The integration measure for optical beam variables $\mathrm{d_b}k$ is defined in (2.105).

8.2.4.2 Fixed reference frame

The transverse Fourier transform is performed at the z-location of the field. As a result, the spectrum at each plane is referenced to the z-value of that plane. Instead, we can specify a fixed reference frame for the definition of all the spectra that are computed at arbitrary z-locations. This redefinition is given by

$$F(\mathbf{K},\omega,z) \to G(\mathbf{K},\omega,z)\exp\left(-\frac{\mathrm{i}z}{2k}|\mathbf{K}|^2\right), \tag{8.32}$$

for the reference frame located at $z = 0$. By fixing the reference frame to a common point along the z-axis, we also remove the linear free space propagation term. The effect of this *fixed reference frame* representation is that the spectrum evolves due to the nonlinear process as a function of z, but represents a configuration-space field (via an inverse transverse Fourier transform) that would be located at the z-location of the reference frame. To reproduce the configuration-space field at any other z-location, a free space propagation is performed from the reference frame to that z-location. In terms of the redefined scalar spectral fields, the equation becomes

$$\begin{aligned}-\mathrm{i}2k\partial_z G(\mathbf{K},\omega,z) = k_0^2\frac{d_{\text{pmc}}}{\epsilon_0}\int\frac{\omega'}{\omega}&\exp(-\mathrm{i}\Delta k_z z)G(\mathbf{K}-\mathbf{K}',\omega-\omega',z)\\ &\times G(\mathbf{K}',\omega',z)\,\mathrm{d}_\mathrm{b}k'\end{aligned} \tag{8.33}$$

where Δk_z is formed by combining Δk with the z-dependent phase factors that come from the fixed reference frame representation. Using the paraxial expansion for the z-component, given by

$$k_z(\mathbf{K},\omega) = \sqrt{k^2 - |\mathbf{K}|^2} \approx k - \frac{|\mathbf{K}|^2}{2k}, \tag{8.34}$$

we obtain

$$\begin{aligned}\Delta k_z &= k_z(\mathbf{K},\omega) - k_z(\mathbf{K}',\omega') - k_z(\mathbf{K}-\mathbf{K}',\omega-\omega')\\ &\approx k - k_1 - k_2 - \frac{|\mathbf{K}|^2}{2k} + \frac{|\mathbf{K}'|^2}{2k_1} + \frac{|\mathbf{K}-\mathbf{K}'|^2}{2k_2}.\end{aligned} \tag{8.35}$$

Note that, in the *non-collinear case*, where the phase-matching conditions produce two down-converted photons that do not propagate in the same direction, the paraxial expansions of the fields under the integral are done relative to beam axes that make nonzero angles with respect to the z-axis. As a result, the paraxial expansions of these k_z's contain transverse parts of the beam axes (see Section 8.6.2).

8.2.4.3 Positive angular frequencies

The *angular frequency* is considered to be a positive quantity. Instead of having the integral running over all angular frequencies from $-\infty$ to ∞, we convert the expression so that the angular frequencies are positive. For this purpose, we divide the integration over the angular frequency into three parts

$$\int_{-\infty}^{\infty}\cdots\,\mathrm{d}\omega' = \int_{-\infty}^{0}\cdots\,\mathrm{d}\omega' + \int_{0}^{\omega}\cdots\,\mathrm{d}\omega' + \int_{\omega}^{\infty}\cdots\,\mathrm{d}\omega', \tag{8.36}$$

where ω is the angular frequency of the field external to the integral. Then we convert the angular frequencies in the arguments of the G's to positive angular frequencies with suitable redefinitions of the integration variables and by using the property

$$G(-\mathbf{K},-\omega,z)\exp(\mathrm{i}kz) = G^*(\mathbf{K},\omega,z)\exp(-\mathrm{i}kz). \tag{8.37}$$

We also symmetrize some of the terms to obtain simpler expressions. For simplicity, we express the optical beam variables $(\mathbf{K},\omega)$ as three-dimensional wavevectors $\mathbf{k}$ and remove all other dependencies. Then we have

$$\begin{aligned}
&\int_{-\infty}^{\infty}\frac{\omega'}{\omega}G(\mathbf{k}')G(\mathbf{k}-\mathbf{k}')\,\mathrm{d}\omega' \\
&= \int_{-\infty}^{0}\frac{\omega'}{\omega}G(\mathbf{k}')G(\mathbf{k}-\mathbf{k}')\,\mathrm{d}\omega' + \int_{0}^{\omega}\frac{\omega'}{\omega}G(\mathbf{k}')G(\mathbf{k}-\mathbf{k}')\,\mathrm{d}\omega' \\
&+ \int_{\omega}^{\infty}\frac{\omega'}{\omega}G(\mathbf{k}')G(\mathbf{k}-\mathbf{k}')\,\mathrm{d}\omega' \\
&= \frac{1}{2}\int_{0}^{\omega}G(\mathbf{k}')G(\mathbf{k}-\mathbf{k}')\,\mathrm{d}\omega' + \int_{0}^{\infty}G^*(\mathbf{k}')G(\mathbf{k}+\mathbf{k}')\,\mathrm{d}\omega.
\end{aligned} \tag{8.38}$$

Hence, the complete expression in optical beam variables becomes

$$\begin{aligned}
-\mathrm{i}2k\partial_z G(\mathbf{K},\omega,z) = {}& k_0^2\frac{d_{\mathrm{pmc}}}{2\epsilon_0}\int\int_0^{\omega}\exp(-\mathrm{i}\Delta k_z^{(1)}z)G(\mathbf{K}',\omega',z) \\
&\times G(\mathbf{K}-\mathbf{K}',\omega-\omega',z)\,\frac{\mathrm{d}\omega'}{2\pi}\frac{\mathrm{d}^2k'}{(2\pi)^2} \\
&+ k_0^2\frac{d_{\mathrm{pmc}}}{\epsilon_0}\int\int_0^{\infty}\exp(-\mathrm{i}\Delta k_z^{(2)}z)G^*(\mathbf{K}',\omega',z) \\
&\times G(\mathbf{K}+\mathbf{K}',\omega+\omega',z)\,\frac{\mathrm{d}\omega'}{2\pi}\frac{\mathrm{d}^2k'}{(2\pi)^2},
\end{aligned} \tag{8.39}$$

where

$$\begin{aligned}
\Delta k_z^{(1)} &= k_z(\mathbf{k}) - k_z(\mathbf{k}') - k_z(\mathbf{k}-\mathbf{k}'), \\
\Delta k_z^{(2)} &= k_z(\mathbf{k}) + k_z(\mathbf{k}') - k_z(\mathbf{k}+\mathbf{k}').
\end{aligned} \tag{8.40}$$

Exercise 8.1. Show that the final result in (8.38) follows from the three separate terms.

8.2.4.4 Separate equations

The two nonlinear terms in (8.39) represent different scenarios. In the first nonlinear term, the angular frequencies in the arguments of the G's add up to ω and they are thus smaller than ω. This situation is what we expect for the two *down-converted fields* with the incoming field being the *pump field*. In the second nonlinear term, the angular frequency of the field that is not complex conjugated is equal to the sum of the angular frequency of the complex conjugated field and that of the incoming field. This field thus represents an *upconverted field*. One can separate the equation into separate equations for the two scenarios, given by

$$\begin{aligned} -\mathrm{i}2k_{\mathrm{p}}\partial_z G_{\mathrm{p}}(\mathbf{K},\omega,z) &= k_0^2 \frac{d_{\mathrm{pmc}}}{2\epsilon_0} \int\int_0^{\omega} \exp(-\mathrm{i}\Delta k_z^{(1)} z) G_{\mathrm{d}}(\mathbf{K}',\omega',z) \\ &\quad \times G_{\mathrm{d}}(\mathbf{K}-\mathbf{K}',\omega-\omega',z)\, \frac{\mathrm{d}\omega'}{2\pi} \frac{\mathrm{d}^2k'}{(2\pi)^2}, \\ -\mathrm{i}2k_{\mathrm{d}}\partial_z G_{\mathrm{d}}(\mathbf{K},\omega,z) &= k_0^2 \frac{d_{\mathrm{pmc}}}{\epsilon_0} \int\int_0^{\infty} \exp(-\mathrm{i}\Delta k_z^{(2)} z) G_{\mathrm{d}}^*(\mathbf{K}',\omega',z) \\ &\quad \times G_{\mathrm{p}}(\mathbf{K}+\mathbf{K}',\omega+\omega',z)\, \frac{\mathrm{d}\omega'}{2\pi} \frac{\mathrm{d}^2k'}{(2\pi)^2}, \end{aligned} \tag{8.41}$$

where we indicate the pump and down-converted fields by subscripts p and d, respectively, and k_{p} and k_{d} are the wavenumbers for the pump and down-converted fields, respectively. They are related to the vacuum wavenumber by $k_{\mathrm{p}} = n_{\mathrm{p}}k_0$ and $k_{\mathrm{d}} = n_{\mathrm{d}}k_0$, where n_{p} and n_{d} are the refractive indices for the pump and down-converted fields, respectively. We can distinguish between the ordinary and extraordinary fields, as required for the phase-matching conditions. For example, in the case of type I phase-matching conditions, we have $n_{\mathrm{p}} = n_{\mathrm{eff}}(\omega_{\mathrm{p}})$ and $n_{\mathrm{d}} = n_{\mathrm{o}}(\omega_{\mathrm{d}})$. Henceforth, type I phase-matching conditions are assumed. The nonlinear coefficient then becomes $d_{\mathrm{pmc}} = d_{\mathrm{I}}$.

8.2.4.5 Combined calculation

The expression for the scalar classical field is obtained from the original electric field by

$$G(\mathbf{K},\omega,z) = \int \vec{\eta}^* \cdot \mathbf{E}(\mathbf{x},t) \exp\left(\mathrm{i}\omega t - \mathrm{i}\mathbf{K}\cdot\mathbf{X} - \mathrm{i}kz + \frac{\mathrm{i}z}{2k}|\mathbf{K}|^2\right)\, \mathrm{d}t\, \mathrm{d}^2x. \tag{8.42}$$

The z-dependent terms in the exponent represent the paraxial expansion of k_z. We can recombine them into k_z and represent the full expression as

$$G(\mathbf{K},\omega,z) = \int \vec{\eta}^* \cdot \mathbf{E}(\mathbf{x},t) \exp\left(\mathrm{i}\omega t - \mathrm{i}\mathbf{k}\cdot\mathbf{x}\right)\, \mathrm{d}t\, \mathrm{d}^2x. \tag{8.43}$$

It is the three-dimensional Fourier transform of a given polarization component of the electric field over time and the transverse spatial coordinates to produce an angular spectrum in optical beam variables as a function of z.

8.3 Propagation operator

The quantum nonlinear wave equations are obtained from the classical nonlinear equations in (8.41) by replacing the classical fields by equivalent quantum field operators, as was done in Section 6.4.8. The equivalent field operators are defined in (6.152) for a generic ladder operator commutation relation given in (6.150). These field operators are defined in the (spatial) *interaction picture*, which is equivalent to the (spatial) *Heisenberg picture* for the non-interacting theory. The resulting field operators are obtained by applying (8.43) to the quantized electric field in (4.46), but for the *commutation relation* given in (6.150).

Replacing $G(\mathbf{k}, z) \to \hat{G}(\mathbf{k}, z)$ in (8.41), we obtain

$$
\begin{aligned}
-\mathrm{i}2k_{\mathrm{p}}\partial_z\hat{G}_{\mathrm{p}}(\mathbf{K}, \omega, z) = k_0^2\frac{d_{\mathrm{I}}}{2\epsilon_0}\int\int_0^{\omega} &\exp(-\mathrm{i}\Delta k_z^{(1)}z)\hat{G}_{\mathrm{d}}(\mathbf{K}', \omega', z) \\
&\times \hat{G}_{\mathrm{d}}(\mathbf{K}-\mathbf{K}', \omega-\omega', z)\,\frac{\mathrm{d}\omega'}{2\pi}\,\frac{\mathrm{d}^2k'}{(2\pi)^2},
\end{aligned} \tag{8.44}
$$

and

$$
\begin{aligned}
-\mathrm{i}2k_{\mathrm{d}}\partial_z\hat{G}_{\mathrm{d}}(\mathbf{K}, \omega, z) = k_0^2\frac{d_{\mathrm{I}}}{\epsilon_0}\int\int_0^{\infty} &\exp(-\mathrm{i}\Delta k_z^{(2)}z)\hat{G}_{\mathrm{d}}^{*}(\mathbf{K}', \omega', z) \\
&\times \hat{G}_{\mathrm{p}}(\mathbf{K}+\mathbf{K}', \omega+\omega', z)\,\frac{\mathrm{d}\omega'}{2\pi}\,\frac{\mathrm{d}^2k'}{(2\pi)^2}.
\end{aligned} \tag{8.45}
$$

These equations are now compared with the quantum propagation equation for operators in (3.237) into which we substituted a suitable ansatz for the propagation operator. The resulting commutation is evaluated using (6.153), but with ϵ_0 replaced by ϵ for the appropriate medium. Note that

$$
\frac{k_0^2}{\epsilon_0} = \frac{k_{\mathrm{p}}^2}{\epsilon_{\mathrm{p}}} = \frac{k_{\mathrm{d}}^2}{\epsilon_{\mathrm{d}}}, \tag{8.46}
$$

which follows by multiplying above and below by the square of the appropriate refractive index. The comparisons between these equations then reveal the expression for the propagation operator.

For the sake of simpler expressions, the optical beam variables $(\mathbf{K}, \omega)$ are henceforth often represented by $\mathbf{k}$. The measure $\mathrm{d}_{\mathrm{b}}k$, defined in (2.105), serves as a reminder that the quantities are functions of the optical beam variables. The integral signs are also combined into one integral sign without integration boundaries. It is henceforth understood that the integrations over ω have different ranges, as shown in (8.44) and (8.45), depending on whether it is associated with down-converted fields or pump fields.

The ansatz for the propagation operator in the fixed reference frame only consists of the *vertex rule* (for parametric down-conversion) plus its Hermitian conjugate. It reads

$$\begin{aligned}\hat{P}_{\Delta} = \int \Big[&\hat{G}_{\mathrm{d}}^{\dagger}(\mathbf{k}_1, z)\hat{G}_{\mathrm{d}}^{\dagger}(\mathbf{k}_2, z)V(\mathbf{k}, \mathbf{k}_1, \mathbf{k}_2, z)\hat{G}_{\mathrm{p}}(\mathbf{k}, z) \\ &+ \hat{G}_{\mathrm{p}}^{\dagger}(\mathbf{k}, z)V^*(\mathbf{k}, \mathbf{k}_1, \mathbf{k}_2, z)\hat{G}_{\mathrm{d}}(\mathbf{k}_1, z)\hat{G}_{\mathrm{d}}(\mathbf{k}_2, z)\Big]\, \mathrm{d}_{\mathrm{b}}k\, \mathrm{d}_{\mathrm{b}}k_1\, \mathrm{d}_{\mathrm{b}}k_2,\end{aligned} \tag{8.47}$$

where V and its complex conjugate are *vertex kernels* for the nonlinear process.

We now substitute (8.47) into the quantum propagation equation for the field operators in (3.237). The commutation relation in (6.153), with $\epsilon_0 \to \epsilon_{\mathrm{p}}$ due to the dielectric medium, leads to

$$\begin{aligned}\partial_z\hat{G}_{\mathrm{p}}(\mathbf{k}, z) &= \frac{\mathrm{i}}{\hbar}\left[\hat{G}_{\mathrm{p}}(\mathbf{k}, z), \hat{P}_{\Delta}\right] \\ &= \frac{\mathrm{i}k}{2\epsilon_{\mathrm{p}}}\int V^*(\mathbf{k}, \mathbf{k}_1, \mathbf{k}_2, z)\hat{G}_{\mathrm{d}}(\mathbf{k}_1, z)\hat{G}_{\mathrm{d}}(\mathbf{k}_2, z)\, \mathrm{d}_{\mathrm{b}}k_1\, \mathrm{d}_{\mathrm{b}}k_2,\end{aligned} \tag{8.48}$$

under the paraxial approximation ($k_z = k$). In a similar way, the commutation with $\hat{G}_{\mathrm{d}}(\mathbf{K}, \omega, z)$ produces

$$\begin{aligned}\partial_z\hat{G}_{\mathrm{d}}(\mathbf{k}_1, z) &= \frac{\mathrm{i}}{\hbar}\left[\hat{G}_{\mathrm{d}}(\mathbf{k}_1, z), \hat{P}_{\Delta}\right] \\ &= \frac{\mathrm{i}k_1^2}{\epsilon_{\mathrm{d}}k_{1z}}\int V(\mathbf{k}, \mathbf{k}_1, \mathbf{k}_2, z)\hat{G}_{\mathrm{d}}^{\dagger}(\mathbf{k}_2, z)\hat{G}_{\mathrm{p}}(\mathbf{k}, z)\, \mathrm{d}_{\mathrm{b}}k\, \mathrm{d}_{\mathrm{b}}k_2,\end{aligned} \tag{8.49}$$

where we combined the two nonlinear terms that are produced, using the symmetry of the vertex with respect to an interchange of the variables of the two down-converted fields. The expressions for the vertex kernels, following from the comparison with the equivalent equations in (8.44) and (8.45), are given by

$$\begin{aligned}V(\mathbf{k}, \mathbf{k}_1, \mathbf{k}_2, z) = &(2\pi)^3\delta(\mathbf{K} - \mathbf{K}_1 - \mathbf{K}_2)\delta(\omega - \omega_1 - \omega_2) \\ &\times \frac{1}{2}d_{\mathrm{I}}\exp[\mathrm{i}(k_z - k_{1z} - k_{2z})z],\end{aligned} \tag{8.50}$$

and its complex conjugate. It is noted that the dispersive refractive indices dropped out of the expressions for the vertex kernel. As a result, the expressions for both commutations are consistently related by complex conjugation. The vertex kernel and its complex conjugate can now be substituted into the ansatz for the propagation operator in (8.47).

8.4 Wigner functional evolution equation

Apart from the definition of the propagation operator, the derivation of the *Wigner functional evolution equation*[c] follows more or less the same steps employed in Section 6.4.8. The expression of the evolution equation is obtained by representing the Wigner functional for the propagation operator in terms of a construction process applied to a generating functional, as described in Section 6.1.3.

Before performing these steps, we specify the factor C that appears in the commutation relation for the ladder operators in (6.150), as well as in the quantized field operator in (6.152). In Section 4.1, the commutation relations are defined to be Lorentz covariant. However, in the context of propagation in a nonlinear medium, together with the assumed paraxial condition of the beams, both of which break Lorentz covariance, it is not reasonable to maintain consistency with the Lorentz covariant formulation. If we want to maintain consistency, we would set $C = k_z$. For the sake of a slightly simpler formulation, and because we have tacitly already made this choice in Section 8.3, we set $C = 1$ in this chapter. It means that the *commutation relation* for the ladder operators in terms of optical beam variables is here given by

$$\left[\hat{a}_r(\mathbf{K},\omega), \hat{a}_s^{\dagger}(\mathbf{K}',\omega')\right] = (2\pi)^3 \delta_{r,s}\delta(\mathbf{K}-\mathbf{K}')\delta(\omega-\omega'), \tag{8.51}$$

instead of (4.42). The measure associated with the $\diamond$-contractions is now given by $\mathrm{d}_{\mathrm{b}}k$, defined in (2.105), noting that for $C = 1$, we have $\mathrm{d}_C k \equiv \mathrm{d}_{\mathrm{b}}k$.

When a field operator (in the Schrödinger picture) is applied to a coherent state, the result is given in (6.161). For $C = 1$ and under the paraxial approximation, it becomes

$$\hat{G}(\mathbf{K},\omega)\,|\alpha\rangle = \mathrm{i}\,|\alpha\rangle\sqrt{\frac{\hbar k}{2\epsilon}}\,\alpha(\mathbf{K},\omega), \tag{8.52}$$

where the wavenumber k and the permittivity ϵ represent the appropriate quantities for a particular dielectric medium.

8.4.1 Wigner functional of the propagation operator

Next, we compute the Wigner functional for the propagation operator using the coherent-state-assisted approach of Section 6.1.3. It requires the overlap of the propagation operator by coherent states for the pump (with β_1 and β_2) and for the down-converted fields (with α_1 and α_2), leading to

c If Wigner functionals could be identified as probability distributions, their evolution equations would be *Fokker–Planck equations*. However, since Wigner functionals are not probability distributions, we do not refer to their evolution equations as Fokker–Planck equations.

$$\begin{aligned}
&\langle\alpha_1|\langle\beta_1|\hat{P}_\Delta|\alpha_2\rangle|\beta_2\rangle \\
&= \hbar\exp\left(-\frac{1}{2}\|\alpha_1\|^2-\frac{1}{2}\|\alpha_2\|^2-\frac{1}{2}\|\beta_1\|^2-\frac{1}{2}\|\beta_2\|^2+\alpha_1^*\diamond\alpha_2+\beta_1^*\diamond\beta_2\right) \\
&\quad\times\left(\beta_1^*\diamond T^*\diamond\diamond\alpha_2\alpha_2+\alpha_1^*\alpha_1^*\diamond\diamond T\diamond\beta_2\right).
\end{aligned} \tag{8.53}$$

The new vertex kernel T is defined as

$$\begin{aligned}
T(\mathbf{k}_1,\mathbf{k}_2,\mathbf{k},z) &= \sqrt{\frac{\hbar}{(2\epsilon_0 c)^3}}\sqrt{\frac{\omega\omega_1\omega_2}{n_{\text{eff}}(\omega)n_o(\omega_1)n_o(\omega_2)}}V(\mathbf{k},\mathbf{k}_1,\mathbf{k}_2,z) \\
&= \frac{\sigma_\text{I}}{c^2}\exp(\text{i}\Delta k_z z)\sqrt{\omega\omega_1\omega_2}(2\pi)^3\delta(\mathbf{K}-\mathbf{K}_1-\mathbf{K}_2)\delta(\omega-\omega_1-\omega_2),
\end{aligned} \tag{8.54}$$

where

$$\Delta k_z = k_z(\mathbf{K},\omega)-k_z(\mathbf{K}_1,\omega_1)-k_z(\mathbf{K}_2,\omega_2), \tag{8.55}$$

and

$$\sigma_\text{I} = \frac{1}{2}\sqrt{\frac{c\hbar}{2\epsilon_0}\frac{\chi_{abc}^{(2)}\eta_a(\vec{K}_1)\eta_b(\vec{K}_2)\eta_c^*(\vec{K})}{\sqrt{n_o(\omega_1)n_o(\omega_2)n_{\text{eff}}(\omega)}}}, \tag{8.56}$$

is the *effective nonlinear coefficient* represented as a scattering cross-section with the units of an area. Here, $n_{\text{eff}}(\omega)$ represents the *effective refractive index* and $n_o(\omega)$ is the *ordinary refractive index* (see Appendix B). Although it depends on the optical beam variables of the pump and down-converted field, the effective nonlinear coefficient is slowly varying so that we can consider it as a constant under monochromatic paraxial conditions. It is henceforth treated as a constant. The third arguments in both $T(\mathbf{k}_1,\mathbf{k}_2,\mathbf{k}_3,z)$ and its complex conjugate $T^*(\mathbf{k}_1,\mathbf{k}_2,\mathbf{k}_3,z)$ are always associated with the pump field.

Why does the vertex $T(\mathbf{k}_1,\mathbf{k}_2,\mathbf{k}_3,z)$ depend on z if the analysis is performed in the spatial Schrödinger picture? Is the full z-dependence not supposed to be carried by the state? The reason lies with the fixed reference frame. It removes the z-dependence due to the *linear evolution* from the state and transfers it to the vertex. The z-dependence of the state thus only represents the *nonlinear evolution*. In a certain sense, the fixed reference frame combines with the spatial Schrödinger picture to produce a situation similar to the (spatial) interaction picture found in quantum field theory [32].

In (8.53), a factor of $\hbar$ is pulled out of the vertex so that it can be cancelled, as was done in (6.168). However, while the cancellation removed all the $\hbar$'s from (6.168), in this case there is still a square root of $\hbar$ in the definition of T in (8.54). It is incorporated into the definition of the effective nonlinear coefficient in (8.56). The presence of the Planck constant (albeit hidden inside σ_I) in the vertex of parametric down-conversion identifies it as a *quantum interaction*.

The Wigner functional for the propagation operator is now obtained by substituting (8.53) into the coherent-state-assisted integral expression and evaluating the functional integrals. As in Sections 6.4.8, the integration process is alleviated with the aid of a generating functional and a construction operation. Here, we have two functional phase spaces. So, the generating functional

$$\begin{aligned}\mathcal{G} = \exp\Big(&-\frac{1}{2}\|\alpha_1\|^2 - \frac{1}{2}\|\alpha_2\|^2 - \frac{1}{2}\|\beta_1\|^2 - \frac{1}{2}\|\beta_2\|^2 + \alpha_1^* \diamond \alpha_2 + \beta_1^* \diamond \beta_2 \\ &+ \alpha_1^* \diamond \mu_1 + \mu_2^* \diamond \alpha_2 + \beta_1^* \diamond \eta_1 + \eta_2^* \diamond \beta_2\Big), \end{aligned} \tag{8.57}$$

is a duplicated version of the one in (6.164). The associated construction process is defined in terms of *functional derivatives* contracted with the vertex kernels:

$$C_\delta = \frac{\delta}{\delta\eta_1} \diamond T^* \diamond\diamond \frac{\delta}{\delta\mu_2^*}\frac{\delta}{\delta\mu_2^*} + \frac{\delta}{\delta\mu_1}\frac{\delta}{\delta\mu_1} \diamond\diamond T \diamond \frac{\delta}{\delta\eta_2^*}. \tag{8.58}$$

Applying this construction process on the generating functional, we reproduce the overlap obtained in (8.53).

To complete the calculation of the Wigner functional with the coherent-state-assisted approach, we substitute $\mathcal{G}$ into the coherent-state-assisted functional integral in (6.20). The resulting generating functional for the Wigner functional of the propagation operator reads

$$\mathcal{W} = \exp\left(\mu_2^* \diamond \alpha + \alpha^* \diamond \mu_1 - \frac{1}{2}\mu_2^* \diamond \mu_1 + \eta_2^* \diamond \beta + \beta^* \diamond \eta_1 - \frac{1}{2}\eta_2^* \diamond \eta_1\right). \tag{8.59}$$

It is the duplicated version of (6.167). The Wigner functional for the propagation operator can now be obtained by computing

$$W_{\hat{P}}[\alpha, \beta] = \hbar\, C_\delta\{\mathcal{W}\}|_{\eta_1=\eta_2^*=\mu_1=\mu_2^*=0}\,. \tag{8.60}$$

8.4.2 Evolution equation

The representation of the Wigner functional for the propagation operator in terms of the construction operation and the generating functional is used to produce the expression of the evolution equation. Apart from having two field variables instead of one, we still have the same expression for the evolution equation as found in (6.168). Here, it becomes

$$-\mathrm{i}\partial_z W_{\hat{P}} = C_\delta\left\{\mathcal{W} \star W_{\hat{P}} - W_{\hat{P}} \star \mathcal{W}\right\}\big|_{\eta_1=\eta_2^*=\mu_1=\mu_2^*=0}\,. \tag{8.61}$$

after cancelling a factor of Planck's constant on both sides, where $\star$ represents the star product. The calculation of the star products are the same as in Section 6.4.8 apart from the duplication of the phase space. The results are

$$
\begin{aligned}
\mathcal{W} \star W_{\hat{\rho}} = {} & \exp\left[\mu_2^* \diamond \alpha + \alpha^* \diamond \mu_1 - \frac{1}{2}\mu_2^* \diamond \mu_1 + \eta_2^* \diamond \beta + \beta^* \diamond \eta_1 - \frac{1}{2}\eta_2^* \diamond \eta_1\right] \\
& \times W_{\hat{\rho}}\left[\alpha^* + \frac{1}{2}\mu_2^*, \alpha - \frac{1}{2}\mu_1, \beta^* + \frac{1}{2}\eta_2^*, \beta - \frac{1}{2}\eta_1\right], \\
W_{\hat{\rho}} \star \mathcal{W} = {} & \exp\left[\mu_2^* \diamond \alpha + \alpha^* \diamond \mu_1 - \frac{1}{2}\mu_2^* \diamond \mu_1 + \eta_2^* \diamond \beta + \beta^* \diamond \eta_1 - \frac{1}{2}\eta_2^* \diamond \eta_1\right] \\
& \times W_{\hat{\rho}}\left[\alpha^* - \frac{1}{2}\mu_2^*, \alpha + \frac{1}{2}\mu_1, \beta^* - \frac{1}{2}\eta_2^*, \beta + \frac{1}{2}\eta_1\right].
\end{aligned}
\tag{8.62}
$$

We now substitute these two terms into (8.61), apply the construction operation, and set the auxiliary fields to zero. The resulting equation reads

$$
\begin{aligned}
-\mathrm{i}\partial_z W_{\hat{\rho}} = {} & \int 2\alpha(\mathbf{k}_1)\frac{\delta W_{\hat{\rho}}}{\delta\alpha^*(\mathbf{k}_2)}T^*(\mathbf{k}_1, \mathbf{k}_2, \mathbf{k}_3, z)\beta^*(\mathbf{k}_3) - 2\frac{\delta W_{\hat{\rho}}}{\delta\alpha(\mathbf{k}_1)}\alpha^*(\mathbf{k}_2) \\
& \times T(\mathbf{k}_1, \mathbf{k}_2, \mathbf{k}_3, z)\beta(\mathbf{k}_3) - \alpha(\mathbf{k}_1)\alpha(\mathbf{k}_2)T^*(\mathbf{k}_1, \mathbf{k}_2, \mathbf{k}_3, z)\frac{\delta W_{\hat{\rho}}}{\delta\beta(\mathbf{k}_3)} \\
& + \alpha^*(\mathbf{k}_1)\alpha^*(\mathbf{k}_2)T(\mathbf{k}_1, \mathbf{k}_2, \mathbf{k}_3, z)\frac{\delta W_{\hat{\rho}}}{\delta\beta^*(\mathbf{k}_3)} - \frac{1}{4}T^*(\mathbf{k}_1, \mathbf{k}_2, \mathbf{k}_3, z) \\
& \times \frac{\delta^3 W_{\hat{\rho}}}{\delta\beta(\mathbf{k}_3)\delta\alpha^*(\mathbf{k}_1)\delta\alpha^*(\mathbf{k}_2)} + \frac{1}{4}T(\mathbf{k}_1, \mathbf{k}_2, \mathbf{k}_3, z) \\
& \times \frac{\delta^3 W_{\hat{\rho}}}{\delta\alpha(\mathbf{k}_1)\delta\alpha(\mathbf{k}_2)\delta\beta^*(\mathbf{k}_3)}\, \mathrm{d_b}k_1\, \mathrm{d_b}k_2\, \mathrm{d_b}k_3,
\end{aligned}
\tag{8.63}
$$

where we expressed the optical beam variables compactly as wavevectors.

The resulting *evolution equation* describes the evolution of the full state (pump and down-converted field) under fairly general conditions. Apart from assuming type I phase-matching conditions and that the pump beam is *monochromatic* and *paraxial*, we did not make any further assumptions or approximations in deriving this equation. Other phase-matching conditions produce the same equation, but with different expressions for the vertex kernels.

8.5 Simplifications

The evolution equation in (8.63) is rather complicated, making it challenging to find a general solution. Since the equation can handle situations where the pump is any kind of state, it includes situations where the pump can become entangled with the down-converted light during the down-conversion process. The fact that the equation is linear in the Wigner functional implies that the solution can be expressed in exponential form. However, the polynomial terms in the exponent do not close. At every order, the equation generates higher-order terms. A general solution therefore tends to be an exponential with a polynomial of infinite order in its argument.

Fortunately, the typical experimental conditions that are employed for parametric down-conversion allow us to introduce simplifications, making it easier to find solutions. Here, we consider such simplifications.

8.5.1 Coherent state pump

Usually, the *pump* that is used in parametric down-conversion is a coherent state. Here, we assume that it remains a coherent state during the down-conversion process and does not become entangled with the down-converted light. This assumption is justified by the fact that a coherent state remains a coherent state when it suffers a loss, as shown in Section 6.7.1. Only the pump's parameter function $\zeta(z)$ can potentially change as a function of z. So, we substitute

$$W_{\hat{\rho}} \rightarrow \mathcal{N}_0 \exp\left[-2\|\beta - \zeta(z)\|^2\right] W_{\hat{\sigma}} \tag{8.64}$$

into (8.63), where $W_{\hat{\sigma}}$ is the Wigner functional for the down-converted part only.

After the substitution, the functional derivatives with respect to β are evaluated and the pump field is factored out and removed. The result is

$$\begin{aligned}
-\mathrm{i}\partial_z W_{\hat{\rho}} = {} & \mathrm{i}2\left[(\beta^* - \zeta^*) \diamond (\partial_z\zeta) + (\partial_z\zeta^*) \diamond (\beta - \zeta)\right] W_{\hat{\sigma}} \\
& - 2\int \alpha^*(\mathbf{k}_1)\alpha^*(\mathbf{k}_2)T(\mathbf{k}_1, \mathbf{k}_2, \mathbf{k}_3, z)\left[\beta(\mathbf{k}_3) - \zeta(\mathbf{k}_3, z)\right] W_{\hat{\sigma}} \\
& - \left[\beta^*(\mathbf{k}_3) - \zeta^*(\mathbf{k}_3, z)\right] T^*(\mathbf{k}_1, \mathbf{k}_2, \mathbf{k}_3, z)\alpha(\mathbf{k}_1)\alpha(\mathbf{k}_2)W_{\hat{\sigma}} \\
& + \frac{1}{4}\frac{\delta^2 W_{\hat{\sigma}}}{\delta\alpha(\mathbf{k}_1)\delta\alpha(\mathbf{k}_2)} T(\mathbf{k}_1, \mathbf{k}_2, \mathbf{k}_3, z)\left[\beta(\mathbf{k}_3) - \zeta(\mathbf{k}_3, z)\right] \\
& - \frac{1}{4}\left[\beta^*(\mathbf{k}_3) - \zeta^*(\mathbf{k}_3, z)\right] T^*(\mathbf{k}_1, \mathbf{k}_2, \mathbf{k}_3, z)\frac{\delta^2 W_{\hat{\sigma}}}{\delta\alpha^*(\mathbf{k}_1)\delta\alpha^*(\mathbf{k}_2)} \\
& + \frac{\delta W_{\hat{\sigma}}}{\delta\alpha(\mathbf{k}_1)}\alpha^*(\mathbf{k}_2)T(\mathbf{k}_1, \mathbf{k}_2, \mathbf{k}_3, z)\beta(\mathbf{k}_3) - \alpha(\mathbf{k}_1)\frac{\delta W_{\hat{\sigma}}}{\delta\alpha^*(\mathbf{k}_2)} \\
& \times T^*(\mathbf{k}_1, \mathbf{k}_2, \mathbf{k}_3, z)\beta^*(\mathbf{k}_3)\, \mathrm{d_b}k_1\, \mathrm{d_b}k_2\, \mathrm{d_b}k_3.
\end{aligned} \tag{8.65}$$

8.5.2 Strong-field perturbation theory

The effective nonlinear coefficient (8.56) obtained from the second-order susceptibility (without bosonic enhancement) is extremely weak. On its own, it allows us to use a perturbative approach to perform calculations. However, the *bosonic enhancement* that is introduced by the large number of photons in the *pump* increases the efficiency of the process. The expansion parameter for the perturbative approach becomes the product of the effective nonlinear coefficient and the magnitude of the *pump parameter function*, which is usually quite large. As a result, the perturbative approach can break down, especially for practical scenarios where the efficiency of the parametric down-conversion process is significantly increased to produce severely squeezed states.

Fortunately, there is a different perturbative approach that becomes more accurate when the strength of the pump field is increased. In fact, it relies on the strength of the pump field for its validity. The strength (or brightness) of the pump field is given by the magnitude of its parameter function $\|\zeta\|$. We refer to this approach as *strong-field perturbation theory*. Our initial assumption that the pump is (and remains) a coherent state, as given in (8.64), is a prerequisite for this approach.

We redefine the pump's field variable relative to its parameter function by replacing $\beta(\mathbf{k}) \to \zeta(\mathbf{k})+\eta(\mathbf{k})$. The new field variable $\eta(\mathbf{k})$ remains small compared to $\zeta(\mathbf{k})$ over the region in phase space where the Wigner functional of the pump's coherent state is significant. After replacing the field variable $\beta(\mathbf{k})$ by this redefinition in the evolution equation (8.65), we expand it in terms of $\eta(\mathbf{k})$. The higher-order terms are severely suppressed because the vertices that are contracted to $\eta(\mathbf{k})$ instead of $\zeta(\mathbf{k})$ are unenhanced. For a strong enough pump field, we can discard all the terms that contain $\eta(\mathbf{k})$. Effectively, we just replaced $\beta(\mathbf{k}) \to \zeta(\mathbf{k})$, which implies a *semiclassical approximation*.

8.5.3 Semiclassical approximation

By setting $\beta(\mathbf{k}) = \zeta(\mathbf{k})$ in (8.65) under the semiclassical approximation, we reduce the equation to

$$
\begin{aligned}
-\mathrm{i}\partial_z W_{\hat{\sigma}} = 2\int \Bigg[& \alpha(\mathbf{k}_1)\frac{\delta W_{\hat{\sigma}}}{\delta\alpha^*(\mathbf{k}_2)}T^*(\mathbf{k}_1,\mathbf{k}_2,\mathbf{k}_3,z)\zeta^*(\mathbf{k}_3) \\
& - \frac{\delta W_{\hat{\sigma}}}{\delta\alpha(\mathbf{k}_1)}\alpha^*(\mathbf{k}_2)T(\mathbf{k}_1,\mathbf{k}_2,\mathbf{k}_3,z)\zeta(\mathbf{k}_3)\Bigg]\,\mathrm{d}_\mathrm{b}k_1\,\mathrm{d}_\mathrm{b}k_2\,\mathrm{d}_\mathrm{b}k_3.
\end{aligned} \tag{8.66}
$$

It can be expressed with the simpler notation in terms of $\diamond$-contractions. The *evolution equation* in the *semiclassical approximation* thus reads

$$
\partial_z W_{\hat{\sigma}} = \frac{1}{2}\frac{\delta W_{\hat{\sigma}}}{\delta\alpha^*}\diamond H^*\diamond\alpha + \frac{1}{2}\alpha^*\diamond H\diamond\frac{\delta W_{\hat{\sigma}}}{\delta\alpha}, \tag{8.67}
$$

where the *bilinear vertex kernels* are defined as

$$
H(\mathbf{k}_1,\mathbf{k}_2,z) \triangleq -\mathrm{i}4\int T(\mathbf{k}_1,\mathbf{k}_2,\mathbf{k},z)\zeta(\mathbf{k})\,\mathrm{d}_\mathrm{b}k, \tag{8.68}
$$

and its complex conjugate. The contraction of the vertex kernel by the parameter function provides the *bosonic enhancement* of the bilinear vertex kernel.

8.5.3.1 Solutions of the semiclassical evolution equation

A solution for (8.67) is obtained by introducing an ansatz in the form of a squeezed vacuum state, as given in (6.337). Here, it is assumed that the kernels $A(z)$, $B(z)$, and $B^*(z)$, are functions of the propagation distance z in the nonlinear medium. (These kernels are still defined in the *fixed reference frame* discussed in Section 8.2.4.) The squeezed vacuum state is a pure state. Therefore, normalization requires that (6.346) applies, following from (C.29).

After substituting (6.337) into (8.67), we separate out three equations for the three kernels by taking the appropriate combinations of functional derivatives with respect to α and α^*. The resulting equations for the kernels are

$$\begin{aligned}
\partial_z A(z) &= \frac{1}{2}B(z)\diamond H^*(z) + \frac{1}{2}H(z)\diamond B^*(z),\\
\partial_z B(z) &= \frac{1}{2}A(z)\diamond H(z) + \frac{1}{2}H(z)\diamond A^*(z),\\
\partial_z B^*(z) &= \frac{1}{2}A^*(z)\diamond H^*(z) + \frac{1}{2}H^*(z)\diamond A(z).
\end{aligned} \tag{8.69}$$

The appearance of A^* follows from the transpose since A is Hermitian.

To solve these equations, we integrate them with respect to z and perform progressive back substitutions to obtain expansions in terms of integrals of contracted H-kernels. The initial conditions for the expansions are $A(z_0) = \mathbf{1}$ and $B(z_0) = 0$, which give the Wigner functional of the vacuum state. The resulting expressions of the kernels that satisfy the equations in (8.69) are

$$\begin{aligned}
A(z) &= \mathbf{1} + \int_{z_0}^{z}\int_{z_0}^{z_1} \mathcal{Z}\{H(z_1)\diamond H^*(z_2)\}\, \mathrm{d}z_2\, \mathrm{d}z_1 + \int_{z_0}^{z}\int_{z_0}^{z_1}\int_{z_0}^{z_2}\int_{z_0}^{z_3} \mathcal{Z}\{H(z_1)\\
&\quad \diamond H^*(z_2)\diamond H(z_3)\diamond H^*(z_4)\}\, \mathrm{d}z_4\, \mathrm{d}z_3\, \mathrm{d}z_2\, \mathrm{d}z_1 + \cdots,\\
B(z) &= \int_{z_0}^{z} H(z_1)\, \mathrm{d}z_1 + \int_{z_0}^{z}\int_{z_0}^{z_1}\int_{z_0}^{z_2} \mathcal{Z}\{H(z_1)\diamond H^*(z_2)\diamond H(z_3)\}\, \mathrm{d}z_3\, \mathrm{d}z_2\, \mathrm{d}z_1\\
&\quad + \int_{z_0}^{z}\int_{z_0}^{z_1}\int_{z_0}^{z_2}\int_{z_0}^{z_3}\int_{z_0}^{z_4} \mathcal{Z}\{H(z_1)\diamond H^*(z_2)\diamond H(z_3)\diamond H^*(z_4)\\
&\quad \diamond H(z_5)\}\, \mathrm{d}z_5\, \mathrm{d}z_4\, \mathrm{d}z_3\, \mathrm{d}z_2\, \mathrm{d}z_1 + \cdots ,
\end{aligned} \tag{8.70}$$

where $\mathcal{Z}\{\cdot\}$ is a *symmetrization operation* recursively defined by

$$\begin{aligned}
\mathcal{Z}\{f_1(z_1)\diamond\cdots\diamond f_n(z_n)\} &= \frac{1}{2}\left[f_1(z_1)\diamond \mathcal{Z}\{f_2(z_2)\diamond\cdots\diamond f_n(z_n)\}\right.\\
&\quad \left. + \mathcal{Z}\{f_1(z_2)\diamond\cdots\diamond f_{n-1}(z_n)\}\diamond f_n(z_1)\right],
\end{aligned} \tag{8.71}$$

with $\mathcal{Z}\{f_1(z_1)\} = f_1(z_1)$.

While B does not necessarily have an inverse, A does. In some situations, the inverse A^{-1} can appear in calculations. Therefore, we need a way to calculate it. Separating the higher-order terms in the definition of A in (8.70) from the leading-order term, we express it as $A = \mathbf{1} + E$. Then we use the fact that $A^{-1} \diamond A = \mathbf{1}$ to write the inverse as $A^{-1} = \mathbf{1} - A^{-1} \diamond E$ in terms of itself. Repeated back substitutions then lead to

$$A^{-1} = \mathbf{1} - E + E \diamond E - E \diamond E \diamond E + \cdots = \mathbf{1} + \sum_{n=1}^{\infty} (-E)^{\diamond n}. \tag{8.72}$$

Each of the contractions involves an infinite sequence of terms when E is replaced by the sum over higher-order terms from (8.70).

8.5.4 Pump evolution beyond the semiclassical approximation

Although a strong enough *pump field* justifies the *semiclassical approximation*, one can consider the higher-order terms that are produced from (8.65) in *strong-field perturbation theory* beyond the semiclassical approximation. The sub-leading-order terms that contain one contraction with the new field variable $\eta(\mathbf{k})$ or $\eta^*(\mathbf{k})$ represent evolution equations for the *pump parameter function*. For example, by taking a functional derivative with respect to $\eta^*(\mathbf{k}_0)$, we obtain

$$\begin{aligned} 0 = {} & \mathrm{i}2\partial_z \zeta(\mathbf{k}_0, z) W_{\hat{\sigma}} + 2 \int T^*(\mathbf{k}_1, \mathbf{k}_2, \mathbf{k}_0, z) \alpha(\mathbf{k}_1) \alpha(\mathbf{k}_2) W_{\hat{\sigma}} \\ & + \frac{1}{4} T^*(\mathbf{k}_1, \mathbf{k}_2, \mathbf{k}_0, z) \frac{\delta^2 W_{\hat{\sigma}}}{\delta \alpha^*(\mathbf{k}_1) \delta \alpha^*(\mathbf{k}_2)} \\ & + \alpha(\mathbf{k}_1) \frac{\delta W_{\hat{\sigma}}}{\delta \alpha^*(\mathbf{k}_2)} T^*(\mathbf{k}_1, \mathbf{k}_2, \mathbf{k}_0, z) \, \mathrm{d_b} k_1 \, \mathrm{d_b} k_2. \end{aligned} \tag{8.73}$$

Substituting the Wigner functional obtained from the semiclassical approximation as the solution from the first-order perturbation, we get an equation that contains new terms with α^*'s and α's, representing corrections to the semiclassical kernels A, B and B^*. The remaining terms without α's represent an equation for the evolution of the pump parameter function. It reads

$$\mathrm{i}\partial_z \zeta(\mathbf{k}_0, z) = \frac{1}{2} \int T^*(\mathbf{k}_1, \mathbf{k}_2, \mathbf{k}_0, z) B(\mathbf{k}_1, \mathbf{k}_2) \, \mathrm{d_b} k_1 \, \mathrm{d_b} k_2. \tag{8.74}$$

Although B contains the bilinear kernel functions, which are enhanced, the additional T-vertex is unenhanced. It requires a strong bosonic enhancement within B to overcome this suppression. Therefore, the evolution represented by this equation is usually weak. We do not consider this contribution further here.

8.6 Expressions of the kernel functions

8.6.1 Bilinear vertex kernel

The contraction of the vertex kernel T with the parameter function of the pump to form the bilinear kernel H implies that the latter depends on the beam profile of the pump. The use of a Gaussian beam for the pump in parametric down-conversion scenarios is quite common. Therefore, it is worth the effort to consider the case of a Gaussian pump beam in detail, and thus provide explicit expressions for the bilinear vertex kernel and the squeezed vacuum state kernels. In Section 9.5.1, we consider the case where the pump parameter function is given by a Laguerre-Gauss mode.

The Gaussian *pump parameter function* is modelled by

$$\zeta(\mathbf{k}) = \sqrt{2}\pi\zeta_0 w_\mathrm{p} h(\omega - \omega_\mathrm{p}, \delta_\mathrm{p}) \exp\left(-\frac{1}{4}w_\mathrm{p}^2|\mathbf{K}|^2\right), \tag{8.75}$$

where ζ_0 is a complex constant, w_p is the beam waist radius, $h(\omega)$ is a normalized real-valued spectral function with δ_p and ω_p being the pump bandwidth and the centre frequency of the pump spectrum, respectively. The squared magnitude of the pump profile function $\|\zeta(\mathbf{k})\|^2 = |\zeta_0|^2$ is the average number of photons in the pump field, providing the *bosonic enhancement* of the process. The spectral function is normalized so that

$$\int h^2(\omega - \omega_\mathrm{p}) \frac{\mathrm{d}\omega}{2\pi} = 1. \tag{8.76}$$

Under the *monochromatic approximation*, $h(\omega - \omega_\mathrm{p})$ is a narrow function centred at ω_p, allowing us to assume that $h^2(\omega - \omega_\mathrm{p}) = 2\pi\delta(\omega - \omega_\mathrm{p})$.

We now substitute (8.54) and (8.75) into (8.68) and evaluate the integrals. The resulting bilinear kernel reads

$$H(\mathbf{k}_1, \mathbf{k}_2, z) = -\mathrm{i}\Omega_0\sqrt{\omega_1\omega_2}h(\omega_1 + \omega_2 - \omega_\mathrm{p}, \delta_\mathrm{p}) \exp\left(-\frac{1}{4}w_\mathrm{p}^2|\mathbf{K}_1 + \mathbf{K}_2|^2 + \mathrm{i}\Delta k_z z\right), \tag{8.77}$$

where

$$\Omega_0 = 4\sqrt{2\pi\omega_\mathrm{p}}\frac{\zeta_0\sigma_\mathrm{I} w_\mathrm{p}}{c^2}, \tag{8.78}$$

assuming that δ_p is small enough so that we can replace $\omega_1 + \omega_2 \to \omega_\mathrm{p}$. The effect of the Dirac delta functions in (8.54) on Δk_z is discussed below.

8.6.2 Phase-matching conditions

The efficiency of the down-conversion process discussed above is determined by the extent to which the three different fields that interact during the process remain in phase. When Δk_z is multiplied by z, as in the argument of the exponential function in the bilinear kernel in (8.77), it produces a phase, leading to the z-dependent phase factor $\exp(i\Delta k_z z)$. Integrated over z, this phase factor plays a significant role in determining the efficiency of the down-conversion process. For $\Delta k_z = 0$, the phase is zero and remains zero for all z. (Here, we ignore the effects of dispersion, which are discussed in Section 8.6.8 below.) The integrated phase factor then has its maximum value, giving the process a high efficiency. The efficiency of the down-conversion process thus depends on the extent to which the different beams produce a coherent addition inside the crystal so that Δk_z is as close to zero as possible. To achieve such a situation, one minimizes the phase mismatch of the z-components of the wavevectors. The conditions that ensure that this minimization is satisfied are called *phase-matching conditions*. They play a central role in the parametric process. These conditions are determined by different combinations of refractive indices due to the different states of polarization for the three beams to achieve phase matching.

We distinguish among different types of phase-matching conditions. These conditions use different combinations of states of polarization (spin) and propagation directions for the three beams in association with the properties of the nonlinear crystal for phase matching. For *critical phase-matching conditions*, the average wavevectors of the three beams add up to zero inside the medium, as shown in Figure 8.1.

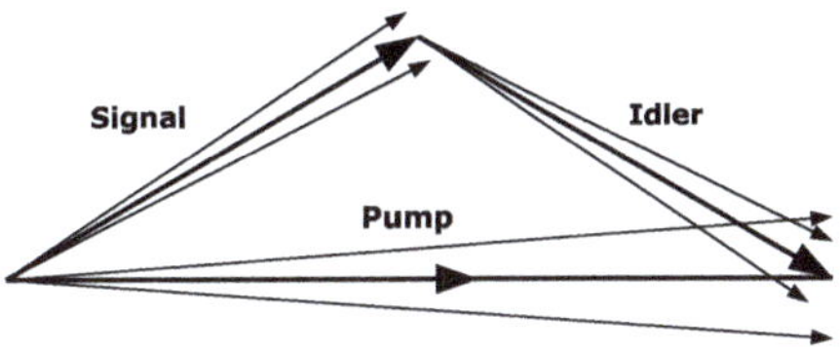

Figure 8.1: Non-collinear phase-matching under paraxial conditions.

With a negative uniaxial crystal for which the *extraordinary* refractive index n_e is smaller than the *ordinary* refractive index n_o (see Appendix B), the critical phase-matching conditions include the *type I* and *type II* phase-matching conditions. In the derivation of the evolution equation for parametric down-conversion, we assume type I phase-matching conditions. For *quasi-phase-matching conditions* (also called type 0 phase-matching conditions), the crystal is *periodically poled*, switching the sign of the phase evolution to maintain phase matching. First, we consider the *type I phase-matching conditions*.

8.6.2.1 Wavevector mismatch under the paraxial approximation

The z-component of the mismatch in the wavevectors Δk_z plays a significant role in the expression of the phase-matching function discussed below. Here, we consider the expression for Δk_z under the *paraxial approximation*. To derive the expression, we impose the paraxial condition for the three optical beams without restricting the *down-conversion angle* θ to be small.[d] For collinear down-conversion where $\theta = 0$, the effect of the paraxial condition on the expression of Δk_z is readily obtained, but when $\theta \neq 0$ the derivation becomes more complicated. The reason is that θ may be larger than the *beam divergence angles* θ_B of the respective beams. In other words, for the general propagation direction of the pump given by the z-direction, we cannot assume that the signal and idler beams propagate close to the z-axis. Each of them propagates close to a direction that makes an angle of θ with respect to the z-axis, as shown in Figure 8.1. Therefore, we cannot assume that the magnitude of the transverse part of the wavevector is small compared to the wavenumber. We need to impose the paraxial condition on each of the beams separately, even for arbitrary large θ.

The z-dependent part of the argument of the exponential function in (8.77) only contains the z-components of the wavevectors in Δk_z. They are given in terms of the transverse parts of the wavevectors via the dispersion relation

$$k_{zm} = \sqrt{k_m^2 - |\mathbf{K}_m|^2}, \tag{8.79}$$

where $m = 1, 2, 3$ labels the three beams, and k_m represents the wavenumbers of the beams in the nonlinear medium. The wavenumbers only depend on the angular frequencies. When these angular frequencies are fixed by monochromatic conditions, the wavenumbers become constants. In that case, we denote them as $k_{\rm s}$, $k_{\rm i}$, and $k_{\rm p}$, for $m = 1, 2, 3$, respectively, where the first two represent the *signal* and *idler* down-converted beams and the last one represents the pump.

For a beam propagating along the z-axis, the magnitude of the transverse wavevector $|\mathbf{K}_m|$ is small. On the other hand, if the beam does not propagate along the z-axis, then $|\mathbf{K}_m|$ is not in general small. We need to subtract the transverse component of the beam axis, which means that $|\mathbf{K}_m|^2 - k_m^2 \sin^2(\theta_m)$ is small, where m represents either the signal or the idler. The *down-conversion angle* θ_m is the angle between the signal or idler beam axis and the z-axis (the assumed propagation direction of the pump). They are different for these two beams under non-degenerate conditions. To perform a paraxial expansion for such a situation, we use *mathematical tagging* whereby we label the part of the expression that is small under the paraxial approximation. In this way, we perform the following replacement:

$$|\mathbf{K}_m|^2 \to k_m^2 \sin^2(\theta_m) + \left[|\mathbf{K}_m|^2 - k_m^2 \sin^2(\theta_m)\right] \epsilon. \tag{8.80}$$

d Strictly speaking, only the pump beam can be assumed to obey paraxial conditions, since the other two beams are produced in a superposition of all possible combinations of wavevectors that satisfy the energy-momentum conservation conditions. Here, we assume that there are line filters imposing monochromatic conditions on the down-converted light leading to the necessary restrictions on the down-converted beams.

The parameter ϵ tags the part of the expression that is small under the paraxial approximation. Setting $\epsilon = 1$, we recover the original expression. On the other hand, for $\epsilon = 0$, we obtain the contribution of the nominal wavevector of the beam along the z-direction. In this way, ϵ serves as an expansion parameter. We make an expansion up to sub-leading order in ϵ for the paraxial expansion.

We substitute (8.80) into (8.79) for $m = 1, 2$, but for $m = 3$, we insert a factor of ϵ into the transverse part inside (8.79), because it represents the pump beam propagating along z. The Dirac delta functions in (8.54) lead to the replacement $\mathbf{K}_3 \to \mathbf{K}_1 + \mathbf{K}_2$ in the latter. The three resulting expressions are then substituted into (8.55), leading to

$$\begin{aligned}\Delta k_z = {} & \sqrt{k_3^2 - |\mathbf{K}_1 + \mathbf{K}_2|^2 \epsilon} - \sqrt{C_1^2 k_1^2 - (|\mathbf{K}_1|^2 - S_1^2 k_1^2)\epsilon} \\ & - \sqrt{C_2^2 k_2^2 - (|\mathbf{K}_2|^2 - S_2^2 k_2^2)\epsilon},\end{aligned} \tag{8.81}$$

where k_m with $m = 1, 2, 3$ is the wavenumber of the respective beams. We also defined $C_m \triangleq \cos(\theta_m)$ and $S_m \triangleq \sin(\theta_m)$, with the down-conversion angles θ_m for $m = 1, 2$ in non-collinear (non-degenerate) phase matching.

For $\epsilon = 0$, we get the leading-order terms

$$\Delta k_z = k_3 - k_1 \cos(\theta_1) - k_2 \cos(\theta_2). \tag{8.82}$$

These terms must add up to zero for *critical phase matching*, so that

$$k_3 = k_1 \cos(\theta_1) + k_2 \cos(\theta_2). \tag{8.83}$$

Expanding (8.81) to sub-leading order in ϵ, while applying critical phase matching (to remove the 0-th-order term) and setting $\epsilon = 1$, we obtain

$$\Delta k_z = \frac{k_{1z} k_{2z}}{2k_3} \left| \frac{\mathbf{K}_1}{k_{1z}} - \frac{\mathbf{K}_2}{k_{2z}} \right|^2 - \frac{k_{1t}^2}{2k_{1z}} - \frac{k_{2t}^2}{2k_{2z}}, \tag{8.84}$$

where the longitudinal and transverse components of the beam axes are

$$k_{mz} \triangleq k_m \cos(\theta_m) \quad \text{and} \quad k_{mt} \triangleq k_m \sin(\theta_m), \tag{8.85}$$

for $m = 1, 2$. In this result, Δk_z contains a weighted difference between the two transverse wavevectors. The additional terms represent shifts in the down-converted field for the non-collinear case.

We consider two special cases in the following exercises. They lead to simplified versions of (8.84).

Exercise 8.2. Consider the *degenerate non-collinear* case, where it is assumed that $\omega_1 = \omega_2 = \frac{1}{2}\omega_3$, so that $\theta_1 = \theta_2 = \theta_d$. Since the wavelengths are those associated with the medium, we have $\lambda_1 = \lambda_2 = \lambda_d \neq 2\lambda_3$. Show that

$$\Delta k_z = \frac{|\mathbf{K}_1 - \mathbf{K}_2|^2}{2k_3} - \frac{k_d \sin^2(\theta_d)}{\cos(\theta_d)}. \tag{8.86}$$

Exercise 8.3. Consider the *collinear non-degenerate* case where the down-conversion angles of the signal and idler beams are zero $\theta_1 = \theta_2 = 0$, but $\omega_1 \neq \omega_2$. Show that

$$\Delta k_z = \frac{k_1 k_2}{2k_3}\left|\frac{\mathbf{K}_1}{k_1} - \frac{\mathbf{K}_2}{k_2}\right|^2 = \frac{|k_2\mathbf{K}_1 - k_1\mathbf{K}_2|^2}{2k_1 k_2 k_3}. \tag{8.87}$$

8.6.2.2 Type II phase matching

For the case of type II phase matching, which is a type of critical phase matching, the only difference compared to type I phase matching is that either the signal beam or idler beam has an extraordinary state of polarization, instead of both having ordinary states of polarization. As a consequence, the down-converted light does not form a single cone for a given angular frequency. Instead, two separate cones are produced for the signal and idler beams, respectively.

The expression for Δk_z is given by (8.84) in Section 8.6.2 for the non-degenerate case. Even for the degenerate case, the wavenumbers of the signal and idler beams are not equal because their associated refractive indices are different. Hence, $k_1 \neq k_2$.

The expression for the *effective nonlinear coefficients* in the type II phase-matching conditions is similar to that of the type I phase-matching conditions, given in (8.56). For the type II phase-matching conditions, it becomes

$$\sigma_{\text{II}} = \frac{1}{2}\sqrt{\frac{c\hbar}{2\epsilon_0}\frac{\chi^{(2)}_{abc}\eta_a(\vec{K}_1)\eta_b(\vec{K}_2)\eta_c^*(\vec{K})}{\sqrt{n_o(\omega_1)n_{\text{eff}}(\omega_2)n_{\text{eff}}(\omega)}}}. \tag{8.88}$$

The *effective refractive index* $n_{\text{eff}}(\omega)$ and the ordinary and extraordinary refractive indices $n_o(\omega)$ and $n_e(\omega)$ are discussed in Appendix B.

8.6.2.3 Quasi-phase matching

A drawback of critical phase-matching conditions is that the components in the susceptibility tensor producing the nonlinear coefficient are not the strongest components that the medium can provide. The strongest component in the susceptibility tensor of a medium is usually one where the polarization vectors of all three fields are aligned. To benefit from such a larger associated conversion efficiency, we need a different scheme to accomplish phase matching.

Quasi-phase matching, which is also referred to as the *type 0* phase-matching conditions, is such a scheme. In this case, the sign of the nonlinear coefficient is periodically inverting to compensate for the mismatch in the wavevectors. This *periodic poling* can be achieved thanks to the asymmetric structure of crystals with nonzero second-order nonlinear susceptibility tensors. These crystals are ferroelectric, which allows one to invert the sign of the coefficient permanently by applying a strong electric field.

An implication of quasi-phase matching is the fact that the phase-matching function and the expression for Δk_z are different from those for critical phase matching. Due to the periodic poling, the optimal conversion efficiency is determined for specific (collinear or non-collinear) down-conversion angles *by design*. Instead of the critical phase-matching condition in (8.83), we have

$$k_3 - k_1 \cos(\theta_1) - k_2 \cos(\theta_2) = \frac{\pi}{\Lambda}, \tag{8.89}$$

where Λ is the half-period of the periodic poling that achieves quasi-phase matching for the specific down-conversion angles. As a result, the leading-order terms obtained from (8.81) are replaced by π/Λ. The remaining sub-leading-order wavevector-dependent terms do not factorize, as in (8.84). Instead, we have

$$\Delta k_z = \frac{\pi}{\Lambda} + \frac{|\mathbf{K}_1|^2}{2k_{1z}} + \frac{|\mathbf{K}_2|^2}{2k_{2z}} - \frac{|\mathbf{K}_1 + \mathbf{K}_2|^2}{2k_3} - \frac{k_{1t}^2}{2k_{1z}} - \frac{k_{2t}^2}{2k_{2z}}, \tag{8.90}$$

where the longitudinal and transverse components of the beam axes k_z and k_t are defined in (8.85).

Exercise 8.4. Consider the *degenerate non-collinear* case, where it is assumed that $\omega_1 = \omega_2 = \frac{1}{2}\omega_3$, so that $\theta_1 = \theta_2 = \theta_d$. Show that

$$\Delta k_z = \frac{\pi}{\Lambda} + \frac{\left(|\mathbf{K}_1|^2 + |\mathbf{K}_2|^2\right)}{2k_d \cos(\theta_d)} - \frac{|\mathbf{K}_1 + \mathbf{K}_2|^2}{2k_3} - \frac{k_d \sin^2(\theta_d)}{\cos(\theta_d)}, \tag{8.91}$$

and that

$$\Lambda = \frac{\lambda_3 \lambda_d}{2\lambda_d - 4\lambda_3 \cos(\theta_d)}. \tag{8.92}$$

Exercise 8.5. Consider the *collinear non-degenerate* case, where $\theta_1 = \theta_2 = 0$. Show that

$$\Delta k_z = \frac{\pi}{\Lambda} + \frac{\left(\lambda_1 |\mathbf{K}_1|^2 + \lambda_2 |\mathbf{K}_2|^2\right)}{4\pi} - \frac{\lambda_3 |\mathbf{K}_1 + \mathbf{K}_2|^2}{4\pi} \tag{8.93}$$

where

$$\Lambda = \frac{\lambda_1 \lambda_2 \lambda_3}{2(\lambda_1 \lambda_2 - \lambda_2 \lambda_3 - \lambda_1 \lambda_3)}. \tag{8.94}$$

8.6.3 Phase-matching function

The leading-order term for B in (8.70) plays a significant role in experiments where spontaneous parametric down-conversion is used to produce *biphotons*. In such experiments, the higher-order terms in the state that contain more than two photons are unwelcome. For that reason, the efficiency of the down-conversion process is reduced to suppress these higher-order terms, leading to *weak squeezing conditions*. Therefore, the leading-order term in B represents the dominant part of the state (apart from the vacuum term, which we discard).

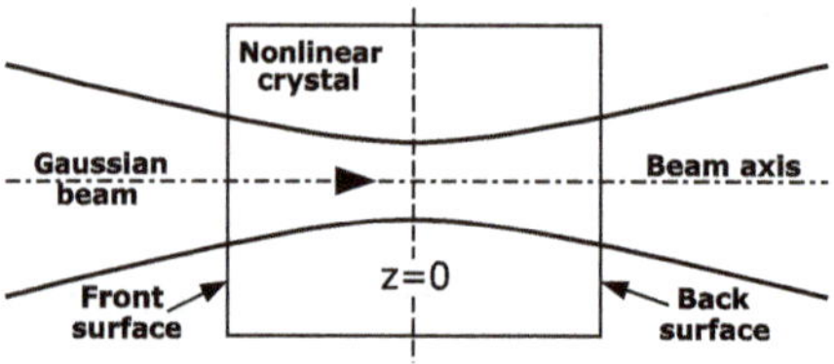

Figure 8.2: Gaussian pump beam propagating through the nonlinear crystal showing the reference plane at $z = 0$ in the centre of the crystal.

After evaluating the z integration, using (8.77) with degenerate non-collinear conditions, we obtain the leading-order term in B given by

$$B \approx \int_{-L/2}^{L/2} H(z)\,dz = -i\Omega_1 \exp\left(-\frac{1}{4}w_p^2|\mathbf{K}_1 + \mathbf{K}_2|^2\right) \operatorname{sinc}\left(\frac{L|\mathbf{K}_1 - \mathbf{K}_2|^2}{4k_p} - L\chi\right), \tag{8.95}$$

where the sinc function is defined as

$$\operatorname{sinc}(x) \triangleq \frac{\sin(x)}{x}, \tag{8.96}$$

L is the length of the crystal, Ω_1 is a *nonlinear coefficient* that represents the efficiency of the down-conversion process, and

$$\chi = \frac{k_d \sin^2(\theta_d)}{2\cos(\theta_d)}, \tag{8.97}$$

is a *non-collinear phase-shift parameter* with $\theta_d = \theta(\omega_d)$ being the down-conversion angle. To compute the nonlinear coefficient Ω_1, we set $\omega_1 = \omega_2 = \omega_d$, and assume a Gaussian shape for the temporal spectrum so that

$$h(0, \delta_p) = \sqrt{\frac{2\sqrt{\pi}}{\delta_p}}. \tag{8.98}$$

The nonlinear coefficient then becomes

$$\Omega_1 = \Omega_0 L \omega_{\rm d} \frac{\sqrt{2}\pi^{1/4}}{\sqrt{\delta_{\rm p}}} = \frac{4\pi^{3/4}\zeta_0\sigma_{\rm I} L w_{\rm p} \omega_{\rm p}^{3/2}}{c^2\sqrt{\delta_{\rm p}}}. \tag{8.99}$$

For the z-integration to produce (8.95), the point where $z = 0$ (for the *fixed reference frame*) is defined in the centre of the crystal, as shown in Figure 8.2. It avoids an unwanted phase factor.

The function in (8.95) is called the *phase-matching function* or the phase-matching kernel. It imposes the phase-matching conditions on the down-converted state.

Due to the sinc function, calculations involving the phase-matching function are often challenging. In those cases where the scale of the phase-matching region is the only relevant aspect, one can replace the sinc function with a more tractable function such as a Gaussian function with the same width, leading to the *Gaussian approximation*. However, the sinc function contains the square of the magnitude of the wavevector difference in its argument. Therefore, its sub-leading-order term is fourth order in the magnitude of the wavevector difference, as opposed to the Gaussian equivalent, which is only second order in the magnitude of the wavevector difference. Its shape differs significantly from that of a Gaussian function, as shown in Figure 8.3. Therefore, the Gaussian approximation does not produce reliable results when the contributions of the sub-leading order and beyond are significant. It is not a *controlled approximation*.

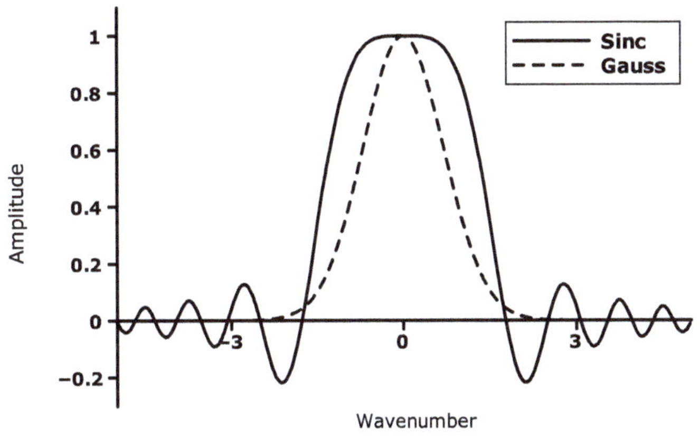

Figure 8.3: Comparison of the shapes of a sinc function with a squared argument and a Gaussian function.

An alternative approach is to leave the z-integral leading to the sinc function in (8.95) unevaluated. Then all wavevector integrals in subsequent calculations are evaluated first, before the z-integral is evaluated. Even then, the final z-integration is often challenging, requiring further approximations (see Section 8.6.4).

8.6.3.1 For quasi-phase matching

The phase-matching function for quasi-phase matching is not obtained by integrating the phase factor over the length of the crystal, as shown in (8.95). Instead, the integration is performed separately for each half-period and the results are added with the appropriate signs. In other words, we have

$$LS_{\mathrm{qpm}}(\Delta k_z, \Lambda, L) = \sum_{n=0}^{N-1} (-1)^n \int_{n\Lambda}^{(n+1)\Lambda} \exp(i\Delta k_z z)\,\mathrm{d}z, \tag{8.100}$$

where N represents the number of half-periods, so that $L = N\Lambda$ is the full length of the crystal. The half-period Λ is related to the average mismatch in the wavevectors Δk_z by

$$\Lambda = \frac{\pi}{\langle \Delta k_z \rangle}. \tag{8.101}$$

It leads to (8.89). Evaluating the integral and the summation, assuming without loss of generality that $N = L/\Lambda$ is an even integer, we obtain

$$S_{\mathrm{qpm}}(\Delta k_z, \Lambda, L) = \frac{2\sin\left(\frac{1}{2}\Delta k_z L\right)\sin\left(\frac{1}{2}\Delta k_z \Lambda\right)}{\Delta k_z L\cos\left(\frac{1}{2}\Delta k_z \Lambda\right)}, \tag{8.102}$$

where we discarded a global phase factor. This expression is not convenient for further calculations. It can be approximated by a shifted sinc function, given by

$$S_{\mathrm{qpm}}(\Delta k_z, \Lambda, L) \approx \frac{2}{\pi}\,\mathrm{sinc}\left(\frac{1}{2}L\Delta k_z - \frac{\pi}{2}N\right). \tag{8.103}$$

The shift removes the first term in (8.91) or (8.93). Without the shift, the expression in (8.103) is the same as the phase-matching function in (8.95), apart from a factor of $2/\pi$ that represents the reduction in the efficiency due to the quasi-phase matching.

8.6.4 Thin-crystal and plane wave approximations

To alleviate the complexity of calculations associated with parametric down-conversion, we exploit the experimental conditions to implement certain approximations. These approximations use the scales associated with the experimental parameters to determine which terms in the expression (typically in the exponent) can be discarded.

There are two such general approaches. One, which is called the *thin-crystal approximation*, is based on the relative scales of the thickness of the nonlinear medium and the Rayleigh range of the pump beam. Considering the phase-matching kernel in (8.95), we replace the transverse wavevectors $\mathbf{K}$ with dimensionless wavevectors $\mathbf{a}$ by absorbing w_{p} into them. Hence, we replace

$$\mathbf{K}_{1,2} \to \frac{2\mathbf{a}_{1,2}}{w_{\mathrm{p}}}, \tag{8.104}$$

in (8.95). The first term in the argument of the sinc function then becomes

$$\frac{L|\mathbf{K}_1 - \mathbf{K}_2|^2}{4k_{\mathrm{p}}} \to \frac{L}{k_{\mathrm{p}} w_{\mathrm{p}}^2} |\mathbf{a}_1 - \mathbf{a}_2|^2 = \frac{L}{2z_R} |\mathbf{a}_1 - \mathbf{a}_2|^2, \tag{8.105}$$

where we used (2.111). It contains the ratio of the thickness of the nonlinear medium and the Rayleigh range of the pump beam. In typical experiments where down-conversion is used to produce biphoton states, the Rayleigh range of the pump is much larger than the thickness of the nonlinear medium $z_R \gg L$. This situation is called the *thin-crystal condition*. Under such a condition, we can implement the *thin-crystal approximation* that sets $L/z_R \to 0$, so that the wavevector dependent part of the argument of the sinc function drops out, leading to

$$B \approx -\mathrm{i}\Omega_1 \exp\left(-\frac{1}{4} w_{\mathrm{p}}^2 |\mathbf{K}_1 + \mathbf{K}_2|^2\right) \operatorname{sinc}(L\chi). \tag{8.106}$$

The remaining sinc function only depends on the down-conversion angle. For the collinear case, its argument becomes zero so that $\operatorname{sinc}(0) = 1$.

The other approach is called the *plane wave approximation*. It treats the *pump beam* as if it is a plane wave. This approach is valid when the beam radius of the pump is much larger than the other transverse scale parameters in the experimental setup. In the Fourier domain, the angular spectrum of the pump beam then becomes a Dirac delta function. Based on the limit process in (2.17), such a Dirac delta function can be produced by a pump beam with a Gaussian angular spectrum in the limit of a large beam radius:

$$\lim_{w_{\mathrm{p}} \to \infty} \exp\left(-\frac{1}{4} w_{\mathrm{p}}^2 |\mathbf{k}|^2\right) = \frac{1}{\pi w_{\mathrm{p}}^2} \delta(\mathbf{k}). \tag{8.107}$$

When we apply the plane wave approximation to (8.95), it becomes

$$\begin{aligned} B \approx \int_{-L/2}^{L/2} H(z)\,\mathrm{d}z &= -\mathrm{i}\frac{\Omega_1}{\pi w_{\mathrm{p}}^2} \delta(\mathbf{K}_1 + \mathbf{K}_2) \operatorname{sinc}\left(\frac{L|\mathbf{K}_1 - \mathbf{K}_2|^2}{4k_{\mathrm{p}}} - L\chi\right) \\ &= -\mathrm{i}\frac{\Omega_1}{\pi w_{\mathrm{p}}^2} \delta(\mathbf{K}_1 + \mathbf{K}_2) \operatorname{sinc}\left(\frac{L|\mathbf{K}_1|^2}{k_{\mathrm{p}}} - L\chi\right). \end{aligned} \tag{8.108}$$

The transverse wavevectors now have equal magnitudes with opposite directions.

These two approximations are complementary. They do not represent opposite situations, but are valid simultaneously. As a result, we can use the two approximations to focus on different complementary aspects of the results.

8.6.5 Leading-order down-converted intensity

In (7.175) of Section 7.6.6, it is shown that the intensity of the down-converted field involves the kernel $A - \mathbf{1}$. Since the leading-order term of A in (8.70) is $\mathbf{1}$, the leading order term of the kernel for the intensity measurement is the second-order term for A in (8.70). Under weak squeezing conditions, the dominant part in the intensity measurement is

$$A - \mathbf{1} \approx \frac{1}{2} \int_{z_0}^{z} \int_{z_0}^{z_1} H(z_1) \diamond H^*(z_2) + H(z_2) \diamond H^*(z_1)\, \mathrm{d}z_2\, \mathrm{d}z_1. \tag{8.109}$$

When we substitute the bilinear kernel in (8.77) and its complex conjugate into this expression and evaluate the $\diamond$-contractions, it becomes

$$\begin{aligned} A - \mathbf{1} \approx & \int_{z_0}^{z} \int_{z_0}^{z_1} \frac{\Omega_2 \omega_1 (\omega_\mathrm{p} - \omega_1)}{\rho(\omega_1, z_1, z_2)} h(\omega_1 - \omega_2, \sqrt{2}\delta_\mathrm{p}) \\ & \times \exp\left[-\frac{1}{8} w_\mathrm{p}^2 |\mathbf{K}_1 - \mathbf{K}_2|^2 + \mathrm{i}(z_1 - z_2) \frac{k_z(\omega_\mathrm{p} - \omega_1)}{8k(\omega_\mathrm{p}) k_z(\omega_1)} |\mathbf{K}_1 - \mathbf{K}_2|^2 \right. \\ & + \mathrm{i} \frac{(z_1 + z_2)|\mathbf{K}_1|^2 - |\mathbf{K}_2|^2}{4k_z(\omega_1)\rho(\omega_1, z_1, z_2)} + \mathrm{i} \frac{(z_1 - z_2) k(\omega_\mathrm{p}) |\mathbf{K}_1 + \mathbf{K}_2|^2}{8k_z(\omega_1) k_z(\omega_\mathrm{p} - \omega_1) \rho(\omega_1, z_1, z_2)} \\ & \left. - \frac{(z_1 + z_2)^2 |\mathbf{K}_1 - \mathbf{K}_2|^2}{8 w_\mathrm{p}^2 k^2(\omega_\mathrm{p}) \rho(\omega_1, z_1, z_2)} - \mathrm{i}(z_1 - z_2)\chi \right] + (z_1 \leftrightarrow z_2)\, \mathrm{d}z_2\, \mathrm{d}z_1. \end{aligned} \tag{8.110}$$

where we left ω_1 as a variable, and defined

$$\begin{aligned} \Omega_2 &= \frac{4|\zeta_0|^2 \sigma_\mathrm{I}^2 2^{3/4} \omega_\mathrm{p} \sqrt{\delta_\mathrm{p}}}{\pi^{1/4} c^4}, \\ \rho(\omega_1, z_1, z_2) &= 1 - \mathrm{i} \frac{(z_1 - z_2) k_z(\omega_1)}{w_\mathrm{p}^2 k(\omega_\mathrm{p}) k_z(\omega_\mathrm{p} - \omega_1)}, \\ \chi &= \frac{k^2(\omega_1)}{2k_z(\omega_1)} + \frac{k^2(\omega_\mathrm{p} - \omega_1)}{2k_z(\omega_\mathrm{p} - \omega_1)} - \frac{1}{2} k(\omega_\mathrm{p}). \end{aligned} \tag{8.111}$$

By setting $\omega_1 = \omega_\mathrm{d} = \frac{1}{2}\omega_\mathrm{p}$, when the outputs are filtered to retain degenerate angular frequencies, we obtain

$$\begin{aligned} A - \mathbf{1} \approx & \int_{z_0}^{z} \int_{z_0}^{z_1} \frac{\Omega_2'}{\rho(\omega_\mathrm{d}, z_1, z_2)} \exp\left[-\frac{1}{8} w_\mathrm{p}^2 |\mathbf{K}_1 - \mathbf{K}_2|^2 + \mathrm{i}(z_1 - z_2) \frac{|\mathbf{K}_1 - \mathbf{K}_2|^2}{8k(\omega_\mathrm{p})} \right. \\ & + \mathrm{i} \frac{(z_1 + z_2)|\mathbf{K}_1|^2 - |\mathbf{K}_2|^2}{2k(\omega_\mathrm{p})\rho(\omega_\mathrm{d}, z_1, z_2)} + \mathrm{i} \frac{(z_1 - z_2)|\mathbf{K}_1 + \mathbf{K}_2|^2}{2k(\omega_\mathrm{p})\rho(\omega_\mathrm{d}, z_1, z_2)} \end{aligned}$$

$$
- \frac{(z_1 + z_2)^2 |\mathbf{K}_1 - \mathbf{K}_2|^2}{8 w_\mathrm{p}^2 k^2(\omega_\mathrm{p}) \rho(\omega_\mathrm{d}, z_1, z_2)} - \mathrm{i}(z_1 - z_2)\chi \Bigg] + (z_1 \leftrightarrow z_2)\, \mathrm{d}z_2\, \mathrm{d}z_1, \tag{8.112}
$$

where χ reverts to its definition in (8.97),

$$
\begin{aligned}
\Omega_2' = \frac{1}{4}\Omega_2 \omega_\mathrm{p}^2 h(0, \sqrt{2}\delta_\mathrm{p}) &= \frac{2|\zeta_0|^2 \sigma_\mathrm{I}^2 \omega_\mathrm{p}^3}{c^4}, \\
\rho(\omega_\mathrm{d}, z_1, z_2) &= 1 - \mathrm{i}\frac{(z_1 - z_2)}{w_\mathrm{p}^2 k(\omega_\mathrm{p})},
\end{aligned} \tag{8.113}
$$

and we replaced $k_z(\omega_\mathrm{d}) \to \frac{1}{2}k(\omega_\mathrm{p})$ based on the phase-matching condition.

We see that the expression of the resulting second-order kernel is still significantly more complicated than the first-order kernel given in (8.95). As we calculate the higher-order terms of these squeezed vacuum state kernels, we obtain progressively more complicated expressions. There does not seem to be a pattern for the expressions of the higher orders. In the next section, we show that the approximations of Section 8.6.4 allow us to obtain simpler expressions for these orders. Moreover, under these approximations, these orders produce a pattern that allows us to provide general expressions that are valid for all orders.

8.6.6 Higher-order terms

The kernels for the squeezed vacuum state A and B are composed of the bilinear kernel in (8.77) according to the expansions in (8.70). These integrals become progressively more complicated for the higher-order terms. While scenarios involving only the first few terms are relatively tractable, calculations involving the higher-order terms generally prove to be severely challenging.

Fortunately, even when these higher-order terms become significant, the typical experimental conditions allow us to employ the approximations that are discussed in Section 8.6.4. The Rayleigh range of the pump beam z_R is usually much longer than the length of the nonlinear crystal L. Therefore, under such conditions, we can use the *thin-crystal approximation* where we consider expansions in terms of the dimensionless *thin-crystal parameter*

$$
\frac{L}{z_R} = \frac{n_\mathrm{p}\lambda_\mathrm{p} L}{\pi w_\mathrm{p}^2}. \tag{8.114}
$$

We can also use the *plane wave approximation*, where applicable.

8.6.6.1 Extreme thin-crystal limit

By retaining only the leading-order terms in an expansion in terms of the thin-crystal parameter, we obtain simpler closed-form expressions. As in (8.106), the thin-crystal approximation removes all the transverse wavevector dependencies from Δk_z, leaving only a z-dependent phase factor. For *collinear* scenarios, the z-dependence completely drops out of the bilinear kernel H in (8.77). In such a case, we can compute all the $\diamond$-contractions for the higher-order terms in (8.70) for the A and B kernels.

From its definition in (8.77), we represent H without the z-dependence, as

$$H(\mathbf{k}_1,\mathbf{k}_2) = -\mathrm{i}\exp(\mathrm{i}\varphi)H_0(\mathbf{k}_1,\mathbf{k}_2), \tag{8.115}$$

where φ is a global phase and $H_0(\mathbf{k}_1,\mathbf{k}_2)$ is a real-valued kernel. The expressions for the kernels can then be easily integrated over the z's. The different orders in the expansions are represented as m $\diamond$-contracted bilinear kernels, represented by $(H_0)^{m\diamond}$. Since the expansions always involve alternating $\diamond$-contractions of H and H^* in sequence, the phase factors cancel, leaving only one factor of the phase in the case where an odd number of bilinear kernels are contracted. Formally, these kernels are represented as

$$\begin{aligned} A &= \mathbf{1} + \frac{L^2}{2!}H\diamond H^* + \frac{L^4}{4!}H\diamond H^*\diamond H\diamond H^*\cdots = \mathbf{1} + \sum_{n=1}^{\infty}\frac{L^{2n}(H_0)^{2n\diamond}}{(2n)!} = \cosh_\diamond(LH_0), \\ B &= LH + \frac{L^3}{3!}H\diamond H^*\diamond H + \frac{L^5}{5!}H\diamond H^*\diamond H\diamond H^*\diamond H\cdots \\ &= -\mathrm{i}\exp(\mathrm{i}\varphi)\sum_{n=1}^{\infty}\frac{L^{2n-1}(H_0)^{(2n-1)\diamond}}{(2n-1)!} = -\mathrm{i}\exp(\mathrm{i}\varphi)\sinh_\diamond(LH_0). \end{aligned} \tag{8.116}$$

The phase φ comes from the *pump parameter function*, where $\zeta_0 = |\zeta_0|\exp(\mathrm{i}\varphi)$. Here, we assume that it is a constant global phase. For more complicated pump parameter functions where the phase becomes a phase function, $\diamond$-contractions are required.

All the $\diamond$-contractions in (8.116) can be readily evaluated. As a result, we obtain closed-form expressions for all the orders, having different forms depending on whether it contains an even or odd number of contractions. The kernels for the squeezed vacuum state are now given by

$$\begin{aligned} A(\mathbf{k}_1,\mathbf{k}_2) &= \mathbf{1} + \sum_{n=1}^{\infty} H_{2n}^{(\mathrm{e})}(\mathbf{k}_1,\mathbf{k}_2), \\ B(\mathbf{k}_1,\mathbf{k}_2) &= -\mathrm{i}\exp(\mathrm{i}\varphi)\sum_{n=1}^{\infty} H_{2n-1}^{(\mathrm{o})}(\mathbf{k}_1,\mathbf{k}_2). \end{aligned} \tag{8.117}$$

The expressions for the even orders (as found in A) are

$$\begin{aligned} H_m^{(\mathrm{e})}(\mathbf{k}_1,\mathbf{k}_2) = {} & \frac{\Lambda_0\Lambda_1^m L^m}{m^{5/4}m!}\left[(\omega_\mathrm{p}-\omega_1)\omega_1\right]^{m/2} h(\omega_1-\omega_2,\sqrt{m}\delta_\mathrm{p}) \\ & \times\exp\left(-\frac{1}{4m}w_\mathrm{p}^2|\mathbf{K}_1-\mathbf{K}_2|^2\right), \end{aligned} \tag{8.118}$$

where m is an even integer. For the odd orders (as found for B and B^*), they are

$$\begin{aligned} H_m^{(o)}(\mathbf{k}_1,\mathbf{k}_2) = {} & \frac{\Lambda_0\Lambda_1^m L^m}{m^{5/4}m!}\left[(\omega_p-\omega_1)\omega_1\right]^{m/2} h(\omega_1+\omega_2-\omega_p,\sqrt{m}\delta_p) \\ & \times \exp\left(-\frac{1}{4m}w_p^2|\mathbf{K}_1+\mathbf{K}_2|^2\right), \end{aligned} \tag{8.119}$$

where m is an odd integer. The two quantities in the prefactors are given by

$$\Lambda_0 \triangleq \frac{\pi^{5/4}w_p^2}{\sqrt{\delta_p}} \quad \text{and} \quad \Lambda_1 \triangleq \frac{4\sqrt{2}|\zeta_0|\sigma_I\sqrt{\omega_p\delta_p}}{\pi^{3/4}c^2 w_p}. \tag{8.120}$$

The above implementation of the thin-crystal approximation is rather crude. We call it the *extreme thin-crystal approximation*. There are a number of issues that are neglected. First, the assumption of collinear down-conversion that removes the χ-term is in general not valid. Non-collinear phase-matching conditions are often used in experimental scenarios. Moreover, by removing the transverse wavevector dependent part of Δk_z, we remove the variables for directions orthogonal to the remaining terms in the exponent. The lack of these terms lead to divergences in some calculations. Although the extreme thin-crystal approximation provides a relatively easy way to consider some applications, it cannot be applied in all cases.

We can also perform a plane wave approximation in a way similar to how we implemented the thin-crystal approximation here. The resulting kernels then contain Dirac delta functions along the wavevector directions associated with the pump beam. The remaining exponential function is a phase function with an imaginary z-dependent argument. These kernels are rather crude in a similar sense to the extreme thin-crystal approximation. Therefore, we can regard them as the *extreme plane wave approximation*. We do not use the kernels under this approximation.

8.6.6.2 Combined rational approximation

To address the shortcomings of the extreme thin-crystal approximation and the extreme plane wave approximation, we consider the implementation of these approximations more carefully. The argument for the removal of the z-dependencies is based on a comparison of scales. The pump beam width introduces a scale w_p while the sinc function in the phase-matching function has a scale given by $(\lambda_p z)^{1/2}$. If the latter is much smaller than the former (as for thin-crystal conditions), then those terms with a z can be neglected. However, the space is multidimensional and these scales are not valid along all the directions in the space. The scale w_p is only relevant along $\mathbf{K}_1+\mathbf{K}_2$ or $\mathbf{K}_1-\mathbf{K}_2$ for the odd or even orders, respectively. As a result, the directions that are orthogonal to these directions are not governed by w_p. Along those directions, the z-dependent scale is applicable. Therefore, those terms need to be retained, which means that some z-dependencies remain in the argument of the exponential function.

As an example, consider the argument of the second-order term in (8.110). The terms that contain $|\mathbf{K}_1-\mathbf{K}_2|^2$ have a combined coefficient given by

$$
-\frac{1}{8}w_{\mathrm{p}}^2 + \mathrm{i}(z_1 - z_2)\frac{k_z(\omega_{\mathrm{p}} - \omega_1)}{8k(\omega_{\mathrm{p}})k_z(\omega_1)} - \frac{(z_1 + z_2)^2}{8w_{\mathrm{p}}^2 k^2(\omega_{\mathrm{p}})\rho(\omega_1, z_1, z_2)}. \tag{8.121}
$$

We see that the second and third terms are suppressed relative to the first term under thin-crystal conditions and can therefore be discarded. On the other hand, the coefficient of the term that contains $|\mathbf{K}_1 + \mathbf{K}_2|^2$ is just

$$
\mathrm{i}\frac{(z_1 - z_2)k(\omega_{\mathrm{p}})}{8k_z(\omega_1)k_z(\omega_{\mathrm{p}} - \omega_1)\rho(\omega_1, z_1, z_2)}. \tag{8.122}
$$

It is not suppressed relative to any other similar terms and thus needs to be retained. The term that contains $|\mathbf{K}_1|^2 - |\mathbf{K}_2|^2$ represents both the previous two directions. It is suppressed relative to the leading order under the thin-crystal approximation and it falls away in the plane wave approximation. Therefore, we discard it.[e] The wavevector-independent χ-term also remains. As a result, z-dependencies are retained in the argument of the exponent and all the wavevector directions are represented so that divergences in calculations are avoided. In effect, the resulting kernels are combinations of those obtained from the extreme thin-crystal approximation and the extreme plane wave approximation. It is referred to as the *combined rational approximation.*

The denominators in the exponents may also contain z-dependent polynomials, such as $\rho(\omega_1, z_1, z_2)$ in (8.111). For these polynomials, we also discard terms that are suppressed by factors of the thin-crystal parameter. Since all the z-dependent terms in these polynomials are suppressed they drop away, leaving the denominators independent of the z's. The same applies for the denominators of the prefactors.

Based on these considerations, we obtain the general expressions for the odd and even orders, equivalent to those in (8.118) and (8.119), but where the z-integrations have not yet been done. For even orders, we have

$$
\begin{aligned}
H_m^{(\mathrm{e})} = {} & \frac{\Lambda_0\Lambda_1^m}{m^{5/4}}\left[(\omega_{\mathrm{p}} - \omega_1)\omega_1\right]^{m/2} h(\omega_1 - \omega_2, \sqrt{m}\delta_{\mathrm{p}}) \\
& \times \exp\left[-\frac{w_{\mathrm{p}}^2}{4m}|\mathbf{K}_1 - \mathbf{K}_2|^2 + \mathrm{i}\frac{Z_m k(\omega_{\mathrm{p}})|\mathbf{K}_1 + \mathbf{K}_2|^2}{8k_z(\omega_1)k_z(\omega_{\mathrm{p}} - \omega_1)} - \mathrm{i}Z_m\chi(\omega_1)\right],
\end{aligned} \tag{8.123}
$$

where m is an even integer, and for the odd orders, we have

$$
\begin{aligned}
H_m^{(\mathrm{o})} = {} & \mathrm{i}\frac{\Lambda_0\Lambda_1^m}{m^{5/4}}\left[(\omega_{\mathrm{p}} - \omega_1)\omega_1\right]^{m/2} h(\omega_1 + \omega_2 - \omega_{\mathrm{p}}, \sqrt{m}\delta_{\mathrm{p}}) \\
& \times \exp\left[-\frac{w_{\mathrm{p}}^2}{4m}|\mathbf{K}_1 + \mathbf{K}_2|^2 + \mathrm{i}\frac{Z_m k(\omega_{\mathrm{p}})|\mathbf{K}_1 - \mathbf{K}_2|^2}{8k_z(\omega_1)k_z(\omega_{\mathrm{p}} - \omega_1)} - \mathrm{i}Z_m\chi(\omega_1)\right],
\end{aligned} \tag{8.124}
$$

---

e The implication is that, without this term, the z-symmetrization causes A to become real and symmetric, so that $A = A^* = A^T = A^\dagger$.

where m is an odd integer. The quantities Λ_0 and Λ_1 are the same as those in (8.125), and

$$Z_m = \sum_{n=1}^{m} (-1)^{n+1} z_n. \tag{8.125}$$

The prefactors in (8.123) and (8.124) lack factors of $L^m/m!$ compared to those in (8.118) and (8.119), which are produced by the z-integrations. The expressions for the squeezed vacuum state kernels now also incorporate the z-integrations and z-symmetrization. Instead of (8.117), they now become

$$\begin{aligned} A(\mathbf{k}_1, \mathbf{k}_2) &= \mathbf{1} + \sum_{n=1}^{\infty} \int H_{2n}^{(e)}(\mathbf{k}_1, \mathbf{k}_2, \mathcal{Z}\{z\})\, \mathrm{d}\{z\}, \\ B(\mathbf{k}_1, \mathbf{k}_2) &= -\mathrm{i} \exp(\mathrm{i}\varphi) \sum_{n=1}^{\infty} \int H_{2n-1}^{(o)}(\mathbf{k}_1, \mathbf{k}_2, \mathcal{Z}\{z\})\, \mathrm{d}\{z\}, \end{aligned} \tag{8.126}$$

where $\{z\}$ includes all the z's in Z_m, as defined in (8.125), and $\mathcal{Z}\{\cdot\}$ represents a z-symmetrization. The expressions need to be z-symmetrized before the z-integrations are performed. The z-symmetrization of the list of z's is recursively defined by

$$f(\mathcal{Z}\{z_1, \ldots, z_n\}) = \frac{1}{2} f(z_1, \mathcal{Z}\{z_2, \ldots, z_n\}) + \frac{1}{2} f(\mathcal{Z}\{z_2, \ldots, z_n\}, z_1), \tag{8.127}$$

with $\mathcal{Z}\{z_1\} = z_1$. The z-integrals are given with the sequence of integration boundaries in (8.70). The z-symmetrization ensures that the even kernels are real-valued. If the z-integrations are done first, they would usually render the wavevector integrations intractable. Therefore, it is often better to leave the z-integrations till after all the wavevector integrations have been performed.

8.6.6.3 Monochromatic approximation

Sometimes when studying a parametric down-conversion system, the only spatiotemporal degrees of freedom that we are interested in are the spatial degrees of freedom. The system or experiment may enforce monochromatic conditions on all the fields, allowing us to discard the temporal dependencies. As a result, we can simplify the kernels by setting the angular frequencies equal to their centre angular frequencies and integrating out the angular frequencies.

Under such conditions both the pump beam and the down-converted fields are *monochromatic*, as imposed by line filters in the experimental setup. Moreover, the down-converted fields are *degenerate* so that $\omega_p = 2\omega_d$. The assumed type I phase-matching conditions then lead to $\theta_1 = \theta_2 = \theta_d$ and $k_1 = k_2 = k_d$. Furthermore, non-collinear conditions imply that $k_p = 2k_d \cos(\theta_d)$ or $n_p = n_d \cos(\theta_d)$, where $n_p = n_{\text{eff}}(\omega_p)$ and $n_d = n_o(\omega_d)$.

Starting from the expression in (8.123), we then obtain for the even kernels

$$H_m^{(e)} = \frac{\pi w_p^2 \Xi^m}{mL^m} \exp\left(-\frac{w_p^2}{4m}|\mathbf{K}_1 - \mathbf{K}_2|^2 + i\frac{Z_m}{2k_p}|\mathbf{K}_1 + \mathbf{K}_2|^2 - iZ_m\chi_d \right), \tag{8.128}$$

where $\chi_d \triangleq \chi(\omega_p)$. The odd kernels in (8.124) become

$$H_m^{(o)} = i\frac{\pi w_p^2 \Xi^m}{mL^m} \exp\left(-\frac{w_p^2}{4m}|\mathbf{K}_1 + \mathbf{K}_2|^2 + i\frac{Z_m}{2k_p}|\mathbf{K}_1 - \mathbf{K}_2|^2 - iZ_m\chi_d \right). \tag{8.129}$$

In these expressions, we replace Λ_1 defined in (8.120) in terms of the *effective squeezing parameter* Ξ, as determined by the experimental conditions, in anticipation of its definition in (9.13) in Section 9.1. The squeezed state kernels A and B are now defined by (8.126) in terms of (8.128) and (8.129), respectively.

8.6.7 Kernels with walk-off

The nonlinear media that mediate the process of parametric down-conversion are in general *anisotropic*. This anisotropy comes in handy for setting up phase-matching conditions without which efficient parametric down-conversion would not be possible. However, the anisotropy also have detrimental effects, such as *walk-off*. The basic mechanism for walk-off, as experienced by the extraordinary fields in anisotropic media, is discussed in Section 2.9.4. The nonlinear evolution of the ordinary and extraordinary fields in the classical nonlinear equation is derived in Section 8.2. Here, we address the effect of walk-off by incorporating it into the kernels that are derived above.

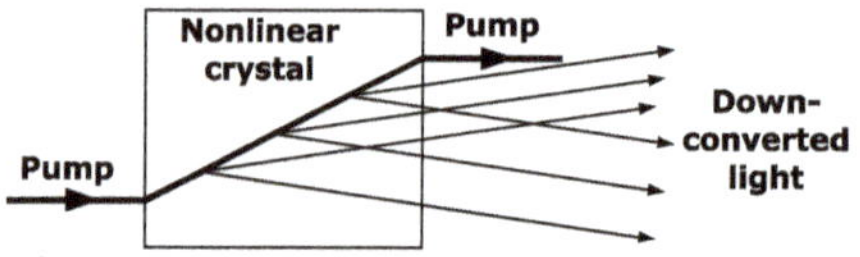

Figure 8.4: Diagrammatic representation of the walk-off effect in parametric down conversion.

The walk-off effect, discussed in Section 2.9.4, here manifests as a z-dependent lateral shift of the *pump beam* as it propagates through the nonlinear crystal due to its extraordinary state of polarization. The down-converted light that is produced while the pump beam experiences this lateral shift is therefore smeared along the direction of the shift. This process is illustrated in Figure 8.4.

We model this shift with the aid of the *parameter function* of the pump's coherent state. It becomes a tilted phase factor in the Fourier domain. Under the semiclassical approximation, the parameter function of the pump is contracted to the vertex kernel to produce the bilinear kernel. As a result, the effect of walk-off implies that the bilinear kernel in (8.77) is multiplied by an appropriate tilted phase factor. The result reads

$$H(\mathbf{k}_1, \mathbf{k}_2, z) = -\mathrm{i}\Omega_0 \sqrt{\omega_1\omega_2}\, h(\omega_1 + \omega_2 - \omega_\mathrm{p}, \delta_\mathrm{p}) \times \exp\left[-\frac{1}{4}w_\mathrm{p}^2 |\mathbf{K}_1 + \mathbf{K}_2|^2 + \mathrm{i}\Delta k_z z + \mathrm{i}z\mathbf{w}\cdot(\mathbf{K}_1 + \mathbf{K}_2)\right], \tag{8.130}$$

where $\mathbf{w}$ is a *walk-off vector*, representing the direction and magnitude of the walk-off, as determined by the walk-off angle given in (2.220). For $\mathbf{w} = 0$, we recover the original expression in (8.77).

The bilinear kernel in (8.130) can be used to compute the kernels A and B in terms of their expansion in (8.70). The walk-off effect leads to additional terms in the expressions for the different orders under the thin-crystal approximation. We do not consider this walk-off effect any further in subsequent calculations.

8.6.8 Dispersion effects

In the derivation presented in this chapter, the frequency spectra are assumed to be narrow enough to replace the variable angular frequencies by the centre angular frequencies of the spectra. Under such conditions, any dispersion effects produced by the medium are removed. In physical systems, the bandwidth is always finite and may be broad enough (as in femtosecond pulsed lasers) to make *dispersion effects* significant. In Section 8.6.2, it is assumed that the phase-matching conditions satisfy *critical phase matching*, which removes the leading-order terms in the phase mismatch Δk_z, as shown in (8.82). In the presence of dispersion, there are additional terms produced at leading order that do not in general cancel. These terms produce phase factors that can reduce the efficiency of the down-conversion process.

Here, we reconsider the derivation of the phase-matching condition in the presence of dispersion. Although the critical phase-matching condition can always be satisfied for a given set of angular frequencies, there are still contributions due to variations in these angular frequencies around the values that satisfy the critical phase-matching condition. These variations give rise to dispersion effects. Here, we fix the angles to those that are required for critical phase matching for a given set of angular frequencies, then we vary the angular frequencies around these values.

Expanding (8.81) to second order in ϵ, without assuming critical phase matching to remove the zeroth-order terms, we obtain

$$\Delta k_z \approx k_3(\omega_3) - k_1(\omega_1)C_1 - k_2(\omega_2)C_2 + \left[\frac{|\mathbf{K}_1|^2}{2k_1(\omega_1)C_1} + \frac{|\mathbf{K}_2|^2}{2k_1(\omega_2)C_2} - \frac{|\mathbf{K}_1 + \mathbf{K}_2|^2}{2k_3(\omega_3)} - \frac{k_1(\omega_1)S_1^2}{2C_1} - \frac{k_2(\omega_2)S_2^2}{2C_2}\right]\epsilon^2, \tag{8.131}$$

showing the *angular frequency* dependencies of the wavenumbers. The numerical subscripts 1, 2, 3 represent the signal, idler, and pump, respectively.

The wavenumber in the medium $k(\omega)$ is a function of the angular frequency ω that can be represented as

$$k(\omega) = \frac{\omega n(\omega)}{c}, \tag{8.132}$$

where $n(\omega)$ is the dispersive refractive index. Since the wavevectors are those in the medium (as opposed to those in vacuum), the transverse components k_x and k_y are independent variables, unaffected by the dispersion properties of the medium. Therefore, $k(\omega)$ carries all the information about the dispersion properties of the medium. The type of phase-matching condition does affect the value of $k(\omega)$. Therefore, we need to distinguish them based on the polarization direction of their beams. For the ordinary polarization, we denote them by $k_o(\omega)$ and for the extraordinary polarization, we denote them by $k_e(\omega)$. The wavenumber of the pump is $k_{\text{eff}}(\omega)$.

Now, we consider an expansion of (8.131) with respect to the angular frequency relative to the centre angular frequencies. Although the bandwidths of the fields are finite, they are still small compared to the centre angular frequencies. For the pump, it is ω_p. The signal and idler beams also have centre angular frequencies together with down-conversion angles as determined by the phase-matching conditions. In the subsequent discussion, we assume degenerate conditions so that the centre angular frequencies of both the signal and idler beams are given by $\omega_d = \frac{1}{2}\omega_p$. (It also means that $C_1 = C_2 = C$.) Therefore, each of the wavenumbers can be expanded relative to these centre angular frequencies. The resulting expansions to second order have the form

$$k(\omega) = k(\omega_c) + \frac{\omega - \omega_c}{v_g(\omega_c)} + \frac{1}{2}(\omega - \omega_c)^2 D_g(\omega_c), \tag{8.133}$$

where ω_c is the generic centre angular frequency, $v_g(\omega_c)$ is the *group velocity* and $D_g(\omega_c)$ is the *group velocity dispersion*, as discussed in Appendix B.3. The latter two are

$$v_g(\omega) \triangleq \left[\frac{\partial k(\omega)}{\partial \omega}\right]^{-1} \quad \text{and} \quad D_g(\omega) \triangleq \frac{\partial^2 k(\omega)}{\partial \omega^2}. \tag{8.134}$$

We use (8.133) for the expansion of (8.131). However, since the second-order terms in (8.131) are already suppressed by the paraxial condition, further suppressions caused by factors of $\Delta\omega/\omega_c$ can be neglected. As a result, the angular frequencies in the second-order terms in (8.131) are replaced by their centre angular frequencies only. The second-order terms in (8.131) thus lead to the expressions of the phase-matching conditions obtained in Section 8.6.2. The dominant dispersion effects are those produced by the leading-order terms.

Expanding the leading-order terms in (8.131) using (8.133), we get

$$\Delta k_z = k_3(\omega_p) + \frac{\omega_3 - \omega_p}{v_g(\omega_p)} + \frac{1}{2}(\omega_3 - \omega_p)^2 D_g(\omega_p)$$
$$- k_1(\omega_d)C - \frac{\omega_1 - \omega_d}{v_g(\omega_d)}C - \frac{1}{2}C(\omega_1 - \omega_d)^2 D_g(\omega_d)$$
$$- k_2(\omega_d)C - \frac{\omega_2 - \omega_d}{v_g(\omega_d)}C - \frac{1}{2}C(\omega_2 - \omega_d)^2 D_g(\omega_d) + O\{\epsilon^2\}. \tag{8.135}$$

The *critical phase-matching conditions*, which are valid among the centre angular frequencies, now implies that

$$k_3(\omega_p) - k_1(\omega_d)C - k_2(\omega_d)C = 0, \tag{8.136}$$

so that these terms are removed from Δk_z.

The additional leading-order terms that remain after we imposed the critical phase-matching conditions produce additional z-dependent phase terms in the exponent of the bilinear kernel. When this modified bilinear kernel is used in subsequent calculations involving integrals over the angular frequencies, these additional z-dependent phase terms lead to a z-dependent profile function with an approximate Gaussian shape. The width of this profile function can be smaller than the length of the nonlinear crystal under certain experimental conditions. The location of the Gaussian profile function can be shifted along the propagation direction, with an additional z-dependent phase tilt. When the profile function is narrower than the crystal, its location within the nonlinear crystal leads to uncorrelated down-conversions from the different points along the propagation inside the crystal under certain experimental conditions. Depending on the application, such a situation is considered undesirable. Therefore, the thickness of the crystal is usually reduced to become comparable to, or smaller than, the width of the Gaussian profile function caused by the dispersion effect.

Note that the phase-matching function in (8.95) is obtained under the degenerate condition by setting $\omega_1 = \omega_2 = \omega_d$, which removes any dispersion terms. In general, a finite bandwidth can provide the phase-matching function with dependencies on ω_1 and ω_2, which can then lead to dispersion effects.

In subsequent calculations, we usually assume that the thickness of the crystal is always small enough to avoid dispersion effects, so that we can use the kernels without dispersion. Nevertheless, in Sections 9.1.5 and 9.2.3, the effect of dispersion is illustrated for the even and odd kernels, respectively, in the context of measurements of the photon-number distribution and photon correlations.

Exercise 8.6. Use the expressions in Appendix B to show that the *group index* associated with the effective refractive index is given by

$$n_{\mathrm{g,eff}} = n_{\mathrm{eff}}^3 \left[\frac{\cos^2(\theta_X) n_{\mathrm{g,o}}}{n_{\mathrm{o}}^3} + \frac{\sin^2(\theta_X) n_{\mathrm{g,e}}}{n_{\mathrm{e}}^3} \right], \tag{8.137}$$

where $n_{\mathrm{g,o}}$ and $n_{\mathrm{g,e}}$ are the group indices for the ordinary and extraordinary refractive indices, respectively.

Exercise 8.7. Use the expressions in Appendix B to show that the *group index dispersion* associated with the effective refractive index is given by

$$\begin{aligned} \Delta n_{\mathrm{g,eff}} &= \frac{\lambda^3}{2\pi} \frac{\partial^2 n_{\mathrm{eff}}(\lambda)}{\partial \lambda^2} \\ &= 3 n_{\mathrm{eff}}^5 \frac{\lambda}{2\pi} \left[\frac{\cos^2(\theta_X) n_{\mathrm{g,o}}}{n_{\mathrm{o}}^3} + \frac{\sin^2(\theta_X) n_{\mathrm{g,e}}}{n_{\mathrm{e}}^3} \right]^2 \\ &\quad - 3 n_{\mathrm{eff}}^3 \frac{\lambda}{2\pi} \left[\frac{\cos^2(\theta_X) n_{\mathrm{g,o}}^2}{n_{\mathrm{o}}^4} + \frac{\sin^2(\theta_X) n_{\mathrm{g,e}}^2}{n_{\mathrm{e}}^4} \right] \\ &\quad + n_{\mathrm{eff}}^3 \left[\frac{\cos^2(\theta_X) \Delta n_{\mathrm{g,o}}}{n_{\mathrm{o}}^3} + \frac{\sin^2(\theta_X) \Delta n_{\mathrm{g,e}}}{n_{\mathrm{e}}^3} \right], \end{aligned} \tag{8.138}$$

where $\Delta n_{\mathrm{g,o}}$ and $\Delta n_{\mathrm{g,e}}$ are the group index dispersions for the ordinary and extraordinary refractive indices, respectively.

8.7 Stimulated parametric down-conversion

In Section 8.5.3, we use the semiclassical approximation to derive the evolution equation in (8.67) for parametric down-conversion. The solution for *spontaneous* parametric down-conversion is obtained from this semiclassical evolution equation with the aid of an ansatz in the form of the squeezed vacuum state.

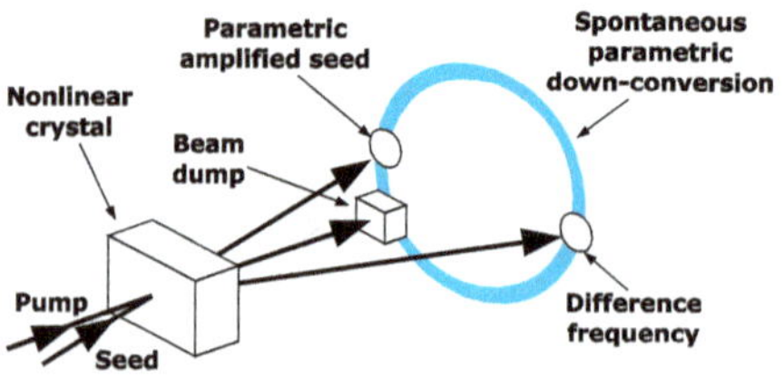

Figure 8.5: Diagrammatic represented of the setup for stimulated parametric down conversion.

Here, we turn our attention to the case where a *seed field* enters the nonlinear crystal together with the *pump*, thus producing *stimulated* parametric down-conversion. Such a stimulated process is also associated with *difference-frequency generation* and *parametric amplification*. The setup for the stimulated process is shown in Figure 8.5.

An arbitrary quantum state can serve as the initial seed field, apart from the fact that its frequency must be lower than that of the pump (typically half the frequency of the pump for degenerate phase matching). In Figure 8.5, the seed and the pump enter the crystal with a nonzero angle between them, calling for non-collinear phase-matching conditions. It is possible to combine the seed and the pump beams so that they enter the crystal with a zero relative angle, allowing collinear phase-matching conditions, by using a *dichroic mirror*, which reflects light of a specific wavelength only. Here, we consider only the non-collinear case.

We obtain a solution for stimulated parametric down-conversion with an arbitrary initial seed state by using the semiclassical evolution equation in (8.67). The approach is to use the fact that the parametric down-conversion process can be represented as a *Bogoliubov transformation* of the initial state, as revealed by the result in Section 6.8.2. The solution is thus obtained by solving for the Bogoliubov kernels. We use the results obtained in Section 6.8.3 for this purpose.

8.7.1 Equations for Bogoliubov kernels

Given that $W_{\hat{\sigma}}[\alpha](z_0)$ represents the Wigner functional of the initial seed field before it enters the crystal at $z = z_0$, the state at $z > z_0$ is assumed to be represented by a Bogoliubov transformation of this initial seed field. Based on the general Bogoliubov transformation of the ladder operators in (6.330), we formulate the equivalent Bogoliubov transformation of the field variables of a Wigner functional. The Bogoliubov transformation changes the arguments of the Wigner functional by

$$\begin{aligned} \alpha &\to U \diamond \alpha + V \diamond \alpha^* \triangleq \bar{\alpha}, \\ \alpha^* &\to \alpha^* \diamond U^\dagger + \alpha \diamond V^\dagger \triangleq \bar{\alpha}^*, \end{aligned} \tag{8.139}$$

where U and V are the Bogoliubov kernels. In (6.331), we derived the conditions that these Bogoliubov kernels satisfy. Here, we summarize them as

$$U \diamond V^T = \left(U \diamond V^T\right)^T \equiv V \diamond U^T \quad \text{and} \quad U \diamond U^\dagger - V \diamond V^\dagger = \mathbf{1}. \tag{8.140}$$

Applied to the initial seed field, the Bogoliubov transformation produces

$$W_{\hat{\sigma}}[\alpha^*, \alpha](z) = W_{\hat{\sigma}}[\alpha^* \diamond U^\dagger(z) + \alpha \diamond V^\dagger(z), U(z) \diamond \alpha + V(z) \diamond \alpha^*](z_0). \tag{8.141}$$

The transformed seed field of (8.141) is substituted into the semiclassical evolution equation in (8.67). The z-derivative produces functional derivatives with respect to the barred field variables via the chain rule, which are equivalent to the functional derivatives produced on the right-hand side. It leads to

$$
\begin{aligned}
\partial_z W_{\hat{\sigma}} &= \left(\alpha^* \diamond \partial_z U^\dagger(z) + \alpha \diamond \partial_z V^\dagger(z)\right) \diamond \frac{\delta W_{\hat{\sigma}}[\bar{\alpha}]}{\delta \bar{\alpha}^*} \\
&\quad + \frac{\delta W_{\hat{\sigma}}[\bar{\alpha}]}{\delta \bar{\alpha}} \diamond \left(\partial_z U(z) \diamond \alpha + \partial_z V(z) \diamond \alpha^*\right) \\
&= \frac{1}{2}\left\{\alpha^* \diamond H(z) \diamond V^\dagger(z) \diamond \frac{\delta W_{\hat{\sigma}}[\bar{\alpha}]}{\delta \bar{\alpha}^*} + \alpha^* \diamond H(z) \diamond U^T(z) \diamond \frac{\delta W_{\hat{\sigma}}[\bar{\alpha}]}{\delta \bar{\alpha}}\right. \\
&\quad \left. + \frac{\delta W_{\hat{\sigma}}[\bar{\alpha}]}{\delta \bar{\alpha}^*} \diamond U^*(z) \diamond H^*(z) \diamond \alpha + \frac{\delta W_{\hat{\sigma}}[\bar{\alpha}]}{\delta \bar{\alpha}} \diamond V(z) \diamond H^*(z) \diamond \alpha\right\}. \qquad (8.142)
\end{aligned}
$$

By comparing the field variable-dependent parts of the different terms in this equation, we can identify four separate equations:

$$
\begin{aligned}
\partial_z U^\dagger(z) &= \frac{1}{2} H(z) \diamond V^\dagger(z), \\
\partial_z V^T(z) &= \frac{1}{2} H(z) \diamond U^T(z), \\
\partial_z V^*(z) &= \frac{1}{2} U^*(z) \diamond H^*(z), \\
\partial_z U(z) &= \frac{1}{2} V(z) \diamond H^*(z), \qquad (8.143)
\end{aligned}
$$

Two of these equations are the Hermitian adjoints of the other two. Hence, they can be reduced to the following two equations:

$$
\begin{aligned}
\partial_z U(z) &= \frac{1}{2} V(z) \diamond H^*(z), \\
\partial_z V(z) &= \frac{1}{2} U(z) \diamond H(z). \qquad (8.144)
\end{aligned}
$$

If we assume that U is Hermitian and that V is symmetric, as found for a simple squeezing process in (6.336), the transpose, complex conjugate, and Hermitian adjoint of these equations would imply that[f]

$$
\begin{aligned}
V(z) \diamond H^*(z) &= H(z) \diamond V^*(z), \\
U(z) \diamond H(z) &= H(z) \diamond U^*(z). \qquad (8.145)
\end{aligned}
$$

Under such conditions, we can represent the differential equations in a symmetrized fashion given by

$$
\begin{aligned}
\partial_z U(z) &= \frac{1}{4} V(z) \diamond H^*(z) + \frac{1}{4} H(z) \diamond V^*(z), \\
\partial_z V(z) &= \frac{1}{4} U(z) \diamond H(z) + \frac{1}{4} H(z) \diamond U^*(z). \qquad (8.146)
\end{aligned}
$$

f It follows from (8.143). The left-hand sides of the first and last equations in (8.143), as well as those of the second and third equations, would be equal.

Note that these equations are of the same form as those in (8.69). By replacing $U \to A$, $V \to B, H \to 2H$ in (8.146), we reproduce those in (8.69). Unless stated otherwise, we do not assume that the conditions in (6.336) apply.

Exercise 8.8. Use (6.338), (8.140) and (8.145) to show that $A \diamond H = H \diamond A^*$ and $B \diamond H^* = H \diamond B^*$.

8.7.2 Solutions of the Bogoliubov kernels

The expressions for the Bogoliubov kernels are obtained by solving the equations in (8.144) or (8.146), depending on their properties. Since the equations in (8.146) are of the same form as those in (8.69), their solutions are of the same form as those in (8.70). The expressions for these Bogoliubov kernels are readily obtained by replacing $A \to U$, $B \to V, H \to \frac{1}{2}H$ in (8.70).

On the other hand, the solutions of the Bogoliubov kernels for the equations in (8.144) are obtained in a similar way by integrating them over z and performing repeated back substitutions. The results read

$$\begin{aligned}
U(z) = \mathbf{1} + \frac{1}{4}\int_{z_0}^{z}\int_{z_0}^{z_1} H(z_2) \diamond H^*(z_1)\,\mathrm{d}z_2\,\mathrm{d}z_1 + \frac{1}{16}\int_{z_0}^{z}\int_{z_0}^{z_1}\int_{z_0}^{z_2}\int_{z_0}^{z_3} H(z_4) \diamond H^*(z_3) \\
\diamond H(z_2) \diamond H^*(z_1)\,\mathrm{d}z_4\,\mathrm{d}z_3\,\mathrm{d}z_2\,\mathrm{d}z_1 + \cdots, \\
V(z) = \frac{1}{2}\int_{z_0}^{z} H(z_1)\,\mathrm{d}z_1 + \frac{1}{8}\int_{z_0}^{z}\int_{z_0}^{z_1}\int_{z_0}^{z_2} H(z_3) \diamond H^*(z_2) \diamond H(z_1)\,\mathrm{d}z_3\,\mathrm{d}z_2\,\mathrm{d}z_1 \\
+ \frac{1}{32}\int_{z_0}^{z}\int_{z_0}^{z_1}\int_{z_0}^{z_2}\int_{z_0}^{z_3}\int_{z_0}^{z_4} H(z_5) \diamond H^*(z_4) \diamond H(z_3) \\
\diamond H^*(z_2) \diamond H(z_1)\,\mathrm{d}z_5\,\mathrm{d}z_4\,\mathrm{d}z_3\,\mathrm{d}z_2\,\mathrm{d}z_1 + \cdots,
\end{aligned} \tag{8.147}$$

similar to those obtained from (8.70), but without the z-symmetrizations.

The contractions in (8.147) are the same as those in (8.70). Therefore, the expression for the different orders of these kernels are the same as those in Section 8.6.6 for the combined rational approximation. They are combined to form the Bogoliubov kernels as in (8.126), but without the z-symmetrizations and with the replacement $H \to \frac{1}{2}H$, leading to an additional factor of 2^{-m} for the m-th-order kernel.

The two equations in (8.144) can be combined into a matrix-vector equation having a form that allows one to formulate the solution in terms of a *Magnus expansion* [37]. However, the result is much more complex than what we obtain in (8.147). Therefore, we don't use the Magnus expansion here.

8.7.3 Consistency with respect to the spontaneous process

If a Bogoliubov transformed initial state is a solution for the stimulated parametric down-conversion process, then a Bogoliubov transformed vacuum state should be the solution for the spontaneous parametric down-conversion process. As a result, the transformation kernels U and V must combine to produce the A and B kernels of the squeezed vacuum state. It implies that we should be able to use the differential equations for U and V in (8.144) to derive the differential equations for A and B given in (8.69).

The transformed state that we obtain by performing a Bogoliubov transformation on the Wigner functional of a vacuum state is the squeezed vacuum state, as obtained in (6.337). By comparing the terms in the exponents with the same field variables, we obtain the relationships between the kernels of the squeezed vacuum state and the Bogoliubov kernels given in (6.338). The rest of the derivation is provided as an exercise.

Exercise 8.9. Assume that all the kernels in (6.338) are functions of z, compute the derivatives of these equations with respect to z, substitute (8.143) into the result, and show that the resulting equations are identical to those in (8.69).

8.8 Upconversion

While *parametric down-conversion* is a process whereby a pump photon is converted into two photons with lower frequencies, *parametric upconversion* is the opposite process in which two photons with lower frequencies are combined to produce a photon with a higher frequency. There are different upconversion processes. In *second-harmonic generation*, the lower frequency photons are all coming from the same input field. *Sum-frequency generation*, on the other hand, involves two different input fields, with one photon from each being combined to produce the higher frequency photon.

Various aspects of second-harmonic generation and sum-frequency generation have been investigated [38]. Here, we are specifically interested in the effect of the spatiotemporal degrees of freedom. It is shown below how these scenarios can be formulated with the aid of the Wigner functional formalism, leading to differential equations for the parameter functions of the fields involved, under the assumption that the input fields are coherent states. These differential equations are much harder to solve than their counterparts without the full spatiotemporal degrees of freedom, so much so that the general solutions are not known. It is not our intention here to provide solutions. Instead, we show how the problem can be formulated so that all the spatiotemporal degrees of freedom are taken into account, demonstrating the challenges thus imposed.

8.8.1 Second-harmonic generation

When the upconversion process is performed with a single input beam, so that pairs of photons from the same input field are converted into single upconverted photons with double the frequency, we call the process *second-harmonic generation* [38]. Under the condition where the input state is a coherent state, second-harmonic generation produces a coherent state in terms of the upconverted field. The process then involves the coupled evolution of these two parameter functions, leading to equations that are surprisingly complicated.

8.8.1.1 Evolution equation

To investigate the process of second-harmonic generation, we again develop an *evolution equation*, starting with (8.63). It is assumed that the input state is a known coherent state represented by its Wigner functional in the down-converted field variables, while the Wigner functional in terms of the upconverted field variables (here referred to as the *antipump*) is unknown. The input state's Wigner functional is therefore given by

$$W_{\hat{p}}[\alpha,\beta] \to \mathcal{N}_0 \exp\left[-2\|\alpha-\xi(z)\|^2\right] W_{\hat{\sigma}}[\beta], \tag{8.148}$$

where $\xi(z)$ is the z-dependent parameter function of the input coherent state and $W_{\hat{\sigma}}[\beta]$ is the Wigner functional for the upconverted antipump. We substitute this Wigner functional into (8.63), evaluate the functional derivatives with respect to α, factor out the exponential part of the Wigner functional that depends on α and remove it. The resulting evolution equation reads

$$\begin{aligned}
&-\mathrm{i}\partial_z W_{\hat{\sigma}} - \mathrm{i}2(\alpha^* - \xi^*) \diamond (\partial_z\xi) W_{\hat{\sigma}} - \mathrm{i}2(\partial_z\xi^*) \diamond (\alpha-\xi) W_{\hat{\sigma}} \\
&= \int 4\alpha^*(\mathbf{k}_1)\left[\alpha^*(\mathbf{k}_2)-\xi^*(\mathbf{k}_2)\right] T(\mathbf{k}_1,\mathbf{k}_2,\mathbf{k}_3,z)\beta(\mathbf{k}_3)W_{\hat{\sigma}} \\
&\quad + \left[\alpha^*(\mathbf{k}_1)-\xi^*(\mathbf{k}_1)\right]\left[\alpha^*(\mathbf{k}_2)-\xi^*(\mathbf{k}_2)\right] T(\mathbf{k}_1,\mathbf{k}_2,\mathbf{k}_3,z)\frac{\delta W_{\hat{\sigma}}}{\delta\beta^*(\mathbf{k}_3)} \\
&\quad + \alpha^*(\mathbf{k}_1)\alpha^*(\mathbf{k}_2)T(\mathbf{k}_1,\mathbf{k}_2,\mathbf{k}_3,z)\frac{\delta W_{\hat{\sigma}}}{\delta\beta^*(\mathbf{k}_3)}\,\mathrm{d}_\mathrm{b}k_1\,\mathrm{d}_\mathrm{b}k_2\,\mathrm{d}_\mathrm{b}k_3 - \mathrm{c.\,c.}\,,
\end{aligned} \tag{8.149}$$

where c. c. represents the complex conjugate of the preceding integral.

Assuming that the parameter function of the input coherent state is (and remains) strong, we may employ the *semiclassical approximation*, by setting $\alpha \to \xi(z)$. Then the equation simplifies to

$$\begin{aligned}
-\mathrm{i}\partial_z W_{\hat{\sigma}} = &\int \xi^*(\mathbf{k}_1,z)\xi^*(\mathbf{k}_2,z)T(\mathbf{k}_1,\mathbf{k}_2,\mathbf{k}_3,z)\frac{\delta W_{\hat{\sigma}}}{\delta\beta^*(\mathbf{k}_3)} \\
&- \frac{\delta W_{\hat{\sigma}}}{\delta\beta(\mathbf{k}_3)}T^*(\mathbf{k}_1,\mathbf{k}_2,\mathbf{k}_3,z)\xi(\mathbf{k}_1,z)\xi(\mathbf{k}_2,z)\,\mathrm{d}_\mathrm{b}k_1\,\mathrm{d}_\mathrm{b}k_2\,\mathrm{d}_\mathrm{b}k_3.
\end{aligned} \tag{8.150}$$

For a highly efficient conversion, the input field may become depleted, in which case the parameter function of the input coherent state may not remain strong. In such a case, the equations for the evolution of the input field's parameter function can be obtained by using *strong-field perturbative theory*, developed in Section 8.5.2. The required evolution equations then follow from the first-order perturbation, obtained by replacing $\alpha \to \xi(z) + \epsilon$ and its complex conjugate into (8.149) and extracting the terms that are proportional to either ϵ or ϵ^*. These equations are

$$\partial_z \xi(\mathbf{k}, z) = \mathrm{i} \int \xi^*(\mathbf{k}_2, z) T(\mathbf{k}, \mathbf{k}_2, \mathbf{k}_3, z) \left[\frac{\delta F_{\hat{\sigma}}}{\delta \beta^*(\mathbf{k}_3)} + 2\beta(\mathbf{k}_3) \right] \mathrm{d_b} k_2 \, \mathrm{d_b} k_3, \tag{8.151}$$

and its complex conjugate, where we defined $W_{\hat{\sigma}}[\beta] \triangleq \exp(F_{\hat{\sigma}}[\beta])$, with $F_{\hat{\sigma}}[\beta]$ being a functional polynomial in β and β^*. While the right-hand side is independent of β, the left-hand side contains β. It implies that β must cancel on the left, which in turn forces a specific form on the expression of $F_{\hat{\sigma}}$. The Wigner functional of the upconverted antipump field must have the general form

$$W_{\hat{\sigma}}[\beta] = \exp(F_{\hat{\sigma}}[\beta]) = \exp\left[-2\|\beta - \zeta(z)\|^2\right], \tag{8.152}$$

where $\zeta(z)$ is an unknown function. Hence, the upconverted state $W_{\hat{\sigma}}$ is again a coherent state. We can conclude that when the input state is a coherent state, the upconversion process produces a coherent state as the upconverted antipump state. The relationship between their parameter functions still needs to be determined. In terms of (8.152), the equation in (8.151) becomes

$$\partial_z \xi(\mathbf{k}, z) = \mathrm{i}2 \int \xi^*(\mathbf{k}_2, z) T(\mathbf{k}, \mathbf{k}_2, \mathbf{k}_3, z) \zeta(\mathbf{k}_3, z) \, \mathrm{d_b} k_2 \, \mathrm{d_b} k_3, \tag{8.153}$$

providing an evolution equation for the parameter function of the input state.

We briefly mention a different approach. If the parameter function of the input field can become depleted, then strictly speaking the strong-field assumption that we use in *strong-field perturbative theory* is not valid. An alternative approach is to derive these evolution equations as *equations of moments*. The idea is to multiply the evolution equation for the Wigner functional of the complete state by powers of the field variables and perform functional integrations over all field variables. For example, multiplying the equation by β and integrating over all field variables, we get an equation for the parameter function of the antipump as the first moment of its Wigner functional. In the same way, we can multiply the equation by α and evaluate the integrals to get an equation for the input field's parameter function. All such moments are produced with the aid of the characteristic functional of the state. So, to obtain all such equations of the moments, we can convert the evolution equation into an equivalent evolution equation for the characteristic functional. Usually, the resulting equations of the moments are the same as those obtained from strong-field perturbation theory. If not, it follows that the perturbation theory broke down and that the equations of the moments represent the correct equations. The equations we obtain here are the same for both approaches.

8.8.1.2 Undepleted solution

First, we consider the case where the input field remains strong. The input field's parameter function is considered to be independent of z under such conditions.

Based on (8.150), the evolution equation has the form

$$\partial_z W_{\hat{\sigma}} = \mathrm{i}\Gamma^* \diamond \frac{\delta W_{\hat{\sigma}}}{\delta \beta^*} - \mathrm{i}\frac{\delta W_{\hat{\sigma}}}{\delta \beta} \diamond \Gamma. \tag{8.154}$$

An evolution equation of this form represents an *evolution equation for displacement*. Its solutions are given by displacements of the arguments of an arbitrary initial state. (In the current scenario, the initial state is known to be a coherent state, but the procedure to solve this equation works for any arbitrary initial state.) For an initial state $W_{\text{in}}[\beta^*, \beta]$ at $z = 0$, we have

$$W_{\hat{\sigma}}(z) = W_{\text{in}}[\beta^* - \zeta^*(z), \beta - \zeta(z)]. \tag{8.155}$$

It then follows that

$$\partial_z \zeta(z) = \mathrm{i}\Gamma \quad \text{and} \quad \partial_z \zeta^*(z) = -\mathrm{i}\Gamma^*. \tag{8.156}$$

The initial conditions for the current context require that $\zeta(0) = 0$ and that $W_{\hat{\sigma}}$ becomes a vacuum state for $z = 0$. Hence, the only solution for the antipump state that is consistent with these conditions is the coherent state, in agreement with (8.152).

With the expression for the antipump in (8.152), the evolution equation in (8.150) leads to evolution equations for its parameter function, given by

$$\partial_z \zeta(\mathbf{k}, z) = \mathrm{i} \int \xi(\mathbf{k}_1)\xi(\mathbf{k}_2) T^*(\mathbf{k}_1, \mathbf{k}_2, \mathbf{k}, z)\, \mathrm{d}_{\mathrm{b}}k_1\, \mathrm{d}_{\mathrm{b}}k_2, \tag{8.157}$$

and its complex conjugate. Since we assume that the input state remains undepleted, its parameter function does not carry a z-dependence. The expression for the antipump's parameter function then reads

$$\zeta(\mathbf{k}, L) = \mathrm{i} \int \xi(\mathbf{k}_1)\xi(\mathbf{k}_2) \int_{-L/2}^{L/2} T^*(\mathbf{k}_1, \mathbf{k}_2, \mathbf{k}, z)\, \mathrm{d}z\, \mathrm{d}_{\mathrm{b}}k_1\, \mathrm{d}_{\mathrm{b}}k_2, \tag{8.158}$$

where the reference plane is chosen in the centre of the crystal. It is usually advisable to perform the wavevector integrations before the z-integration.

Exercise 8.10. Use the expression for $T^*(\mathbf{k}_1, \mathbf{k}_2, \mathbf{k}, z)$ given in (8.54) and assume the expression for the input state's parameter function is given by

$$\xi(\mathbf{k}) = \sqrt{2\pi}\xi_0 w_0 h(\omega - \omega_0, \delta_0) \exp\left(-\frac{1}{4}w_0^2|\mathbf{K}|^2\right), \tag{8.159}$$

where ξ_0 is the complex amplitude, w_0 is the beam width, ω_0 is the centre angular frequency, and δ_0 is the bandwidth. Compute the parameter function for the antipump $\zeta(\mathbf{k}, L)$ using (8.158). Show that this parameter function is given by the product $\zeta(\mathbf{k}, L) = \zeta_0(L)G(\mathbf{k})$ where $\zeta_0(L)$ is an amplitude function, and $G(\mathbf{k})$ is a normalized shape function independent of L given by

$$G(\mathbf{k}) = \sqrt{\pi}w_0 h(\omega - 2\omega_0, \sqrt{2}\delta_0) \exp\left(-\frac{1}{8}w_0^2|\mathbf{K}|^2\right). \tag{8.160}$$

8.8.1.3 Depletable solution

If the efficiency of the process is significant, the power in the input state could be significantly reduced during the upconversion process. It may even become depleted. It is also necessary to consider the possibility that the shape of the input state's parameter function is modified during the process. The angular spectrum representing the input's state parameter function becomes z-dependent, affecting the antipump's parameter function.

Exercise 8.11. Use (8.153), with $\xi^*(\mathbf{k})$ obtained from (8.159), $T(\mathbf{k}_1, \mathbf{k}_2, \mathbf{k}, z)$ given in (8.54), and $\zeta(\mathbf{K})$ given by

$$\zeta(\mathbf{k}) = \sqrt{2\pi}\zeta_0 w_1 h(\omega - \omega_1, \delta_1) \exp\left(-\frac{1}{4}w_1^2|\mathbf{K}|^2\right), \tag{8.161}$$

to compute an evolving parameter function $\xi(\mathbf{k}, z)$ for the input state and demonstrate its z-dependence.

To investigate the z-evolution of both parameter functions, we use the equation obtained in (8.153) from the first-order strong-field perturbation, and the equation obtained in (8.157) in which ξ becomes z-dependent. In summary, the full set of equations are

$$\begin{aligned}
\partial_z\zeta(\mathbf{k}, z) &= \mathrm{i}\int \xi(\mathbf{k}_1, z)\xi(\mathbf{k}_2, z)T^*(\mathbf{k}_1, \mathbf{k}_2, \mathbf{k}, z)\, \mathrm{d}_b k_1\, \mathrm{d}_b k_2, \\
\partial_z\xi(\mathbf{k}, z) &= \mathrm{i}2\int \xi^*(\mathbf{k}_1, z)T(\mathbf{k}_1, \mathbf{k}, \mathbf{k}_3, z)\zeta(\mathbf{k}_3, z)\, \mathrm{d}_b k_1\, \mathrm{d}_b k_3,
\end{aligned} \tag{8.162}$$

and their complex conjugates. It then follows that

$$\begin{aligned}
\partial_z\|\xi(z)\|^2 = \mathrm{i}2\int & \xi^*(\mathbf{k}_1, z)\xi^*(\mathbf{k}_2, z)T(\mathbf{k}_1, \mathbf{k}_2, \mathbf{k}_3, z)\zeta(\mathbf{k}_3, z) - \xi(\mathbf{k}_1, z) \\
& \times \xi(\mathbf{k}_2, z)T^*(\mathbf{k}_1, \mathbf{k}_2, \mathbf{k}_3, z)\zeta^*(\mathbf{k}_3, z)\, \mathrm{d}_b k_1\, \mathrm{d}_b k_2\, \mathrm{d}_b k_3, \\
\partial_z\|\zeta(z)\|^2 = \mathrm{i}\int & \xi(\mathbf{k}_1, z)\xi(\mathbf{k}_2, z)T^*(\mathbf{k}_1, \mathbf{k}_2, \mathbf{k}_3, z)\zeta^*(\mathbf{k}_3, z) - \xi^*(\mathbf{k}_1, z) \\
& \times \xi^*(\mathbf{k}_2, z)T(\mathbf{k}_1, \mathbf{k}_2, \mathbf{k}_3, z)\zeta(\mathbf{k}_3, z)\, \mathrm{d}_b k_1\, \mathrm{d}_b k_2\, \mathrm{d}_b k_3.
\end{aligned} \tag{8.163}$$

The right-hand sides of these two equations are the same apart from a sign change and a factor of 2 for the first equation. As a result, we can show that

$$\partial_z \left[\|\xi(z)\|^2 + 2\|\zeta(z)\|^2\right] = 0. \tag{8.164}$$

The quantity

$$\mathcal{E} \triangleq \|\xi(z)\|^2 + 2\|\zeta(z)\|^2, \tag{8.165}$$

is proportional to the total energy in the system, because the upconverted photons have twice the amount of energy as the input photons. The fact that the z-derivative vanishes indicates that the *energy is conserved*. The magnitudes of the parameter functions are thus constrained by

$$0 \leq \|\xi(z)\| \leq \sqrt{\mathcal{E}} \quad \text{and} \quad 0 \leq \|\zeta(z)\| \leq \sqrt{\frac{\mathcal{E}}{2}}. \tag{8.166}$$

To simplify the subsequent analysis, we represent the parameter functions in terms of $\mathcal{E}$, using real-valued z-dependent amplitude functions and normalizable complex-valued shape functions. Hence,

$$\xi(\mathbf{k}, z) = \sqrt{\mathcal{E}}a(z)F(\mathbf{k}, z) \quad \text{and} \quad \zeta(\mathbf{k}, z) = \sqrt{\frac{\mathcal{E}}{2}}b(z)G(\mathbf{k}, z). \tag{8.167}$$

The amplitude functions are constrained to $0 \leq a(z), b(z) \leq 1$. The initial conditions are $a(0) = 1$ and $b(0) = 0$. Moreover, $a^2(z) + b^2(z) = 1$. Therefore, we can always represent one in terms of the other. Although the amplitude functions can always be defined as being real valued in this way, the upconversion process also develops an unknown z-dependent phase that accompanies the upconverted antipump field. For the moment, this phase is absorbed into the shape functions. In terms of these definitions, the magnitudes of the fields are

$$\begin{aligned} \|\xi(z)\|^2 &= \int |\xi(\mathbf{k}, z)|^2 \, \mathrm{d}_b k = \mathcal{E}a^2(z), \\ \|\zeta(z)\|^2 &= \int |\zeta(\mathbf{k}, z)|^2 \, \mathrm{d}_b k = \frac{1}{2}\mathcal{E}b^2(z). \end{aligned} \tag{8.168}$$

The equations in (8.163) then become

$$\partial_z a(z) = -a(z)b(z)\partial_z \gamma(z) \quad \text{and} \quad \partial_z b(z) = a^2(z)\partial_z \gamma(z), \tag{8.169}$$

where

$$\partial_z \gamma(z) = \mathrm{i}\sqrt{\frac{\mathcal{E}}{2}} \left[K^*(z) - K(z)\right] \tag{8.170}$$

is the derivative of a dimensionless *gain function*, defined in terms of

$$K(z) = \int F^*(\mathbf{k}_1, z) F^*(\mathbf{k}_2, z) T(\mathbf{k}_1, \mathbf{k}_2, \mathbf{k}_3, z) G(\mathbf{k}_3, z)\, \mathrm{d}_\mathrm{b}k_1\, \mathrm{d}_\mathrm{b}k_2\, \mathrm{d}_\mathrm{b}k_3, \tag{8.171}$$

and its complex conjugate.

Using $a^2(z) = 1 - b^2(z)$, we obtain a differential equation for $b(z)$, given by

$$\partial_z b(z) = \left[1 - b^2(z)\right] \partial_z \gamma(z), \tag{8.172}$$

in terms of the gain function. The solution of this equation is

$$b(z) = \tanh\left[\gamma(z)\right]. \tag{8.173}$$

It implies that

$$a(z) = \sqrt{1 - b^2(z)} = \operatorname{sech}\left[\gamma(z)\right]. \tag{8.174}$$

Hence, the evolution of the amplitudes of the input and antipump fields is governed by the gain function. Knowing the gain function, we can solve the z-evolution of the two parameter functions. As a first attempt, we may assume that the gain is linear in z, which directly leads to the evolution of the amplitudes as given above. However, when we include the spatiotemporal degrees of freedom below, it becomes clear that z-dependence of the gain function is not so simple. Based on (8.170) and (8.171), the gain function is given by the overlap of the vertex kernel by the shape functions. Its z-dependence thus depends on the z-dependencies of the shape functions and the vertex kernel.

8.8.1.4 Constant input shape functions

First, we consider a simplified scenario by assuming that the shapes of the parameter functions remain constant, even when their magnitudes (the average photon numbers) change as a function of z. Although such an assumption is consistent with the result obtained with (8.158), the result from (8.153) produces a z-dependent shape function, as shown in the related exercises. Therefore, the assumption of constant shape functions is a rather crude approximation. However, it is constructive to consider such a case.

In addition, the parameter functions also have z-dependent phases. Hence, we express the parameter functions as

$$\begin{aligned} \xi(\mathbf{k}, z) &= \sqrt{\mathcal{E}}\, a(z) \exp[\mathrm{i}\varphi(z)] F(\mathbf{k}), \\ \zeta(\mathbf{k}, z) &= \sqrt{\frac{\mathcal{E}}{2}}\, b(z) \exp[\mathrm{i}\varrho(z)] G(\mathbf{k}), \end{aligned} \tag{8.175}$$

which differ from those in (8.167) in that the shape functions $F(\mathbf{k})$ and $G(\mathbf{k})$ are independent of z and we introduce phases $\varphi(z)$ and $\varrho(z)$ associated with $\xi(\mathbf{k}, z)$ and $\zeta(\mathbf{k}, z)$, respectively. These expressions are substituted into the equations for the parameter functions in (8.162). Each is then contracted with the missing shape function to produce

$$K_0(z) = \int F^*(\mathbf{k}_1)F^*(\mathbf{k}_2)T(\mathbf{k}_1,\mathbf{k}_2,\mathbf{k}_3,z)G(\mathbf{k}_3)\,\mathrm{d_b}k_1\,\mathrm{d_b}k_2\,\mathrm{d_b}k_3$$
$$\equiv \exp[\mathrm{i}2\varphi(z) - \mathrm{i}\varrho(z)]K(z), \tag{8.176}$$

or its complex conjugate on the right-hand sides. Their z-dependencies are produced by that of the vertex kernel in (8.171). The real parts of the resulting equations then produce

$$\partial_z a(z) = -a(z)b(z)g(z)\sin[\Phi(z)] \quad \text{and} \quad \partial_z b(z) = a^2(z)g(z)\sin[\Phi(z)], \tag{8.177}$$

in accordance with (8.169), where we replaced

$$K_0(z) = \frac{g(z)}{\sqrt{2\mathcal{E}}}\exp[\mathrm{i}\nu(z)], \tag{8.178}$$

and combined all the phases into

$$\Phi(z) = \nu(z) + \varrho(z) - 2\varphi(z). \tag{8.179}$$

Based on the similarity between (8.177) and (8.169), we conclude that the equations in (8.177) produce solutions similar to those in (8.173) and (8.174), with

$$\partial_z \gamma(z) = g(z)\sin[\Phi(z)]. \tag{8.180}$$

In addition, we also have equations for the phases, given by the imaginary parts of the equations from which (8.177) followed. The additional equations are

$$a(z)\partial_z\varphi(z) = a(z)b(z)g(z)\cos[\Phi(z)],$$
$$b(z)\partial_z\varrho(z) = a^2(z)g(z)\cos[\Phi(z)]. \tag{8.181}$$

A differential equation for $\Phi(z)$ follows from (8.179), with the aid of (8.177), (8.181), and

$$L(z) = \ln\left[b(z)a^2(z)\right], \tag{8.182}$$

leading to

$$\partial_z\Theta(z) = \partial_z\nu(z) + \cot[\Theta(z)]\partial_z L(z). \tag{8.183}$$

Assuming that the phase of $K(z)$ is constant, so that $\partial_z\nu(z) = 0$, we obtain a solution for $\Theta(z)$, given by

$$\Theta(z) = \arccos\left[\frac{\Gamma}{b(z)a^2(z)}\right], \tag{8.184}$$

where Γ is a constant, independent of z. Since $b(z_0) = 0$, it follows that $\Gamma = 0$, leading to $\Theta = \frac{\pi}{2}$. Then we get

$$\partial_z a(z) = -a(z)b(z)g(z),$$
$$\partial_z b(z) = a^2(z)g(z),$$
$$\partial_z \varphi(z) = \partial_z \varrho(z) = 0. \tag{8.185}$$

Although we are able to obtain explicit solutions, they come at the price of crude assumptions. The shape functions are assumed to remain constant and the overlap function $K_0(z)$ is replaced by a real-valued function $g(z)$ times a constant phase factor. Based on (8.176) and the z-dependent vertex kernel given in (8.54), it is unlikely that the phase of $K_0(z)$ would be constant regardless of the shapes of the parameter functions. To demonstrate this issue, we consider the case where the shape functions are Gaussian functions.

8.8.1.5 Constant Gaussian shape functions

The two relationships for $K_0(z)$ in (8.176) and (8.178) provide us with a way to compute the z-dependent magnitude and phase functions $g(z)$ and $\nu(z)$. However, an explicit calculation requires knowledge of the shape functions. Here, we consider the case where the input shape function is a Gaussian function, as given in (8.159), and the output shape function is the Gaussian function obtained in (8.160). The Gaussian beam is the most common beam shape found in practical second-harmonic generation experiments. When we substitute these shape functions into (8.176), we obtain

$$g(z) = \frac{\Xi_0}{\sqrt{w_0^4 k(\omega_0)^2 + 16z^2}} \quad \text{and} \quad \exp[i\nu(z)] = \sqrt{\frac{w_0^2 k(\omega_0) + i4z}{w_0^2 k(\omega_0) - i4z}}, \tag{8.186}$$

where

$$\Xi_0 = \frac{\sqrt{\mathcal{E}}\sigma w_0 k(\omega_0)\omega_0^{3/2}\sqrt{\delta_0}}{2^{3/4}\pi^{3/4}c^2}. \tag{8.187}$$

Since $w_0^2 k(\omega_0)$ is proportional to the Rayleigh range (2.111), it thus follows that $\nu(z)$ is the Gouy phase and $g(z)$ is proportional to the magnitude of the complex prefactor that contains the Gouy phase. Therefore, its derivative with respect to z is not zero. Instead

$$\partial_z \nu(z) = \frac{\Xi_0^2}{4w_0^2 k(\omega_0)g^2(z)}. \tag{8.188}$$

For $\partial_z \nu(z) \neq 0$, the differential equation in (8.183) is not readily solvable. As a result, the phases of the two fields are not easily obtained.

Clearly, the full analysis of the second-harmonic generation process with all the spatiotemporal degrees of freedom incorporated is challenging. We leave this matter here as an open question.

8.8.2 Sum-frequency generation

When upconversion is performed with two different input states entering the nonlinear medium, the process is called *sum-frequency generation*. A photon from each of the two input states is combined in the nonlinear interaction to produce one upconverted photon. Here, we consider the case where both input states are coherent states, focusing on the spatiotemporal effects of the process.

In such a scenario, one can consider the possibility that the two photons are taken from the same input state, leading to second-harmonic generation from the two respective input states, which are then produced together with the sum-frequency generation process involving both input states. In general, the parameter functions of the two input states can differ in terms of their properties, such as the frequency, the propagation direction, and the shape of the modes. Such differences cause the phase-matching conditions to favour one of these processes over the other. Since we are interested in sum-frequency generation, we assume that the phase-matching conditions are only satisfied for this process and that the second-harmonic generation processes can be ignored.

To obtain equations for the evolution of the parameter functions, we derive an *evolution equation* for the state similar to what we have in (8.63), but where the two down-converted fields are associated with different phase spaces. For this purpose, we modify (8.62) to have two different down-converted field variables, each with its own set of auxiliary field variables. We also modify the construction process in (8.58) so that the functional derivatives associated with the down-converted fields at each of the two vertices are taken with respect to different auxiliary field variables. The evolution equation thus obtained reads

$$
\begin{aligned}
\partial_z W_{\hat{\rho}} = {} & \mathrm{i} \int \alpha(\mathbf{k}_1) \frac{\delta W_{\hat{\rho}}}{\delta \eta^*(\mathbf{k}_2)} T^*(\mathbf{k}_1, \mathbf{k}_2, \mathbf{k}_3) \beta^*(\mathbf{k}_3) - \frac{\delta W_{\hat{\rho}}}{\delta \eta(\mathbf{k}_2)} \alpha^*(\mathbf{k}_1) T(\mathbf{k}_1, \mathbf{k}_2, \mathbf{k}_3) \\
& \times \beta(\mathbf{k}_3) + \frac{\delta W_{\hat{\rho}}}{\delta \alpha^*(\mathbf{k}_1)} \eta(\mathbf{k}_2) T^*(\mathbf{k}_1, \mathbf{k}_2, \mathbf{k}_3) \beta^*(\mathbf{k}_3) - \frac{\delta W_{\hat{\rho}}}{\delta \alpha(\mathbf{k}_1)} \eta^*(\mathbf{k}_2) \beta(\mathbf{k}_3) \\
& \times T(\mathbf{k}_1, \mathbf{k}_2, \mathbf{k}_3) - \alpha(\mathbf{k}_1) \eta(\mathbf{k}_2) T^*(\mathbf{k}_1, \mathbf{k}_2, \mathbf{k}_3) \frac{\delta W_{\hat{\rho}}}{\delta \beta(\mathbf{k}_3)} + \alpha^*(\mathbf{k}_1) \eta^*(\mathbf{k}_2) \\
& \times T(\mathbf{k}_1, \mathbf{k}_2, \mathbf{k}_3) \frac{\delta W_{\hat{\rho}}}{\delta \beta^*(\mathbf{k}_3)} - \frac{1}{4} T^*(\mathbf{k}_1, \mathbf{k}_2, \mathbf{k}_3) \frac{\delta^3 W_{\hat{\rho}}}{\delta \beta(\mathbf{k}_3) \delta \alpha^*(\mathbf{k}_1) \delta \eta^*(\mathbf{k}_2)} \\
& + \frac{1}{4} T(\mathbf{k}_1, \mathbf{k}_2, \mathbf{k}_3) \frac{\delta^3 W_{\hat{\rho}}}{\delta \alpha(\mathbf{k}_1) \delta \eta(\mathbf{k}_2) \delta \beta^*(\mathbf{k}_3)} \, \mathrm{d_b} k_1 \, \mathrm{d_b} k_2 \, \mathrm{d_b} k_3,
\end{aligned}
\tag{8.189}
$$

where $\eta(\mathbf{k})$ is the extra field variable associated with the other input field.

The derivation of the evolution equations for the parameter functions then follows the same procedure that we used for second-harmonic generation. The two input states are replaced by coherent state Wigner functionals and the functional derivatives with respect to α and η are evaluated. Under the assumption that these input fields are strong, we use the *semiclassical approximation* for both the input states, replacing their field variables by $\alpha \to \xi_1$ and $\eta \to \xi_2$, where $\xi_1(\mathbf{k})$ and $\xi_2(\mathbf{k})$ are the two parameter functions for the two coherent states. The resulting evolution equation is

$$\partial_z W_{\hat{\sigma}} = \mathrm{i}\int \xi_1^*(\mathbf{k}_1)\xi_2^*(\mathbf{k}_2)T(\mathbf{k}_1,\mathbf{k}_2,\mathbf{k}_3)\frac{\delta W_{\hat{\rho}}}{\delta\beta^*(\mathbf{k}_3)} - \xi_1(\mathbf{k}_1)\xi_2(\mathbf{k}_2)T^*(\mathbf{k}_1,\mathbf{k}_2,\mathbf{k}_3)\frac{\delta W_{\hat{\rho}}}{\delta\beta(\mathbf{k}_3)}\,\mathrm{d}_\mathrm{b}k_1\,\mathrm{d}_\mathrm{b}k_2\,\mathrm{d}_\mathrm{b}k_3. \tag{8.190}$$

For the sub-leading-order equations in the strong-field perturbative expansion, we substitute $\alpha \to \xi_1 + \epsilon_1$ and $\eta \to \xi_2 + \epsilon_2$ into the equation with the coherent states for the input fields, where $\epsilon_1(\mathbf{k})$ and $\epsilon_2(\mathbf{k})$ are the two new field variables. Then we extract four equations associated with ϵ_1, ϵ_2, and their complex conjugates. These equations are

$$\begin{aligned}
\partial_z\xi_1 &= \mathrm{i}\int \xi_2^*(\mathbf{k}_2)T(\mathbf{k}_1,\mathbf{k}_2,\mathbf{k}_3)\left[\beta(\mathbf{k}_3) + \frac{1}{2}\frac{\delta W_{\hat{\rho}}}{\delta\beta^*(\mathbf{k}_3)}\right]\mathrm{d}_\mathrm{b}k_2\,\mathrm{d}_\mathrm{b}k_3,\\
\partial_z\xi_2 &= \mathrm{i}\int \xi_1^*(\mathbf{k}_1)T(\mathbf{k}_1,\mathbf{k}_2,\mathbf{k}_3)\left[\beta(\mathbf{k}_3) + \frac{1}{2}\frac{\delta W_{\hat{\rho}}}{\delta\beta^*(\mathbf{k}_3)}\right]\mathrm{d}_\mathrm{b}k_1\,\mathrm{d}_\mathrm{b}k_3,
\end{aligned} \tag{8.191}$$

and their complex conjugates. Again, the β-dependencies in these equations imply that the upconverted state must be a coherent state. When we replace the upconverted state in the leading-order and sub-leading-order equations by a coherent state with an unknown parameter function ζ, they become

$$\begin{aligned}
\partial_z\zeta &= \mathrm{i}\int \xi_1(\mathbf{k}_1)\xi_2(\mathbf{k}_2)T^*(\mathbf{k}_1,\mathbf{k}_2,\mathbf{k})\,\mathrm{d}_\mathrm{b}k_1\,\mathrm{d}_\mathrm{b}k_2,\\
\partial_z\xi_1 &= \mathrm{i}\int \xi_2^*(\mathbf{k}_2)T(\mathbf{k},\mathbf{k}_2,\mathbf{k}_3)\zeta(\mathbf{k}_3)\,\mathrm{d}_\mathrm{b}k_2\,\mathrm{d}_\mathrm{b}k_3,\\
\partial_z\xi_2 &= \mathrm{i}\int \xi_1^*(\mathbf{k}_1)T(\mathbf{k}_1,\mathbf{k},\mathbf{k}_3)\zeta(\mathbf{k}_3)\,\mathrm{d}_\mathrm{b}k_1\,\mathrm{d}_\mathrm{b}k_3,
\end{aligned} \tag{8.192}$$

and their complex conjugates.

Following the same steps used for second-harmonic generation, we obtain

$$\begin{aligned}
\partial_z\|\xi_1(z)\|^2 &= \mathrm{i}\int \xi_1^*(\mathbf{k}_1,z)\xi_2^*(\mathbf{k}_2,z)T(\mathbf{k}_1,\mathbf{k}_2,\mathbf{k}_3,z)\zeta(\mathbf{k}_3,z)\\
&\quad - \xi_1(\mathbf{k}_1,z)\xi_2(\mathbf{k}_2,z)T^*(\mathbf{k}_1,\mathbf{k}_2,\mathbf{k}_3,z)\zeta^*(\mathbf{k}_3,z)\,\mathrm{d}_\mathrm{b}k_1\,\mathrm{d}_\mathrm{b}k_2\,\mathrm{d}_\mathrm{b}k_3,\\
\partial_z\|\xi_2(z)\|^2 &= \mathrm{i}\int \xi_1^*(\mathbf{k}_1,z)\xi_2^*(\mathbf{k}_2,z)T(\mathbf{k}_1,\mathbf{k}_2,\mathbf{k}_3,z)\zeta(\mathbf{k}_3,z)\\
&\quad - \xi_1(\mathbf{k}_1,z)\xi_2(\mathbf{k}_2,z)T^*(\mathbf{k}_1,\mathbf{k}_2,\mathbf{k}_3,z)\zeta^*(\mathbf{k}_3,z)\,\mathrm{d}_\mathrm{b}k_1\,\mathrm{d}_\mathrm{b}k_2\,\mathrm{d}_\mathrm{b}k_3,\\
\partial_z\|\zeta(z)\|^2 &= \mathrm{i}\int \xi_1(\mathbf{k}_1,z)\xi_2(\mathbf{k}_2,z)T^*(\mathbf{k}_1,\mathbf{k}_2,\mathbf{k}_3,z)\zeta^*(\mathbf{k}_3,z)\\
&\quad - \xi_1^*(\mathbf{k}_1,z)\xi_2^*(\mathbf{k}_2,z)T(\mathbf{k}_1,\mathbf{k}_2,\mathbf{k}_3,z)\\
&\quad \times\zeta(\mathbf{k}_3,z)\,\mathrm{d}_\mathrm{b}k_1\,\mathrm{d}_\mathrm{b}k_2\,\mathrm{d}_\mathrm{b}k_3.
\end{aligned} \tag{8.193}$$

In this case, we have two *conserved quantities*

$$\partial_z\left[\|\xi_1(z)\|^2+\|\zeta(z)\|^2\right]=\partial_z\left[\|\xi_2(z)\|^2+\|\zeta(z)\|^2\right]=0. \tag{8.194}$$

These conserved quantities are defined as

$$\mathcal{E}_1\triangleq\|\xi_1\|^2+\|\zeta\|^2 \quad\text{and}\quad \mathcal{E}_2\triangleq\|\xi_2\|^2+\|\zeta\|^2, \tag{8.195}$$

representing the initial average number of photons in the two respective input fields. The total energy in the system is proportional to

$$\mathcal{E}=\mathcal{E}_1+\mathcal{E}_2=\|\xi_1\|^2+\|\xi_2\|^2+2\|\zeta\|^2. \tag{8.196}$$

As with second-harmonic generation, the upconverted field starts from zero. During the conversion process, photons from each of the two input fields combine to form an upconverted photon. Therefore, the process stops when either of the two input fields becomes exhausted. As a result, the maximum average number of photons in the upconverted field is equal to the initial average number of photons in the smaller input field. We use this information to specify bounds for the average number of photons of the three fields during this process. Assuming that $\mathcal{E}_1>\mathcal{E}_2$, the bounds are

$$\mathcal{E}_1-\mathcal{E}_2\le\|\xi_1\|^2\le\mathcal{E}_1,\quad 0\le\|\xi_2\|^2\le\mathcal{E}_2,\quad\text{and}\quad 0\le\|\zeta\|^2\le\mathcal{E}_2. \tag{8.197}$$

Using these bounds, we parameterize the fields as

$$\begin{aligned}\xi_1(\mathbf{k},z)&=\sqrt{\mathcal{E}_1}a_1(z)F_1(\mathbf{k},z),\\ \xi_2(\mathbf{k},z)&=\sqrt{\mathcal{E}_2}a_2(z)F_2(\mathbf{k},z),\\ \zeta(\mathbf{k},z)&=\sqrt{\mathcal{E}_2}b(z)G(\mathbf{k},z).\end{aligned} \tag{8.198}$$

in terms of amplitude and shape functions. Moreover,

$$\|\xi_1\|^2=\mathcal{E}_1a_1^2(z),\quad \|\xi_2\|^2=\mathcal{E}_2a_2^2(z),\quad\text{and}\quad \|\zeta\|^2=\mathcal{E}_2b^2(z). \tag{8.199}$$

All the normalized amplitude functions can be expressed in terms of $b(z)$ as

$$a_1^2(z)=1-\mathcal{R}b^2(z)\quad\text{and}\quad a_2^2(z)=1-b^2(z), \tag{8.200}$$

where

$$\mathcal{R}\triangleq\frac{\mathcal{E}_2}{\mathcal{E}_1}. \tag{8.201}$$

Similar to what we did for second-harmonic generation to obtain (8.169), we derive three differential equations for the three magnitude functions. They are given by

$$\begin{aligned}\partial_z a_1(z) &= -\mathcal{R}a_2(z)b(z)\partial_z\gamma(z),\\ \partial_z a_2(z) &= -a_1(z)b(z)\partial_z\gamma(z),\\ \partial_z b(z) &= a_1(z)a_2(z)\partial_z\gamma(z).\end{aligned} \tag{8.202}$$

where $\gamma(z)$ is given by

$$\partial_z\gamma(z) = \mathrm{i}\frac{1}{2}\sqrt{\mathcal{E}_1}\left[K^*(z) - K(z)\right], \tag{8.203}$$

with $K(z)$ defined by

$$K(z) = \int F_1^*(\mathbf{k}_1, z)F_2^*(\mathbf{k}_2, z)T(\mathbf{k}_1, \mathbf{k}_2, \mathbf{k}_3, z)G(\mathbf{k}_3, z)\,\mathrm{d_b}k_1\,\mathrm{d_b}k_2\,\mathrm{d_b}k_3. \tag{8.204}$$

Note that the definitions of $\gamma(z)$ and $K(z)$ in the current context differ from those in (8.170) and (8.171).

Using (8.200), we convert the differential equation for $b(z)$ into one that only contains $b(z)$ and the gain function $\gamma(z)$. It reads

$$\partial_z b(z) = \sqrt{1 - \mathcal{R}b^2(z)}\sqrt{1 - b^2(z)}\partial_z\gamma(z). \tag{8.205}$$

Solving this equation for $b(z)$ in terms of the gain function, we obtain

$$b(z) = \mathrm{SN}\left(\gamma - \gamma_0, \sqrt{\mathcal{R}}\right), \tag{8.206}$$

where $\mathrm{SN}(x, k)$ is the Jacobi elliptic function sn [39] and γ_0 is the initial value of the gain. It then follows that

$$\begin{aligned}a_1(z) &= \sqrt{1 - \mathcal{R}\,\mathrm{SN}^2\left(\gamma - \gamma_0, \sqrt{\mathcal{R}}\right)},\\ a_2(z) &= \mathrm{CN}\left(\gamma - \gamma_0, \sqrt{\mathcal{R}}\right),\end{aligned} \tag{8.207}$$

where $\mathrm{CN}(x, k)$ is the Jacobi elliptic function cn [39]. As with the second-harmonic generation, we obtain solutions for the magnitude functions in terms of the gain function $\gamma(z)$, which in turn depends on the overlap of the vertex kernel by the shape functions. The solution of the gain function is now even more challenging than for second-harmonic generation, because there are three independent shape functions instead of just two. Apart from this issue, the problem introduced by the spatiotemporal degrees of freedom is similar to what we found with second-harmonic generation.

In Section 9.7, we return to the process of sum-frequency generation. There we consider it in the application of upconversion teleportation.

Bibliography

[1] H. H. Arnaut and G. A. Barbosa. Orbital and intrinsic angular momentum of single photons and entangled pairs of photons generated by parametric down-conversion. *Phys. Rev. Lett.*, 85:286–289, 2000.

[2] A. Mair, A. Vaziri, G. Weihs, and A. Zeilinger. Entanglement of the orbital angular momentum states of photons. *Nature*, 412:313–316, 2001.

[3] J. P. Torres, A. Alexandrescu, and L. Torner. Quantum spiral bandwidth of entangled two-photon states. *Phys. Rev. A*, 68:050301, 2003.

[4] G. Gibson, J. Courtial, M. J. Padgett, M. Vasnetsov, V. Pas'ko, S. M. Barnett, and S. Franke-Arnold. Free-space information transfer using light beams carrying orbital angular momentum. *Opt. Express*, 12:5448–5456, 2004.

[5] S. Gröblacher, T. Jennewein, A. Vaziri, G. Weihs, and A. Zeilinger. Experimental quantum cryptography with qutrits. *New J. Phys.*, 8:75, 2006.

[6] S. P. Walborn, D. S. Lemelle, M. P. Almeida, and P. H. Souto Ribeiro. Quantum key distribution with higher-order alphabets using spatially encoded qudits. *Phys. Rev. Lett.*, 96:090501, 2006.

[7] G. Vallone, V. D'Ambrosio, A. Sponselli, S. Slussarenko, L. Marrucci, F. Sciarrino, and P. Villoresi. Free-space quantum key distribution by rotation-invariant twisted photons. *Phys. Rev. Lett.*, 113:060503, 2014.

[8] S. K. Goyal, A. Hamadou Ibrahim, F. S. Roux, T. Konrad, and A. Forbes. The effect of turbulence on entanglement-based free-space quantum key distribution with photonic orbital angular momentum. *J. Opt.*, 18:064002, 2016.

[9] S. Goyal, P. E. Boukama-Dzoussi, S. Ghosh, F. S. Roux, and T. Konrad. Qudit-teleportation for photons with linear optics. *Sci. Rep.*, 4:4543, 2014.

[10] D. Bouwmeester, J.-W. Pan, K. Mattle, M. Eibl, H. Weinfurter, and A. Zeilinger. Experimental quantum teleportation. *Nature*, 390:575–579, 1997.

[11] D. V. Strekalov, A. V. Sergienko, D. N. Klyshko, and Y. H. Shih. Observation of two-photon "ghost" interference and diffraction. *Phys. Rev. Lett.*, 74:3600–3603, 1995.

[12] J. H. Shapiro and R. W. Boyd. The physics of ghost imaging. *Quantum Inf. Process.*, 11:949–993, 2012.

[13] V. Giovannetti, S. Lloyd, and L. Maccone. Quantum-enhanced positioning and clock synchronization. *Nature*, 412:417–419, 2001.

[14] R. Quan, Y. Zhai, M. Wang, F. Hou, S. Wang, X. Xiang, T. Liu, S. Zhang, and R. Dong. Demonstration of quantum synchronization based on second-order quantum coherence of entangled photons. *Sci. Rep.*, 6:30453, 2016.

[15] R. Loudon and P. L. Knight. Squeezed light. *J. Mod. Opt.*, 34:709–759, 1987.

[16] T. S. Iskhakov, I. N. Agafonov, M. V. Chekhova, and G. Leuchs. Polarization-entangled light pulses of 10^5 photons. *Phys. Rev. Lett.*, 109:150502, 2012.

[17] M. V. Chekhova, G. Leuchs, and M. Żukowski. Bright squeezed vacuum: Entanglement of macroscopic light beams. *Opt. Commun.*, 337:27–43, 2015.

[18] U. L. Andersen, T. Gehring, C. Marquardt, and G. Leuchs. 30 years of squeezed light generation. *Phys. Scr.*, 91:053001, 2016.

[19] T. C. Zhang, K. W. Goh, C. W. Chou, P. Lodahl, and H. J. Kimble. Quantum teleportation of light beams. *Phys. Rev. A*, 67:033802, 2003.

[20] S. Pirandola, J. Eisert, C. Weedbrook, A. Furusawa, and S. L. Braunstein. Advances in quantum teleportation. *Nat. Photonics*, 9:641–652, 2015.

[21] A. Gatti, L. A. Lugiato, G.-L. Oppo, R. Martin, P. Di Trapani, and A. Berzanskis. From quantum to classical images. *Opt. Express*, 1:21–30, 1997.

[22] A. Gatti, E. Brambilla, and L. A. Lugiato. Entangled imaging and wave-particle duality: from the microscopic to the macroscopic realm. *Phys. Rev. Lett.*, 90:133603, 2003.

[23] S. Lloyd. Enhanced sensitivity of photodetection via quantum illumination. *Science*, 321:1463–1465, 2008.
[24] F. Dell'Anno, S. De Siena, and F. Illuminati. Multiphoton quantum optics and quantum state engineering. *Phys. Rep.*, 428:53–168, 2006.
[25] V. Giovannetti, S. Lloyd, and L. Maccone. Advances in quantum metrology. *Nat. Photonics*, 5:222–229, 2011.
[26] P. R. Sharapova, G. Frascella, M. Riabinin, A. M. Pérez, O. V. Tikhonova, S. Lemieux, R. W. Boyd, G. Leuchs, and M. V. Chekhova. Properties of bright squeezed vacuum at increasing brightness. *Phys. Rev. Res.*, 2:013371, 2020.
[27] S. Mrowczynski and B. Mueller. Wigner functional approach to quantum field dynamics. *Phys. Rev. D*, 50:7542–7552, 1994.
[28] F. S. Roux. Combining spatiotemporal and particle-number degrees of freedom. *Phys. Rev. A*, 98:043841, 2018.
[29] F. S. Roux. Erratum: Combining spatiotemporal and particle-number degrees of freedom [Phys. Rev. A 98, 043841 (2018)]. *Phys. Rev. A*, 101:019903(E), 2020a.
[30] F. S. Roux. Evolution equation for multi-photon states in turbulence. *J. Phys. A, Math. Theor.*, 52:405301, 2019.
[31] F. S. Roux. Quantifying entanglement of parametric down-converted states in all degrees of freedom. *Phys. Rev. Res.*, 2:023137, 2020b.
[32] M. E. Peskin and D. V. Schroeder. *An Introduction to Quantum Field Theory*. Addison-Wesley Publishing Company, Reading, Massachusetts, USA, 1995.
[33] P. A. Zyla and et al. Particle data group. *Prog. Theor. Exp. Phys.*, 2020:083C01, 2020.
[34] F. S. Roux. Parametric down-conversion beyond the semiclassical approximation. *Phys. Rev. Res.*, 2:033398, 2020c.
[35] P. Sharapova, A. M. Pérez, O. V. Tikhonova, and M. V. Chekhova. Schmidt modes in the angular spectrum of bright squeezed vacuum. *Phys. Rev. A*, 91:043816, 2015.
[36] W. Wasilewski, A. I. Lvovsky, K. Banaszek, and C. Radzewicz. Pulsed squeezed light: Simultaneous squeezing of multiple modes. *Phys. Rev. A*, 73(6):063819, 2006.
[37] W. Magnus. On the exponential solution of differential equations for a linear operator. *Commun. Pure Appl. Math.*, VII:649–673, 1954.
[38] Y. R. Shen. *The Principles of Nonlinear Optics*. John Wiley & Sons, New York, USA, 2003.
[39] M. Abramowitz and I. A. Stegun. *Handbook of Mathematical Functions*. Dover, Toronto, 1972.

9 Applications in parametric down-conversion

The Wigner functional formalism, developed in Chapters 4 to 7, makes it possible to perform a comprehensive analysis of parametric down-conversion, incorporating all the degrees of freedom. Such an analysis is provided in Chapter 8. Since parametric down-conversion is a widely used process in quantum optics [1–17] and photonic quantum information systems [18–24], there are numerous applications that can benefit from that analysis. In this chapter, a few of these applications are discussed.

The state produced by *spontaneous parametric down-conversion* is a *squeezed vacuum state*. Therefore, we use the results provided in Chapter 7 for measurements on squeezed vacuum states. The expressions for the squeezed vacuum state kernels A and B, obtained in Chapter 8, are employed to provide detailed calculations.

Often the calculations in applications of Wigner functional theory can be broken down into different stages. The first stage is the modelling process of the scenario in terms of Wigner functionals, which leads to the evaluation of some functional integrals. The results are usually in the form of contracted kernels and function. The evaluation of the integrals for these contractions is the second stage of the calculation process. It is often more challenging than the first stage. Therefore, it is instructive to consider some examples with challenging wavevector and z-integrations.

In the next two sections, we consider measurements that respectively involve the A and B kernels of the squeezed vacuum state, discussing the calculations involving the wavevector and z-integrations in detail. In the first of these sections, we discuss the measurement of the photon-number distribution (intensity function) produced by the down-conversion process, as governed by A. In the subsequent section, correlation measurements are discussed, which are mediated by B. Various other applications are addressed in the subsequent sections.

9.1 Imaging the down-converted field

To perform the measurement of the far-field intensity of a squeezed vacuum state, as produced by spontaneous parametric down-conversion, we use the setup shown in Figure 9.1. The down-converted light emerging from the nonlinear crystal is Fourier transformed with a $2f$ system, as discussed in Section 7.2, producing the far-field intensity as a function of the transverse coordinates in the back-focal plane of the lens. In this plane, a CCD array is placed. It consists of small detector elements (pixels) located in an array on the output plane. The lens diameter is assumed to be large enough so that all the down-converted light can pass through it unobstructed. We also assume that the detector elements are smaller than the *resolution* of the intensity function in the output plane. So, we ignore the size of the detector elements by assuming them to be infinitely small.

The expression for the intensity of a squeezed vacuum state is given in terms of the photon-number distribution in (7.70). The detector kernel for a far-field single-mode

https://doi.org/10.1515/9783111445342-012

detector system is given in terms of the detector modes provided in (7.27). It is assumed that the experimental conditions for the spontaneous parametric down-conversion process is such that we can use the expressions of the squeezed vacuum state kernels given in (8.123) under the *combined rational approximation*. If we use the kernels with the *extreme thin-crystal approximation*, the result would be a constant without any dependence on the output coordinates.

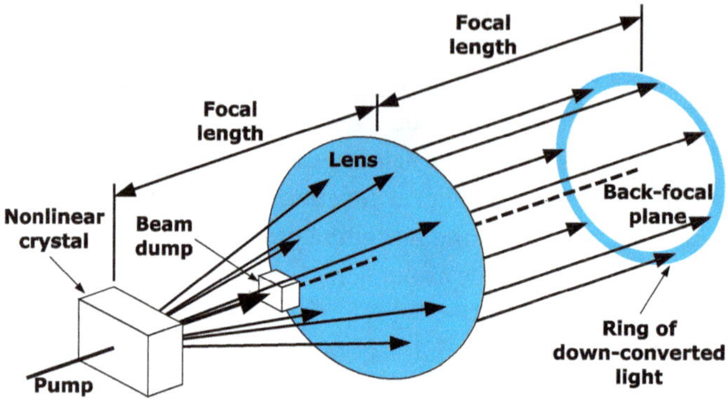

Figure 9.1: Setup for the measurement of the far-field intensity of spontaneously parametric down-converted light.

The detector kernel is modelled as a single-mode kernel with $D = MM^*$, leading to the overlap $\frac{1}{2}M^* \diamond (A - \mathbf{1}) \diamond M$. The detector mode is modelled by the real-valued Gaussian function given in (7.27). Therefore, M^* and M are given by the same function, with different optical beam variables. They also have the same parameter values for the modes size, centre angular frequency, and bandwidths. As a result, the overlap enforces degeneracy in the kernels.

The photon-number distribution (proportional to the intensity) in the far-field is represented by

$$\langle n\rangle_{\text{spdc}} = \frac{1}{2}M^* \diamond (A - \mathbf{1}) \diamond M = \frac{1}{2}\sum_{p=1}^{\infty} M^* \diamond H_{2p}^{(\text{e})} \diamond M. \tag{9.1}$$

The expression for $H_{2p}^{(\text{e})}$ is given in (8.123).

Here, we perform the calculation in the Fourier domain (as usual), but in the plane of the crystal. Since the far-field distribution is related to the crystal plane via a Fourier transform, this calculation is equivalent to performing the calculation in terms of the *configuration space* coordinates in the far field. Assuming a small enough resolution, the far-field intensity function is thus obtained from the Fourier domain calculation in the crystal plane by a direct replacement of the transverse wavevectors in terms of the far-field coordinates.

9.1.1 Integrating over optical beam variables

The integral that needs to be evaluated is

$$
\begin{aligned}
M^* \diamond H_m^{(e)} \diamond M = {} & 2\pi w_0^2 \frac{\Lambda_0 \Lambda_1^m}{m^{5/4}} \int \int_z \mathcal{Z}\left\{[(\omega_p - \omega_1)\omega_1]^{m/2} h(\omega_1 - \omega_2, \sqrt{m}\delta_p)\right. \\
& \times h(\omega_1 - \omega_d, \delta_d) h(\omega_2 - \omega_d, \delta_d) \exp\left(-\frac{1}{4} w_0^2 |\mathbf{K}_1 - \mathbf{K}_0|^2\right) \\
& \times \exp\left(-\frac{1}{4} w_0^2 |\mathbf{K}_2 - \mathbf{K}_0|^2\right) \exp\left[-\frac{w_p^2}{4m} |\mathbf{K}_1 - \mathbf{K}_2|^2 - \mathrm{i} Z_m \chi(\omega_1)\right. \\
& \left.\left. + \mathrm{i} \frac{Z_m k(\omega_p) |\mathbf{K}_1 + \mathbf{K}_2|^2}{8 k_z(\omega_1) k_z(\omega_p - \omega_1)}\right]\right\} \mathrm{d_b} k_1 \, \mathrm{d_b} k_2 \, \mathrm{d}\{z\},
\end{aligned}
\tag{9.2}
$$

where $\mathcal{Z}\{\cdot\}$ is the z-symmetrization defined in (8.71). In (7.29), it is shown how the shift $\mathbf{K}_0$ is related to the pixel location in the detector plane $\mathbf{X}_0 = \{x_0, y_0\}$, where f is the focal length of the lens in the $2f$ system, and $\lambda = \lambda_d$ is the down-converted wavelength. Moreover, the mode size w_0 is related to the pixel size by (7.28). The three h-functions are modelled as normalized Gaussian-shaped spectral functions, distinguished by their parameters. The centre angular frequency of the pump is twice the degenerate down-converted centre angular frequency $\omega_p = 2\omega_d$ and the bandwidths δ_p and δ_d are those for the pump and the down-converted light, respectively. It is assumed that these bandwidths are small enough to consider them as monochromatic so that we can replace all the angular frequencies in the rest of the expression by their associated centre angular frequencies. Then we evaluate the integrals over the *angular frequencies* to produce

$$
\int h(\omega_1 - \omega_2, \sqrt{m}\delta_p) h(\omega_1 - \omega_d, \delta_d) h(\omega_2 - \omega_d, \delta_d) \frac{\mathrm{d}\omega_1}{2\pi} \frac{\mathrm{d}\omega_2}{2\pi} = \frac{\sqrt{2} m^{1/4} \delta_p^{1/2} \delta_d}{\pi^{1/4} \sqrt{2\delta_d^2 + m\delta_p^2}}. \tag{9.3}
$$

The square root factor in the denominator shows how the bandwidth increases for the higher orders. Its m-dependence makes summations complicated. Here, we assume that the bandwidth of the pump is small enough, so that we can take the limit $\delta_p \to 0$. This limit is well-defined thanks to the factor of $\delta_p^{-1/2}$ in Λ_0, given in (8.120). It may seem that for any physical value of the bandwidth there exists a value of m where the term for the pump in the square root starts to dominate. However, the value of m is limited by the point in the expansion of the kernel where the terms start to become smaller. Such a point must exist to ensure convergence. This point happens roughly where the value of m becomes comparable to the squeezing parameter Ξ, which is defined below. Therefore, the monochromatic assumption for the pump requires that $\delta_p^2 \ll 2\delta_d^2/\Xi$. Provided that this condition holds, we can discard the bandwidth factors in the expression.

The Gaussian integrals over the transverse wavevectors in (9.2) are evaluated next, leading to some simplifications. The complete expression thus becomes

$$M^* \diamond H_m^{(e)} \diamond M = \int_z \mathcal{Z}\left\{\frac{\Omega'_m}{w_0^2 k_p + i4Z_m} \exp\left[i\frac{2Z_m w_0^2 |\mathbf{K}_0|^2}{w_0^2 k_p - i4Z_m} - iZ_m \chi(\omega_d)\right]\right\} d\{z\}, \quad (9.4)$$

where $k_p = k(\omega_p)$ is the pump wavenumber, and

$$\Omega'_m = \frac{w_0^2 w_p^2 k_p \Lambda_1^m \omega_d^m}{\left(m w_0^2 + 2 w_p^2\right)}. \quad (9.5)$$

It is interesting to note that, if we used the kernels for the extreme thin-crystal approximation in (8.118), the term with $\mathbf{K}_0$ in the exponent would be absent thus removing the coordinate dependence in the output photon-number distribution.

9.1.2 Integrating over *z*

Before addressing the *z*-integration, we apply a *thin-crystal approximation* on the detector mode size. Since we are not interested in the effect of the size of the pixels, we let it go to zero; by implication $w_0 \to \infty$. Even for a finite pixel size, the size of w_0 is large enough to discard the *z*-dependent part in the denominators, so that $w_0^2 k_p - i4Z_m \to w_0^2 k_p$. Hence,

$$M^* \diamond H_m^{(e)} \diamond M = \Omega_m \int_z \mathcal{Z}\left\{\exp\left[i\frac{2Z_m |\mathbf{K}_0|^2}{k_p} - iZ_m \chi(\omega_d)\right]\right\} d\{z\}, \quad (9.6)$$

where

$$\Omega_m = \frac{w_p^2 \Lambda_1^m \omega_d^m}{m w_0^2} = \frac{\eta_w \Lambda_1^m \omega_d^m}{m}. \quad (9.7)$$

In the last expression, we replaced $w_p^2/w_0^2 \to \eta_w < 1$, which represents a reduced efficiency due to the difference in mode sizes (the pixel size, which is inversely proportional to w_0 is assumed to be much smaller than the pump mode size).

For the *z*-integrations, we expand Z_m in terms of (8.125) and perform the symmetrization. Then we evaluate the *z*-integrals in sequence with their integration boundaries, as given in (8.70). The individual *z*-integrations are quite simple, but in the end we need an expression for arbitrary *m*. With $m = 2p$ for $p > 0$, it is given by

$$\int_z \mathcal{Z}\left\{\exp\left(iZ_{2p}\kappa\right)\right\} d\{z\} = \sum_{n=0}^{\infty} \frac{(-1)^n (n+p-1)!(z-z_0)^{2p+2n}\kappa^{2n}}{n!(p-1)!(2p+2n)!}, \quad (9.8)$$

where z_0 is the lower bound of the *z*-integrations, and

$$\kappa \triangleq \frac{2|\mathbf{K}_0|^2}{k_p} - \chi(\omega_d). \quad (9.9)$$

For z-integration boundaries $-\frac{1}{2}L$ and $\frac{1}{2}L$, the complete expression for the photon-number distribution is then given by

$$\langle n\rangle_{\text{spdc}} = \frac{\eta_w}{4}\sum_{p=1}^{\infty}\sum_{n=0}^{\infty}\frac{(-1)^n(n+p-1)!L^{2p+2n}\kappa^{2n}\Lambda_1^{2p}\omega_{\text{d}}^{2p}}{n!p!(2p+2n)!}, \tag{9.10}$$

where Λ_1 is defined in (8.120).

Exercise 9.1. Compute the z-integral on the left-hand side of (9.8) for $p = 1, 2, 3$ and show that the results agree with the expression on the right-hand side.

9.1.3 Summations

To evaluate the summations, we first redefine the index $n \to q - p$. Then we change the order of summation so that

$$\sum_{p=1}^{\infty}\sum_{q=p}^{\infty} f_{p,q} = \sum_{q=1}^{\infty}\sum_{p=1}^{q} f_{p,q}. \tag{9.11}$$

Now, we can evaluate the summation over p from 1 to q. The result is

$$\langle n\rangle_{\text{spdc}} = \frac{\eta_w}{4}\sum_{q=1}^{\infty}\frac{\left[\Xi^2 - \rho^2(\mathbf{K}_0)\right]^q - (-1)^q\rho^{2q}(\mathbf{K}_0)}{q(2q)!}, \tag{9.12}$$

where

$$\Xi \triangleq \Lambda_1 L\omega_{\text{d}} = \frac{2\sqrt{2}L|\zeta_0|\sigma_{\text{I}}\omega_{\text{p}}^{3/2}\delta_{\text{p}}^{1/2}}{\pi^{3/4}c^2 w_{\text{p}}}, \tag{9.13}$$

represents the *effective squeezing parameter*, and

$$\rho(\mathbf{K}_0) \triangleq L\kappa = \frac{2L|\mathbf{K}_0|^2}{k_{\text{p}}} - L\chi(\omega_{\text{d}}), \tag{9.14}$$

in terms of (9.9).

The summation over q is complicated by the factor of q^{-1}. To address this challenge, we replace this factor by an auxiliary integral

$$\frac{1}{q} \to \int_0^1 2y^{2q-1}\,\text{d}y. \tag{9.15}$$

After evaluating the summation over q, we obtain an expression for which we can then evaluate the integral over y. Hence,

$$\begin{aligned}\langle n\rangle_{\text{spdc}} &= \frac{\eta_w}{2}\int_0^1 \frac{1}{y}\cosh\left(y\sqrt{\Xi^2-\rho^2}\right)-\frac{1}{y}\cos(y\rho)\ \mathrm{d}y \\ &= \frac{\eta_w}{2}\left[\mathrm{Chi}\left(\sqrt{\Xi^2-\rho^2}\right)-\mathrm{Ci}(\rho)-\frac{1}{2}\ln\left(\Xi^2-\rho^2\right)+\ln(\rho)\right],\end{aligned} \tag{9.16}$$

for $\Xi^2 \geq \rho^2$, and

$$\begin{aligned}\langle n\rangle_{\text{spdc}} &= \frac{\eta_w}{2}\int_0^1 \frac{1}{y}\cos\left(y\sqrt{\rho^2-\Xi^2}\right)-\frac{1}{y}\cos(y\rho)\ \mathrm{d}y \\ &= \frac{\eta_w}{2}\left[\mathrm{Ci}\left(\sqrt{\rho^2-\Xi^2}\right)-\mathrm{Ci}(\rho)-\frac{1}{2}\ln\left(\rho^2-\Xi^2\right)+\ln(\rho)\right],\end{aligned} \tag{9.17}$$

for $\Xi^2 < \rho^2$, where the Chi- and Ci-functions are defined as [25],

$$\begin{aligned}\mathrm{Ci}(z) &= \gamma + \ln(z) + \int_0^z \frac{\cos(t-1)}{t}\,\mathrm{d}t, \\ \mathrm{Chi}(z) &= \gamma + \ln(z) + \int_0^z \frac{\cosh(t-1)}{t}\,\mathrm{d}t,\end{aligned} \tag{9.18}$$

for $\mathrm{Re}\{z\} > 0$.

9.1.4 Parameters and variables

The variables and parameters in the arguments of the results in (9.16) and (9.17) can be simplified. In the degenerate case (as imposed by the single-mode detector kernel), we have $\omega_{\mathrm{p}} = 2\omega_{\mathrm{d}}$ or $\lambda_{\mathrm{d}} = 2\lambda_{\mathrm{p}}$. Then, from (8.97), we get

$$\chi(\omega_{\mathrm{d}}) = \frac{k_{\mathrm{d}}\sin^2(\theta_{\mathrm{d}})}{\cos(\theta_{\mathrm{d}})} = \frac{2\pi n_{\mathrm{o}}(\omega_{\mathrm{d}})}{\lambda_{\mathrm{d}}}\frac{\sin^2(\theta_{\mathrm{d}})}{\cos(\theta_{\mathrm{d}})} = \frac{\pi n_{\mathrm{p}}}{\lambda_{\mathrm{p}}}\tan^2(\theta_{\mathrm{d}}). \tag{9.19}$$

The degenerate phase-matching conditions imply that $k_z(\omega_{\mathrm{d}}) = \frac{1}{2}k_{\mathrm{p}}$, which also means that $n_{\mathrm{p}} = n_{\mathrm{eff}}(\omega_{\mathrm{p}}) = n_{\mathrm{o}}(\omega_{\mathrm{d}})\cos(\theta_{\mathrm{d}})$. Together with (7.29), it follows that ρ in (9.14) is

$$\rho = \frac{\pi L}{n_{\mathrm{p}}\lambda_{\mathrm{p}}f^2}|\mathbf{X}_0|^2 - \frac{\pi L n_{\mathrm{p}}}{\lambda_{\mathrm{p}}}\tan^2(\theta_{\mathrm{d}}) = \frac{r^2-r_0^2}{R^2}, \tag{9.20}$$

where $r^2 = |\mathbf{X}_0|^2$,

$$R = f\sqrt{\frac{n_\mathrm{p}\lambda_\mathrm{p}}{\pi L}} \quad \text{and} \quad r_0 = fn_\mathrm{p}\tan(\theta_\mathrm{d}). \tag{9.21}$$

The function in (9.16) thus becomes

$$\begin{aligned}\langle n\rangle_\mathrm{spdc} = \frac{w_\mathrm{p}^2}{2w_0^2}\Bigg[&\mathrm{Chi}\left(\sqrt{\Xi^2 - \frac{(r^2-r_0^2)^2}{R^4}}\right) - \mathrm{Ci}\left(\frac{r^2-r_0^2}{R^2}\right) \\ &- \frac{1}{2}\ln\left(\Xi^2 - \frac{(r^2-r_0^2)^2}{R^4}\right) + \ln\left(\frac{r^2-r_0^2}{R^2}\right)\Bigg],\end{aligned} \tag{9.22}$$

and similar for (9.17).

A three-dimensional plot of the spontaneous parametric down-converted photon-number distribution in the far field (in the output plane of a $2f$ system), as given in (9.22), is shown in Figure 9.2 for weak squeezing with $\Xi = 1$. The ripples are an indication that the squeezing parameter is weak.

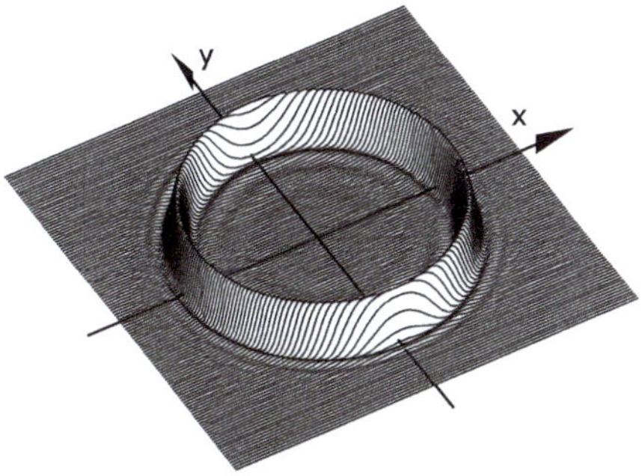

Figure 9.2: Output photon-number distribution after a $2f$ system produced from non-collinear spontaneous parametric down-conversion with $\Xi = 1$ and $r_0 = 5R$.

When the squeezing becomes stronger, the ring in Figure 9.2 becomes broader and the ripples disappear. When the down-conversion angle (represented by r_0) becomes small compared to the ring width the interior of the ring starts to fill up, and eventually forms a blob at the origin in the collinear case.

9.1.5 With dispersion

In Section 8.6.8, the effects of dispersion in parametric down-conversion is briefly discussed. Here, we illustrate these effects with a detailed calculation in the context of the intensity measurements discussed above. The origin of these effects is an additional phase factor produced by the dispersion in the expansion of Δk_z. Based on (8.135), this phase has the form $\psi_{\text{disp}} = zk_{\text{disp}}(\omega_1, \omega_2)$, where

$$\begin{aligned} k_{\text{disp}}(\omega_1, \omega_2) = {} & \frac{\omega_1 + \omega_2 - \omega_{\text{p}}}{v_{\text{g}}(\omega_{\text{p}})} - \frac{\omega_1 - \omega_{\text{d}}}{v_{\text{g}}(\omega_{\text{d}})}C - \frac{\omega_2 - \omega_{\text{d}}}{v_{\text{g}}(\omega_{\text{d}})}C \\ & + \frac{1}{2}(\omega_1 + \omega_2 - \omega_{\text{p}})^2 D_{\text{g}}(\omega_{\text{p}}) - \frac{1}{2}C(\omega_1 - \omega_{\text{d}})^2 D_{\text{g}}(\omega_{\text{d}}) \\ & - \frac{1}{2}C(\omega_2 - \omega_{\text{d}})^2 D_{\text{g}}(\omega_{\text{d}}), \end{aligned} \tag{9.23}$$

after we replaced $\omega_3 \to \omega_1 + \omega_2$, based on the phase-matching conditions. Since it depends on the *angular frequencies*, this phase affects the integration represented in (9.3) over these angular frequencies. Here, the result becomes

$$\begin{aligned} h_{\text{disp}}(Z_m) = {} & \int \exp[\text{i}Z_m k_{\text{disp}}(\omega_1, \omega_2)] h(\omega_1 - \omega_2, \sqrt{m}\delta_{\text{p}}) h(\omega_1 - \omega_{\text{d}}, \delta_{\text{d}}) \\ & \times h(\omega_2 - \omega_{\text{d}}, \delta_{\text{d}}) \frac{\text{d}\omega_1}{2\pi} \frac{\text{d}\omega_2}{2\pi} \\ = {} & \frac{A_0}{\sqrt{1 + \text{i}Z_m A_1}\sqrt{1 - \text{i}Z_m A_2}} \exp\left[-\frac{Z_m^2}{(1 + \text{i}Z_m A_1)w_z^2}\right], \end{aligned} \tag{9.24}$$

where $C = \cos(\theta_{\text{d}}) = n_{\text{eff}}(\omega_{\text{p}})/n_{\text{o}}(\omega_{\text{d}})$, and

$$\begin{aligned} A_0 &= \frac{4\pi^{7/4}\delta_{\text{d}}\sqrt{2\delta_{\text{p}}}}{\sqrt{\delta_{\text{p}}^2 + 2\delta_{\text{d}}^2}}, \quad & A_1 &= [2D_{\text{g}}(\omega_{\text{p}}) - D_{\text{g}}(\omega_{\text{d}})C]\delta_{\text{d}}^2, \\ A_2 &= \frac{D_{\text{g}}(\omega_{\text{d}})C\delta_{\text{p}}^2\delta_{\text{d}}^2}{\delta_{\text{p}}^2 + 2\delta_{\text{d}}^2}, \quad & w_z &= \frac{v_{\text{g}}(\omega_{\text{p}})v_{\text{g}}(\omega_{\text{d}})}{|v_{\text{g}}(\omega_{\text{d}}) - v_{\text{g}}(\omega_{\text{p}})C|\delta_{\text{d}}}. \end{aligned} \tag{9.25}$$

To facilitate numerical evaluation of these quantities, we express the group velocity and the group velocity dispersion as

$$v_{\text{g}} = \frac{c}{n_{\text{g}}} \quad \text{and} \quad D_{\text{g}} = \frac{\Delta n_{\text{g}}}{c^2}, \tag{9.26}$$

respectively, using the expressions in Appendix B.3. We also express the angular frequency bandwidths in terms of wavelength bandwidth by

$$\delta_{\text{p}} = \frac{2\pi c}{\lambda_{\text{p}}^2}\Delta\lambda_{\text{p}} \quad \text{and} \quad \delta_{\text{d}} = \frac{2\pi c}{\lambda_{\text{d}}^2}\Delta\lambda_{\text{d}} = \frac{\pi c}{2\lambda_{\text{p}}^2}\Delta\lambda_{\text{d}}. \tag{9.27}$$

The quantities then become

$$
\begin{aligned}
A_0 &= \frac{4\pi^{9/4}\Delta\lambda_\mathrm{d}}{\lambda_\mathrm{p}}\sqrt{\frac{2c\Delta\lambda_\mathrm{p}}{8\Delta\lambda_\mathrm{p}^2+\Delta\lambda_\mathrm{d}^2}},\\
A_1 &= \frac{\pi^2\Delta\lambda_\mathrm{d}^2}{4n_\mathrm{o}(\omega_\mathrm{d})\lambda_\mathrm{p}^4}[2n_\mathrm{o}(\omega_\mathrm{d})\Delta n_\mathrm{g}(\omega_\mathrm{p})-n_\mathrm{eff}(\omega_\mathrm{p})\Delta n_\mathrm{g}(\omega_\mathrm{d})],\\
A_2 &= \frac{2\pi^2 n_\mathrm{eff}(\omega_\mathrm{p})\Delta n_\mathrm{g}(\omega_\mathrm{d})\Delta\lambda_\mathrm{p}^2\Delta\lambda_\mathrm{d}^2}{(\Delta\lambda_\mathrm{p}^2+2\Delta\lambda_\mathrm{d}^2)n_\mathrm{o}(\omega_\mathrm{d})\lambda_\mathrm{p}^4},\\
w_z &= \frac{2n_\mathrm{o}(\omega_\mathrm{d})\lambda_\mathrm{p}^2}{[n_\mathrm{o}(\omega_\mathrm{d})n_\mathrm{g}(\omega_\mathrm{p})-n_\mathrm{eff}(\omega_\mathrm{p})n_\mathrm{g}(\omega_\mathrm{d})]\pi\Delta\lambda_\mathrm{d}}.
\end{aligned}
\tag{9.28}
$$

In Figure B.3, we see that Δn_g, which has the units of a distance, is much smaller than the wavelength. In addition, the fractional bandwidth $\Delta\lambda/\lambda$ is also very small. As a result, it follows that $A_1 L \ll 1$ and $A_2 L \ll 1$. Therefore, we can discard the z-dependent terms in the denominators in (9.24), leading to the simpler expression

$$
h_\mathrm{disp}(Z_m) \approx A_0 \exp\left(-\frac{Z_m^2}{w_z^2}\right). \tag{9.29}
$$

The resulting function is a Gaussian envelope as a function of Z_m. The integrand of the z-integrals in (9.8) can thus be replaced by

$$
\exp(\mathrm{i}Z_{2r}\kappa) \to \exp\left(-\frac{Z_{2r}^2}{w_z^2}+\mathrm{i}Z_{2r}\kappa\right). \tag{9.30}
$$

Since w_z is inversely proportional to $\Delta\lambda_\mathrm{d}$, we see that the width of this envelope depends on the width of the observed down-converted spectrum. For $L \gg w_z$, we can extend parts of these z-integrals to infinity. The result can therefore be approximate by

$$
M^* \diamond H_m^{(\mathrm{e})} \diamond M \approx \mathcal{N}_m(L,w_z)\exp\left(-\frac{1}{4}w_z^2\kappa^2\right), \tag{9.31}
$$

where $\mathcal{N}_m(L,w_z)$ is a constant factor that depends on the experimental parameters, but not on the output coordinates. It shows that the part of the result that depends on the output plane coordinates is a Gaussian function that is independent of L and m. So, this Gaussian function factors out of the summation over m, leaving the latter to produce a constant independent of the output plane coordinates. Retaining only the dominant term at each order, we get

$$
M^* \diamond H_m^{(\mathrm{e})} \diamond M \approx \frac{\sqrt{\pi}w_z L^{m-1}}{(m-1)!}\exp\left(-\frac{1}{4}w_z^2\kappa^2\right). \tag{9.32}
$$

When this result is inserted into the summation over m, it evaluates to

$$
\langle n\rangle_\mathrm{spdc} \approx \frac{\sqrt{\pi}\eta_w w_z}{2L}[\cosh(\Xi)-1]\exp\left(-\frac{1}{4}w_z^2\kappa^2\right). \tag{9.33}
$$

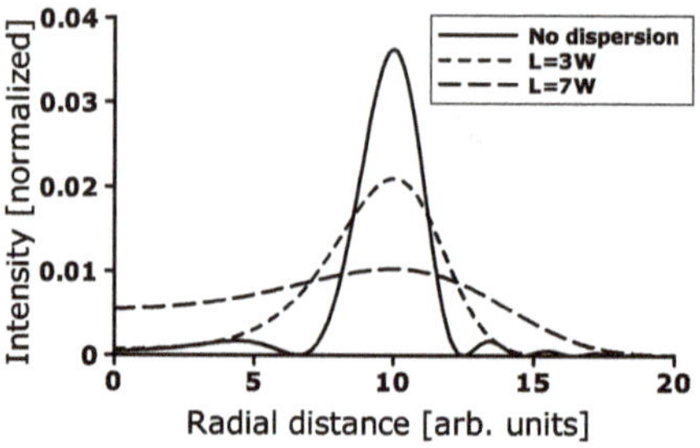

Figure 9.3: Comparison of the shapes of the intensity profiles of the down-converted light in the weak squeezing limit with different amounts of dispersion as functions of the radial coordinate.

In Figure 9.3, the effect of this dispersion is shown as a comparison among the cases without dispersion and with dispersion for $L = 3w_z$ and $L = 7w_z$ in the weak squeezing limit. It shows that the dispersion causes the ring that is produced in non-collinear phase-matching to become broader. When squeezing becomes stronger, the width of the ring is increased further.

In the opposite limit where $L \ll w_z$, the result of Section 9.1.3 is recovered. The latter situation represents the preferred experimental conditions.

Exercise 9.2. Compute the numerical values of A_1 and A_2, using the expressions in Appendices B.2 and B.3, for the case where $\lambda_p = 0.4\,\mu\text{m}$ in a BBO crystal with a down-conversion angle of 10 degrees. Show that $A_1L \ll 1$ and $A_2L \ll 1$ for a crystal thickness of $L = 1$ mm.

9.2 Second-order correlations

In Section 7.6.6, we consider the basic measurements of second-order correlations in squeezed vacuum states. Different approaches for these calculations are presented there. With the benefit of having derived the expressions for the squeezed vacuum state kernels produced in spontaneous parametric down-conversion in Chapter 8, we can proceed to compute the *probability* for the detection of two (signal and idler) photons in coincidence. For this purpose, we combine the conditions for single-mode detection (7.196) and weak squeezing (7.208), which gives the second-order correlation as

$$\Gamma^{(2)} = \frac{1}{4}|M_s \diamond B^* \diamond M_i|^2. \tag{9.34}$$

The detector modes M_s and M_i differ only in their locations, as represented by $\mathbf{K}_0$ in (7.27). We assume thin-crystal conditions, for which the squeezed vacuum state kernel B^* can be represented in terms of the complex conjugates of (8.124) and (8.126). A similar setup to the one used for imaging in Section 9.1 is used here for correlation measurements. This setup, which incorporates a $2f$ system, is shown in Figure 9.4.

The overlap that we need to calculate in full is represented by β_{si}, as defined in (7.180). Here, we use (6.340) so that we can represent the *crossed overlap* by

$$\beta_{\text{si}} = M_{\text{s}} \diamond B^* \diamond M_{\text{i}} = \sum_{p=0}^{\infty} M_{\text{s}} \diamond H_{2p+1}^{(\text{o})} \diamond M_{\text{i}} \triangleq \sum_{p=0}^{\infty} \beta_{2p+1}. \tag{9.35}$$

It differs from the overlap considered for the intensity in that it involves the odd kernel B^*, given in (8.124). Moreover, the two modes are located at different points $\mathbf{x}_1$ and $\mathbf{x}_2$ in the detector plane. Here, we use the *combined rational approximation* (as we did for the intensity measurement) and not the extreme thin-crystal approximation, because the information about the phase-matching condition is lost in the latter.

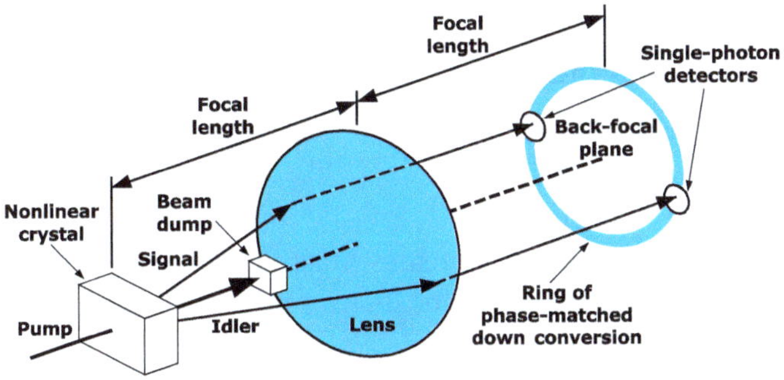

Figure 9.4: Optical $2f$ system for correlation measurements on a squeezed vacuum state produced by spontaneous parametric down-conversion.

9.2.1 Integrating over optical beam variables

The integral expression for the current overlap reads

$$\begin{aligned}\beta_m \triangleq M_{\text{s}} \diamond H_m^{(\text{o})} \diamond M_{\text{i}} = \text{i}\frac{\Lambda_0\Lambda_1^m}{m^{5/4}} \int\int_z \mathcal{Z}\Big\{ [(\omega_{\text{p}} - \omega_1)\omega_1]^{m/2}\, h(\omega_1 + \omega_2 - \omega_{\text{p}}, \sqrt{m}\delta_{\text{p}}) \\ \times h(\omega_1 - \omega_{\text{d}}, \delta_{\text{d}}) h(\omega_2 - \omega_{\text{d}}, \delta_{\text{d}}) 2\pi w_0^2 \exp\left(-\frac{1}{4} w_0^2 |\mathbf{K}_1 - \mathbf{K}_{\text{s}}|^2\right) \\ \times \exp\left(-\frac{1}{4} w_0^2 |\mathbf{K}_2 - \mathbf{K}_{\text{i}}|^2\right) \exp\Bigg[-\frac{w_{\text{p}}^2}{4m} |\mathbf{K}_1 + \mathbf{K}_2|^2 - \text{i} Z_m \chi(\omega_1) \\ + \text{i}\frac{Z_m k(\omega_{\text{p}}) |\mathbf{K}_1 - \mathbf{K}_2|^2}{8 k_z(\omega_1) k_z(\omega_{\text{p}} - \omega_1)} \Bigg] \Big\}\, \text{d}_{\text{b}}k_1\, \text{d}_{\text{b}}k_2\, \text{d}\{z\},\end{aligned} \tag{9.36}$$

where $\mathcal{Z}\{\cdot\}$ is the z-symmetrization defined in (8.71). Although we now have two detectors instead of the one we had for intensity measurements, we still assume degeneracy so that $\omega_p = 2\omega_d$. Based on (7.29), the shifts of the angular spectra of the signal and idler detector modes are related to the pixel positions on these two detector arrays by

$$\mathbf{K}_{s,i} = \frac{k_d \mathbf{X}_{s,i}}{f} = \frac{\pi \mathbf{X}_{s,i}}{\lambda_p f}, \tag{9.37}$$

where k_d is the free space wavenumber of the down-converted light and f is the focal length of the $2f$ system.

The integrations over the angular frequencies produce the same expression as in (9.3), with the only difference being that m represents an odd integer as opposed to an even integer. We again assume a monochromatic pump and discard the bandwidth factors. The subsequent wavevector integrations then produce

$$\beta_m = \int_z \mathcal{Z}\left\{\frac{-i2\Omega'_m}{w_0^2 k(\omega_p) + i4Z_m} \exp\left[-\frac{1}{4}\frac{w_0^2 w_p |\mathbf{K}_s + \mathbf{K}_i|^2}{m w_0^2 + 2w_p^2} - i\frac{1}{2}\frac{Z_m w_0^2 |\mathbf{K}_s - \mathbf{K}_i|^2}{w_0^2 k(\omega_p) + i4Z_m} + iZ_m \chi(\omega_d)\right]\right\} d\{z\}, \tag{9.38}$$

where Ω'_m is the same as in (9.5), but with m now representing odd integers. The thin-crystal approximation in which the pixel size on the detector plane is assumed to be small is applied, which simplifies the expression to

$$\beta_m = -i2\Omega_m \int_z \mathcal{Z}\left\{\exp\left[-\frac{w_p}{4m}|\mathbf{K}_s + \mathbf{K}_i|^2 - i\frac{Z_m}{2k_p}|\mathbf{K}_s - \mathbf{K}_i|^2 + iZ_m \chi(\omega_d)\right]\right\} d\{z\}, \tag{9.39}$$

with Ω_m given in (9.7) and $k_p = k(\omega_p)$.

9.2.2 Summations and z-integrations

For the z-integrations of the odd powers to all orders, we use the results for even powers with one extra z-integration. The last z-integration for odd powers always comes as a result from the z-integrations of an even power, times a factor of $\exp(-i\kappa z)$.

Exercise 9.3. Show that the z-symmetrizations for odd powers $m = 2p + 1$ with $p = 1, 2, 3$ produce a factor of $\exp(-i\kappa z)$ with the same negative sign in the exponent for the final integration over z.

We follow the same steps as before, but leave the last z-integration unevaluated. The integrand for an odd number of z's has the same form as in (9.8), with $m = 2p + 1$ and

$$\kappa = \frac{|\mathbf{K}_s - \mathbf{K}_i|^2}{2k_p} - \chi(\omega_d). \tag{9.40}$$

Without the last integration, the resulting expression can be obtained from the result in (9.8), by replacing $z_0 \to -\frac{1}{2}L$. Thus, we have

$$\int_z \mathcal{Z}\{\exp(-iZ_{2p+1}\kappa)\}\, d\{z\} = \int_{-L/2}^{L/2} \exp(-i\kappa z) \sum_{n=0}^{\infty} \left(z + \frac{L}{2}\right)^{2p+2n} \times \frac{(-1)^n (n+p-1)!\kappa^{2n}}{n!(p-1)!(2p+2n)!}\, dz, \tag{9.41}$$

for $p > 0$. For $p = 0$ ($m = 1$), we just have

$$\begin{aligned} \beta_1 = M_s \diamond H_1^{(0)} \diamond M_i &= -i2\frac{\eta\Xi}{L} \exp(-\rho_\Sigma) \int_{-L/2}^{L/2} \exp(-iz\kappa)\, dz \\ &= -i2\eta\Xi \exp(-\rho_\Sigma) \operatorname{sinc}\left(\frac{1}{2}L\kappa\right), \end{aligned} \tag{9.42}$$

where we used (9.7) and (9.13), and defined $\operatorname{sinc}(x) \triangleq \sin(x)/x$ and

$$\rho_\Sigma \triangleq \frac{1}{4} w_p^2 |\mathbf{K}_s + \mathbf{K}_i|^2 = \frac{|\mathbf{X}_s + \mathbf{X}_i|^2}{w_c^2}. \tag{9.43}$$

The last expression is represented in terms of the pixel locations with the aid of (9.37), where we define the pump mode size in the detector plane as

$$w_c \triangleq \frac{\lambda_d f}{\pi w_p} = \frac{2\lambda_p f}{\pi w_p}, \tag{9.44}$$

according to (7.28), with $\lambda_d = 2\lambda_p$ being the free space wavelengths.

To deal with the m in the denominator of the exponent in (9.39), we use

$$\frac{1}{a^{s+1}} = \frac{1}{s!} \int_0^\infty y^s \exp(-ay)\, dy, \tag{9.45}$$

to express

$$\frac{1}{2p+1} \exp\left(-\frac{\rho_\Sigma}{2p+1}\right) = \sum_{s=0}^{\infty} \frac{(-\rho_\Sigma)^s}{s!(2p+1)^{s+1}} = \sum_{s=0}^{\infty} \frac{(-\rho_\Sigma)^s}{(s!)^2} \int_0^\infty y^s \exp[-(2p+1)y]\, dy. \tag{9.46}$$

With these substitutions, the expression for the crossed overlap becomes

$$\beta_{\rm si} = \sum_{p=0}^{\infty} M_{\rm s} \diamond H_{2p+1}^{(0)} \diamond M_{\rm i}$$
$$= \beta_1 - {\rm i}2\eta_w \sum_{p=1}^{\infty} \sum_{n,s=0}^{\infty} (\Lambda_1\omega_{\rm d})^{2p+1} \frac{(-1)^{n+s}(n+p-1)!\kappa^{2n}\rho_\Sigma^s}{n!(p-1)!(2r+2n)!(s!)^2}$$
$$\times \int_{-L/2}^{L/2} (z + \frac{1}{2}L)^{2p+2n} \exp(-{\rm i}\kappa z)\, {\rm d}z \int_0^{\infty} y^s \exp[-(2p+1)y]\, {\rm d}y, \tag{9.47}$$

where η_w is defined beneath (9.7).

Now we are ready to evaluate the summations. First, we replace $n \to q - p$ and change the order of the summations. Then we sum over p from 1 to q, after which we sum over q from 1 to ∞, and finally over s from 0 to ∞. The result is

$$\beta_{\rm si} = \beta_1 - {\rm i}2\eta_w \frac{\Xi^3}{L} \int_0^{\infty} \int_{-L/2}^{L/2} \frac{\cosh\left[\frac{1}{L}(z+\frac{1}{2}L)\sqrt{\Xi^2\exp(-2y)-\rho_\Delta^2}\right] - 1}{\Xi^2\exp(-2y)-\rho_\Delta^2}$$
$$\times J_0\left(2\sqrt{y\rho_\Sigma}\right) \exp(-3y) \exp(-{\rm i}\kappa z)\, {\rm d}z\, {\rm d}y, \tag{9.48}$$

where $J_0(\cdot)$ is the Bessel function of the first kind [25], we replaced Λ_1 in terms of the squeezing parameter Ξ as defined in (9.13), and defined

$$\rho_\Delta \triangleq L\kappa = \frac{L|\mathbf{K}_{\rm s} - \mathbf{K}_{\rm i}|^2}{2k_{\rm p}} - L\chi(\omega_{\rm d}) = \frac{r_s^2 - 4r_0^2}{4R^2}, \tag{9.49}$$

with r_0 and R given in (9.21), and $r_s = |\mathbf{X}_{\rm s} - \mathbf{X}_{\rm i}|$. Here, $\mathbf{X}_{\rm s}$ and $\mathbf{X}_{\rm i}$ are respectively related to $\mathbf{K}_{\rm s}$ and $\mathbf{K}_{\rm i}$ by (7.29).

At this point, we can evaluate the remaining z-integral. It produces

$$\beta_{\rm si} = \beta_1 + 2\Xi\eta_w \int_0^{\infty} J_0\left(2\sqrt{y\rho_\Sigma}\right) \exp(-y) \left\{ {\rm i}\frac{2\sin(\frac{1}{2}\rho_\Delta)}{\rho_\Delta} \right.$$
$$+ \frac{\rho_\Delta \exp(-{\rm i}\frac{1}{2}\rho_\Delta)}{\Xi^2\exp(-2y)-\rho_\Delta^2} \left[\cosh\left(\sqrt{\Xi^2\exp(-2y)-\rho_\Delta^2}\right) - 1\right]$$
$$\left. - {\rm i}\frac{\exp(-{\rm i}\frac{1}{2}\rho_\Delta)}{\sqrt{\Xi^2\exp(-2y)-\rho_\Delta^2}} \sinh\left(\sqrt{\Xi^2\exp(-2y)-\rho_\Delta^2}\right) \right\} {\rm d}y. \tag{9.50}$$

The way that ρ_Δ is combined with Ξ in (9.50) is reminiscent of the way ρ is combined with Ξ in (9.16) and (9.17).

The integration over y is more challenging. We can interpret it as an averaging of the expression over different values of Ξ, ranging from the nominal value of Ξ at $y = 0$ to zero for $y \to \infty$. However, the averaging is weighed by the Bessel function and exponential factor $J_0\left(2\sqrt{y\rho_\Sigma}\right)\exp(-y)$, which has its largest value at $y = 0$. Without the y-dependencies inside the curly brackets, the integral over y then only involves the Bessel function and one exponential factor. This integral produces

$$\int_0^\infty J_0(2\sqrt{y\rho_\Sigma})\exp(-y)\,dy = \exp(-\rho_\Sigma). \tag{9.51}$$

However, the y-dependencies inside the curly brackets cannot simply be discarded for the general case without reasonable consideration.

Exercise 9.4. Make the assumption that the y-dependencies inside the curly brackets can be discarded. Derive a simplified expression from (9.50) under this assumption.

Here, we take advantage of the experimental conditions that are normally employed for correlation experiments to simplify the expression. Since such correlation measurements need to avoid the probability for accidental correlations, the down-conversion efficiency is usually reduced to suppress the probability for multiple photon pairs. In other words, the correlations represented by β_{si} are usually measured under *weak squeezing conditions* where the biphoton component dominates. We can therefore restrict our attention to the leading-order term provided in (9.42). The crossed overlap β_{si} is then expressed as

$$\beta_{si} = -i4\eta_w\Xi\exp(-\rho_\Sigma)\operatorname{sinc}\left(\frac{1}{2}\rho_\Delta\right) + O\{\Xi^3\} \approx \beta_1. \tag{9.52}$$

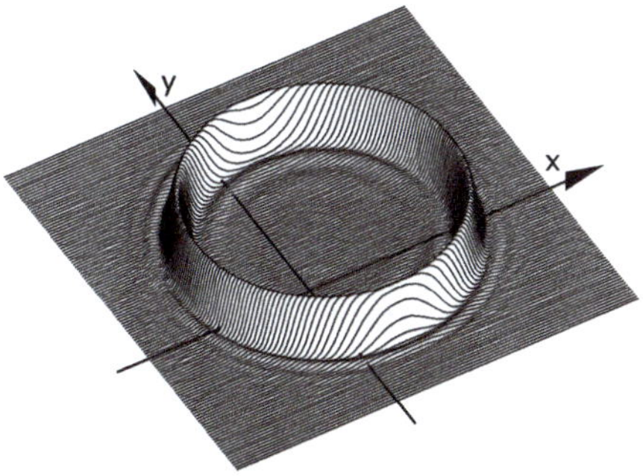

Figure 9.5: Three-dimensional plot of the modulus square of the ρ_Δ-dependent part of (9.53).

The full expression for β_{si} is complex valued. However, it appears as the modulus square $|\beta_{si}|^2$ in expressions for the measured correlation, which becomes

$$|\beta_{si}|^2 \approx 16\eta_w^2\Xi^2\exp(-2\rho_\Sigma)\operatorname{sinc}^2\left(\frac{1}{2}\rho_\Delta\right) \tag{9.53}$$

in the weak squeezing limit. The correlation is therefore the result of the product of two functions, respectively dependent on ρ_Σ and ρ_Δ.

Based on (9.49), the sinc function with ρ_Δ describes a ring with twice the radius compared to that of ρ given in (9.20): the radius r_0 is multiplied by a factor of 2, which implies that the ring represented by $r_s^2 - 4r_0^2$ is scaled by a factor of 2 on the detector plane, compared to the ring of the intensity in (9.22). The ring is also shifted to be centred at $\mathbf{X}_s$, the location of the detector for the signal beam. This ring is plotted in Figure 9.5 for the weak squeezing limit and is shown as the large ring in Figure 9.6. Apart from having twice the radius and being shifted, it is qualitatively the same form as the expression in (9.22), as shown in Figure 9.2.

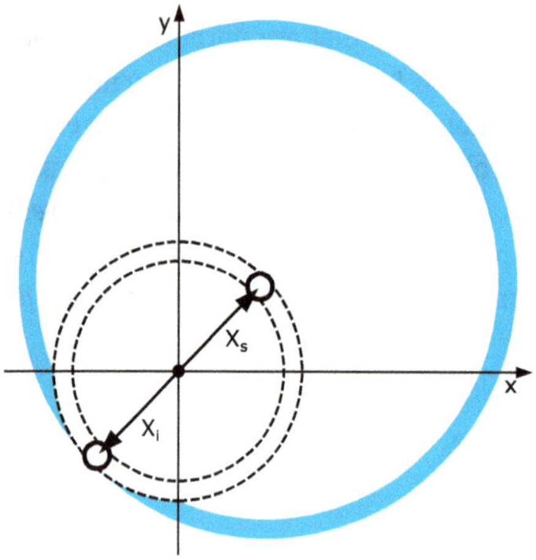

Figure 9.6: Diagram of the correlation measurements on the output plane, showing the rings.

We also need to take the effect of ρ_Σ into accounts. The exponential factor that contains ρ_Σ represents a beam profile with a size given by w_c, which is the pump beam's transformed beam size in the output plane (i. e., the detector plane). It is usually small, especially under thin-crystal experimental conditions. Therefore, the Gaussian exponential factor produces a tiny spot on the detector plane. The location of the spot is dictated by r_s. While the large ring in Figure 9.6 is centred at $\mathbf{X}_s$, the Gaussian spot in (9.50) is shifted in the opposite direction as a function of $\mathbf{X}_i$ to ensure that the argument of the exponential represented by ρ_Σ is small. For a successful correlated detection at the two detectors, the Gaussian spot must lie on the large ring. Therefore, the magnitudes of the two position vectors $\mathbf{X}_s$ and $\mathbf{X}_i$ must add up to $2r_0$, the radius of the large ring. So, each of the two position vectors has a length of approximately r_0. The collection of all such locations describes a ring of radius r_0, shown as the ring with the dashed lines in Figure 9.6. The locations of correlated measurements always lie on opposite points on this ring. It is equivalent to the ring that we obtained for the intensity measurement in (9.22), as shown in Figure 9.2.

9.2.3 With dispersion

The calculation of the second-order correlation with dispersion is similar to the calculation in Section 9.1.5, but with some differences. In this case, the integration over the *angular frequencies* is given by

$$\begin{aligned} h_{\text{disp}}(Z_m) &= \int \exp[\mathrm{i}Z_m k_{\text{disp}}(\omega_1,\omega_2)]h(\omega_1+\omega_2-\omega_{\text{p}},\sqrt{m}\delta_{\text{p}})h(\omega_1-\omega_{\text{d}},\delta_{\text{d}}) \\ &\quad \times h(\omega_2-\omega_{\text{d}},\delta_{\text{d}})\frac{\mathrm{d}\omega_1}{2\pi}\frac{\mathrm{d}\omega_2}{2\pi} \\ &= \frac{A_0}{\sqrt{1+\mathrm{i}Z_m A_1}\sqrt{1-\mathrm{i}Z_m A_2}}\exp\left[-\frac{Z_m^2}{(1+\mathrm{i}Z_m A_1)w_z^2}\right], \end{aligned} \tag{9.54}$$

where k_{disp} is given in (9.23). The result has the same form as in Section 9.1.5, but some of the quantities are different. While A_0 remains the same, the other two become

$$\begin{aligned} A_1 &= \frac{\delta_{\text{p}}^2\delta_{\text{d}}^2}{\delta_{\text{p}}^2+2\delta_{\text{d}}^2}[2D_{\text{g}}(\omega_{\text{p}})-D_{\text{g}}(\omega_{\text{d}})C] \\ &= \frac{2\pi^2\Delta\lambda_{\text{p}}^2\Delta\lambda_{\text{d}}^2}{n_{\text{o}}(\omega_{\text{d}})\lambda_{\text{p}}^4(8\Delta\lambda_{\text{p}}^2+\Delta\lambda_{\text{d}}^2)}[2n_{\text{o}}(\omega_{\text{d}})\Delta n_{\text{g}}(\omega_{\text{p}})-n_{\text{eff}}(\omega_{\text{p}})\Delta n_{\text{g}}(\omega_{\text{d}})], \\ A_2 &= D_{\text{g}}(\omega_{\text{d}})C\delta_{\text{d}}^2 = \frac{\pi^2 n_{\text{eff}}(\omega_{\text{p}})\Delta n_{\text{g}}(\omega_{\text{d}})\Delta\lambda_{\text{d}}^2}{4n_{\text{o}}(\omega_{\text{d}})\lambda_{\text{p}}^4}. \end{aligned} \tag{9.55}$$

We also provide the expressions for the quantities when the replacements in (9.26) and (9.27) are applied. As in Section 9.1.5, the numerical values of A_1 and A_2 are small enough be discarded in the denominators. The width of the remaining Gaussian function is

$$\begin{aligned} w_z &= \frac{v_{\text{g}}(\omega_{\text{p}})v_{\text{g}}(\omega_{\text{d}})\sqrt{\delta_{\text{p}}^2+2\delta_{\text{d}}^2}}{|v_{\text{g}}(\omega_{\text{d}})-v_{\text{g}}(\omega_{\text{p}})C|\delta_{\text{p}}\delta_{\text{d}}} \\ &= \frac{n_{\text{o}}(\omega_{\text{d}})\lambda_{\text{p}}^2\sqrt{8\Delta\lambda_{\text{p}}^2+\Delta\lambda_{\text{d}}^2}}{\sqrt{2}\pi[n_{\text{o}}(\omega_{\text{d}})n_{\text{g}}(\omega_{\text{p}})-n_{\text{eff}}(\omega_{\text{p}})n_{\text{g}}(\omega_{\text{d}})]\Delta\lambda_{\text{p}}\Delta\lambda_{\text{d}}}, \end{aligned} \tag{9.56}$$

which is now governed by both $\Delta\lambda_{\text{p}}$ and $\Delta\lambda_{\text{d}}$.

The z-integrations become more complicated. We cannot use the same approach of Section 9.2.2, which represents them as an additional z-integration applied to the even case times a phase factor. Instead we repeat the calculation provided in Section 9.1.5, for an odd number of z-integrations. In this case, the final expression does not in general separate into a Gaussian that depends on the output plane coordinates and a factor that depends on the squeezing parameter. However, we can assume that $w_z \ll L$ and only consider the leading order in the expansion of w_z/L. In that case, the expression is

$$M^* \diamond H_m^{(e)} \diamond M \approx \frac{\sqrt{\pi} w_z L^{m-1}}{(m-1)!} \exp\left(-\frac{1}{4} w_z^2 \kappa^2\right), \tag{9.57}$$

as found in Section 9.1.5, but m is an odd integer and w_z is given by (9.56).

Using this result in the summations performed in Section 9.2.2, we obtain

$$\beta_{si} = -i \frac{2\sqrt{\pi}\eta_w w_z}{L} \Xi \cosh(\Xi) \exp(-\rho_\Sigma) \exp\left(-\frac{1}{4}\kappa^2 w_z^2\right). \tag{9.58}$$

Although it is qualitatively similar to the result without dispersion, it produces a broader ring, depending on the value of w_z compared to L.

9.3 Parametric amplified coherent state

The output intensity function of a squeezed vacuum state that we calculated in Section 9.1 is obtained from *spontaneous parametric down-conversion*. Here, we compute the output intensity function obtained for *stimulated parametric down-conversion*, where the initial state is a coherent state. For this purpose, we use the results of Section 8.7, where we derived the state obtained from stimulated parametric down-conversion for an arbitrary initial state with the aid of a *Bogoliubov transformation*.

To see what happens when we send a coherent state as seed field, together with the pump, into a nonlinear crystal, we apply the Bogoliubov transformation of (8.139) to the arguments of the Wigner functional for a coherent state, using β as the new field variable. In addition, the parameter function also needs to be represented in terms of the transformation. The result is

$$\begin{aligned} W_{\text{coh}}[\alpha] &= \mathcal{N}_0 \exp\left(-2\|\alpha - \xi\|^2\right) \\ &\rightarrow \mathcal{N}_0 \exp\left(-2\|U \diamond \beta + V \diamond \beta^* - \xi\|^2\right) \\ &= \mathcal{N}_0 \exp\left[-2(\beta^* - \zeta^*) \diamond A \diamond (\beta - \zeta)\right. \\ &\quad \left. - (\beta^* - \zeta^*) \diamond B \diamond (\beta^* - \zeta^*) - (\beta - \zeta) \diamond B^* \diamond (\beta - \zeta)\right]. \end{aligned} \tag{9.59}$$

The final expression is expressed with shifted field variables. The Bogoliubov kernels U and V are combined and represented in terms of the squeezed vacuum state kernels according to their definitions in (6.338). The result is a *displaced squeezed vacuum state*, with the displacement given by the *inversely Bogoliubov transformed* version of the parameter function of the initial coherent state, given by

$$\zeta \triangleq U^\dagger \diamond \xi - V^T \diamond \xi^*, \tag{9.60}$$

based on (6.335). The displaced squeezed vacuum state is the most general form for a pure Gaussian state. The expression for a general mixed Gaussian state is the same but in that case the kernels A and B do not satisfy the identities in Section 6.8.5.

9.3.1 Intensity measurement

Our aim is to perform a measurement of the photon-number distribution (proportional to the intensity function) on the parametric amplified coherent state, which is given by (9.59) in terms of the transformed parameter function given in (9.60). The equivalent calculation in Section 9.1 is done with the aid of the photon-number distribution in (7.70), which is derived for a squeezed vacuum state. Here, we derive an expression for the photon-number distribution that is suitable for a parametric amplified coherent state. The photon-number distribution is calculated with the aid of a localized number operator given in (7.17) applied on this state. The kernel of the imaging detector after a $2f$ system is discussed in Section 7.2.

Following the approach presented in Section 7.3.3, we place the Wigner functional for the localized number operator into the exponent of (9.59) multiplied by an auxiliary variable J. The resulting generating function reads

$$\begin{aligned}\mathcal{W}(J) &= \mathcal{N}_0 \int \exp\left[-2(\beta^* - \zeta^*)\diamond A\diamond(\beta-\zeta) - (\beta^* - \zeta^*)\diamond B\diamond(\beta^* - \zeta^*)\right.\\ &\quad \left. - (\beta-\zeta)\diamond B^*\diamond(\beta-\zeta) + J\beta^*\diamond D\diamond\beta - \frac{1}{2}J\,\mathrm{tr}\{D\}\right]\mathcal{D}^\circ[\beta]\\ &= \mathcal{N}_0 \int \exp\left[-2\alpha^*\diamond A\diamond\alpha - \alpha^*\diamond B\diamond\alpha^* - \alpha\diamond B^*\diamond\alpha\right.\\ &\quad \left. + J(\alpha^* + \zeta^*)\diamond D\diamond(\alpha+\zeta) - \frac{1}{2}J\,\mathrm{tr}\{D\}\right]\mathcal{D}^\circ[\alpha],\end{aligned} \tag{9.61}$$

where we shifted $\beta \to \alpha + \zeta$. The generic result of an anisotropic functional integration of this form is provided in (C.30). For the integral in (9.61), we replace $A \to A - \frac{1}{2}JD$, $F^* \to -J\zeta^*\diamond D$, and $F \to -JD\diamond\zeta$ in (C.30), and multiply it by $\exp(J\zeta^*\diamond D\diamond\zeta - \frac{1}{2}J\,\mathrm{tr}\{D\})$. The result reads

$$\begin{aligned}\mathcal{W}(J) &= \mathcal{N}_0 \int \exp\left[-2\alpha^*\diamond\left(A - \frac{1}{2}JD\right)\diamond\alpha - \alpha^*\diamond B\diamond\alpha^* - \alpha\diamond B^*\diamond\alpha\right.\\ &\quad \left. + J\alpha^*\diamond D\diamond\zeta + J\zeta^*\diamond D\diamond\alpha + J\zeta^*\diamond D\diamond\zeta - \frac{1}{2}J\,\mathrm{tr}\{D\}\right]\mathcal{D}^\circ[\alpha]\\ &= \frac{\exp\left([J^2\text{-terms}] + J\zeta^*\diamond D\diamond\zeta - \frac{1}{2}J\,\mathrm{tr}\{D\}\right)}{\sqrt{\det\{A - \frac{1}{2}JD\}\det\{A - \frac{1}{2}JD - B\diamond(A^* - \frac{1}{2}JD^*)^{-1}\diamond B^*\}}}.\end{aligned} \tag{9.62}$$

We can ignore the J^2-terms, because only one derivative is evaluated in J, which is then set equal to zero. The resulting expression for the average number of photons, which is proportional to the intensity, is given by

$$\langle n \rangle = \partial_J \mathcal{W}(J)\big|_{J=0} = \zeta^* \diamond D \diamond \zeta + \frac{1}{2}\operatorname{tr}\{(A - \mathbf{1}) \diamond D\}, \tag{9.63}$$

where we used the identities in Section 6.8.5. The first term represents the part produced by *stimulated parametric down-conversion*, where ζ is given in (9.60) in terms of the initial parameter function of the seed field. The second term represents the spontaneous parametric down-converted light, as given by (7.70), which has already been considered in detail in Section 9.1. The spontaneous parametric down-converted light overlaps the stimulated parametric down-converted light because they are both governed by the same phase-matching conditions. As a result, the spontaneous parametric down-converted light contributes as "noise" to the "signal" represented by the stimulated parametric down-converted light. However, for a strong enough seed, the latter is much larger than the former.

Unlike the photon-number distribution of spontaneous parametric down-converted light, which involves the trace of the detector kernel and the even kernel of the squeezed state, the photon-number distribution for stimulated parametric down-converted light involves the overlap of the detector kernel by the parameter functions representing the displacement of the squeezed vacuum state. These parameter functions are produced by Bogoliubov transformations. So, the calculation of the photon-number distribution for stimulated parametric down-converted light is significantly different from what we have in Section 9.1.

9.3.2 Detector kernel for infinite resolution

To compute the first terms in (9.63), we need to specify the detector kernel, which depends on the details of the experimental setup. In case we want to obtain an image of the output intensity function, we need a CCD array. It can be modelled as a small detector located at a position $\mathbf{X}_0 = \{x_0, y_0\}$, which is varied over the output plane, as discussed in Section 7.2.1. If we are interested in the far-field intensity function of the down-converted field, the output plane is located in the far-field, or in the back-focal plane of a $2f$ system, as shown in Figure 9.7.

As in Section 9.1, we assume that the resolution is better than what is necessary to reproduce the photon-number distribution in the far field accurately. Having performed the integrations for the Gaussian modes of the detectors in Section 9.1 as an exercise, we now simplify the calculation by assuming that the size of the detector elements (pixels) of the CCD is infinitely small. By taking the limit where $w_{\text{pixel}} \to 0$, we convert the detector mode into a Dirac delta function. It leads to the *infinite resolution detector kernel* considered in Section 7.2.5.

When we use this detector kernel in the stimulated parametric down-conversion term in (9.63), we get

$$\zeta^* \diamond D \diamond \zeta \propto \left|\zeta\left(\frac{k_d}{f}\mathbf{X}_0\right)\right|^2 = \left|\zeta_1\left(\frac{k_d}{f}\mathbf{X}_0\right)\right|^2 + \left|\zeta_2\left(\frac{k_d}{f}\mathbf{X}_0\right)\right|^2, \tag{9.64}$$

where we assume that the two terms that are produced by the inverse Bogoliubov transformed parameter function

$$\zeta_1 = U^\dagger \diamond \xi \quad \text{and} \quad \zeta_2 = V^T \diamond \xi^*, \tag{9.65}$$

do not overlap. So, we only need to calculate the two transformed parameter functions in (9.65) to determine the output photon-number distribution due to the stimulated down-conversion. It is considered in the next section.

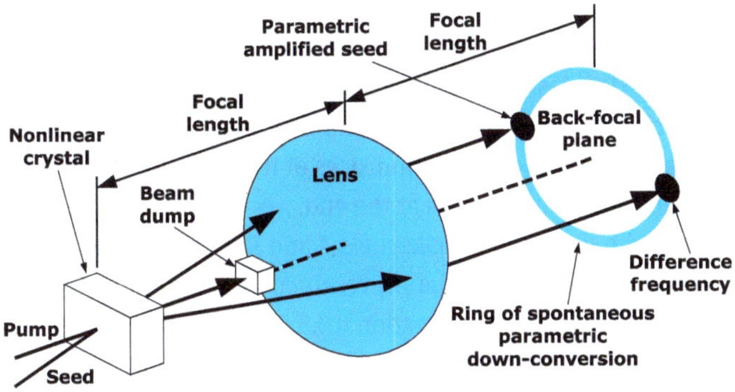

Figure 9.7: Diagrammatic represented of a $2f$ system to produce the far-field distribution with stimulated parametric down-conversion.

9.4 Mode transformations

In Sections 7.8.3 and 7.8.4, we arrived at the expressions for the Wigner functionals of photon-subtracted and photon-added states, respectively, in terms of modes given by the detector modes after being transformed by the kernels of the squeezed vacuum state and the Bogoliubov transformations, respectively. For photon-subtracted states, the transformed detector modes are

$$M_E = (A - \mathbf{1}) \diamond M = E \diamond M \quad \text{and} \quad M_B = B \diamond M^*, \tag{9.66}$$

where M is the detector mode and A and B are the kernels of the squeezed vacuum state. For photon-added states, the transformed detector mode is given by $M_V = V \diamond M^*$, where V is the odd Bogoliubov kernel. Then in Section 9.3, we investigated the output photon-number distribution produced by a parametric amplified coherent state and found that it is given by the inverse Bogoliubov transformation of the coherent state parameter function. These transformed parameter functions are represented by those in (9.65).

In all these cases, the kernels that perform the transformations are in some way associated with the squeezing process imposed by parametric down-conversion. Mode transformations are therefore quite common in systems involving parametric down-conversion. When measurements involve these modes, it is necessary to know how the transformations affect their shapes and sizes.

We are specifically interested in the effect of the transformation on the spatial properties of the modes. Since we have considered the angular frequency integration in detail in Sections 9.1 and 9.2, we now ignore the angular frequency dependencies in the modes, parameter functions, and also in the kernels. The assumption is that these spectra are monochromatic. So, we use the monochromatic versions of the kernels, as provided in (8.128) and (8.129) for the even and odd kernels, respectively.

Since the kernels that perform the transformations are not unitary,[a] the transformed modes are not normalized. The magnitudes of the modes contain information about losses due to reduced efficiencies and the magnification effect of the nonlinear process. Here, we are not specifically interested in such information. Instead, we focus on the effect of the transformations on the shapes and sizes of these modes. Therefore, when necessary, we normalize the resulting mode at the end.

Whether they are given by the squeezed state kernels A and B or by the Bogoliubov kernels U and V, the transforming kernels can be represented under suitable experimental conditions by the expressions derived in Section 8.6.6. In these calculations, we model the original detector modes or parameter functions as Gaussian functions in a $2f$ system, as given by (7.27). It represents the angular spectrum in the input plane of the $2f$ system. The mode size w_0 may in some cases be related to a small detector in the output plane, in which case it can be assumed that w_0 is large. Sometimes, the application involves collinear phase-matching conditions, for which we set $\mathbf{K}_0 = 0$. There are also cases where the phase-matching conditions need to be non-collinear. Therefore, we consider both scenarios, starting with collinear phase-matching conditions, as found for the transformed mode produced by the even kernel in heralded photon subtraction.

9.4.1 Transformed by the even kernel

There are two transformed detector modes found in heralded photon subtraction from a squeezed vacuum state discussed in Sections 7.8.3. One of them is produced by the even kernel A. The transformed mode is given by

$$M_E = E \diamond M = (A - \mathbf{1}) \diamond M = \sum_{p=1}^{\infty} H_{2p}^{(e)} \diamond M. \tag{9.67}$$

a While a Bogoliubov transformation is unitary, the individual Bogoliubov kernels are not.

With the leading-order term **1** being removed from the even intensity kernel A, only the higher-order terms in this kernel $H_{2p}^{(e)}$ contribute. For the current calculation, they are represented under the monochromatic approximation provided in (8.128). The generic integral expression for the different orders, with the angular frequency set equal to the centre frequency, is then given by

$$H_m^{(e)} \diamond M = \frac{\Xi^m}{mL^m} \int_z \int M(\mathbf{K}_2) \mathcal{Z} \left\{ \exp\left[-\frac{w_p^2}{4m} |\mathbf{K}_1 - \mathbf{K}_2|^2 + i\frac{Z_m |\mathbf{K}_1 + \mathbf{K}_2|^2}{2k_p} - iZ_m \chi(\omega_d) \right] \right\} \frac{d^2 k_2}{(2\pi)^2} \, d\{z\}. \tag{9.68}$$

Here, m is an even integer, Ξ is the effective squeezing parameter in (9.13), $M(\mathbf{K})$ is the two-dimensional angular spectrum of the mode to be transformed, $\mathcal{Z}\{\cdot\}$ is the z-symmetrization defined in (8.71), and Z_m is defined in (8.125). We discarded a global constant factor and substituted $k_z(\omega_d) = \frac{1}{2}k_p$, based on the degenerate *critical phase-matching condition.*

Since it is natural to perform heralded photon subtraction with collinear phase matching to allow the down-converted light easy passage through the beamsplitter, we can discard the χ-term in the exponent. However, for the sake of other similar transformations that are considered later, we keep it. The optical system between the beamsplitter and the heralding detector can either be a Fourier transforming $2f$ system or an imaging system such as the $4f$ system. As a result, the transformed detector mode size in the crystal plane can be much larger than that of the pump beam, or it can be of the same size (it would not make sense to be much smaller due to the implied loss). Therefore, we do not make assumptions about the relative size of w_0 in the current case. We can either use the thin-crystal approximation, or the plane wave approximation to simplify the analysis, because they are both valid in this scenario. These two approximations reveal different, complementary aspects of the system. So, we consider both and combine them into a rational approximation, as in Section 8.6.6.

Starting with the thin-crystal approximation, we consider the transformed mode in the spatial domain of the input plane (the crystal plane). Therefore, we compute the inverse two-dimensional Fourier transform with respect to the kernel's output wavevector, which is not contracted with the original mode's angular spectrum. After performing the inverse Fourier transform and applying the thin-crystal approximation, we obtain

$$\mathcal{F}^{-1}\{H_m^{(e)} \diamond M\}\big|_{\text{tc}} = \frac{\Xi^m}{L^m} \int_z \int M(\mathbf{K}) \exp\left(i\mathbf{X}\cdot\mathbf{K} - \frac{m|\mathbf{X}|^2}{w_p^2} \right) \times \mathcal{Z}\{\exp[-iZ_m \chi(\omega_d)]\} \frac{d^2 k}{(2\pi)^2} \, d\{z\}. \tag{9.69}$$

The remaining two-dimensional wavevector integration over **K** represents an inverse Fourier transform of the original mode with respect to its spatial degrees of freedom. It produces the original mode on the spatial domain $M(\mathbf{X})$.

When we use the plane wave approximation, instead of the thin-crystal approximation the expression becomes

$$\mathcal{F}^{-1}\{H_m^{(e)} \diamond M\}\big|_{\text{pw}} = \int\int_z \mathcal{Z}\left\{\exp\left[\mathrm{i}\mathbf{X}\cdot\mathbf{K} + \mathrm{i}\frac{2Z_m|\mathbf{K}|^2}{k_\mathrm{p}} - \mathrm{i}Z_m\chi(\omega_\mathrm{d})\right]\right\}$$
$$\times \frac{\Xi^m}{L^m} M(\mathbf{K}) \frac{\mathrm{d}^2k}{(2\pi)^2}\,\mathrm{d}\{z\}. \tag{9.70}$$

This time the integral over **K** does not represent a simple inverse Fourier transform of the original mode. Instead, it is better to compute the Fourier transform of the expression to remove the **X**-dependence and express the profile function in the Fourier domain, multiplied by the angular spectrum of the original mode. The expression in (9.70) only differs from (9.69) due to the quadratic phase term that contains $|\mathbf{K}|^2$ in the exponent instead of the Gaussian term with $|\mathbf{X}|^2$.

As in Section 8.6.6, we consider the different kinds of terms in the exponent after the inverse Fourier transform and expand them separately to leading order in their respective dimension parameter. The result is a combination of the previous two cases

$$\mathcal{F}^{-1}\{H_m^{(e)} \diamond M\}\big|_{\text{cr}} = \frac{\Xi^m}{L^m}\int\int_z \mathcal{Z}\left\{\exp\left[\mathrm{i}\frac{2Z_m|\mathbf{K}|^2}{k_\mathrm{p}} - \mathrm{i}Z_m\chi(\omega_\mathrm{d})\right]\right\}\,\mathrm{d}\{z\}$$
$$\times M(\mathbf{K})\exp\left(\mathrm{i}\mathbf{X}\cdot\mathbf{K} - \frac{m|\mathbf{X}|^2}{w_\mathrm{p}^2}\right)\frac{\mathrm{d}^2k}{(2\pi)^2}. \tag{9.71}$$

It represents the combined rational approximation for the current scenario. We use this expression for the subsequent calculation, but eventually we return to the two approximations to extract complementary information about the configuration space and Fourier domain functions, respectively.

9.4.1.1 Integrations of the z's

Next, we consider the z-integrations. Based on (9.8), we get

$$\mathcal{F}^{-1}\{H_{2p}^{(e)} \diamond M\}\big|_{\text{cr}} = \Xi^{2p}\int M(\mathbf{K})\exp(\mathrm{i}\mathbf{X}\cdot\mathbf{K})\exp\left(-\frac{2pr^2}{w_\mathrm{p}^2}\right)$$
$$\times \sum_{n=0}^{\infty}\frac{(-1)^n(n+p-1)!(L\kappa)^{2n}}{n!(p-1)!(2n+2p)!}\frac{\mathrm{d}^2k}{(2\pi)^2}, \tag{9.72}$$

where $r = |\mathbf{X}|$, $m = 2p$, and κ is defined in (9.9), but with **K** instead of $\mathbf{K}_0$.

9.4.1.2 Summations

The summations are now evaluated over both n and p, following the same process used in Section 9.1.3. The result is

$$\begin{aligned} M_E(\mathbf{X}) &= \sum_{p=1}^{\infty} \mathcal{F}^{-1}\{H_{2p}^{(e)} \diamond M\}\Big|_{\text{cr}} \\ &= \int \frac{\Xi^2(r)}{\Xi^2(r) - \rho^2(\mathbf{K})} \left\{\cosh\left[\sqrt{\Xi^2(r) - \rho^2(\mathbf{K})}\right] - 1\right\} \\ &\quad \times M(\mathbf{K}) \exp(\mathrm{i}\mathbf{X}\cdot\mathbf{K}) \frac{\mathrm{d}^2 k}{(2\pi)^2}, \end{aligned} \tag{9.73}$$

where

$$\Xi(r) \triangleq \Xi \exp\left(-\frac{r^2}{w_{\mathrm{p}}^2}\right) \quad \text{and} \quad \rho(\mathbf{K}) \triangleq L\kappa = \frac{2L}{k_{\mathrm{p}}}|\mathbf{K}|^2 - \rho_0. \tag{9.74}$$

In the last expression

$$\rho_0 \triangleq L\chi(\omega_{\mathrm{d}}) = \frac{r_0^2}{R^2}, \tag{9.75}$$

where r_0 and R are given in (9.21). The transformed mode is a complicated combination of configuration space and Fourier domain properties. However, by considering the two approximations separately, we get a better picture of the effects of the transformation on the respective domains. In both cases, the effect of the mode transformation is represented in terms of a profile function that is multiplied with the original mode, either in configuration space or in the Fourier domain, depending on whether we used the thin-crystal or the plane wave approximation. We consider both these approximations.

9.4.1.3 Thin-crystal approximation

Under the thin-crystal approximation, the $\mathbf{K}$-dependent term in $\rho(\mathbf{K})$ is removed, leaving only ρ_0. The integration over $\mathbf{K}$ then becomes an inverse Fourier transform of the initial mode's angular spectrum, leading to

$$M_E(\mathbf{X}) = M(\mathbf{X}) F_{\mathrm{tc}E}(r) \tag{9.76}$$

where $F_{\mathrm{tc}E}(r)$ is a radially symmetric profile function of $r = |\mathbf{X}|$,

$$F_{\mathrm{tc}E}(r) = \frac{\Xi^2(r)}{\Xi^2(r) - \rho_0^2} \left\{\cosh\left[\sqrt{\Xi^2(r) - \rho_0^2}\right] - 1\right\}, \tag{9.77}$$

associated with the transformation of the even kernel $E = A - \mathbf{1}$, under the thin-crystal approximation. While the detector mode $M(\mathbf{X})$ is normalized, the profile function $F_{\mathrm{tc}E}(r)$

is not. Its magnitude grows quickly with increasing values of the squeezing parameter Ξ. We are more interested in the shape of the profile function. Despite the nonlinear nature of the hyperbolic trigonometric function, the shape of the profile function is close to that of the Gaussian function of the pump beam in its argument. Its peak remains at the origin, regardless of the value of ρ_0.

The width of the profile function changes significantly as a function of the squeezing parameter Ξ. An estimate of this width is obtained from the coefficient of the sub-leading order of the expansion of the normalized profile function. It can be represented as

$$\frac{F_{\text{tc}E}(r)}{F_{\text{tc}E}(0)} = 1 - \frac{r^2}{w_{\text{eff}}^2} + \mathrm{O}\{r^4\}, \tag{9.78}$$

where w_{eff} represents an *effective width* of the profile function. This effective width is determined by computing the *curvature* of the normalized profile function at the origin for collinear phase matching ($\rho_0 = 0$), which is the expected phase-matching conditions for the mode transformation in Sections 7.8.3. The result

$$w_{\text{eff}}(\Xi) = \left[-\frac{1}{2}\partial_r^2 \frac{F_{\text{tc}E}(r)}{F_{\text{tc}E}(0)}\bigg|_{r=0} \right]^{-1/2} = w_{\text{p}} \sqrt{\frac{\cosh(\Xi) - 1}{\Xi \sinh(\Xi)}}, \tag{9.79}$$

is a modification of the beam width of the pump beam by a Ξ-dependent factor. The factor $w_{\text{eff}}(\Xi)/w_{\text{eff}}(0)$ is plotted in Figure 9.8 as a function of Ξ, in comparison with the other factors that are calculated for the other cases. It is evident that the effect of the transformation, as imposed by the even kernel A of the squeezed vacuum state, becomes more pronounced for larger values of the squeezing parameter. This effect plays a significant role when the state's Wigner functional is considered for subsequent processing.

In the limit of weak squeezing, the profile function has the same width as the square of the pump beam mode. For larger values of the squeezing parameter, the width decreases, reaching roughly half the width of the squared pump mode at a moderately strong value of $\Xi = 10$. It means that the transformed mode in terms of spatial coordinates becomes narrower in the crystal plane. If the original mode is significantly smaller than the profile function, it would remain more or less undistorted. However, if not, the profile function would produce a distortion of the original mode. The angular spectrum of the transformed mode is produced by the convolution of the angular spectrum of the original mode and that of the profile function, causing a broadening of the mode together with some smearing. With a $2f$ system after the nonlinear crystal, the configuration space mode in the crystal plane becomes the angular spectrum in the far field (at the output of the $2f$ system), and vice versa. For an imaging system, such as a $4f$ system, the output mode is the same as the transformed mode in the crystal plane.

9.4.1.4 Plane wave approximation

Under the plane wave approximation, the Gaussian function that comes from the pump beam is removed from the squeezing parameter $\Xi(r) \to \Xi$. The only $\mathbf{X}$-dependence is now found in the inverse Fourier kernel, which is readily removed by computing the Fourier transform of the output mode. As a result, we obtain

$$M_E(\mathbf{K}) = M(\mathbf{K})F_{\text{pw}E}(\mathbf{K}), \tag{9.80}$$

where $F_{\text{pw}E}(\mathbf{K})$ is a radially symmetric profile function on the Fourier domain, associated with the transformation of the even kernel $E = A - \mathbf{1}$, under the plane wave approximation. It reads

$$F_{\text{pw}E}(\mathbf{K}) = \frac{\Xi^2}{\Xi^2 - \rho^2(\mathbf{K})} \left\{\cosh\left[\sqrt{\Xi^2 - \rho^2(\mathbf{K})}\right] - 1\right\}. \tag{9.81}$$

with $\rho(\mathbf{K})$ given in (9.74). In this case, it is the angular spectrum that is deformed by the profile function. In the collinear case ($\rho_0 = 0$), we have

$$F_{\text{pw}E}(k) = \left[\cosh\left(\sqrt{\Xi^2 - \frac{4L^2}{k_\text{p}^2}k^4}\right) - 1\right]\left(1 - \frac{4L^2}{\Xi^2 k_\text{p}^2}k^4\right)^{-1}, \tag{9.82}$$

where $k = |\mathbf{K}|$. In the limit of weak squeezing, it becomes the squared sinc function with the modulus square of the transverse wavevector in its argument.

Again, the magnitude of this profile function grows with increasing Ξ. The width of this profile function on the Fourier domain also increases with increasing Ξ. The profile function now depends on k^4, where k is the magnitude of the transverse wavevector. The expansion of the normalized profile function is therefore represented as

$$\frac{F_{\text{pw}E}(k)}{F_{\text{pw}E}(0)} = 1 - \frac{k^4}{w_\text{eff}^4} + \text{O}\{k^8\}, \tag{9.83}$$

where w_eff now represents the *effective width* of the Fourier domain profile function. The curvature (second-order derivative with respect to k) is zero. Instead, an estimate of the width of this profile function follows from the fourth-order derivative with respect to k, giving the coefficient of the sub-leading-order term in the expansion of the normalized profile function. We thus extract the effective width, which reads

$$w_\text{eff}(\Xi) = \left[-\frac{1}{4!}\partial_k^4 \frac{F_{\text{pw}E}(k)}{F_{\text{pw}E}(0)}\bigg|_{k=0}\right]^{-1/4} = \sqrt{\frac{\Xi k_\text{p}}{2L}}\left[\frac{\cosh(\Xi) - 1}{1 - \cosh(\Xi) + \frac{1}{2}\Xi\sinh(\Xi)}\right]^{1/4}. \tag{9.84}$$

In Figure 9.8, it is shown that the Ξ-dependent factor $w_\text{eff}(\Xi)/w_\text{eff}(0)$ increases with Ξ. Therefore, while the width of the spatial profile function decreases, the width of the Fourier domain profile function increases as a function of the squeezing parameter.

9.4.1.5 Profiles in other planes

The above discussion only concerns the configuration space or Fourier domain profile functions in the input plane (crystal plane). When the plane of interest is produced in the output (back focal) plane of a $2f$ system, the thin-crystal profile function in (9.77) becomes a Fourier domain profile function, and the plane wave profile function in (9.81) becomes a spatial domain profile function. For an imaging system such the $4f$ system, the profile functions are again those obtained in the input plane, apart from magnifications.

9.4.2 Transformed by the odd kernel

The other mode transformation found in the heralded photon subtraction process discussed in Sections 7.8.3, is produced by the odd kernel B. It reads

$$M_B = B \diamond M^* = \sum_{p=0}^{\infty} H_{2p+1}^{(o)} \diamond M^*, \tag{9.85}$$

where the different orders of the kernel $H_{2p+1}^{(o)}$ are provided in (8.129) under the monochromatic approximation. The calculation proceeds in a similar fashion as in the calculation for the transformation by the even kernel with the inverse Fourier transform and the two approximations. Again, we consider the expression at the centre angular frequency and discard a global constant factor. There are a few differences in the results due to the odd orders. Therefore, we review the calculation in the current case, glossing over those parts that are already discussed in Sections 9.4.1.

The expressions for the current case, as obtained from the two approximations prior to the z-integration are similar to those in (9.69) and (9.70). The one obtained from the combined rational approximation is

$$\begin{aligned}\mathcal{F}^{-1}\{H_m^{(o)} \diamond M^*\}\big|_{\text{cr}} = \mathrm{i}\frac{\Xi^m}{L^m} \int \int_z \mathcal{Z}\left\{\exp\left[\mathrm{i}\frac{2Z_m|\mathbf{K}|^2}{k_\mathrm{p}} - \mathrm{i}Z_m\chi(\omega_\mathrm{d})\right]\right\} \mathrm{d}\{z\} \\ \times M^*(\mathbf{K}) \exp\left(-\mathrm{i}\mathbf{X}\cdot\mathbf{K} - \frac{m|\mathbf{X}|^2}{w_\mathrm{p}^2}\right) \frac{\mathrm{d}^2k}{(2\pi)^2}.\end{aligned} \tag{9.86}$$

The pertinent difference is that m represents an odd integer. It leads to significant differences in the result of the z-integrations and the subsequent summations, similar to the situation in Section 9.2.

The z-integrations are treated according to the approach in Section 9.2.2. It leads to

$$\mathcal{F}^{-1}\{H_1^{(o)} \diamond M^*\}\big|_{\text{cr}} = \mathrm{i} \int M^*(\mathbf{K}) \exp(-\mathrm{i}\mathbf{X}\cdot\mathbf{K}) \frac{2\Xi \sin(\frac{1}{2}L\kappa)}{L\kappa} \frac{\mathrm{d}^2k}{(2\pi)^2}, \tag{9.87}$$

for $m = 1$, where κ is given in (9.9), but with $\mathbf{K}$ instead of $\mathbf{K}_0$, and to

$$\mathcal{F}^{-1}\{H_{2p+1}^{(o)} \diamond M^*\}\big|_{\text{cr}} = \mathrm{i}\int M^*(\mathbf{K}) \exp(-\mathrm{i}\mathbf{X}\cdot\mathbf{K}) \left[\frac{\Xi}{L}\exp\left(-\frac{|\mathbf{X}|^2}{w_{\mathrm{p}}^2}\right)\right]^{2p+1}$$
$$\times \int_{-L/2}^{L/2} \exp(\mathrm{i}\kappa z) \sum_{n=0}^{\infty} (z+\frac{1}{2}L)^{2p+2n} \frac{(-1)^n (n+p-1)!\kappa^{2n}}{n!(p-1)!(2p+2n)!}\,\mathrm{d}z, \quad (9.88)$$

for $m > 1$ ($p > 0$), where the final z-integration is left unevaluated till after the summations. When the summations are evaluated, as in Section 9.2.2, and the final z-integration is performed, the expression for $m = 1$ is added to it, cancelling one of the terms. The result then reads

$$M_B(\mathbf{X}) = \sum_{p=0}^{\infty} \mathcal{F}^{-1}\{H_{2p+1}^{(o)} \diamond M^*\}\big|_{\text{cr}}$$
$$= \int \Xi(r) \exp\left[\mathrm{i}\frac{1}{2}\rho(\mathbf{K})\right] \left\{\rho(\mathbf{K}) \frac{\cosh\left[\sqrt{\Xi^2(r)-\rho^2(\mathbf{K})}\right]-1}{\Xi^2(r)-\rho^2(\mathbf{K})}\right.$$
$$\left. + \mathrm{i}\frac{\sinh\left[\sqrt{\Xi^2(r)-\rho^2(\mathbf{K})}\right]}{\sqrt{\Xi^2(r)-\rho^2(\mathbf{K})}}\right\} M^*(\mathbf{K}) \exp(-\mathrm{i}\mathbf{X}\cdot\mathbf{K}) \frac{\mathrm{d}^2k}{(2\pi)^2}, \quad (9.89)$$

where $\Xi(r)$ and $\rho(\mathbf{K})$ are defined in (9.74).

We now restrict this general expression for the odd kernel transformation in (9.89) to either the thin-crystal approximation or the plane wave approximation, to obtain profile functions for the configuration space and the Fourier domain, respectively. For the thin-crystal approximation, the **K**-dependence in $\rho(\mathbf{K})$ is removed, so that

$$M_B(\mathbf{X}) = M^*(\mathbf{X}) F_{\text{tc}B}(r), \quad (9.90)$$

where $F_{\text{tc}B}(r)$ is a complex profile function associated with the odd kernel B^* under the thin-crystal approximation. The complex conjugate of the initial mode in configuration space is obtained because the integration over **K** represents the complex conjugate of the inverse Fourier transform applied to the complex conjugate of the initial mode's angular spectrum. The profile function is a radially symmetric function of the radial coordinate $r = |\mathbf{X}|$, given by

$$F_{\text{tc}B}(r) = \Xi(r) \exp\left(\mathrm{i}\frac{1}{2}\rho_0\right) \left\{\rho_0 \frac{\cosh\left[\sqrt{\Xi^2(r)-\rho_0^2}\right]-1}{\Xi^2(r)-\rho_0^2} + \mathrm{i}\frac{\sinh\left[\sqrt{\Xi^2(r)-\rho_0^2}\right]}{\sqrt{\Xi^2(r)-\rho_0^2}}\right\}. \quad (9.91)$$

The typical scenario for photon subtraction, as discussed in Section 7.8.3, calls for collinear phase-matching conditions, with $\chi(\omega_{\mathrm{d}}) = 0$, so that $\rho_0 = 0$. In that case, the profile function simplifies to

$$F_{\text{tc}B}(r) = \mathrm{i}\sinh\left[\Xi\exp\left(-\frac{r^2}{w_\text{p}^2}\right)\right]. \tag{9.92}$$

Its magnitude grows from zero, becoming larger with increasing values of Ξ.

For the plane wave approximation, we have $\Xi(r) \to \Xi$ as the Gaussian pump mode drops out, which allows us to remove the remaining inverse Fourier kernel by applying a Fourier transform of the output mode. The output transformed mode is then expressed in terms of its angular spectrum by

$$M_B(\mathbf{K}) = M^*(\mathbf{K})F_{\text{pw}B}(k), \tag{9.93}$$

where $F_{\text{pw}B}(k)$ is a radially symmetric profile function of $k = |\mathbf{K}|$ on the Fourier domain,

$$\begin{aligned} F_{\text{pw}B}(k) = \Xi\exp\left[\mathrm{i}\frac{1}{2}\rho(k)\right]\Bigg\{&\rho(k)\frac{\cosh\left[\sqrt{\Xi^2-\rho^2(k)}\right]-1}{\Xi^2-\rho^2(k)} \\ &+\mathrm{i}\frac{\sinh\left[\sqrt{\Xi^2-\rho^2(k)}\right]}{\sqrt{\Xi^2-\rho^2(k)}}\Bigg\}, \end{aligned} \tag{9.94}$$

under the plane wave approximation, for the transformation by the odd kernel B.

Apart from the differences between these expressions and those for the even kernel, the qualitative behaviour is the same. The profile function under the thin-crystal approximation, which deforms the initial mode in configuration space, decreases its width as a function of Ξ, but retains its Gaussian shape, assuming the pump mode is Gaussian. Under the plane wave approximation, the deformation of the initial mode is performed on the Fourier domain. In this case, the curvature of the Fourier domain profile function is not zero, but it is imaginary, indicating a pure phase curvature. The magnitude of the profile function is again a function of k^4, with a width that increases as a function of Ξ. The effective widths of these profile functions under the thin-crystal approximation and the plane wave approximation are respectively given by

$$w_\text{eff}(\Xi) = \left[-\frac{1}{2}\partial_r^2\frac{F_{\text{tc}E}(r)}{F_{\text{tc}E}(0)}\bigg|_{r=0}\right]^{-1/2} = w_\text{p}\sqrt{\frac{\sinh(\Xi)}{\Xi\cosh(\Xi)}}, \tag{9.95}$$

and

$$\begin{aligned} w_\text{eff}(\Xi) &= \left[-\frac{1}{4!}\partial_k^4\frac{F_{\text{pw}E}(k)}{F_{\text{pw}E}(0)}\bigg|_{k=0}\right]^{-1/4} \\ &= \sqrt{\frac{\Xi k_\text{p}}{2L}}\left[\frac{2\sinh(\Xi)}{1-\sinh(\Xi)+\frac{1}{4}\Xi\sinh(\Xi)}\right]^{1/4}. \end{aligned} \tag{9.96}$$

The Ξ-dependent factors $w_{\text{eff}}(\Xi)/w_{\text{eff}}(0)$ in the expressions of these effective widths are plotted in comparison with their equivalents from the transformation induced by the even kernel in Figure 9.8 as functions of the squeezing parameter Ξ. The effective widths for both cases under the thin-crystal approximation decrease as a function of Ξ, with the curve for the odd kernel decreasing faster than the one for the even kernel. For strong squeezing, the configuration space profile functions become narrower. On the other hand, the effective widths under the plane wave approximation increase as a function of Ξ, showing that the Fourier domain profile functions become broader with stronger squeezing. The one for the odd kernel becomes saturated at a value of $3^{1/4}$.

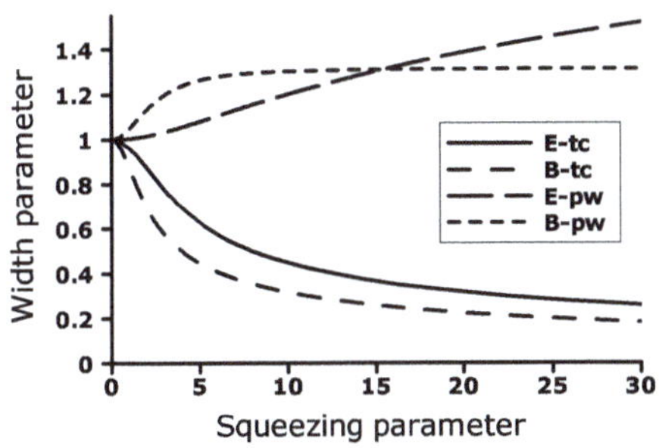

Figure 9.8: Curves as functions of the squeezing parameter of the width parameters in the profile functions produced by the even and odd kernels *E* and *B*, in both the thin-crystal (tc), and plane wave (pw) approximations.

The observed increase in the widths of the profile functions in the Fourier domain corresponds to a decrease of the transformed mode's width in configuration space. It follows from the fact the widths of functions in configuration space are inversely related to the widths of their spectra. Therefore, both approximations for a given case represent decreases in widths in configuration space as a function of an increasing squeezing. We can understand this effect as the result of the local bosonic enhancement leading to higher efficiency where the mode has a larger amplitude. It causes the function to become narrower. However, the curves of the widths for the two approximations for a given scenario are not directly related via the Fourier transform. It demonstrates the fact that the two approximations provide complementary information about the process, that is different even though their trends are qualitatively the same.

9.4.3 Transformed parameter function

It is shown in Section 9.3 that a parametric amplified coherent state produces a displaced squeezed vacuum state. The displacement is given by an inverse Bogoliubov transformed version of the initial coherent state's parameter function. Since (inverse)

Bogoliubov transformations are represented by two Bogoliubov kernels U and V, it produces two transformed versions of the initial parameter function, given by

$$\zeta_1 = U^\dagger \diamond \xi \quad \text{and} \quad \zeta_2 = V^T \diamond \xi^*, \tag{9.97}$$

where ξ is the original parameter function. The complete transformed parameter function is given by $\zeta = \zeta_1 - \zeta_2$.

Here, the transformation of the parameter function is discussed in more detail. We can use the results of the calculations in Sections 9.4.1 and 9.4.2, because U is an even kernel corresponding to A and V is an odd kernel corresponding to B. If we assume that U is Hermitian and V is symmetric, they would be directly obtained by replacing the squeezing parameter in A and B by half the squeezing parameter $\Xi \to \frac{1}{2}\Xi$. Here we consider the case where the Bogoliubov kernels do not have these additional properties.

In general, implementations of stimulated parametric down-conversion implies non-collinear phase-matching conditions (the pump and the seed enter the nonlinear crystal with a nonzero angle between them), unlike the collinear phase-matching conditions assumed for the final results in Sections 9.4.1 and 9.4.2. The initial parameter function is shifted in the Fourier domain to represent the angle that the seed beam makes with respect to the pump beam as they enter the crystal to match the non-collinear phase-matching conditions. Fortunately, the calculations in Sections 9.4.1 and 9.4.2 include the possibility of non-collinear phase-matching conditions. Therefore, we can use those results in the current context. Due to the off-axis nature of non-collinear phase-matching conditions, the effect of the distortion by the profile function is not isotropic as found for the collinear case.

9.4.3.1 Transformation by the even Bogoliubov kernel

First, we consider $\zeta_1 = U^\dagger \diamond \xi$ by using the results in Section 9.4.1. One difference is that, while the leading-order term (the identity) is removed from the even kernel A in the mode transformation in (9.67), the leading-order identity is still present in the transformation of the parameter function by the even Bogoliubov kernel. This leading-order term represents the original photons from the seed that pass through the nonlinear crystal. Although they stimulate the process, they are not affected (assuming we can ignore the losses). Another difference is that we do not have the z-symmetrization for the general Bogoliubov kernels. The result of the transformation is

$$\begin{aligned}\zeta_1(\mathbf{X}) &= \mathcal{F}^{-1}\{\xi\} + \sum_{p=1}^{\infty} \frac{1}{4^p} \mathcal{F}^{-1}\{H_{2p}^{(e)} \diamond \xi\}\Big|_{\text{cr}} \\ &= \int \left\{ \cosh\left[\frac{1}{2}\sqrt{\Xi^2(r) - \rho^2(\mathbf{K})}\right] - \frac{i\rho(\mathbf{K})}{\sqrt{\Xi^2(r) - \rho^2(\mathbf{K})}} \sinh\left[\frac{1}{2}\sqrt{\Xi^2(r) - \rho^2(\mathbf{K})}\right] \right\} \\ &\quad \times \exp\left[i\frac{1}{2}\rho(\mathbf{K})\right] \xi(\mathbf{K} - \mathbf{K}_0) \exp(i\mathbf{X}\cdot\mathbf{K}) \frac{d^2k}{(2\pi)^2},\end{aligned} \tag{9.98}$$

where $\Xi(r)$ and $\rho(\mathbf{K})$ are defined in (9.74), and $\mathbf{K}_0$ represents the shift of the angular spectrum of the seed beam in the Fourier domain due to its angle of incidence.

The current situation is complicated by the non-collinear phase-matching conditions. The distorting part of (9.98) produces a ring in the Fourier domain (leading to a ring in the output plane of a $2f$ system after the nonlinear crystal). The angular spectrum of the seed only overlaps this ring on one side, provided that its angle of incidence satisfies the non-collinear phase-matching condition.

To investigate this scenario, we use the plane wave approximation, which assumes that the spatial extent of the pump parameter function is much larger than that of the seed field in the plane of the crystal. (The thin-crystal approximation is not suitable for this scenario, because it removes the information about the ring.) We replace the parameter function of the pump by 1. The inverse Fourier transform in (9.98) is removed, so that the transformed parameter function is represented in the Fourier domain as

$$\zeta_1(\mathbf{K}) = F_{\mathrm{pw}U}(\mathbf{K})\xi(\mathbf{K} - \mathbf{K}_0), \tag{9.99}$$

where the profile function is

$$\begin{aligned} F_{\mathrm{pw}U}(\mathbf{K}) = & \left\{\cosh\left[\frac{1}{2}\sqrt{\Xi^2(r) - \rho^2(\mathbf{K})}\right] - \frac{\mathrm{i}\rho(\mathbf{K})}{\sqrt{\Xi^2(r) - \rho^2(\mathbf{K})}} \sinh\left[\frac{1}{2}\sqrt{\Xi^2(r) - \rho^2(\mathbf{K})}\right]\right\} \\ & \times \exp\left[\mathrm{i}\frac{1}{2}\rho(\mathbf{K})\right], \end{aligned} \tag{9.100}$$

with $\rho(\mathbf{K})$ given in (9.74). In the output plane of a $2f$ system after the nonlinear crystal, (9.99) becomes a spatial function by the replacement $\mathbf{K} \to \pi\mathbf{X}/\lambda f$.

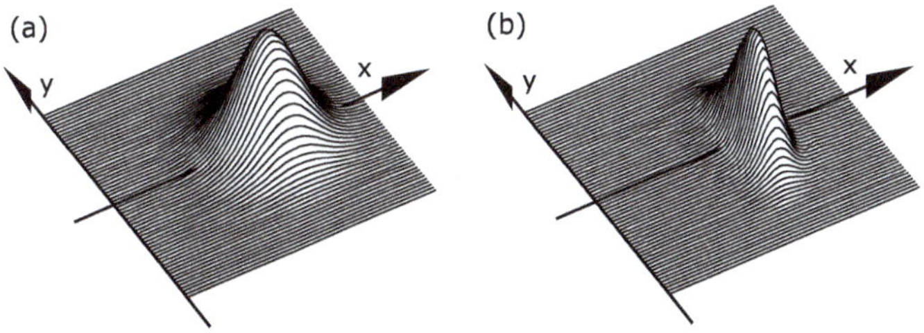

Figure 9.9: Comparison of (a) the undistorted Gaussian parameter function and (b) the distorted Gaussian parameter function of the seed. Here, the radius of the ring is $r_0 = 8$, the initial Gaussian width w is equal to the ring width of $w = R = 2.8$ in arbitrary units, and the squeezing parameter is $\Xi = 7$.

Due to the off-axis operation under non-collinear phase-matching conditions, the distortion produced by this profile function is anisotropic. The widths of the transformed parameter function are different along the radial and azimuthal directions, respectively. The widths of this profile functions along the two directions are obtained from a similar procedure to what is used in Section 9.4.1. The curvatures are evaluated at a point on the ring, for example, at $x = r_0, y = 0$, where $\{x, y\}$ are the output coordinates.

To illustrate the anisotropic effect of the distortion, we consider a seed with a Gaussian parameter function. In Figure 9.9, the effect of the distortion is demonstrated with a comparison between the intensity functions of the distorted and undistorted Gaussian parameter functions. For this purpose, the initial width w of the parameter function is chosen to be comparable to the ring width of the profile function. The parameters are $R = w = 2.8$ and $r_0 = 8$ all in arbitrary units, and the squeezing parameter is $\Xi = 7$.

Exercise 9.5. Compute the widths along the x- and y-directions, respectively, for the profile function in (9.100) at $x = r_0, y = 0$, using the procedures of Sections 9.4.1 and 9.4.2.

9.4.3.2 Transformation by the odd Bogoliubov kernel

For the transformed parameter function given by $\zeta_2 = V^T \diamond \xi^*$, we are aided by the results in Section 9.4.2. First, we point out a few differences between the expressions for ζ_1 and ζ_2. While ζ_1 contains the original parameter function ξ as a first term, ζ_2 does not contain such a term. The idler ζ_2 contains the complex conjugate of the parameter function ξ, thus representing the *phase conjugated parameter function.*

The two beams associated with ζ_1 and ζ_2, are shifted in the opposite directions, because the expressions for kernel functions $U(\mathbf{k}_1, \mathbf{k}_2)$ and $V(\mathbf{k}_1, \mathbf{k}_2)$ contain opposite signs when combining the transverse wavevectors in the angular spectra. The widths and heights of the different terms in the two beams are different due to the odd and even numbered terms.

The expression for the transformation of the parameter function produced by V only differs from (9.89) in that $\Xi \to \frac{1}{2}\Xi$. Hence,

$$\begin{aligned}\zeta_2(\mathbf{X}) &= \sum_{p=0}^{\infty} \frac{1}{2^{2p+1}} \mathcal{F}^{-1}\{H_{2p+1}^{(o)} \diamond \xi^*\}\big|_{\text{cr}} \\ &= \int \frac{\Xi(r)}{\sqrt{\Xi^2(r) - \rho^2(\mathbf{K})}} \sinh\left[\tfrac{1}{2}\sqrt{\Xi^2(r) - \rho^2(\mathbf{K})}\right] \\ &\quad \times \xi^*(\mathbf{K} - \mathbf{K}_0) \exp(-i\mathbf{X} \cdot \mathbf{K}) \frac{d^2k}{(2\pi)^2},\end{aligned} \tag{9.101}$$

where $\Xi(r)$ and $\rho(\mathbf{K})$ are provided in (9.74).

Again, we use the plane wave approximation to express the transformed parameter function in the Fourier domain as

$$\zeta_2(\mathbf{K}) = F_{\text{pw}V}(\mathbf{K})\xi^*(\mathbf{K} - \mathbf{K}_0), \tag{9.102}$$

where the profile function is

$$F_{\text{pw}V}(\mathbf{K}) = \frac{\Xi(r)}{\sqrt{\Xi^2(r) - \rho^2(\mathbf{K})}} \sinh\left[\tfrac{1}{2}\sqrt{\Xi^2(r) - \rho^2(\mathbf{K})}\right]. \tag{9.103}$$

Although the expression in (9.103) is quantitatively different, it gives the same qualitative effect as (9.100). The non-collinear phase-matching conditions produce a ring in the Fourier domain, with the angular spectrum of the seed overlapping it at a specific point on one side of the ring. The resulting overlap leads to an anisotropic distortion of the seed's angular spectrum, similar to what is shown in Figure 9.9. To avoid such a distortion, the initial angular spectrum of the seed must be much narrower than the width of the ring. The ring width is determined by applying the procedure in Section 9.4.1 to the modulus of the profile function in (9.103).

9.4.3.3 Mode transformation for photon addition

Another mode transformation that we encountered is the transformation of the detector mode in heralded photon addition, discussed in Section 7.8.4. The transformation is provided by the odd Bogoliubov kernel V. Therefore, we can use the results of Section 9.4.3 above. Instead of the parameter function, we have a detector mode that is aligned with the difference-frequency beam. Due to the non-collinear phase-matching conditions, this detector mode is shifted off-axis in the detector plane. The scenario is similar to the transformation by the odd Bogoliubov kernel, considered above.

The application of photon addition usually requires weak squeezing. Therefore, we consider the limit where $\Xi \to 0$ in the above analysis. As a result, the transformed detector mode is given by the leading-order term

$$\begin{aligned} \zeta_2(\mathbf{X}) &= \frac{1}{2}\,\mathcal{F}^{-1}\{H_1^{(0)} \diamond M^*\}\Big|_{\text{cr}} \\ &= \mathrm{i}\int \Xi(r)\,\mathrm{sinc}\left[\frac{1}{2}\rho(\mathbf{K})\right]\xi^*(\mathbf{K})\exp(-\mathrm{i}\mathbf{X}\cdot\mathbf{K})\,\frac{\mathrm{d}^2k}{(2\pi)^2}, \end{aligned} \tag{9.104}$$

with $\Xi(r)$ and $\rho(\mathbf{K})$ are provided in (9.74). Here, the sinc function is responsible for producing the ring on the Fourier domain. Qualitatively, the distortion effect is similar to the case of the previous section.

9.4.3.4 General mode transformations

The different scenarios for mode transformations that we consider here show how diverse the situations are in which it can occur. It also demonstrates that such mode transformations can be quite common in applications based on parametric down-conversion. Generally, such mode transformations are unwanted. To avoid them, one can either work in the weak squeezing limit or use modes that are smaller than the distorting profile functions. However, one does not usually have the freedom to make such choices. For this reason, these calculations play a significant role in the design of such systems.

9.5 Pump shaping

In all the calculations so far, we have assumed that the *pump state* for the parametric down-conversion process is a coherent state with a Gaussian parameter function. However, the parameter function of the pump provides a way to control some of the properties of the down-converted state. In particular, the spectrum of eigenstates that are obtained by diagonalizing the kernel can be modified. In such applications, the pump beam is modulated to have a specific angular spectrum. One of the interesting cases is where the pump has a nonzero *angular momentum*. Then the signal and idler photons carry different amounts of angular momenta.

9.5.1 Laguerre–Gauss modes

Here, we consider the case where the pump is still a coherent state, but its parameter function is not a simple Gaussian function. For this purpose, the parameter function is given by a Laguerre–Gauss mode (see Section 2.5.2), with the radial index set to zero. Such Laguerre–Gauss modes with zero radial index are called *helical modes*. The *pump parameter function* can then be represented by the generating function in (2.134) with $v = 0$ and $z = 0$. Since the temporal frequency does not play much of a role here, we ignore the temporal frequency spectrum in this discussion. The parameter function is then represented in part as a generating function, given by

$$\mathcal{L}(\mathbf{K};\mu,\sigma) = \sqrt{2\pi}\zeta_0 w_{\mathrm{p}} \mathcal{N}_\ell \exp\left(-\mathrm{i}\mu w_{\mathrm{p}} \Gamma\cdot\mathbf{K} - \frac{1}{4}w_{\mathrm{p}}^2|\mathbf{K}|^2\right), \tag{9.105}$$

where ζ_0 is the complex amplitude of the parameter function, μ is the generating parameter, σ is the sign of the *azimuthal index* ℓ, and $\Gamma = \frac{1}{2}(\vec{x} + \mathrm{i}\sigma\vec{y})$, so that $\Gamma\cdot\mathbf{K} = \frac{1}{2}(k_x + \mathrm{i}\sigma k_y)$. The normalization constant is given by

$$\mathcal{N}_\ell = \sqrt{\frac{2^{|\ell|}}{|\ell|!}}, \tag{9.106}$$

for an azimuthal index ℓ.

First, we repeat the calculation of the bilinear vertex kernel, as defined in (8.68). Here, it becomes a generating function, given by

$$\mathcal{H}(\mathbf{K}_1,\mathbf{K}_2,z;\mu,\sigma) = -\mathrm{i}\Omega_0\mathcal{N}_\ell \exp\left[-\mathrm{i}\mu w_{\mathrm{p}}\Gamma\cdot(\mathbf{K}_1+\mathbf{K}_2) - \frac{1}{4}w_{\mathrm{p}}^2|\mathbf{K}_1+\mathbf{K}_2|^2 + \mathrm{i}\Delta k_z z\right], \tag{9.107}$$

where Ω_0 is a product of dimension parameters and constants, including the nonlinear coefficient, and the phase-matching condition is represented by Δk_z, given by the expression in (8.86) for degenerate non-collinear type I phase matching. We represent the kernel with $\mathcal{H}$ to remind us that it is a generating function. It differs from the bilinear vertex kernel in (8.77) due to the additional linear term with the generating parameter μ in the exponent and the normalization constant for different values of the azimuthal index ℓ. Since the bilinear vertex kernel is a generating function, any calculation in which we need a product of different bilinear vertex kernels, we need to use a different generating parameter in each of them.

Exercise 9.6. Provide modified versions of (9.105) and (9.107) for the case where $\nu \neq 0$.

Calculating the different terms in the expansion of the kernels in Section 8.6.6, we find that the helical modes produce additional factors that are multiplied with the expressions of the original kernels. The additional factors are functions that depend on all the generating parameters. We denote these sets of generating parameters by $\{\mu\} = \{\mu_1, \mu_2, \ldots\}$, numbering them in sequence for the contracted kernels from right to left. The new kernels are therefore given by

$$\mathcal{H}_m^{(\mathrm{x})}(\mathbf{K}_1, \mathbf{K}_2, \{\mu\}) = \mathcal{G}_m^{(\mathrm{x})}(\mathbf{K}_1, \mathbf{K}_2, \{\mu\}) H_m^{(\mathrm{x})}(\mathbf{K}_1, \mathbf{K}_2), \tag{9.108}$$

where x is either o or e for the odd or even kernels, respectively, and $H_m^{(\mathrm{x})}$ represents the expression for the original kernel (without the temporal frequency part). The factors are given for the even and odd kernels by

$$\begin{aligned}
\mathcal{G}_m^{(\mathrm{e})}(\mathbf{K}_1, \mathbf{K}_2, \{\mu\}) &= \mathcal{N}_\ell^m \exp\left[-\mathrm{i}\frac{w_\mathrm{p}}{m} \left(\Sigma_\mu^{(\mathrm{o})} \kappa_- + \Sigma_\mu^{(\mathrm{e})} \kappa_-^* \right) + \frac{\Sigma_\mu^{(\mathrm{o})} \Sigma_\mu^{(\mathrm{e})}}{m} \right], \\
\mathcal{G}_m^{(\mathrm{o})}(\mathbf{K}_1, \mathbf{K}_2, \{\mu\}) &= \mathcal{N}_\ell^m \exp\left[-\mathrm{i}\frac{w_\mathrm{p}}{m} \left(\Sigma_\mu^{(\mathrm{o})} \kappa_+ + \Sigma_\mu^{(\mathrm{e})} \kappa_+^* \right) + \frac{\Sigma_\mu^{(\mathrm{o})} \Sigma_\mu^{(\mathrm{e})}}{m} \right],
\end{aligned} \tag{9.109}$$

respectively, where

$$\Sigma_\mu^{(\mathrm{e})} = \sum_{p=1}^{m \,\mathrm{Mod}\, 2} \mu_{2p} \quad \text{and} \quad \Sigma_\mu^{(\mathrm{o})} = \sum_{p=0}^{m \,\mathrm{Mod}\, 2} \mu_{2p+1}, \tag{9.110}$$

are the sums of the even and odd numbered generating parameters, and

$$\begin{aligned}
\kappa_\pm &\triangleq \Gamma \cdot (\mathbf{K}_1 \pm \mathbf{K}_2) = \frac{1}{2}(k_{1x} \pm k_{2x}) + \mathrm{i}\sigma\frac{1}{2}(k_{1y} \pm k_{2y}), \\
\kappa_\pm^* &\triangleq \Gamma^* \cdot (\mathbf{K}_1 \pm \mathbf{K}_2) = \frac{1}{2}(k_{1x} \pm k_{2x}) - \mathrm{i}\sigma\frac{1}{2}(k_{1y} \pm k_{2y}),
\end{aligned} \tag{9.111}$$

are complex helical wavevector variables that we define to simplify the expressions. It follows that

$$|\kappa_\pm|^2 = \frac{1}{4}|\mathbf{K}_1 \pm \mathbf{K}_2|^2, \tag{9.112}$$

which cannot contribute a nonzero *topological charge*. The generating functions, represented by $\mathcal{G}_m^{(\mathrm{e})}(\{\mu\})$ and $\mathcal{G}_m^{(\mathrm{o})}(\{\mu\})$ produce polynomial factors with which the Gaussian kernels are multiplied.

The number of odd-numbered generating parameters always exceeds the number of even-numbered generating parameters by one for odd kernels. On the other hand, the number of odd-numbered generating parameters equals the number of even-numbered generating parameters for even kernels. When we apply the derivatives to generate the different terms in the expansions for a specific helical mode, we always have to apply equal numbers of derivatives to all generating parameters, because the same helical mode appears in all the bilinear kernels. As a result, the terms in the expansions of the even kernel (A or U) always have a zero topological charge, whereas the terms in the expansion of odd kernels (B or V) always carry the same topological charge, equal to that of the pump parameter function.

9.5.2 Demonstration

As a demonstration, we consider the case with $\ell = 2$ and compute the polynomial prefactors up to $m = 4$. The expansion of the odd kernel has the form

$$B = \mathcal{N}_2 P_1 H_1 + \mathcal{N}_2 P_3 H_3 + \cdots , \tag{9.113}$$

where $\mathcal{N}_2 = \sqrt{2}$ is the normalization constant from (9.106), H_1 and H_3 represent the original odd kernel orders, and P_1 and P_3 are the prefactors given by

$$\begin{aligned} P_1 &= \left.\frac{\partial^2 \mathcal{G}_1^{(\mathrm{o})}}{\partial \mu_1^2}\right|_{\mu_1=0} = \left.\frac{\partial^2}{\partial \mu_1^2} \exp\left(-\mathrm{i}\mu_1 \kappa_+ w_{\mathrm{p}}\right)\right|_{\mu_1=0} , \\ P_3 &= \left.\frac{\partial^6 \mathcal{G}_3^{(\mathrm{o})}}{\partial \mu_1^2 \partial \mu_2^2 \partial \mu_3^2}\right|_{\{\mu\}=0} = \frac{\partial^2}{\partial \mu_1^2}\frac{\partial^2}{\partial \mu_2^2}\frac{\partial^2}{\partial \mu_3^2} \exp\left[-\mathrm{i}\frac{1}{3}\left(\mu_1\kappa_+ + \mu_2\kappa_+^* + \mu_3\kappa_+\right) w_{\mathrm{p}} \right. \\ &\qquad \left.\left. + \frac{1}{3}(\mu_1+\mu_3)\mu_2\right]\right|_{\{\mu\}=0} . \end{aligned} \tag{9.114}$$

Likewise for the even kernel, we have

$$A = \mathcal{N}_2 P_2 H_2 + \mathcal{N}_2 P_4 H_4 + \cdots , \tag{9.115}$$

with the same normalization constant $\mathcal{N}_2 = \sqrt{2}$, the original even kernel orders H_2 and H_4, and the prefactors P_2 and P_4 given by

$$P_2 = \left.\frac{\partial^4 \mathcal{G}_2^{(e)}}{\partial\mu_1^2\partial\mu_2^2}\right|_{\{\mu\}=0} = \frac{\partial^2}{\partial\mu_1^2}\frac{\partial^2}{\partial\mu_2^2}\exp\left[-i\frac{1}{2}\left(\mu_1\kappa_- + \mu_2\kappa_-^*\right)w_p + \frac{1}{2}\mu_1\mu_2\right]\Big|_{\{\mu\}=0},$$

$$\begin{aligned} P_4 &= \left.\frac{\partial^8 \mathcal{G}_4^{(e)}}{\partial\mu_1^2\partial\mu_2^2\partial\mu_3^2\partial\mu_4^2}\right|_{\{\mu\}=0} \\ &= \frac{\partial^2}{\partial\mu_1^2}\frac{\partial^2}{\partial\mu_2^2}\frac{\partial^2}{\partial\mu_3^2}\frac{\partial^2}{\partial\mu_4^2}\exp\left[-i\frac{1}{4}\left(\mu_1\kappa_- + \mu_2\kappa_-^* + \mu_3\kappa_- + \mu_4\kappa_-^*\right)w_p\right. \\ &\quad \left. + \frac{1}{4}(\mu_1+\mu_3)(\mu_2+\mu_4)\right]\Big|_{\{\mu\}=0}. \end{aligned} \tag{9.116}$$

Working out all the prefactors, we get

$$\begin{aligned} P_1 &= -w_p^2\kappa_+^2, \\ P_2 &= \frac{1}{2} - \frac{1}{2}w_p^2|\kappa_-|^2 + \frac{1}{16}w_p^4|\kappa_-|^4, \\ P_3 &= -\frac{4}{27}w_p^2\kappa_+^2\left(1-\frac{1}{6}w_p^2|\kappa_+|^2\right)\left(1-\frac{1}{18}w_p^2|\kappa_+|^2\right), \\ P_4 &= \frac{3}{32} - \frac{3}{32}w_p^2|\kappa_-|^2 + \frac{9}{512}w_p^4|\kappa_-|^4 - \frac{1}{1024}w_p^6|\kappa_-|^6 + \frac{1}{65536}w_p^8|\kappa_-|^8. \end{aligned} \tag{9.117}$$

The prefactors of the odd orders always have an overall factor of κ_+^2, which carries a topological charge of 2. In addition, they may also contain a real-valued polynomial in $|\kappa_+|^2 = \frac{1}{4}|\mathbf{K}_1 + \mathbf{K}_2|^2$. The prefactors of the even orders are real-valued polynomials of $|\kappa_-|^2 = \frac{1}{4}|\mathbf{K}_1 - \mathbf{K}_2|^2$. Their topological charges are 0. The odd orders combine to produce the odd kernel, carrying the same topological charge as the pump parameter function. On the other hand, the even kernel, produced from the even orders, has a zero topological charge regardless of the topological charge of the pump parameter function. The polynomials in the different orders become progressively more complicated. Since we set the radial index to zero, these polynomials are not to be confused with the Laguerre polynomials that are produced for nonzero radial indices. By implication, the nonzero topological charge of the pump parameter function is enough to produce very complicated kernels in terms of their *spatiotemporal degrees of freedom*. The situation only becomes even more complex when the pump parameter function is given by a Laguerre–Gauss mode with a nonzero radial index.

In the case of weak squeezing, the higher orders are suppressed. Therefore, the complexity in the spatiotemporal degrees of freedom presented in these higher orders is not observed in such weakly squeezed applications.

9.6 Arbitrary pump state

While *pump shaping* is concerned with the effect of the spatiotemporal degrees of freedom of the pump on the down-converted field, the *state of the pump* also determines how its *particle-number degree of freedom* affects the down-converted state. Here, we use an *evolution equation*, derived from (8.63), to investigate the down-conversion process for a pump with an arbitrary Gaussian state.

9.6.1 Semiclassical approximation

We represent the Gaussian state of the pump as a displaced squeezed thermal state,

$$W_{\text{pump}}[\beta] = \mathcal{N} \exp\left[-2(\beta^* - \zeta^*) \diamond C \diamond (\beta - \zeta) \right. \\ \left. - (\beta^* - \zeta^*) \diamond S \diamond (\beta^* - \zeta^*) - (\beta - \zeta) \diamond S^* \diamond (\beta - \zeta)\right], \tag{9.118}$$

where ζ is a parameter function, C and S are kernels, and $\mathcal{N}$ is a normalization constant. The kernels and parameter function are unspecified at this stage. However, we distinguish between cases where the parameter function is strong enough to allow a *semiclassical approximation*, and those where it is not valid.

Exercise 9.7. What does it require for the parameter function to be *strong enough* to allow a semiclassical approximation? Consider, for example, the case where the pump state (prior to displacement) is a hot thermal state (a thermal state with a larger average number of photons).

For the semiclassical approximation, we assume that the complete state (pump and down-converted light) is represented as a Gaussian state without terms in the exponent that couple the down-converted field variables to the pump field variables. It means that the down-converted state and the pump state are separable (not entangled). So, we substitute $W_{\hat{\rho}}[\alpha,\beta] \rightarrow W_{\hat{\sigma}}[\alpha] W_{\text{pump}}[\beta]$, where $W_{\hat{\sigma}}[\alpha](z)$ is the down-converted state, into the evolution equation in (8.63). Then we evaluate the *functional derivatives* with respect to β, factor out $W_{\text{pump}}[\beta]$, and remove it. The resulting equation becomes

$$\begin{aligned} -\mathrm{i}\partial_z W_{\hat{\sigma}} = \int & 2[\beta^*(\mathbf{k}_0) - \zeta^*(\mathbf{k}_0)] C(\mathbf{k}_0,\mathbf{k}_3) T^*(\mathbf{k}_1,\mathbf{k}_2,\mathbf{k}_3,z) \alpha(\mathbf{k}_1)\alpha(\mathbf{k}_2) W_{\hat{\sigma}} \\ & + 2[\beta(\mathbf{k}_0) - \zeta(\mathbf{k}_0)] S^*(\mathbf{k}_0,\mathbf{k}_3) T^*(\mathbf{k}_1,\mathbf{k}_2,\mathbf{k}_3,z) \alpha(\mathbf{k}_1)\alpha(\mathbf{k}_2) W_{\hat{\sigma}} \\ & + \frac{1}{2}[\beta^*(\mathbf{k}_0) - \zeta^*(\mathbf{k}_0)] C(\mathbf{k}_0,\mathbf{k}_3) T^*(\mathbf{k}_1,\mathbf{k}_2,\mathbf{k}_3,z) \frac{\delta^2 W_{\hat{\sigma}}}{\delta\alpha^*(\mathbf{k}_1)\delta\alpha^*(\mathbf{k}_2)} \\ & + \frac{1}{2}[\beta(\mathbf{k}_0) - \zeta(\mathbf{k}_0)] S^*(\mathbf{k}_0,\mathbf{k}_3) T^*(\mathbf{k}_1,\mathbf{k}_2,\mathbf{k}_3,z) \frac{\delta^2 W_{\hat{\sigma}}}{\delta\alpha^*(\mathbf{k}_1)\delta\alpha^*(\mathbf{k}_2)} \\ & + 2\beta^*(\mathbf{k}_3) T^*(\mathbf{k}_1,\mathbf{k}_2,\mathbf{k}_3,z) \alpha(\mathbf{k}_1) \frac{\delta W_{\hat{\sigma}}}{\delta\alpha^*(\mathbf{k}_2)} \, \mathrm{d}_\mathrm{b}k_1 \, \mathrm{d}_\mathrm{b}k_2 \, \mathrm{d}_\mathrm{b}k_3 - \text{c. c.}, \end{aligned} \tag{9.119}$$

where c. c. represents the complex conjugate of the previous terms. There are no terms that involve only one α (unless such terms are bootstrapped in some way). Therefore, the process does not generate a displacement in α. There are also no terms that contain one α and one β. Therefore, if we can get rid of the higher-order terms, the pump and the down-converted state would remain separable, consistent with our assumption. The only terms that need to be present are those associated with the squeezed thermal state.

Assuming that the parameter function of the *pump* is strong enough to allow the semiclassical approximation, we substitute $\beta = \zeta$. As a result, all but two of the terms fall away, leaving

$$-\mathrm{i}\partial_z W_{\hat{\sigma}} = \int 2\zeta^*(\mathbf{k}_3)T^*(\mathbf{k}_1,\mathbf{k}_2,\mathbf{k}_3,z)\alpha(\mathbf{k}_1)\frac{\delta W_{\hat{\sigma}}}{\delta\alpha^*(\mathbf{k}_2)} - 2\frac{\delta W_{\hat{\sigma}}}{\delta\alpha(\mathbf{k}_1)}\alpha^*(\mathbf{k}_2)T(\mathbf{k}_1,\mathbf{k}_2,\mathbf{k}_3,z)\zeta(\mathbf{k}_3)\, \mathrm{d_b}k_1\, \mathrm{d_b}k_2\, \mathrm{d_b}k_3, \tag{9.120}$$

which is the same equation that is obtained for a coherent state pump in (8.66). Under the semiclassical approximation, the kernels C and S do not play any role in the state of the down-converted light. Provided that $W_{\hat{\sigma}}$ is a Gaussian state, all the terms in the equation have only two field variables, maintaining the state's Gaussian nature.

9.6.2 Without the semiclassical approximation

When the semiclassical approximation cannot be applied, as for a pump state located at the origin of phase space, the evolution equation can produce terms that have more than two field variables. Unless these terms cancel among themselves, they have to be represented by equivalent terms in the exponent of the state. It implies that the pump state and the down-converted state are not separable. However, it does not necessarily mean that they are entangled. When there are terms in the exponent of the state consisting of more than two field variables, the resulting state is not a Gaussian state, which usually leads to intractable calculations.

In those cases where the *pump state* does not have any displacement, all the terms on the right-hand side of the evolution equation contain β. As a result, the vacuum term for α is the only term in the argument of the exponential for the state that only contains α. It is unlikely that a cancellation occurs among the higher-order terms in such a case.

9.6.2.1 Example: Thermal state

To consider an example, we assume that the pump is a thermal state. For this purpose, we can set $C = \theta$ in (9.119), representing the kernel of the thermal state. In addition, we also set $S = S^* = 0$ (no squeezing) and $\zeta = 0$ (no displacement). It effectively assumes that the complete state produced by the down-conversion process has the form

$$W_{\hat{\rho}}[\alpha,\beta] = \mathcal{N}\exp\left(-2\beta^*\diamond\theta\diamond\beta - 2\alpha^*\diamond\alpha + F[\alpha,\beta]\right), \tag{9.121}$$

where $F[\alpha,\beta]$ is a Hermitian polynomial in all the field variables. The first two terms are all that is present prior to the down-conversion process, while $F[\alpha,\beta]$ is produced during the down-conversion process. It also means that all the terms in $F[\alpha,\beta]$ contain the unenhanced vertex T, and are therefore suppressed.

To determine what $F[\alpha,\beta]$ looks like, we follow an iterative approach, starting with the first two terms that are initially present. Without any z-dependent kernels, the resulting equation has the form

$$0 = 4\alpha^*\alpha^* \diamond \diamond T \diamond (\mathbf{1} - \theta) \diamond \beta - 4\beta^* \diamond (\mathbf{1} - \theta) \diamond T^* \diamond \diamond \alpha\alpha. \tag{9.122}$$

Both these terms are first order in the unenhanced vertex T. The thermal state kernel θ cannot be cancelled by the identity (unless $\theta = \mathbf{1}$, in which case the thermal state is the vacuum state, giving a trivial result). Therefore, these terms must be balanced by new z-dependent terms in the exponent with the same field variables.

Hence, for our next iteration, the ansatz for the evolved state is

$$\begin{aligned} W_{\hat{\rho}}[\alpha,\beta] = \mathcal{N} \exp\big[&-2\beta^* \diamond \theta \diamond \beta - 2\alpha^* \diamond \alpha \\ &+ \alpha^*\alpha^* \diamond \diamond \kappa(z) \diamond \beta + \beta^* \diamond \kappa^*(z) \diamond \diamond \alpha\alpha\big]. \end{aligned} \tag{9.123}$$

The equation then produces the z-derivatives of the new kernels $\kappa(z)$ and $\kappa^*(z)$ on the left-hand side to match the terms that are produced on the right-hand side in the previous round. Although the new kernels produce new terms on the right-hand side of the evolution equation, each of them contains both the T-vertex and a new kernel, both of which produce suppressions. These new terms are thus doubly suppressed, which allows us to discard them. The evolution equation now leads to differential equations for the new kernels, given by

$$\begin{aligned} \partial_z\kappa(z) &= \mathrm{i}T \diamond (\mathbf{1} - \theta), \\ \partial_z\kappa^*(z) &= -\mathrm{i}(\mathbf{1} - \theta) \diamond T^*, \end{aligned} \tag{9.124}$$

showing their dependences on the vertex kernel.

At this point, we trace out the pump degrees of freedom, leading to an expression that contains a term in the exponent that is fourth order in α:

$$\mathrm{tr}_\beta\{W_{\hat{\rho}}\} = \mathcal{N}' \exp\left(-2\alpha^* \diamond \alpha + \frac{1}{2}\alpha^*\alpha^* \diamond \diamond \kappa \diamond \theta^{-1} \diamond \kappa^* \diamond \diamond \alpha\alpha\right), \tag{9.125}$$

where $\mathcal{N}'$ is a modified normalization constant. Thanks to the double suppression produced by the two κ's, we can treat the expression perturbatively, by expanding the part of the exponential that contains the higher-order term. We follow an approach similar to what we used in Section 6.4.8, where the resulting polynomial is converted into a construction process, operating on source terms added in the remaining part of the exponential. This exponential function, serving as a generating function for the construction process, is in Gaussian form and can be used in subsequent calculations.

Although the thermal state does not provide a displacement that gives a *bosonic enhancement* due to the strength of the parameter function, there is a bosonic enhancement of the process due to the average number of photons in the thermal state. This enhancement is demonstrated in the following exercise.

Exercise 9.8. Use the single-mode thermal state kernel in (6.204) to show that the higher-order term in the exponent in (9.125) receives a bosonic enhancement from the average number of photons in the single-mode thermal state.

Here, we use the Wigner functional formalism to perform the analysis. In this scenario, it may be more convenient to use Glauber–Sudarshan *P*-functionals, because *P*-functionals provide a more convenient way to represent mixed states such as thermal states. Such *P*-functionals can be converted to Wigner functionals with a convolution, as shown in (5.90).

9.7 Upconversion teleportation

Quantum teleportation is a process in which a state that exists in one optical field is transferred to another optical field, without bringing these two optical fields into contact with each other. The process is accomplished with the aid of entanglement. It facilitates the transfer of the information via quantum correlations as mediated by a *joint measurement*. As such, quantum teleportation is a *heralded process*, like those discussed in Section 7.8.

To consider teleportation in the context of functional phase spaces, we need at least three separate phase space field variables. The optical field with the initial quantum state that is to be transferred (the *sender*) is represented in terms of a field variable γ. The optical field that is to receive the quantum state (the *receiver*) is represented by a field variable β. It is entangled with a third optical field (the *mediator*), which we represent with a field variable α.

The teleportation process is now accomplished by performing a *joint measurement* on the optical fields represented in terms of α and γ (the mediator and the sender). What the joint measurement does is to extract the coefficients of terms in an expansion of the tensor product state (mediator $\otimes$ sender) in terms of a complete basis of entangled states. For example, if the states are all just qubits, then the tensor product space is four-dimensional and the four Bell states in (4.98) would then represent a complete basis of entangled states in terms of which any such tensor products of qubits can be expanded. The joint measurement causes a projection of a specific state at the receiver that is related to the state of the sender via a unitary transformation. The required unitary transformation is determined by the specific result of the joint measurement. By sending the measurement result via a classical communication channel to the receiver, the required inverse unitary transformation can be performed to convert the projected state of the receiver into the same state that the sender had prior to the measurement. The process is shown in the diagram in Figure 9.10.

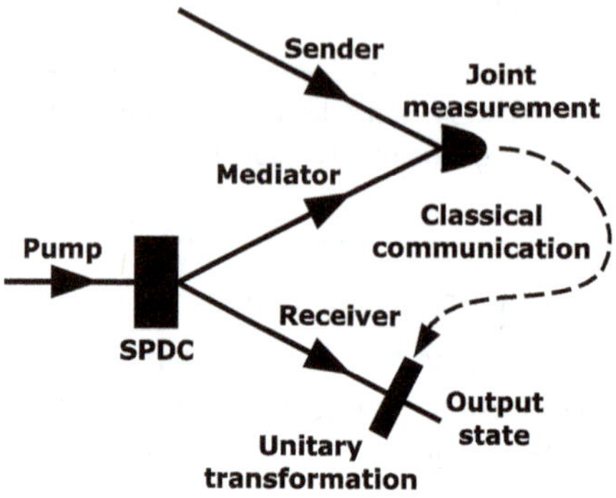

Figure 9.10: Diagrammatic represented of quantum teleportation.

There are different ways in which quantum teleportation can be implemented in practical systems. We can distinguish among different implementations based on the nature of the joint measurement that heralds the teleportation process.

The simplest implementation is *qubit teleportation* [18, 26–28] where the joint measurement is performed with the aid of the *Hong–Ou–Mandel effect*, using a beamsplitter as described in Section 7.6.7. This measurement can only observe one of the Bell states, namely Ψ_{-}. Therefore, it is applicable for the teleportation of an arbitrary qubit thus representing a *discrete variable teleportation process*. The unitary transformation associated with this Bell state is trivially given by the identity. Therefore, the joint measurement is equivalent to a pure *heralding* measurement, as discussed in Section 7.8. Such a system can only transfer the state of a single qubit. The particle-number degree of freedom of the teleported state is typically just a single photon. Generalizations of such a system to larger discrete variable states is possible, but at the cost of significant system complexity and generally significantly lower heralding probabilities.

Another way to implement teleportation is by using a *continuous variable teleportation process* [19, 20]. In this case, the state to be transferred is represented in terms of the particle-number degree of freedom. The joint measurement is done with the aid of *homodyning*, discussed in Section 7.9. The measured quadrature values are communicated to the receiver where a unitary transformation is performed in the form of a displacement on phase space. The spatiotemporal degrees of freedom do not play a role in this scenario. It is possible to generalize the system to transfer some spatiotemporal information with the particle-number degree of freedom. However, such generalizations again tend to increase system complexity severely, and the spatiotemporal information is generally in the form of a few discrete modes.

The scenario that we investigate here is where the joint measurement is performed with the aid of *upconversion* [29], as discussed in Section 8.8, through a process based on sum-frequency generation. The optical fields of the mediator and the sender (represented in terms of α and γ, respectively) are sent into a nonlinear crystal where they combine to form upconverted photons. The measurement of such an upconverted photon then *heralds* the teleportation of the spatiotemporal properties of the state of the sender. Since the nonlinear process is very weak, one can use bosonic enhancement to increase the efficiency. Therefore, the sender is prepared in the form of a strong coherent state. Its parameter function represents the spatiotemporal properties of the state to be teleported. Although all the photons in the coherent state aid the *bosonic enhancement* of the process, only one of these photons combine with a mediator photon to produce the upconverted photon that is detected to herald the teleportation of the spatiotemporal properties to the photon at the receiver that is entangled with this mediator photon.

Quantum teleportation protocols feature prominently in quantum information technology, especially in secure quantum communication systems. In those cases, additional constraints are added for quantum teleportation to ensure that the transfer is done securely. Here, we are not specifically interested in such quantum information systems. Our interest in quantum teleportation is focussed on the physics of the process, with the potential application of communicating spatiotemporal information. In that context, the upconversion teleportation process that we consider here is in principle able to transfer spatiotemporal information of a state without restrictions. However, any practical implementation imposes bandwidth limitations. Moreover, the scenario does not transform the particle-number degree of freedom of the state. As shown below, the nature of the state at the receiver is instead determined by the nature of the joint measurement. Nevertheless, the particle-number degree of freedom plays a significant role in the measurement process. Hence, the reason why we investigate this particular implementation of quantum teleportation.

Our aim is to use the functional formalism to model the practical system that implements upconversion teleportation so that we can investigate any distortions that the spatiotemporal information may incur during the teleportation process. The analysis reveals the nature of the spatiotemporal properties that are obtained at the receiver in the form of the transformed parameter function of the sender coherent state. We use the results of the mode transformations considered in Section 9.4 for this purpose. In this way, the fidelity of the spatiotemporal information at the receiver is given in terms of the experimental parameters.

9.7.1 Input state

Our first order of business is to obtain an expression for the complete state of the system. For this purpose, we use the appropriate *evolution equation*. It allows us to determine the complete state within the appropriate approximations. There are two approximations that are relevant for the current scenario, based on the relevant experimental conditions. One is the fact that the sender coherent state is strong, which allows us to use *strong-field perturbation theory*, leading to the *semiclassical approximation*. The other is the fact that the entangled state comprising the mediator and the receiver operates in terms of single biphotons. Therefore, we can employ *weak squeezing conditions*. We apply these approximations when they become necessary.

The input state consisting of the tensor product of a coherent state and a twin-beam squeezed vacuum state is represented by the product of their Wigner functionals with different field variables. Hence,

$$W_{\text{in}}[\alpha,\beta,\gamma] = \mathcal{N}_0^3 \exp[-2\alpha^* \diamond A \diamond \alpha - 2\beta^* \diamond A \diamond \beta - 2\alpha^* \diamond B \diamond \beta^* \\ - 2\beta \diamond B^* \diamond \alpha - 2(\gamma^* - \gamma_0^*) \diamond (\gamma - \gamma_0)] , \quad (9.126)$$

where α and β are the field variables of the twin-beam squeezed vacuum state for the mediator and the receiver, respectively, and γ is the field variable of the sender coherent state with γ_0 being its parameter function. The process produces an upconverted field, which is represented in terms of the field variable η.

9.7.1.1 Complete state from the evolution equation

To see what terms we need to include in the exponent for the upconverted degrees of freedom, we use an iterative approach similar to the approach in Section 9.6. Substituting the input state, together with a vacuum state for the upconverted field into the evolution equation given in (8.63), we see what kind of terms the evolution equation generates. Similar terms are added in the exponent of the complete state. When the modified state is substituted into the evolution equation, it generates more terms. These terms are either higher-order corrections to existing terms or they require new terms in the exponent of the complete state. In this way, we determine what the necessary terms in the expression of the complete state are.

Unfortunately, it often happens that this iteration process never ends. Typically, the evolution equation produces terms that contain both input and upconverted field variables. Eventually these terms lead to higher-order terms with more than two field variables. As a result, the state becomes a super-Gaussian state. Since functional integrals cannot be evaluated for such super-Gaussian states, calculations become intractable.

Fortunately, each iteration produces suppression factors in the form of the nonlinear coefficient of the upconversion process. The nonlinear coefficient, together with the magnitudes of displacement parameter functions and other dimension parameters produce the squeezing parameter of the process. Under weak squeezing conditions, the squeezing parameter is used as an expansion parameter for the exponent and allows us to discard terms beyond a certain level. In this way, the iteration process is truncated at a given level and the Wigner functional retains its Gaussian form.

Starting with the input state in (9.126), together with a vacuum state for the upconverted field, we obtain an evolution equation with the form

$$\begin{aligned}
\mathrm{i}\partial_z W_{\hat{\rho}} = & -2W_{\hat{\rho}} \int \alpha^*(\mathbf{k}_1)\gamma_0^*(\mathbf{k}_2)T(\mathbf{k}_1,\mathbf{k}_2,\mathbf{k}_3,z)\eta(\mathbf{k}_3) - \eta^*(\mathbf{k}_3)T^*(\mathbf{k}_1,\mathbf{k}_2,\mathbf{k}_3,z) \\
& \times \gamma_0(\mathbf{k}_1)\alpha(\mathbf{k}_2)\, \mathrm{d}k_1\, \mathrm{d}k_2\, \mathrm{d}k_3 - 2W_{\hat{\rho}} \int \eta^*(\mathbf{k}_3)T^*(\mathbf{k}_1,\mathbf{k}_2,\mathbf{k}_3,z)\gamma_0(\mathbf{k}_1) \\
& \times A(\mathbf{k}_2,\mathbf{k}_4)\alpha(\mathbf{k}_4) - \alpha^*(\mathbf{k}_4)A(\mathbf{k}_4,\mathbf{k}_1)\gamma_0^*(\mathbf{k}_2)T(\mathbf{k}_1,\mathbf{k}_2,\mathbf{k}_3,z)\eta(\mathbf{k}_3) \\
& + \eta^*(\mathbf{k}_3)T^*(\mathbf{k}_1,\mathbf{k}_2,\mathbf{k}_3,z)B(\mathbf{k}_1,\mathbf{k}_4)\beta^*(\mathbf{k}_4)\gamma_0(\mathbf{k}_2) - \beta(\mathbf{k}_4)B^*(\mathbf{k}_4,\mathbf{k}_1) \\
& \times \gamma_0^*(\mathbf{k}_2)T(\mathbf{k}_1,\mathbf{k}_2,\mathbf{k}_3,z)\eta(\mathbf{k}_3)\, \mathrm{d}k_1\, \mathrm{d}k_2\, \mathrm{d}k_3\, \mathrm{d}k_4,
\end{aligned} \tag{9.127}$$

All the γ field variables cancel at this level of the iteration. This cancellation happens because the upconverted field at this level of the iteration is still a vacuum state. It happens regardless of the strength of the parameter function and does not require the semiclassical approximation.

Thanks to the cancellation, the vertex T in all the terms is contracted to the parameter function γ_0. We therefore define bilinear kernels

$$\begin{aligned}
K(\mathbf{k}_1,\mathbf{k}_3,z) &= \mathrm{i} \int \gamma_0^*(\mathbf{k}_2)T(\mathbf{k}_1,\mathbf{k}_2,\mathbf{k}_3,z)\, \mathrm{d}k_2, \\
K^*(\mathbf{k}_3,\mathbf{k}_1,z) &= -\mathrm{i} \int \gamma_0(\mathbf{k}_2)T^*(\mathbf{k}_1,\mathbf{k}_2,\mathbf{k}_3,z)\, \mathrm{d}k_2.
\end{aligned} \tag{9.128}$$

Note that the kernels K and K^* always link a down-converted field variable with an upconverted field variable (unless other kernels are involved), one being complex conjugated and the other one not. Since they are all bilinear kernels, we use $\diamond$ to represent contractions involving down-converted degrees of freedom and $\bullet$ for the contractions with the upconverted degrees of freedom. As far as possible, we always put the complex conjugated field variable on the left-hand side of K or K^*, with the other (not-complex-conjugated) field variable on the right-hand side. In some cases, we may be forced to deviate from this practice. Since we use different symbols for contractions with the different field variables, such inversions are easy to identify. Note that we do not represent the kernel in such inverted cases with a transpose, even though the kernel is not symmetric with respect to such inversions.

The first iteration thus introduces four additional terms. We represent their contribution to the state as a Gaussian Wigner functional

$$\begin{aligned}
W_{\hat{\sigma}} = \mathcal{N} \exp [& -2\alpha^* \diamond N(z) \bullet \eta - 2\eta^* \bullet N^*(z) \diamond \alpha \\
& - 2\beta \diamond G(z) \bullet \eta - 2\eta^* \bullet G^*(z) \diamond \beta^*],
\end{aligned} \tag{9.129}$$

that is multiplied with the initial state, so that $W_{\hat{\rho}} = W_{\text{in}} W_{\hat{\sigma}}$. The differential equations that link the z-derivatives of these terms with the terms in the evolution equation after the first iteration are

$$\partial_z N(z) = E \diamond K(z), \quad \partial_z N^*(z) = K^*(z) \diamond E,$$
$$\partial_z G(z) = B^* \diamond K(z), \quad \partial_z G^*(z) = K^*(z) \diamond B, \tag{9.130}$$

where $E = A - \mathbf{1}$. These equations lead to kernels given by

$$N(z) = E \diamond V(z), \quad N^*(z) = V^*(z) \diamond E,$$
$$G(z) = B^* \diamond V(z), \quad G^*(z) = V^*(z) \diamond B, \tag{9.131}$$

where

$$V(\mathbf{k}_1, \mathbf{k}_2, z) = \int_{z_0}^{z} K(\mathbf{k}_1, \mathbf{k}_2, z')\, dz',$$
$$V^*(\mathbf{k}_3, \mathbf{k}_1, z) = \int_{z_0}^{z} K^*(\mathbf{k}_3, \mathbf{k}_1, z')\, dz'. \tag{9.132}$$

The next iteration produces additional terms when the functional derivatives operate on the new terms in the exponent, making the evolution equation significantly more complicated. In addition to those terms produced in (9.127), it now also generates terms that are third order in field variables, causing the Wigner functional of the state to be super-Gaussian. All these additional terms are at least second order in the vertex T. Therefore, they are suppressed by the upconversion efficiency. In a sense, the suppression at this second-order level is more severe than the suppression in the first iteration because all the T-vertices in the first iteration are enhanced by the parameter function γ_0 through a stimulated process. By contrast, most of the T-vertices produced by the second iteration do not contain such enhancements. The exceptions include terms produced by the triple functional derivatives leading to constant terms with doubly contracted kernels independent of field variables. Such terms only affect the normalization of the state. There are also terms that are third order in the enhanced vertex K, representing higher-order corrections to the kernels that we already have.

The enhancement produced by γ_0 assumes a strong input field provided by the coherent state. Therefore, we can impose the *semiclassical approximation*, provided that γ_0 is large. The semiclassical approximation is implemented by substituting $\gamma \to \gamma_0$. The second iteration thus produces the following new terms in the evolution equation:

$$\begin{aligned} \mathrm{i}\partial_z W_{\hat{\rho}} = (&\text{old terms} - \mathrm{i}2\eta^* \bullet N^* \diamond K \bullet \eta - \mathrm{i}2\eta^* \bullet K^* \diamond N \bullet \eta \\ &+ \mathrm{i}2\alpha^* \diamond N \bullet K^* \diamond \alpha + \mathrm{i}2\alpha^* \diamond K \bullet N^* \diamond \alpha \\ &+ \mathrm{i}2\beta \diamond G \bullet K^* \diamond \alpha + \mathrm{i}2\alpha^* \diamond K \bullet G^* \diamond \beta^*)W_{\hat{\rho}}. \end{aligned} \tag{9.133}$$

There terms represent modifications for the kernels (A_0, B_0 and B_0^*) of the input squeezed state that involves α and for the vacuum state term in the upconverted degrees of freedom represented by η, with a kernel represented by A_u. These corrections are all second order in the enhanced vertex K. The differential equations for these kernels with the terms in the evolution equation after the second iteration are

$$\begin{aligned}
\partial_z A_0(z) &= -K(z) \bullet N^*(z) - N(z) \bullet K^*(z), \\
\partial_z B_0(z) &= -K(z) \bullet G^*(z), \quad \partial_z B_0^*(z) = -G(z) \bullet K^*(z), \\
\partial_z A_u(z) &= K^*(z) \diamond N(z) + N^*(z) \diamond K(z).
\end{aligned} \tag{9.134}$$

With the appropriate initial conditions, these equations lead to kernels given by

$$\begin{aligned}
A_0(z) &= A - \int_{z_0}^{z} N(z') \bullet K^*(z') + K(z') \bullet N^*(z') \, dz' \\
&= A - E \diamond Y(z) - Y^*(z) \diamond E, \\
B_0(z) &= B - \int_{z_0}^{z} K(z') \bullet G^*(z') \, dz' = B - Y^*(z) \diamond B, \\
B_0^*(z) &= B^* - \int_{z_0}^{z} G(z') \bullet K^*(z') \, dz' = B^* - B^* \diamond Y(z), \\
A_u(z) &= \mathbf{1} + \int_{z_0}^{z} K^*(z') \diamond N(z') + N^*(z') \diamond K(z') \, dz' \\
&= \mathbf{1} + R(z),
\end{aligned} \tag{9.135}$$

where we used the expressions in (9.131), and defined

$$\begin{aligned}
Y(z) &= \int_{z_0}^{z} V(z') \bullet K^*(z') \, dz, \quad Y^*(z) = \int_{z_0}^{z} K(z') \bullet V^*(z') \, dz', \\
R(z) &= \int_{z_0}^{z} K^*(z') \diamond E \diamond V(z') + V^*(z') \diamond E \diamond K(z') \, dz' \\
&= V^*(z) \diamond E \diamond V(z),
\end{aligned} \tag{9.136}$$

in the final expressions. We terminate the iterative process at this point.

Exercise 9.9. Show that $R(z) = V^*(z) \diamond E \diamond V(z)$.
Hint: Invert the order of integration in one of the two terms.

Incorporating all the terms from the original input state in (9.126), the first iteration in (9.127) and the second iteration in (9.133), we obtain an expression for the Wigner functional of the complete state given by

$$\begin{aligned}W_{\text{full}} = \mathcal{N} \exp\big[&-2(\gamma^* - \gamma_0^*) \diamond (\gamma - \gamma_0) - 2\alpha^* \diamond A_0(z) \diamond \alpha - 2\beta^* \diamond A \diamond \beta \\ &- 2\alpha^* \diamond B_0(z) \diamond \beta^* - 2\beta \diamond B_0^*(z) \diamond \alpha - 2\alpha^* \diamond N(z) \bullet \eta - 2\eta^* \bullet N^*(z) \diamond \alpha \\ &- 2\beta \diamond G(z) \bullet \eta - 2\eta^* \bullet G^*(z) \diamond \beta^* - 2\eta^* \bullet A_u(z) \bullet \eta\big],\end{aligned} \tag{9.137}$$

where $\mathcal{N}$ is the normalization constant. Thus, we have obtained the complete state produced by the upconversion process, under the given conditions. However, there are some simplifications that can be made.

9.7.1.2 Simplifications

Thanks to the semiclassical approximation, the field variable γ is not contracted with any other field variable and does not take part in any measurement. So, we can trace it out. The resulting expression is the same as (9.137), but without the γ-term.

Another field variable that does not take part in any measurement is α. However, the integration over α has a more significant effect on the expression. The result of this integration is represented by

$$\begin{aligned}W_{\text{full}} = \mathcal{N}' \exp\big[&-2\beta^* \diamond A_b \diamond \beta - 2\eta^* \bullet A_u'(z) \bullet \eta \\ &- 2\beta \diamond G'(z) \bullet \eta - 2\eta^* \bullet G'^*(z) \diamond \beta^*\big],\end{aligned} \tag{9.138}$$

where the effect on the normalization is absorbed into $\mathcal{N}'$, and

$$\begin{aligned}A_b(z) &= A - B_0(z) \diamond A_0^{*-1}(z) \diamond B_0^*(z), \\ A_u'(z) &= A_u(z) - N^*(z) \diamond A_0^{-1}(z) \diamond N(z), \\ G'(z) &= G(z) - B_0^*(z) \diamond A_0^{-1}(z) \diamond N(z), \\ G'^*(z) &= G^*(z) - N^*(z) \diamond A_0^{-1}(z) \diamond B_0(z).\end{aligned} \tag{9.139}$$

The scenario under consideration calls for the squeezed part of the input state to be a weakly squeezed state. This condition is necessary to reduce the probability for the simultaneous production of two or more photon pairs. The weak squeezing condition allows us to discard the second terms of A_u', G', and G'^*, thanks to the expressions in (9.131), in which $E = A - \mathbf{1}$ is second order in the squeezing parameter. The expression of A_b is left as it is for the time being. The upconversion process, on the other hand, is assumed to be strong. Therefore, the kernels K and K^* do not produce the same suppressions as those of the weakly squeezed input state. So, the complete state becomes

$$\begin{aligned}W_{\text{full}} = \mathcal{N}' \exp\big[&-2\beta^* \diamond A_b \diamond \beta - 2\eta^* \bullet A_u(z) \bullet \eta \\ &- 2\beta \diamond G(z) \bullet \eta - 2\eta^* \bullet G^*(z) \diamond \beta^*\big].\end{aligned} \tag{9.140}$$

The resulting expression in (9.140) has the form of a twin-beam squeezed vacuum state. It represents entanglement between the upconverted degrees of freedom, represented by η, and the teleported degrees of freedom at the receiver, represented by β.

Therefore, measurements that are performed on the upconverted degrees of freedom affect the state obtained in the teleported degrees of freedom.

9.7.2 Joint measurement—single photons

A measurement in the upconverted degrees of freedom serves as the joint measurement heralding the teleportation process. The teleportation of the state is to be heralded by the successful detection of a single photon in the upconverted field. The calculation of this measurement is done with the aid of the generating function in (7.87), for which we use a single-mode detector kernel $D_u(\mathbf{k}_1,\mathbf{k}_2) = M(\mathbf{k}_1)M^*(\mathbf{k}_2)$, where $M(\mathbf{k})$ is a normalized spectral function, so that $\operatorname{tr}\{D_u\} = 1$. This detector mode serves a similar, but conjugate, role to that of the pump in the down-conversion process. We can therefore think of the detector mode as the parameter function of an *antipump* state. A significant difference is that the detector mode is normalized, whereas the larger than unity magnitude of the *pump parameter function* in a down-conversion process serves to produce a *bosonic enhancement* of the process.

We multiply (9.140) with the generating function in (7.87) as a functional of η, and integrate over η. It leads to

$$\begin{aligned}
\mathcal{W}_{\text{tele}} =& \frac{2\mathcal{N}'\mathcal{N}_0}{1+J}\int \exp\left[-2\beta^*\diamond A_b\diamond\beta-2\beta\diamond G(z)\bullet\eta-2\eta^*\bullet G^*(z)\diamond\beta^*\right.\\
&\left.-2\eta^*\bullet A_u(z)\bullet\eta-2\mathcal{J}\eta^*\bullet MM^*\bullet\eta\right]\mathcal{D}^\circ[\eta]\\
=&\frac{2\mathcal{N}_0\exp\left(-2\beta^*\diamond A_b\diamond\beta\right)}{(1+J)\det\{A_u+\mathcal{J}MM^*\}}\exp\left[2\beta\diamond G\bullet\left(A_u+\mathcal{J}MM^*\right)^{-1}\bullet G^*\diamond\beta^*\right]\\
=&\frac{2\mathcal{N}'}{1+J+(1-J)\tau}\exp\left(-2\beta^*\diamond A_b\diamond\beta+2\beta\diamond G\bullet A_u^{-1}\bullet G^*\diamond\beta^*\right)\\
&\times\exp\left[-\frac{2(1-J)\beta\diamond G\bullet A_u^{-1}\bullet MM^*\bullet A_u^{-1}\bullet G^*\diamond\beta^*}{1+J+(1-J)\tau}\right],
\end{aligned} \tag{9.141}$$

where

$$\tau = M^*\bullet A_u^{-1}\bullet M, \tag{9.142}$$

and $\mathcal{J}$ is given in (7.88). For the simplified final expression, we used the identities in Appendix D, which lead to

$$\begin{aligned}
\left(A_u+\mathcal{J}MM^*\right)^{-1} &= A_u^{-1}-\frac{\mathcal{J}}{1+\mathcal{J}\tau}A_u^{-1}\bullet MM^*\bullet A_u^{-1},\\
\det\left\{A_u+\mathcal{J}MM^*\right\} &= \det\{A_u\}\left(1+\mathcal{J}\tau\right).
\end{aligned} \tag{9.143}$$

The term in the exponent of (9.141) with the generating parameter is second order in the weak squeezing parameter. Those without the generating parameter can be combined to give a term that looks like that of a thermal state. However, it differs from the identity by terms that are all second order in weak squeezing parameter making it a very cold thermal state. Therefore, we discard these second-order corrections, but retain the second-order term with the generating parameter. When we apply the derivative with respect to the generating parameter, it produces only terms that are second order in the weak squeezing parameter. This suppression reflects the fact that the probability for a successful upconversion is low due to the weakness of the initial squeezed state. Since the heralding process performs post-selection, the resulting heralded state needs to be normalized again, which then removes the suppression.

Before we apply the derivative, some further simplifications can be made. Based on (9.135), the inverse of A_u^{-1} can be represented by

$$A_u^{-1} \approx \mathbf{1} - R(z), \tag{9.144}$$

where the correction is second order in the weak squeezing parameter. Inserting the inverse into the expression for the state, we drop the second-order correction, but retain the τ. The reason is that

$$\tau = M^* \bullet A_u^{-1} \bullet M = 1 - M^* \bullet R \bullet M \triangleq 1 - \delta, \tag{9.145}$$

which implies that $1 + J + (1 - J)\tau = 2 - (1 - J)\delta$. Since δ is associated with the generating parameter, it still plays a role. So, the expression becomes

$$\mathcal{W}_{\text{tele}} = \frac{\mathcal{N}' \exp(-2\beta^* \diamond \beta)}{1 - \frac{1}{2}(1 - J)\delta} \exp\left[-\frac{1 - J}{1 - \frac{1}{2}(1 - J)\delta} \beta \diamond G \bullet MM^* \bullet G^* \diamond \beta^*\right]. \tag{9.146}$$

At this point, we remind ourselves that the spatiotemporal properties that we strive to teleport, are represented by the parameter function γ_0. It is contracted with the vertex kernel, sitting inside $V(z)$, which in turn forms part of $G(z)$. To make the presence of γ_0 apparent, we use the definitions in (9.128) and (9.132) to redefine

$$\begin{aligned} V(z) \bullet M &= \mathrm{i} \int_{z_0}^{z} \gamma_0^* \diamond T(z') \bullet M \, \mathrm{d}z' \triangleq \gamma_0^* \diamond B_u(z), \\ M^* \bullet V^*(z) &= -\mathrm{i} \int_{z_0}^{z} M^* \bullet T^*(z') \diamond \gamma_0 \, \mathrm{d}z' \triangleq B_u^*(z) \diamond \gamma_0, \end{aligned} \tag{9.147}$$

which is analogues to the integrated version of the H's in (8.68), with M taking over the role of the pump parameter function. In other words, B_u is analogous to B for the upconversion process in the weak squeezing limit. We also represent the original B's as B_d to emphasize the fact that it is associated with the down-conversion process that produced the initial weakly squeezed vacuum state. However, since M is normalized, it does not enhance the vertex in B_u as the pump parameter function does in B_d. Instead, the magnitude of γ_0 provides the enhancement for B_u. It then follows that

$$G \bullet M = \gamma_0^* \diamond B_u \diamond B_d^* \triangleq \gamma_1^* \quad \text{and} \quad M^* \bullet G^* = B_d \diamond B_u^* \diamond \gamma_0 \triangleq \gamma_1, \tag{9.148}$$

where γ_1 now represents a transformed or distorted version of the original parameter function. We also make the parameter function γ_0 apparent in the definition of δ in (9.145). Using (9.136) and (9.147), it becomes

$$\begin{aligned} \delta &= M^* \bullet R \bullet M = \gamma_0 \diamond B_u^* \diamond E \diamond B_u \diamond \gamma_0^* \\ &= \frac{1}{2}\gamma_0 \diamond B_u^* \diamond B_d \diamond B_d^* \diamond B_u \diamond \gamma_0^* = \frac{1}{2}\|\gamma_1\|^2, \end{aligned} \tag{9.149}$$

where we used (7.206) to replace $E \to \frac{1}{2}B_d \diamond B_d^*$.

After these simplifications, the generating function for the heralded state at the receiver becomes

$$\mathcal{W}_{\text{tele}} = \frac{\mathcal{N}\exp(-2\beta^* \diamond \beta)}{1-\frac{1}{4}(1-J)\|\gamma_1\|^2} \exp\left[-\frac{(1-J)\beta^* \diamond \gamma_1\gamma_1^* \diamond \beta}{1-\frac{1}{4}(1-J)\|\gamma_1\|^2}\right]. \tag{9.150}$$

The normalization constant $\mathcal{N}$ is to be computed for each heralded state, because the measurement of a single photon is not trace preserving. It is obtained by computing the trace of the generating function. Thus, a generating function for the inverse of the normalization constant is produced, which reads

$$\mathcal{W}_{\mathcal{N}^{-1}} = \frac{1}{\mathcal{N}_0\left[1+\frac{1}{4}(1-J)\|\gamma_1\|^2\right]}. \tag{9.151}$$

Using these generating functions, we calculate the normalized heralded state obtained from a single-photon detection. It is

$$W_{\text{tele-1}}[\beta] = \frac{\partial_J \mathcal{W}_{\text{tele}}|_{J=0}}{\partial_J \mathcal{W}_{\mathcal{N}^{-1}}|_{J=0}} = (4\beta^* \diamond FF^* \diamond \beta - 1)\exp(-2\beta^* \diamond \beta), \tag{9.152}$$

where we removed all sub-leading-order terms in the exponent and the polynomial, and defined the normalized transformed parameter function

$$F \triangleq \frac{\gamma_1}{\|\gamma_1\|}. \tag{9.153}$$

The resulting Wigner functional is that of a single-photon Fock state parameterized by the normalized transformed parameter function F. At the origin where $\beta = 0$, the Wigner functional is negative.

The input state that we wished to teleport is a coherent state. Yet, what we obtained is a Fock state. It shows that the scheme does not teleport the particle-number degree of freedom, but only the spatiotemporal degrees of freedom (with some distortions). Instead, the particle-number degree of freedom of the state at the receiver is imposed by the nature of the joint measurements. Here, it is a single-photon measurement in the upconverted degrees of freedom, which leads to the teleported state being a Fock state.

9.7.3 Teleportation channel limitations

The kernel contractions in (9.148) represent the transformation that converts the initial parameter function at the sender into the parameter function at the receiver. In this sense, we regard the two contracted odd kernels $B_d \diamond B_u^*$ as a representation of the *teleportation channel*, which is the communication channel that transfers the spatiotemporal information of the state from the sender to the receiver. In practical applications, this channel must transfer the information as accurately as possible. Therefore, we need to understand the limitations that exist in this channel.

In physical implementations of classical communication channels, the limitations are often represented by the (temporal or spatial) bandwidth, as well as the limitations in the duration or spatial extent of the *signal* carrying the information. All these limitations are combined into the *time-bandwidth product* or the *space-bandwidth product* of the system, the latter being more relevant for our current application. The space-bandwidth product of a communication system is a finite-dimensionless value, representing the average number of modes that can pass through the communication system.

For physical implementations of quantum systems, similar limitations exist. Although quantum teleportation is a truly quantum process, the final expression for the transformation kernel representing the teleportation channel is formally similar to the transformation or distortion of the modes of a classical optical field. So, the space-bandwidth product is relevant for the current case.

To compute the space-bandwidth product for this process, we use the equivalence between a *quantum channel* and a *bipartite quantum state*, as given by the Choi–Jamiołkowski isomorphism [30]. The quantum channel, given by

$$\hat{C} = \sum_{m,n} |m\rangle\, H_{mn}\, \langle n| \,, \tag{9.154}$$

is thus associated with a bipartite quantum state that reads

$$|\psi\rangle = \sum_{m,n} |m\rangle\, |n\rangle\, H_{mn}. \tag{9.155}$$

In Section 4.3.3, we show that any pure bipartite state can be represented in terms of a *Schmidt decomposition*. The *Schmidt number* then gives the effective number of Schmidt coefficients. It thus also indicates the average number of modes that can pass through the associated channel, similar to the space-bandwidth product. By computing the

Schmidt number for the quantum channel, using the Choi–Jamiołkowski isomorphism, we get an estimate of the space-bandwidth product of the teleportation channel.

In the current case, the calculation of the Schmidt number produces

$$\kappa = \frac{\operatorname{tr}\{H \diamond H^{\dagger}\}^2}{\operatorname{tr}\{H \diamond H^{\dagger} \diamond H \diamond H^{\dagger}\}}, \tag{9.156}$$

where $H = B_d \diamond B_u^*$. The resulting integrals are in general rather complicated, unless some approximations are employed to simplify them.

Fortunately, we already investigated in Section 9.4 how these kernels place restrictions in configuration space and in the Fourier domain, as represented by the effective widths of the profile functions on these respective domains. Here, we have two consecutive odd kernels, each imposing restrictions. Those imposed by the combined kernel are therefore determined by the smaller effective widths. For the odd kernels, the effective widths are given in (9.95) and (9.96) as a function of the squeezing parameter. In configuration space, the smallest width is given by the Ξ-dependent width in (9.95) associated with B_u. On the other hand, the smallest width in the Fourier domain is obtained for the smallest value of the squeezing parameter. Therefore, we need to use the width provided in (9.96) associated with B_d in the weak squeezing limit $\Xi \to 0$. The space-bandwidth product is given by the product of these two widths. It reads

$$\text{SBWP} = 24^{1/4}\sqrt{\frac{\sinh(\Xi)}{\beta\Xi\cosh(\Xi)}}, \tag{9.157}$$

where

$$\beta = \frac{2L}{k_{\text{p}} w_{\text{p}}^2}, \tag{9.158}$$

is the dimensionless expansion parameter for the thin-crystal approximation. Therefore, thin-crystal conditions lead to better space-bandwidth products. Considering the effect of the squeezing parameter, we see that the space-bandwidth products decreases for large values of the squeezing parameter.

Bibliography

[1] D. F. Walls and G. J. Milburn. *Quantum Optics*. Springer, Berlyn, 1995.

[2] C. C. Gerry and P. L. Knight. *Optical Coherence and Quantum Optics*. Cambridge University Press, New York, 2005.

[3] C. K. Hong and L. Mandel. Theory of parametric frequency down conversion of light. *Phys. Rev. A*, 31:2409, 1985.

[4] G. J. Milburn. Multimode minimum uncertainty squeezed states. *J. Phys. A, Math. Gen.*, 17:737–745, 1984.

[5] H. H. Arnaut and G. A. Barbosa. Orbital and intrinsic angular momentum of single photons and entangled pairs of photons generated by parametric down-conversion. *Phys. Rev. Lett.*, 85:286–289, 2000.

[6] R. S. Bennink and R. W. Boyd. Improved measurement of multimode squeezed light via an eigenmode approach. *Phys. Rev. A*, 66:053815, 2002.

[7] C. K. Law and J. H. Eberly. Analysis and interpretation of high transverse entanglement in optical parametric down conversion. *Phys. Rev. Lett.*, 92:127903, 2004.

[8] W. Wasilewski, A. I. Lvovsky, K. Banaszek, and C. Radzewicz. Pulsed squeezed light: Simultaneous squeezing of multiple modes. *Phys. Rev. A*, 73(6):063819, 2006.

[9] A. Biswas and G. S. Agarwal. Nonclassicality and decoherence of photon-subtracted squeezed states. *Phys. Rev. A*, 75:032104, 2007.

[10] E. Brambilla, L. Caspani, L. A. Lugiato, and A. Gatti. Spatiotemporal structure of biphoton entanglement in type-II parametric down-conversion. *Phys. Rev. A*, 82:013835, 2010.

[11] D. Li, C.-H. Yuan, Z. Y. Ou, and W. Zhang. The phase sensitivity of an SU(1, 1) interferometer with coherent and squeezed-vacuum light. *New J. Phys.*, 16:073020, 2014.

[12] H. Vahlbruch, M. Mehmet, K. Danzmann, and R. Schnabel. Detection of 15 dB squeezed states of light and their application for the absolute calibration of photoelectric quantum efficiency. *Phys. Rev. Lett.*, 117:110801, 2016.

[13] P. Sharapova, A. M. Pérez, O. V. Tikhonova, and M. V. Chekhova. Schmidt modes in the angular spectrum of bright squeezed vacuum. *Phys. Rev. A*, 91:043816, 2015.

[14] F. S. Roux. Quantifying entanglement of parametric down-converted states in all degrees of freedom. *Phys. Rev. Res.*, 2:023137, 2020a.

[15] P. R. Sharapova, G. Frascella, M. Riabinin, A. M. Pérez, O. V. Tikhonova, S. Lemieux, R. W. Boyd, G. Leuchs, and M. V. Chekhova. Properties of bright squeezed vacuum at increasing brightness. *Phys. Rev. Res.*, 2:013371, 2020.

[16] F. S. Roux. Parametric down-conversion beyond the semiclassical approximation. *Phys. Rev. Res.*, 2:033398, 2020b.

[17] F. S. Roux. Stimulated parametric down-conversion for spatiotemporal metrology. *Phys. Rev. A*, 104:043514, 2021.

[18] D. Bouwmeester, J.-W. Pan, K. Mattle, M. Eibl, H. Weinfurter, and A. Zeilinger. Experimental quantum teleportation. *Nature*, 390:575–579, 1997.

[19] T. C. Zhang, K. W. Goh, C. W. Chou, P. Lodahl, and H. J. Kimble. Quantum teleportation of light beams. *Phys. Rev. A*, 67:033802, 2003.

[20] S. Pirandola, J. Eisert, C. Weedbrook, A. Furusawa, and S. L. Braunstein. Advances in quantum teleportation. *Nat. Photonics*, 9:641–652, 2015.

[21] B. Brecht and C. Silberhorn. Characterizing entanglement in pulsed parametric down-conversion using chronocyclic Wigner functions. *Phys. Rev. A*, 87:053810, 2013.

[22] S. L. Braunstein. Squeezing as an irreducible resource. *Phys. Rev. A*, 71:055801, 2005.

[23] F. Hudelist, J. Kong, C. Liu, J. Jing, Z. Y. Ou, and W. Zhang. Quantum metrology with parametric amplifier-based photon correlation interferometers. *Nat. Commun.*, 5:3049, 2014.

[24] Z. Qin, M. Gessner, Z. Ren, X. Deng, D. Han, W. Li, X. Su, A. Smerzi, and K. Peng. Characterizing the multipartite continuous-variable entanglement structure from squeezing coefficients and the Fisher information. *npj Quantum Inf.*, 5:1–6, 2019.

[25] M. Abramowitz and I. A. Stegun. *Handbook of Mathematical Functions*. Dover, Toronto, 1972.

[26] J. Yin, J.-G. Ren, H. Lu, Y. Cao, H.-L. Yong, Y.-P. Wu, C. Liu, S.-K. Liao, F. Zhou, Y. Jiang, and et al. Quantum teleportation and entanglement distribution over 100-kilometre free-space channels. *Nature*, 488:185–188, 2012.

[27] X. S. Ma, T. Herbst, T. Scheidl, D. Q. Wang, S. Kropatschek, W. Naylor, B. Wittmann, A. Mech, J. Kofler, E. Anisimova, V. Makarov, T. Jennewein, R. Ursin, and A. Zeilinger. Quantum teleportation over 143 kilometres using active feed-forward. *Nature*, 489:269–273, 2012.

[28] S. Goyal, P. E. Boukama-Dzoussi, S. Ghosh, F. S. Roux, and T. Konrad. Qudit-teleportation for photons with linear optics. *Sci. Rep.*, 4:4543, 2014.
[29] B. Sephton, A. Vallés, I. Nape, M. A. Cox, F. Steinlechner, T. Konrad, J. P. Torres, F. S. Roux, and A. Forbes. Quantum transport of high-dimensional spatial information with a nonlinear detector. *Nat. Commun.*, 14:8243, 2023.
[30] M. Jiang, S. Luo, and S. Fu. Channel-state duality. *Phys. Rev. A*, 87:022310, 2013.

Part IV: **Epilogue**

10 Summary and outlook

10.1 The aim

Photonic quantum information systems, based on quantum optics, are often of such a complex nature that the design and analysis of these systems become extremely challenging. Existing mathematical formalisms, such as traditional operator-based quantum mechanics, or even the continuous variable formalism for quantum optics, cannot always model such systems adequately due to intrinsic limitations in these formalisms. Such limitations are typically imposed by the spatiotemporal degrees of freedom in the system. While there are many quantum optical systems that can be adequately modelled with a finite-dimensional formulation, some widely used systems are intrinsically infinite dimensional. To meet this challenge, a suitable formalism that can model arbitrary quantum optical systems and scenarios without any limitations is required.

To illustrate the issue, consider the process of an arbitrary quantum optical system as a transformation of the parameter function of some input state. The infinite-dimensional nature of the process is revealed by the kernel implementing the transformation when it has an infinite spectrum. Assuming we know how to diagonalize such a kernel $K(\mathbf{k}_1, \mathbf{k}_2)$, it would be given by an infinite sum

$$K(\mathbf{k}_1, \mathbf{k}_2) = \sum_{n=0}^{\infty} \lambda_n \phi(\mathbf{k}_1) \phi^*(\mathbf{k}_2). \tag{10.1}$$

Here, $\phi(\mathbf{k})$ and λ_n denote the eigenfunctions and their associated eigenvalues. The eigenfunctions span an infinite-dimensional space of functions. The transformation of an arbitrary parameter function can produce any function in such an infinite-dimensional space, representing the spatiotemporal properties of the photons in the states that are found in such systems. Moreover, operations on such states thus represent mappings between states consisting of photons with arbitrary spatiotemporal properties.

Obviously, any finite-dimensional representation of such a system introduces errors due to the truncation that it imposes. If the system is already well understood, one may be able to guess how to introduce the truncation to minimize the errors. Unfortunately, such knowledge of the system is seldom available. Even if the kernels are known, the diagonalized form of such kernels are generally not known. To avoid truncation errors, the formalism needs to accommodate infinite-dimensional representations of systems without having to diagonalize the kernels.

By implication, in such a formalism, states and operations need to be represented as *functionals* (functions of functions). Due to the functional nature of the state representations, it naturally leads to the development of a *functional phase space (Moyal) formalism*. The states and operations are represented as Wigner functionals that are defined on the functional phase space.

https://doi.org/10.1515/9783111445342-014

Most calculations therefore involve functional integrals. Since the list of tractable functional integrals are relatively short—those that lead to Dirac delta functionals and those that involve Gaussian functionals—special methods are developed to convert expressions that are not represented in this short list into equivalent expressions that are. These methods are generally based on *generating functions or functionals*. If an expression does not have a Gaussian form, but a polynomial form or a polynomial Gaussian form, it can be represented in terms of a generating function(al) that is in pure Gaussian form. Once the functional calculations are completed, some (functional) derivatives are applied to produce the result for the original non-Gaussian input. Such methods work for most cases. Scenarios that cannot be addressed in this way include those states or operations that require expressions in super-Gaussian form or in terms of some other transcendental functional of the field variables. Fortunately, such cases are not often encountered. It is not to say that methods cannot be developed to deal with such cases.

It is imperative that the resulting formalism is readily utilizable. It would be pointless to develop a formalism that ends up being too cumbersome to be useful. The main difference between a functional formalism and the more traditional existing formalisms is that the variables in traditional formalisms are replaced by *field variables*. Such field variables are *contracted* (integrated) as opposed to being merely multiplied and summed. A functional formalism also tends to involve various kernels contracted to these field variables. In the end, the functional formalism leads to complicated expressions in terms of such kernels. The results are often in the form of functional determinants and inverses.

It is therefore necessary, not only to develop the functional formalism, but also to develop a *calculation technology* that can deal with the complexities of the resulting expressions. For this reason, a large part of the development is to formulate various methodologies addressing the challenges in the simplification of expressions that are produced by functional integrations.

10.2 Development

The crucial step in the development of any infinite-dimensional formalism is to find a way to represent the states and operations. From a functional analysis point of view, it implies that we need to identify a basis that can represent all such entities. Various candidates are considered, including Fock states and coherent states, but the preferred bases are those that facilitate convenience in calculations. Such bases are those that are complete and orthogonal. The fixed-spectrum Fock states can be formulated as a complete orthogonal basis for an infinite-dimensional stratified Hilbert space, but they are not convenient to use due to their discrete nature. Coherent states are more convenient, but they are not mutually orthogonal.

The bases that turn out to be complete, orthogonal, and convenient to use for calculations are the quadrature bases into which all the spatiotemporal degrees of freedom are incorporated in such a way that their orthogonality extends over the entire space. A formulation of quadrature bases in terms of the fixed-spectrum Fock bases, in direct analogy with the particle-number-degree-of-freedom-only approach, does not work because the stratification of the Hilbert space frustrates the extension of their orthogonality to the entire space. Instead, the quadrature bases are obtained as eigenvectors of the wavevector-dependent quadrature operators without stratification. These eigenvectors are parameterized by the eigenvalue functions in an infinite-dimensional functional space. With the suitable bases being the quadrature bases, the functional space naturally forms a phase space, leading to an infinite-dimensional functional *Moyal formalism*. The parameter functions of the quadrature bases thus become *quadrature field variables* that parameterize the functional phase space. States and operations become functionals on this phase space. We choose the Wigner functionals as the preferred phase space representations. If we decided to use either Q- or P-functionals, we would have ended up having to use both of them. For example, if the states are represented as Q-functionals, the observables need to be represented as P-functionals. The latter are often severely singular and, therefore, not convenient for complicated calculations. Wigner functionals generally do not present such challenges.

Apart from the complexities introduced by the functional nature of this phase space formulation, it mirrors the particle-number-only version exactly. One can define characteristic functionals in terms of functional symplectic Fourier transforms of the Wigner functionals and vice versa. The mapping between operators on Hilbert space and Wigner functionals on phase space is invertible, as shown by the functional Weyl transform. Products of operators on Hilbert space become *star products* of Wigner functionals on phase space. By implication, there is nothing that can be calculated on an infinite-dimensional Hilbert space that does not have an equivalent representation on functional phase space.

10.2.1 Outstanding issues

While there are many topics that have not been considered due to limited space, most of them can be addressed in terms of the approaches and techniques that have been discussed. However, there are some aspects of the material as it is presented that still needs some further consideration.

One such issue is the question about the conditions for Wigner functionals to be valid representations of states. In Section 5.7, this matter is partially addressed in terms of covariance kernels of such Wigner functionals. However, it does not exhaustively define such valid Wigner functionals in the context of an infinite number of degrees of freedom. It is merely assumed that the conditions for valid covariance kernels with a finite number of degrees of freedom can be generalized to an infinite number of degrees of freedom.

Another issue is the inverse of the even squeezed vacuum state kernel A. The expression for this inverse provided in (8.72) implies an infinite sequence of contractions among infinite sequences. Currently, there is no known simplification for this expression. In those cases where calculations involve this inverse, as in Section 9.2, we avoided the issue by considering the weak squeezing limit, or we left it unevaluated, as in the definitions of the overlaps in Section 7.6.6. While these situations are justified within their context, it would be beneficial to have a simpler way to treat the inverse A^{-1} when it appears in quantities that require explicit calculations.

In Chapter 6, we discussed squeezing in the context of a *squeezing operator*, showing that it leads to the Bogoliubov transformation of the arguments of the Wigner functional of a state on which squeezing is applied. When applied to a vacuum state, it produces the *squeezed vacuum state* parameterized in terms of kernels that are related to the Bogoliubov kernels, as shown in (6.338). The squeezed vacuum state is also obtained as a solution of the evolution equation for parametric down-conversion in Chapter 8. However, the kernels of this squeezed vacuum state are then represented by the expansions shown in (8.70). Although these kernels can be related to Bogoliubov kernels that are produced in a similar evolution process, there is currently no clear way to express the result from the evolution process as a once-off squeezing process in terms of a squeezing operator as presented in Chapter 6. This issue is partly the result of additional constraints that the direct squeezing process impose on the Bogoliubov kernels requiring them to be Hermitian and symmetric, which is not required for general Bogoliubov transformations. The question is whether these constraints are in some way also imposed by the down-conversion process. In the context of the current analysis it does not seem to be the case.

10.3 Applications

Several applications are addressed in the book. There are many more. It stands to reason that not all quantum systems need a functional formalism. To understand which systems would benefit from the functional formalism, we provide a classification of quantum systems in terms of the nature of their degrees of freedom.

Often such systems are distinguished based on how the particle-number degree of freedom is handled. If the system produces single photons or a finite number of discrete photons in terms of which states are defined and processing is performed, the system is said to work with *discrete variables*. If, on the other hand, the system uses states with multiple photons that are only specified by an average, such as with coherent states or squeezed states, the system is said to work with *continuous variables*. This distinction only addresses the particle-number degree of freedom. One can extend this distinction also to the spatiotemporal degrees of freedom. When a system treats the spatiotemporal degrees of freedom in terms of a finite number of discrete modes, we refer to such a system as having *discrete spatiotemporal variables*. When it deals with the spatiotemporal

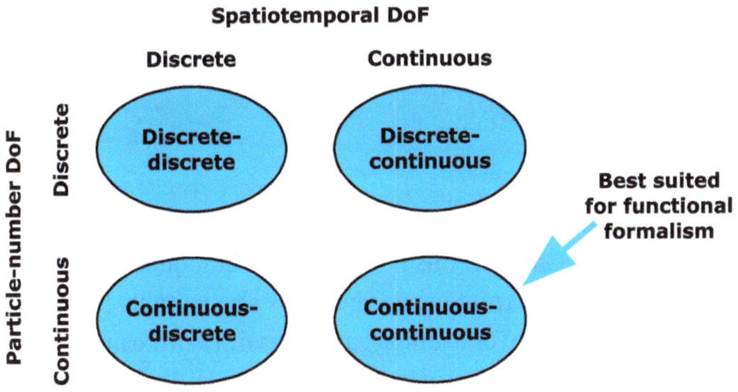

Figure 10.1: Classification of quantum optical systems.

degrees of freedom as an infinite-dimensional space of mode functions, it can be said to have *continuous spatiotemporal variables*.

These distinctions allow us to specify four categories of quantum optical systems: *discrete-discrete* systems, *discrete-continuous* systems, *continuous-discrete* systems, and *continuous-continuous* systems. They are shown in Figure 10.1.

The functional formalism is best suited for the last category where both the particle-number degree of freedom and the spatiotemporal degrees of freedom are represented by continuous variables. There are various systems that naturally fall into this category. Although there may be many analyses of these systems in terms of approaches from some of the other categories, such analyses generally fall short of providing thorough exhaustive representations of such systems due to their intrinsic dimensional limitations.

In this book, systems that involve the parametric down-conversion process are discussed extensively, as examples of continuous-continuous systems requiring analyses in terms of the functional formalism. The states that are produced by this process contain multiple photons of an undetermined number. Moreover, the kernels that mediate this process cannot be accurately represented in terms of finite-dimensional matrices. The functional formalism facilitates the derivation of a functional *evolution equation* that governs the behaviour of any system based on this process.

Other examples of continuous-continuous systems that require the functional formalism for their analyses include: propagation of light through random media such as a turbulent atmosphere; propagation of spatiotemporally varying light through periodic media such as photonic crystals; and the process of lasing induced by a spatiotemporally varying pump. Even when the number of dimensions in the spatiotemporal degrees of freedom is finite, it may be beneficial to treat the spatiotemporal degrees of freedom of the system as being infinite dimensional. For example, if the number of dimensions is large, it can be cumbersome to calculate the behaviour of such a system. Often such calculations are relegated to numerical analyses, making it challenging to extract analytical properties of the system, especially when the parameter space is large. In such cases, a functional approach, based on modelling the system as being continuous, could be beneficial.

10.3.1 Beyond quantum optics

While the functional formulation presented in this book is focused on quantum optics for applications in photonic quantum information systems, it is also suitable for other physical scenarios involving bosonic fields. For example, a lattice with a finite number of trapped bosonic particles (ions or atoms) can be treated as an infinite lattice illuminated by an optical beam. Likewise, there are various scenarios in solid state physics that can be treated in a similar way.

The bosonic functional phase space approach can be augmented with an equivalent fermionic approach, based on a functional Grassmann phase space [1]. These two functional phase space formalisms can be combined to allow analyses of arbitrary quantum systems in the continuous-continuous category.

Further afield, the functional formalism developed here may also be beneficial in fundamental physics. It is worth pointing out that, while quantum field theory serves as the *formalism* in terms of which the theories in the standard model of particle physics are modelled, it does not *dictate* these theories. In other words, it provides a powerful yet flexible tool for the modelling of fundamental dynamics. However, it does not allow models for gravity.

A similar approach may serve the development of a comprehensive theory that also incorporates gravity. Instead of building the theory already into the formalism, it may be more beneficial to develop a formalism that is powerful enough to model all the dynamics from particle physics as well as gravity, yet remains agnostic about the specific theory that is used to model them.

One way to develop such a powerful formalism is to use a functional phase space as point of departure. The notion of a $\diamond$-contraction provides a principle that can then be developed into a *background independent* formulation. In the preceding chapters, we often used this $\diamond$-contraction without specifying whether we consider the Lorentz covariant formulation or not. This flexibility demonstrates the possibility for a background independent formulation.

Moreover, a functional phase space may provide some added benefits in that it comes with a ready-made capability to represent arbitrary states and operations. Quantum field theory is by contrast more focussed on modelling the fundamental dynamics and is not well-suited for modelling exotic states. The latter may be an essential ingredient for successful models of gravity. The formal correspondence between the functional path integral approach in quantum field theory and the functional phase space approach may facilitate the incorporation of fundamental dynamics in the latter without sacrificing its powerful state representation capability. Obviously, the actual development of such a formalism would be far more daunting than these few optimistic remarks may indicate.

10.4 Mathematical foundation

The development and application of the functional phase space formalism is done in a mathematical language that is familiar to most physics and engineers. It is not the intention to represent it in terms of formal abstract mathematics. Nevertheless, it may be beneficial to consider a rigorous mathematical foundation for the functional phase space formalism presented here.

The functional formalism follows from the solutions of the eigenvalue equation for quadrature operators with the spatiotemporal and spin degrees of freedom incorporated, as discussed in Section 4.5. The subsequent development is guided to a large extent by the results of that analysis. Yet, the formal foundation based on these results is lacking.

We review the currently implied foundation briefly. The starting point is a mathematical formalism for quantum theory in terms of the particle-number degree of freedom only. It is derived from Fourier theory at the beginning of Chapter 3, where it serves to produce the basic operator formalism for quantum mechanics. This approach pins down the mathematical properties of the functions representing the particle-number degree of freedom for states in quantum optics. Part of this development requires the introduction of *coordinate bases*, which converts the real line into a continuous set of coordinate basis elements with the necessary orthogonality and completeness conditions. However, these properties of the coordinate bases have no bearing on the properties of the states, because their definitions given in (3.22) show that these states inherit their properties from the space of functions (Schwartz space) whose properties are dictated by Fourier theory. The only purpose of the coordinate bases is to provide the mathematical mechanism (a form of *mathematical tagging*) for the inner products and to set up unitary invariance, not applicable for Fourier transformable functions.

The next step in the development is the incorporation of the spatiotemporal degrees of freedom. It involves the conversion of the particle-number-degrees-of-freedom-only ladder operators to those at the beginning of Chapter 4 defined over wavevector space. The space on which they operate is a combination of the space of functions for the particle-number degree of freedom together with the space of functions defined on the three-dimensional space of all wavevectors. The latter is also restricted by their Fourier relationship to fields in configuration space. However, the situation becomes more complicated because every quantum can carry a full set of the spatiotemporal and spin degrees of freedom. The search for a suitable way to represent such states is the topic of Chapter 4. While it provides the opportunity to discuss *fixed-spectrum Fock states* and *fixed-spectrum coherent states* that are relevant for their own reasons, this search eventually culminates in the formulation of the quadrature bases of Section 4.5 (starting from

Section 4.5.2). According to the orthogonality condition derived in Section 4.5.3, the parameter functions (eigenvalue functions) that distinguish the elements of these quadrature bases are functions with finite Euclidean norms (Schwartz functions). Although one can define an inner product for these functions, it is not used in the formalism. The orthogonality among quadrature bases elements is related to the *metric* (the distance) between their parameter functions. Any two quadrature bases elements with a nonzero metric between their parameter functions are orthogonal.

The situation at this point is analogous to what the coordinate bases in Chapter 3 present. Although the span of these quadrature bases imply some vast non-separable Hilbert space, the space of states that are defined in terms of them is a much smaller proper subset of this vast Hilbert space. Eventually, the state space is not defined in terms of a Hilbert space. Instead, it is given by all the Wigner functionals for states on the functional phase space that is formed in terms of the parameter functions of the quadrature bases acting as field variables, as discussed in Chapter 5.

10.4.1 Outstanding issues

There are two challenges for the mathematical formulation of Wigner functional theory on this functional phase space. The first is to define the space of all valid Wigner functionals. Some aspects of the validity of Wigner functionals are discussed in Section 5.7, based on the properties of the covariance kernel. It is currently based on the assumption that the finite-dimensional analysis [2] can be extended to infinite dimensions.

The other challenge is to define functional integrations over this phase space. The notion of functional integration is first encountered in the context of the Dirac delta functional in Section 4.5.2 where certain assumed properties of the functional integration measure are used to demonstrate the result of the orthogonality relation in terms of the Dirac delta functional. It is again used for the completeness conditions in Section 4.5.5 leading to the implied representation of Dirac delta functionals in terms of functional integrals. Ultimately, the defining property of these functional integrals is the assumed result of the functional integration of a Gaussian functional, as presented in Appendix C.

10.4.2 Fundamental mathematical properties

As for the fundamental mathematical properties of quantum systems with infinitely many degrees of freedom, one can study different cases of such systems. Two examples from historic investigations that are relevant within quantum optics are the case where a free Hamiltonian is augmented by a source term (the van Hove Hamiltonian [3]) and the case where the two quadrature operators associated with a free Hamiltonian are scaled by opposite factors so that their commutation relation remains the same [4].

In the former case, the addition of the source terms gives rise to displaced eigenstates that are related to those of the original free Hamiltonian by unitary displacement operators provided that the parameter function of the source term has a finite Euclidean norm. In the case of a point source, this requirement is not satisfied, leading to infinite displacements for which the eigenstates of the augmented system become orthogonal to those of the original free system. Such infinite displacements produce states with infinite average numbers of particles, indicating that such a system is unphysical. The modelling of a point source thus needs more careful considerations.

In the latter case, the scaling of the quadrature operators lead to a Bogoliubov transformation of the ladder operators. Such a Bogoliubov transformation can be produced with the application of squeezing operators on the ladder operators. Such squeezing operators are unitary, provided that the squeezing kernel in terms of which they are defined has a finite trace. If the squeezing kernel is proportional to the identity, its trace is not finite, leading to eigenstates with infinite average numbers of particles. Hence, again, such a system is not physical.

In both of these examples, the distinction between the physically realistic scenario with eigenstates having a finite average number of particles and the unphysical scenario with unphysical eigenstates having infinite average numbers of particles, is clearly made. In all cases, the physically realistic scenario is related to the original free system by valid unitary transformations. Stated differently, there are no unitarily inequivalent physical quantum optical systems in these quadratic examples. The functional formalism provides a powerful way to investigate the nature of more complicated interacting theories along similar lines.

Bibliography

[1] F. S. Roux. Fermion quadrature bases for Wigner functionals. *J. Phys. A, Math. Gen.*, 57:225302, 2024.

[2] R. Simon, E. C. G. Sudarshan, and N. Mukunda. Gaussian–Wigner distributions in quantum mechanics and optics. *Phys. Rev. A*, 36:3868–3880, 1987.

[3] L. van Hove. Les dificultés de divergences pour un modelle particulier de champ quantifié. *Physica*, 18:145–159, 1952.

[4] R. Haag. On quantum field theories. *Danske Videnskabernes Selskab Matematisk-Fysiske Meddelelser*, 29:1–37, 1955.

A Generating functions

At various stages during derivations or analyses, we are dealing with infinite sequences of functions or operators. In those scenarios, we often use *generating functions* or a *generating functionals* to alleviate the calculations. Such scenarios are encountered in numerous sections. Therefore, they play significant roles in these derivations and calculations. Here, we discuss what they are and how they are used.

A.1 Mathematical tagging

A generating function(al) represents a form of *mathematical tagging*, where multiple mathematical entities are combined into one entity in which all the different entities are tagged in some way. These tags enable the extraction of any specific entity from the whole. Usually such tags cannot be evaluated in terms of numerical values, and even in those cases where some numerical value can be assigned to such a tag, its numerical value does not have any physical meaning.

For a simple example of mathematical tagging, consider the three-dimensional vector $\mathbf{v} = a\vec{x}+b\vec{y}+a\vec{z}$. Here, the unit vectors $\vec{x}, \vec{y}$, and $\vec{z}$ serve as the tags, and the components a, b, and c are the entities. The vector $\mathbf{v}$ is a linear combination of the entities, each multiplied by its unique tag. To extract a specific component, we perform the dot-product of the vector with the tag of that component: $\mathbf{v} \cdot \vec{y} = b$.

The coordinate bases defined in Section 3.1.3 also represent examples of mathematical tagging. Every function value associated with a given coordinate point is tagged by an element from that basis, denoted by a ket. The tags can be transformed to different bases, which can then be extracted with inner products using the associated bras.

There are various other forms of mathematical tagging found in various fields of mathematics, some of which can be rather sophisticated and even esoteric. Those cases do not concern us here.

A.2 Basics

In the case of a generating function, the tagging is done with the aid of powers of an auxiliary parameter that we call the *generating parameter*. As an example, we can consider an infinite discrete set of functions $\{g_n(\mathbf{x})\}$. A generating function for this set of functions is given by

$$\mathcal{G}(\mu) = \sum_{n=0}^{\infty} \mu^n g_n(\mathbf{x}), \tag{A.1}$$

https://doi.org/10.1515/9783111445342-015

where μ is the generating parameter. The n-th function is thus tagged by μ^n. The individual functions are extracted with the aid of a differential operation

$$\frac{1}{n!}\,\partial_\mu^n \mathcal{G}(\mu)\Big|_{\mu=0} = g_n(\mathbf{x}). \tag{A.2}$$

A.3 Generating functionals

Sometimes it is necessary to transfer some values to the arguments of the selected function in the generating function. In that case, we use a *field variable* to act as the auxiliary parameter. Then the generating function becomes a *generating functional*. Consider, for example, a finite discrete set of spectral functions $\{G_n(\mathbf{k})\}$ for $n = 0 \ldots N$. The generating functional for this set of functions is then given by

$$\mathcal{G}(\{\xi\}) = \sum_{n=0}^{N} \xi_n \diamond G_n, \tag{A.3}$$

where ξ_n is a set of *generating field variables*. In this case, the individual functions are extracted with the aid of a *functional derivative*

$$\frac{\delta}{\delta \xi_n(\mathbf{k}_0)}\mathcal{G}(\{\xi\})\Big|_{\{\xi\}=0} = G_n(\mathbf{k}_0). \tag{A.4}$$

Note that the wavevector of the functional derivative is transferred to the argument of the extracted spectral function. Such generating functionals are used in the procedure to compute the Wigner functionals of polynomial operators, as shown toward the end of Section 6.1.3; they are also used for Fock states with multiple parameter functions in Section 6.3.3. The *characteristic functional*, discussed in Section 5.4, is a generating functional for the moments of Wigner functionals.

A.4 Applications

The power of generating function(al)s is partly due to their linearity. Any linear operation that is performed on the generating function is in effect performed on all the individual functions in its expansion

$$\mathcal{L}\{\mathcal{G}(\mu)\} = \sum_{n=0}^{\infty} \mu^n \mathcal{L}\{g_n(\mathbf{x})\}. \tag{A.5}$$

The result of such a linear operation becomes a generating function for the results of applying the linear operation on the individual functions.

We can generalize this property of generating functions to bilinear operations among the functions. In this case, different generating parameters must be used for the two generating functions. The result is

$$\mathcal{L}\{\mathcal{G}(\mu),\mathcal{G}(\nu)\} = \sum_{m,n=0}^{\infty} \mu^m \nu^n \mathcal{L}\{g_m(\mathbf{x}),g_n(\mathbf{x})\}. \tag{A.6}$$

An example of such a bilinear operation is where we investigate the orthogonality of a set of functions, such as the paraxial modes found in Section 2.5.2. For this purpose, we compute the inner product between a generating function and itself with different generating parameters

$$\langle \mathcal{G}(\mu),\mathcal{G}(\nu)\rangle = \sum_{m,n=0}^{\infty} \mu^m \nu^n \langle g_m, g_n\rangle. \tag{A.7}$$

Note that such an inner product usually requires the complex conjugate of one of the entities. The complex conjugate of a generating function produces a generating for the complex conjugates of the functions. If the functions are orthogonal, then

$$\langle g_m, g_n\rangle = \lambda_n \delta_{m,n}, \tag{A.8}$$

where λ_n is the *orthogonality constant* and $\delta_{m,n}$ is the *Kronecker delta function* defined in (2.6). The inner product then becomes

$$\langle \mathcal{G}(\mu),\mathcal{G}(\nu)\rangle = \sum_{n=0}^{\infty} (\mu\nu)^n \lambda_n, \tag{A.9}$$

which is a function of only the product of the two generating parameters. Therefore, when the result of such an inner product between generating functions only depends on the product of the two generating parameters, we know that the individual functions are mutually orthogonal.

Generating functions can also be used to represent infinite sets of operators, such as the products of annihilation operators, considered in Section 3.1.8. When such a generating function for products of operators are applied to eigenstates of these operators, the result produces a generating function for the products of the eigenvalues.

A.5 Statistical calculations

One can represent statistical distributions in terms of generating functions. An example is the Poisson distribution presented in Section 3.2.4. In Section 7.4, we develop a generating function for the measurement of the photon statistics in an arbitrary state. The

Wigner functional of a state is multiplied by this generating function and the result is integrated. The result then represents a generating function for the statistical distribution of the number of photons in that state.

Generating functions for statistical distributions provide powerful methods to perform statistical calculations. The generating function for such a statistical distribution is represented by

$$\mathcal{G}(J) = \sum_{n=0}^{\infty} J^n P_n, \tag{A.10}$$

where P_n represents the individual probabilities in the distribution. The average value of the distribution (its first moment) is readily computed with the following operation:

$$\partial_J \mathcal{G}(J)\big|_{J=1} = \sum_{n=0}^{\infty} n J^{n-1} P_n \bigg|_{J=1} = \sum_{n=0}^{\infty} n P_n = \langle n \rangle. \tag{A.11}$$

The second moment can then be computed as

$$\begin{aligned} \partial_J \left[J \partial_J \mathcal{G}(J)\right]\big|_{J=1} &= \partial_J \left[\sum_{n=0}^{\infty} n J^n P_n \right] \bigg|_{J=1} \\ &= \sum_{n=0}^{\infty} n^2 J^{n-1} P_n \bigg|_{J=1} = \sum_{n=0}^{\infty} n^2 P_n = \langle n^2 \rangle. \end{aligned} \tag{A.12}$$

The variance is then obtained from these results by

$$\sigma^2 = \langle n^2 \rangle - \langle n \rangle^2. \tag{A.13}$$

In Section 7.6, we show how two generating functions for the measurement of the photon statistics, developed in Section 7.4, can be combined to produce a generating function for the statistical distribution of correlation measurements. A more sophisticated generating function for the correlation measurement in homodyne tomography is derived in Section 7.9.2.

A.6 Auxiliary representations

One can also use generating functions for the Wigner functionals of states such as those for the Fock states in Section 6.3. It allows us to overcome a limitation of functional integration, which is that they can only be applied for a limited number of types of functionals (see Appendix C). These generating functions allow us to represent polynomial functionals or polynomial-Gaussian functionals in terms of Gaussian functionals, which can then be integrated. When necessary, we also convert second-order polynomial functionals into Gaussian generating functions, although their only purpose is to be differentiated once after evaluating the functional integration. It is, for example, done with the Wigner functional of the number operator, as in Section 7.3.3. Polynomial Wigner functionals are often encountered due to the polynomial nature of operators, such as the free Hamiltonian operator in Section 6.5.2, or the operators associated with the dynamics of certain processes, such as free space propagation in Section 6.4.8, or measurement operators, such as the homodyne operator in Section 7.9.1.

Another way to represent polynomial functionals or polynomial-Gaussian functionals is in terms of a *construction operation* applied to a generic generating function(al). This procedure is developed in Section 6.1.3 and discussed in Section 7.1.2.

B Dispersive birefringence

A *birefringent* (uniaxial anisotropic) medium is a dielectric medium where the refractive index for light that is linearly polarized along a specific direction, defined as the *crystal axis*, differs from the refractive index for light that is linearly polarized perpendicular to that direction. The refractive index along the special direction is referred to as the *extraordinary refractive index* and for the perpendicular directions we have the *ordinary refractive index*. In birefringent media, the refractive indices are often dispersive—they depend on the wavelength or *angular frequency* of the light.

B.1 Effective refractive index

The *effective refractive index* denoted by n_{eff} is the refractive index experienced by an electric field propagating through a birefringent medium while its polarization vector lies in the plane defined by the crystal (or optic) axis and the propagation vector. It therefore depends on the angle between the propagation vector and the crystal axis θ_X. The expression for the effective refractive index is given by

$$\frac{1}{n_{\text{eff}}^2} = \frac{\cos^2(\theta_X)}{n_{\text{o}}^2} + \frac{\sin^2(\theta_X)}{n_{\text{e}}^2}, \tag{B.1}$$

where n_{o} is the *ordinary refractive index* and n_{e} is the *extraordinary refractive index*.

B.2 Sellmeier equations

The refractive indices of nonlinear crystals (n_{o} and n_{e} in a birefringent medium) are in general dispersive. These crystals are also birefringent (uniaxial) or even biaxial, which means that the refractive indices depend on the polarization of the light. The frequency dependences of these refractive indices can be modelled by *Sellmeier equations*. Here, we provide examples of such Sellmeier equations for two nonlinear crystals (BBO and KTP) often found in quantum optical experiments.

B.2.1 Barium borate

For BBO (barium borate BaB_2O_4), the Sellmeier equations are [1]

$$\begin{aligned} n_{\text{o}}^2 &= 1 + \frac{0.90291\lambda^2}{\lambda^2 - 0.003926} + \frac{0.83155\lambda^2}{\lambda^2 - 0.018786} + \frac{0.76536\lambda^2}{\lambda^2 - 60.01}, \\ n_{\text{e}}^2 &= 1 + \frac{1.151075\lambda^2}{\lambda^2 - 0.007142} + \frac{0.21803\lambda^2}{\lambda^2 - 0.02259} + \frac{0.656\lambda^2}{\lambda^2 - 263}, \end{aligned} \tag{B.2}$$

https://doi.org/10.1515/9783111445342-016

where λ is the wavelength in µm. It is a birefringent crystal. These equations are valid within the range: $0.188\,\mu m < \lambda < 5.2\,\mu m$. The curves for the ordinary refractive index and the extraordinary refractive index of BBO are shown in Figure B.1 over the visible wavelengths.

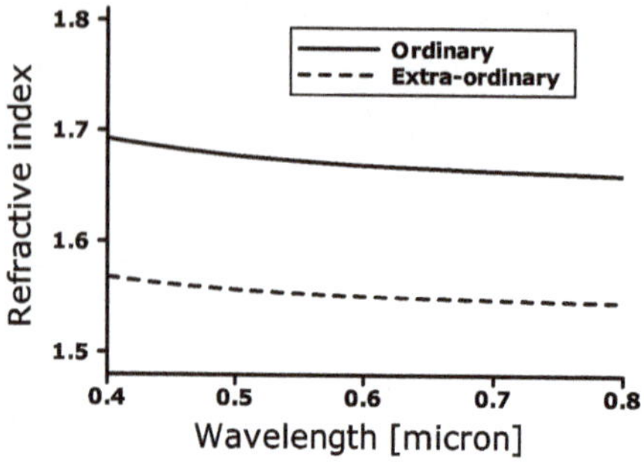

Figure B.1: Curves of the ordinary and extraordinary refractive indices of BBO over the range of wavelengths for visible light.

B.2.2 KTP

For KTP (KTiOPO4), which is a biaxial crystal, the Sellmeier equations are given by [2]

$$\begin{aligned} n_1^2 &= 3.29100 + \frac{0.04140}{\lambda^2 - 0.03978} + \frac{9.35522}{\lambda^2 - 31.45571}, \\ n_2^2 &= 3.45018 + \frac{0.04341}{\lambda^2 - 0.04597} + \frac{16.98825}{\lambda^2 - 39.43799}, \\ n_3^2 &= 4.59423 + \frac{0.06206}{\lambda^2 - 0.04763} + \frac{110.80672}{\lambda^2 - 86.12171}, \end{aligned} \tag{B.3}$$

where λ is the wavelength in µm. They are valid within the range: $0.43\,\mu m < \lambda < 3.54\,\mu m$.

B.3 Dispersion

In free space, the wavenumber is proportional to the *angular frequency* via the speed of light $ck = \omega$. In a dielectric medium, the relationship between the wavenumber and the angular frequency also involves the refractive index. It is given by the *phase velocity*

$$v_p \triangleq \frac{\omega}{k(\omega)} = \frac{c}{n(\lambda)}, \tag{B.4}$$

where $n(\lambda)$ is the dispersive refractive index as a function of the wavelength. When the medium is *dispersive*, instead of the phase velocity, the effective propagation velocity of an electromagnetic field is given by the *group velocity*,

$$v_g(\omega) \triangleq \left[\frac{\partial k(\omega)}{\partial \omega}\right]^{-1} = c\left[n(\lambda) - \lambda\frac{\partial n(\lambda)}{\partial \lambda}\right]^{-1}. \tag{B.5}$$

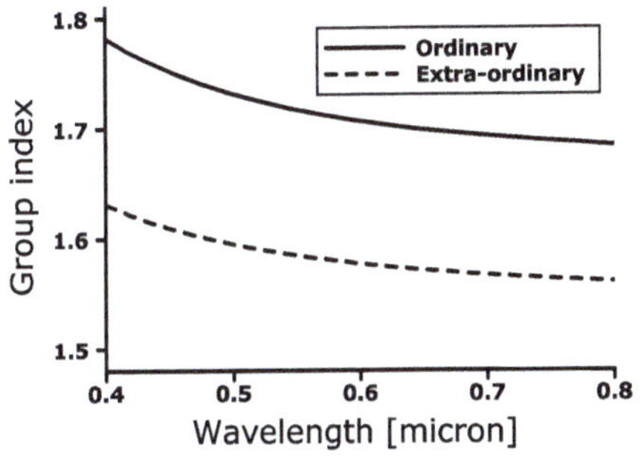

Figure B.2: Curves of the ordinary and extraordinary group indices of BBO over the range of wavelengths for visible light.

In analogy to the refractive index $n(\lambda)$ associated with the phase velocity, we define the group velocity in terms of a *group index*

$$v_g = \frac{c}{n_g}, \tag{B.6}$$

where the *group index* is given by

$$n_g \triangleq \frac{c}{v_g(\omega)} = c\frac{\partial k(\omega)}{\partial \omega} = n(\lambda) - \lambda\frac{\partial n(\lambda)}{\partial \lambda}. \tag{B.7}$$

In Figure B.2, curves are provided for the ordinary and extraordinary group indices for BBO, plotted as functions of the wavelength over the visible range. These group indices are larger than the corresponding refractive indices, shown in Figure B.1, indicating that the group velocity is always smaller than its corresponding phase velocity.

The dispersion of a medium is quantified by the *group velocity dispersion*, given by

$$D_g(\omega) \triangleq \frac{\partial^2 k(\omega)}{\partial \omega^2} = \frac{\lambda^3}{2\pi c^2}\frac{\partial^2 n(\lambda)}{\partial \lambda^2}. \tag{B.8}$$

We define the *group index dispersion* by

$$\Delta n_g(\omega) = c^2 D_g(\omega) = \frac{\lambda^3}{2\pi}\frac{\partial^2 n(\lambda)}{\partial \lambda^2}, \tag{B.9}$$

which has the units of a distance. Curves for the ordinary and extraordinary group index dispersions for BBO are plotted in Figure B.3 as functions of the wavelength over the visible range.

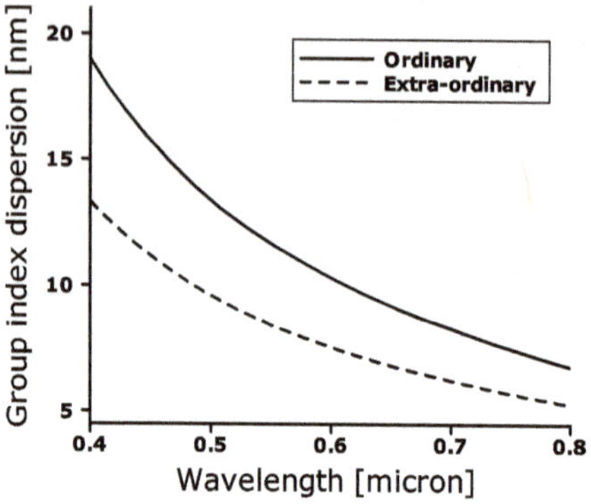

Figure B.3: Curves of the ordinary and extraordinary group index dispersions of BBO over the range of wavelengths for visible light.

References

[1] G. Tamošauskas, G. Beresnevičius, D. Gadonas, and A. Dubietis. Transmittance and phase matching of BBO crystal in the 3-5 μm range and its application for the characterization of mid-infrared laser pulses. *Opt. Mater. Express*, 8:1410–1418, 2018.

[2] K. Kato and E. Takaoka. Sellmeier and thermo-optic dispersion formulas for KTP. *Appl. Opt.*, 41:5040–5044, 2002.

C Functional integration

A functional formalism, as presented in this book, would not be of much use if one cannot evaluate functional integrals. The notion of an infinite-dimensional phase space, parameterized in terms of *field variables* that are functions themselves, may sound conceptually daunting. However, treated as a purely formal device to perform calculations, much in the same way as it is done for *path integrals* in quantum field theory, the concept of a functional phase space integral becomes less daunting. The requirements for such a formal procedure is consistency: any such calculation must produce an unambiguous result. Moreover, if the result represents a quantity that can be measured, then it must be finite. An inevitable consequence of a functional formalism, representing an infinite-dimensional space, is the appearance of quantities that are formally divergent ("infinities"). It is imperative that such divergent quantities all cancel in calculations leading to measurable quantities.

There are only a few forms of functional integrals that we can evaluate. However, with the aid of generating functions (or generating functionals), as discussed in Appendix A, they are powerful enough to allow the analysis of most physical scenarios where Wigner functional theory is relevant.

The first time we use a functional integral in this book is in (4.171) in Section 4.5.3, where it serves to test the validity of the identification of the Dirac delta functionals. Later, in Section 4.5.5, we use functional integrals to discuss the completeness of the quadrature bases. In those cases, the functional integrations are used in a general fashion without the need to evaluate specific integrals. In contrast, calculations involving Wigner functionals that are found in later chapters often require the evaluation of such functional integrals. Here, we discuss functional integrals and how they are evaluated.

C.1 Dirac delta functionals

The simplest non-trivial functional integrals that we can evaluate are those that lead to Dirac delta functionals. The derivation of this type of functional integral is provided in (5.4). We summarize those results for integrations over the two respective quadrature field variables as

$$\begin{aligned}\int \exp(iq_0 \diamond p)\, \mathcal{D}^{\circ}[p] &= \delta[q_0],\\ \int \exp(-iq \diamond p_0)\, \mathcal{D}[q] &= (2\pi)^{\Omega}\delta[p_0],\end{aligned} \tag{C.1}$$

where $\delta[\cdot]$ represents the *Dirac delta functional*, and q_0 and p_0 are (parameter) functions, but can also represent polynomials in terms of other field variables.

Combining the two integrals in (C.1), we obtain an integral in terms of *complex field variables*. The expression then becomes

https://doi.org/10.1515/9783111445342-017

$$\int \exp\left(\alpha_0^* \diamond \alpha - \alpha^* \diamond \alpha_0\right) \mathcal{D}^\circ[\alpha] = (2\pi)^\Omega \delta[\alpha_0], \tag{C.2}$$

where we defined

$$\begin{aligned} \alpha &\triangleq \frac{1}{\sqrt{2}}(q + \mathrm{i}p), \quad \alpha_0 \triangleq \frac{1}{\sqrt{2}}(q_0 + \mathrm{i}p_0), \\ \mathcal{D}^\circ[\alpha] &\triangleq \mathcal{D}[q]\,\mathcal{D}^\circ[p], \quad \delta[\alpha_0] \triangleq \delta[p_0]\delta[q_0]. \end{aligned} \tag{C.3}$$

As an example, consider the integral

$$\begin{aligned} \int W[\alpha_0] \exp\left(\alpha_0^* \diamond \alpha - \alpha^* \diamond \alpha_0\right) \mathcal{D}^\circ[\alpha, \alpha_0] &= (2\pi)^\Omega \int W[\alpha_0]\delta[\alpha_0]\,\mathcal{D}^\circ[\alpha_0] \\ &= W[0]. \end{aligned} \tag{C.4}$$

Note that the factor of $(2\pi)^\Omega$, produced by the integral over α, as in (C.2) to produce the Dirac delta functional, eventually cancels the factor of $(2\pi)^{-\Omega}$ in the measure $\mathcal{D}^\circ[\alpha_0]$ of the final integral, as defined in (5.29).

C.2 Change of integration field variables

The simplest change of variables is a shift $\alpha \to \beta = \alpha + \alpha_0$ where α_0 is any field (function) in the functional domain. The resulting field variable β is still an element of the functional domain. Since all functional integrals in this book are evaluated over the entire functional domain, such a shift does not affect any integration boundaries. Therefore, the functional integration is invariant with respect to such a shift. This property has already been tacitly assumed in the discussion beneath (4.172) in Section 4.5.3.

Although we usually represent the functionals as $W[\alpha]$, they are in fact functionals of both α and α^* acting as independent variables, so that we can represent them as $W[\alpha^*, \alpha]$. The same is true for the integration measure. Although we denote them as $\mathcal{D}^\circ[\alpha]$ for simplicity, we are in fact integrating over both α and α^*, so that the measure is in fact given by $\mathcal{D}^\circ[\alpha^*, \alpha]$.

The reason why this viewpoint is significant is because we need to think of the measure in this way when we perform a change of integration variables. For instance, when we change from integrating over α back into integrating over p and q, we need to consider the change with respect to α^* as well. So, we have

$$\mathcal{D}[\alpha^*, \alpha] \to |\det\{\mathcal{J}\}|\mathcal{D}[q, p], \tag{C.5}$$

where $\mathcal{J}$ is the Jacobian

$$\mathcal{J} \triangleq \begin{bmatrix} \partial_q \alpha^* & \partial_q \alpha \\ \partial_p \alpha^* & \partial_p \alpha \end{bmatrix} = \begin{bmatrix} \frac{1}{\sqrt{2}} & \frac{1}{\sqrt{2}} \\ -\mathrm{i}\frac{1}{\sqrt{2}} & \mathrm{i}\frac{1}{\sqrt{2}} \end{bmatrix}, \tag{C.6}$$

implying that $|\det\{\mathcal{J}\}| = 1$.

When the integration field variables are changed by a transformation involving a kernel $\alpha \to K \diamond \beta$, then we need to remember that we also have $\alpha^* \to \beta^* \diamond K^{\dagger}$. As a result, the Jacobian involves both transformations, leading to

$$\mathcal{D}[\alpha^*, \alpha] \to |\det\{K\}\det\{K^{\dagger}\}|\mathcal{D}[\beta^*, \beta], \tag{C.7}$$

or, as it is usually represented

$$\mathcal{D}[\alpha] \to |\det\{K\}|^2 \mathcal{D}[\beta], \tag{C.8}$$

because $\det\{K^{\dagger}\} = \det\{K^*\}$.

Occasionally, a change of variables is done with transformations involving both the field variable and its complex conjugate. The generic form of such a transformation is

$$\alpha \to A_{ab} \diamond \beta + A_{ac} \diamond \beta^* \quad \text{and} \quad \alpha^* \to A_{cb} \diamond \beta + A_{cc} \diamond \beta^*. \tag{C.9}$$

The field variables and their complex conjugates are treated as independent variables. Such a transformation is converted into two separate transformations performed consecutively. The first transformation only transforms one field variable, leaving the other as it is. The second transformation then transforms the remaining field variable. If we transform α first, the two transformations are

$$\alpha \to \left(A_{ab} - A_{ac} \diamond A_{cc}^{-1} \diamond A_{cb}\right) \diamond \beta + A_{ac} \diamond A_{cc}^{-1} \diamond \alpha^*, \tag{C.10}$$

which is then followed by

$$\alpha^* \to A_{cb} \diamond \beta + A_{cc} \diamond \beta^*. \tag{C.11}$$

The sequence of two separate transformations allows us to compute the combined Jacobian as a product of the Jacobians for the separate transformations. Hence,

$$\begin{aligned} \mathcal{D}[\alpha] &\to \left|\det\left\{A_{ab} - A_{ac} \diamond A_{cc}^{-1} \diamond A_{cb}\right\}\det\{A_{cc}\}\right| \mathcal{D}[\beta] \\ &= \left|\det\left\{A_{ab} \diamond A_{cc} - A_{ac} \diamond A_{cc}^{-1} \diamond A_{cb} \diamond A_{cc}\right\}\right| \mathcal{D}[\beta]. \end{aligned} \tag{C.12}$$

In the last expression, we could also have performed the contraction with the transpose A_{cc}^T because $\det\{A\} = \det\{A^T\}$.

C.3 Isotropic Gaussian functionals

The most general form of an isotropic Gaussian functional in terms of complex field variables is

$$W[\alpha] = \mathcal{N} \exp\left(-2\alpha^* \diamond K \diamond \alpha - \alpha^* \diamond F_1 - F_2^* \diamond \alpha\right), \tag{C.13}$$

where $\mathcal{N}$ is the normalization constant, K is an invertible kernel with $\det\{K\} \geq 0$, and F and F^* are arbitrary complex parameter functions with finite norms. These functionals are isotropic in the sense that they are rotationally symmetric with respect to rotations around their centres (with appropriate redefinitions of the complex parameter functions). The coherent states are the only pure states with isotropic Gaussian Wigner functionals. (See Section 6.1.) Thermal states are mixed states with isotropic Gaussian Wigner functionals located at the origin.

The functional integration of an isotropic Gaussian functional is by analogy based on the integration of a multivariate Gaussian function. In the latter case, we have

$$\int \exp\left(-\mathbf{a}^T M \mathbf{a} - \mathbf{a}^T \mathbf{b}\right) \, \mathrm{d}^n a = \frac{\pi^{n/2} \exp\left(\frac{1}{4} \mathbf{b}^T M^{-1} \mathbf{b}\right)}{\sqrt{\det\{M\}}}, \tag{C.14}$$

where n is the number of variables, $\mathbf{a}$ is a vector of the variables, M is an $n \times n$ matrix, $\mathbf{b}$ is a constant vector, and $\det\{M\}$ is the determinant of the matrix.

By analogy, the equivalent functional integral evaluates to

$$\int \exp(-q \diamond K \diamond q - q \diamond F) \, \mathcal{D}[q] = \frac{\pi^{\Omega/2} \exp\left(\frac{1}{4} F \diamond K^{-1} \diamond F\right)}{\sqrt{\det\{K\}}}, \tag{C.15}$$

where K represents an invertible kernel, q is a real-valued *integration field variable*, F is a parameter function, Ω is the divergent constant defined in (4.154), and $\det\{K\}$ represents a *functional determinant* of the kernel. For such a result to be well-defined, the kernel K must be positive definite (all its eigenvalues must be larger than zero). The functional determinant can be defined in terms of the zeta function,[a] but we prefer the equivalent definition

$$\det\{K\} \triangleq \exp[\operatorname{tr}\{\ln_\diamond(K)\}]. \tag{C.16}$$

Here, $\operatorname{tr}\{\cdot\}$ is the *kernel trace operation* over all the degrees of freedom of the kernel (as opposed to the *operator trace*), and $\ln_\diamond(\cdot)$ is a *logarithmic mapping* from kernels to kernels. It is defined as the inverse map of the *exponential mapping* from kernels to kernels $\exp_\diamond(\cdot)$, which in turn is defined in terms of its Taylor expansion where all multiplications are represented by $\diamond$-contractions and the first term in the expansion is the identity kernel $\mathbf{1}$, defined in (4.54) for the Lorentz invariant measure. Hence, for a kernel A, it is

$$\exp_\diamond(A) \triangleq \mathbf{1} + A + \frac{1}{2} A \diamond A + \cdots = \mathbf{1} + \sum_{m=1}^{\infty} \frac{A^{\diamond m}}{m!}. \tag{C.17}$$

a The functional determinant is $\det\{K\} = \exp[-\zeta_K'(0)]$, where $\zeta_K(z) = \operatorname{tr}\{K\}^{-z}$ is the zeta function.

The kernel traces are distinguished from *operator traces* in that they contain kernels in their arguments instead of operators. For a kernel $K(\mathbf{k}_1, \mathbf{k}_2)$ the kernel trace is formally evaluated by

$$\mathrm{tr}\{K\} = \int K(\mathbf{k}, \mathbf{k}) \, \mathrm{d}_{\mathcal{C}} k, \tag{C.18}$$

where $\mathrm{d}_{\mathcal{C}} k$ is a generic measure, as defined in (6.156).

The functional integral in (C.15) is used to define functional integration in terms of complex-valued integration field variables. First, we replace the kernel in (C.15) by the identity kernels $K(\mathbf{k}_1, \mathbf{k}_2) \rightarrow \mathbf{1}(\mathbf{k}_1, \mathbf{k}_2)$, so that

$$\int \exp(-q \diamond q - q \diamond F) \, \mathcal{D}[q] = \pi^{\Omega/2} \exp\left(\frac{1}{4} F \diamond F\right), \tag{C.19}$$

where we use the fact that $\det\{\mathbf{1}\} = 1$, as shown in Appendix D. We multiply this integral by another version of itself in which q is replaced by p and F is replaced by another arbitrary function G. Then we perform a change of integration field variables to convert the two real-valued integration field variables to a complex-valued integration field variables as in (C.3), which gives

$$\int \exp(-2\alpha^* \diamond \alpha - \alpha^* \diamond F_a - F_b^* \diamond \alpha) \, \mathcal{D}[\alpha] = \pi^{\Omega} \exp\left(\frac{1}{2} F_b^* \diamond F_a\right), \tag{C.20}$$

where $F_a = (F + \mathrm{i}G)/\sqrt{2}$ and $F_b^* = (F - \mathrm{i}G)/\sqrt{2}$. Finally, we do another change of integration field variables to introduce an arbitrary positive definite Hermitian kernel $\alpha \rightarrow Q \diamond \alpha$ and $\alpha^* \rightarrow \alpha^* \diamond Q$, leading to

$$\begin{aligned} &\int \exp(-2\alpha^* \diamond Q^2 \diamond \alpha - \alpha^* \diamond Q \diamond F_a - F_b^* \diamond Q \diamond \alpha) \det\{Q\}^2 \, \mathcal{D}[\alpha] \\ &= \pi^{\Omega} \exp\left(\frac{1}{2} F_b^* \diamond Q^{-2} \diamond F_a\right), \end{aligned} \tag{C.21}$$

which becomes

$$\int \exp(-2\alpha^* \diamond K \diamond \alpha - \alpha^* \diamond F_1 - F_2^* \diamond \alpha) \, \mathcal{D}^{\circ}[\alpha] = \frac{\exp(\frac{1}{2} F_2^* \diamond K^{-1} \diamond F_1)}{\mathcal{N}_0 \det\{K\}}, \tag{C.22}$$

where we divided by $(2\pi)^{\Omega} \det\{K\}$, and defined $F_1 \triangleq Q \diamond F_a$, $F_2^* \triangleq F_b^* \diamond Q$, $K \triangleq Q^2$, and $\mathcal{N}_0 \triangleq 2^{\Omega}$, as in (5.20).

Special cases, such as when there are no linear terms, or where the kernel is the identity kernel, can be readily obtained from the general expression. If the kernel is just a constant C times the identity, then the determinant becomes $\det\{K\} = C^{\Omega}$, as discussed in more detail in Appendix D.

C.4 Anisotropic Gaussian functionals

The anisotropic Gaussian functionals are those Gaussian functionals that are not rotationally symmetric with respect to rotations around their centres. They represent the most general Gaussian functionals. Those that are centred at the origin of phase space can be represented as

$$W[\alpha] = \mathcal{N} \exp\left(-2\alpha^* \diamond A \diamond \alpha - \alpha^* \diamond B \diamond \alpha^* - \alpha \diamond B^* \diamond \alpha\right), \tag{C.23}$$

where A and B are Hermitian $A^*(\mathbf{k}_1, \mathbf{k}_2) = A(\mathbf{k}_2, \mathbf{k}_1)$ and symmetric $B(\mathbf{k}_1, \mathbf{k}_2) = B(\mathbf{k}_2, \mathbf{k}_1)$ kernels, respectively. Such anisotropic states are called *squeezed vacuum states*, as discussed in Section 6.8.

To evaluate integrals of such anisotropic Gaussian functionals,

$$\kappa = \int \exp\left(-2\alpha^* \diamond A \diamond \alpha - \alpha^* \diamond B \diamond \alpha^* - \alpha \diamond B^* \diamond \alpha\right) \mathcal{D}^\circ[\alpha], \tag{C.24}$$

we use a trick.[b] Consider the square of the integral

$$\begin{aligned} \kappa^2 = \int \exp\big(&-2\alpha_1^* \diamond A \diamond \alpha_1 - \alpha_1^* \diamond B \diamond \alpha_1^* - \alpha_1 \diamond B^* \diamond \alpha_1 \\ &- 2\alpha_2^* \diamond A \diamond \alpha_2 - \alpha_2^* \diamond B \diamond \alpha_2^* - \alpha_2 \diamond B^* \diamond \alpha_2\big) \mathcal{D}^\circ[\alpha_1, \alpha_2], \end{aligned} \tag{C.25}$$

where α_1 and α_2 are the field variables for the two separate functional integrals. Next, we perform a change of integration field variables in which the two field variables are transformed into the sum and difference of two new field variables:

$$\alpha_1 \to \frac{1}{\sqrt{2}}(\alpha + \beta) \quad \text{and} \quad \alpha_2 \to \mathrm{i}\frac{1}{\sqrt{2}}(\alpha - \beta). \tag{C.26}$$

Since the Jacobian is 1, it does not affect the integration result. The change of integration field variables produces

$$\begin{aligned} \kappa^2 = \int \exp\big(&-2\alpha^* \diamond A \diamond \alpha - 2\beta^* \diamond A \diamond \beta \\ &- 2\alpha^* \diamond B \diamond \beta^* - 2\alpha \diamond B^* \diamond \beta\big) \mathcal{D}^\circ[\alpha, \beta], \end{aligned} \tag{C.27}$$

which is equivalent to the Wigner functional for a *twin-beam squeezed vacuum state*, discussed in Section 6.8.8. Applying the isotropic Gaussian integration in (C.22) twice on (C.27), we now get

$$\kappa^2 = \frac{1}{\mathcal{N}_0^2 \det\{A\} \det\{A^* - B^* \diamond A^{-1} \diamond B\}}. \tag{C.28}$$

---

b Apparently, this trick was first used by Richard Feynman for the integration of a Gaussian function.

The integration of an anisotropic Gaussian functional at the origin thus produces

$$
\begin{aligned}
\kappa &= \int \exp\left(-2\alpha^* \diamond A \diamond \alpha - \alpha^* \diamond B \diamond \alpha^* - \alpha \diamond B^* \diamond \alpha\right) \mathcal{D}^\circ[\alpha] \\
&= \frac{1}{\mathcal{N}_0 \sqrt{\det\{A\} \det\{A^* - B^* \diamond A^{-1} \diamond B\}}},
\end{aligned} \tag{C.29}
$$

assuming the determinants are positive definite, as required for Wigner functionals of physical states. When linear terms are added to the exponent, we need to complete the square. The functional integration then produces

$$
\begin{aligned}
&\int \exp\left(-2\alpha^* \diamond A \diamond \alpha - \alpha^* \diamond B \diamond \alpha^* - \alpha \diamond B^* \diamond \alpha - \alpha^* \diamond F - F^* \diamond \alpha\right) \mathcal{D}^\circ[\alpha] \\
&= \frac{\exp\left(\frac{1}{4} F^* \diamond A^{-1} \diamond F\right)}{\mathcal{N}_0 \sqrt{\det\{A\} \det\{A^* - B^* \diamond A^{-1} \diamond B\}}} \exp\left[\frac{1}{4}\left(F - F^* \diamond A^{-1} \diamond B\right)\right. \\
&\quad \left.\diamond \left(A^* - B^* \diamond A^{-1} \diamond B\right)^{-1} \diamond \left(F^* - B^* \diamond A^{-1} \diamond F\right)\right].
\end{aligned} \tag{C.30}
$$

If the anisotropic Gaussian functional represents a pure state, we can use the identities in Section 6.8.5 to get a simpler expression for the functional integral:

$$
\begin{aligned}
&\int \exp\left(-2\alpha^* \diamond A \diamond \alpha - \alpha^* \diamond B \diamond \alpha^* - \alpha \diamond B^* \diamond \alpha - \alpha^* \diamond F - F^* \diamond \alpha\right) \mathcal{D}^\circ[\alpha] \\
&= \frac{1}{\mathcal{N}_0} \exp\left(\frac{1}{2} F^* \diamond A \diamond F - \frac{1}{4} F \diamond A^* \diamond B^* \diamond A^{-1} \diamond F - \frac{1}{4} F^* \diamond A^{-1} \diamond B \diamond A^* \diamond F^*\right) \\
&= \frac{1}{\mathcal{N}_0} \exp\left(\frac{1}{2} F^* \diamond A \diamond F - \frac{1}{4} F \diamond B^* \diamond F - \frac{1}{4} F^* \diamond B \diamond F^*\right),
\end{aligned} \tag{C.31}
$$

where we used (6.340) for the last expression, assuming the kernels have the necessary properties.

D Inverses and determinants

Functional integrations often lead to expressions with *functional determinants* and inverses that contain contractions of kernels. In their most general form, these functional determinants and inverses are not tractable. However, one often finds functional determinants and inverses that have expressions that can be simplified. Here, we discuss various such cases.

D.1 One idempotent kernel

Consider a kernel that has the form

$$K = \mathbf{1} + aP, \tag{D.1}$$

where $\mathbf{1}$ is the identity kernel, a is a constant, and P is an *idempotent kernel* $P \diamond P = P$. The inverse of such a kernel has the form

$$K^{-1} = \mathbf{1} - bP, \quad \text{with} \quad b = \frac{a}{1+a}. \tag{D.2}$$

It can be confirmed by showing that

$$K \diamond K^{-1} = (\mathbf{1} + aP) \diamond (\mathbf{1} - bP) = \mathbf{1}. \tag{D.3}$$

For the determinant of K, we assume that $a < 1$ to ensure convergence. Using the expression of a *functional determinant* in (C.16), we then obtain

$$\begin{aligned} \det\{K\} &= \exp\left[\operatorname{tr}\left\{\ln_{\diamond}(\mathbf{1} + aP)\right\}\right] = \exp\left[\operatorname{tr}\left\{\sum_{m=1}^{\infty} \frac{a^m}{m} P^{m\diamond}\right\}\right] \\ &= \exp\left[\operatorname{tr}\left\{\sum_{m=1}^{\infty} \frac{a^m}{m} P\right\}\right] = \exp\left[\sum_{m=1}^{\infty} \frac{a^m}{m} \operatorname{tr}\{P\}\right] \\ &= \exp\left[\ln(1+a)\operatorname{tr}\{P\}\right] = (1+a)^{\operatorname{tr}\{P\}}. \end{aligned} \tag{D.4}$$

Note that this result is only valid when a is independent of the wavevector or the optical beam variables. If a is a function of such variables, it cannot be removed from the traces, so that $\operatorname{tr}\{a^m P\}$ needs to be evaluated for all m.

If P is a *single-mode projection kernel*,

$$P(\mathbf{k}_1, \mathbf{k}_2) = M(\mathbf{k}_1) M^*(\mathbf{k}_2), \tag{D.5}$$

where $M(\mathbf{k})$ is a normalized function. Then $\operatorname{tr}\{P\} = M^* \diamond M = 1$, so that

$$\det\{\mathbf{1} + aMM^*\} = 1 + a. \tag{D.6}$$

https://doi.org/10.1515/9783111445342-018

D.2 Inverses with multimode kernels

More complicated scenarios are often encountered where kernels consist of linear combinations of mode pairs. Generically, they have the form

$$K(\mathbf{k}_1,\mathbf{k}_2)=\mathbf{1}(\mathbf{k}_1,\mathbf{k}_2)+\sum_{n=1}^{N}F_n(\mathbf{k}_1)G_n^*(\mathbf{k}_2), \tag{D.7}$$

where the modes represented by F_n are linearly independent, but not necessarily mutually orthogonal or normalized, and the same for G_n^*. Moreover, the two sets of modes F_n and G_n^* are not necessarily the same, but they are also not orthogonal. The coefficients for the expansion are absorbed into F_n.

We'll refer to the terms consisting of the two modes as *mode transcriptions*, because while the one performs an inner product, the resulting complex constant becomes a coefficient for the other. If the two modes of a mode transcription are the same (one being the complex conjugate of the other), then the term is proportional to a *single-mode projection kernel*.

A procedure to compute the inverse of such a multimode kernel can be formulated as an iterated process, where a kernel with one mode transcription is removed and inverted consecutively. The iterative process proceeds as follows:

$$\begin{aligned}K^{-1}&=\left(\mathbf{1}+\sum_{n=1}^{N}F_nG_n^*\right)^{-1}=\left(\mathbf{1}+\sum_{n=2}^{N}M_1^{-1}\diamond F_nG_n^*\right)^{-1}\diamond M_1^{-1}\\&=\left(\mathbf{1}+\sum_{n=3}^{N}M_2^{-1}\diamond M_1^{-1}\diamond F_nG_n^*\right)^{-1}\diamond M_2^{-1}\diamond M_1^{-1},\end{aligned} \tag{D.8}$$

and so forth, where

$$\begin{aligned}M_1&=\mathbf{1}+F_1G_1^*,\\M_2&=\mathbf{1}+M_1^{-1}\diamond F_2G_2^*,\\M_3&=\mathbf{1}+M_2^{-1}\diamond M_1^{-1}\diamond F_3G_3^*.\end{aligned} \tag{D.9}$$

The respective inverses are obtained as in Section D.1. For M_1, we have

$$\begin{aligned}M_1^{-1}\diamond M_1&=(\mathbf{1}-bF_1G_1^*)\diamond(\mathbf{1}+F_1G_1^*)\\&=\mathbf{1}-bF_1G_1^*+F_1G_1^*-bF_1G_1^*\diamond F_1G_1^*=\mathbf{1}.\end{aligned} \tag{D.10}$$

It means that

$$b=\frac{1}{1+G_1^*\diamond F_1}, \tag{D.11}$$

so that

$$M_1^{-1} = \mathbf{1} - \frac{F_1 G_1^*}{1 + G_1^* \diamond F_1}. \tag{D.12}$$

For M_2, we then have

$$\begin{aligned} M_2 &= \mathbf{1} + M_1^{-1} \diamond F_2 G_2^* = \mathbf{1} + \left(\mathbf{1} - \frac{F_1 G_1^*}{1 + G_1^* \diamond F_1}\right) \diamond F_2 G_2^* \\ &= \mathbf{1} + \left(F_2 - F_1 \frac{G_1^* \diamond F_2}{1 + G_1^* \diamond F_1}\right) G_2^*. \end{aligned} \tag{D.13}$$

It is of the same form as M_1 and we can therefore represent its inverse with the expression for M_1^{-1} by replacing $G_1^* \to G_2^*$ and

$$F_1 \to F_2 - F_1 \frac{G_1^* \diamond F_2}{1 + G_1^* \diamond F_1}. \tag{D.14}$$

Hence,

$$M_2^{-1} = \mathbf{1} - \frac{(1 + G_1^* \diamond F_1) F_2 G_2^* - (G_1^* \diamond F_2) F_1 G_2^*}{(1 + G_1^* \diamond F_1)(1 + G_2^* \diamond F_2) - (G_2^* \diamond F_1)(G_1^* \diamond F_2)}. \tag{D.15}$$

The inverse of a kernel with two mode transcriptions is obtained by combining (D.12) and (D.15),

$$\begin{aligned} (\mathbf{1} + F_1 G_1^* + F_2 G_2^*)^{-1} &= M_2^{-1} \diamond M_1^{-1} \\ &= \mathbf{1} - \frac{(1 + \tau_{2,2}) F_1 G_1^* - \tau_{1,2} F_1 G_2^* - \tau_{2,1} F_2 G_1^* + (1 + \tau_{1,1}) F_2 G_2^*}{(1 + \tau_{1,1})(1 + \tau_{2,2}) - \tau_{2,1}\tau_{1,2}}, \end{aligned} \tag{D.16}$$

where

$$\tau_{m,n} \triangleq G_m^* \diamond F_n. \tag{D.17}$$

In the same way, we find the result for three mode transcriptions to be

$$K_3^{-1} = (\mathbf{1} + F_1 G_1^* + F_2 G_2^* + F_3 G_3^*)^{-1} = \mathbf{1} - \sum_{m,n=1}^{3} \frac{a_{m,n}}{d_0} F_m G_n^*, \tag{D.18}$$

where d_0 is the determinant given by

$$\begin{aligned} d_0 = {} & (1 + \tau_{1,1})(1 + \tau_{2,2})(1 + \tau_{3,3}) - (1 + \tau_{1,1})\tau_{2,3}\tau_{3,2} - (1 + \tau_{2,2})\tau_{1,3}\tau_{3,1} \\ & - (1 + \tau_{3,3})\tau_{2,1}\tau_{1,2} + (\tau_{1,2})\tau_{2,3}\tau_{3,1} + (\tau_{3,2})\tau_{2,1}\tau_{1,3}, \end{aligned} \tag{D.19}$$

and

$$\begin{aligned}
a_{1,1} &= (1+\tau_{2,2})(1+\tau_{3,3}) - \tau_{2,3}\tau_{3,2},\\
a_{2,2} &= (1+\tau_{1,1})(1+\tau_{3,3}) - \tau_{1,3}\tau_{3,1},\\
a_{3,3} &= (1+\tau_{2,2})(1+\tau_{1,1}) - \tau_{2,1}\tau_{1,2},\\
a_{1,2} &= \tau_{1,3}\tau_{3,2} - \tau_{1,2}(1+\tau_{3,3}), \quad a_{2,3} = \tau_{2,1}\tau_{1,3} - \tau_{2,3}(1+\tau_{1,1}),\\
a_{3,1} &= \tau_{3,2}\tau_{2,1} - \tau_{3,1}(1+\tau_{2,2}), \quad a_{1,3} = \tau_{1,2}\tau_{2,3} - \tau_{1,3}(1+\tau_{2,2}),\\
a_{2,1} &= \tau_{2,3}\tau_{3,1} - \tau_{2,1}(1+\tau_{3,3}), \quad a_{3,2} = \tau_{3,1}\tau_{1,2} - \tau_{3,2}(1+\tau_{1,1}).
\end{aligned} \tag{D.20}$$

In effect, the coefficients $a_{m,n}/d_0$ are the elements of the inverse of the matrix

$$A = \begin{pmatrix} 1+\tau_{1,1} & \tau_{1,2} & \tau_{1,3} \\ \tau_{2,1} & 1+\tau_{2,2} & \tau_{2,3} \\ \tau_{3,1} & \tau_{3,2} & 1+\tau_{3,3} \end{pmatrix}. \tag{D.21}$$

Sometimes the identity kernel in (D.7) is replaced by an arbitrary invertible kernel H. For a single-mode transcription, we then have

$$K = H + F_1 G_1^*. \tag{D.22}$$

Its inverse is obtained from the previous result

$$\begin{aligned}
K^{-1} &= (H + F_1 G_1^*)^{-1} = (\mathbf{1} + H^{-1} \diamond F_1 G_1^*)^{-1} \diamond H^{-1} \\
&= H^{-1} - \frac{H^{-1} \diamond F_1 G_1^* \diamond H^{-1}}{1 + G_1^* \diamond H^{-1} \diamond F_1}.
\end{aligned} \tag{D.23}$$

For two mode transcriptions, it then follows that

$$\begin{aligned}
K^{-1} &= (H + F_1 G_1^* + F_2 G_2^*)^{-1} \\
&= H^{-1} - \big[(1+\kappa_{2,2})H^{-1} \diamond F_1 G_1^* \diamond H^{-1} - \kappa_{1,2} H^{-1} \diamond F_1 G_2^* \diamond H^{-1} \\
&\quad - \kappa_{2,1} H^{-1} \diamond F_2 G_1^* \diamond H^{-1} + (1+\kappa_{1,1}) H^{-1} \diamond F_2 G_2^* \diamond H^{-1}\big] \\
&\quad \times \big[(1+\kappa_{1,1})(1+\kappa_{2,2}) - \kappa_{2,1}\kappa_{1,2}\big]^{-1}.
\end{aligned} \tag{D.24}$$

where

$$\kappa_{m,n} = G_m^* \diamond H^{-1} \diamond F_n. \tag{D.25}$$

It this way, the result can be generalized for arbitrary numbers of transcriptions, but the expressions quickly become rather complex.

D.3 Determinants with multimode kernels

To compute the *functional determinants* of the multimode kernels given in (D.7), we also follow an iterated process, similar to the one for inverses. At each step, a kernel with one mode transcription is pulled out to become a separate determinant, leaving behind the inverted kernel. The consecutive steps are

$$\begin{aligned}
\det\{K\} &= \det\left\{\mathbf{1} + \sum_{n=1}^{N} F_n G_n^*\right\} \\
&= \det\left\{\mathbf{1} + \sum_{n=2}^{N} M_1^{-1} \diamond F_n G_n^*\right\} \det\{M_1\} \\
&= \det\left\{\mathbf{1} + \sum_{n=3}^{N} M_2^{-1} \diamond M_1^{-1} \diamond F_n G_n^*\right\} \det\{M_2\} \det\{M_1\} \\
&= \left(1 + G_N^* \diamond M_{N-1}^{-1} \diamond \cdots \diamond M_1^{-1} \diamond F_N\right) \\
&\quad \times \det\{M_{N-1}\} \cdots \det\{M_1\}.
\end{aligned} \tag{D.26}$$

The individual determinants are obtained from (D.6), leading to

$$\begin{aligned}
\det\{M_1\} &= \det\{\mathbf{1} + F_1 G_1^*\} = 1 + \tau_{1,1}, \\
\det\{M_2\} &= \det\left\{\mathbf{1} + \left(F_2 - F_1 \frac{\tau_{1,2}}{1 + \tau_{1,1}}\right) G_2^*\right\} \\
&= (1 + \tau_{2,2}) - \frac{\tau_{2,1}\tau_{1,2}}{1 + \tau_{1,1}}, \\
\det\{\mathbf{1} + F_1 G_1^* + F_2 G_2^*\} &= \det\{M_2\} \det\{M_1\} \\
&= (1 + \tau_{1,1})(1 + \tau_{2,2}) - \tau_{2,1}\tau_{1,2}.
\end{aligned} \tag{D.27}$$

Note that the denominator of the second term in the inverse of a multimode kernel is the same as the determinant for that kernel.

When the identity **1** in the determinant of a kernel with one mode transcription is replaced by an invertible kernel H, we have

$$\begin{aligned}
\det\{H + F_1 G_1^*\} &= \det\{H\}(1 + G_1^* \diamond H^{-1} \diamond F_1) \\
&= \det\{H\}(1 + \kappa_{1,1}).
\end{aligned} \tag{D.28}$$

The cases with more transcriptions is obtained by the iterated process provided above.

D.4 Determinants of diagonal kernels

Some kernels are diagonal, so that $K(\mathbf{k}_1, \mathbf{k}_2) = 0$ for $\mathbf{k}_1 \neq \mathbf{k}_2$. Then, either the kernel can be represented as $K(\mathbf{k}_1, \mathbf{k}_2) = G(\mathbf{k}_1)\mathbf{1}(\mathbf{k}_1, \mathbf{k}_2)$ or it is a function of measure zero. In the former case, $\diamond$-contractions are replaced by products of the diagonal function. For example, $K \diamond K = G^2\mathbf{1}$. It implies that $f_\diamond(K) = f(G)\mathbf{1}$. By implication $\ln_\diamond(K) = \ln(G)\mathbf{1}$, which is relevant for the determinant. More explicitly, we have

$$\det\{G\mathbf{1}\} = \exp[\mathrm{tr}\{\ln_\diamond(G\mathbf{1})\}] = \exp\left\{\int \ln[G(\mathbf{k})]\delta(0)\,\mathrm{d}^3k\right\}. \tag{D.29}$$

For the case where $G(\mathbf{k})$ is a constant C, it becomes

$$\det\{C\mathbf{1}\} = \exp\left[\ln(C)\int \delta(0)\,\mathrm{d}^3k\right] = \exp\left[\ln(C)\Omega\right] = C^\Omega, \tag{D.30}$$

which is consistent with the fact that $\det\{\mathbf{1}\} = 1$.

D.5 Derivatives of determinants and inverse

When functional integrations involve generating functions, the determinants and inverses that they produce usually contain generating parameters. In other words, the generating functions produced by such functional integrations contain determinants and inverses that are functions of the generating parameters, and which are then subjected to derivatives with respect to these generating parameters. Here, we apply the derivatives to determinants and inverses without any simplifying assumptions based on the properties of the kernels.

For the inverse of a kernel, we consider the derivative of that kernel contracted to its inverse. The kernel depends on some parameter x with respect to which a derivative is applied. It leads to

$$\partial_x\left[K^{-1}(x)\diamond K(x)\right] = \left[\partial_x K^{-1}(x)\right]\diamond K(x) + K^{-1}(x)\diamond[\partial_x K(x)] = 0. \tag{D.31}$$

Hence, the derivatives of the inverse of a kernel is given by

$$\partial_x K^{-1}(x) = -K^{-1}(x)\diamond\partial_x K(x)\diamond K^{-1}(x). \tag{D.32}$$

The derivative of a *functional determinant* leads to the derivative of the functional natural logarithm

$$\begin{aligned}\partial_x \det\{K(x)\} &= \exp[\mathrm{tr}\{\ln_\diamond(K)\}]\,\mathrm{tr}\{\partial_x \ln_\diamond[K(x)]\} \\ &= \det\{K(x)\}\,\mathrm{tr}\{K^{-1}(x)\diamond K(x)\diamond\partial_x \ln_\diamond[K(x)]\} \\ &= \det\{K(x)\}\,\mathrm{tr}\{K^{-1}(x)\diamond \exp_\diamond(\ln_\diamond[K(x)])\diamond\partial_x \ln_\diamond[K(x)]\} \\ &= \det\{K(x)\}\,\mathrm{tr}\{K^{-1}(x)\diamond\partial_x \exp_\diamond(\ln_\diamond[K(x)])\}.\end{aligned} \tag{D.33}$$

Here, we first inserted the contraction of the kernel with its inverse into the trace, and then converted the kernel into an exponential applied to the logarithm of the kernel.

The derivative of the logarithm is expressed as a derivative applied to the exponential. In the end, the exponential of the logarithm of the kernel is converted back to just being the kernel so that the expression for the derivative of a determinant becomes

$$\partial_x \det\{K(x)\} = \det\{K(x)\} \operatorname{tr}\left\{K^{-1}(x) \diamond \partial_x K(x)\right\} . \tag{D.34}$$

Index

https://doi.org/10.1515/9783111445342-019

www.ingramcontent.com/pod-product-compliance
Lightning Source LLC
Chambersburg PA
CBHW060942210326
41598CB00031B/4699